INDUSTRIAL
PLASTICS
Theory and Applications

6th Edition

INDUSTRIAL PLASTICS

Theory and Applications

6th Edition

Erik Lokensgard, Ph.D.
Eastern Michigan University
Ypsilanti, Michigan

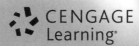

CENGAGE
Learning®

Australia • Brazil • Mexico • Singapore • United Kingdom • United States

Industrial Plastics: Theory and Applications, Sixth Edition
Erik Lokensgard

SVP, GM Skills & Global Product Management: Dawn Gerrain

Product Director: Matthew Seeley

Associate Product Manager: Nicole Robinson

Senior Director, Development: Marah Bellegarde

Senior Product Development Manager: Larry Main

Senior Content Developer: Meaghan Tomaso

Product Assistant: Maria Garguilo

Vice President, Marketing Services: Jennifer Ann Baker

Marketing Manager: Jonathon Sheehan

Senior Production Director: Wendy Troeger

Production Director: Andrew Crouth

Senior Content Project Manager: Betsy Hough

Senior Art Director: Benjamin Gleeksman

Cover and Interior Design Image: © Miles Studio/Shutterstock

For product information and technology assistance, contact us at
Cengage Learning Customer & Sales Support, 1-800-354-9706

For permission to use material from this text or product, submit all requests online at **www.cengage.com/permissions**.

Further permissions questions can be e-mailed to **permissionrequest@cengage.com**

Library of Congress Control Number: 2015943890

ISBN: 978-1-2850-6123-8

Cengage Learning
20 Channel Center Street
Boston, MA 02210
USA

Cengage Learning is a leading provider of customized learning solutions with employees residing in nearly 40 different countries and sales in more than 125 countries around the world. Find your local representative at **www.cengage.com.**

Cengage Learning products are represented in Canada by Nelson Education, Ltd.

To learn more about Cengage Learning, visit **www.cengage.com**

Purchase any of our products at your local college store or at our preferred online store **www.cengagebrain.com**

Notice to the Reader
Publisher does not warrant or guarantee any of the products described herein or perform any independent analysis in connection with any of the product information contained herein. Publisher does not assume, and expressly disclaims, any obligation to obtain and include information other than that provided to it by the manufacturer. The reader is expressly warned to consider and adopt all safety precautions that might be indicated by the activities described herein and to avoid all potential hazards. By following the instructions contained herein, the reader willingly assumes all risks in connection with such instructions. The publisher makes no representations or warranties of any kind, including but not limited to, the warranties of fitness for particular purpose or merchantability, nor are any such representations implied with respect to the material set forth herein, and the publisher takes no responsibility with respect to such material. The publisher shall not be liable for any special, consequential, or exemplary damages resulting, in whole or part, from the readers' use of, or reliance upon, this material.

Printed in the United States of America
Print Number: 01   Print Year: 2015

CONTENTS

INTENDED USE

Industrial Plastics: Theory and Applications, Sixth Edition, covers all facets of industrial plastics technology as well as major manufacturing processes, serving as an indispensable resource for those individuals enrolled in polymer technology or plastics technology programs at community colleges, technical colleges, and universities. Comprehensive in nature, this text will also be useful to professionals who wish to review the basics and remain up to date on the latest technology in plastics manufacturing.

TEXT LAYOUT

Presented in a logical sequence, *Industrial Plastics* builds topics from the ground up—covering everything from the history of plastics to the running of a successful plastics business.

CHAPTER 1, provides a historical introduction to plastics.

CHAPTER 2, includes extensive updates on the current status of the plastics industry. It features US consumption of major plastics materials, recycling, disposal, and significant organizations within the industry.

CHAPTER 3, treats elementary polymer chemistry. It attempts to present basics about plastics and polymer chemistry in a practical context.

CHAPTER 4, on health and safety, reflects the organization in material safety data sheets. The intent of this chapter, which has been updated to reflect current standards, is to assist students in becoming adept at reading and understanding MSDSs for plastics.

CHAPTER 5, on elementary statistics, relies on graphical techniques rather than hypothesis testing.

CHAPTER 6, on properties and tests, has been updated to show current varieties of testing equipment.

CHAPTER 7, on ingredients of plastics, includes a new section on nanocomposites.

CHAPTER 8, on the selection of plastics for specific applications, has the intent of explaining the differences between various grades of plastics.

CHAPTER 9, on machining and finishing, treats common processes for shaping and polishing plastics products.

CHAPTER 10, on molding processes, includes a new section on all-electric and hybrid injection-molding machines. It continues to feature thorough treatment of injection-molding safety.

CHAPTER 11, on extrusion, now includes several recent photos of multilayer blow-molding equipment and blown film equipment.

CHAPTER 12, on laminating processes, discusses layers of plastics, paper, glass fibers, and metal.

CHAPTER 13, on reinforcing processes, includes numerous processes to create a matrix of fibrous reinforcements and plastics.

CHAPTER 14, on casting processes, includes several new photographs of large rotational molding equipment.

CHAPTER 15, on thermoforming, treats the major methods to form sheet materials with vacuum, pressure, and mechanical forces.

CHAPTER 16, on expansion processes, discusses techniques to create foamed materials. It includes several new photographs of synthetic turf.

CHAPTER 17, on coating processes, concerns the application of coatings onto plastics substrates and the application of plastics onto non-polymeric substrates.

CHAPTER 18, on fabrication, treats both mechanical and chemical techniques.

CHAPTER 19, on decoration processes, includes updated information on hot-foil stamping, as well as other decorating techniques.

CHAPTER 20, on radiation processes, treats growth in the use of radiation processing.

CHAPTER 21, on design, includes a new section on stereolithography.

CHAPTER 22, on tooling and mold making, covers major machining techniques.

CHAPTER 23, on commercial considerations, provides updated coverage of auxiliary equipment.

APPENDIX A, the glossary, provides definitions of terms.

APPENDIX B, on abbreviations, includes chemical or generic names for an updated list of abbreviations.

APPENDIX C, on trade names, provides trade names, the corresponding name of the plastics, and the manufacturer.

APPENDIX D, on material identification, offers several methods to identify unknown plastics.

APPENDIX E, on thermoplastics, contains extensive material and a thorough list of thermoplastics.

APPENDIX F, on thermosets, treats most major thermoset materials.

APPENDIX G, provides useful tables showing conversions of various units.

APPENDIX H, provides contacts for many organizations and also a selected bibliography.

The new and updated sixth edition of *Industrial Plastics* will further enhance the ease of use and the depth of content.

NEW TO THIS EDITION
The Latest Technology

The sixth edition of *Industrial Plastics: Theory and Applications* provides updated materials on websites in all chapters, current data on the plastics industry, and expanded treatments of emerging technologies, in particular, nanoplastics, and bioplastics.

Related Internet Sites

A list of Internet sites at the end of each chapter guides students to find more information on specific topics. Each site relates to the material found in the chapter and further enhances learning for the reader. Many of these companies provide extensive discussions of their materials, processes, and products on their websites. The more elaborate related Internet sites also include photos, video, and audio presentations.

OTHER FEATURES

Lab Activities

Where applicable, lab activities are included at the end of the chapter. The philosophy embedded in the laboratory activities is that practical applications are essential for thorough understanding of many theoretical concepts. The activities contain tried approaches but also include suggestions for further investigations. It is hoped that students and instructors will build on the laboratory activities and customize them for available equipment and materials.

*The confidential and proprietary details outlined or other information provided are intended only as a guide. They are not to be taken as a license under which to operate or as a recommendation to infringe on any patents.

Chapter Review

All chapters provide vocabulary lists and review questions for students to use as a self-study guide and to test their knowledge of important concepts.

SUPPLEMENTS

Instructor Resources

The Instructor Resources, now available online, contain a thorough classroom guide for instructors, including answers to all end-of-chapter questions, chapter tests powered by Cognero with hundreds of test questions, and an image gallery with all photos and illustrations from the text.

To access these Instructor Resources, go to login.cengagebrain.com, and create an account or log into your existing account.

ABOUT THE AUTHOR

Erik Lokensgard is a professor at Eastern Michigan University in the School of Engineering Technology. In addition to teaching, he has been active with the Detroit Section of the Society of Plastics Engineers and has provided training for several plastics companies in the Detroit area.

ACKNOWLEDGMENTS

The author and publisher wish to thank the following individuals, whose technical expertise and thorough review of the manuscript contributed to the development of the revised text:

Dan Burklo, Northwest State Community College, Archibold, OH

Barry David, Millersville University, Millersville, PA

George Comber, Weber State University, Ogden, UT

David Meyer, Sinclair Community College, Dayton, OH

Matthew Meyer, Ashville-Buncombe Technical College, Ashville, NC

Dan Ralph, Hennepin Technical College, Brooklyn Park, MN

Mike Ryan, University of Buffalo, Buffalo, NY

Neil Thomas, Ivy Tech State College, Evansville, IN

HISTORICAL INTRODUCTION TO PLASTICS

CHAPTER 1

INTRODUCTION

Life without plastics is rather hard to imagine. In everyday activities, we rely on plastic items such as milk jugs, eyeglasses, telephones, nylons, automobiles, and videotapes. However, not much more than one hundred years ago, the plastics taken for granted today did not exist. Long before the development of commercial plastics, some existing materials displayed unique features. Although they were strong, translucent, lightweight, and moldable, only a few substances combined these qualities. Today, these materials have the name natural plastics. They provide the starting point for a brief history of plastics materials.

This chapter will provide information about advantages of early plastics and the difficulties encountered during their manufacturing. It sets modern materials and processes in a historical context and demonstrates the powerful influence of pioneers in the plastics industry. The topics included are listed here:

   I.   **Natural plastics**
        A. **Horn**
        B. **Shellac**
        C. **Gutta percha**
   II. **Early modified natural materials**
        A. **Rubber**
        B. **Celluloid**
   III. **Early synthetic plastics**
   IV. **Commercial synthetic plastics**

NATURAL PLASTICS

The starting point for this section is set in Medieval England. In medieval times, English surnames—last names—indicated professions. Some of these professions are easily recognized today. Occupational references for names such as Smith, Baker, Carpenter, Weaver, Taylor, Cartwright, Barber, Farmer, and Hunter are obvious. Occupational origins of other names, such as Fuller, Tucker, Cooper, and Horner, are less familiar.

> Little Jack Horner
> Sat in a corner
> Eating his Christmas pie;
> He put in his thumb
> And pulled out a plum,
> And said, "What a good boy am I."

This rhyme implies that Jack was not hungry or poor and did not have to share his Christmas treat with other family members. He enjoyed a special treat alone. Apparently, Jack's father had a comfortable income. What did Jack's father, or perhaps grandfather, do? He was a horner—a man who made small items from horns, hooves, and occasionally from tortoiseshells.

A typical response to horn working is to dismiss it as quaint, irrelevant, or disgusting. The horner's craft was smelly and often unpleasant. Today, horners can only be found in rare historically oriented craft museums. However, horn working is not irrelevant with relation to the plastics industry. The unique properties of horn inspired a search for substitutes. The quest for synthetic horn led to the production of early plastics and the beginnings of the modern plastics industry.

Horn

Spoons, combs, and lantern windows were common products made by horners in England and Europe during the Middle Ages. Horn spoons were strong and lightweight. They did not rust, corrode, or give an undesirable taste to food. Horn combs were flexible, smooth, glossy, and often decorative. As seen in Figure 1-1, lantern windows exploited the translucent quality of horn. They also flexed without shattering and withstood some impact. No other material provided this combination of properties.

Making utilitarian objects from natural polymers did not begin in the Middle Ages. One of the oldest known uses of horn dates from the times of the pharaohs of Egypt. Approximately 2000 BC, ancient Egyptian craftsmen formed ornaments and food utensils by softening tortoiseshells in hot oils. When the shell was sufficiently pliable, they pressed it into the desired shape. They trimmed any rough shapes, scraped, sanded, and finally polished them to a high luster with fine powders.

Little Jack Horner's forefather worked in a manner similar to the ancient Egyptians. He softened pieces of cow horn by either boiling them in water or soaking them in alkaline solutions and pressing the pieces flat. Some horns were delaminated along growth lines, which yielded thin sheets. If thicker pieces were needed, several thinner sheets were welded together. After creating the desired thickness, horn pieces were squeezed into molds to create a useful shape. Sometimes, horners dyed the pieces to make them look like expensive tortoiseshell.

Two items are particularly important for this history because they are made using different techniques: combs and buttons.

Combs. Some English horners emigrated to the American colonies and established small businesses. By 1760, horn workers were well established in Massachusetts. Leominster, Massachusetts, became a center for the comb business and earned the name "Comb City."

In comb factories, craftsmen sawed flattened horn pieces to size, cut in teeth with fine saws, smoothed rough edges, colored, and polished the combs. The final operation was called bending. A contoured wooden form imparted a curve to a softened comb and maintained the shape as the comb cooled.

Figure 1-2 shows a photograph of a comb made from tortoiseshell. Notice that several teeth are slightly warped. Even in the most carefully made combs, thin teeth were easily broken. Notice also that the comb is generally uniform in cross section. Usually, combs were not embossed with raised artistic motifs because shell and horn do not flow easily.

Although comb makers in Massachusetts developed machines to mechanize manufacturing, they could not establish stable production. This was not the fault of the machines, but of the material. Clamping fixtures and cutter movements required uniform, flat workpieces. Horn was neither flat nor uniform in size and flexibility.

The lack of dimensional consistency, low "flow ability," and inherent waste caused by the shape of horn encouraged comb manufacturers to search for substitutes.

Buttons. Horn button makers faced a different set of problems. Flat, utilitarian buttons were molded from horn pieces, cut out in pre-sized blanks, and then pressed into heated molds. However, customers also wanted decorative buttons to complement fine garments. Hand-carved buttons of ivory

From the Collections of the Henry Ford Museum and Greenfield Village

Figure 1-1. This candle lantern displays horn windows.

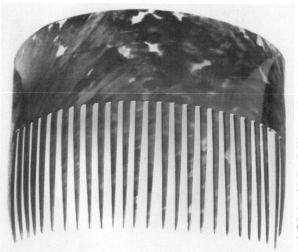

From the Collections of the Henry Ford Museum and Greenfield Village

Figure 1-2. This well-preserved tortoiseshell comb shows only one broken tooth.

had been available for centuries, but they were expensive and one of a kind. In order to make embossed and raised motifs, the molding material had to flow easily in the mold. To achieve this, the horners developed molding powders of ground horn. Horn buttons often consisted of ground cow hooves colored with a water solution. The horn powder was poured into molds and compressed or rolled out into sheets. The sheets were cut into small blanks with tools similar to small cookie cutters. The blanks were then compressed into a mold to achieve three-dimensional surfaces. Figure 1-3 shows two horn buttons—one with a prominent relief.

Buttons required rather undemanding physical properties. They were thick enough to be strong and devoid of fragile teeth. The impetus to seek alternatives came from the actual working of the horn. Removing the tissue mass and cleaning the slimy membrane from the inside of the horn was dirty work accompanied by strong odors from boiled horns. When shellac became readily available, horners carefully evaluated its qualities.

Shellac

In about 1290, when Marco Polo returned to Europe from his travels in Asia, he brought back shellac. He had found shellac in India, where people had been using it for centuries. They had discovered the unique properties of a natural polymer that came from insects rather than cow horn.

The insect that produces a polymer is a small bug called the lac, native to India and southeast Asia. A female lac inserts a stinger-like proboscis into the twig or small branch of a tree. She lives off the sap drawn from the host plant, and exudes a thick liquid, which dries slowly. As the deposit of hardened liquid grows, the insect becomes immobilized. After the male lac fertilizes the female, she increases the juice excretions and is totally covered. Inside this deposit, she lays hundreds of eggs and eventually dies. When the eggs hatch, the young insects eat their way out of the covering and go off to repeat the cycle.

The hardened excretion has unique properties. When cleaned, dissolved in alcohol, and applied to a surface, it makes a shiny, almost transparent coating. The name shellac was descriptive because it came from the *shell* of the *lac*. In addition to being used as a protective coating for furniture and floors, solid shellac was moldable.

Under heat and pressure, shellac will flow into the recesses of intricate and detailed molds. Because pure shellac is brittle and weak, compounds containing various fibers were developed to give the moldings some strength. An early product made from molded shellac was the daguerreotype case, seen in Figure 1-4. Their manufacture in the United States began about 1852.

In addition to such cases, shellac was molded into buttons, knobs, and electrical insulators. By 1870, the shellac molding business was well established. The business got a big boost when phonograph records were made from shellac. Shellac molding materials could accurately reproduce the intricate detail needed for sound. Molded shellac parts maintained a niche in the growing plastics industry until the 1930s, when synthetic plastics finally surpassed their qualities.

Several undesirable traits offset the desirable characteristics of this material. The amount and quality of the lac harvest were affected by predator insects, insufficient rain, wide temperature variations, hot winds, and the geographic regions of India. In a drought, farmers harvested twigs hosting live lac and eggs. They stored the lac brood in pits and kept the sticks and twigs wet with cool water. If they did not continue this burdensome task, the death of the lac brood stock would result.

Under normal conditions, the farmers collected the encrusted twigs after the larva left the sheltering deposit. Next, they then scraped the hardened residue off and cleaned it. Cleaning was

From the collection of Evelyn Gibbons

Figure 1-3. These black horn buttons show the three-dimensional relief possible with horn-molding compounds.

Figure 1-4. This daguerreotype case, molded in about 1855, contains shellac and wood flour. The detail is remarkable.

not a simple process, due to sand, dirt, dead lac bodies, leaves, and wood fibers.

After the shellac was ready for use as a coating or a molding powder, problems still persisted. The largest problem was moisture absorption. When a shellac molding or coating gets wet, it absorbs water. If soaked for 48 hours, it will absorb up to 20% water and change to a whitish color. Antique furniture suffered from water rings caused by condensation on containers of ice water. Shellac also takes on moisture from the atmosphere. In high-humidity environments, enough water will be absorbed to whiten shellac finishes. In moldings, moisture absorption could lead to cracking. Even such stable forms as buttons cracked due to moisture absorption.

The color of shellac was not consistent. The most common colors—yellow and orange—depended on the type of tree the lac infested. To create white shellac, chlorine bleaches were used to lighten the natural color. However, the bleaching process also affected its solubility in alcohol. Bleached shellac often coalesced into a gummy, worthless lump.

Another problem involved aging. Shellac finishes and moldings darkened with age. Old shellac became insoluble in alcohol. Shellac finishes stored in steel cans also absorbed iron, which caused the finish to turn gray or black.

These problems caused manufacturers to seek alternatives. During the 1920s and 1930s, new plastics began replacing shellac. In response, shellac producers tried to improve its qualities. Because shellac contained several polymers, they hoped to separate the most desirable portion by fractional distillation. However, this effort did not result in a material that could withstand the competition from the synthetic plastics.

Gutta Percha

Gutta percha is a natural polymer with remarkable properties. It is produced by the Palaquium gutta trees indigenous to the Malay peninsula. In 1843, William Montgomerie reported that in Malaya, gutta percha was used to make knife handles.

The material was softened in hot water and pressed by hand into a desired shape. His report stirred interest in the material and led to the formation of the Gutta Percha Company, which remained active until 1930. This company manufactured molded items.

The characteristics of gutta percha are unusual. At room temperature, it is a solid. It can be dented but does not break easily. When heated, it can be drawn out into long strips that will not rebound like rubber. Gutta percha is highly inert and resists vulcanization. Its resistance to chemical attack made it an excellent insulator for electric wires and cable. When long strips of extended gutta percha were wound tightly around a wire, the resulting cable was flexible, waterproof, and impervious to chemical attack.

The first underwater telegraph cable ran across the English Channel from Dover to Calais. Its success was due to gutta percha insulation. In the United States, the Morse Telegraph Company laid a cable insulated with gutta percha across the Hudson River in 1849. Gutta percha also protected the first transatlantic cable laid in 1866. Figure 1-5 shows the use of gutta percha in the first transatlantic cable.

Like other natural materials, gutta percha was inconsistent. Contamination created regions in the insulation that were low in resistance to electricity. These areas eventually lost the ability to insulate, which led to the shorting out of electric circuits. Despite these problems, it remained unsurpassed as an insulator until the development of synthetic plastics in the 1920s and 1930s. Only then did gutta percha become less important in electrical applications.

EARLY MODIFIED NATURAL MATERIALS

It was difficult to harvest, gather, or purify natural plastics. Using these materials in manufacturing processes was arduous. Virtually any material that held potential as a substitute for horn and shellac received attention. Many materials were

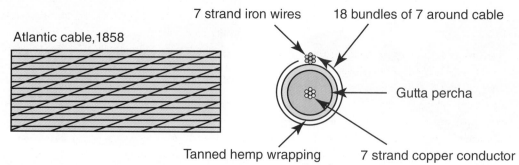

Figure 1-5. The first transatlantic cable had an overall diameter of 0.62 inches and contained 1 pound of gutta percha in every 23 feet of cable. The amount of gutta percha used for the entire cable was over 260 tons.

complete failures. Others failed in their natural condition but became useful when chemically altered.

Casein, a material made from milk curd, appeared to have some value as artificial horn. Dried milk curds were ground into powder and plasticated with water. The resulting dough was molded into various shapes. This attempt met failure because the molded items dissolved when wet. Casein held no significance as a rival to horn until 1897. In that year, a German printer, Adolf Spitteler, learned how to harden the casein dough with formaldehyde. The hardened casein was called Galalith, which means "milkstone." It was a moldable plastic used for buttons, umbrella handles, and other small items.

Galalith is important because it exemplifies a group of materials that originate in nature and become useful for manufacturing only after chemical modification. One of the earliest and most important materials in this category is rubber.

Rubber

Natural rubber, also called gum rubber, is a natural latex. It is found in the sap or juice of many plants and trees. The white, sticky juice of the milkweed plant is rich in latex. Several trees also produce natural latex in great quantities. For example, the rubber tree (*Hevea brasiliensis*), a prolific producer of latex, was cultivated in large plantations in India.

Compared to gutta percha, natural rubber had little industrial significance. Natural rubber is extremely sensitive to temperature. When the weather is hot, it becomes very soft. When the ambient temperature is cool or cold, it becomes stiff. One of the first uses of gum rubber was to make cloth waterproof.

In 1823, Charles Mackintosh got a patent for waterproof cloth. By pressing a layer of rubber between two pieces of cloth, he solved one problem. In comfortable temperatures, gum rubber becomes tacky. However, when rubber is placed between two pieces of fabric, it no longer has a tacky feeling. Some waterproof jackets called Mackintoshes were manufactured, but they had all the problems of gum rubber. In cold weather, the jackets were stiff and frequently cracked. When it was hot, the jackets melted. In addition to getting sticky in warm weather, gum rubber decomposed easily, creating a strong, foul odor.

In 1839, Charles Goodyear discovered that kneading powdered sulfur into rubber tremendously improved its characteristics. His finding did not happen easily. Goodyear spent years trying to alter gum rubber. He tried blending it with ink, castor oil, soup, and even cream cheese. He finally mixed gum rubber with powdered sulfur and heated the blend. The resulting rubber was stronger, tougher, less sensitive to temperature, and more resilient than before. He had learned how to vulcanize the gum rubber. Kneading in small amounts of sulfur produced flexible rubber. Large amounts of sulfur—up to 50%—yielded ebonite, a rubber so hard it could shatter like glass.

In 1844, Goodyear received an American patent for his discovery. He hoped that a large display at the London Exhibition of 1851 would send him on the road to riches. Goodyear put considerable effort into his display, which was called the Vulcanite Court. Its walls, roof, and furniture were made of rubber. His display included combs, buttons, canes, and knife handles molded of hard rubber. His products of flexible rubber featured large rubber balloons and a rubber raft. Goodyear also set up an exhibit in 1855 at the Paris Exposition. Here he featured electric wires insulated with hard rubber, toys, sporting equipment, dental plates, telegraph equipment, and fountain pens.

These displays convinced many people that vulcanized rubber held immense commercial potential. However, before Goodyear could win personal wealth from his idea, he died in 1860. He did not live to see the rise of the rubber industry, which became significant during the Civil War. During that period, the Union Army purchased rubber products valuing $27 million. The Goodyear Company moved into the forefront of the new rubber industry.

Horners were particularly interested in hard rubber as a substitute for horn. In England, comb makers purchased tons of hard rubber. They preferred it to both horn and tortoiseshell because it reduced waste.

Although rubber was less wasteful, it did not have an advantage in appearance. The highly sulfur-loaded material was usually black or dark brown. It could not replace the many horn products that imitated tortoiseshell or ivory. This appearance limitation prevented ebonite from sweeping other materials aside.

Vulcanized rubber was one of the first modified natural polymers. Without vulcanization, gum rubber was of limited utility. Vulcanized rubber was both flexible and hard, making it a very significant industrial material.

Celluloid

To produce celluloid, cellulose, in the form of cotton linters, underwent a series of chemical modifications. One alteration was the conversion of cotton into nitrocellulose. In 1846, a Swiss chemist, C. F. Schönbein, discovered that a combination of nitric acid and sulfuric acid transformed cotton into a high explosive. Explosive nitrocellulose is highly nitrated. Moderately nitrated cellulose is not explosive but is useful in other ways.

Moderately nitrated cellulose is called pyroxylin, a material that dissolves in several organic solvents. When applied to a surface, the solvents evaporate and leave behind a thin, transparent film. This film was named collodion. Collodion found widespread use as a carrier for photosensitive materials. Anyone familiar with the photographic processes common in the 1850s and 1860s observed dried collodion. When a thick layer of collodion dried, the resulting material was hard, water resistant, somewhat elastic, and very similar to horn.

Alexander Parkes, a British businessman, decided to focus his efforts on developing collodion into an industrial material. Parkes lived in Birmingham, England, and had considerable experience in working with natural polymers. He had worked with gum rubber, gutta percha, and chemically treated gum rubber. He understood the qualities of natural plastics and their limitations. In 1862, he announced a new material, which he called Parkesine.

He claimed that Parkesine was a substance "partaking in a large degree of the properties of ivory, tortoiseshell, horn, hardwood, India rubber, gutta percha, etc., and which will . . . to a considerable extent, replace such materials." In 1866, he founded a company that would sell his new material, but his expectations did not match reality. When mixing pyroxylin with various stiff oils, he used several solvents. When the solvents evaporated, the new plastic shrunk excessively. Combs became so warped and twisted that they were useless. Parkes did not find buyers flocking to his door to purchase his material, and his company failed in two years.

Parkes's failure did not turn others away from the effort to convert hardened collodion into an industrial material. An American, John W. Hyatt, also turned his attention to the problem. In 1863, he decided to try for a $10,000 reward promised to anyone who could find a substitute for ivory billiard balls. Hyatt made a few billiard balls out of shellac and wood pulp, similar to the

material used for daguerreotype cases. However, these were poor substitutes because they lacked the elasticity of ivory.

Hyatt then set out to make a solid material from pyroxylin. In 1870, he received a patent on the process for making a new material he called celluloid. He mixed powdered pyroxylin with pulverized gum camphor. Hyatt wet the mixture to evenly disperse the powders. He then removed the water by pressing it with blotting paper. The material, by then a fragile block, was placed in a mold, heated, and pressed. This resulted in a block of material that was uniform throughout. This block could be used as a molding compound but was usually shaved into sheets that needed seasoning to remove residual water. Figure 1-6 shows the shaving of a large block of celluloid into sheets.

John and his brother, Isaiah S. Hyatt, established a few companies to use their new material. The first was the Albany Dental Plate Company, established in 1870. Next, they formed the Albany Billiard Ball Company. In both cases, the applications they selected for celluloid were failures. Dental plates were a very poor choice because they tasted of camphor. Some dental plates softened, warped, or flaked. They were not nearly as good as dental plates made with hard rubber and never seriously competed with them for the market. The celluloid billiard balls, pictured in Figure 1-7, had the same problems as shellac balls. Hyatt's company abandoned dental plates and billiard balls, and focused on horn products.

Celluloid was a very good substitute for horn. It easily imitated ivory, tortoiseshell, and horn. Celluloid became a commercial success, and the Celluloid Manufacturing Company brought the Hyatts substantial profits. By 1874, celluloid combs and mirrors

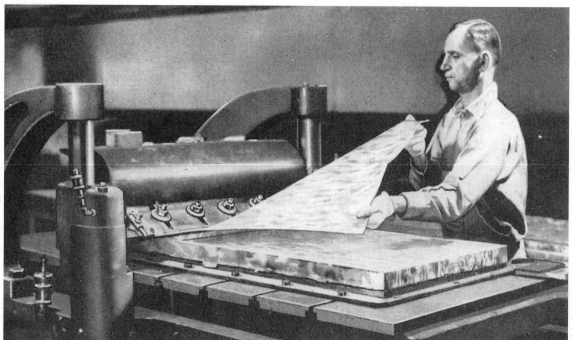

Figure 1-6. This machine shaved sheets of celluloid from large blocks. The block shown contains spots of various colors to make the celluloid appear like tortoiseshell.

Figure 1-7. The Hyatt billiard ball, an early celluloid product.

were readily available. Between 1890 and 1910, comb manufacturers in Leominster, Massachusetts, switched almost completely to celluloid. Figure 1-8 shows these products.

Instead of trying to monopolize celluloid manufacturing, the Hyatts licensed a number of companies to use their material. Between 1873 and 1880, they began affiliations with the Celluloid Harness Trimming Company, the Celluloid Novelty Company, the Celluloid Waterproof Cuff and Collar Company, the Celluloid Fancy Goods Company, the Celluloid Piano Key Company, and the Celluloid Surgical Instrument Company. As this list indicates, celluloid products were generally small items related to clothing or novelties.

Celluloid was not adequate for most industrial applications. One example of its failure in the engineering materials market was safety glass. Layers of celluloid were placed between two pieces of glass to make safety glass for automobiles. The

Figure 1-8. A comb and brush produced from celluloid in 1880.

problem was that exposure to sunlight caused yellowing and deterioration. Celluloid did meet the needs of one major application that could never have been filled by ivory, tortoiseshell, horn, or hard rubber—it was used for photographic film.

By 1895, motion pictures based on celluloid roll film were available to some audiences. Celluloid made possible the early silent films and famous celluloid personalities. The biggest problem with celluloid film was flammability. Carbon arcs provided light for projection, but when films jammed in the projector, the intense heat caused the ignition of film. Hundreds of people lost their lives in disastrous theater fires. However, those deaths did not limit the use of celluloid for motion pictures. No other material could perform like celluloid. Not until safety film was invented in the 1930s was a photographic base available that eliminated the fire hazard.

The consumption of celluloid rose until the mid-1920s. It did not mold easily and was used mainly as a fabricating material. Therefore, moldings of plastics continued to be dominated by shellac. Starting in the 1920s, more robust synthetic polymers took over celluloid applications. Table tennis balls are one of the few remaining products made of celluloid.

EARLY SYNTHETIC PLASTICS

Dr. Leo H. Baekeland was a research chemist who searched for a substitute for shellac and varnish. In June 1907, when working with the chemical reaction of phenol and formaldehyde, he discovered a plastic material that he named Bakelite. Phenol and formaldehyde came from chemical companies rather than nature. This marked a major difference between Bakelite and modified natural plastics.

In his notebook, Baekeland wrote that, with some improvements, his material might be "a substitute for celluloid and hard rubber." In 1909, he reported his finding to the New York section of the American Chemical Society. He claimed that Bakelite made excellent billiard balls because its elasticity was very similar to that of ivory.

The General Bakelite Company was established in 1911. Bakelite use grew rapidly. In contrast to celluloid, it found applications beyond the area of novelties and fashion apparel. Other companies began production of phenolics, plastics very similar to Bakelite. By 1912, the Albany Billiard Ball Company, a company founded by J. W. Hyatt, adopted Bakelite for billiard balls. In 1914, Western Electric began to use phenolic resins for the telephone earpiece. That same year, Kodak cameras used phenolic for end panels.

In 1916, Delco began to use phenolics for molded insulation in automotive electrical systems. By 1918, molded phenolic resins appeared in dozens of automotive parts. During World War I, aircraft and communication systems increasingly used molded phenolics parts. Phenolics are not obsolete. In 1991, the US plastics industry used 165 million pounds of phenolics.

With Bakelite, a new era began in plastics. Previously, plastics were either natural or chemical modifications of natural materials. The production of Bakelite proved that it was possible to do in a laboratory or factory what the lac insects and rubber trees did in nature. In fact, the controlled conditions of a factory allowed the production of purer and more uniform materials than those produced from trees, insects, or horns.

COMMERCIAL SYNTHETIC PLASTICS

Bakelite was the first in a long, continuing stream of new plastics. Pioneers in the development of early commercial synthetic plastics struggled with two basic problems—one theoretical and one practical. The theoretical problem was that they did not have a clear understanding of the chemical and structural nature of plastics. This confusion continued until 1924, when Herman Staudinger claimed that polymers were long linear molecules consisting of many small units held together by chemical bonds. This idea served as the starting point for the development of many plastics. The practical problem involved the purity of the chemicals required for sustained chemical reactions in making plastics. After many failed attempts, chemists understood that the purity requirements far exceeded their expectations. Consequently, polymer-grade chemicals became synonymous with the highest purity commercially available.

During the 1930s, the solutions to these two problems became rather clear. The needs of World War II also contributed to a surge in the development of new plastics. Table 1-1 provides a partial chronology of the growth of plastics. The period between 1935 and 1945 stands out because many of the materials developed at that time received wide use and continue today as major industrial materials. Details about many of the materials in this list are available in Appendixes E and F.

Table 1-1. Chronology of Plastics

Date	Material	Example
1868	Cellulose nitrate	Eyeglass frames
1909	Phenol-formaldehyde	Telephone handset
1909	Cold molded	Knobs and handles
1919	Casein	Knitting needles
1926	Alkyd	Electrical bases
1926	Analine-formaldehyde	Terminal boards
1927	Cellulose acetate	Toothbrushes, packaging
1927	Polyvinyl chloride	Raincoats
1929	Urea-formaldehyde	Lighting fixtures
1935	Ethyl cellulose	Flashlight cases
1936	Acrylic	Brush backs, displays
1936	Polyvinyl acetate	Flashbulb lining
1938	Cellulose acetate butyrate	Irrigation pipe
1938	Polystyrene or styrene	Kitchen housewares
1938	Nylon (polyamide)	Gears
1938	Polyvinyl acetal	Safety glass interlayer
1939	Polyvinylidene chloride	Auto seat covers
1939	Melamine formaldehyde	Tableware
1942	Polyester	Boat hulls
1942	Polyethylene	Squeezable bottles
1943	Fluorocarbon	Industrial gaskets
1943	Silicone	Motor insulation
1945	Cellulose propionate	Automatic pens and pencils
1947	Epoxy	Tools and jigs
1948	Acrylonitrile-butadiene styrene	Luggage
1949	Allylic	Electrical connectors

1954	Polyurethane or urethane	Foam cushions
1956	Acetal	Automotive parts
1957	Polypropylene	Safety helmets
1957	Polycarbonate	Appliance parts
1959	Chlorinated polyether	Valves and fittings
1962	Phenoxy	Bottles
1962	Polyallomer	Typewriter cases
1964	Ionomer	Skin packages
1964	Polyphenylene oxide	Battery cases
1964	Polyimde	Bearings
1964	Ethylene-vinyl acetate	Heavy-gauge flexible sheeting
1965	Parylene	Insulating coatings
1965	Polysulfone	Electrical and electronic parts
1965	Polymethylpentene	Food bags
1970	Poly(amide-imide)	Films
1970	Thermoplastic polyester	Electrical and electronic parts
1972	Thermoplastic polyimides	Valve seats
1972	Perfluoroalkoxy	Coatings
1972	Polyaryl ether	Recreation helmets
1973	Polyethersulfone	Oven windows
1974	Aromatic polyesters	Circuit boards
1974	Polybutylene	Pipes
1975	Nitrile barrier resins	Packaging
1976	Polyphenylsulfone	Aerospace components
1978	Bismaleimide	Circuit boards
1982	Polyetherimide	Ovenable containers
1983	Polyetheretherketone	Wire coating
1983	Interpenetrating Networks (IPN)	Shower stalls
1983	Polyarylsulfone	Lamp housings
1984	Polyimidesulfone	Convey links
1985	Polyketone	Automotive engine parts
1985	Polyether sulfonamide	Cams
1985	Liquid crystal polymers	Electronic Components
1986	Polycarbonate/ABS blends	Automotive parts
1987	High-purity polyacetylene	Electrical conductors
1991	Metallocene polyethylene	Packaging films
1992	Linear low density polyethylene	Packaging films
1992	Syndiotactic polystyrene	Thinwall electrical parts
1992	Syndiotactic polypropylene	Automotive interior parts
1992	Cyclic olefin copolymers	Appliance parts
1998	Ethylene-styrene copolymers	Toys
1998	Nanocomposites	Truck body parts
2001	Transparent polyester/polycarbonate alloys	Sunglasses
2005	Polylactic Acid	Compostable packaging

SUMMARY

For centuries, natural plastics combined light weight, strength, water resistance, translucency, and moldability. Their potential was obvious, but the materials were difficult to gather, or they were only available in limited volumes or sizes. All over the world, people tried to improve natural plastics or find substitutes.

The manufacture of modified natural plastics converted natural raw materials, such as cotton linters or gum rubber, into new and better forms. Celluloid had many qualities that surpassed horn. However, modified materials still relied on natural sources for their prime ingredient. Not until the development of Bakelite was it possible to create a material that rivaled nature in a factory. Bakelite opened the door to the development of a host of synthetic polymers tailored to meet specific requirements.

The search for improved materials continues today. Many modern fibers are a result of attempts to create artificial silk. Composite materials are now taking over applications previously reserved for metals. The possibilities for new substitutes seem endless. Leo Baekeland saw boundless potential in phenolic plastics and used the infinity symbol to represent its uses. That symbol applies today to the unlimited future facing those who strive to find and use new polymers.

RELATED INTERNET SITES

- **www.americanchemistry.com.** The American Chemistry Council's home page lists "Products and Technology" as a main selection. Under that section, choosing "Plastics" leads to a list containing "Education and Resources." Within "Education and Resources," there are a number of documents, including "Plastics 101" and "Hands on Plastics."

- **http://museo.cannon.com.** The Sandretto Plastics Museum in Italy offers a guided tour of the museum, which is dedicated to plastics products. The site also includes an extensive section on the history of plastics.

VOCABULARY

The following vocabulary words are found in this chapter. Use the glossary in Appendix A to look up the definitions of any of these words you do not understand as they apply to plastics.

Bakelite
casein
celluloid
collodion
ebonite
Galalith
gum rubber
gutta percha
lac
natural plastics
natural rubber
nitrocellulose
Parkesine
phenolics
pyroxylin
shellac
vulcanize

QUESTIONS

1-1. Why was machine manufacture of horn combs often unsuccessful?

1-2. Why weren't daguerreotype cases made from horn or horn powders?

1-3. Casein comes from _____.

1-4. Moderately nitrated nitrocellulose is called _____.

1-5. What is the difference between collodion and moderately nitrated nitrocellulose?

1-6. What is the difference between Parkesine and celluloid?

1-7. Bakelite comes from a chemical reaction between _____ and _____.

ACTIVITIES

1-1. Write a report tracing the development of one piece of sporting equipment that has a long history. Examples include golf clubs, tennis rackets, tennis balls, snow skis, tennis shoes, billiard balls, and fishing rods. Try to gather information on the manufacturing processes involved.

As an example of the changes in sports equipment, the following section begins to treat golf balls:

Golf balls began as ovals or spheres of wood, ivory, or iron. Featheries, which were leather pouches tightly stuffed with feathers, replaced the solid balls. Featheries could not withstand moisture and prevented any play on wet grass or in rain.

Between 1846 and 1848, balls of solid gutta percha began to replace featheries. They were fine in the rain but fractured on cold days. However, the pieces could be remolded into a usable ball. Figure 1-9 shows a mold for balls of gutta percha.

In about 1899, a new ball made gutta percha balls obsolete. It consisted of stretched elastic thread wound into a sphere and covered with gutta percha or *balata*, a natural rubber that could be vulcanized. These balls required high-quality elastic thread and the

Figure 1-9. This mold shaped gutta percha into golf balls.

Solid horn

Note: *This activity does not reflect current practices in the plastics industry. It does demonstrate the qualities of molded horn.*

a. Acquire cow horns from your local pet store. If they are fresh, boil them for about 30 minutes to prevent decay. Saw the horns and remove the tissue mass. Scrape the membrane from the inside of the horn pieces. When finished, the half horn will appear as shown in Figure 1-10.

b. Cut a small piece, about 1 inch square. Measure length, width, and thickness. Soak in boiling water for about 15 minutes. Place between polished plates and squeeze in a platen press. If heated, keep heat about 250°F (121°C). Keep pressed and let cool. Examine the resulting horn piece. If the original piece appears off-white in color, the flattened and thinned piece should be quite transparent. Figure 1-11 shows some rather transparent pieces of flattened horn. The relative smoothness of the press plates will affect the transparency of the horn.

c. Measure the flattened horn.

d. Resoak in boiling water.

e. Remeasure. Did the horn return to its original dimensions?

f. Put food coloring in the water to see how readily the horn takes dye.

g. Examine the eggcup spoons made from horn, available at some kitchen utensil retailers. Make a wooden form for a similar spoon. You may have seen a shoehorn like the one shown in Figure 1-12.

equipment to tightly and uniformly wind the balls. The new balls were much more elastic and could be hit further.

In 1966, the balls of wound elastic thread became outdated when a ball molded of a solid synthetic type of rubber came to the market. It was much tougher than the previous balls and depended on new materials that were just being developed by chemical companies. Since then, new outer coverings have been developed that are almost impossible to cut with a golf club.*

1-2. Trace the development of a plastics manufacturing company that has a long history. Companies like DuPont, Celanese, and Monsanto have ties to companies that used or manufactured celluloid. Companies that had factories in Leominster, Massachusetts, often have ties to horn comb making. For example, the Foster Grant company, famous for its sunglasses, began operations in Leominster to make plastic combs. Many of its early employees had likely been horn workers.

1-3. Investigate natural plastics.

Equipment. Heated platen glass or unheated press, saws, polished plates, or stainless steel mirror stock.

CAUTION

Perform activities in a laboratory environment under supervision.

Figure 1-10. This half horn is ready for cutting and flattening.

Figure 1-11. This photograph demonstrates the transparent qualities of horn.

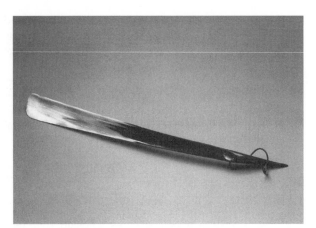

Figure 1-12. Molding altered the shape and surface finish of this shoehorn.

Powdered horn or hoof. Cow hooves are available in pet stores for dog chews.

  a. Make powder by filing with a coarse file.

  b. Compact with a plunger in a cylinder, as seen in Figure 1-13. The rectangular piece of material is a rubber bumper.

  c. Explore the surface detail by pressing it against a coin. See how well the detail transfers.

  d. Experiment with colors.

Shellac. It is available in stick form for furniture scratch repair or as flake from fine woodworking supply stores (see Figure 1-14).

  a. Do simple moldings using cylinder, plunger, and a coin. Figure 1-15 shows a shellac molding using a quarter as the impression. Note the reproduction of details. Shellac will flow at a temperature just above 212°F (100°C). Without a device to apply pressure to the cylinder, the shellac will raise the cylinder and flash excessively. The rubber "bumper" keeps pressure on the cylinder during compression.

Figure 1-14. Shellac in flake form is rather rare, but it is still available.

Figure 1-15. This shellac molding, made with a quarter as the impression piece, indicates the fine detail possible with molded shellac.

Figure 1-13. This type of simple cylinder and plunger can mold shellac and horn powders.

  b. Compare powdered hoof moldings to shellac moldings.

  c. Investigate the effect of fibers.
    (1) Create a sandwich of paper towels and shellac.
    (2) Press the sandwich in a heated platen press.

(3) Does shellac flow through paper? How much does paper improve the strength of shellac? Figure 1-16 shows a thin laminate, containing two layers of paper toweling and a small amount of shellac.

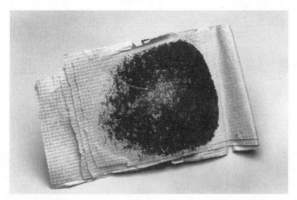

Figure 1-16. This thin laminate is significantly more flexible than a similar sheet of unreinforced shellac.

(4) Create a molding compound using powdered shellac and glass or cotton fibers. How greatly do the fibers improve the physical properties of the moldings?

Casein. Casein, a natural plastic made from milk curds, was molded into various shapes in the 1800s. Because the finished objects would dissolve when wet, casein did not gain industrial importance. To make casein:

a. Slowly warm ½ cup of heavy cream (or milk). Do not boil.

b. When the cream begins simmering, stir in a few spoonfuls of vinegar or lemon juice.

c. Continue to slowly add vinegar until it begins to gel.

d. Cool the mixture and remove the "blob."

e. Wash the "blob" with water, revealing little plastic curds.

f. Mold the material into a shape and let it solidify for a few hours.

CURRENT STATUS OF THE PLASTICS INDUSTRY

CHAPTER 2

INTRODUCTION

Chapter 1 used the words *polymers, rubber,* and *plastics* without providing thorough definitions. Polymers are natural or synthetic organic compounds. Natural polymers include horn, shellac, gutta percha, and gum rubber. Synthetic polymers appear in thousands of plastics products, clothing, automobile parts, finishes, and cosmetics. Whether natural or synthetic, polymers have chemical structures characterized by repeating small units called *mers*. In order for a compound to be a polymer, it should have at least 100 mers. Many polymers found in plastics products have 600 to 1000 mers.

The word plastics comes from the Greek word *plastikos*, which means "to form or fit for molding." A more definitive explanation comes from The Society of the Plastics Industry. It identifies plastics as follows:

> Any one of a large and varied group of materials
> consisting wholly or in part of combinations of carbon
> with oxygen, nitrogen, hydrogen, and other organic or
> inorganic elements which, while solid in the finished
> state, at some stage in its manufacture is made liquid,
> and thus capable of being formed into various shapes,
> most usually through the application, either singly or
> together, of heat and pressure.

In this book, the word *plastics* will always end with an *s* when it refers to a material. The word *plastic*, without an *s*, will be considered an adjective meaning formable. Because plastics are closely related to resins, the two are often confused. Resins are gumlike solid or semisolid substances used in making such products as paints, varnishes, and plastics. A resin is not a plastics unless it has become a "solid in the finished state."

The English chemist Joseph Priestley coined the word rubber after he noticed that a piece of natural latex was good to rub out pencil marks. Natural rubber is one material in a group called elastomers. Elastomers are natural or synthetic polymeric materials that can be stretched to at least 200% of their original length and, at room temperature, return quickly to approximately their original length.

Although elastomers and plastics have been considered as separate categories of materials, the distinction between them has eroded significantly. In the early development of plastics and rubber materials, the plastics tended to be stiff, whereas the rubbers tended to be flexible. Now many plastics exhibit characteristics traditionally available only in rubber. Although rubbers still have unique characteristics— especially the ability to retract rapidly—the categories now partially overlap.

In recent years, a family of materials called thermoplastic elastomers (TPEs) have partially bridged the gap between traditional rubbers and plastics. Within the TPEs, subsets include thermoplastic elastomers based on urethanes, polyesters, styrenics, and olefins. By far the most dominant are the thermoplastic olefin elastomers (TPOs). They have replaced many grades of rubber, particularly in automotive parts. The TPEs have supplanted many traditional rubber products because of their ease of processing (see Figure 2-1).

Plastics and elastomer products do not contain 100% polymer. They generally consist of one or more polymers plus various additives (see Chapter 7 for discussion of additives and their effects). The relationship between these basic terms appears in Figure 2-2. Note that polymers is the umbrella category. With the presence of additives, some polymers become plastics or elastomers.

Figure 2-1. TPO bumper fascia must withstand sunlight, heat, and impact from stones.

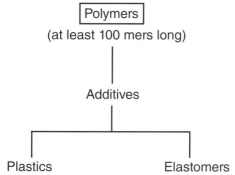

Figure 2-2. Plastics and elastomers contain additives to make them reliable in a wide variety of environments. Even natural-color plastics contain additives.

This chapter presents the current status of the plastics industry, with emphasis on the United States. The content outline is presented here:

I. **Major plastics materials**

II. **Recycling of plastics**
   A. Bottle deposit laws and their effect
   B. Curbside recycling
   C. Recycling of postconsumer recycled HDPE
   D. Automotive recycling
   E. Chemical recycling
   F. Recycling in Germany

III. **Disposal by incineration or degradation**
   A. History of incineration in the United States
   B. Advantages of incineration
   C. Disadvantages of incineration
   D. Degradable plastics

IV. **Organizations in the plastics industry**
   A. Publications for the plastics industry
   B. Trade newspapers

MAJOR PLASTICS MATERIALS

The plastics industry plays a major role in the economy of the United States. In 2010, the US plastics industry provided employment to about 900,000 people. This was lower than the 1.1 million employed in 2005, when the industry ranked as the third largest manufacturing industry in the country.

Although the production of plastics in the United States grew steadily in the 1980s and 1990s, since 2000 the growth has been less steady. A number of factors contribute to that development. The destruction of the World Trade Center on September 11, 2001; the wars in Afghanistan and Iraq; problems with some mortgage financing; and growing global demand for crude oil have all influenced the plastics industry. The financial crisis that started in 2008 led to a steep drop in plastics production. As seen in Figure 2-3, the US production is now growing slowly after the lowest point in 2009. Although the financial crisis hurt the plastics industry, much

US Plastics Production

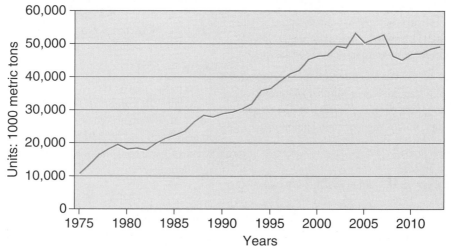

Figure 2-3. US production of plastics is recovering from a sharp drop in 2008 and 2009.

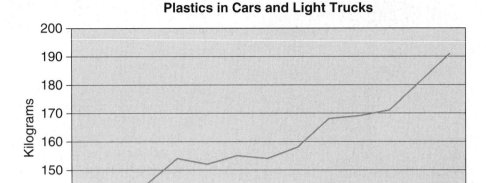

Figure 2-4. Plastics steady growth in cars and light trucks.

of the decrease was not inherent to the industry, but derived from national and international markets. The use of plastics in cars and light trucks continued to increase slowly during the financial crisis, as shown in Figure 2-4. The projections for 2020 have been based on growth of plastics in the automotive industry.

The commodity thermoplastics, which includes six plastics, accounted for 57% of the total production:

Polypropylene (PP)	7065
High-density polyethylene (HDPE)	6458
Polyvinyl chloride (PVC)	4024
Linear low-density polyethylene (LLDPE)	4586
Low-density polyethylene (LDPE)	2473
Polystyrene (PS)	2093
Total	26,699
Units: 1000 metric tons	

Figure 2-5 shows the percentages of the total production by material type.

Figures 2-6 through 2-11 provide graphic representation of the expected utilization of six thermoplastics in the year 2012.

The overall growth of plastics sales reflects the ability of plastics products to fulfill an increasing number of consumer demands. The increasing utilization of plastics has also caused concern for the role of plastics in environmental pollution.

RECYCLING OF PLASTICS

The first Earth Day, which occurred in 1970, signaled the development of a new level of awareness and concern about the environment. During the 1970s, a number of antilitter campaigns sprang up. In 1976, the federal government passed

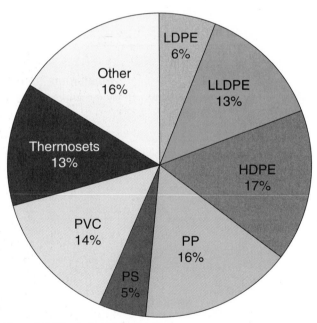

Figure 2-5. US production by types of plastics.

the **Resource Conservation and Recovery Act (RCRA)**. It promoted reuse, reduction, incineration, and recycling of materials. The combined effects of public concern and legislation created major changes in two arenas: hazardous waste management and recycling of nonhazardous materials. Treatment of hazardous materials in the plastics industry appears in Chapter 4.

Recycling is a term generally reserved for postconsumer waste materials. In contrast, reuse or reprocessing usually handles waste materials generated during manufacturing. Plastics industries have for decades reused the materials in defective parts, trimmings, and other manufacturing scrap. This usage

Projected PP Consumption 2020

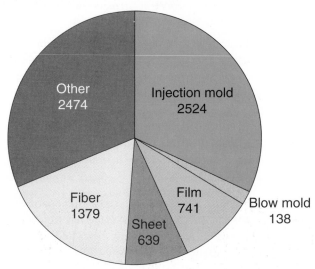

Figure 2-6. Expected polypropylene (PP) usage for the year 2020.

Projected HDPE Consumption 2020

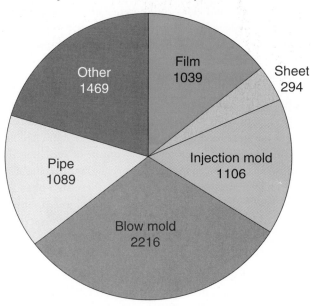

Figure 2-7. Expected HDPE usage for the year 2020.

Projected PVC Consumption 2020

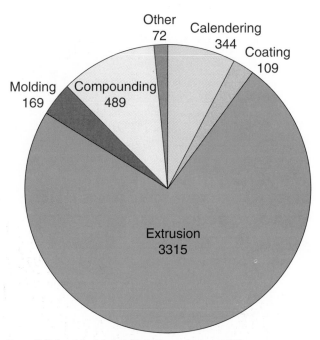

Figure 2-8. Expected polyvinyl chloride (PVC) usage for the year 2020.

Projected LLDPE Consumption 2020

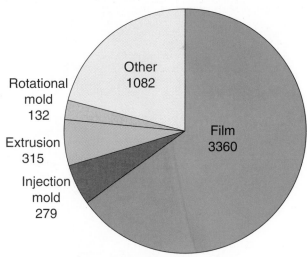

Figure 2-9. Expected LLDPE usage for the year 2020.

ranges from very small-scale reprocessing in small companies to huge programs that generate thousands of tons of reprocessed materials. Systematic recycling of postconsumer plastics occurs through three main channels: bottle returns for deposits, drop-off stations, and curbside pickup programs.

Bottle Deposit Laws and Their Effect

Efforts to encourage or force recycling of bottles have involved state laws, federal proposals, and state mandates. The recycling of PET (polyethylene terephthalate) containers reflects the effectiveness of these efforts.

State legislation. During the 1970s, five states enacted bottle deposit laws. In chronological sequence, they were Oregon, Vermont, Maine, Michigan, and Iowa. In the 1980s, five more states required deposits: Connecticut, Delaware, Massachusetts, New York, and California. The first deposit law took effect in 1972 in Oregon, and the latest occurred in 1987

Projected LDPE Consumption 2020

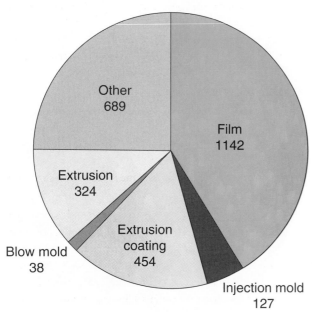

Figure 2-10. Expected LDPE usage for the year 2020.

Projected PS Consumption 2020

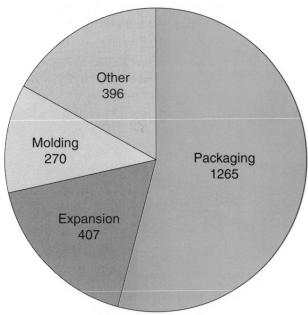

Figure 2-11. Expected polystyrene usage for the year 2020.

in California. The deposit law states require 5 cents deposit per bottle or can, except Michigan, which requires 10 cents.

Since 1987, bottle deposit legislation has been introduced annually in about 25 states, but no additional states have enacted deposit laws. This is explained by the powerful opposition of major soft drink manufacturers such as Coca-Cola® and Pepsi-Cola® companies. They view bottle deposits as an unfair

tax and a hardship for beverage retailers. In 2015, 11 states have bottle deposit laws, and the other states remain without deposit laws.

Because Michigan was the only state to charge more than cents, it has held a rather unique position in bottle recycling efforts. The Michigan Department of Natural Resources indicated that the deposit law caused a 90% reduction in litter along highways and parks. In addition, the deposit law led to the annual recovery of about 18,000 metric tons of plastics—almost entirely PET. In 2011, Michigan recovered about 97% of the plastics bottles covered by this law. In contrast, states with deposits of five cents recover about 85%. States with no bottle deposit law recover about 20% of the containers.

Federal legislation. Concurrent with the initiation of state laws, federal legislators and officials investigated deposit laws. In 1976, a Federal Energy Administration study recommended a nationwide deposit law. In 1977, the General Accounting Office favored deposits. In 1978, the Office of Technological Assessment issued a report favoring deposits. In 1981, Senate Bill 709 for deposits received hearings—but no action. In 1983, Senate Bill 1247 and House of Representatives Bill 2960 proposed a 5-cent deposit on carbonated beverages. No action was taken. In 2007, H.R. 4238, Bottle Recycling Climate Act, was introduced but did not pass. Consequently, bottle deposit laws have remained at the state level.

State recycled content mandates for bottles. Because of pressure from the federal government and concern at the state level, by the end of 1994, 40 states had established legislated goals for litter control or recycling. In an effort to force more thorough recycling, some states also passed recycling content mandate laws. These laws often replaced laws enacted in the early 1990s that subjected various plastics to bans. For example, some states placed polystyrene foam packaging under a ban. In many cases, these laws were not enforced.

A differing legislative approach was to mandate recycling rates or percentages of postconsumer recycled (PCR) content used in "new" containers. These laws usually specify the percentage of PCR of various types of containers. Starting in 1994, Florida sought a rate of 25% PCR in bottles and jars and planned to require wholesale distributors to pay an advanced deposit fee if containers did not contain this percentage. Oregon had a similar rule, expecting 25% PCR in rigid containers.

These regulations did not lead to compliance. The history of California reflects some of the problems with such mandates. In California, a law offered three alternatives. First, rigid plastic containers were to contain 25% PCR by 1995. Second, containers that could be reused or refilled five times were exempt. Third, the recycling rate of containers had to be at least 45%. If the recycling rate fell below 25%, California was to begin vigorous enforcement.

The beverage industry did not meet these mandates. Therefore, in 1997 the State of California decided not to enforce its laws and to give the industry time to comply voluntarily. One major problem concerned the cost and availability of PCR supplies. In some cases, recycled material became more expensive than new plastics. Some of the PET collected in California did not go into reprocessing in the United States. It was sold to Asia because shipping costs were lower to Asia than to the East Coast of the United States.

However, in July 2001, the state recycling board declared the recycling rate for rigid plastic containers was below 25% in 1999 and 2000. If California chose to enforce its laws, significant changes would be required by the beverage manufacturers.

Several groups have tried to push Coca-Cola and Pepsi-Cola to use more aggressive actions to promote recycling. In April 2001, Coca-Cola announced plans to include 10% recycled content in all its PET bottles by 2005. This announcement was a partial answer to a group of shareholders who supported a resolution asking Coca-Cola to set a rate of 25% recycled content and a target of 80% recycling rate. The problem with this announcement was that in the early 1990s, Coca-Cola had announced its intention to use 25% recycled content in its bottles and later backed away from that pledge. In 2012, a state in Australia instituted a 10-cent bottle deposit program. Coca-Cola has sued to stop the program. An environmental organization, SumOfUs.org, has created a petition campaign to get Coca-Cola to end its opposition to public recycling programs.

PET bottle recycling. In states with bottle deposits, consumers return bottles to the stores, often using reverse vending machines to claim their refund. Stores sort the bottles for pickup by the distributors. The distributors then sell the materials to the recycling market. To make shipping more cost-effective, densification equipment squeezes the bottles into bales, which are sold to recycling companies. The recyclers unbale the containers, chop them into flakes, clean, wash, dry, and in some cases reprocess the materials. Table 2–1 shows the growth of PET recycling. In 2011, the recycling rate reached a new high, in part due to increases in collection in California and a ban on plastic bottles in landfills in North Carolina. Budget concerns caused decreases in collections by discontinuing or curtailing some collection programs. However, these decreases did not match the increases.

Finding uses for the recycled PET has not always been an easy task. For years, the US Food and Drug Administration (FDA) did not allow recycled materials to be used in food-contact applications. Consequently, PET had to be used in nonfood applications. A major use for recycled PET was as fiber. For example, 35 soft drink bottles provide enough material for the fiberfill used in one sleeping bag. Other significant uses are the production of strapping, film/sheet, beverage containers, and polyester fabrics for clothing.

Another possibility was to make multilayer containers, with an inner layer of virgin material, a middle layer of recycled material, and an outer layer of virgin material. Figure 2-12 shows a sketch of this style container. This allows use of recycled material, yet provides control of the food-contact layer and the outer or show surface.

Yet another possibility was to chemically depolymerize the PET, and then use the resulting materials for polymerization into "new" PET. This option has found limited success because it produces PET that is more expensive than newly manufactured PET. According to one estimate in 1994, recycled PET made by depolymerization cost 20 to 30 cents per pound more than virgin material.

In an important decision made in August 1994, the US Food and Drug Administration approved the use of 100% recycled PET for food-contact packaging. This was the first time the FDA approved 100% recycled content in food and beverage packaging. That means that PET soft drink bottles could be reprocessed into new food-use bottles.

To win this approval, a recycling facility in Michigan had to develop new ways of thoroughly cleaning the recycled material. The new process features high-intensity washing, temperatures of about 500°F (260°C), and other cleaning techniques. Since the first approval of bottle-to-bottle recycling, several other companies have developed comparable systems. In 2001, a company in Holland, Ohio, received FDA approval to produce

Table 2-1. Recycling of PET

1982	1989	1993	1999	2006	2011	2013
18	88.5	203.5	336	578	729	817

Units: 1000 metric tons

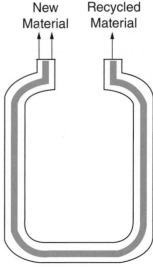

Figure 2-12. An inner layer of recycled material is sandwiched between two layers of prime material.

100% recycled PET bottles from curbside collection. Another company, which owns about 10% of the worldwide PET production capacity, opened a recycling facility in West Virginia in 2001. The company hopes to process more than 10,000 metric tons of PCR PET per year.

In 1993, the recycling rate for all PET packaging was approximately 30%. In states with bottle deposit laws, the containers covered by deposits are returned at a rate of about 95%. Because deposit laws cover 18% of the American population, that means that the one-fifth of the population under deposit laws accounts for about 60% of the total recycled PET bottles. Although these figures might indicate support for bottle deposits, opponents assert that comprehensive waste-management programs are far more effective than forced deposits.

The Association of Postconsumer Plastic Recyclers (APR) is an organization whose members represent more than 90% of the companies and organizations involved in recycling in North America. In 2006, the APR issued a position statement supporting the expansion of bottle deposit laws to include noncarbonated as well as carbonated beverage bottles. That move is partially a reflection of the growth in sales of bottled water and juices. The New York Times estimated that Americans consumed 30 billion single-serving bottles of water in 2007. Because the large beverage companies continue to oppose bottle deposits in any form, the APR realizes that new and expanded bottle deposit laws will be difficult to enact.

The statistics about recycling indicate the magnitude of the problem. The 2005 Report on Postconsumer PET Container Recycling Activity from the National Association for PET Container Resources (NAPCOR) and the APR indicates that in 2005, 5.075 billion pounds of PET were made into bottles in the United States. Of that amount, only 1.17 billion pounds came into recycling facilities. That amounts to a gross recycling rate of 23.1%. Although the total number of pounds of recycled PET has grown every year since 1997, the gross recycling rate declined from 27.1% in 1997 to 23.1% in 2005. Since 2005, the gross recycling rate has grown slightly, and in 2010 reached 29%. The lowest percent was in 2003, with 19.6% as the gross recycling rate. For several years, the largest uses for the recycled PET were for fiber, strapping, and food and beverage bottles.

Curbside Recycling

During the early 1990s, the number of US communities that offered drop-off stations or curbside recycling grew significantly. Since 1997, the growth of these opportunities has been very slow. By the year 2000, over 7000 communities offered curbside recycling to about 52% of the US population. By 2009, 9000 programs existed. Over 80% of the population has access to drop-off stations or curbside recycling. In small communities, the programs tend to be managed by companies or agencies established to fill that need. Many medium to large communities

hire a nationwide solid-waste management company to organize the collection, establish and maintain facilities, and locate buyers for recycled materials. Although curbside recycling is helping in the reclamation of plastics, in 2009, 2.12 million tons of plastics was recycled. However, that represented only 7% of the total production.

Identification coding. Unlike bottle deposit materials, in which the type of plastics was clearly known, it was difficult to distinguish the types of plastics obtained in the curbside recycling programs. In some cases, mistaken identity of a few containers could ruin a large amount of otherwise useful material. For example, PET bottles and PVC bottles are often impossible to distinguish by appearance. If a small amount of PVC is mixed into a large batch of PET, the PET will be ruined.

To avoid this and other similar problems, the Plastic Bottle Institute of the Society of the Plastics Industry established a system for identifying plastics containers in 1988. Each code has a number in the triangular symbol and an abbreviation below it, as shown in Figure 2-13.

The "chasing arrows" symbol has come to imply recycling. Some individuals and organizations feel that the recycling symbol is deceptive, for when they tried to recycle containers marked 3 or 6, they frequently found no one who would accept those materials. They thought the symbol implied recycling, not just the capability of the material to be recycled. However, systematic and widespread recycling of plastics will require

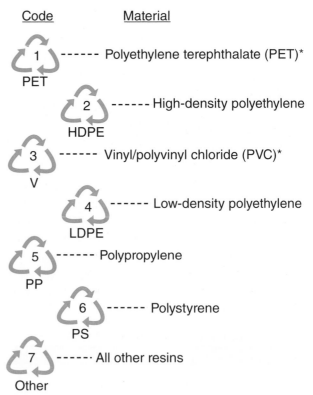

Code	Material
1 PET	------ Polyethylene terephthalate (PET)*
2 HDPE	------ High-density polyethylene
3 V	------ Vinyl/polyvinyl chloride (PVC)*
4 LDPE	------ Low-density polyethylene
5 PP	------ Polypropylene
6 PS	------ Polystyrene
7 Other	------ All other resins

Figure 2-13. These codes were recommended by the Plastic Bottle Institute.

some identification system, with or without the chasing arrows symbol.

Collection. Most recycling programs accept metals, plastic, and paper/cardboard. Some programs require residents to extensively separate the materials. For the plastics, the residents look at the recycling codes and sort plastics into selected groups. During pickup, the materials go into bins or containers to preserve the sorting. Most programs accept plastics number 1 (PET) and 2 (HDPE). Requiring residents to sort the plastics containers has two major drawbacks. First, they often make mistakes in sorting. Second, and often more important, the pickup drivers need more time to place the sorted materials into appropriate containers. Some cities have found that a central sortation line is much more efficient. One very basic manual sorting line requires six employees working only 3.5 hours to process the containers from 5000 houses.

Because of the efficiency of centralized sorting, many large communities accept commingled materials. Residents in many cities throw plastics 1, 2, and 6, plus aluminum and steel cans, newsprint, cardboard, scrap metal, and other paper, into one large container. The importance of compaction affects the decision to accept commingled materials. Uncompacted materials can occupy 10 cubic yards, but if the items are compacted, they need only 3 cubic yards. Compaction in the collection truck can result in savings, primarily in transportation costs. Some trucks can compact paper and cardboard—leaving glass, metal, and plastics loose.

The American Plastics Council promoted "all-bottle" recycling as an alternative to bottle deposits. It conducted studies that indicate an all-bottle approach can boost the amount of containers collected by about 12%. In 2001, only 10% of US communities participated in all-bottle collections. Although most material recovery facilities are not equipped to handle all types of bottles, Waste Management, Inc. (WMI) plans to create a facility near Raleigh, North Carolina, that will allow it to gather and sort plastic containers from all resin types. The biggest concern for the company is whether recycling PVC, LDPE, PS, and PP will be economical.

Figure 2-14 shows a collection truck. This style features two large bins, one for paper and cardboard, the other for commingled containers. Figure 2-15 shows the dumping of the commingled containers.

Because the plastics containers account for the most volume, some trucks are outfitted with special compacting equipment. Figure 2-16 shows a small compaction bin for crushing plastics containers, particularly milk jugs. Without the ability to compact milk jugs, the drivers would have to make more trips to unload.

Sortation. The collection trucks deliver materials to a **materials recovery facility (MRF)**. In 1995, about 750 MRFs

Figure 2-14. The bins in this style truck allow separation of paper from mixed containers.

Figure 2-15. The plastics containers consume far greater volume than steel or glass containers.

Figure 2-16. The compaction bin reduces the volume of collected plastics bottles.

were operating in the United States. Only about 60 had automatic sortation lines, and almost 700 had mostly manual procedures. Between 1995 and 2000, some MRFs closed and diverted materials to larger facilities. In 1999, 480 MRFs were operating in the United States, with a daily throughput of 50,000 metric tons per day. Although there is interest in automatic sortation, many new facilities have manual techniques. Even low-level automatic systems, which can handle only a limited stream of waste plastics, cost about $100,000.

Trucks tip the collected materials into initial sorting equipment. Figure 2-17 shows a conveyor that delivers the commingled containers into the sorting equipment. Many of the manual MRFs can magnetically separate all ferrous containers. Manual sortation can be a very simple process that involves depositing various materials into bins or tubs. Manual sortation can also occur on a picking line. Picking lines feature conveyor belts that move recycled materials past employees who sort them into various categories.

Some recycling facilities accept foamed polystyrene. Foamed PS presents several problems. Because its bulk density is so low, the initial storage point must be rather large. Figure 2-18 shows an 8-foot-tall metal storage container. A full semitrailer load of PS foam for recycling will weigh about 1500 lb. Even bales of

Figure 2-17. To remove the steel from the mixed container stream, a magnetic drum pulls cans out of the main flow.

Courtesy of the City of Ann Arbor, Solid Waste Dept.

Figure 2-18. This enclosed bin protects scrap PS foam from the elements. When full, a truck delivers the foam to an MRF for baling.

Figure 2-19. This tightly packed bale of PS foam came from a vertical-style baler.

foamed PS are very light. Figure 2-19 shows a bale of PS foam. These bales weigh from 80 to 90 lb. In contrast, a similarly sized bale of HDPE weighs about 450 lb.

Recycling of PCR HDPE

Because HDPE containers do not fall under bottle deposit laws, the effectiveness of recycling depends on curbside recycling and drop-off stations. HDPE bottles were recycled at a rate of about 38% in the year 2000. By 1999, HDPE recycling reached 346,000 metric tons. Table 2-2 shows the growth of HDPE recycling.

The recovery rate on natural HDPE bottles (predominantly 1-gallon and 1-half-gallon milk jugs) was slightly less than 25% in 1993. In contrast, the rate for all HDPE packaging was about 10%. The total packaging sales in 1993 was 1,929,000 metric tons. The total sales of HDPE for the same year was 4,820,000 metric tons, which indicated that packaging of the type that appears in the postconsumer waste stream is only 25% of the total sales. The other products eventually find their way to landfills or incinerators.

In 1995, the demand for recycled materials was up, and the recycling companies were making profits. The recycled HDPE, number 2, had a demand of 500 million pounds in 1994, which was about one-third more than what came through recycle streams. Because of this high demand, a number of major manufacturers began to market materials containing recycled postconsumer plastics. Among these companies were Dow, Eastman Chemical, and Ticona.

Table 2-2. Recycling of HDPE

	1982	1989	1993	1999	2005	2011
HDPE	–	59	216.4	346.0	449	443
Units: 1000 metric tons

Figure 2-20. When the bin is full, this pile of mixed-color HDPE will go to the baler.

Figure 2-21. The conveyor lifts the containers to the top of the baling chamber.

Figure 2-22. This photo shows a bale of PET exiting the baling machine.

Although only a limited number of MRFs handle polystyrene foams, almost all accept PET, natural HDPE, and mixed colored HDPE. Picking lines and automatic sorting equipment separate natural from colored HDPE. When the volume of either type is great enough, the stored containers will go to a baling machine. Figure 2-20 shows a view of a large bin of mixed color HDPE.

Balers are of two major types, horizontal and vertical. The distinction refers to the direction of movement of the ram, which compacts the containers. Vertical balers may be the simplest type that requires hand loading. Because vertical balers force the containers into a closed rectangular space, the bales so produced are often very tight.

Horizontal balers raise the containers up a long conveyor and drop them into a chamber. Figure 2-21 shows the feed conveyor of a horizontal baler. The containers fall into a chamber, and a ram presses them into a bale. Wire-tying equipment finishes the bales. In contrast to many vertical balers, which often make tight bales, some horizontal balers produce bales which are not tightly packed. This occurs because the ram on some horizontal balers pushes against the resistance provided by previously generated bales. Figure 2-22 shows bales exiting the machine. The difficulty with loose bales is that they may break

up during handling with fork lifts. Figure 2-23 shows two bales of HDPE on a fork lift.

Because most MRFs do not have equipment to reprocess the plastics, they sell the bales to a reprocessing company. Transportation from the MRF to the reprocessing company generally occurs in semitrailers.

Because a reprocessing facility acquires baled HDPE from several sources, the quality differences in various types of baling equipment are apparent. Figure 2-24 shows tightly packed bales. In contrast, Figure 2-25 shows a loose bale—one in danger of breaking up before reaching the reprocessing equipment.

The first step in reprocessing is to unbale the milk jugs and feed them into a chopper. This provides an opportunity for some control on the materials going into the reprocessing system. The flakes coming from the chopper contain irregular shapes and considerable contamination. Figure 2-26 shows unwashed chopped flakes. A blower system then separates the fine pieces out, because they are too light to fall through a controlled updraft. This process is called elutriation. Elutriation

Figure 2-23. A forklift carries bales to a waiting semitrailer.

Figure 2-24. These large, tight bales promote easy handling.

Figure 2-25. This loose bale may break when moved with a forklift.

Figure 2-26. Unwashed flakes contain many types of contamination, including paper, dirt, pebbles, and undesired plastics.

refers to a purification by straining, washing, or decanting. In this context, the purification is done with air. Figure 2-27 shows the types of fines, paper, and dirt removed during the first elutriation process.

Storage bins hold the flake until it enters a washer. Figure 2-28 shows a washer for HDPE flake. After a pre-weighed charge of flakes drops into the washer, a determined amount of water enters, and a vigorous washing cycle begins. Some systems utilize detergents to assist in cleaning. Others rely on the abrasive characteristics of the flakes to scrub each other.

The washer dumps the load of water and flakes out an outlet pipe, as seen in Figure 2-29. The charge of flakes then enters a **flotation tank**. Slow-moving paddles move the flakes, which float in water, through the tank. The paddles also agitate the flakes and cause heavy particles to fall to the bottom of the tank. In the tank, dirt, sand, and most plastics, other than polyethylene and polypropylene, settle out.

The paddles of the flotation tank lift the flakes to an exit chute, which delivers them to a centrifugal dewatering device.

Figure 2-27. This photo reveals dust, lint, paper, and plastic films. The particles were light enough for separation by an elutriation system.

Figure 2-28. This cylindrical machine is a high-intensity washer.

Figure 2-30. This photo shows the blower unit that separates out the fines and lightweight contaminants.

Figure 2-29. The washer dumps a load of scrubbed flakes and water into the flotation tank.

It spins the water out of the flakes and then delivers the flakes to a rapid drying treatment. After drying, a second elutriation system creates a controlled updraft, which separates fines from the washed flakes. Figure 2-30 shows a blower used for this separation. This elutriation yields mostly label films. Figure 2-31 shows the type of fines collected by the second air sorting. Blowers deliver the clean flakes to storage bins, such as the one seen in Figure 2-32. If the storage bins are full, flakes can be stored in **gaylords**, which are large boxes containing approximately 1 cubic yard of material.

The clean flakes (as seen in Figure 2-33) then enter the feel throat of an extruder. The extruder melts the flakes and forces the melted material through a die. Figure 2-34 shows an extruder prepared for this operation. This extruder has a water-wall-type pelletizer, which chops strands of extruded material just after they exit the die. The particles are thrown into a cylindrically shaped wall of water. Figure 2-35 shows this type of pelletizer head. The pellets cool rapidly. They are then dewatered and dumped into a gaylord for shipment to a processing facility, which will make new products.

Figure 2-31. This group of fines consists primarily of films used in labels on HDPE bottles.

Because the HDPE found in milk jugs is blow-molding grade material, it frequently goes into additional blow-molded products.

Automatic sortation. Manual sortation has two serious limitations. It is not effective in distinguishing PVC from PET, and it becomes unworkable when the total volume

Courtesy of Michigan Polymer Reclaim

Figure 2-32. Below the storage bin is a gaylord, which can store flakes if the bin fills up. Notice the stockings hanging around the bin. The air used to deliver the flakes into the bin must escape. The stockings catch any fines in this exhaust air stream.

Courtesy of Michigan Polymer Reclaim

Figure 2-33. These are clean flakes. Notice the one dark piece. It is a portion of a milk bottle top and consists of blue polypropylene. It is a contaminant, but because PP and HDPE have similar densities, it cannot be excluded by a flotation system.

Courtesy of Michigan Polymer Reclaim

Figure 2-34. Clean flakes drop into the feed hopper at the rear of this extruder.

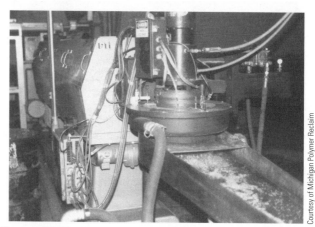

Courtesy of Michigan Polymer Reclaim

Figure 2-35. This photo shows a water-well-type pelletizer. It has the advantage of keeping the die hot and dry, yet cooling hot pieces of plastics in a water bath.

of materials is very high. To develop faster systems, companies have developed various types of automatic sorting devices. Many automatic systems include similar equipment to prepare the containers for identification. They all use bale loaders, unbaling machines, and screens to remove rocks, dirt, and extremely large or small containers. Great diversity occurs in the technical method used to identify various plastics. Identification systems have two major components: a method to separate and convey the containers and a method to identify the plastics.

The separation and transportation usually involve high-speed conveyors. If there is a single detector in the system, the conveyors must singulate the containers and bring them past the detector one at a time. Frequently, an air blast is used to separate one container from the next. In addition to air blasts, vibratory conveyors also aid in singulation. If a single detector fails to recognize the container, or if the container is improperly positioned for optimal recognition, it may be unidentified. To increase identification rates, some systems have multiple detectors with a singulating conveyor system. Some systems are capable of multiple containers and require multiple detectors for this application. Figure 2-36 shows several possible configurations for automatic identification.

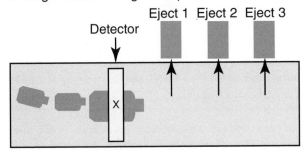

1. Single detector/single sample

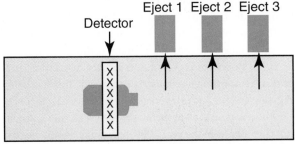

2. Multiple detector/single sample

Figure 2-36. Singulating conveying systems present containers to the detection equipment one at a time.

The general purpose of the detector is to determine the chemical makeup of a container. Once that is complete, computer-based systems must track the location of the known container and then activate the air jet to eject to the appropriate takeoff conveyor. The four major types of detectors are optical, X-ray, single wavelength infrared (IR), and multiple wavelength IR.

Optical systems rely on vision systems that determine the color of a container; X-ray sensors can distinguish PET from PVC by sensing the presence of chlorine atoms in the PVC. Chemical detection can occur rapidly—with some systems requiring less than 20 milliseconds. Single wavelength IR systems can determine opacity and, based on the results, sort containers into clear, translucent, and opaque streams. Multiple wavelength IR systems can determine the chemical constitution of a container by comparing its results to a known standard. Compared to the single wavelength IR systems, the multiple wavelength systems require more time for identification.

Automated sorting systems can be modest, with equipment to recognize only a few materials. A basic system can distinguish three main classes: natural HDPE; PP, PET, and PVC; and mixed-color HDPE. A slightly more powerful system would distinguish PET from PVC.

Some color systems claim the ability to distinguish millions of shades of color. IR systems extend the capabilities further. After identification is complete, the selected containers are blown by air jets onto appropriate conveyors or hoppers. These automated systems can process containers at a rate of two to three containers per second or about 1500 lb per hour. To improve on that rate, multiple lines are needed. However, the costs of such systems can reach $1 million.

Chopped commingled materials. Elimination of all sorting can simplify the systems in MRFs. However, the chopping of commingled plastics results in mixed chopped flakes. Sorting the flakes into appropriate material streams can be done in various ways.

Flotation systems can distinguish materials based on differences in density. Froth flotation separates plastics by differing surface-wetting potentials.

Systems based on an optical sorting technology bring a stream of chopped flakes past detectors. If the detectors indicate the presence of an unwanted flake, air blasts remove it from the stream. The efficiency and speed of these systems is still under development. Currently, one pass through the system can remove about 98% of contaminates; however, to clean the stream to the level of 10 parts per million (ppm), several additional passes are required. Similar systems based on magnetic separation can distinguish PVC from PET.

Automotive Recycling

Approximately 10 million automobiles are discarded every year in the United States. They arrive first at automotive dismantlers. After removing parts useful for resale, the dismantlers hand the remains over to shredders. There are approximately 180 shredders in the United States. After turning a car into pieces, the facilities sort ferrous from nonferrous metals and feed these materials to foundries and steel mills for reprocessing. About 75% of the materials in automobiles are recycled. The shredder residue, sometimes called shredder fluff, contains plastics, glass, fabrics, adhesives, paint, and rubber. The automotive shredder residue (ASR) accounts for 3 to 4 million metric tons per year. Between 20 and 30% of the ASR is plastics materials, most of which goes into landfills.

In an effort to reduce the shredder residue, the major automobile companies have established procedures and guidelines for recycling. To make recycling work throughout the automotive industry, several major automobile companies have created the Vehicle Recycling Partnership (VRP). This partnership is currently attached to the United States Council for Automotive Research (USCAR). As part of the efforts, they established a code for making plastics—SAE code J1344—to aid in identification during disassembly. This code utilizes the ISO (International Organization for Standardization) designations for plastics. (For information on ISO tests, see Chapter 6.)

In addition to efforts aimed at the entire automobile industry, major companies have established internal guidelines for recycling. For example, Ford Motor Company established recycling guidelines in 1993. These guidelines promote the use of recycled materials and encourage the reduction of painted plastics. Molded-in color reduces volatile emissions from painting operations and makes recycling easier. The guidelines also recommend the use of a limited number of plastics and suggest the use of polypropylene (PP), acrylonitrile-butadiene-styrene (ABS), polyethylene (PE), polyamide (PA), polymethyl methacrylate (PMMA), and polylcarbonate (PC). PVC should be used only where separation and recycling techniques are established. In addition, suppliers were urged to recover and recycle materials at the end of the useful life of vehicles. In 2003, Ford started a recycling program called Core Recovery Program. By 2013, the company estimated that it had diverted 120 million pounds, including plastic bumpers, from landfills.

Chemical Recycling

Chemical recycling involves two levels of depolymerization. The original polymerization of some plastics is reversible. Depolymerization yields monomers, which can be used to make new polymers. Other plastics do not depolymerize to immediately useful monomers.

The type of depolymerization that produces useful monomers is called hydrolysis. It is beyond the scope of this discussion to thoroughly explain hydrolysis and the chemical reactions it utilizes. It is, however, a technique that is feasible for polyesters, polyamides, and polyurethanes. The chemical techniques vary depending on the type of plastics. For years, companies making PET have used a form of hydrolysis to convert manufacturing waste PET to monomer.

Some plastics can be converted to monomers with thermal depolymerization. This process involves heating the plastics in the absence of oxygen. This process is also labeled pyrolysis. Acrylic, polystyrene, and some grades of acetal yield monomers under these conditions.

Other plastics arose through irreversible reactions and consequently cannot be depolymerized into monomers. They can be depolymerized into useful petrochemical materials by pyrolytic liquefaction. Materials appropriate for this treatment are HDPE, PP, and PVC.

In simple pyrolytic liquefaction, pieces of plastics enter a tube that is heated to about 1000°F (538°C). The plastics melt and decompose into vapors, and the vapors condense into liquids. Liquids are used as feed stocks in petrochemical plants. Some facilities hope to make gasoline from the plastics.

Recycling in Germany

In the United States, some packaging and container manufacturers now face various recycling laws and mandates. Such laws do not currently extend to producers of appliances, automobiles, and electronics. In contrast, Germany has placed full responsibility for recovering, reprocessing, or disposal of packaging on the manufacturers.

To meet this challenge, over 600 manufacturers and distributors combined their forces and created a system called Duales System Deutschland (DSD). DSD was founded in 1990 and functioned as a private company until 1997, when it was converted into a public company with about 600 shareholders. DSD collects, sorts, and arranges for reprocessing of all packaging materials. During 1993, DSD collected 360,000 metric tons of material. In the year 2000, it collected over 5.6 million metric tons of material. The materials recovered include glass, plastics, and metals. Before 2006, stores were required to take back only the containers they sold, but since May 2006, stores are required to take back any and all packages made from the same materials they sell. This has helped make the system more efficient. In 2006, DSD recovered 5.2 million metric tons of materials. Although that volume is about the same as in the year 2000, packaging companies have continuously optimized package designs. As a consequence, the annual consumption of sales packaging has been regularly reduced. In 2012, DSD received 100% of the packaging containers that participate in the system. That meant that none of those containers went to landfills.

DISPOSAL BY INCINERATION OR DEGRADATION

An alterantive to landfills has been to burn waste materials. Burning reduces the volume of the materials and yields considerable heat. Plastics in solid waste are an energy-rich fuel. However, incineration also generates emissions, sometime toxic. Degradable plastics, although still small in total volume, avoid these undesirable emissions.

History of Incineration in the United States

Burning of solid wastes in open dumping areas was common for centuries; however, the concerns for environmental protection closed open dumping and burning areas and replaced them with landfills. Some communities constructed special incinerators to dispose of the solid wastes. By 1960, approximately 30% of municipal solid waste (MSW) was handled in incinerators without any attempt to recover or use the heat. This method grew until the early 1970s. The Clean Air Act, which was passed in 1970, closed about half the incinerators because the retrofills for pollution control were considered too expensive. To make incineration viable, some method to pay for the cost of equipping a facility was required. One possible solution was to use the solid waste as fuel in the generation of electrical energy or steam.

From the late 1970s through 1980, there was a tremendous increase in waste-to-energy (WTE) facilities. By 1990, about 15% of the MSW was incinerated. These waste-to-energy facilities were expensive; a well-equipped facility required at least $50 million to construct. In 1991, there were 168 active incinerators in the United States. In comparison, Japan had about 1900 and Western Europe had over 500 incinerators. Because of strict regulations on incineration, a number of US incinerators closed in the mid-1990s. In 1999, the United States had only 102 combustion facilities with a capacity to burn up to 87,000 metric tons per day. In 2010, the United States had 87 plants for its 300 million people, and Denmark had 29 plants for its 5.5 million.

In Japan, construction began on a unique WTE facility in 2001. The $81 million power plant was designed to generate 74,000 kW and to burn only plastic waste. The plant's capacity was 680 metric tons per day, and it should emit fewer pollutants than most oil-fired power stations.

Advantages of Incineration

One of the advantages of incineration is that it does not require sorting of the solid wastes. The entire collection of paper, plastics, and other materials can go into the incinerator. It provides an 80% to 90% reduction in the volume of the solid waste, turning many cubic yards of material into a few pounds of ash. WTE plants incur an operating cost of about $10 to $20 per ton. To offset the cost, the WTEs have two sources of revenue. They charge city solid-waste organizations to deliver truckloads of compacted MSW. Incinerator facilities receive approximately $50 per ton for accepting the MSW. In addition, the incinerators sell electricity. The cost of electricity from these facilities is comparable to those of other electric generating plants—generally not significantly lower.

For many cities, the alternative to incineration is landfilling. However, landfill costs are high in highly populated regions of the country. In New York and Massachusetts, landfill costs can exceed $50 to $60 per ton. In some areas of California, landfill costs have reached as much as $85 per ton.

Disadvantages of Incineration

When new incinerators are proposed, some citizen groups vehemently oppose them. The grounds for trying to stop new incinerators involve two factors: first, the incinerator ash, and second, the emissions from the incineration process.

Incinerators generate two types of ash: bottom ash and fly ash. Bottom ash comes from the bottom of the incineration chamber and contains noncombustible materials. Mass burn incinerators deliver the full contents of the compacted load of waste into the burning chamber. Bricks, rocks, steel, iron, and glass go into the bottom ash, along with the residues of combustion. The fly ash is material collected from the smokestack gases by pollution control equipment.

Fly ash often contains relatively high concentrations of heavy metals and some hazardous chemicals. In contrast, the bottom ash usually contains less toxic materials. Some incinerator facilities mix the fly ash and bottom ash together.

A point of conflict concerns the danger to citizens from the ashes and the proper disposal method for the ashes. If the ash is considered a toxic material, then it must be landfilled in a special landfill designed for toxic materials. That raises the cost tremendously compared to regular landfills.

The emissions from incinerators often contain various levels of furans, dioxins, arsenic, cadmium, and chromium. These are all very toxic materials, and many people fear potential adverse health effects. Proponents of incineration argue for comparisons between incinerators and power plants that burn pulverized coal. Environmentalists argue that the emissions are potentially carcinogenic and should be banned immediately.

A particular concern for the plastics industry is dioxin. When compounds that contain chlorine, such as PVC and paper whitened by chlorine bleaching, are incinerated at high temperatures, various chlorinated chemicals are produced. When the gases containing these chemicals cool to about 300°C (572°F), dioxin forms. The Environmental Protection Agency (EPA) views dioxin as a probable human carcinogen and a dangerous noncarcinogenic health threat.

An EPA report published in 1995 cites medical waste incinerators as the biggest source of dioxin in the United States. The second largest source is municipal solid-waste incinerators. Although the medical waste incinerators burn small volumes compared to MSW incinerators, the medical waste stream has a high PVC content. In addition, there are many more medical waste incinerators than MSW incineration facilities. In 1994, over 6700 medical waste incinerators were operational.

Two approaches have been proposed: first, the elimination of all chlorinated wastes from incinerators, and second, the improvement of emissions controls for incinerators. Reducing dioxin emissions will require significant investments in pollution-control equipment, perhaps involving at least 60% of existing incinerators. If the new air quality regulations are enforced, about 80% of the existing medical incinerators will probably cease operation. To handle medical waste, a small number of extremely well-controlled incinerators may be able to meet the standards. Hauling the waste to a few incinerators will also add to the cost of handling medical waste.

Another possibility is to deliver medical waste to huge autoclaves, which heat the materials to a temperature that kills biohazards. After autoclaving, the waste materials then go into a traditional landfill.

Degradable Plastics

Considerable controversy surrounds biodegradable plastics. A focus for the controversy was the rings or yokes used to hold six-packs and eight-packs of beer or soft drink cans or bottles together. It was proven by various environmental groups that some animals, particularly sea birds, were getting stuck or trapped in the rings. The response to this was legislation requiring that such rings, yokes, or other devices be degradable.

Raw material suppliers introduced photodegradable and biodegradable grades of material for these applications. The photodegradable materials contain chemicals that are sensitive to sunlight and cause the rings to disintegrate. Biodegradable grades often contain cornstarch or other starches, which are attacked by microorganisms in water or soil. The rings also disintegrate over time. In all cases, the thinner the packaging material, the quicker the physical deterioration.

Currently, 28 states have laws requiring degradable connecting devices. Michigan requires degradable within 360 days; Florida allows only 120 days. Other states do not specify the time lapse permitted.

This did not end the controversy—environmental groups attacked the degradable materials. They argued that bio- or photodegradable plastics can contaminate otherwise useful recycled plastics. They also view the degraded materials as a potential threat to water purity. After sufficient bio- or photodegradation, the rings break down into small pieces. However, these small chips are often rather chemically stable. For example, a grocery sack made of polyethylene with cornstarch as an additive will disintegrate into tiny particles of polyethylene. It will appear to be gone from the roadside, but the tiny pieces are still there. They are likely to remain intact for long periods of time.

Since 2000, a number of companies have begun selling bioplastics. Bioplastics and biopolymers are polymer materials that derive from renewable resources. In contrast to petroleum-based plastics, these materials begin as corn or other plants. Many bioplastics are also biodegradable, but some are not. In a major development, Braskem company, a Brazilian-based chemical company, started making conventional polyethylene using sugar instead of oil or gas as the raw material. In 2010, Braskem started a plant with an annual capacity of 200,00 metric tons. The company claims its polyethylene is a sustainable product,

made with a lower carbon footprint than polyethylene from oil or gas. By 2011, the global production of bioplastics reached 1161 thousand metric tons, divided into nondegradable materials and degradable materials. The leaders among the nondegradable materials were PET and polytheylene. The leader in degradable materials was PLA. Although the usage of biopolymers is limited, they offer a potential to compete with conventional polymers if oil prices continue to rise and production of bioplastics becomes more cost-effective. Polyactic acid (PLA) usage represents about 40% of the total of nonpetroleum-based plastics.

The "green" plastic that has received the greatest attention is polyactic acid, which is based on starch from corn, wheat, and other grains. The chemical name for PLA is polyhydroxyalkanoate, a natural polyester that has been studied for many years. In the past, it was expensive and had very limited properties and processing ranges. The process involves fermentation of cornstarch to yield lactic acid. The lactic acid then goes through a catalytic ring-opening polymerization to generate PLA. Even before it was ready for the market, big agricultural companies were interested in the potential of PLA.

In the United States, both ADM (Archer Daniels Midlands) and Cargill have entered the market. During the 1990s, Metabolix, Inc., developed techniques to make PLA easier to process and at reduced costs. ADM and Metabolix established a joint venture and created a new company, Telles™. Cargill created a new unit, NatureWorks®. Telles™ offers three injection-molding grades of its PLA, named Mirel™. Telles™ also sells two grades of Mvrea, a compostable material intended for grocery bags and other film applications. The unit of Cargill now sells Natureworks polymer and Ingeo® extruded fibers for clothing. In 2012, Natureworks had a capacity of 140,000 tons (US) of Ingeo®. Because of 25% to 30% annual increases in the early 2010s, NatureWorks had instituted efforts to build a new processing plant in Thailand.

PLA continues to have various limitations. It is difficult to foam. It has relatively low thermal stability and needs careful drying to remove moisture. It is also prone to color change due to lower thermal resistance. Some applications also require special equipment.

A German company makes Bio-Flex® mulch, a biodegradable film for agricultural use. It is intended to replace traditional PE agricultural films. After use, it can be plowed under, eliminating the time and cost of removing traditional films. Some automobile companies are considering PLA for components in order to promote the claim of environmental concern. Toyota has established a pilot plant for making PLA. In 2009, Toyota announced that it had developed an alloy of PLA and PP that was intended for use in injection molded parts.

Some of the most visible applications of PLA have been in food and beverage packaging. BIOTA, a company based in Colorado, sells spring water in bottles made from PLA. They claim that the bottle will completely degrade in 80 days in commercial composting. Excellent Packaging and Supply (EPS) is a distributor for SpudWare™, a line of biodegradable cutlery and plates made in China from a bioplastic that is 80% vegetable starch and 20% vegetable oil. Some organic fruits and vegetables are packaged in biodegradable films.

A few organizations now claim the promotion of bioplastics as their mission. European Bioplastics, an association begun in 2006 to represent users of bioplastics and biodegradable polymers (BDP), states its mission is to support the use of renewable raw materials. It also encourages compliance with the provisions of EN 13432, a standard for compostable polymers that is valid in all European Union (EU) member states. International Biodegradable Polymers Association & Working Groups is an organization based in Germany. In addition to promoting biodegradable plastics, the organization has assisted in the development of a system for the certification and labeling of biodegradable materials and compostable plastics products. This system identifies biodegradable and compostable plastics so they will not enter the recycling stream with conventional plastics.

The concern with recycling may hinder the growth of PLA, because a small amount of PLA can contaminate the PET recycling stream, which is the largest of all recycling flows. Most bioplastics are labeled as "7," the recycling code for "other" materials, but beverage bottles such as the ones offered by BIOTA could easily end up in recycling bins.

ORGANIZATIONS IN THE PLASTICS INDUSTRY

The plastics industry supports a large number of organizations, ranging from international, all-encompassing societies to small, highly specific groups. Some of the larger groups will receive attention; many of the smaller organizations will receive mention only.

The Society of the Plastics Industry (SPI)

The SPI was established in 1937 to serve as "the voice of the plastics industry." According to its mission statement, the SPI seeks to "promote the development of the plastics industry and enhance public understanding of its contributions while meeting the needs of society." The Society has a structure composed of three major councils and one special interest group. The councils are Equipment, Material Suppliers, and Processors. The Equipment council concerns machine safety, standards, equipment statistics, and programs/conferences. The Material Suppliers council consists of fluoropolymer manufacturers, organic peroxide producers, vinyl producers, and programs/conferences. The Processors council includes film and bag manufacturers, thermoformers, medical products, emerging issues, statistics/benchmarks, and programs/conferences.

One of its most familiar activities is organizing the National Plastics Exposition (NPE) and International Plastics Exposition every three years. This exposition began in 1946 and now draws visitors from 75 countries. The exposition brings together equipment manufacturers, raw materials, suppliers, laminators, fabricators, mold makers, and manufacturers of plastics.

The SPI maintains a publications service. To locate the various publications, visit the Plastics E-Store on the SPI's website: http://www.plasticsindustry.org.

Society of Plastics Engineers (SPE)

In 1942, 60 salesmen and engineers met near Detroit, Michigan, and began the SPE "to promote scientific and engineering knowledge related to plastics." In 1993, its members numbered over 37,800—divided among 91 sections in 19 countries. Large local sections hold monthly meetings, which provide social as well as technical benefits to members. The SPE maintains its world headquarters in Brookfield, Connecticut, and recently opened a European office in Brussels, Belgium.

The SPE supports and promotes formal education related to plastics. It encourages students to engage in research projects on plastics materials and processes, and provides scholarships for both undergraduate and graduate study. Students can join SPE student chapters, which number 32 throughout the world.

The SPE provides continuing education for its members through a wide range of seminars and conferences. Regional and national technical conferences provide members with access to the most current research in the field. The largest conference is the Annual Technical Conference and Exhibition (ANTEC). In 1943, the SPE held its first Annual Technical Conference, which had 50 exhibitors and fewer than 2000 visitors. In 1993, ANTEC drew about 5000 members and hosted over 650 technical presentations. In 2001, ANTEC was held in Dallas, Texas, and featured over 700 technical papers, 20 seminars, and 130 exhibitors. The 2012 ANTEC was held in Orlando, Florida, in conjunction with the NPE. The advantage for the SPE members was the ability to visit the extensive exhibitions, featuring equipment in 13 international pavilions, and over 400 domestic companies.

Every month, the SPE publishes *Plastics Engineering*, a magazine containing articles on current developments in plastics. In addition to *Plastics Engineering*, the Society also publishes three technical journals that include articles and papers of scientific and scholarly merit: *Journal of Vinyl Technology*, published four times a year; *Polymer Engineering & Science*, published 24 issues per year; and *Polymer Composites*, with six issues per year.

The SPE publishes a catalog annually, containing a variety of books from world-renowned publishers. The books cover two broad areas: one concerning engineering and processing of plastics materials, and the other on polymer science. The catalog also offers the proceedings of regional technical conferences and the ANTEC proceedings and subscriptions to the SPE journals.

The SPE holds the position as the world's largest technical organization for scientific and engineering knowledge relating to plastics. For further information or membership applications, visit the website or write to:

Society of Plastics Engineers
14 Fairfield Drive
Brookfield, CT 06804-0403
www.4spe.org

The Plastics Institute of America (PIA)

The PIA is now a not-for-profit educational and research group. Scholarships are offered to students seeking careers in the plastics industry, and employee training is provided in collaboration with the University of Massachusetts Lowell. To contact PIA, visit the website or write to:

The Plastics Institute of America, Inc.
UMass-Lowell Campus, Wannalancit Center
600 Suffolk Street
CVIP, 2nd Fl South
Lowell, MA 01854
www.plasticsinstitute.org

Society for the Advancement of Material and Process Engineering (SAMPE)

This society has members who are involved in the development of materials and process—chiefly materials and process engineers. It publishes the *SAMPE Journal* bimonthly and the *Journal of Advanced Materials*. To contact SAMPE, visit the website or write to:

SAMPE Headquarters
1161 Park View Drive, Suite 200
Covina, CA 91724-3751
www.sampe.org

Other Organizations

International Association of Plastics Distributors (IAPD)
www.iapd.org

The Association of Postconsumer Plastic Recyclers (APR)
www.plasticsrecycling.org

The National Association for PET Container Resources (NAPCOR)
www.napcor.com

Association of Rotational Molders International (ARM)
www.rotomolding.org

Plastic Lumber Trade Association
www.plasticlumber.org

Plastics Foodservice Packaging Group (PFPG), formerly Polystyrene Packaging Council
www.polystyrene.org

Polyurethane Foam Association
www.pfa.org

Polyurethane Manufacturers Association
www.pmahome.org

Publications for the Plastics Industry

Plastics Technology is published monthly by Bill Communications, Inc. It features articles for plastics processors and a pricing update. Plastics Technology also publishes *PLASPEC*, which provides information on various grades of materials. For more information:

Plastics Technology
www.ptonline.com

Trade Newspapers

Plastics News is a weekly publication from Crain Communications, Inc. It includes short articles on business topics and reports on sales and acquisitions of processing plants, new materials and designs, seminars, and resin prices. For subscription information:

www.plasticsnews.com

RELATED INTERNET SITES

- **www.braskem.com.** Braskem is a major, global chemical company. On the home page, select "Sustainable Chemistry" to access "Green Products." This section contains information on the production of "Green PE" from sugar.

- **www.cereplast.com.** Cereplast, Inc., sells biopolymers. Selecting "Products" on the home page leads to a discussion of compostable plastics, which will completely degrade, and sustainable materials, which replace up to 95% of the petroleum content with materials from renewable sources.

- **www.epa.gov.** The Environmental Protection Agency (EPA) has an extensive site. For a summary of facts about recycling of municipal waste, contact **www.epa.gov/wastes/nonhaz/municipal/**.

- **www.european-bioplastics.org.** This association began in 2006 to represent users of bioplastics and biodegradable polymers (BDP). Its mission is to support the use of renewable raw materials. Selecting "Standards" on the home page leads to choice of "Standardization." That section contains information about EN13432: 2000 Packaging, and EN14995: 2006 Plastics. These two standards define the specifications a bioplastic must meet to be considered compostable.

- **www.excellentpackaging.com.** Excellent Packaging and Supply (EPS) includes a division, BioMass Packaging. This division distributes a number of biobased packaging materials and features, SpudWare™, Biobag®, ECOSAFE®, and Greenware®.

- **www.natureworksllc.com.** NatureWorks LLC sells bioplastics that have gloss and clarity like polystyrene and strength like PET. The company claims that this material has a much lower adverse environmental impact than conventional packaging materials.

VOCABULARY

The following vocabulary words are found in this chapter. Use the glossary in Appendix A to look up the definitions of any of these words you do not understand as they apply to plastics.

automotive shredder residue (ASR)
biodegradable
bottom ash
elastomers
elutriation
flotation tank
fly ash
gaylords
hydrolysis
infrared (IR)
materials recovery facility (MRF)
municipal solid waste (MSW)
photodegradable
picking lines
plastic
plastics
polymers
pyrolysis
resins
Resource Conservation and Recovery Act (RCRA)
rubber
SAE code J1344
thermoplastic elastomer (TPE)
thermoplastic olefin elastomer (TPO)
Vehicle Recycling Partnership (VRP)
waste-to-energy (WTE)

QUESTIONS

2-1. Explain the difference between recycling and reprocessing.

2-2. What is pyrolysis?

2-3. How large is a gaylord?

2-4. How do plastics and elastomers differ?

2-5. How are resins different from plastics?

2-6. Approximately what percentage of the US workforce is directly involved in plastics manufacturing?

2-7. How effective are bottle deposit laws?

2-8. Containers marked with the number 5 contain what type of plastics?

2-9. What does the acronym MRF stand for?

2-10. What are the four major types of detectors used in automatic sorting systems for plastics?

2-11. Explain hydrolysis.

2-12. What conditions are required for the formation of dioxins during incineration?

ACTIVITIES

Recycling HDPE

Schools, businesses, and communities often have containers for the recycling of various bottles. Among bottles, one of the easiest material to recycle is high-density polyethylene.

Introduction. Recycling of postconsumer HDPE is one of the biggest and most successful plastics recycling efforts. The bulk of the recycled HDPE comes through curbside pickup of HDPE bottles and containers—particularly milk jugs. Some companies advertise their use of recycled materials to demonstrate their environmental concern. Figure 2-37 shows a tool for removing bicycle tires from rims.

Procedure

2-1. If a curbside recycling program exists nearby, determine which plastics are acceptable. If the program accepts both PET and PVC, how do they guarantee that the PVC does not contaminate the PET?

2-2. Does the program expect the participants to sort the materials? Is the collection put into a truck with specific compartments, or is the material comingled?

2-3. Is there an MRF nearby? If so, how does it sort the plastics? Does it have any automated sorting equipment?

2-4. Does the program distinguish between natural HDPE and colored HDPE? If yes, how does this sorting occur?

2-5. How are the sorted HDPE containers handled? Is a baler used? Is the material chopped into flakes?

2-6. Who buys the HDPE from the recycling organization? What is the current price for baled material, dirty chopped flakes, clean chopped flakes, and reprocessed pellets?

2-7. How much difference is there between the price of recycled materials and virgin materials?

2-8. Is there a mandate on the percentage of postconsumer recycled materials in packaging? If yes, what percentage must be PCR?

2-9. What technologies are required for a company to turn bales for crushed HDPE containers into clean pellets or flakes?

2-10. Write a report summarizing your findings.

Additional recycling activity. This activity is not feasible without adequate equipment. If processing equipment and choppers are available, try to determine the difference between recycled material and virgin material.

! CAUTION

Use processing equipment only under the guidance of trained supervisors.

Procedure

2-1. Acquire milk jugs from a local recycling company. Generally, such companies are pleased to donate some HDPE milk jugs for students learning about recycling.

2-2. List the decisions needed. For example, how should labels be treated? How should cleaning occur? If a chopper is small and cannot handle complete bottles, how should they be reduced in size? Figure 2-38 shows pieces of milk jugs cut into a size that fits into a small chopper, along with chopped flakes. Should contaminated bottles be rejected? What determines the difference between contaminated and not contaminated? What should be done with screw-on or pop-on closures?

2-3. Determine a procedure to follow, and process enough jugs to yield 3 to 5 lb of chopped flakes. Should the flakes be washed? If so, how will this occur? Figure 2-38 also shows flakes produced from the milk jug pieces.

2-4. Extrude the clean flakes to produce pellets. Is there any odor of sour milk attached to the pellets?

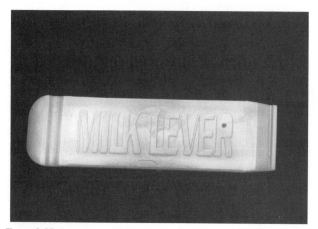

Figure 2-37. The marketing of this bike tire removal tool appeals to the environmental concerns of customers.

2-5. Is there a color difference between virgin pellets and recycled pellets?

2-6. One way to examine the recycled material for contaminants is to make it into film. If blown film equipment is available, try to process the recycled material. If not, squeeze out the thinnest film possible using a heated platen press. Are impurities visible in the recycled materials? A microscope or magnifying glass will help answer the question.

2-7. Stretch, tear, and bend film samples of virgin and recycled materials. Do they appear to possess differing physical properties?

2-8. Write a report summarizing the findings.

Figure 2-38. These milk jug pieces were an appropriate size for a small chopper. The fines were not removed from the flakes.

ELEMENTARY POLYMER CHEMISTRY

CHAPTER 3

INTRODUCTION

Many people have no difficulty recognizing common metals such as copper, aluminum, lead, iron, and steel. They also know the difference between oak, pine, walnut, and cherry. In addition to identifying several metals and woods, many people also understand some physical qualities of these materials. For example, they know that steel is harder and stronger than copper.

In contrast, significant characteristics and names of major types of plastics are often unknown. This chapter discusses several commercial plastics, acquainting the reader with polymer names and chemical structures. To do this, some knowledge of basic chemistry is required. This chapter assumes the reader has an understanding of basic chemistry, elements, the periodic table, and some chemical structures. In this and following chapters, every reference to an atom also includes its chemical symbol. The outline for this chapter follows:

REVIEW OF BASIC CHEMISTRY

Understanding basic chemistry is crucial to learning about plastics. Chemical terms are used to explain the names and properties of polymers. Chemical structures determine the unique characteristics of polymers as well as their limitations. It is convenient to begin this section with a review of molecules and chemical bonds.

Molecules

Molecules occur when two or more atoms combine. The properties of molecules stem from three major factors: the elements involved, the number of atoms joined together, and the type of chemical bonds present. The number of atoms joined determines the size of the molecule, and the bonds determine its strength.

For example, water (H_2O) is made up of hydrogen atoms chemically combined with oxygen. Water consists of two hydrogen atoms and one oxygen atom—three atoms in total. Because only three atoms are involved, water is a very small molecule. The molecular mass of water is 18 because the number of atomic mass units (amus) is 16 for oxygen and 1 for each hydrogen.

Chemical bonding concerns the ways atoms can attach to each other. There are three basic categories of primary chemical bonds: metallic bonds, ionic bonds, and covalent bonds. Of these three, covalent bonding is the most important for plastics. Usually, covalent chemical bonding involves the sharing of electrons between two atoms. The exact mechanism of this sharing is beyond the scope of this chapter. However, it is important to know that covalent bonds have known strengths and lengths that depend on the atoms combined. Covalent bonds can involve varying numbers of electrons, as shown in Figure 3-1. The bond

$$C : C$$
or
$$C—C$$
Single bonds

$$C : : C$$
or
$$C = C$$
Double bonds

$$C \vdots C$$
or
$$C \equiv C$$
Triple bonds

Figure 3-1. Types of covalent bonds.

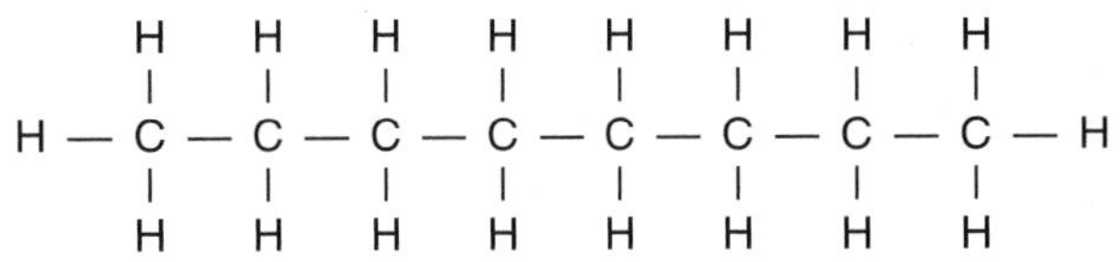

Figure 3-2. Chemical structure of octane.

that consists of the least number of electrons (2) is called a single covalent bond. If more electrons are involved (4 or 6), the bonds are called double or triple covalent bonds, respectively.

Unless otherwise specified, comments about covalent bonding will refer only to single covalent bonds.

HYDROCARBON MOLECULES

Hydrocarbons are materials that consist mainly of carbon and hydrogen. Pure hydrocarbons contain only carbon and hydrogen. When hydrocarbon molecules have only single covalent bonds, they are considered saturated. The word *saturated* implies that the bonding cites are fully "loaded up." In contrast, unsaturated molecules contain some double bonds. Because double bonds are more chemically reactive than single bonds, the saturated molecules tend to be more stable than unsaturated molecules. Table 3-1 lists saturated hydrocarbon molecules.

Table 3-1. Saturated Hydrocarbon Molecules with Melting and Boiling Points

Formula	Name	Melting Point, °C	Boiling Point, °C
CH_4	Methane	−182.5	−161.5
C_2H_6	Ethane	−183.3	−88.6
C_3H_8	Propane	−187.7	−42.1
C_4H_{10}	Butane	−138.4	−0.5
C_5H_{12}	Pentane	−129.7	+36.1
C_6H_{14}	Hexane	−95.3	68.7
C_7H_{16}	Heptane	−90.6	98.4
C_8H_{18}	Octane	−56.8	125.7
C_9H_{20}	Nonane	−53.5	150.8
$C_{10}H_{22}$	Decane	−30	174
$C_{11}H_{24}$	Undecane	−26	196
$C_{12}H_{26}$	Dodecane	−10	216
$C_{15}H_{32}$	Pentadecane	+10	270
$C_{20}H_{42}$	Eicosane	36	345
$C_{30}H_{62}$	Triacontane	66	distilled at reduced pressure to avoid decomposition
$C_{40}H_{82}$	Tetracontane	81	
$C_{50}H_{102}$	Pentacontane	92	
$C_{60}H_{122}$	Hexacontane	99	
$C_{70}H_{142}$	Heptacontane	105	

As indicated in Table 3-1, molecules with one, two, three, or four carbon atoms have boiling points below 0°C (32°F). This means that they exist as gases at room temperature. Molecules from 5 to 10 carbons are rather volatile liquids at room temperature. When the number of carbons increases beyond 20, the materials become solid.

The hydrocarbon molecule with eight carbons is octane, which is familiar because of its use as automotive fuel (Figure 3-2). The molecular mass of octane is 114 amus, and here is its chemical structure:

Note: *Carbon has 12 amus and hydrogen 1.*

$$(8 \times 12) + (18 \times 1) = 114$$

Notice that the hydrogens are connected (with single covalent bonds) only to carbons. There are no bonds between the hydrogens. This means that the structural integrity of the molecule is proved by bonds between the carbons. If we imagine the hydrogen removed, what would remain is a row of eight carbon atoms. This line is called the backbone of the molecule. The strength of the molecule is substantially determined by the strength of the carbon-to-carbon single bonds.

As the number of carbons in the backbone increases, the molecules get longer and longer. Light liquids give way to viscous liquids or oils. Oils become greases. Greases turn to waxes, which are weak solids. Weak solids become flexible solids. Flexible solids turn to rigid solids. Eventually, the molecules get so long that they become rigid and strong at room temperature. Figure 3-3 graphically portrays this sequence of changes.

MACROMOLECULES

The simplest plastic structure is polyethylene, a saturated hydrocarbon with the abbreviation PE. A common PE molecule contains approximately 1000 carbon atoms in its backbone, also called a carbon chain. Molecules of plastics materials are often called macromolecules because of their large size.

Although these molecules are very large, they cannot easily be seen. A common PE molecule is approximately 0.0025 mm stretched out. The thickness of a sheet of typing paper is about 0.076 mm. It would require 30 stretched-out molecules placed end to end to reach from the top side to the bottom side of the paper sheet.

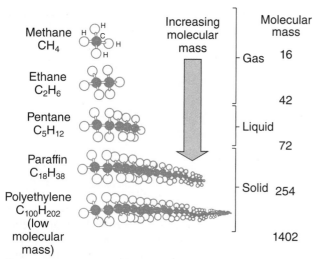

Figure 3-3. Growing chain length (molecular mass) from a molecule of gas to a solid plastics.

Because macromolecules are so long, chemists don't show their extended chemical structures. An abbreviated method is used. The smallest repeating structure is named the **mer**. It is drawn in brackets as follows:

$$\left[\begin{array}{cc} H & H \\ | & | \\ -C\!-\!C- \\ | & | \\ H & H \end{array}\right]_n$$

The little *n* is called the subscript. It stands for the degree of polymerization. **Polymerization** means joining many mers together. (Chapter 8 will discuss various polymerization techniques.) The **degree of polymerization (DP)** represents the number of mers joined in one molecule. If the DP is 500, then 500 of the repeating units are linked together. The molecule would be PE, 1000 carbons long, with a molecular mass of 14,000 amus. The molecular mass is calculated by multiplying the DP (in this case, 500) by the molecular mass of the mer (in this case, 28). If the DP were 9, the material would be paraffin, a hydrocarbon wax with a backbone 18 carbons long.

Frequently, as the degree of polymerization increases, the mechanical properties also increase. This becomes a problem because plastics also become more difficult to process as the degree of polymerization rises. For example, some high-molecular-weight polyethylene grades will not melt in conventional processing equipment.

Carbon Chain Polymers

There are literally thousands of plastics, and polymer chemists are constantly developing new polymers. However, most industrial plastics contain a rather limited number of elements.

A large number of plastics consist of carbon (C), hydrogen (H), and the following atoms:

oxygen (O)   fluorine (F)
nitrogen (N)   sulfur (S)
chlorine (Cl)

Homopolymers. The simplest chemical plastics are called **homopolymers** because they contain only one basic structure. A convenient approach to understanding several plastics is to rewrite the structural formula of PE as follows:

$$\left[\begin{array}{cc} H & H \\ | & | \\ -C\!-\!C- \\ | & | \\ H & X \end{array}\right]_n$$

If a hydrogen (H) goes in the *X* position, the material becomes a PE. If a chlorine (Cl) goes in the *X* position, the material becomes a polyvinyl chloride, PVC. Table 3-2 lists several other plastics that have this form.

In some cases, 2 hydrogen (H) atoms are replaced. Again, it is convenient to rewrite the PE structure as follows:

$$\left[\begin{array}{cc} H & Y \\ | & | \\ -C\!-\!C- \\ | & | \\ H & X \end{array}\right]_n$$

Table 3-2. Plastics Involving Single Substitutions

X Position	Material Name	Abbreviation
H	Polyethylene	PE
Cl	Polyvinyl chloride	PVC
Methyl group	Polypropylene	PP
Benzene ring	Polystyrene	PS
CN	Polyacrylonitrile	PAN
OOCCH$_3$	Polyvinyl acetate	PvaC
OH	Polyvinyl alcohol	PVA
COOCH$_3$	Polymethyl acrylate	PMA
F	Polyvinyl fluoride	PVF

Methyl group:

$$H-\overset{\displaystyle |}{\underset{\displaystyle |}{C}}-H$$
$$H$$

Benzene ring:

Table 3-3. Plastics Involving Two Substitutions

X Position	Y Position	Material Name	Abbrev.
F	F	Polyvinylidene fluoride	PVDF
Cl	Cl	Polyvinyl dichloride	PVDC
COOCH$_3$	CH$_3$	Polymethyl methacrylate	PMMA
CH$_3$	CH$_3$	Polyisobutylene	PIBSA

Table 3-3 includes a number of common plastics that exhibit the double substitution.

If three or more hydrogens are replaced, the new atoms are often fluorine, and the resulting plastics are fluoroplastics. If all four hydrogens are replaced with fluorine, the material becomes a PTFE, polytetrafluoroethylene, which is sold under the trade name Teflon®.

Copolymers. Up to this point, all the plastics mentioned contain only one type of functional group. The structural formulas require only one bracket of the [H H] type.

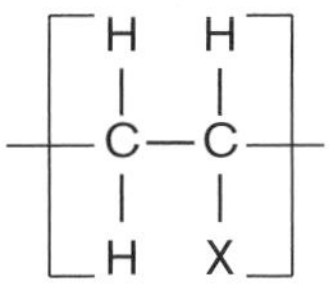

These materials are homopolymers. However, some plastics combine two or more different functional groups—two or more different mers. If only two different mers are involved, the material is called a copolymer.

An example of a copolymer is styrene-acrylonitrile (SAN). Using the information in Table 3-1, we can draw its chemical structure. First, here is the styrene structure:

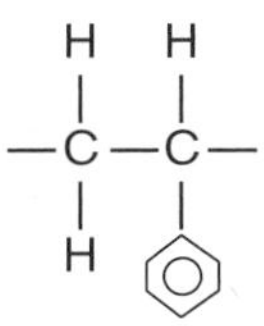

Next, the acrylonitrile structure:

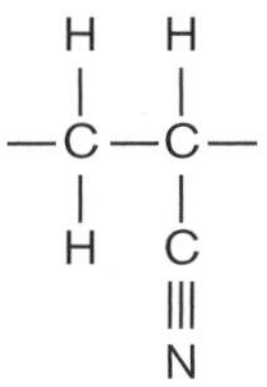

To complete the chemical structure of the copolymer, place the two structures side by side. The resulting structure represents SAN, styrene-acrylonitrile:

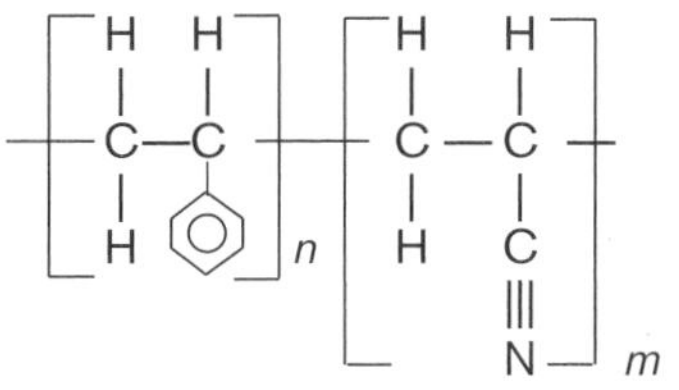

When two different mers are involved, four possible combinations arise: (1) If the mers alternate, ABABABABAB, the material is an alternating copolymer; (2) if the mers join up in a haphazard or random fashion, ABBAAAABAABBBABBAB-BBBAA, the material is a random copolymer; (3) if the mers join up in chunks of like mers, AAAABBBAAAAABBBAAAAABB-BAAAABBB, the material is a block copolymer; and (4) if the backbone remains made up of one mer, and joining side groups of the second mer, the material is called a graft copolymer. Its structure follows:

```
AAAAAAAAAAAAAAAAAAAAAAAAA
B                       B
B                       B
B                       B
```

Printed chemical structures do not clearly indicate whether a copolymer is alternating, random, or block. Printed structures identify the mers but are too short to show larger arrangements. Do not assume a material is an alternating copolymer because the structure shows two mers side by side.

All the information needed for the chemical structure of SAN is provided in Table 3-2. Other copolymers are also specified in this table. For example, ethylene-vinyl acetate (PEVA) and polyethylene methyl acrylate (EMAC) are two more copolymers based on the materials listed in Table 3-2.

Terpolymers. If three separate mers combine to form a material, it is identified as a terpolymer. Terpolymers can also exhibit alternating, random, block, or branch structures. An example of a terpolymer built from the materials in Table 3-1 is acrylic-styrene-acrylonitrile (ASA).

The acrylic structure:

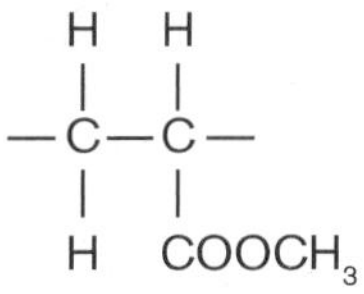

The styrene structure:

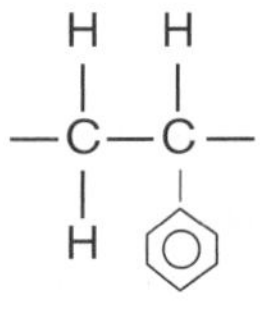

The acrylonitrile structure:

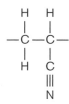

Putting these three pieces together results in the chemical structure for ASA.

Carbon and Other Elements in the Backbone

The types of macromolecules discussed so far all have carbon backbones. This is not true of all common macromolecules. Many include oxygen, nitrogen, sulfur, or benzene rings in the main chain. Most of these materials are homopolymers with decreasing counts of co- and terpolymers. These materials tend to be unique in chemical structure and are not easily categorized. Specific details about the structures of these materials will appear in Chapter 8.

MOLECULAR ORGANIZATION

Molecular organization deals with the arrangement of molecules rather than the details of elements and chemical bonding (which is termed *molecular structure*). This section discusses the major categories of molecular arrangement and the effects of arrangement on selected properties.

Amorphous and Crystalline Polymers

Plastics exhibit two basic types of molecular arrangements: amorphous and crystalline. In amorphous plastics, the molecular chains have no order. They are randomly twisted, kinked, and coiled, as shown in Figure 3-4.

Amorphous plastics can be rather easily identified because they are transparent as long as no fillers or color pigments are present. Department store display cases are often acrylic because of the high clarity of this amorphous polymer.

Some plastics develop crystalline regions. In those regions, the molecules take on a highly ordered structure. The way in which this occurs is beyond the scope of this chapter. However, it is generally accepted that the polymer chains fold back and forth, producing highly ordered crystalline regions, as seen in Figure 3-5.

Plastics do not fully crystallize like metals. Crystalline plastics are more accurately termed *semicrystalline* materials. This means that they consist of crystalline regions surrounded by noncrystalline, amorphous areas (see Figure 3-6).

Although plastics have carbon backbones, some of them crystallize and others remain amorphous. Regularity and flexibility of the polymer chain are major factors that account for this difference.

Because hydrogen (H) is the smallest atom, any atom that replaces it will be larger. Single atom replacements create small "bumps" on a polymer chain. Small groups, such as methyl (1 carbon, 3 hydrogens), or ethyl (2 carbons, 5 hydrogens), represent medium-sized bumps on the chain. Groups containing roughly 10 or more atoms, such as a benzene ring (6 carbons, 6 hydrogens), are large bumps on a molecule.

A group of atoms attached to a backbone is often called a side group, especially if it is chemically different from the main chain. If the chemical structures attached to a backbone are identical to it, the molecule is often considered a

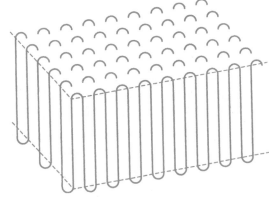

Figure 3-5. A crystalline region.

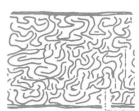

Figure 3-4. Amorphous arrangement.

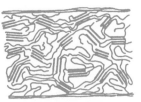

Figure 3-6. Mixture of amorphous and crystalline regions.

branched structure. Chain branching concerns both the size (usually the length) of the branch and the frequency of branching. When the branches are long and/or numerous, they prevent molecules from getting very close. When the branches are small or infrequent, the molecules can "snuggle up" and form denser solids.

Drawings that try to depict the shape and size of molecules come in a variety of types. The simplest graphic model is a line formula, which shows only the bonds—not the atoms. A line formula for PE is shown in Figure 3-7. Notice that the carbon atoms create a zigzag pattern because the bonds are slightly angular. The bond angles between two carbon atoms in a carbon backbone polymer is 109.5°.

When chains fold and coil, they twist or rotate between carbons. Figure 3-8 shows how a coiled PE molecule in a line formula might appear.

Keep in mind that a coiled molecule is a three-dimensional structure. Therefore, it is not well represented by a two-dimensional drawing. Figure 3-9 refers to an attempt to draw a kinked backbone. For clarity, all atoms connected to the carbon backbone were omitted.

A space-filling model is one that tries to show a three-dimensional structure but straightens and flattens out the backbone. Figure 3-10 shows a space-filling drawing of a polyethylene molecule.

If it shows the regularity of "smoothness" of the PE molecule, this image will be successfully conveyed. Because it is so even, it readily crystallizes.

In PVC, a chlorine replaces one hydrogen per mer. A model of this molecule is shown in Figure 3-11. The bumps caused by the chlorine atoms affect the ability of the molecules to crystallize. It is partially crystalline, but less so than PE.

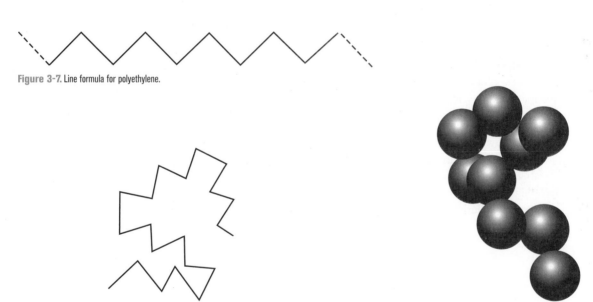

Figure 3-7. Line formula for polyethylene.

Figure 3-8. Line formula of a coiled molecular chain.

Figure 3-9. Graphic representation of a kinked backbone.

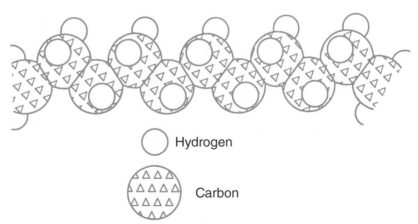

◯ Hydrogen

◯ Carbon

Figure 3-10. Model of a section of a PE molecule.

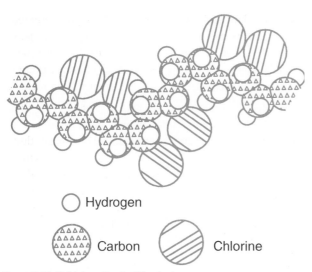

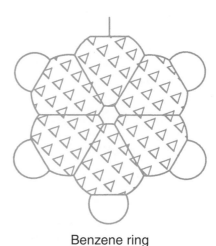

Benzene ring

Figure 3-11. Model of a section of a PVC molecule.

Figure 3-12. Model of a benzene ring.

In PS, a **benzene ring** replaces one hydrogen per mer. The resulting structure is very bumpy. A drawing of a benzene ring appears in Figure 3-12. To reduce confusion, the zigzag line in Figure 3-13, a drawing of a section of a PS molecule, indicates the backbone. The benzene rings prevent crystallization and cause PS to be amorphous throughout.

Optical effects of crystallinity. Amorphous materials are transparent because the haphazard arrangement of the chains does not uniformly disrupt light. In contrast, semicrystalline polymers have highly ordered crystalline regions. These crystalline regions significantly deflect light. This results in semicrystalline materials that are usually translucent or opaque.

This difference is made apparent through a short demonstration. When sufficiently heated, semicrystalline polymers lose their crystalline regions and become completely amorphous. A piece of a plastic milk jug (high-density polyethylene, HDPE) is readily available and contains only PE, with no fibers or colorants.

At room temperature, a piece of HDPE is translucent. If heated on a hot plate or flame, the regions that get hot enough will become transparent.

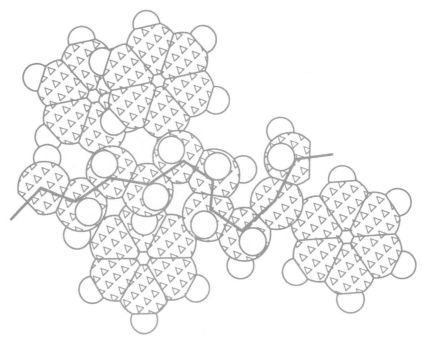

Figure 3-13. Model of a section of a PS molecule.

The crystalline regions are disrupted by heat, causing the material to become amorphous throughout. Upon cooling, some regions become highly ordered, whereas others become disordered. The ordered regions cause light beams to be diffracted rather than to pass through with little disturbance.

Dimensional effects of crystallinity. Besides the optical differences, when melted plastics cool to a solid state, the crystalline materials shrink more than amorphous ones. This is because when crystal regions form, those regions require less volume than they did when amorphous, due to the closeness of the folded chains. The result is greater shrinkage. The simplest, hand-operated injection-molding machine can demonstrate this difference. Using the same mold, inject a natural crystalline polymer such as HDPE. After purging out the HDPE, inject a natural amorphous polymer like PS. The difference in length is readily noticeable, especially if the part is at least 2 inches long.

Melt characteristics of crystallinity. The amount of crystallinity also influences how a plastic melts. Water melts from solid ice to liquid at exactly 0°C [32°F]. In a somewhat similar manner, highly crystalline plastics go through the change from solid to melted over a narrow temperature range. When glass is heated, it begins to soften and becomes pliable. With sufficient heat, it will melt into a viscous liquid. Similarly, amorphous plastics change from a solid to a leathery consistency, then to pliable, and finally to melted over a rather wide range of temperatures. Two temperature notations correspond to these differences. The melt temperature (T_m) usually refers to crystalline materials, whereas the glass transition temperature (T_g) refers to amorphous materials.

It is important to be aware of how these melt characteristics influence processing. Most thermoforming is done with amorphous plastics so that the processing temperature range is large. Thermoforming of highly crystalline materials occurs only in a narrow temperature "window" and is consequently much more difficult.

INTERMOLECULAR FORCES

The size of atoms and side groups affects the crystallinity of plastics. The amount of crystallinity affects optical characteristics and shrinkage. However, there are major differences in some physical properties that are not explained by crystallinity. Two such properties are melting point and tensile strength. Two types of polyamide (nylon) provide a good example. Nylon 6 has a melting point of 220°C (428°F) and a tensile strength of 78 megapascals (MPa) (11,000 pounds per square inch [psi]). In contrast, nylon 12 has a melting point of 175°C (347°F) and a tensile strength of 50 MPa (7100 [psi]).

Intermolecular interactions are a major factor in these differences. **Intermolecular interactions** are attractions between molecules or atoms of different molecules. These forces are much weaker than chemical bonds. However, these intermolecular forces influence the amount of energy needed to disrupt (melt, break, stretch, dissolve) the materials. There are three types of important interactions in plastics: **Van der Waals forces, dipole interactions**, and **hydrogen bonds**. Van der Waals forces occur between all molecules. They contribute only slightly to differences between various polymers. Dipole interactions occur when molecules or portions of molecules exhibit **polarity**, or unbalanced electrical charges. Hydrogen bonding is a special case of dipole interaction and requires a bond between hydrogen and oxygen or hydrogen and nitrogen. Hydrogen bonds are the strongest of the intermolecular forces.

Hydrogen bonds are very important in the production of plastics. To better understand the effect of hydrogen bonds, think about the physical properties of water and methane.

	Molecular Weight	Melting Point	Boiling Point
H_2O	18	0°C	100°C
CH_4	16	−183°C (361°F)	−162°C (−259–F)

These two molecules are of similar size. Yet why are these properties so different? The answer lies in hydrogen bonds. Water is abundant in hydrogen bonds, and methane has no hydrogen bonds. In Figure 3-14, the dotted line represents the attraction between oxygen in one molecule and hydrogen in a neighboring molecule.

Nylon, a plastic that contains nitrogen in the backbone, is an excellent example of hydrogen bonding. Nylon fibers are elastic because the hydrogen bonds act like springs. Fibers spring back after stretching because of the force of hydrogen bonds. If it were not for these bonds, nylon stockings would sag.

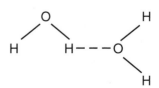

Figure 3-14. Hydrogen bonding in water.

To know whether a plastic has dipoles or hydrogen bonds, it is not necessary to memorize the bonding characteristics of various plastics. A simple set of conditions can signal the presence of dipoles and/or hydrogen bonds.

The following combinations of atoms signal a permanent dipole. Dipoles occur if we find one of the following:

- A carbon-chlorine single bond
- A carbon-fluorine single bond
- A carbon = oxygen double bond

The following combinations signal a hydrogen bond:

- A carbon-OH single bond
- A nitrogen-hydrogen single bond

When hydrogen bonds or permanent dipoles are present, the physical properties change. An explanation of the differences between nylon 6 and nylon 12 can be derived when analyzing the secondary bonding forces.

Nylon 6 has one dipole and one hydrogen bond every six carbons in the main chain. Nylon 12 has one dipole and one hydrogen bond every 12 carbons in the main chain. The additional secondary bonds in nylon 6 cause it to be stronger and have a higher melting point.

MOLECULAR ORIENTATION

Under normal conditions, amorphous macromolecules are not straight. Instead, they are kinked, coiled, twisted, and wrapped around one another. When they are melted, the molecules retain much of their intertwining. When crystalline plastics are melted, the crystalline regions unfold, and the entire structure becomes amorphous. When a melted plastic moves or flows, some of the molecules stretch out. If the flow rate is high, the molecules will stretch almost straight out. This is called orientation.

If they have a chance, when highly oriented amorphous plastics cool, the molecules will return to the coiled, kinked state. This depends on the cooling rate. If cooling is slow, molecules will have the time to rearrange and will coil. If the cooling time is very short, then the stretched out molecules will freeze before coiling.

If oriented semicrystalline plastics cool rapidly, some of the orientation will be locked in. In addition, the cooling rate will also influence the degree of crystallinity. Slow cooling allows a greater degree of crystallinity, whereas rapid cooling inhibits some crystal formation. Because changes in the degree of crystallinity can alter the dimensions of parts, controlling the cooling rate is of great practical importance.

When molecules freeze in a stretched out condition, they don't "like" it. The molecules are stressed and would "like" to get into a less stressed state. The stresses locked in by rapid freezing are called residual stresses. Given the chance, the molecules will coil. In many materials, this will happen very slowly over time but will also occur rapidly if the material gets hot enough. In this case, the material will change shape and generally become unfit for use.

If a plastic part cannot be allowed to change slowly over time or must withstand spikes of high heat, then stress relief must be introduced to the manufacturing process. This is called annealing and usually involves the controlled heating of parts. After the parts change shape, they are sometimes machined or pressed to achieve desired dimensions.

Frequently, relatively thick plastic objects will cool slowly enough that the center portions are not oriented. However, very thin objects may exhibit high orientation throughout. A good example is a thin-wall drinking glass injection-molded with PS. When these containers break, the fracture lines show directionality, as indicated in the following sketch (see Figure 3-15).

These glasses will break easily only in one direction because the molecules are oriented in one direction. The technical term for this is uniaxial orientation.

Uniaxial Orientation

Because highly oriented materials often appear identical to materials with low residual stresses, simple tests can help identify orientation. These tests involve fracture, stretch, and tear characteristics.

Some thin sheets used in thermoforming processes are highly oriented. They will crack easily in one direction, but not perpendicularly. Pieces of these sheets will curl up when heated. The direction of curl indicates the direction of orientation.

Teflon tape, sold in rolls for sealing pipe fittings, shows orientation in its stretch differences. A piece of this tape will stretch easily in the crosswise direction, but not so readily lengthwise.

Figure 3-15. Directional fractures in a molded PS container.

Some labels on 2-liter pop bottles are thin films of plastic stretched around the bottles. This film may tear easily in one direction, but not perpendicularly.

A dramatic demonstration of orientation occurs when test bars of HDPE are pulled at a low rate of less than 1 inch per minute. They will stretch out several hundred percent and become fibrous, allowing them to be separated by hand.

Biaxial Orientation

Biaxial orientation means that a plastic object or sheet contains molecules that are stretched in two directions, usually perpendicular to each other. When heated, biaxially oriented materials shrink in two directions. In contrast, uniaxially oriented materials shrink dramatically in one direction and may even increase in length in the other direction.

Some materials are intentionally biaxially oriented, such as shrink-wrap and window-covering materials that shrink when heated with a hair dryer. A similar product is Shrinky-Dinks®, which is used as a handicraft. These materials are heated, then stretched in two directions prior to rapid cooling. Later, heating produces high shrinkage in two directions.

A familiar product, the 2-liter pop bottle also has a biaxial orientation. They are blown up like balloons, stretching the molecules in two directions. Pouring very hot water into a 2-liter bottle causes vertical and circumferential shrinkage. When trying this, be sure to place the bottle in a sink so that when the bottle shrinks, excess water is safely contained.

THERMOSETS

So far, all the materials discussed in this chapter have been long-chain hydrocarbons. Even though the molecules are very long, they do have ends. The long chains are not connected to each other. Because the molecules are not chemically tied together, they can slide past each other when pulled. If the chains are smooth, the molecules can slide significantly. When heated, the molecules can move, and the materials will soften or melt if heated sufficiently.

Plastics that consist of disconnected chains are called thermoplastics. As the word implies, when they are heated (thermo), they become soft and formable (plastic). In contrast, some plastics are referred to as thermoset. When heated, they do not soften or become pliable. Most thermosets cure into nonmelting, insoluble solids. The chemical basis for this characteristic is that molecules in thermosets are chemically linked to each other. The chemical bonds between molecules are cross-links (Figure 3-16). It is theoretically possible that large thermoset objects, such as the hoods of diesel trucks, are actually one immense molecule.

Figure 3-16. Cross-linked molecules.

The image of a bowl of spaghetti can also help explain the difference between thermosets and thermoplastics. Tying one strand of spaghetti to another is like creating one cross-link. If many knots tied many strands to many neighboring strands, then the bowl would have cured.

Thermosets fall into two large categories: rigid and flexible. The rigid thermosets often find applications in high-heat environments. They don't soften under heat and will char at high temperatures. Because thermosets have a tight chemical bond, they also tend to resist attack by solvents.

Flexible thermosets have a long history. In Chapter 1, Charles Goodyear and the vulcanization of rubber received attention. Chemically speaking, Goodyear found a way to cause the formation of cross-links between rubber molecules. Another group of flexible thermosets is based on urethane. Foams for car seats, sofas, furniture, and beds come from polyurethane.

Both rigid and flexible thermosets cannot be recycled or reprocessed like thermoplastics. Rubber tires cannot be chopped up and used again to make new tires. Current recycling programs for consumer items focus on thermoplastic materials that are reprocessable. For more information on efforts to recycle thermosets, see Chapter 2.

SUMMARY

Although many systems can be used to classify plastics, this chapter emphasizes using differences in chemical structures to group various plastics. The first level of distinction is between cross-linked thermosets and those which are not cross-linked.

Within the thermoset group, two subcategories extend the classification system. One subcategory contains highly cross-linked materials, which tend to be rigid and often brittle. The other is for slightly cross-linked materials, which tend to be elastic and soft. If the materials are sufficiently elastic, they also warrant an elastomer description. The second subcategory distinguishes between amorphous thermoplastics and semicrystalline thermoplastics. Figure 3-17 presents this classification in graphic form.

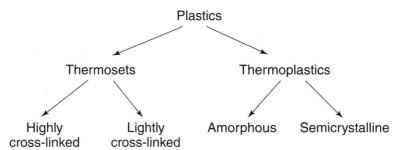

Figure 3-17. A classification system for plastics.

RELATED INTERNET SITES

- **www.acs.org.** The site of the American Chemical Society includes information about polymer chemistry. Selecting "Education" on the home page leads to a listing of the ACS educational sites. The section titled "Student" includes materials for high school, undergraduate, and graduate/postdoctorate.

- **www.americanchemistry.com.** A number of excellent resources for teachers are available here. To find these materials, select "Products and Technology," then select "Plastics," and then "Education and Resources." Under "Education and Resources," a list of selections contains "Hands On Plastics." Don't choose "Hands On Plastics 2" or "Plastics 101." Within "Hands on Plastics," find "Introduction to Plastics." That will lead to a section containing a link to "Background Information for Teachers."

VOCABULARY

The following vocabulary words are found in this chapter. Use the glossary in Appendix A to look up the definitions of any of these words you do not understand as they apply to plastics.

alternating copolymer
amorphous
annealing
atomic mass units (amus)
backbone
benzene ring
block copolymer
branched

copolymer
covalent bonds
    double covalent bond
    triple covalent bond
cross-links
crystalline
degree of polymerization (DP)
dipole interactions
graft copolymer
homopolymer
hydrocarbons
hydrogen bonds
intermolecular interaction
ionic bonds
macromolecules
mer
metallic bonds
molecular mass
molecules
orientation
    uniaxial
    biaxial
polarity
polymerization
primary chemical bonds
random copolymer
residual stresses
saturated/unsaturated
side group
single covalent bonds
terpolymer
thermoplastics
thermoset
Van der Waals forces

QUESTIONS

3-1. A bond between two carbon atoms is a _____ bond.

3-2. A CH_3 group is called a _____ group.

3-3. A dash between atoms indicates a _____ bond.

3-4. The small repeating units that make up a plastic molecule are called _____.

3-5. What does poly mean?

3-6. Three types of intermolecular forces found in plastics are _____, _____, and _____.

3-7. Crystalline plastics are often more rigid and not as transparent as _____ plastics.

3-8. A _____ material may be softened repeatedly when heated and will harden when cooled.

3-9. The term used to describe the tying together of adjacent polymer chains is _____.

3-10. If two different mers make up the composition of a polymer, it is called a _____.

3-11. A hydrocarbon molecule that contains some double bonds is called _____.

3-12. Is PVC a copolymer? Explain.

3-13. What is the general structure of an alternating copolymer?

3-14. Should a polyethylene molecule be considered branched if the branches have the same structure as the backbone?

3-15. If a material is transparent, is it crystalline?

3-16. What effect do dipoles and hydrogen bonds have on the melting point of polymers?

3-17. Is residual stress synonymous with orientation?

ACTIVITIES

Oriented Thermoforming Sheet

Equipment. Toaster oven or heat lamp, tongs, calipers, oriented thermoforming sheet.

Procedure

3-1. Locate some highly oriented thermoforming sheets. To tell whether the material is highly oriented, bend it to see whether it fractures easily in one direction but bends without fracture in the other direction. This phenomenon appears most clearly in rather thin sheets.

3-2. Cut out squares approximately 75 mm by 75 mm. Carefully measure length, width, and thickness. Mark the pieces for identification of length and width.

3-3. Heat until the stress relief occurs. A toaster oven fitted with hardware mesh can provide a convenient source of heat (Figure 3-18).

Without mesh, samples may fall through the toaster tray. When the sample is sufficiently hot, the material may curl up. Quickly remove it before it melts, and flatten immediately.

! CAUTION

Use proper protective clothing and equipment to avoid burns. Conduct this activity in a well-ventilated area.

Figure 3-18. A toaster oven for stress relief of small samples.

3-4. When cool, measure length, width, and thickness. Record the change in dimensions as a percentage.

3-5. If the molecules have disoriented completely, the samples should bend with equal toughness in both directions.

3-6. Repeat the test on other samples. Injection-molded PS drinking glasses/cups also show a huge change when heated.

Orientation of HDPE

Equipment. Tensile tester, HDPE tensile bars.

Procedure

3-1. Slowly pull a tensile test bar of HDPE. Do not exceed 1 inch per minute of cross-head speed. The sample should extend several hundred percent. Figure 3-19 shows a sample before and after slow extension.

3-2. Calculate the ultimate strength of the sample.

3-3. Cut a section of the thinned-out region. Mount this piece in a tensile tester, and pull it to failure. Calculate the ultimate strength of the oriented strip. Compare the strength resulting from step 2 with the strength resulting from step 3.

3-4. Pull off "threads" of oriented polyethylene, using only finger strength. Figure 3-20 shows the threads that formed when the sample failed. These threads can be further separated with fingers.

3-5. Stress-relieve a piece of the thinned-out region. After heating, was it thicker than before?

Storm Window Covering

Obtain a piece of shrink film for window covering. Measure it, then stress-relieve it with heat. Measure the annealed piece. Calculate the percentage reduction in length and width. Was the amount of stress in the original film equal in each direction?

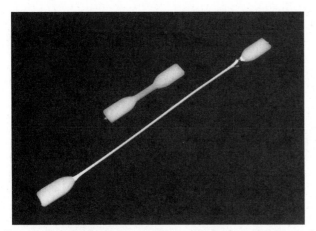

Figure 3-19. The extreme elongation of HDPE, when pulled slowly, causes a high degree of molecular orientation.

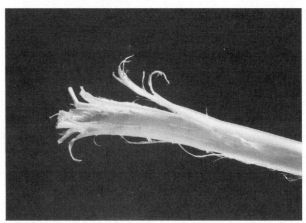

Figure 3-20. The tendency of this sample to fray or "thread" demonstrates an effect of molecular orientation.

HEALTH AND SAFETY

CHAPTER 4

INTRODUCTION

Most industries inform their workers on a wide range of potential health and safety hazards. One method of categorizing these hazards is to divide them into three types: physical, biomechanical, and chemical. The outline for this chapter follows:

PHYSICAL HAZARDS

Physical hazards include machine motions, electrical systems, hydraulic and pneumatic pressure systems, noise, heat, vibrations, and other potential dangers. Physical hazards also encompass ionizing, ultraviolet, microwave, and thermal radiation.

To guard against physical hazards, industrial safety personnel must guarantee that machines have proper safety devices, guards, and warning systems. Protection devices against noise, vibrations, and radiation may include safety glasses, ear plugs, and various shields. Two-hand controls may be installed to ensure that the operator's hands are not near cutting blades, moving shafts, shearing blades, or hot surfaces.

BIOMECHANICAL HAZARDS

Biomechanical hazards are often related to repetitive motions. Ergonomics deals with those types of actions that are not of immediate danger but may lead to injury if repeated over days, weeks, and months. Reducing or eliminating these hazards requires proper hand-tool and machine design, good visual conditions, and air quality. Fatigue generated by poor conditions may lead to accidents. In addition to physical injuries, psychological and mental problems can also arise. Some problems possibly linked with poor conditions are anxiety and irritability, substance abuse, sleeping difficulties, neuroses, headaches, and gastrointestinal symptoms.

CHEMICAL HAZARDS

Although the plastics industry has its share of physical and biomechanical hazards, the largest hazard is chemical. Many of the compounds and processes used in the plastics industry are potentially dangerous.

The **inhalation** and absorption through the lungs of toxic substances accounts for nearly 90% of toxicity cases in the plastics industry. In some cases, workers expose themselves to danger because they are unaware of the hazards. To increase employee awareness, many reputable companies carry on safety education programs intended to protect and inform workers. This chapter discusses chemical health and safety hazards and their correction and prevention.

SOURCES OF CHEMICAL HAZARDS

The overall consumption of plastics in the United States is likely to rise at a rather steady rate, assuming trends continue without major upsets. Until 2012, the following plastics—listed here in order of sales volume—should dominate sales:

1. Polypropylene
2. Polyvinyl chloride
3. High-density polyethylene
4. Linear low-density polyethylene
5. Low-density polyethylene
6. Polyester, thermoplastic
7. Polystyrene
8. Phenolic

In order of volume processed, the major techniques for converting these materials into products will likely be as follows:

1. Extrusion
2. Injection molding
3. Blow molding
4. Polyurethane foaming
5. Phenolic adhesive application
6. Polystyrene expanding

The dominating materials and processes indicate that the major form is solid pelletized plastics. Consequently, health and safety issues for pellets receive primary attention. Powders and liquids will receive comment when they are especially hazardous. The polyolefins on the list above, namely high- and low-density polyethylene and polypropylene, pose low hazard during processing. Thermoplastic polyester also raises minor hazards. Concerns about these materials focus on additives and their possible toxic effects. For more information about additives, see Chapter 7.

The two thermoset materials—polyurethane and phenolic—potentially expose humans to hazardous byproducts of polymerization. A discussion of these hazards appears later in this chapter.

READING AND UNDERSTANDING MSDSs

Hazardous methods and materials are common in the plastics industry. In the United States, a **material safety data sheet (MSDS)** accompanies any purchase of hazardous industrial raw materials. Federal Standard 313B, which provides guidelines for the preparation of an MSDS, defines what the term *hazardous* means. The definition is broad and covers plastics because in the course of normal use plastics "may produce dusts, gases, fumes, vapors, mists, or smokes" that are dangerous.

A similar definition of the word *hazardous* appears in the Code of Federal Regulations (CFR). Hazardous substances are those that have been found capable of posing an unreasonable risk to health, safety, and property.

When a customer repeatedly purchases the same material, an MSDS is sent with the first order every year. Although each producer of raw materials is responsible for creating the MSDS, the guidelines require certain categories of information. A thorough understanding of MSDSs as they relate to the plastics industry can promote safety for all personnel.

New regulations specify the essential information that must be in an MSDS. Because the format is left to the supplier, MSDSs differ in format, but not in information. Many MSDSs use the following order of sections:

1. Product and company identification
2. Composition/information on ingredients
3. Hazards identification
4. First-aid measures
5. Firefighting measures
6. Accidental release measures
7. Handling and storage
8. Exposure controls and personal protection
9. Physical and chemical properties
10. Stability and reactivity data
11. Toxicological information
12. Ecological information
13. Disposal considerations
14. Transport information
15. Regulatory information
16. Other information

Section 1: Product and Company Identification

Section 1 contains information about the product name and the manufacturer's identity. This section usually includes emergency telephone numbers and the trade and chemical family name of the material. A product identification section indicates the chemical name of the product.

For example, Lexan®, manufactured by GE Plastics, is a type of polycarbonate appropriate for injection molding. Its chemical name is Poly (Bisphenol-A carbonate). Every chemical has a CAS Registry number. CAS stands for Chemical Abstracts Services. In addition to cataloging chemical substances, the CAS Registry provides an unambiguous identification of materials. Chemical companies promote their materials with trade names, such as Lexan. To know whether Lexan is chemically identical to Makrolon®—a polycarbonate made by Bayer Company—compare the CAS numbers. The number for both brands is 25971-65-5.

Section 2: Composition

Section 2 contains information on hazardous ingredients. Because polycarbonate is not a controlled product, the MSDS provides no additional data in this section. However, in addition to the major constituents of a material, all hazardous additives, fillers, or colorants must appear in this section. An MSDS for a grade of ABS lists the following ingredients:

CAS #	Chemical Name	OSHA PEL Units	ACGIH TLV Units
7631-86-9	Silica	0.05 mg/m³	0.05 mg/m³
100-42-5	Styrene	50.0 ppm	50.0 ppm
1333-86-4	Carbon black	3.5 mg/m³	3.5 mg/m³

It is very important to understand in detail what this information means. The acronym OSHA stands for the Occupational Safety and Health Administration. ACGIH stands for the American Conference of Governmental Industrial Hygienists. Both these organizations publish standards for exposure to various industrial materials.

OSHA utilizes a measure called the permissible exposure limit (PEL). A PEL indicates the amount and time of exposure permissible. Thus the PEL is a time-weighted average (TWA). A TWA represents the exposure level considered acceptable for an 8-hour day as part of a 40-hour week. In addition to the PEL values, OSHA reports a recommended exposure limit (REL), and a short-term exposure limit (STEL).

In this example, the permissible exposure limit for silica is 0.05 mg/m³. The unit, mg/m³, is appropriate for dusts, powders, or fibers.

The abbreviation TLV means threshold limit value. This is the value recommended by the American Congress of Governmental Industrial Hygienists. The TLV is also a time-weighted average, acceptable for 8 hours a day as part of a 40-hour week.

The ACGIH also has two additional categories of TLVs. The first category is TLV-STEL. It indicates acceptable exposure for 15 minutes and should not be exceeded at any time during an 8-hour day, even if the TWA for the day is within the limits.

The TWA is generally lower than the STEL. The procedure for exposures, which is above the TWA yet below the STEL, is clearly specified. Such exposures should be:

- No longer than 15 minutes
- No more than 4 times per day

The second additional category of TWA is the ceiling value. This value should not be exceeded during any part of a workday.

In this example, silica and carbon black are powders or dusts. Styrene is as hazardous as a gas. Consequently, the PEL or TLV units are ppm (parts per million).

It is important to clearly distinguish between PEL and TLV levels and the percentage of an ingredient by weight. The acrylonitrile-butadiene-styrene (ABS) used as an example contained 3% of carbon black and 0.2% of residual styrene monomer. Residual styrene monomer is a material that did not combine to produce polymer molecules, but remained trapped in the polymer. The silica was present at 5% by weight. Because carbon black and silica are powdered solid materials, they are encapsulated by the major plastics and are very unlikely to exist in a free form. Consequently, the TLV and PEL values have little practical application. However, the monomeric styrene can escape as a gas when the material is at processing temperatures. Its TLV and PEL are of practical value.

Styrene as a healthy hazard. Styrene is so important to the plastics industry that it warrants special attention. Styrene holds a crucial position in the plastics industry because it is a building block for thermoplastic styrenics, which include polystyrene, impact polystyrene, SAN, ABS, and others. In addition, styrene appears in polyester casting resins.

In thermoplastic styrenic resins, monomeric styrene is a very minor ingredient. Some ABS plastics contain less than 0.2% of monomeric styrene. In addition to the residual monomer, commercial styrenic plastics yield styrene during thermo-oxidative degradation. These sources can combine to emit styrene into the air during thermoplastic molding and forming processes. One study found a range of 1 to 7 ppm of styrene in the atmosphere of a plant using injection-molding polystyrene.

The potential hazards of styrene from thermoplastics are tiny compared to the danger from polyester resins. In particular, open-mold operations are common for the production of boats, yacht hulls, large tanks or pipes, tub and shower stalls, and truck or tractor tops. Polyester resins provide a matrix for reinforcing glass fibers—either as cloth, mat, or chopped strand. Common processes are *chop-and-spray* and *hand-layup*. In the chop-and-spray method, catalyzed resin and glass strands meet in a chop/mix head. Operators direct the coated glass fibers into a form. In the hand-layup method, operators position reinforcing glass layers in molds and apply resin, sometimes with airless spray guns.

Whatever the process, open-mold work exposes operators to styrene vapors. Polyester resins contain about 35% styrene by weight. Although MSDS recommends that personnel "avoid breathing vapors," this is possible only if operators wear self-contained breathing apparatus. Some companies making large yacht hulls do fit their operators with such equipment. However, many manufacturers rely on ventilation systems.

Ventilation systems have varying efficiencies. If the ventilation systems are not adequate, the concentrations may exceed the current TLV of 20 ppm. One study found styrene concentrations of 109 ppm when an open mold had a surface area of 1.3 m³. An 8.3 m³ surface yielded 123 ppm. Another study found 120 ppm for a chop-and-spray operation and 86 ppm for a roll-out station.

However, if the ventilation system is well placed and powerful, the concentrations drop. One yacht company consumes 8000 lb of polyester resin per week. Its plant has local ventilation at each hull workstation, capable of exhausting 17,000 cubic feet per minute (cfm). In addition, the system for the entire building guarantees 10–15 air changes per hour. Under these circumstances, the hull lamination workers experience exposure rates of 17–25 ppm. According to another study, the average exposure for yacht companies is 37 ppm, whereas small boat companies showed an average of 82 ppm. These exposure readings were probably due to less efficient ventilation systems.

Even lower exposures occur in closed-mold or press-mold operations. Press-mold companies had exposures between 11 and 26 ppm. In order to achieve lower exposures, some companies are converting some or all operations to closed molds. If the ACGIH lowers the TLV for styrene, the incentive to eliminate open molds will be even greater.

Table 4-1 lists the limits currently recommended by ACGIH. These recommendations are available in 2007 TLVs® and BEIs® based on the documentation of the TLVs for chemical substances and physical agents and biological exposure indices. The ACGIH regularly publishes updates to these values in order to keep the documentation up to date.

The ACGIH also rates materials as carcinogens, which means cancer-causing agents. A1 is the rating for confirmed human carcinogens. A2 is used for suspected human *carcinogens*, and A3 for animal carcinogens.

Table 4-1. Threshold Limit Values for Selected Chemicals

Material	Flash Point (°C)	TLV (ppm)	TLV (mg/m³)	Health Hazard
Acetaldehyde A3		25 ceiling	180	Animal carcinogen
Acetone (dimethyl ketone)	−18	500	1,186	Skin irritation, moderate narcosis from inhalation
Acrylonitrile (vinyl cyanide) A3	5	2	4.3	Absorbed through skin, inhalation carcinogenic
Ammonia		25	17	
Asbestos A1 (as amosite)			0.1 fibers/cc	Respiratory (inhalation) diseases, carcinogenic
Benzene (benzol) A1	11	0.5	1.6	Poisoning by inhalation, carcinogenic skin irritation and burns
Bisphenol A				Skin and nasal irritation
Boron fibers				Irritant, respiratory discomfort
Carbon dioxide	gas	5,000	9,000	Asphyxiation possible, chronic (inhalation) poisoning in small amounts
Carbon monoxide	gas	25	29	Asphyxiation
Carbon tetrachloride A2		5	31.5	Inhaled and absorbed, chronic poisoning in small amounts, carcinogenic A2
Chlorine A3		0.5	1.5	Bronchial distress, poisoning, and chronic effects
Chlorobenzene A3 (phenyl chloride)	29	10	46	Absorbed and inhaled, paralyzing in acute poisoning
Cobalt A3 (as metal dust and fume)			0.02	Possible (inhalation) pneumoconiosis and dermatitis
Cyclohexane	−20	300	1,030	Liver and kidney damage, inhalation
Cyclohexanol	66	50	206	Inhaled and absorbed, possible organ damage
O-Dichlorobenzene	66	25	150	Possible liver damage, inhalation, percutaneous

1,2-Dichloroethane (ethylene dichloride)	13	200	793	Anesthetic and narcotic, inhalation, possible nerve damage
Epichlorohydrin A3	95	0.5	1.9	Highly irritating to eyes, percutaneous and respiratory tract, carcinogenic
Ethanol (ethyl alcohol)	13	1,000	1,880	Possible liver damage, narcotic effects
Fluorine		1	1.6	Respiratory distress, acute in high concentrations
Formaldehyde A2		0.3 ceiling		Skin and bronchial irritations, inhalation, carcinogenic
Glass fiber		10		Irritant
Hydrogen chloride A4		2 ceiling	3 ceiling	Eye, skin, and mucous membrane irritant
Hydrogen fluoride		2 ceiling	1.6 ceiling	
Isopropyl alcohol		200	491	Narcotic, irritation to respiratory tract, dermatitis
Methanol (methyl alcohol)	11	200	262	Chronic if inhaled, poisoning may cause blindness
Methyl acrylate	3	2	7	Inhaled and absorbed, liver, kidney, and intestinal damage
Methyl chloride A4	gas	50	174	Possible liver damage, narcotic, severe skin irritation, moderate narcosis through inhalation, ingestion, carcinogenic
Methyl ethyl ketone	−6	200	590	Slightly toxic, effects disappearing after 48 h
Mica			3	Pneumoconiosis, respiratory discomfort
Nickel			1.5	Chronic eczema, carcinogenic
Phenol (carbolic acid)	80	5	19	Inhaled, absorbed through skin, narcotic, tissue damage, skin irritation
Phosgene		0.1	0.4	Lung damage
Pyridine A3	20	1	3.2	Liver and kidney damage
Silane				Inhalation, organ damage
Silica (fused)			0.1	Respiratory discomfort, silicosis, potential carcinogen
Styrene		20	85	Eye and mucous membrane irritant
Toluene A4	4	20	75	Similar to benzene, possible liver damage
Vinyl acetate A3	8	10	35	Inhalation, irritant
Vinyl chloride A1		1	2.6	Carcinogenic

Source: Adapted from 2007 TLVs and BEIs Based on the Documentation of the Threshold Limit Values for Chemical Substances and Physical Agents & Biological Exposure Indices, American Conference of Governmental Industrial Hygienists, Inc., Cincinnati, OH.

Section 3: Hazards Identification

This section deals with possible entry routes of toxic substances into humans. The most common routes are ingestion, inhalation, skin, and eyes. In addition to toxicity, this section reports on chronic effects and carcinogenicity.

Ingestion. Ingestion of pellets is rather unlikely, and some companies state, "Not a probable route of exposure." Other companies are more cautious and provide statements such as "The oral LD-50 in rats is in excess of 1000 milligrams per kilogram of body weight. Two-week feeding tests

with dogs and rats showed no evidence for gross pathological change."

Toxicity ratings utilize the term LD_{50} or LD-50. LD stands for "lethal dose," and the subscript 50 means that this dose is capable of killing 50% of a population of experimental animals. Many pelletized plastics are rather inert, and experimental animals can eat large quantities with little effect.

Discussions of toxicity levels often rely on a few general categories. When lethal doses are less than 1 mg or 10 ppm, the material is considered extremely toxic. If the lethal range is less than 100 ppm or 50 mg, the material is highly toxic. Moderate toxicity refers to doses less than 1000 ppm or

500 mg. Toxicity ratings over 1000 ppm or 500 mg should be considered *slightly toxic.*

Toxicity ratings refer to the total weight of the subject. An MSDS might contain a rating similar to the following:

$$\text{ORALLD}_{50} \qquad 265\text{mg/Kg}$$

In other words, this moderately toxic substance kills 50% of experimental guinea pig populations. If the toxicity to humans is identical to the response shown by guinea pigs, then the lethal oral dose for a human with a body weight of 70 kg would be 264 g $\times$ 70, which yields 18,480 mg, or 18.48 g.

Inhalation. Some MSDSs view inhalation of pelletized plastics as unlikely because of their physical form. Others provide the following data:

$$\text{INHALATION LC}_{50} \quad \text{NO DATA AVAILABLE}$$

The abbreviation LC stands for "lethal concentration." It normally refers to a vapor or gas. The units for numeric values for vapors and gases are ppm at standard temperature and pressure.

Although inhalation of pellets is unlikely, inhalation of gases and vapors is a major concern. Many studies have measured the odor response of humans in order to determine how effectively odor warns humans of possible dangers. Many gases are easily detected. An example is acetaldehyde, a chemical used in the production of some phenolic resins. Overheated PET also releases small amounts of acetaldehyde. Acetaldehyde has a STEL of 25 ppm, but its air odor threshold is 0.050 ppm. Because it is easily detected at concentrations well below the TLV, odor can provide a warning about exposure to this material.

However, some gases provide no odor warning. Vinyl chloride is the primary monomer used in the polymerization of PVC. It has a TLV of 1 ppm and an odor threshold of 3000 ppm. It is also classed as A1—"known human carcinogen." Odor provides no warning of exposure to this dangerous material.

Table 4-2 lists chemicals' selected uses in the production of plastics. Some are also decomposition products that are formed when selected plastics are overheated.

The inhalation of isocyanates is a hazard associated with the manufacture of polyurethane products. Polyurethane involves the polymerization of TDI (toluene diisocyanate) or MDI (methylene diisocyanate).

This polymerization routinely occurs during the manufacture of polyurethane foams and reaction-injection molding of polyurethane products. In some construction projects, foamed polyurethane is sprayed on the inside of walls and roofs. According to the ACGIH, the TLV for both isocyanates is 0.005 ppm, with the additional restriction that TDI has a STEL of 0.02 ppm. These levels are very low and therefore require significant effort to achieve and maintain.

Table 4-2. Comparison of TLVs to Odor Thresholds for Selected Chemicals

Name	TLV, ppm	Air Odor Threshold, ppm
Acetaldehyde	25 ceiling	0.05
Acrylonitrile	2	17
Ammonia	25	5.2
1,3 Butadiene A2	2	1.6
Formaldehyde A2	0.3 ceiling	0.83
Hydrogen chloride	5 ceiling	0.77
Hydrogen cyanide	4.7 ceiling	0.58
Methyl methacryiate	50	0.083
Phenol	5	0.040
Styrene	20	0.32
Tetrahydrofuran	200	2.0
Vinyl acetate A3	10	0.5
Vinyl chloride A1	1	3000

Dermal. A *dermal* exposure is one in which a substance comes in contact with the skin. Skin is an effective barrier against some chemicals that pose no hazard through dermal exposure. Other chemicals and materials may irritate the surface of the skin. Their danger is limited. Some materials can penetrate the skin and cause sensitization. Epoxies may cause sensitization after repeated exposure. The most severe hazard is the penetration of a chemical into the skin that enters the blood stream and acts directly on bodily systems. This is called a *systemic hazard.*

Pelletized plastics cannot penetrate the skin but may cause skin irritation or dermatitis. This is particularly true if the material contains abrasive fibers such as glass. Skin contact with molten plastics can cause severe burns.

Liquid materials may pose a severe dermal hazard. A catalyst used to harden an epoxy resin has the following data in its MSDS:

Diethylene Triamine	CAS #11140-0
ACGIH	
TLV	STEL
1 ppm (skin)	NE
OSHA	
PEL	STEL
1 ppm (skin)	NE

This substance will enter through the skin and has a low TLV and PEL. This particular material is very dangerous and requires special precautions. If this substance received laboratory testing on animals, it might have a lethal dose (LD) value.

Eyes. If pellets get into a person's eyes, the eyes may be injured mechanically. Liquids and gases may cause severe eye damage.

For example, methylenedianiline is an ingredient in a catalyst for hardening an epoxy resin. One MSDS indicates that methylenedianiline causes irreversible blindness in cats and visual impairment in cattle.

Carcinogenicity. Solid plastics in pelletized form are often not regulated as carcinogenic. However, residual monomers may have links to cancer. For example, an MSDS of ethylene-vinyl acetate (EVA) lists vinyl acetate as a hazardous ingredient present at a maximum of 0.3%. When EVA is polymerized, a small amount of the vinyl acetate monomer remains. Extensive exposure of test animals to vinyl acetate monomer at 600 ppm caused some carcinomas in the nose and air passages of some animals. Consequently, it bears the A3 notation.

Section 4: First-Aid Measures

This section gives you specific instruction on first-aid treatment for the different routes of exposure along with special notes to physicians.

Section 5: Firefighting Measures

Because most pelletized plastics are not explosive, this section generally concentrates on firefighting. In the presence of sufficient heat and oxygen, most plastics will burn and yield carbon dioxide and water vapor. Many plastics can be made self-extinguishing or fire retardant. All thermosets are self-extinguishing. Glass and other inorganic reinforcement may reduce flammability. Most flame-retardant additives act by interfering chemically with flame reactions.

Many MSDSs recommend water as the best medium for extinguishing fires. They also warn of hazardous chemicals produced during combustion, such as dense black smoke, carbon monoxide, hydrogen cyanide, and ammonia.

There are several factors that cause deaths due to fire, including direct burns, oxygen deficiency, and exposure to toxic chemicals. One of the greatest dangers in a fire is carbon monoxide. It is a colorless, odorless gas that can cause unconsciousness in less than 3 minutes.

Combustion often produces toxic by-products. Table 4-3 gives the relative fire toxicity of selected polymers and fibers. The values in Table 4-3 come from experiments using test animals exposed to gases generated by pyrolysis of selected materials. Such data do not directly predict fire toxicity in humans. Note that wool and silk are the most toxic, with wool rated higher than many polymers.

Many commercial plastics have **flash points** so high that MSDS report them as "Not Applicable." On a grade of nylon, one MSDS reported a flash point of 400°C (752°F) as determined by the American Society for Testing and Materials using

Table 4-3. Relative Fire Toxicity of Selected Polymers and Fibers

Material	Approximate Time to Death, min	Approximate Time to Incapacitate, min
Acrylonitrile-butadiene-styrene	12	11
Bisphenol A polycarbonate	20	15
Chlorinated polyethylene	26	9
Cotton fiber, 100%	13	8
Polyamide	14	12
Polyaryl sulfone	13	10
Polyester fiber, 100%	11	8
Polyether sulfone	12	11
Polyethylene	17	11
Polyisocyanurate rigid foam	22	19
Polymethyl methacrylate	16	13
Polyphenyl sulfone	15	13
Polyphenylene oxide	20	9
Polyphenylene sulfide	13	11
Polystyrene	23	17
Polyurethane flexible foam	14	10
Polyurethane rigid foam	15	12
Polyvinyl chloride	17	9
Polyvinyl fluoride	21	17
Polyvinylidene fluoride	16	7
Silk fiber, 100%	9	7
Wood	14	10
Wool fiber, 100%	8	5

Note: Information from Fire Safety Center of the University of San Francisco with the support of the National Aeronautics and Space Administration. All data modified to show approximate values.

the open-cup method—ASTM D-56. In the open-cup method, a material is heated in an open container. The flash point is the lowest temperature at which enough vapors are given off to form an ignitable mixture of vapor and air immediately above the surface of the melt or liquid.

In contrast to the flash points of commercial plastics, the flash points of many liquids have practical implications. According to the Occupational Safety and Health Administration (OSHA) and the National Fire Protection Association (NFPA), a **flammable** liquid is any material having a flash point below 38°C (100°F). **Combustible** liquids are those with flash points at or above 38°C. Table 4-1 lists the flash points of selected materials. Note that many hydrocarbons often have low flash points of below 0°C (32°F). These flammable materials pose a severe fire hazard if they are not properly handled.

Section 6: Accidental Release Measures

Section 6 gives spill or leak procedures. In the case of pelletized materials, such procedures amount to sweeping up spilled pellets. For liquids, much more elaborate actions must be taken.

Section 7: Handling and Storage

This section addresses the areas, methods, and precautions needed for storage of the material.

Section 8: Exposure Controls and Personal Protection

This section lists protective measures for individuals and their working space.

Workplace protection. Adequate ventilation is needed in processing areas. The ACGIH has prepared guidelines for industrial ventilation that are available from the ACGIH Committee on Industrial Ventilation, P.O. Box 116153, Lansing, MI 48901.

Personal protective devices include safety glasses, eye goggles, and face protection. Skin protection consists of gloves, long sleeves, and face shields. Ear plugs serve as hearing protection, and respiratory protection is achieved through using a respirator whenever processing fumes are out of control. Respirators should also be worn when secondary operations such as grinding, sanding, or sawing create excessive dust.

The ACGIH has published *Guidelines for the Selection of Chemical Protective Clothing*, and the American National Standards Institute (ANSI) and OSHA address eye and face protection. Protective barrier cream systems may be used to protect workers from minor skin irritants and can curb the incidence of dermatitis.

Allergic-reactive people should be warned of possible bronchial or skin irritants. Any irritation to the skin, eyes, nose, or throat should be dealt with promptly.

Section 9: Physical and Chemical Properties

This section does not treat ingredients separately. Instead, it considers the material to be one substance. Examples of data are evaporation rate, melting point, boiling point, specific gravity, solubility in water, and form. For pelletized plastics, some of these characteristics are not established (NE).

Section 10: Stability and Reactivity Data

Because most pelletized plastics are very stable, they are not reactive under normal conditions. Some materials will react with strong acids and oxidizing agents, whereas many remain inert.

However, most plastics degrade when sufficiently hot. The degradation of plastics is considered thermooxidative degradation, which is thermal degradation in the presence of oxygen. A few plastics begin to degrade at normal processing temperatures, and others begin to degrade when heated beyond normal processing ranges. In either case, hazardous gases and vapors enter the air and can enter humans through inhalation.

The following list specifies examples of decomposition products:

- At 230°C (446°F), POM releases formaldehyde.
- At 100°C (212°F), PVC releases HCl.
- At 300°C (572°F), PET releases acetaldehyde.
- At 300°C (572°F), nylons release carbon monoxide and ammonia.
- At 340°C (644°F), nylon 6 releases e-caprolactam.
- At 250°C (482°F), fluoroplastics release HF. Inhaling fumes containing decomposition products of fluoroplastics may cause influenza-like symptoms. This is sometimes called "polymer-fume fever" and includes fever, cough, and malaise.
- At 100°C (212°F), PMMA releases MMA.

Thermal degradation of PVC. The potential decomposition of PVC is a serious problem. In both extrusion and injection-molding processes, PVC can decompose catastrophically if overheated and held too long at processing temperatures in the barrel of a machine.

Figure 4-1 shows decomposed polymer that was removed from the barrel and nozzle of an injection-molding machine. Notice the tightly packed carbon that remains after decomposition. When packed into the nozzle and barrel of an

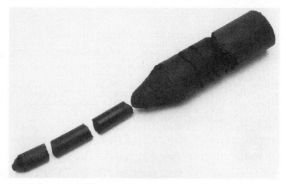

Figure 4-1. Decomposed PVC.

injection-molding machine, this carbon prevents purging of the PVC that remains in the machine.

The sequence of PVC degradation often involves initial discoloration of the PVC, and perhaps the beginnings of small black specks in the parts. If degradation continues, dust and fumes will "spit" out of the nozzle. The fumes will contain high concentrations of hydrogen chloride, which is highly toxic. If such an event should occur, all personnel should immediately evacuate, and anyone near the molding machine should wear a respirator appropriate for organic vapors and acids.

If the nozzle and barrel contain carbon powder, attempting to purge the remaining PVC may prove futile. Because this material will not purge, it may be best to shut down the machine and return after the machine is cool. The nozzle and end cap must come off in order to remove the carbon powder.

The addition of flame retardants, especially flame retardants that contain zinc, may increase the potential of PVC to degrade in this manner. Manufacturers of flame retardants containing zinc urge customers to use caution to avoid "catastrophic zinc failure."

Thermal degradation of POM. *Polyacetal*, also called POM or *polyoxymethylene*, is an engineering thermoplastic. It thermally degrades by depolymerization and releases formaldehyde into the air. Although antioxidants can retard depolymerization, they do not prevent the chemical breakdown. Also, ventilation systems fail to remove all formaldehyde. Consequently, the air near injection-molding machines and extruders may contain excess amounts of formaldehyde. One MSDS indicates that heating above 230°C (446°F) causes the formation of formaldehyde.

Molding technicians may be overexposed to formaldehyde, particularly when purging molding machines. Some technicians have claimed that fumes from acetal purgings can "knock you down." Such statements indicate exposures to formaldehyde well over the TLV of 0.3 ppm ceiling.

Studies of four injection-molding plants found formaldehyde concentrations of 0.05 ppm to 0.19 ppm. In these plants, molding machines had local exhaust systems. The study did not measure the effect of purgings.

Thermal degradation of phenolics. Phenolic resins find a major use in adhesive applications, particularly in the manufacture of plywood and particleboard. Phenolic molding compounds are common in compression and transfer moldings. One MSDS warns that processing phenolics may release small amounts of ammonia, formaldehyde, and phenol. Phenol has a TLV of 5 ppm, a LD_{50} (for a rat) of 414 mg/kg, and a LC_{50} (inhalation rat) of 821 ppm. Formaldehyde has a ceiling of 0.3 ppm. One MSDS asserts that the symptoms of exposure to formaldehyde include eye, nose, throat, and upper respiratory irritation; tearing; and nose stuffiness. These symptoms

often occur at concentrations ranging from 0.2 to 1.0 ppm and become more severe above 1 ppm.

Thermal degradation of nylon 6. Nylon 6 thermally degrades into the monomer from which it is formed, namely e-caprolactam. In addition, many grades of nylon often contain residual caprolactam of less than 1% by weight. The ACGIH has established 5 ppm as the TLV for caprolactam vapor. Caprolactam is a toxic material. The LD_{50} (rat) is 2.14 mg/kg.

Normal injection-molding operations yield some caprolactam vapor. In addition, purgings release greater amounts, and extrusion operations add a steady amount into the workplace atmosphere.

One study involving two injection-molding plants and one extrusion operation found concentrations of 0.01 to 0.03 ppm for e-caprolactam. This is well below the TLV of 5 ppm. However, the study did not consider the purging of molding machines.

Thermal degradation of PMMA. Polymethyl methacrylate (PMMA) is often called acrylic. Acrylic is familiar in sheet form and is sold under the trade name Plexiglas®. It is molded, extruded, and thermofounded into numerous products. During processing, PMMA thermally degrades into methyl methacrylate (MMA). Besides the degradation products, most PMMA contains a small amount of residual monomer. One MSDS reports that acrylic molding pellets contain less than 0.5% by weight of methyl methacrylate.

A study involving one injection-molding plant, two thermoforming operations, and one extrusion location indicated concentrations ranging from a low of 0.06 mg/m³ for injection molding to a high of 4.6 mg/m³ for thermoforming at 160°C (320°F). These are also well below the TLV for MMA, which is 205 mg/m³ (50 ppm). If the processing temperature rises dramatically due to an uncontrolled, or "runaway," heater, the release of MMA could also rise considerably.

Section 11: Toxicological Information

This section is concerned with the material as a potential poison and its effects.

Section 12: Ecological Information

This section covers ecotoxicity, which refers to the ill effects on wildlife and the environment.

Section 13: Disposal Considerations

This section addresses recycling in accordance with federal, state, and local regulations.

Section 14: Transport Information

Section 14 concerns transportation. Most pelletized plastics have no special restrictions for transportation and are not regulated.

Section 15: Regulatory Information

This section ensures that all ingredients comply with the Toxic Substances Control Act of 1976 (TSCA). The Superfund Amendments and Reauthorization Act (SARA) also established various regulations. This section requires companies to state whether their materials are in compliance with regulations and that they do not contain any chemicals that need to be reported per the requirements of section 313 of the Emergency Planning and Community Right-to-Know Act of 1986.

Section 16: Other Information

Many MSDSs provide a separate section on handling and storage. Hazards in handling concern typical secondary operations of grinding, sanding, and sawing. These operations produce dusts that are potentially explosive. Table 4-4 lists the explosion characteristic of dusts found in the plastics industry.

RELATED INTERNET SITES

- **www.acgih.org.** The American Conference of Governmental Industrial Hygienists, Inc., provides information about the activities, publications, and educational programs offered by the ACGIH. It also publishes the journal *Applied Occupational and Educational Hygiene.*

Table 4-4. Explosion Characteristics of Selected Dusts Used in the Plastics Industry

Type of Dust	Ignition Temperature, °C [°F]		Explosibility	Ignition Sensitivity
Cornstarch	400	[752]	Severe	Strong
Wood flour, white pine	470	[878]	Strong	Strong
Acetal, linear	440	[824]	Severe	Severe
Methyl methacrylate polymer	480	[896]	Strong	Severe
Methyl methacrylate-ethyl acrylate-styrene copolymer	440	[824]	Severe	Severe
Methyl methacrylate-styrene-butadiene-acrylonitrile copolymer	480	[896]	Severe	Severe
Acrylonitrile polymer	500	[932]	Severe	Severe
Acrylonitrile-vinyl pyridine copolymer	510	[950]	Severe	Severe
Cellulose acetate	420	[788]	Severe	Severe
Cellulose triacetate	430	[806]	Strong	Strong
Cellulose acetate butyrate	410	[770]	Strong	Strong
Cellulose propionate	460	[860]	Strong	Strong
Chlorinated polyether alcohol	460	[860]	Moderate	Moderate
Tetrafluoroethylene polymer	670	[1,238]	Moderate	Weak
Nylon polymer	500	[932]	Severe	Severe
Polycarbonate	710	[1,310]	Strong	Strong
Polyethylene, high-pressure process	450	[842]	Severe	Severe
Carboxy polymethylene	520	[968]	Weak	Weak
Polypropylene	420	[788]	Severe	Severe
Polystyrene molding compound	560	[1,040]	Severe	Severe
Styrene-acrylonitrile copolymer	500	[932]	Strong	Strong
Polyvinyl acetate	550	[1,022]	Moderate	Moderate
Polyvinyl butyral	390	[734]	Severe	Severe
Polyvinyl chloride, fine	660	[1,220]	Moderate	Weak
Vinylidene chloride polymer, molding compound	900	[1,652]	Moderate	Weak

Alkyd molding compound	500	[932]	Weak	Moderate
Melamine-formaldehyde	810	[1,490]	Weak	Weak
Urea-formaldehyde molding compound	460	[860]	Moderate	Moderate
Epoxy, no catalyst	540	[1,004]	Severe	Severe
Phenol formaldehyde	580	[1,076]	Severe	Severe
Polyethylene terephthalate	500	[932]	Strong	Strong
Styrene-modified polyester-glass-fiber mix	440	[824]	Strong	Strong
Polyurethane foam	510	[950]	Severe	Severe
Coumarone-indene, hard	550	[1,022]	Severe	Severe
Shellac	400	[752]	Severe	Severe
Rubber, crude	350	[662]	Strong	Strong
Rubber, synthetic, hard	320	[608]	Severe	Severe
Rubber, chlorinated	940	[1,724]	Moderate	Weak

Source: Compiled in part from *The Explosibility of Agricultural Dusts*, R1 5753, and *Explosibility of Dusts Used in the Plastics Industry*, R1 5971, U.S. Department of Interior.

- **www.aiha.org.** The American Industrial Hygiene Association offers current information on industrial hygiene. It also publishes *The American Industrial Hygiene Association Journal*. The journal is available online.

- **www.osha.gov.** OSHA, the Occupational Safety and Health Administration, maintains an extensive site. In addition to articles and reports, the permissible exposure limits (PELs) for various materials are included in OSHA materials.

VOCABULARY

The following vocabulary words are found in this chapter. Use the glossary in Appendix A to look up the definitions of any of these words you do not understand as they apply to plastics.

American Conference of Governmental Industrial Hygienists (ACGIH)

carcinogens
ceiling
Chemical Abstracts Services Registry number (CAS number)
combustible
depolymerization
flammable
flash points
inhalation
LD-50 or LD$_{50}$
material safety data sheet (MSDS)
permissible exposure limit (PEL)
recommended exposure limit (REL)
short-term exposure limit (STEL)
threshold limit value (TLV)
time-weighted average (TWA)

QUESTIONS

4-1. A highly toxic, colorless, odorless gas is named _____.

4-2. Combustible liquids are those with flash points at or above _____ °C.

4-3. Name three natural materials that may be more toxic than plastics.

4-4. Name three broad categories of hazards in working with industrial materials.

4-5. The overheating of _____ plastics may cause polymer-fume fever with flu-like symptoms.

4-6. Liquids that have a flash point below 38°C are called _____.

4-7. What is the ignition temperature of cellulose acetate plastics dust?

4-8. Many plastics degrade when overheated. Name one that releases gaseous hydrochloric acid when heated.

4-9. What time period is attached to a STEL?

4-10. Approximately what percentage of polyester resin is styrene?

4-11. Approximately what percentage of ABS pellets is styrene?

4-12. What is a systemic hazard?

ACTIVITIES

4-1. Investigate CAS numbers by acquiring MSDSs on several grades of the same basic material. For example, find MSDSs for various melt index grades and various colors of polyethylene.

- Does a high-melt-index polyethylene have a different CAS number than a low-melt-index polyethylene?

- Does red polyethylene have a different CAS number than green polyethylene?

4-2. Investigate the percentage of styrene in various brands of polyester resin. Acquire MSDSs on polyester resins from several manufacturers. Do some companies have reduced amounts of monomeric styrene in the resin?

4-3. Investigate the percentage of residual styrene in polystyrene pellets. Acquire MSDSs from several manufacturers of polystyrene. Do some companies offer polystyrene with reduced residual monomer?

ELEMENTARY STATISTICS

CHAPTER 5

INTRODUCTION

Global competition is a fact of life for most industries. In the plastics industry, worldwide competition affects small companies as well as large ones. Many companies have focused their efforts on improving the quality of their products. They hope to withstand competition by offering consumers reliable items of high quality. However, the raw materials and processes must also be high quality in order to achieve such a product.

Buying quality materials is a complex task in itself. A frequent, major problem in acquiring materials is knowing how uniform or consistent they are. Sales agents describe the degree of uniformity in statistical terms. Purchasing agents and others involved in buying raw materials need to understand basic statistical concepts.

Variations in the manufacturing process must be controlled so that products are uniform and consistent. To reduce unwanted variations in products, manufacturing personnel seek to minimize random changes in processing. This effort also relies on statistics that can accurately document the repeatability of production equipment.

Understanding and using statistics is essential for companies as they fight to withstand global competition. This chapter introduces a few basic statistical techniques. It assumes the reader has basic math computational skills and has no previous exposure to statistics. The content outline for this chapter follows:

CALCULATING THE MEAN

Comparisons of size or shape abound in everyday life. When someone says, "Look at that big house," we can assume that an average house is the point of comparison. When a tall man walks by, he stands out because of his difference from an average man.

Most women are rather close to the average height. However, many women are somewhat taller or shorter than average. Few women are much taller or much shorter than average and only a rare few are very much taller or very much shorter than average. When an extremely tall woman walks by, a mental comparison to the average identifies her as tall. A different comparison—not to the average, but to the range of height possibilities—places her as rare or unique. Quantifying the average and the range of possibilities is the topic of the next few sections.

When calculating an average, the first step is to define the type of average. The **mean**, the **median**, and the **mode** are averages, but this book treats only the mean.* To calculate the mean, add the values in a group to create a sum. Then divide the sum by the number of values in the group.

---

* The mode is the most numerous value in a distribution. The median is the middle value, which has equal numbers of values above and below it.

The following example illustrates this procedure:

12
11
10
9
8

The sum of the values $(12 + 11 + 10 + 9 + 8)$ is 50. The number of values in this set is 5. Instead of writing out the phrase "the number of values in a set," statisticians use n as an abbreviation. In this case, $n = 5$. Dividing the sum (50) by n (5) results in 10.

$$50/5 = 10$$

The mean is 10. It is often abbreviated $\bar{x}$, which is pronounced "x-bar."

A **distribution** is a collection of values. The set of numbers used for calculation of the mean is a distribution. Virtually any collection of numbers is a distribution. However, statistical analysis relies on only a few standard distributions to explain the countless groups of collected data. In this book, the only standard distribution treated is the **normal distribution**.

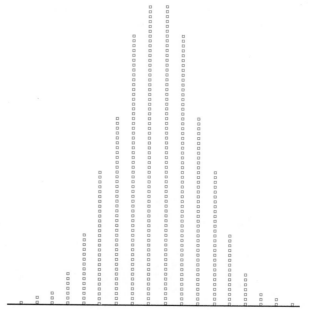

Figure 5-1. A thousand women grouped according to height. (Data adapted from National Center for Health Statistics, Height and Weight of Adults, Ages 18–74 Years, by Socioeconomic and Geographic Variables, United States, 1971–74)

THE NORMAL DISTRIBUTION

For a distribution to be normal, it must exhibit two characteristics. First, it must show **central tendency**. That means that the values must cluster around one central point. Second, it must be rather well centered on the mean. In other words, it must be a **symmetrical distribution**. The next example will help clarify the ideas of central tendency and symmetry.

Imagine 1000 randomly selected women standing on a football field in groups according to their height. The height groups change in 1-inch increments. A person looking down from an overhead blimp would see the shape illustrated in Figure 5-1. Figure 5-1 contains 1000 small circles, one for each person.

If each woman held up a large card, as stadium crowds sometimes do to create words or images, it would look like Figure 5-2.

Figure 5-2 is similar to a **histogram**, which is a vertical bar graph of the frequency in each height group. Converting Figure 5-2 into a histogram requires labeling for the height groups along the x-axis and the frequency count along the y-axis, as seen in Figure 5-3.

If the person at the top of each row grabbed onto a long rope and all the remaining women left the field, the shape shown in Figure 5-4 would appear.

In Figure 5-4, the rope runs as a straight line from one person to the next. Moving the rope into a smooth line results in Figure 5-5.

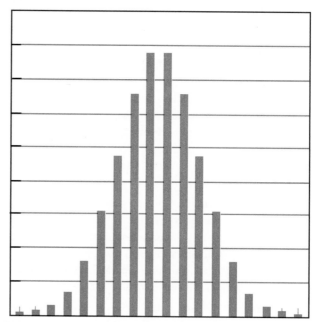

Figure 5-2. Grouped women holding cards.

This shape is called a curve. This particular curve is called the **bell curve** because it resembles the shape of a bell. There are many shapes for bell curves. Figure 5-6 shows different variations of bell curves.

In order to distinguish these curves, a measure of the *spread* of the bell is necessary. The next section presents a method for calculating a numerical measure of this spread.

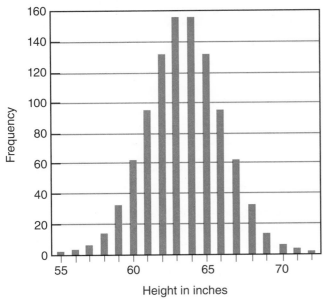

Figure 5-3. Histogram: Women by height.

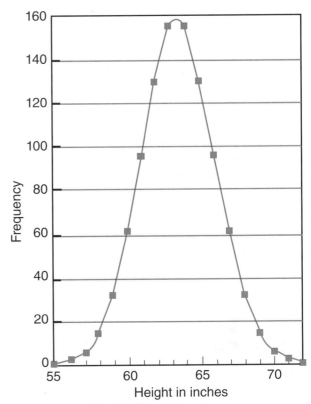

Figure 5-5. Bell curve: Women by height.

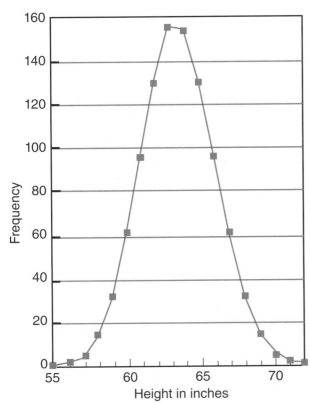

Figure 5-4. Curve composed of straight-line segments: Women by height.

CALCULATING THE STANDARD DEVIATION

Carefully examine the following distributions. Notice that these distributions are not normal distributions. However, they do illustrate an important point.

A	B	C	D	E
12	14	18	11	10
11	12	14	10	10
10	10	10	10	10
9	8	6	10	10
8	6	2	9	10
$\bar{x} = \overline{10}$	$\bar{x} = \overline{10}$	$\bar{x} = \overline{10}$	$\bar{x} = \overline{10}$	$\bar{x} = \overline{10}$

The means of distributions A, B, C, D, and E are all 10. However, these five distributions are not equivalent. E has values that

Figure 5-6. Variations of bell curves.

are the closest together. C has values that are the most spread out. How is it possible to describe the spread numerically?

One approach is to determine the difference between each value and the mean. For distribution A, the largest value is 12. After subtracting the mean (10), the result is a value of 2. Repeating the same procedure for each value yields the results summarized below. The column marked *d* stands for the deviation between each value and the mean.

A	d
12	2
11	1
10	0
9	−1
8	−2

There is a problem with the column of deviations, because their sum is zero. This prevents further calculations to arrive at a numerical measure of the spread. To overcome the difficulties caused by the zero sum, square all the deviations. Squaring will eliminate negative values.

(d)²
4
1
0
1
4
sum = 10

The sum of the squared deviations from the mean is 10. Because the sum is not zero, work can continue. Next, calculate the average deviation by dividing the sum by *n* – 1. In other words, divide 10 by 4.

When calculating the mean, the divisor was *n*. Here, the divisor is 4 (*n* – 1) instead of 5(*n*). The reason for altering the divisor is beyond the scope of this chapter. When *n* is 30 or greater, the difference between *n*(30) and *n* – 1(29) becomes insignificant. However, when the total numbers of values collected is small, the difference between dividing by *n* and dividing by *n* – 1 can be important. To be on the safe side, use *n* – 1.

Squaring the values not only eliminated the zero sum but also introduced squared units. If the original units were pounds, the squared values are pounds squared. To get back to the original units, take the square root of the average squared deviations.

$$10/4 = 2.5$$
$$\sqrt{2.5} = 1.58$$

The final number, 1.58, is the **standard deviation**. The calculation steps for distribution B are identical.

B	d	(d)²	
14	4	16	
12	2	4	40/n = 40/4 = 10
10	0	0	
8	−2	4	√10 = 3.16
6	−4	16	
x̄ = 10		40	

To practice this procedure, calculate the standard deviations for distributions C, D, and E. Here are the results:

Standard deviation for C 5 6.32
Standard deviation for D 5 0.71
Standard deviation for E = 0

If a distribution is normal, the mean and standard deviation describe it completely. By convention, the total area under a bell curve is constant. Changing the standard deviation causes the height and width to change. Narrow distributions are tall, whereas broad distributions are short. Refer to Figure 5-6 to see changes in the curve as the standard deviation increases.

THE STANDARD NORMAL DISTRIBUTION

Mathematicians noticed that bell curves described many physical measurements, including height, weight, length, temperature, and density. They worked to identify a *generic* bell curve that could be used to explain details about unique curves. This generic curve is the **standard normal distribution**. A standard normal distribution has several important characteristics. It has a mean of zero and a standard deviation equal to 1.0. Figure 5-7 shows percentages of the area included in each major section.

GRAPHICAL REPRESENTATION OF HARDNESS TEST RESULTS

To put this statistical information to use, consider experimenting on a piece of clear acrylic material—often called Plexiglas, a popular trade name. The experiment has two goals—to

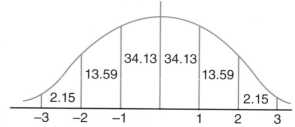
Figure 5-7. Standard normal distribution with area regions identified.

measure the hardness of the piece and determine the uniformity of hardness.

Before beginning any testing activities, state the expected results. In this case, all testing occurs on one piece of material. If the material is not badly flawed, the hardness results should be quite uniform.

To check the reality of these expectations, a Rockwell hardness tester was used, and 100 spots were tested. (For further information on Rockwell hardness tests, see Chapter 6.) A histogram graphically represents the frequency of readings at each degree of hardness.

This histogram (Figure 5-8) shows two important characteristics. First, it shows central tendency because the values cluster around one point. Second, it shows symmetry. Although it is not perfectly symmetrical, it matches the theoretical normal distribution well. A number of statistical techniques can evaluate this match, but they are beyond the scope of this chapter. These graphic techniques rely on an informed judgment. If a distribution of interest shows central tendency and considerable symmetry, consider it normal. If the empirical data are clearly not normal, then stop further analysis. The sketches in Figure 5-9 should help when making a decision.

Because the distribution of the hardness test results is normal, the next step involves calculating the mean and standard deviation.

Mean: 87.7 (Rockwell R scale)
Standard deviation: 1.24

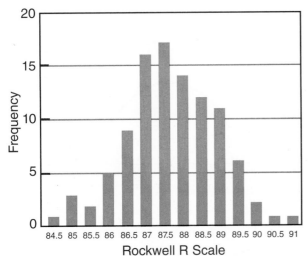

Figure 5-8. Histogram of Rockwell hardnesses for acrylic.

Usually, testing is far too expensive to gather these many results. Common industrial practice is to stop with 5 to 10 tests. However, this raises a problem. If a technician gathers only 10 data points, is it fair to assume that the data follows the normal distribution?

The answer usually relies on past experience. If previous tests indicated normalcy, then it is assumed that future tests of the same type will also be normal. In this case, another 10 points were tested on the same piece of acrylic. Here are the results:

Mean: 87.4
Standard deviation: 1.10

These results clearly go along with the larger sample previously discussed.

An example of the practical use of these statistical procedures concerns the effect of heat on the hardness of acrylic. Suppose acrylic was selected for use in a department store display case. Making the display involves bending acrylic sheets at hot temperatures.

Employees in department stores routinely clean displays by washing with window cleaner and drying with paper towels. Management wanted to know whether this would cause the corners to scratch easily, making the display appear old and worn in a short time.

To answer this question, a test piece of acrylic (of the same type as previously tested) was placed on a strip heater. A band approximately 20 mm (0.78 in.) wide was heated to 150°C (300°F). After the plastic cooled to room temperature, another 10 hardness readings were taken. The results follow:

Mean: 80.3
Standard deviation: 1.68

What happened to the hardness of the acrylic? Observation of the readings indicates that it decreased from 87.4 to 80.3. Is that a big difference? Does that mean that after heating it is a little softer, a lot softer, or extremely soft?

To understand the change in hardness, comparisons must take the standard deviation into account. The standard deviation for the group of 10 readings before being heated was 1.10 and the standard deviation for the group of 10 after heating was 1.68. It is possible to sketch bell curves for each group. However, the two curves would be different in shape because the deviations are not the same. Figure 5-10 shows two very different curves.

Figure 5-9. Normal and non-normal distributions.

To eliminate differing shapes, pool the standard deviations. *Pooling* is simply calculating the mean of the two values. In this case:

$$1.10 + 1.68 = 2.78$$
$$2.78/2 = 1.39$$

Because raw data for the hardness test was accurate to the nearest one-half unit, round the standard deviation and mean to the nearest half unit also. Rounding on the standard deviation must be done after completing the pooling calculation. The **pooled standard deviation** becomes 1.5; the mean unheated becomes 87.5; and the mean heated becomes 80.5.

A problem arises when the deviations of two groups are not similar. Because tests involved the same material, test equipment, time period, and environmental conditions, it is reasonable to expect the random variations in the two groups to be similar. Pooling assumes that differences between the two standard deviations come from random causes. However, sometimes differences are not the same. Statistical techniques are available to determine whether pooling is permitted. However, in the absence of advanced statistical knowledge, you must use the following rules to decide whether pooling is allowed.

1. Create a ratio of two standard deviations with the largest as the numerator and the smallest as the denominator. If the value of that ratio equals 1.5 or less, then pool.
2. If the ratio is larger than 1.5, do not pool. Review the test procedure, looking for a difference in treatment of the groups. If a difference appears, run the test again to see whether deviations between the two groups come closer to each other.

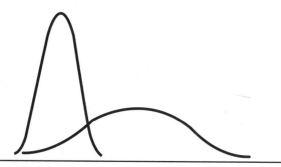

Figure 5-10. Bell curves showing large differences in standard deviations.

SKETCHING GRAPHS

To create an accurately scaled graphic, follow these steps. First, after pooling the deviations, draw a base line (Figure 5-11).

Second, arbitrarily locate and label the mean of one group and make three equally spaced marks on each side of the mean, as shown in Figure 5-12.

Third, label these marks as ± 1, ± 2, and ± 3 standard deviations. Attach the calculated deviations to the appropriate marks (Figure 5-13).

Fourth, sketch a bell curve, making it symmetrical, centered at the mean, and almost touching the base line at ± 3 standard deviations (Figure 5-14).

Fifth, extend and label increments along the base line to reach the mean of the second group. In this example, more increments are needed to the left of -3 (Figure 5-15).

Sixth, locate the mean of the second group on the base line. Notice that the mean of the second group did not line up

Figure 5-11. Base line.

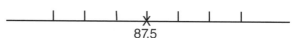

87.5

Figure 5-12. Base line with mean and increments.

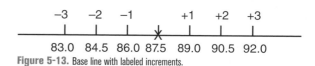

$$-3 \quad -2 \quad -1 \quad \quad +1 \quad +2 \quad +3$$
$$83.0 \quad 84.5 \quad 86.0 \; 87.5 \quad 89.0 \quad 90.5 \quad 92.0$$

Figure 5-13. Base line with labeled increments.

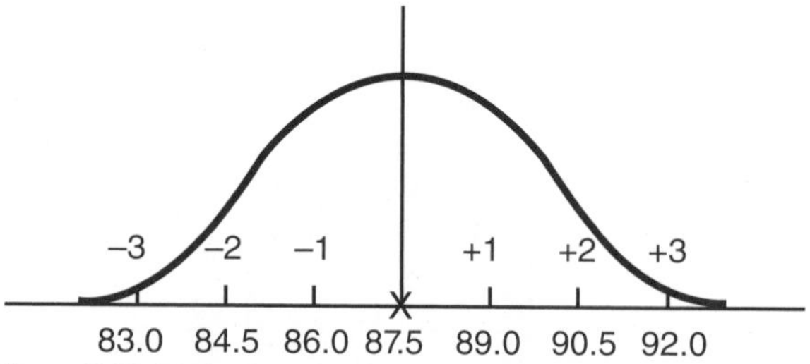

$$-3 \quad -2 \quad -1 \quad \quad +1 \quad +2 \quad +3$$
$$83.0 \quad 84.5 \quad 86.0 \; 87.5 \quad 89.0 \quad 90.5 \quad 92.0$$

Figure 5-14. Sketch of one bell curve.

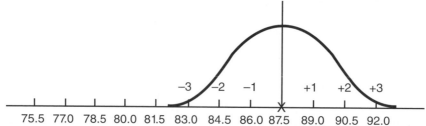

$$-3 \quad -2 \quad -1 \quad \quad +1 \quad +2 \quad +3$$
$$75.5 \quad 77.0 \quad 78.5 \quad 80.0 \quad 81.5 \quad 83.0 \quad 84.5 \quad 86.0 \; 87.5 \quad 89.0 \quad 90.5 \quad 92.0$$

Figure 5-15. Base line extended and labeled.

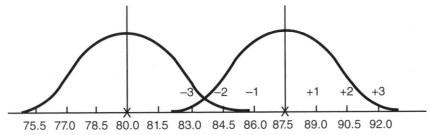

Figure 5-16. Sketch of second bell curve.

exactly with the increments on the base line. Only in rare cases will it align exactly. Sketch in a bell curve identical in height and width to the one already drawn (Figure 5-16).

GRAPHIC COMPARISON OF TWO GROUPS

Look at the curves. Do two distinct bells occur? Do the two bells overlap almost completely? Do they overlap partially? For the purpose of this chapter, evaluations rely on interpreting the graphics and a few simple decision criteria. If the two bells don't touch, the answer is clear—the difference is very significant. If the two bells overlap completely, then there is no difference. Assuming the use of 10 pieces in each sample, the following criteria are appropriate:

1. If the difference between the two means is smaller in size than one standard deviation, there is no difference (Figure 5-17).
2. If the difference between two means is four standard deviations or more, the difference is significant. In Figure 5-16, the difference between the means is slightly less than five standard deviations.
3. If the difference is between one and four standard deviations, graphic analysis is insufficient for a definitive answer (Figure 5-18). A host of statistical techniques are available to determine the significance in cases when the graphics are inconclusive. Without those procedures, report the results as inconclusive or repeat the test to see whether new results will be conclusive.

SUMMARY

This chapter introduced the normal distribution, calculation of the mean and standard deviation, the standard normal distribution, and graphic techniques to compare two samples. Several criteria help determine whether data are normal. These determinations may be made by deciding when it is appropriate to pool the standard deviations and how to accurately sketch bell curves. The procedures reappear in other chapters of this book.

For the most reliable conclusions, be sure to observe these necessary conditions:

1. There should be a normal distribution in each group.
2. The standard deviation for the first group should be equal or almost equal to the deviation in the second group.
3. The sample size should be at least 10.

RELATED INTERNET SITES

- **www.asq.org.** The American Society for Quality (ASQ) is an international organization dedicated to quality and education in quality concepts. Two features of the website are of interest to those with a minimal background in quality measurements. First, *Quality Progress*, the primary journal of the ASQ, is published monthly. Selected articles are available online. Second, ASQ runs a number of educational programs. The Web-based training includes virtual courses, online programs, and webinars. One of the courses is Quality 101, a self-directed, basic-level course.

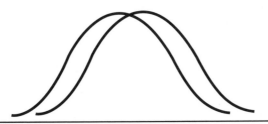

Figure 5-17. Overlapping curves.

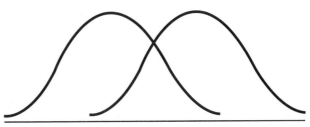

Figure 5-18. Inconclusive curves.

• **www.sae.org.** The Society of Automotive Engineers (SAE) provides a number of educational offerings. Select "Training/Education" on the home page, and then investigate seminars, e-learning, academics, and webinars. The e-learning section lists the titles of all their online courses.

VOCABULARY

The following vocabulary words are found in this chapter. Use the glossary in Appendix A to look up the definitions of any of these words you do not understand as they apply to plastics.

bell curve
central tendency
distribution
histogram
mean, median, mode
n
normal distribution
pooled standard deviation
standard deviation
standard normal distribution
symmetrical distribution

QUESTIONS

5-1. $\bar{x}$ is an abbreviation that stands for _____ .

5-2. A normal distribution must exhibit _____ and _____ .

5-3. A vertical bar graph of frequency and groups is a _____ .

5-4. Squaring the deviations from the mean has the goal of _____ .

5-5. The difference between *n* and *n* – 1 becomes insignificant if *n* is _____ or larger.

5-6. The mean of the standard normal distribution is _____ .

5-7. The percentage of area of the standard normal distribution between the mean and +1 standard deviation is _____ .

5-8. The percentage of area of the standard normal distribution beyond +3 is _____ .

5-9. What is the basis for assuming a small sample (10 pieces or less) is normal?

5-10. What does pooling assume?

5-11. If doubling the smaller standard deviation results in a number less than the larger deviation, is pooling appropriate?

5-12. Find distributions A, B, C, D, and E on page 63. Do these distributions appear normal?

ACTIVITIES

Measuring Standard Deviation

Equipment. Ruler, printed on page 72 (Figure 5-19), pencil, and calculator. (When using a calculator to determine standard deviation, check to see whether it uses *n* or *n* – 1 in the calculation.)

Procedure

5-1. Measure the hand reach of men and women who are at least 18 years old, making sure that only fingers are included, not fingernails. If the number of people measured is at least 30 and the data are normal, the percentages listed in Figure 5-12 will directly apply. Record your findings to the nearest unit of scale. Measure both right

and left hands, and be sure to record whether the subjects are male or female.

5-2. Make a histogram of all results.

5-3. Make two histograms—one for females and another for males.

5-4. Make four histograms—one for right hands of females, a second for left hands of females, and a third and fourth for right hands of males and left hands of males.

5-5. Do the plots indicate normal data?

5-6. Calculate the means and standard deviations for normal data. It is pointless to complete these calculations if the data are not normal.

Weighing Standard Deviation

Equipment. Scale.

Procedure. Exercises based on weight are easy to create. Find some manufactured items intended to be identical. Carefully weigh them and calculate the mean and standard deviation. If virtually identical items are available from competing manufacturers, comparisons are appropriate.

One exercise concerns the weight of plastic pellets. This will require a scale accurate to 0.001 gram. Try to guarantee that the pellets are representative of their bag or box. In industrial sampling of gaylords—1000 lb boxes—a grain sampler guarantees that the sample includes pellets from the top, middle, and bottom of the box.

5-1. Find chopped pellets, identifiable by a cylindrical shape and cut ends.

   a. Weigh 30 chopped pellets. Calculate the mean and standard deviation.

   b. Weigh 30 chopped pellets from another manufacturer.

   c. Are the means identical?

   d. Does one manufacturer have a more consistent chop than the other?

5-2. Find pellets that are oval or spherical in shape. These pellets will show no marks from chopper blades.

   a. Weigh 30 oval/spherical pellets, and calculate the mean and standard deviation.

   b. Are the spherical pellets heavier than the chopped ones?

   c. Is the process for making spherical pellets more consistent than the chopping process?

Representative results:

Chopped PP: mean 0.0174 g   sd. 0.0015 g
Chopped PS: mean 0.0164 g   sd. 0.0024 g
HIPS (spherical): mean 0.0352 g   sd. 0.0050 g

5-3. Weigh individual pellets of the chopped type and the spherical type. Do the data exhibit normalcy?

5-4. Are chopped pellets more uniform or less uniform in weight than the spherical type?

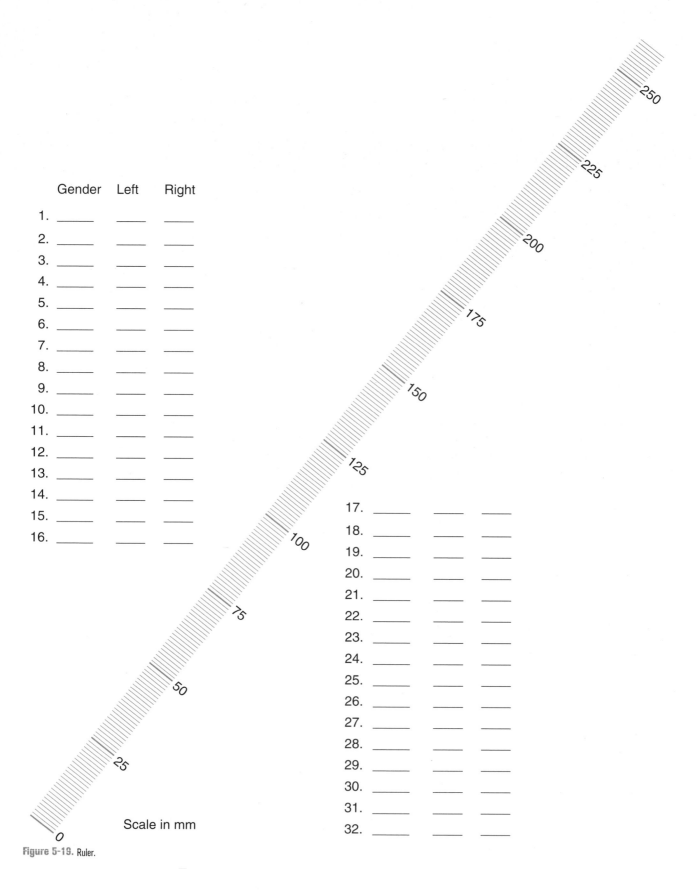

	Gender	Left	Right
1.	____	____	____
2.	____	____	____
3.	____	____	____
4.	____	____	____
5.	____	____	____
6.	____	____	____
7.	____	____	____
8.	____	____	____
9.	____	____	____
10.	____	____	____
11.	____	____	____
12.	____	____	____
13.	____	____	____
14.	____	____	____
15.	____	____	____
16.	____	____	____

17.	____	____	____
18.	____	____	____
19.	____	____	____
20.	____	____	____
21.	____	____	____
22.	____	____	____
23.	____	____	____
24.	____	____	____
25.	____	____	____
26.	____	____	____
27.	____	____	____
28.	____	____	____
29.	____	____	____
30.	____	____	____
31.	____	____	____
32.	____	____	____

Scale in mm

Figure 5-19. Ruler.

PROPERTIES AND TESTS OF SELECTED PLASTICS

CHAPTER 6

INTRODUCTION

Virtually every segment of the plastics industry relies on test data to direct its activities. Raw-materials manufacturers use testing to maintain control of their processes and characterize their products. Designers base their selection of plastics for new products on the results of standard tests. Mold and tool makers depend on shrinkage factors to build molds that will produce finished parts that meet dimensional requirements. Plastics manufacturers use test results to help establish process parameters. Quality-control personnel check to see whether products meet customers' requirements, which often cite standard tests. Therefore, a thorough understanding of testing is essential to many positions within the plastics industry.

This chapter will discuss the most common tests for plastics, which have been grouped into categories. The outline for this chapter follows:

I.   **Testing agencies**
    A. ASTM
    B. ISO
    C. SI units

II. **Mechanical properties**
    A. Stress
    B. Strain
    C. Tensile strength (ISO 527-1, ASTM D-638)
    D. Compressive strength (ISO 75-1 and 75-2, ASTM D-695)
    E. Shear strength (ASTM D-732)
    F. Impact strength
    G. Flexural strength (ISO 178, ASTM D-790 and D-747)
    H. Fatigue and flexing (ISO 3385, ASTM D-430 and D-813)
    I. Damping
    J. Hardness
    K. Abrasion resistance (ASTM D-1044)

III. **Physical properties**
    A. Density and relative density (ISO 1183, ASTM D-792 and D-1505)
    B. Mold shrinkage (ISO 2577, ASTM D-955)
    C. Tensile creep (ISO 899, ASTM D-2990)
    D. Viscosity

IV. **Thermal properties**
    A. Thermal conductivity (ASTM C-177)
    B. Specific heat (heat capacity)
    C. Thermal expansion (ASTM D-696 and D-864)
    D. Deflection temperature (ISO 75, ASTM D-648)
    E. Ablative plastics
    F. Resistance to cold
    G. Flammability (ISO 181, 871, and 1210; ASTM D-635, D-568, and E-84)
    H. Melt index (ISO 1133, ASTM D-1238)
    I. Glass transition temperature
    J. Softening point (ISO 306, ASTM D-1525)

V.   **Environmental properties**
    A. Chemical properties
    B. Weathering (ASTM D-2565, D-4329, G-154, and G-155; ISO 4892)
    C. Ultraviolet resistance (ASTM G-23 and D-2565)
    D. Permeability (ISO 2556, ASTM D-1434 and E-96)

E. Water absorption (ISO 62, 585, and 960; ASTM D-570)

F. Biochemical resistance (ASTM G-21 and G-22)

G. Stress cracking (ISO 4600 and 6252, ASTM D-1693)

VI. Optical properties
A. Specular gloss (ASTM D-2457)
B. Luminous transmittance (ASTM D-1003)
C. Color
D. Index of refraction (ISO 489, ASTM D-542)

VII. Electrical properties
A. Arc resistance (ISO 1325, ASTM D-495)
B. Resistivity (ISO 3915, ASTM D-257)
C. Dielectric strength (ISO 1325 and 3915, ASTM D-149)
D. Dielectric constant (ISO 1325, ASTM D-150)
E. Dissipation factor (ASTM D-150)

TESTING AGENCIES

Several national and international agencies establish and publish testing specifications for industrial materials. In the United States, the standards generally come from the American National Standards Institute, the US military services, and the American Society for Testing and Materials (ASTM). A major international organization similar to the ASTM is the International Organization for Standardization (ISO).

ASTM

The ASTM is an international, nonprofit technical society devoted to "the promotion of knowledge of the materials of engineering, and the standardization of specifications and methods of testing." The ASTM publishes testing for specifications for most industrial materials. Plastics testing comes under the jurisdiction of the ASTM Committee D on plastics. The ASTM annually publishes the *Book of ASTM Standards*, which includes 15 sections covering various materials and industries. In total, there are 76 volumes and over 11,000 standards in the complete ASTM set. Each section contains several volumes of standards. Section 8, which comprises four volumes, is devoted to plastics. Volumes 8.01 through 8.03 provide standards for plastics in general, and volume 8.04 covers plastics pipe and building products. Section 9, which consists of two volumes, covers rubber and rubber products.

ISO

The International Organization for Standardization (ISO) contains national standards organizations from over 90 countries. "The object of ISO is to promote the development of standards in the world with a view to facilitating international exchange of goods and services, and to developing cooperation in the sphere of intellectual, scientific, technological and economic activity." The test procedures for plastics are in ISO catalogue 83. It contains section 83.080.01, Plastics in General, which includes most common physical tests for plastics.

Several US companies that manufacture plastics are adding ISO methods to their testing capabilities. Manufacturers hoping to initiate material sales in Europe and Asia, and those planning on expanding their overseas operations need to meet ISO standards. Some companies supply both ISO and ASTM test results to potential customers. Others are waiting for the introduction of ISO methods.

Table 6-1 identifies a number of common tests for plastics, along with corresponding ISO and ASTM methods.

Although the ASTM specifications use both metric measurements and English units, ISO methods use only *System International d'Unites* (SI), metric units. In the United States, both systems are used. Therefore, students must be able to

Table 6-1. Summary of ISO and ASTM Testing Methods

Property	ISO Method	ASTM Test Method*	SI Units
Apparent density		D-1895	g/cm³
Free flowing	60		
Non-pouring	61		
Arc resistance			
High voltage	1325	D-495	s
Low current			
Brittleness temperature	974	D-746	°C at 50%
Bulk factor	171	D-1895	Dimensionless
Chemical resistance	175	D-543	Changes recorded

Property	Value	ASTM	Units
Compression set	1856	D-395	Pa
Compressive strength	604	D-695	Pa
Conditioning procedure	291	D-618	Metric units
Creep	899	D-2990	Pa
Creep rupture		D-2990	Pa
Deflection temperature	75	D-648	°C at 18.5 MPa
Density	1183	D-1505	g/cm³
Dielectric constant	1325	D-150	Dimensionless
Dissipation factor at 60 Hz, 1 KHz, 1 MHz			
Dielectric strength	3915	D-149	V/mm
Short time			
Step by step			
Dynamic mechanical properties		D-2236	Dimensionless
Logarithmic decrement			
Elastic shear modulus			
Elasticity modulus			
Compressive	4137	D-695	Pa
Tangent, flexural		D-790	Pa
Tensile		D-638	Pa
Elongation	R527	D-638	%
Fatigue Strength	3385	D-671	Number of cycles
Flammability	181, 871, 1210	D-635	cm/min (burning), cm/s
Flexural strength	178	D-790	Pa
Flexural stiffness		D-747	Pa
Flow temperature		D-569	°C
Rossi-Peakes			
Gel time & peak exothermic temperature	2535	D-2471	
Hardness			
Durometer	868	D-2240	Read dial
Rockwell	2037/2	D-785	Read dial
Haze		D-1003	%
Impact resistance			
Dart		D-1709	Pa @ 50% failure
Charpy	179		
Izod	180	D-256	J/m
Indention Hardness		D-2583	Read dial
Barcol Impressor			
Linear coefficient of thermal expansion		D-696	mm/mm/°C
Load deformation		D-621	%
Luminous transmittance		D-1003	%
Melt-flow rate, thermoplastics	1133	D-1238	g/10 min
Melting point	1218, 3146	D-2117	°C

(Continued)

Table 6-1. Summary of ISO and ASTM Testing Methods (*Continued*)

Property	ISO Method	ASTM Test Method*	SI Units
Mold shrinkage	3146	D-955	mm/mm
Molding index		D-731	Pa
Notch sensitivity		D-256	J/m
Oxygen index		D-2863	%
Particle size		D-1921	Micrometer
Physical property changes		D-759	Changes recorded
Subnormal temperature			
Supernormal temperature	1137, 2578		
Refraction index	489	D-542	Dimensionless
Relative density	1183	D-792	Dimensionless
Shear strength		D-732	Pa
Solvent swell		D-471	J
Surface abrasion resistance		D-1044	Changes recorded
Tear resistance		D-624	Pa
Tensile strength	R527	D-638	Pa
Thermal conductivity		C-177	W/K × m
Vicat softening point	306	D-1525	ohm/cm
Volume resistivity 1 min. at 500V		D-257	%
Water absorption			
24-hour immersion	62, 585	D-570	%
Long-term immersion	960		
Water vapor		E-96	g/24 h
Permeability		E-42	
Weathering	45, 85, 877 4582, 4607	D-1435	Changes

Note: *Use the latest version of any ISO and ASTM method referenced.

work quickly and easily in both SI and English units. This chapter follows the publication guidelines of the Society of Plastics Engineers. SI units are always used, occasionally followed by English units in parentheses.

SI Units

The SI metric system consists of seven *base units,* shown in Table 6-2. To simplify large and small numbers, the SI system uses a set of prefixes listed in Table 6-3. When the base units are combined or when additional measures are needed, derived units are used. Table 6-4 lists selected *derived units* frequently used in the plastics industry.

Table 6-2. SI Base Units

Quantity	Unit	Symbol
Length	meter	m
Mass	kilogram	kg
Time	second	s
Thermodynamic temperature	kelvin	K
Electric current	ampere	A
Luminous intensity	candela	cd
Amount of substance	mole	mol

Table 6-3. Prefix and Numerical Expression

Symbol	Prefix	Phonic	Decimal Equivalent	Factor	Prefix Origin	Original Meaning
E	exa	*x'a*	1000000000000000000	10^{18}	Greek	colossal
P	peta	*pet'a*	1000000000000000	10^{15}	Greek	enormous
T	tera	*ter'a*	1000000000000	10^{12}	Greek	monstrous
G	giga	*ji'ga*	1000000000	10^{9}	Greek	gigantic
M	mega	*meg'a*	1000000	10^{6}	Greek	great
k	kilo	*kil'o*	1000	10^{3}	Greek	thousand
h	hecto	*hek'to*	100	10^{2}	Greek	hundred
da	deka	*dek'a*	10	10^{1}	Greek	ten
d	deci	*des'i*	0.1	10^{-1}	Latin	tenth
c	centi	*cen'ti*	0.01	10^{-2}	Latin	hundredth
m	milli	*mill'i*	0.001	10^{-3}	Latin	thousandth
µ	micro	*mi'kro*	0.000001	10^{-6}	Greek	small
n	nano	*nan'o*	0.000000001	10^{-9}	Greek	very small
p	pico	*pe'ko*	0.000000000001	10^{-12}	Spanish	extremely small
f	femto	*fem'to*	0.000000000000001	10^{-15}	Danish	fifteen
a	atto	*at'to*	0.000000000000000001	10^{-18}	Danish	eighteen

Table 6-4. Selected Derived SI Units

Quantity	Unit	Symbol	Formula
Absorbed dose	gray	Gy	J/kg
Area	square meter	m^2	
Volume	cubic meter	m^3	
Frequency	hertz	Hz	s^{-1}
Mass density (density)	kilogram per cubic meter	kg/m^3	
Speed, velocity	meter per second	m/s	
Acceleration	meter per second squared	m/s^2	
Force	newton	N	$kg \cdot m/s^2$
Pressure (mechanical stress)	pascal	Pa	N/m^2
Kinematic viscosity	square meter per second	m^2/s	
Dynamic viscosity	pascal second	Pa · s	$N \cdot s/m^2$
Work, energy, quantity of heat	joule	J	N · m
Power	watt	W	J/s
Quantity of electricity	coulomb	C	A · s
Electric potential, potential difference, electromotive force	volt	V	W/A
Electric field strength	volt per meter	V/m	
Electric resistance	ohm	Ω	V/A

MECHANICAL PROPERTIES

The *mechanical properties* of a material describe how it responds to the application of a force or load. There are only three types of mechanical forces that can affect materials. These are *compression, tension,* and *shear.* Figure 6-1 shows them as push (Figure 6-1A), pull (Figure 6-1B), and opposing pull, which threatens to shear off the bolt in Figure 6-1C. Mechanical testing considers these forces separately and in combinations. Tensile, compression, and shear tests measure only one force. Flexural, impact, and hardness tests involve two or more simultaneous forces.

Brief discussions of selected tests for mechanical properties follow. The tests treated are tensile strength, compressive strength, flexural strength, impact strength, fatigue strength, hardness, and abrasion resistance. Measurements of tensile, compression, flexural, and shear strength depend on a determination of force.

Calculation of force requires the SI base unit for mass and the derived unit of acceleration. By definition:

$$\text{Force} = \text{mass} \times \text{acceleration}$$

The unit of *mass* is the kilogram (kg), and the unit of acceleration is meters per second squared (m/s²). The standard value for acceleration caused by gravity on earth is 9.806 65 meters per second squared. This value, 9.807 m/s², is called the **gravity constant.** The SI unit of force is called the *newton,* which is the force of gravity acting on one kilogram.

$$1\,\text{N} = 1\,\text{kg} \times 9.807\,\text{m/s}^2$$

Stress

Pressure is force applied over an area. The technical term for pressure is **stress.** The metric unit for stress is referred to as the *pascal* (Pa). One pascal equals the force of one newton exerted on the area of one square meter. In the English system, the unit is pounds per square inch (psi). Strength is measured in pascals and is the ratio of the force in newtons and the original cross-sectional area of the sample in square meters.

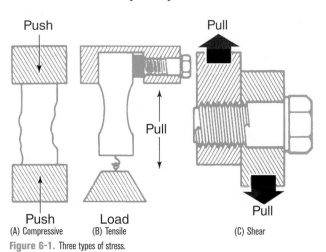

Figure 6-1. Three types of stress.

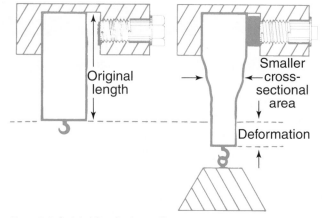

Figure 6-2. Strain is deformation due to pulling stress.

$$\text{Strength (Pa)} = \frac{\text{force (N)}}{\text{cross section (m}^2)}$$

Strain

Stress usually causes material to deform in length. As shown in Figure 6-2, the change in length as it relates to the original length is called **strain.**

Strain is measured in millimeters per millimeter (inches per inch). It can be expressed as a percentage, in which case it is called *percent elongation.* To convert strain in meters per meter to a percent, merely multiply by 100 and report it as a percentage. Strain is apparent when tensile testing plastics deform readily. Figure 6-3 shows typical deformation in an unreinforced plastic.

Stress-Strain Diagrams. The universal testing machine (UTM), shown in Figure 6-4, is used for tensile, compression, flexural, and shear testing. Changing from one test to another

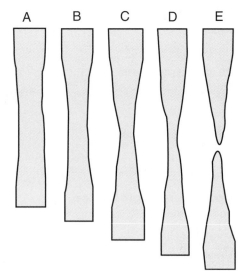

Figure 6-3. Stages of deformation in unreinforced plastics.

Understanding stress-strain curves requires familiarity with a few technical terms. *Yield point A* is the point on the stress-strain curve—also called the load/extension curve—at which the extension increases without an increase in load (stress). Up to the yield point, the resistance of the PC to the applied force was linear. After point *A*, the relation between stress and strain was no longer linear. Calculations can provide the strength and elongation of yield.

At *break point B*, the material failed completely and broke into two pieces. Calculations can readily provide the strength at break and the elongation to break. *Ultimate strength* measures the greatest resistance of the material to stress. On a stress-strain curve, it corresponds to the highest point, *C*.

Figure 6-6 shows a typical stress-strain curve for ABS. This curve shows that the ABS reached its ultimate strength at the yield point (*A* and *C* together).

Figure 6-7 is a typical stress-strain curve for LDPE. This curve exhibits no clear yield point. However, to determine strength or elongation at yield, a yield point must be located. The **offset yield** is used when the curve is not conclusive. This is the point where a line parallel to the linear portion and offset by a specified amount crosses the curve. Figure 6-8 shows the offset line and the location of its intersection with the stress-strain curve (point *A*).

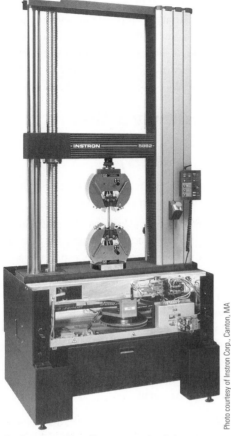

Figure 6-4. Tensile testing machine with covers removed to display components.

Photo courtesy of Instron Corp., Canton, MA

involves altering the direction and speed of the crosshead and switching fixtures that hold the test specimens. The universal testing machine can operate as a stand-alone device or may interface with computers and printers. The testing machine will produce stress-strain diagrams that accurately document the relationship between stress and strain as an increasing load is applied to specimens.

Figure 6-5 shows a stress-strain curve generated by polycarbonate (PC).

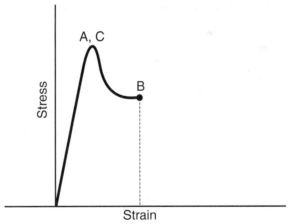

Figure 6-6. Typical stress-strain curve for ABS.

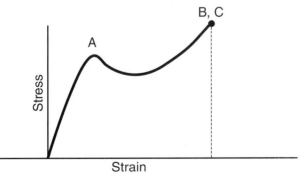

Figure 6-5. Typical stress-strain curve for polycarbonate.

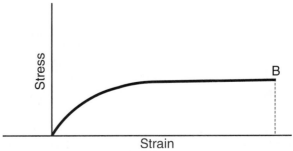

Figure 6-7. Typical stress-strain curve for LDPE.

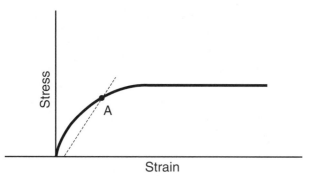

Figure 6-8. Stress-strain curve with offset yield point shown at point A.

Toughness. One generalization about stress-strain curves is that brittle materials are often stronger and less extensible than soft materials. Weaker plastics often exhibit high elongation and low strength. A few materials are both strong and elastic. The area under the curve represents the energy required to break the sample. This area is an approximate measure of **toughness**. In Figure 6-9, the toughest sample has the largest area under the stress-strain curve.

Modulus of Elasticity. The *modulus of elasticity*, also called Young's modulus, is the ratio between the stress applied and the strain within the linear range of the stress-strain curve. Young's modulus has no meaning at stress beyond the yield point. It is calculated by dividing the stress (load) in pascals by the strain (mm/mm). Mathematically speaking, Young's modulus is identical to the slope of the linear portion of the stress-strain curve. When the linear relation remains constant until yield, dividing the yield strength (Pa) by the elongation to yield (mm/mm) results in the modulus of elasticity.

Young's modulus of elasticity = stress (Pa)/strain (m/m)

The ratio of force to elongation is useful in predicting how far a part will stretch under a given load. A large modulus indicates that the plastic is rigid and resists elongation.

Tensile Strength (ISO 527-1, ASTM D-638)

Tensile strength is one of the most important indications of strength in a material. It is the ability of a material to withstand

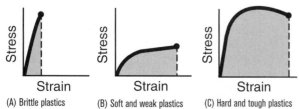

(A) Brittle plastics (B) Soft and weak plastics (C) Hard and tough plastics

Figure 6-9. Toughness is a measure of the amount of energy needed to break a material. It is often defined as the total area under the stress-strain curve.

forces pulling it apart. ISO test specimens are typically 150 mm long, 20 mm wide at the ends, 10 mm wide in the narrow portion, and 3 mm thick. The machine pulls the specimen and records both force and elongation on a stress-strain diagram. The tensile strength is the maximum force divided by the cross-sectional area of the narrow section of the sample.

Compressive Strength (ISO 75-1 and 75-2, ASTM D-695)

Compressive strength is a value that shows how much force is needed to rupture or crush a material.

Compressive strength values may be useful in distinguishing between grades of plastics and comparing plastics to other materials. Compressive strength is especially significant in testing cellular or foamed plastics.

When calculating compressive strength, the units needed are multiples of the pascal, such as kPa, MPa, and GPa. To determine compressive strength, divide the maximum load (force) in newtons by the area of the specimen in square meters. Not all plastics will completely fail in compression. In such a case, the stress can be reported at arbitrary strain, which is usually between 1 and 10 percent.

Compressive strength (Pa) = force (N)/cross-sectional area (m^2)

If 50 kg is required to rupture a 1.0 mm^2 plastics bar,

Force (N) = 50 kg × 9.8 m/s^2

where 9.8 m/s^2 is the gravity constant.

$$\begin{aligned} \text{Compressive strength (Pa)} &= (50 \times 9.8) \text{ N}/1 \text{ mm}^2 \\ &\quad 490 \text{ N}/1 \text{ mm}^2 \\ &= 490 \text{ N}/0.000\,001 \text{ m}^2 \\ &= 490 \text{ MPa or} \\ &\quad 490{,}000 \text{ kPa (71,076 psi)} \end{aligned}$$

Shear Strength (ASTM D-732)

Shear strength is the maximum load (stress) needed to produce a fracture, completely separating the movable part from the stationary by a shearing action. To calculate shear strength, divide the applied force by the area of the sheared edge.

$$\text{Shear strength (Pa)} = \frac{\text{force (N)}}{\text{area of sheared edge (m}^2)}$$

To shear a sample, several methods are common. Figure 6-10 shows three of them.

Impact Strength

Impact strength is not a measure of stress needed to break a sample. However, it does indicate the energy absorbed by

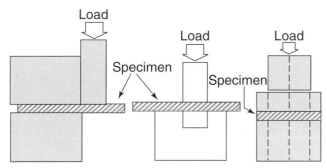

Figure 6-10. Various methods used to test shear strength.

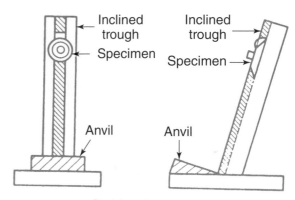

Guided drop test

Figure 6-12. Guided drop test.

the sample prior to its fracture. There are two basic methods for testing impact strength: falling mass tests and pendulum tests.

Falling mass test. (ASTM D-5420 and D-1709). ASTM D-5420 is the impact resistance of a flat rigid plastics specimen to a falling mass. This test is typically performed with a drop impact tester. In such testers, as shown in Figure 6-11A, a weight is guided from various heights by a fixed tube. Containers, dinnerware, and helmets are often tested in this manner. ASTM D-1709 is the impact resistance of a plastics film to a falling dart. This test is typically performed with a falling dart impact tester, where the film is fixed at the base of the tester, and a weighted dart is dropped from varying heights (Figure 6-11B). Sometimes the sample is allowed to slide down a trough and strike a metal anvil (Figure 6-12). This test may be repeated from various heights. If damaged, the sample will show cracks, chips, or other fractures.

Pendulum test. (ISO 179, 180, ASTM D-256, and D-618). Pendulum tests use the energy of a swinging hammer to strike the plastics sample. The result is a measure of energy or work absorbed by the specimen.

Here is the basic formula:

$$\text{Energy (J)} = \text{force (N)} \times \text{distance (m)}$$

The hammers of most plastics-testing machines have a kinetic energy of 2.7–22 J (2–16 ft-lb). Figures 6-13A to 6-13E show impact-testing machines and methods.

In the *Charpy* (simple beam) method, the test piece is supported at both ends, but not held down. The hammer strikes the sample in the center (Figures 6-13A and 6-13B). In the *Izod* (cantilever beam) method, the hammer strikes a specimen supported at one end.

Chip impact tests (ASTM D-4508) also set the hammer to strike one end of the sample, which is often positioned horizontally. Tensile impact tests (ASTM D-1822) hold the specimen in the hammer and swing it against a fixed striker.

Impact tests may specify notched or unnotched samples. In the Charpy test, the notch is on the side away from the striker. In the Izod test, the notch is on the same side as the striker, as shown in Figure 6-13C. In both tests, the depth and radius of the notch can dramatically alter the impact strength, especially if the polymer exhibits notch sensitivity. Figures 6-13D and 6-13E show two other Izod impact-testing machines.

PVC is a rather notch-sensitive material. If prepared with a blunt notch with a radius of 2 mm, PVC has a higher impact strength than ABS. If the samples have sharp notches with a radius of 0.25 mm, then the impact strength of PVC drops below that of ABS. Other materials that are notch brittle are acetals, HDPE, PP, PET, and dry PA.

Moisture in plastics can also influence impact strength. Polyamides (nylons) exhibit a large difference. They have impacts of 5 kJ/m² (50 ft-lbs) when thoroughly dried and strengths over 20 kJ/m² (200 ft-lbs) when containing moisture.

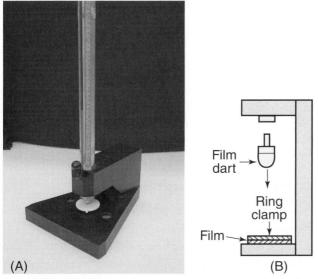

Figure 6-11. Falling mass tests: (A) Drop dart impact tester for rigid materials and (B) Drop dart tester for films.

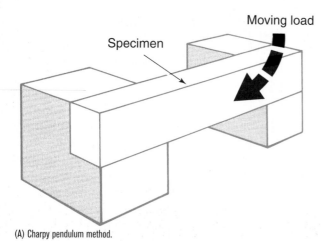

(A) Charpy pendulum method.

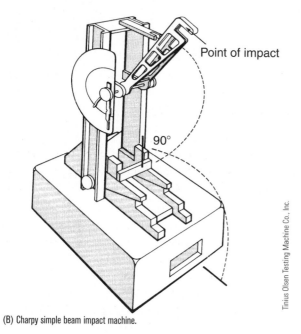

(B) Charpy simple beam impact machine.

Tinius Olsen Testing Machine Co., Inc.

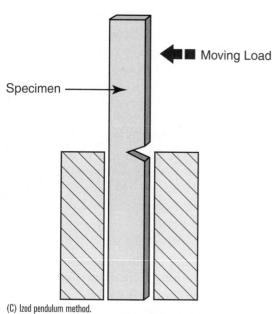

(C) Izod pendulum method.

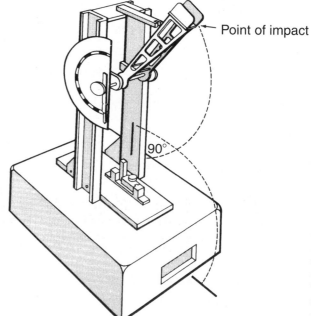

(D) Izod cantilever beam impact machine.

Tinius Olsen Testing Machine Co., Inc.

(E) Impact tester for Izod testing.

Tinius Olsen Testing Machine Co., Inc.

Figure 6-13. Charpy and Izod method testing equipment.

Because impact measurements must take the thickness of samples into account, impact-strength values are expressed in joules per square meter (J/m^2) or foot-pounds (ft-lbs) per inch of notch.

Flexural Strength (ISO 178, ASTM D-790 and D-747)

Flexural strength is the measure of how much stress (load) can be applied to a material before it breaks. Both tensile and compressive stresses are involved in bending the sample. Both the ASTM and ISO samples are supported on test blocks that span a minimum of 16 times the thickness. For example, if a

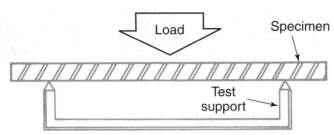

Figure 6-14. A method used to test flexural strength (flexural modulus).

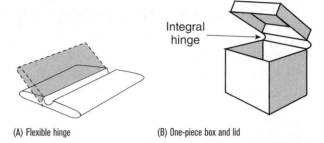

(A) Flexible hinge   (B) One-piece box and lid

specimen is 6 mm thick, the span would be 96 mm. The load is applied in the center (Figure 6-14).

Because most plastics do not break when deflected, the flexural strength at fracture cannot be calculated easily. Using the ASTM method, most thermoplastics and elastomers are measured when 5 percent strain occurs in the samples. This is found by measuring in pascals the load that causes the sample to stretch 5 percent. In the ISO procedure, the force is measured when the deflection equals 1.5 times the thickness of the sample.

Fatigue and Flexing (ISO 3385, ASTM D-430 and D-813)

Fatigue strength is a term used to express the number of cycles a sample can withstand before it fractures. Fatigue fractures are dependent on temperature and stress as well as frequency, amplitude, and mode of stressing.

If the load (stress) does not exceed the yield point, some plastics may be stressed for a great many cycles without failure. In producing integral hinges and one-piece box-and-lid containers, the fatigue characteristics of plastics must be considered. Figure 6-15 shows two integral hinges and apparatus for testing folding endurance.

Damping

Plastics can absorb or dissipate vibrations. This property is called damping. On an average, plastics have 10 times more damping capacity than steel. Gears, bearings, appliance housings, and architectural applications of plastics make effective use of this vibration-reducing property.

Hardness

The term hardness does not describe a definite or single mechanical property of plastics. Scratch, mar, and abrasion resistance are closely related to hardness. Surface wearing of vinyl floor tile and marring of PC optical lenses are affected by several factors. However, a widely accepted definition of hardness is the resistance to compression, indentation, and scratch.

There are several types of instruments used to measure hardness. Because each instrument has its own scale for measurement, the values must also identify the scale used. Two tests of

Photograph courtesy of Tinius Olsen

(C) A folding endurance tester, which records the number of flexings that take place before a plastics sample breaks.

Figure 6-15. Fatigue testing.

rather limited use in testing plastics are the Mohs scale and the scleroscope.

The Mohs hardness scale is used by geologists and mineralogists. It is based on the fact that harder materials scratch softer ones. The scleroscope shown in Figure 6-16

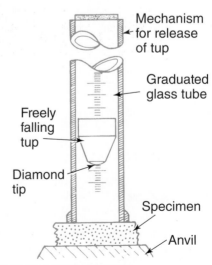

Figure 6-16. Scleroscope for making hardness tests.

is considered a nondestructive hardness test. The instrument measures the rebound height of a free-falling hammer called a *tup*.

Indentation test instruments (ASTM D-2240) are used for more sophisticated quantitative measurements. Rockwell, Wilson, Barcol, Brinell, and Shore are names of well-known testing tools. Figure 6-17 shows the basic differences in hardness tests and scales. Table 6-5 provides details about various hardness scales. In these tests, either depth or area of indentation is used as the measure of hardness.

The Brinell test compares hardness to the area of indentation. Here are typical Brinell numbers for selected plastics: acrylic, 20; polystyrene, 25; polyvinyl chloride, 20; and polyethylene, 2. Figure 6-18 shows a universal hardness tester.

The Rockwell test (ASTM D-785) relates hardness to the difference in depth of penetration of two different loads. The minor load (usually 10 kg) and the major load (from 60–150 kg) are applied to a ball-shaped indentor (Figure 6-19). These

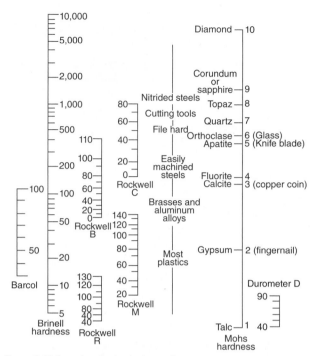

Figure 6-17. Comparison of various hardness scales.

Table 6-5. **Comparison of Selected Hardness Tests**

Instrument	Indentor	Load	Comments
Brinell	Ball, 10 mm diameter	500 kg 3,000 kg	Average out hardness differences in material. Load applied for 15–30 seconds. View through Brinell microscope shows and measures diameter (value, of impression). Not for materials with high creep factors.
Barcol	Sharp-point, rod 26° 0.157 mm flat tip	Spring loaded. Push against specimen with hand 5–7 kg.	Portable. Readings taken after 1 or 10 s.
Rockwell C	Diamond cone	Minor 10 kg Minor 150 kg	Hardest materials, steel. Table model.
Rockwell B	Ball 1.58 mm ($^1/_{16}$ in)	Minor 10 kg Minor 100 kg	Soft metals and filled plastics.
Rockwell R	Ball 12.7 mm ($^1/_2$ in)	Minor 10 kg Minor 60 kg	Within 10 seconds after applying minor load, apply major load. Remove major load 15 s after application. Read hardness scale 15 s after removing major load
Rockwell L	Ball 6.35 mm ($^1/_4$ in)	Minor 10 kg Minor 60 kg	or
Rockwell M	Ball 6.35 mm ($^1/_4$ in)	Minor 10 kg Minor 100 kg	Apply minor load and zero within 10 s. Apply major load immediately after zero adjust. Read number of divisions pointer passed during 15 s of major load.
Rockwell E	Ball 3.175 mm ($^1/_8$ in)	Minor 10 kg Minor 100 kg	Portable. Readings taken in soft plastics after 1 or 10 s.
Shore A	Rod, 1.40 mm diameter, sharpened to 35° 0.79 mm.	Spring loaded. Push against specimen with hand pressure.	Portable. Readings taken in soft plastics after 1 or 10 s.
Shore D	Rod, 1.40 mm diameter. Sharpened to 30° point with 0.100 mm radius.	As above	As above

Figure 6-18. A digital Brinell hardness tester.

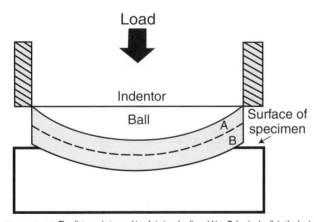

Figure 6-19. The distance between Line A (minor load) and Line B (major load) is the basis for Rockwell hardness readings.

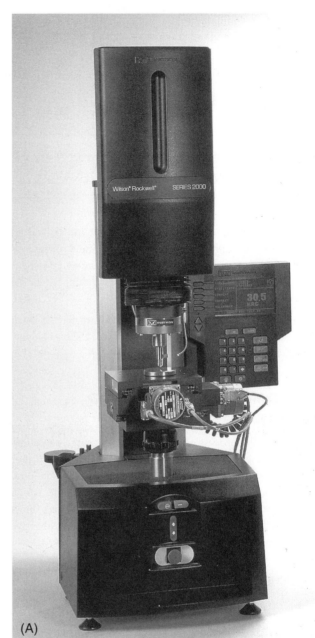

(A)

(B)

Figure 6-20. Rockwell testing. (A) Rockwell hardness tester. (B) A sample is positioned under the penetrator for hardness testing.

are typical Rockwell numbers for certain plastics: acrylic, M 100; polystyrene, M 75; polyvinyl chloride, M 115; and polyethylene, R 15. Figure 6-20 shows Rockwell tests in progress.

For soft or flexible plastics, the Shore durometer instrument (ASTM D-2240, ISO 868) may be used. There are two ranges

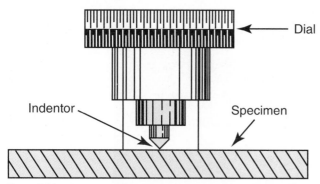

Figure 6-21. A sharp-pointed indentor is used on the Barcol instrument (ASTM D-2583).

of durometer hardness. Type A uses a blunt rod-shaped indentor to test soft plastics. Type D uses a pointed rod-shaped indentor to measure harder materials. The reading or value is taken after 1 or 10 seconds of applying pressure by hand. The scale range is 0 to 100.

The Barcol tester is similar to the Shore durometer, D type. It also uses a sharp-pointed indentor. Figure 6-21 shows a drawing of a Barcol tester.

Abrasion Resistance (ASTM D-1044)

Abrasion is the process of wearing away the surface of a material by using friction. The Williams, Lambourn, and Tabor abraders measure the resistance of plastics materials to abrasion. In each test, an abrader rubs against the sample, which removes some material. The amount of material loss (mass or volume) indicates how well the sample resists the abrasive treatment.

$$\text{Abrasion resistance} = \frac{\text{original mass} - \text{final mass}}{\text{relative density}}$$

PHYSICAL PROPERTIES

Mechanical properties require the basic forces of tension, compression, and shear. However, physical properties of plastics may not involve these forces. The molecular structure of the material often affects the physical properties. A few properties that will receive attention are relative density, mold shrinkage, tensile creep, and viscosity.

Density and Relative Density (ISO 1183, ASTM D-792 and D-1505)

Density is equal to mass per unit volume. The proper SI derived unit of density is kilograms per cubic meter, but it is commonly expressed as grams per cubic centimeter.

Example
Density = mass (kg)/volume (m³)
For PVC:
Density = 1300 kg/1 m³ or 1.3 g/cm³

Relative density is the ratio of the mass of a given volume of material to the mass of an equal volume of water at 23°C (73°F). Relative density is a dimensionless quantity that will remain the same in any measurement system.

Example
Relative density of PVC

$$\text{Relative density of PVC} = \frac{\text{density of PVC}}{\text{density of water}}$$

$$\frac{1300 \text{ kg/m}^3}{1000 \text{ kg/m}^3} = 1.3$$

The relative densities of a number of selected materials are given in Table 6-6. Notice that polyolefins have densities less than 1.0, which means they float in water.

A simple method for determining relative density is to weigh the sample both in air and water (ASTM D-792). A fine wire may be used to suspend the plastics sample in water from a laboratory balance, as shown in Figure 6-22. To measure relative density, divide the mass of a sample in air by the difference between that mass and its mass in water. Thus,

$$\text{Density} = \frac{\text{mass in air}}{\text{mass in air} - \text{mass in water}}$$

Notice that, if a sample floats, the mass of the sample and the wire when immersed will be less than the immersed wire only. The difference will yield a negative number, which indicates buoyancy. Consequently, the denominator of the equation will be larger than the numerator. Calculate the relative density by using the following formula:

$$D = \frac{a - b}{(a - b) - (c - d)}$$

a = mass of specimen and wire in air
b = mass of wire in air
c = mass of wire and specimen immersed in water
d = mass of wire with end immersed in water

Another method, given in ASTM D-1505, is a **density gradient column**. This column is composed of liquid layers that decrease in density from bottom to top. The layer to which the sample sinks shows its density. A density gradient column is rather complex and requires periodic maintenance to clean the column and verify that the layers are of specified density.

A simpler approach would be to create one or more mixtures of known density, as shown in Figure 6-23. For densities greater than water, prepare a solution of distilled water and calcium

Table 6-6. Relative Densities of Selected Materials

Substance	Relative Density
Woods (based on water)	
Ash	0.73
Birch	0.65
Fir	0.57
Hemlock	0.39
Red oak	0.74
Walnut	0.63
Liquids	
Acid, muriatic	1.20
Acid, nitric	1.217
Benzine	0.71
Kerosene	0.80
Turpentine	0.87
Water 20°C	1.00
Metals	
Aluminum	2.67
Brass	8.5
Copper	8.85
Iron, cast	7.20
Iron, wrought	7.7
Steel	7.85
Plastics	
ABS	1.02–1.25
Acetal	1.40–1.45
Acrylic	1.17–1.20
Allyl	1.30–1.40
Aminos	1.47–1.65
Casein	1.35
Celullosics	1.15–1.40
Chlorinated polyesters	1.4
Epoxies	1.11–1.8
Fluoroplastics	2.12–2.2
Ionomers	0.93–0.96
Phenolic	1.25–1.55
Phenylene oxidce	1.06–1.10
Polyamides	1.09–1.14
Polycarbonate	1.2–1.52
Polyester	1.01–1.46
Polyolefins	0.91–0.97
Polystyrene	0.98–1.1
Polysulfone	1.24
Silicones	1.05–1.23
Urethanes	1.15–1.20
Vinyls	1.2–1.55

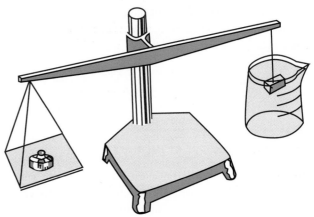

Figure 6-22. An analytical balance is used as shown to determine the relative density of plastics samples.

Relative density 1.20

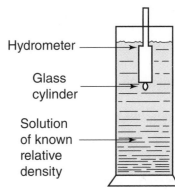

Hydrometer

Glass cylinder

Solution of known relative density

Figure 6-23. An arrangement for measuring density.

nitrate, and measure it with a technical-grade hydrometer. Add calcium nitrate until the desired density is obtained. For densities less than that of water, mix water and isopropyl alcohol to achieve the selected density.

When conducting density tests, remember that dirt, grease, and machining scratches may entrap air on the sample and cause inaccurate results. The presence of fillers, additives, reinforcements, and voids or cells will also alter the relative density.

Mold Shrinkage (ISO 2577, ASTM D-955)

The mold (linear) shrinkage influences the size of molded parts. Typical mold cavities are larger than the desired finished parts. When shrinkage of the parts is complete, they should meet dimensional specifications.

Mold parts shrink when they crystallize, harden, or polymerize in a mold. Shrinkage continues for some time after molding. To allow for complete post-mold shrinkage, do not take measurements until 48 hours have passed.

Mold shrinkage is the ratio of the decrease in length to the original length. The result is reported as mm/mm (in./in.) formula according to this:

$$\text{Mold shrink} = \frac{\text{length of cavity} - \text{length of molded bar}}{\text{length of cavity}}$$

Tensile Creep (ISO 899, ASTM D-2990)

When a mass suspended from a test sample causes the sample to change shape over a period of time, the strain is called creep. When creep occurs at room temperature, it is called cold flow.

Figure 6-24 depicts cold flow. The time duration required to go from the beginning of the test, *A*, to failure of the specimen, *E*, may be well over 1000 hours. Tensile creep test results report the strain in millimeters, as a percentage and as a modulus.

Creep and cold flow are very important properties to consider in the design of pressure vessels, pipes, and beams where a constant load (pressure or stress) may cause deformation or dimensional changes. PVC pipes undergo specialized creep tests to measure their ability to withstand given pressures over time and to determine the burst or rupture strength. Figure 6-25 shows a section of pipe being tested for burst strength. This sample ruptured while under 5.85 MPa (848 psi) of pressure.

Viscosity

The property of a liquid that describes its internal resistance to flow is called viscosity. The more sluggish the liquid, the greater its viscosity. Viscosity is measured in pascal-seconds (Pa × s) or units called poises (see Table 6-7).

Figure 6-25. A large diameter pipe hydrostatic burst unit.

Table 6-7. Viscosity of Selected Materials

Material	Viscosity, Pa × s	Viscosity, centipoises
Water	0.001	1
Kerosene	0.01	10
Motor oil	0.01–1	10–100
Glycerine	1	1000
Corn syrup	10	10000
Molasses	100	100000
Resins	<0.1 to $>10^3$	<100 to >106
Plastics (hot viscoelastic state)	$<10^2$ to $>10^7$	$<10^5$ to $>10^{10}$

Viscosity is an important factor in transporting resins, injecting plastics in a liquid state, and obtaining critical dimensions of extruded shapes. Fillers, solvents, plasticizers, thixotropic agents (materials that are gel-like until shaken), degree of polymerization, and density all may affect viscosity. The viscosity of a resin such as polyester ranges from 1 to 10 Pa 3 s (1000 to 10,000 centipoises). One centipoise equals 0.001 poise. In the metric system, one centipoise equals 0.001 pascal-second. For a complete definition of poise, see any standard physics text or other reference work in which viscosity is described.

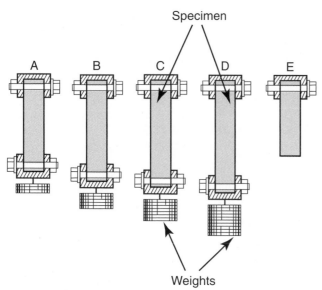

Specimen

A   B   C   D   E

Weights

Figure 6-24. Stages of creep and cold flow.

THERMAL PROPERTIES

Important thermal properties of plastics are thermal conductivity, specific heat, coefficient of thermal expansion, heat deflection, resistance to cold, burning rate, flammability, melt index, glass transition point, and softening point.

As thermoplastics are heated, molecules and atoms within the material begin to oscillate more rapidly. This causes the molecular chains to lengthen. More heat may cause slippage between molecules held by the weaker van der Waals forces. The material may become a viscous liquid. In thermosetting plastics, bonds are not easily set free. They must be broken or decomposed.

Thermal Conductivity (ASTM C-177)

Thermal conductivity is the rate at which heat energy is transferred from one molecule to another. Plastics are thermal insulators for the same molecular reasons that they are electrical insulators.

Thermal conductivity is expressed as a coefficient and is called the k factor. It should not be confused with the symbol K that indicates Kelvin temperature scale. Aluminum has a k factor of 122 W/K $\times$ m. Some foamed or cellular plastics have k values of less than 0.01 W/K $\times$ m (Table 6-8). The k values for most plastics show that they do not conduct heat as well as equal amounts of metal.

The flow of heat energy should be measured in watts, not calories or British thermal units (Btus) per hour. Watt (W) is the same as the joule per second (J/s). It is best to remember that the joule is a unit of energy and a watt is a unit of power.

Specific Heat (Heat Capacity)

Specific heat is the amount of heat required to raise the temperature of a unit of mass by 1 kelvin or 1 degree Celsius (Figure 6-26). It should be expressed in joules per kilogram per kelvin (J/kg $\times$ K). The specific heat at room temperature for ABS is 104 J/kg $\times$ K; for polystyrene, 125 J/kg $\times$ K; and for polyethylene, 209 J/kg $\times$ K. This indicates that it will take more thermal energy to soften crystalline plastics polyethylene than to soften ABS. The values for most plastics indicate that they require a greater amount of heat energy to raise their temperature than water, because the specific heat of water is 1. The amount of heat may be expressed in joules per gram per degree Celsius (J/g $\times$ °C).

Thermal Expansion (ASTM D-696 and D-864)

Plastics expand at a much greater rate than metals, making it difficult to join them. Figure 6-27 shows the differences in coefficients of expansion of selected materials. The coefficient of expansion is used to determine thermal expansion in length, area, or volume per unit of temperature rise. It is expressed as a ratio per degree Celsius. ASTM D-696 provides standards for determining linear thermal expansion, whereas ASTM D-864 involves cubic thermal expansion.

If a PVC rod 2 m in length is heated from –20°C to 50°C (–4 to 122°F), it will change 7 mm (0.27 in.) in length.

Table 6-8. Thermal Conductivity of Selected Materials

Material	Thermal Conductivity (k-factor), W/K × m	Thermal Resistivity (R-factor), K × m/W	Thermal Conductivity, Btu × in/hr × ft² °F
Acrylic	0.18	5.55	1.3
Aluminum (alloyed)	122	0.008	840
Copper (beryllium)	115	0.008	800
Iron	47	0.021	325
Polyamide	0.25	4.00	1.7
Polycarbonate	0.20	5.00	1.4
Steel	44	0.022	310
Window glass	0.86	1.17	6.0
Wood	0.17	5.88	1.2

*R-factor is the reciprocal of the k factor.

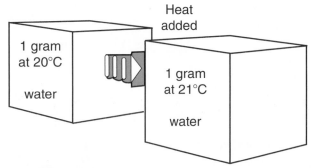

Figure 6-26. How much heat was added?

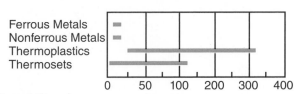

Figure 6-27. Coefficient of expansion (per °C 3 10–6).

Example:

Change in length = coefficient of linear expansion $\times$
original length $\times$ change in temperature

$$= \frac{0.000050}{°C \times 2\,m \times 70°}$$

$$= 0.007\,m, \text{ or } 7\,mm$$

Because area is a product of two lengths, the value of the coefficient must double. Similarly, we must triple the coefficient value to obtain thermal expansion for volume. Table 6-9 shows the thermal expansion of selected materials.

Deflection Temperature (ISO 75, ASTM D-648)

Deflection temperature (formerly called heat distortion) is the highest continuous operating temperature that a material can withstand. In general, plastics are not often used in high-heat environments. However, some special phenolics have been subjected to temperatures as high as 2760°C (4032°F).

A device that provides heat, pressure, linear measurement, and a printout of results is shown in Figure 6-28. In the ASTM test, a specimen (3.175 mm–3.140 mm) is placed on supports 100 mm apart, and a force of 455–1820 kPa is pressed on the sample. The temperature is raised 2°C per minute. The temperature at which the sample will deflect 0.25 mm is reported as the deflection temperature.

In addition to the standard test, a number of nonstandard tests provide information about the heat deflection of various plastics. For instance, materials may be tested in an oven. The temperature is raised until the material chars, blisters, distorts, or loses appreciable strength. Sometimes boiling water provides the heat and temperature level. Figure 6-29 shows a deflection experiment using an infrared radiant heater.

Test bars of glass-reinforced polycarbonate, polysulfone, and thermoplastic polyester were locked in a laboratory vise. Equal 175 g loads were applied. After only 1 minute under 155°C (311°F) heat from an infrared radiant heater, the polycarbonate bar began to deflect. One minute later the polysulfone bar followed suit. The thermoplastic polyester bar was still not bent after 6 minutes at 185°C (365°F).

Table 6-9. Thermal Expansion of Selected Materials

Substance	Coefficient of Linear Expansion $\times 10^{-6}$, mm/mm * °C
Nonplastics	
Aluminum	23.5
Brass	18.8
Brick	5.5
Concrete	14.0
Copper	16.7
Glass	9.3
Granite	8.2
Iron, cast	10.5
Marble	7.2
Steel	10.8
Wood, pine	5.5
Plastics	
Diallyl phthalate	50–80
Epoxy	40–100
Melamine-formaldehyde	20–57
Phenol-formaldehyde	30–45
Polyamide	90–108
Polyethylene	110–250
Polystyrene	60–80
Polyvinylidene chloride	190–200
Polytetrafluoroethylene	50–100
Polymethyl methacrylate	54–110
Silicones	8–50

Figure 6-28. This automatic heat deflection tester can test up to six samples simultaneously.

Photograph courtesy of Tinius Olsen

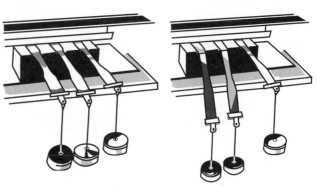

Figure 6-29. Heat deflection testing. (A) Before heat was applied. (B) After two minutes of heating.

Ablative Plastics

Ablative plastics have been used for spacecraft and missiles. Upon reentry, the temperature of the outer surface of a heat shield may be greater than 13,000°C (23,800°F), whereas the inner surface is no more than 95°C (203°F). Ablative plastics may be composed of phenolic or epoxy resin and graphite, asbestos, or silica matrixes.

In ablative materials, heat is absorbed through a process known as pyrolysis. This takes place in the near-surface layer exposed to heat energy. Much of the plastic is consumed and sloughed off as large amounts of heat energy are absorbed.

Resistance to Cold

As a rule, plastics have good resistance to cold. Food packages made of polyethylene routinely withstand temperatures of –60°F (–51°C). Some plastics can withstand extremely low temperatures of –19°F (–196°C) with little loss of physical properties.

Flammability (ISO 181, 871, and 1210; ASTM D-635, D-568, and E-84)

Flammability, also called *flame resistance*, is a term that indicates the measure of the ability of a material to support combustion. Several tests measure this property. In one test, a plastics strip is ignited and the heat source (flame) is removed. The time and amount of material consumed are measured, and the result is expressed in mm/min. Highly combustible plastics such as cellulose nitrate will have high values.

Self-extinguishing is a rather loosely used term related to flammability. It indicates that the material will not continue to burn once the flame is removed. Nearly all plastics may be made self-extinguishing with the proper additives.

Table 6-10 indicates that plastics will burn when exposed to direct flame. To cause self-ignition, the temperature must be higher than the temperature of ignition from a direct flame.

Melt Index (ISO 1133, ASTM D-1238)

Viscosity and flow properties affect both the processing of plastics and the design of molds. Metal viscosity provides the most accurate data, but melt index values are common because they require little time.

Melt flow index is a measure of the amount of material in grams that is extruded through a small orifice at a given pressure and temperature in 10 minutes. A common load is 43.5 psi (300 kPa). The ASTM procedure specifies temperatures of

Table 6-10. Ignition Temperatures and Flammability of Various Materials

Material	Flash-ignition Temperature, °C	Self-ignition Temperature, °C	Burning Ratio, mm/min
Cotton	230–266	254	SB
Paper, newsprint	230	230	SB
Douglas fir	260		SB
Wool	200		SB
Polyethylene	341	349	7.62–30.48
Polypropylen, fiber		570	17.78–40.64
Polytetrofluoro-ethylene		530	NB
Polyvinyl chloride	391	454	SE
Polyvinylidene chloride	532	532	SE
Polystyrene	345–360	488–496	12.70–63.5
Polymethyl methacrylate	280–300	450–462	15.24–40.64
Acrylic, fiber		560	SB
Cellulose nitrate	141	141	Rapid
Cellulose acetate	305	475	12.70–50.80
Cellulose triacetate, fiber	540	SE	
Ethyl cellulose	291	296	27.94
Polyamide (Nylon)	421	424	SE
Nylon 6, 6, fiber		532	SE
Phenolic, glass fiber laminate	520–540	571–580	SE-NB
Melamine, glass fiber laminate	475–500	623–645	SE
Polyester, glass fiber laminate	346–399	483–488	SE
Polyurethane, polyether, rigid foam	310	416	SE
Silicone, glass fiber laminate	490–527	550–564	SE
NB–Nonburning SE–Self-extinguishing SB–Slow burning			

190°C (374°F) for polyethylene and 230°C (446°F) for polypropylene. The ISO method specifies die diameter, temperature, die factor, reference time, and nominal load. Figure 6-30 shows a melt index measuring device.

A high melt index value indicates a low-viscosity material. Usually, low-viscosity plastics have a relatively low molecular mass. In contrast, high molecular mass materials are resistant to flow and have lower melt index values.

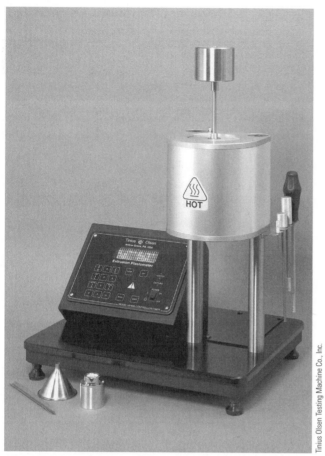

Timius Olsen Testing Machine Co., Inc.

Figure 6-30. This melt index tester includes a microprocessor-based controller and timer.

Table 6-11. Glass Transition of Selected Amorphous Plastics

Plastic	T_g °C
ABS	110
PC	150
PMMA	105
PS	95
PVC	85

Table 6-12. Glass Transition of Selected Crystalline Plastics

Plastics	T_g °C	T_m °C
PA	50	265
PE	−35	130
PETE	65	265
PP	−10	165

Figure 6-31 graphically shows the difference between amorphous and crystalline materials. Note the two points of inflection on the curve for crystalline materials.

Softening Point (ISO 306, ASTM D-1525)

In the **Vicat softening point** test, a sample is heated at a rate of 50°C (120°F) per hour. The temperature at which a needle penetrates the sample—0.039 in. (1 mm)—is the Vicat softening point.

ENVIRONMENTAL PROPERTIES

Plastics are found in nearly every environment. They are used as containers for chemicals, packages for food storage, and as medical implants inside the human body. Before a product is designed, plastics must be tested for endurance under expected environmental extremes. The environmental properties of plastics include chemical resistance, weathering, ultraviolet resistance, permeability, water absorption, biochemical resistance, and stress cracking.

Chemical Properties

The statement that "most plastics resist weak acids, alkalies, moisture, and household chemicals" must be used only as a broad rule. Any statement about the response of plastics to chemical environments must be a generalization only. It is best to test each plastics to determine how it can be specifically applied and which chemicals it is expected to resist.

Glass Transition Temperature

At room temperature, the molecules in amorphous plastics are in rather limited motion. As an amorphous material heats up, the relative motion of the molecules increases. When the material reaches a certain temperature, it loses its rigidity and becomes leathery. That temperature is identified as the **glass transition temperature** (T_g). Often glass transition temperatures are reported as a range of temperatures because the transition does not occur at one specific temperature. The glass transition points for several amorphous plastics are shown in Table 6-11.

Crystalline plastics contain both crystalline and amorphous regions. Consequently, they exhibit two changes upon heating. When the temperature is high enough, the amorphous regions change from glass-like to flexible. As the temperature continues to rise, the energy disrupts the crystalline regions, causing the material to become a viscous liquid throughout. This transition occurs over a limited temperature range. It is identified as the melting temperature (T_m). Table 6-12 shows both the T_g and T_m for selected crystalline plastics.

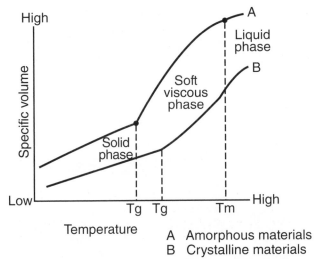

A Amorphous materials
B Crystalline materials

Figure 6-31. This melt index tester includes a microprocessor-based controller and timer.

The chemical resistance of plastics depends to a large degree on the elements combined into molecules and on types and strengths of chemical bonds. Some combinations are very stable, whereas others are quite unstable. Polyolefins are exceptionally inert, nonreactive, and resistant to chemical attack. This is due to the C–C bonds in the backbone of the molecules, which are very stable. In contrast, polyvinyl alcohol contains hydroxyl groups (–OH) attached to the carbon backbone of the molecule. Bonds holding the hydroxyl groups onto the main chain break down in the presence of water.

Table 6-13 lists the chemical resistance of a number of plastics. This table provides information only about natural materials. However, fillers, plasticizers, stabilizers, colorants, and catalysts can affect the chemical resistance of plastics.

The resistance of plastics to organic solvents (**solvent resistance**) can provide information for the identification of unknown materials (see Appendix D on Material Identification). The reactivity of both plastics and organic solvents has been assigned a **solubility parameter**. In principle, a polymer will dissolve in a solvent with a similar or lower solubility parameter. This general principle may not apply in all cases due to crystallization, hydrogen bonding, and other molecular interactions. Table 6-14 contains the solubility parameters of selected solvents and plastics.

Weathering (ASTM D-2565, D-4329, G-154, and G-155; ISO-4892)

Many weathering tests are conducted in Florida, where samples receive considerable exposure to heat, moisture, and sunlight. Exposed samples are rated on color change, gloss change, cracks and crazing, and loss of physical properties. Because weathering tests are done over long periods of

Table 6-13. Chemical Resistance of Selected Plastics at Room Temperature

Plastics	Strong Acids	Strong Alkalies	Organic Solvents
Acetal	Attacked	Resistant	Resistant
Acrylic	Attacked	Slight	Attacked
Cellulose acetate	Affected	Affected	Attacked
Epoxy	Slight	Slight	Slight
Ionomer	Slight	Resistant	Resistant
Melamine	Slight	Slight	Resistant
Phenolic	Resistant	Attacked	Affected
Phenoxy	Resistant	Resistant	Attacked
Pollallomer	Resistant	Resistant	Resistant
Polyamide	Attacked	Slight	Resistant
Polycarbonate	Resistant	Attacked	Attacked
Polychlorotri-fluoroethylene	Resistant	Resistant	Resistant
Polyester	Slight	Affected	Affected
Polyethylene	Resistant	Resistant	Affected
Polyimide	Affected	Attacked	Resistant
Polyphenylene oxide	Resistant	Resistant	Slight
Polypropylene	Resistant	Resistant	Resistant
Polysulfone	Resistant	Resistant	Affected
Polystyrene	Affected	Resistant	Affected
Polytetrafluoro ethylene	Resistant	Resistant	Resistant
Polyurethane	Resistant	Affected	Slight
Polyvinyl chloride	Resistant	Resistant	Affected
Silicone	Slight	Affected	Slight

time, accelerated tests try to provide similar exposure in a shorter amount of time. Figure 6-32 shows accelerated weathering testers. These testers reproduce the damage caused by sunlight, rain, and dew. They test materials by exposing them to alternating cycles of light and moisture at controlled, elevated temperatures. Fluorescent ultraviolet (UV) lamps simulate the effects of sunlight, whereas condensing humidity or water spray simulates the effects of dew or rain. Xenon arc testers provide full spectrum light and consequently are better at determining color changes than the fluorescent testers.

Ultraviolet Resistance (ASTM G-23 and D-2565)

The resistance of plastics to the effects of direct sunlight or artificial weathering devices is linked with weatherability.

Table 6-14. Solubility Parameters of Selected Solvents and Plastics

Solvent	Solubility Parameter
Water	23.4
Methyl alcohol	14.5
Ethyl alcohol	12.7
Isopropyl alcohol	11.5
Phenol	14.5
n-Butyl alcohol	11.4
Ethyl acetate	9.1
Chloroform	9.3
Trichloroethylene	9.3
Methylene chloride	9.7
Ethylene dichloride	9.8
Cyclohexanine	9.9
Acetone	10.0
Isopropyl acetate	8.4
Carbon tetrachloride	8.6
Toluene	9.0
Xylene	8.9
Methyl isopropyl ketone	8.4
Cyclohexane	8.2
Turpentine	8.1
Methyl amyl acetate	8.0
Methyl cyclohexane	7.8
Heptane	7.5

Plastics	Solubility Parameter
Polytetrafluoroethylene	6.2
Polyethylene	7.9–8.1
Polypropylene	7.9
Polystyrene	8.5–9.7
Polyvinyl acetate	9.4
Polymethyl methacrylate	9.0–9.5
Polyvinyl chloride	9.38–9.5
Bisphenol A polycarbonate	9.5
Polyvinylidene chloride	9.8
Polyethylene terephthalate	10.7
Cellulose nitrate	10.56–10.48
Cellulose acetate	11.35
Epoxide	11.0
Polyacetal	11.1
Polyamide 6, 6	13.6
Coumarone indene	8.0–10.6
Alkyd	7.0–11.2

Ultraviolet radiation (combined with water and other environmental oxidizing conditions) may cause color fading, pitting, crumbling, surface cracking, crazing, and brittleness.

(A) This accelerated weathering tester uses a fluorescent ultraviolet light source combined with moisture.

(B) This accelerated weathering tester uses a xenon arc light source, which generates a full spectrum of ultraviolet light.

(C) This device, installed in Florida, tracks the sun across the sky and focuses the light on the test panels. It greatly accelerates outdoor weathering.

Figure 6-32. Accelerated weathering test equipment.

Permeability (ISO 2556, ASTM D-1434 and E-96)

Permeability may be described as the volume or mass of gas or vapor penetrating an area of film in 24 hours. Permeability is an important concept in the food packaging industry. In some applications, a packaging film must allow oxygen to pass through, keeping meats and vegetables looking fresh. Other applications must selectively prevent gases, moisture, and other agents from contaminating the package contents. Packages frequently contain several layers of different materials to achieve the desired control of permeability.

Water Absorption (ISO 62, 585, and 960; ASTM D-570)

Some plastics are hygroscopic. That means that they absorb moisture, usually by taking on water from humid air. Table 6-15 contains water absorption data for selected hygroscopic plastics. These materials must be dry before entering any processes that involve heating or melting. If they are not properly dried, the moisture in these plastics will turn to steam, which may cause surface defects and voids in the material. To verify that drying equipment is functioning properly, many companies periodically test samples for moisture content.

One simple test involves accurately weighing a sample, heating it in an oven for a period of time, and reweighing it to determine the weight loss. Some instruments provide rapid, accurate results based on this thermogravimetric principle.

The thermogravimetric method makes the assumption that all weight loss represents moisture. That assumption is not always accurate because some materials also lose lubricants, oils, and other volatiles when heated. To provide extremely accurate measurements of moisture content, a moisture-specific apparatus is needed. Figure 6-33 shows a moisture meter that heats a sample and draws the evolved gases into an analysis cell that traps only water vapor. The result is a test that accurately measures moisture in a specimen. Two simple, low-cost methods

Figure 6-33. A moisture-specific moisture tester.

of checking moisture content are the Tomasetti's Volatile Indicator (TVI) and the test tube/hot block (TTHB) techniques. Follow the procedure in Figure 6-34 for the TVI technique.

1. Place two glass slides on a hot plate and heat from 1 to 2 minutes at 275°C–348°C (527°F–659°F) (Figure 6-34).
2. Place four pellets or granular plastics samples on one of the glass slides.
3. Place the second hot slide on top of the sample and press the sandwiched pellets to about 10 mm (0.393 in.) diameter.
4. Remove the slides from the hot plate platen and allow them to cool.
5. The number and size of bubbles seen in the plastics samples indicate the percentage of moisture absorbed. Some bubbles may be the result of trapped air. However, numerous bubbles indicate moisture-laden material. There will be a direct correlation between the number of bubbles and moisture content.

The TTHB procedure is shown in Figure 6-35:

1. Heat a hot block with test tube holes to 26°C ± 10°C (500°F ± 50°F).
2. Place 5.0 g of plastics in a 20 × 150 mm Pyrex® test tube.
3. Place a stopper in the test tube, and carefully place the test tube in the hot block.
4. Allow the material to melt (about 7 minutes).
5. Remove the tube and sample from the hot block, and allow it to cool for 10 minutes.
6. Observe and record the correlation of moisture content with the surface area of the condensation in the test tube.

Biochemical Resistance (ASTM G-21 and G-22)

Although most plastics are resistant to bacteria and fungi, some plastics and additives are not. They may not be approved

Table 6-15. Water Absorption

Material	Water Absorbed, percent (24-Hour Immersion)
Polychlorotrifluoroethylene	0.00
Polyethylene	0.01
Polystyrene	0.04
Epoxy	0.10
Polycarbonate	0.30
Polyamide	1.50
Cellulose acetate	3.80

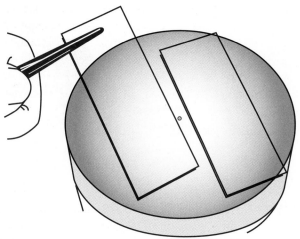

(A) Plug in the hot plate and calibrate it to a surface temperature of 270°C +10°C (518°F). Be sure the surface is clean; place two glass slides on the surface for 1–2 minutes.

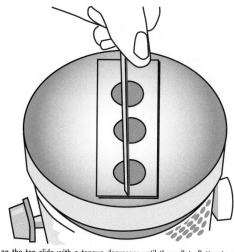

(D) Press on the top slide with a tongue depressor until the pellets flatten to about 10 mm in diameter.

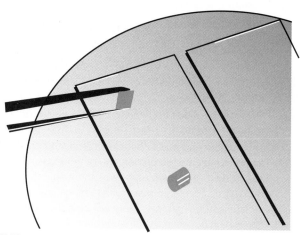

(B) When the glass surface temperature reaches 230°C–°C (446°F–500°F) place four or five pellets on one glass slide, using tweezers.

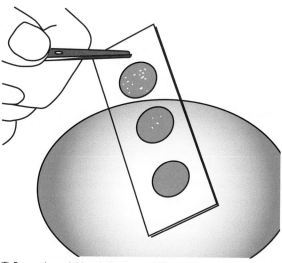

(E) Remove the sandwich and allow it to cool. The amount and size of bubbles indicate the percentage of moisture.

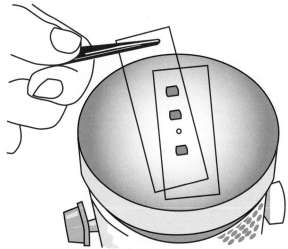

(C) Place the second hot slide over the pellets to form a sandwich.

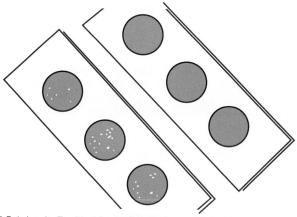

(F) Typical results. The slide at the right indicates dry material; the slide at the left indicates moisture-laden material. One or two bubbles may be only trapped air.

Figure 6-34. The six simple steps of the resin moisture test.

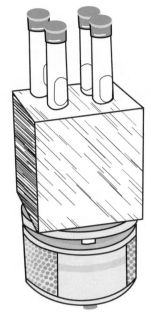

(A) Plastics samples being heated to drive off moisture.

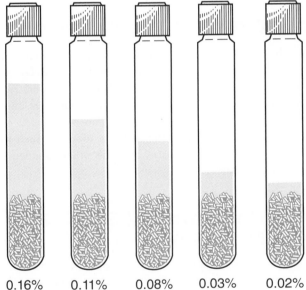

0.16%   0.11%   0.08%   0.03%   0.02%

(B) The area of condensation on the test tube surface shows the percentage of moisture in each plastics sample.

Figure 6-35. The test tube/hot block method of moisture measurement.

by the US Food and Drug Administration (FDA) for use in packaging or as containers for food and drugs. Various preservatives or antimicrobial agents may be added to plastics to make them resistant.

Stress Cracking (ISO 4600 and 6252, ASTM D-1693)

Environmental stress cracking of plastics may be caused by solvents, radiation, or constant strain. There are several tests that expose the sample to a surface agent.

One such test holds various test bars in a flexed position in a fixture. Selected solvents are then sprayed on the samples. Acetone spray has a dramatically different impact on polysulfone than it has on thermoplastic polyester. The polyester is likely to display no response to the acetone spray, whereas the polysulfone frequently cracks instantly into two pieces. Polyester also withstands stresses in the presence of carbon tetrachloride, methyl ethyl ketone, and other aromatic chemicals.

OPTICAL PROPERTIES

Optical properties are closely linked with molecular structure. Therefore, the electrical, thermal, and optical properties of plastics are interrelated. Plastics exhibit many optical properties. Among the most important are gloss, transparency, clarity, haze, color, and refractive index.

Specular Gloss (ASTM D-2457)

Specular gloss is the relative luminous reflectance factor of a plastics sample. A glossmeter directs light onto the sample at incidence angles of 20°, 45°, and 60°. The light that reflects off the surface is collected and measured by a photosensitive device. A perfect mirror is used as a standard and yields values of 1000 for the 20° and 60° incidence angles. Test results on plastics samples provide comparative data that can rate samples and estimate surface flatness. Comparisons should only be made between similar types of samples. For example, opaque films should not be compared to transparent films.

Luminous Transmittance (ASTM D-1003)

A cloudy or milky appearance in plastics is known as haze. A plastics termed *transparent* is one that absorbs very little light in the visible spectrum. *Clarity* is a measure of distortion seen when viewing an object through a transparent plastic. All these terms relate to the test for luminous transmittance.

Luminous transmittance is the ratio of transmitted light to incident light. During this test, a beam of light passes through the air and into a receptor. This measures the incident light. After positioning a sample, the light shines through it and into the receptor. The ratio of the reading through a sample to the reading through air provides a measure of total transmittance.

Unfilled amorphous plastics are the most transparent of the plastics. Even small amounts of fillers, colorants, and other additives interfere with the passage of light.

Color

The perception of color requires three things: a light source, an object, and an observer. When the light reaches the object, some of the rays are reflected and others are absorbed. If the object does not absorb the blue frequencies but absorbs all other rays, the observer will perceive the object as blue.

Many products contain several parts that should match each other in color. To describe color, three components are used. Value, or lightness, is the range from dark to light. Black is one extreme of the value range, white is the other extreme, and gray lies in the middle. Chroma, or saturation, represents the intensity of the color. Hue distinguishes between red, blue, yellow, and green.

Currently, two methods are widely used to measure color. A spectrophotometer is an instrument that compares the color of a standard to the color of a sample. It then assigns numeric results to value, chroma, and hue. Figure 6-36 shows a portable color measurement instrument. Another color measuring method is the human eye. Individuals with acute color discrimination abilities view samples in a light booth under differing light sources. Although the spectrophotometers are valuable tools, the final judgment often depends on visual comparison of samples.

Index of Refraction (ISO 489, ASTM D-542)

When light enters a transparent material, part of the light is reflected and part is refracted (Figure 6-37). The index of

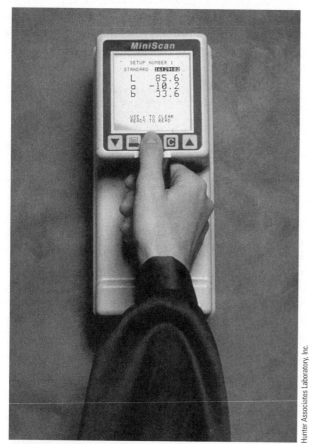

Figure 6-36. A portable color meter.

Hunter Associates Laboratory, Inc.

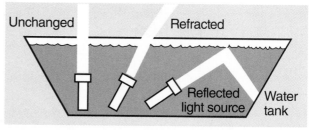

Figure 6-37. Reflection and refraction of light.

refraction (n) may be expressed in terms of the angle of incidence (i) and the angle of refraction (r).

$$n = \frac{\sin i}{\sin r}$$

where i and r are taken relative to the perpendicular to the surface at the point of contact. The index of refraction for most transparent plastics is about 1.5. This does not differ greatly from most window glass. Table 6-16 gives indexes of refraction for selected plastics.

ELECTRICAL PROPERTIES

Five basic properties that describe the electrical behavior of plastics are resistance, insulation resistance, **dielectric strength**, dielectric constant, and dissipation (power) factor. The predominantly covalent bonds of polymers limit their electrical conductivity and cause most plastics to be electrical insulators. With the addition of fillers such as graphite or metals, plastics can be made either conductive or semiconductive.

Arc Resistance (ISO 1325, ASTM D-495)

Arc resistance is a measure of time needed for a given electrical current to render the surface of a plastics conductive due to carbonization. This measurement is reported in seconds. The higher the value, the more resistive the plastics is to arcing. The breakdown in arc resistance may be the result of corrosive

Table 6-16. Optical Properties of Plastics

Material	Refractive Index	Light Transmission, %
Methyl methacrylate	<1.49	94
Cellulose acetate	1.49	87
Polyvinyl chloride acetate	1.52	83
Polycarbonate	1.59	90
Polystyrene	1.60	90

chemicals. Ozone, nitric oxides, or a buildup of moisture or dust may also lower values.

Resistivity (ISO 3915, ASTM D-257)

Insulation resistance is the resistance between two conductors of a circuit or between a conductor and ground when they are separated by an insulator. The insulation resistance is equal to the product of the resistivity of the plastics and the quotient of its length divided by its area.

$$\text{Insulation resistance} = \frac{\text{resistivity} \times (\text{length})}{\text{area}}$$

Resistivity is expressed in ohm-centimeters. Table 6-17 gives resistivities for certain plastics.

Dielectric Strength (ISO 1325 and 3915, ASTM D-149)

Dielectric strength is a measurement of electrical voltage required to break down or arc through a plastics material. The units are represented as volts per millimeter of thickness (V/mm). This electrical property gives an indication of the ability of a plastics to act as an electrical insulator. See Figure 6-38 and Table 6-17.

Dielectric Constant (ISO 1325, ASTM D-150)

The *dielectric constant* of a plastics is a measure of the ability of the plastics to store electrical energy, as shown in Figure 6-39. Plastics are used as dielectrics in the production of capacitors used in radios and other electronic equipment. The dielectric constant is based on air, which has a value of 1.0. Plastics with

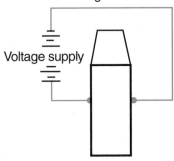

How great can the voltage be before it breaks through the material?

Voltage supply

Figure 6-38. Testing dielectric strength, an important characteristic of plastics for insulating applications.

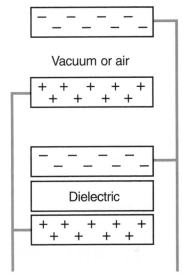

Vacuum or air

Dielectric

Figure 6-39. The dielectric constant is the amount of electricity stored across an insulating material, divided by the amount of electricity stored across air or a vacuum.

a dielectric constant of 5 will have five times the electricity-storing capacity of air or a vacuum.

Nearly all electrical properties of plastics will vary according to time, temperature, or frequency. For example, the values

Table 6-17. Electrical Properties of Selected Plastics

Plastics	Resistivity, ohm-cm	Dielectric Strength, V/mm	Dielectric Constant		Dissipation (Power) Factor	
			At 60 Hz	At 10^6 Hz	At 60 Hz	At 10^6 Hz
Acrylic	10^{16}	15,500–19,500	3.0–4.0	2.2–3.2	0.04–0.06	0.02–0.03
Cellulosic	10^{15}	8,000–23,500	3.0–7.5	2.8–7.0	0.005–0.12	0.01–0.10
Fluoroplastics	10^{18}	10,000–23,500	2.1–8.4	2.1–6.43	0.0002–0.04	0.0003–0.17
Polyamides	10^{15}	12,000–33,000	3.7–5.5	3.2–4.7	0.020–0.014	0.02–0.04
Polycarbonate	10^{16}	13,500–19,500	2.97–3.17	2.96	0.0006–0.0009	0.009–0.010
Polyethylene	10^{16}	17,500–39,000	2.25–4.88	2.25–2.35	<0.0005	<0.0005
Polystyrene	10^{16}	12,000–23,500	2.45–2.75	2.4–3.8	0.0001–0.003	0.0001–0.003
Silicones	10^{15}	8,000–21,500	2.75–3.05	2.6–2.7	0.007–0001	0.001–0.002

may vary as frequency is increased (see Table 6-17 for dielectric constant and dissipation factor).

Dissipation Factor (ASTM D-150)

Dissipation (power) factor, or loss tangent, also varies with frequency. It is a measure of the power (watts) lost in the plastics insulator. A test similar to those used in determining the value of the dielectric constant can measure this power loss. As a rule, measurements are made at 1 million hertz. They indicate the percentage of alternating current lost as heat within the dielectric material. Plastics with low dissipation factors waste little energy and do not become overheated. For some plastics, this is a disadvantage because they cannot be preheated or heat sealed by high-frequency methods of heating (see Table 6-17 for various dissipation factors).

The relationship between heat, current, and resistance is shown in the following power equation:

$$P = I^2 R$$

The power P used to perform wasted work represents lost or dissipated power. In this formula, the amount of power can be decreased by lowering either the current (I) or the resistance (R). In electrical appliances designed to produce heat, a low dissipation factor is not considered desirable.

RELATED INTERNET SITES

- **www.astm.org.** The American Society for Testing and Materials (ASTM) maintains an extensive website. Selection of "Standards" on the home page leads to a "Browse by Interest Area" opportunity. Choosing "Plastics" in the browse list brings up hundreds of standard test methods. A summary is available for each one, and it is possible to purchase the full documents.

- **www.iso.org.** The International Organization for Standardization (ISO) publishes its standards and test methods for plastics and rubber materials. To find these standards, select "Standards" on the home page. Then look for the link "Browse the ISO catalogue." Field 83 represents rubber and plastic industries, with subset 83.080 plastics materials, and 83–080.20 the standards for thermoplastics.

- **www.element.com.** Element is a global network of laboratories involved in material testing. On the home page, select "Materials" to access "Composites," "Polymers," and "Rubber And Elastomers."

Many companies make testing equipment. Here are the sites of some leading companies:

- **www.azic.com.** Arizona Instrument LLC.

- **www.ccsi-inc.com.** CCSi, Inc.

- **www.cwbrabender.com.** C.W. Brabender Instruments, Inc.

- **www.goettfert.com.** Goettfert Inc.

- **www.instron.com.** Instron Corp.

- **www.photovolt.com.** Photovolt Instruments Inc.

- **www.q-panel.com.** Q-Lab Corp.

- **www.tiniusolsen.com.** Tinius Olsen, Inc.

VOCABULARY

The following vocabulary words are found in this chapter. Use the glossary in Appendix A to look up the definitions of any of these words you do not understand as they apply to plastics.

cold flow
compressive strength
creep
damping
density gradient column
dielectric strength
fatigue strength
flexural strength
glass transition temperature
gravity constant
hardness
haze
hygroscopic
impact strength
melt flow index
Mohs scale
offset yield
relative density
scleroscope
shear strength
solubility parameter
solvent resistance
specular gloss
strain
stress
tensile strength
thixotropic agents
toughness
Vicat softening point
viscosity

QUESTIONS

6-1. Name the seven base units of the SI metric system.

6-2. A gigahertz is equal to _____ Hz.

6-3. Identify the SI metric unit for force and its formula.

6-4. Tensile strength, modulus of elasticity, and air pressure are measured in _____.

6-5. In the SI metric system, temperatures are measured in _____.

6-6. Two international technical societies that develop standards and specifications for testing of plastics are _____ and _____.

6-7. True or false: In testing mechanical properties, it is generally important to apply force at a specified rate.

6-8. Young's modulus is the ratio of _____ to _____.

6-9. To choose a tougher plastic, choose one having a _____ area under the stress-strain curve.

6-10. The pendulum test measures _____.

6-11. A plastics hinge relies on which property?

6-12. Resistance to transmitting vibration is called _____.

6-13. Viscosity is defined as a measure of the _____ of a fluid.

6-14. Elongation over time due to a constant force is called _____.

6-15. Plastics for spacecraft heat shields are chosen for their _____ properties.

6-16. As the melt index value of a plastic goes up, the viscosity goes _____.

6-17. Below the glass transition temperature, a plastic becomes _____.

6-18. Name a plastic that is hygroscopic.

6-19. Fillers used to make plastics conduct electricity are _____ and _____.

6-20. In the arc resistance test, the surface of a sample becomes conductive because of _____.

6-21. If the resistivity of a material is high, insulation resistance will be _____.

6-22. Dielectric strength indicates suitability of a plastic for service as _____.

6-23. Plastics used in electrical capacitors must have a high _____.

6-24. For heat sealing a plastic film by high-frequency methods, the _____ must not be low.

6-25. A viscosity of 1 pascal-second is equal to _____ poise.

6-26. Handles of pots and pans are often made of plastics because of the property of low _____.

6-27. The plastics with the lowest self-ignition temperature is _____.

6-28. Stress-cracking tests combine physical stress and _____ stress.

ACTIVITIES

Tensile Testing

Equipment. Constant speed tensile tester, stress-strain plotter, calipers, and tensile test bars. Acquire standard ISO or ASTM test bars if available. Samples can also be cut from sheet materials.

6-1. Acquire or prepare 10 sample pieces and measure:

overall length
gage length
gage width and thickness

(Record dimensions in meters and inches.)

6-2. Pull samples to failure at constant speed. Calculate strength at yield and break, elongation to yield and break, and modulus of elasticity, in SI and English systems.

6-3. Calculate mean and standard deviations on stress (load) and elongation (strain) at yield and break.

6-4. Prepare an additional 10 bars and pull them to failure at a significantly different strain rate than the first group. For example, use 25 mm/min. (1 in./min.) on one group and 500 mm/min. (20 in./min.) on the second group.

6-5. Calculate mean and standard deviations as in step 6–3.

6-6. Sketch bell curves, comparing strengths at yield and elongations at yield.

6-7. How great an effect did changing the strain rate have?

6-8. Write a report summarizing the findings.

Additional activity for tensile testing. If samples are available with different gating arrangements, create groups based on their gate locations. Test to see what effect the gate location has on strength and elongation. Molds that provide bars gated from one end and others gated at both ends are extremely useful. The double gate yields a weld line at the center of the part. This allows comparisons between parts with weld lines and parts with no weld line.

Hardness testing

Equipment. Rockwell hardness tester, strip heater, and temperature measuring device.

Procedure

6-1. Cut a piece of sheet material (acrylic or polycarbonate) into a square 75 mm × 75 mm. The material should be at least 3 mm thick.

6-2. Test for hardness at 10 locations on the sample.

6-3. Place the sample on a strip heater, and heat until soft enough to bend. Measure the highest temperature achieved by the sample. Do not bend, but cool the sample, preserving a flat surface.

6-4. After letting it cool, test 10 locations within the "heat-affected zone."

6-5. Calculate mean and standard deviations for the heated group and the unheated group. Sketch the bell curves.

6-6. How much effect did the heating have on hardness?

6-7. Summarize the findings in a short report.

Additional activities. Systematically alter the temperature achieved in sample pieces. Find the temperature range that causes the largest or smallest change in hardness.

Impact testing

Equipment. Izod or Charpy tester and samples of appropriate dimensions.

Procedure

6-1. Impact 10 pieces and record the results.

6-2. Expose 10 pieces of the same material to cold. Remove the samples one at a time from the cold, and impact as rapidly as possible.

6-3. Calculate means and standard deviations. Sketch the bell curves.

6-4. How great an effect did the cold have on impact strength?

Additional activities. Expose samples to extreme cold. Let them return to room temperature before impact testing. Did the exposure to cold produce a lasting effect?

Linear Thermal Expansion Tests

6-1. If thermal expansion apparatus is available, follow the manufacturer's instructions for use.

6-2. To obtain a relative measurement of thermal expansion, carefully measure the length of a sample. Acquire 1 L of water that has a temperature of 20°C (68°F).

6-3. Place the sample in the water and heat the water to 40°C (104°F). Remove the sample and quickly measure the length.

> **! CAUTION**
>
> Do not exceed 40°C (104°F).

6-4. Calculate the theoretical thermal expansion with the following formula:

Theoretical thermal expansion (mm) = difference in temperature (°C) × coefficient of thermal expansion (1/°C) × original length (mm)

The coefficients for selected plastics are shown in Table 6-10.

6-5. Calculate the observed thermal expansion with the following formula:

Observed thermal expansion (mm) = length hot − length cool.

6-6. Compare the observed and theoretical values.

INGREDIENTS OF PLASTICS

CHAPTER 7

INTRODUCTION

Most plastics products consist of a polymeric material that has been altered in order to change or improve selected properties. This chapter focuses on special ingredients applied to alter and enhance plastics. For information on the processes used to mix these special materials with selected plastics, see Chapter 11 on extrusion. Three large categories of these ingredients are included in the chapter outline below:

I. **Additives**
   A. Antioxidants
   B. Antistatic agents
   C. Colorants
   D. Coupling agents
   E. Curing agents
   F. Flame retardants
   G. Foaming/blowing agents
   H. Heat stabilizers
   I. Impact modifiers
   J. Lubricants
   K. Nucleating agents
   L. Plasticizers
   M. Preservatives
   N. Processing aids
   O. UV stabilizers

II. **Reinforcements**
   A. Lamina
   B. Fibrous reinforcements

III. **Fillers**
   A. Nanocomposites
   B. Large-scale fillers

Here are some of the reasons for including additives, reinforcements, and fillers:

- Improving processability
- Reducing material costs
- Reducing shrinkage
- Permitting higher curing temperatures by reducing or diluting reactive materials
- Improving surface finish
- Changing thermal properties such as expansion coefficient, flammability, and conductivity
- Improving electrical properties, including conductivity or resistance
- Preventing degradation during fabrication and service; providing desirable color or tint
- Improving mechanical properties such as modulus, strength, hardness, abrasion resistance, and toughness
- Lowering the coefficient of friction

A host of chemicals have been used in plastics materials because they produce desired changes in properties. However, some of the most successful chemicals were dangerous and even toxic.

The environmental movement has had a significant effect on chemical usage in the plastics industry. Public concern about water and air pollution has caused many significant changes in plastics materials and manufacturing processes. Agencies overseeing packaging for food, drugs, and cosmetics have sought to eliminate the use of dangerous toxic chemicals. One potent approach used by several agencies is the banning of selected chemicals. This chapter describes the efforts of the plastics industry to comply with environmental regulations.

ADDITIVES

The term *additives* refers to a wide range of chemicals that are *added* to plastics. The major categories of additives are antioxidants, antistatic agents, colorants, coupling agents, curing agents, flame retardants, foaming/blowing agents, heat stabilizers, impact modifiers, lubricants, nucleating agents, plasticizers, preservatives, processing aids, and UV stabilizers.

Typically, most additives are compounded into base material prior to manufacturing. However, colorants and blowing agents can also be added during molding or extrusion processes. Compounding is covered in Chapter 11.

In 2011, additive sales reached $33.4 billion, divided into $22.2 billion in property modifiers and $11.2 billion in property stabilizers. The projected growth is about 5% annually until 2018.

Antioxidants

Oxidation of plastics involves a series of chemical reactions with oxygen that result in the breaking of bonds in polymers. Long-chain molecules are cut into shorter chains. If oxidation continues, the chain cutting (often called *chain scission*) progresses to the point where the material becomes very weak and will disintegrate into a powder (Figure 7-1). At elevated temperatures, oxidation generally occurs much more rapidly than at room temperature. Consequently, tests for oxidation usually expose samples to heat.

To combat oxidation, chemical substances that slow down or stop oxidation are added to plastics. These substances are named **antioxidants**. Because the chemical reactions that occur in oxidation are rather complex, *antioxidant packages*

Figure 7-2. Interior automotive parts require antioxidants to provide resistance to high temperatures and intense sunlight.

Photo courtesy of BASF Corporation

combine two or more chemicals to increase resistance to oxidation. Most antioxidant packages contain a *primary antioxidant* and a *secondary antioxidant*. The primary antioxidant works to stop or terminate oxidative reactions. The secondary antioxidant works to neutralize reactive materials that cause new cycles of oxidation. When properly selected, the primary and secondary antioxidants may work together with a synergistic effect that enhances the results (Figure 7-2).

The major types of antioxidants are listed here:

1. Phenolic
2. Amine
3. Phosphite
4. Thioesters

Phenolic and amine materials are often used as primary antioxidants. The phosphite and thioesters serve as secondary antioxidants.

Some plastics are more susceptible to breakdown through oxidation than others. Polypropylene and polyethylene oxidize readily. Because of this tendency, chemical companies that manufacture polypropylene usually add a small amount of primary antioxidant to prevent oxidation during the extrusion processes needed to pelletize it.

Antistatic Agents

Antistatic agents may be compounded into plastics or applied to the product surface. These agents attract moisture from the air, which makes the surface more conductive and dissipates static charges.

The most common antistatic agents are amines, quaternary ammonium compounds, organic phosphates, and polyethylene

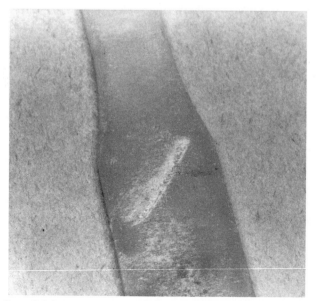

Figure 7-1. Oxidative degradation of unstabilized polypropylene. This damage occurred during 50 hours at 180°C (356°F). A fingernail made the diagonal scratch.

glycol esters. Concentrations of antistatic agents may exceed 2 percent, but application and FDA approval are prime considerations in their use.

Colorants

Plastics can possess a wide range of colors, and plastics designers have exploited this particular property. In fact, some uses for plastics rely completely on the availability of a multitude of colors.

When making colored products, processors use precolor, dry color, liquid color, or color concentrates. *Precolor* is material that has already been compounded to a desired color. *Dry color* is powdered colorant. It is frequently difficult to handle and leads to dust problems. *Liquid color* is a color in a liquid base and requires special pumps. A *color concentrate* is a high loading of colorant carried in a base resin. It comes in pelletized and diced forms.

There are four basic types of colorants used in these various forms:

1. Dyes
2. Organic pigments
3. Inorganic pigments
4. Special-effect pigments

Dyes. *Dyes* are organic colorants. In contrast to pigments, dyes are soluble in plastics and color the material by forming chemical links with molecules. They are often brighter and stronger than inorganic colorants. Dyes are the best choice for a totally transparent product. Although some dyes have poor thermal and light stability, thousands of dyes are currently used in plastics.

Because dyes are soluble in plastics, they may move or migrate. A red dye may migrate into a white part, causing it to turn pink.

Organic pigments. *Pigments* are not soluble in common solvents or in resin. Therefore, they must be mixed and evenly dispersed within the resin. Organic pigments provide the most brilliant opaque colors available. However, the translucent and transparent colors achieved with organic pigments are not as brilliant as those produced with dyes. Organic pigments can be hard to disperse. They tend to form clumps of pigment particles called *agglomerates* that cause spots and specks in products.

Inorganic pigments. Most *inorganic pigments* are based on metals. Oxides and sulfides of titanium, zinc, iron, cadmium, and chromium are common heavy metals that some of the colorants rely on (Figure 7-3).

Environmental agencies have analyzed the health effects of heavy metals and recommend a ban on many of them. In 1992,

Figure 7-3. This inorganic pigment is a powdered form of titanium dioxide. Titanium white is both bright and stable.

the Toxics in Packaging Clearinghouse started, with a mission of reducing the use of heavy metals in packaging. By 1993, 11 states had banned or restricted heavy metals in packaging applications. By 2011, 19 states had such restrictions. The metals of chief concern are lead, mercury, cadmium, and hexavalent chromium. The use of these materials must be regulated to less than 100 parts per million (ppm) within a few years after passing legislation. The EPA has also proposed legislation regulating the amount of cadmium and lead permitted in incinerator ash.

Selected metals in order of weight are listed here:

Metal	Weight in Grams per Mole
Lead	207
Mercury	201
Gold	197
Tungsten	184
Barium	137
Cesium	133
Iodine	127
Tin	119
Cadmium	112
Silver	108
Bromine	80
Chromium	52

The use of some of these metals is restricted. A major concern is that heavy metals may leach out of landfills and enter groundwater, posing a health hazard. Also, when heavy metals are incinerated, the metal residue left in the ash is significant. The incinerator ash cannot go to a traditional landfill, causing its handling, storage, or disposal to be a major problem.

Pigments containing lead, mercury, cadmium, and hexavalent chromium are either banned or under scrutiny. Many companies are developing and marketing heavy-metal-free (HMF) colorants. Some companies expect similar restrictions in the use of barium.

Other inorganic pigments pose no environmental or health danger. They include simple chemicals such as carbon (black), iron oxide (red), and cobalt oxide (blue). Although lead sulfate (white) and cadmium sulfide (yellow) were popular for years, these pigments have recently lost sales.

These metallic oxides are easily dispersed in resin. They do not produce colors as brilliant as those from organic pigments and dyes, but their inorganic structure resists light and heat more effectively. Most inorganic pigments are used in high concentrations to produce opaque-colored plastics. Low concentrations of iron-oxide pigment will produce a translucent color.

Special-effects pigments.

Special-effects pigments may be either organic or inorganic compounds. Colored glass is used in a fine powdered form and is a heat- and light-stable pigment for plastics. Colored glass powder is effective in exterior uses because of its color stability and chemical resistance.

Flakes of aluminum, brass, copper, and gold may be used to produce a striking metallic sheen. Iridescent plastics are used by the automotive industry to produce metallic finishes. When metallic powders are mixed with colored plastics, a finish varying in highlights and reflective hues is produced. Either natural or synthetic, pearl essence may be used to produce a brilliance where pearl luster is desired.

When energy is absorbed by a material, a portion of that energy may be released in the form of light. This light is radiated when the molecules and atoms have their electrons excited to such a state that they begin to lose energy in the form of particles called *photons*. If heat causes electrons to release photons of light energy, the radiation is referred to as *incandescence*.

When chemical, electrical, or light energy excites electrons, the radiation of light is called luminescence. Luminescent materials are often added to plastics to produce special effects. Luminescence is categorized into fluorescence and phosphorescence (Figure 7-4). Fluorescent materials emit light only when their electrons are being excited. These materials cease to emit light when the energy source exciting their electrons is removed. Fluorescent materials are made from sulfides of zinc, calcium, and magnesium. To be environmentally safe, some companies are offering fluorescent colors that are formaldehyde

(A) Illuminated signs.

(B) Nonilluminated signs.

Figure 7-4. Phosphorescent pigments glow in the dark after exposure to light.

free. Fluorescent paint on instrument dials allows a pilot to read instruments with little visible light being emitted. Fluorescent materials are also used in hunting jackets, protective helmets, gloves, life preservers, rain slickers, bicycle stripes, and road warning signs.

Phosphorescent pigments possess an afterglow—that is, they continue to emit light for a limited time after the exciting force has been removed. The most common example of phosphorescence is a television picture tube that emits light when electrical energy excites the phosphorescent materials coating the inside of its face. Phosphorescent pigments used in plastics and paints are made from either calcium sulfide or strontium sulfide.

Mesothorium and radium compounds are radioactive materials sometimes used for special luminescence. Note that there may be harmful effects from prolonged exposure to radioactive materials.

Coupling Agents

Coupling agents are sometimes called promotors (not to be confused with promoters). They are especially important in processing composites. Coupling agents are used as surface treatments to improve the interfacial bond between the matrix and the reinforcements, fillers, or laminates. Without this treatment, many resins and polymers will not adhere to reinforcements or other substrates. Good adhesion is essential in order for the polymer matrix to transfer stress from one fiber, particle, or laminar substrate to the next. Commonly used coupling agents are silane and titanate.

Curing Agents

Curing agents are a group of chemicals that cause cross-linking. These chemicals cause the ends of the monomers to join, forming long polymer chains and cross-linkages.

Because resins may be partially polymerized systems (e.g., B-stage resins), other forms of energy may cause premature polymerization. Inhibitors (stabilizers) may be used to prolong storage and block polymerization.

Catalysts, sometimes called hardeners (more correctly initiators), are chemicals that help the monomers join and/ or cross-link. Organic peroxides are used to polymerize and cross-link thermoplastics (PVC, PS, LDPE, EVA, and HDPE) in addition to thermosetting polyesters.

The most widely used initiators are unstable peroxides called *azo compounds*. Benzoyl peroxide and methyl ethyl ketone peroxide are widely used organic initiators.

Polymerization begins when catalysts are added. Catalysts are influenced little by inhibitors in the resin. As organic peroxides are added to polyester resin, the polymerization reaction begins and yields exothermic heat. This formation of heat further speeds up cross-linking and polymerization. Promoters (or accelerators) are additives that react in a manner opposite that of inhibitors and are often added to resins to aid in polymerization. Promotors react only when a catalyst is added. The reaction that causes polymerization produces heat energy. A common promotor used with the catalyst methyl ethyl ketone peroxide is cobalt naphthanate. All promoters and peroxides should be handled with caution.

! CAUTION

Peroxides may cause skin irritation and acid burns. If promotors and catalysts are added at the same time, a violent reaction may occur. Always thoroughly mix in the promotor, and then add the desired amount of catalyst to the resin. Be sure there is adequate ventilation, and use personal protective wear.

Resins that have not been pre-promoted generally have a longer shelf life. Remember that other forms of energy can cause polymerization. Heat, light, or electrical energy also may initiate this reaction. Always store curing agents at the recommended storage temperature in their original containers.

Flame Retardants

Most commercial flame-retardant chemicals are based on combinations of bromine, chlorine, antimony, boron, and

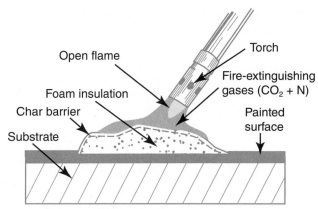

Figure 7-5. This protective finish swells to form an insulating char barrier when heated. It also emits a fire-extinguishing gas to retard burning.

phosphorus. Many of these retardants emit a fire-extinguishing gas (halogen) when heated. Others react by swelling or foaming, which forms an insulation barrier against heat and flame (Figure 7-5).

Some of the most common chemicals used for retarding combustion are alumina trihydrate (ATH), halogenated materials, and phosphorus compounds. In 2006, ATH was still the leader in volume use. ATH cools the flame area by producing water, whereas halogenated materials release inert gases that reduce combustion. Several phosphorus materials form char barriers that insulate combustibles.

Chlorinated and brominated flame retardants have received considerable governmental regulation for decades. In 1977, tris-BP (tris(2, 3, dibromopropyl) phosphate), which was used as a flame retardant in sleepwear for children, was found to cause mutations and was listed as a probable carcinogen. As of 2006, chlorinated flame-retardant compounds (FRs) have varying levels of environmental restriction in Europe, Japan, and North America. China continues to use significant volumes of chlorinated FRs.

Various forms of brominated FRs are also restricted or banned. In 1977, the United States banned polybrominated biphenyls (PBBs). Polybrominated diphenyl ethers (PBDEs) are a group of chemicals. The penta-PDBE form was banned in the EU countries starting in July 2003. The manufacture and import of penta-PDBE and octa-PDBE in the United States stopped in 2004. Deca-PDBE has been under scrutiny because it is claimed to bioaccumulate, and consequently small exposures can mount up over time. In 1999, a study conducted in Sweden found a doubling of PDBEs in a five-year period. In contrast, in 2004 the EU found that deca-PDBEs were not a threat. In January 2014, the EPA put out a final report recommending the use of alternative materials to reduce deca-PDBE use.

In spite of the controversy about the chlorinated and brominated FRs, the need for flame retardants continues to grow. In 2004, global consumption for FRs was 2.9 billion pounds. The expected global demand for FRs is 4.6 billion pounds in 2009, which corresponds to a value of $4.3 billion. In 2010, the global production of FRs reached 1.7 million metric tons.

Foaming/Blowing Agents

The terms *foaming, blowing, frothed, cellular,* and *bubble* are sometimes used to cover a wide variety of compounds and processing techniques to make polymers with a cellular structure (see Chapter 16, Expansion Processes). There are two major types of foaming agents—physical and chemical. *Physical foaming agents* decompose at specific temperatures and release gases. These gases cause cells or voids in the plastics. *Chemical foaming agents* release gases due to a chemical reaction.

Foaming agents are used widely in the manufacture of foamed polyurethane pads and seats for cars, trucks, sofas, and other furniture. Chlorinated fluorocarbons (CFCs) are very efficient physical foaming agents for polyurethane and have been used widely for many years. However, research has indicated that CFCs cause damage to the ozone layer in the upper atmosphere. In 1987, the Montreal Protocol on "Substances that Deplete the Ozone Layer" proposed the reduction and eventual elimination of production and use of ozone-depleting substances (ODS). Twenty-nine countries had ratified the protocol by 1989. Since then, the number has grown to 191 nations. The United States supports the protocol, and the EPA has created several programs to end production of ODS. In 1999, the Beijing Amendment to the Montreal Protocol specified reductions of ODS by January 2004. In developed nations, 90 percent of CFC will be phased out by 2015, and total phaseout achieved by 2020.

The Clean Air Act prohibited the production and import of CFCs starting in 1996. In response to these concerns, many foam makers have switched to hydrochlorofluorocarbon (HCFC), which is much less ozone-depleting than traditional CFCs. Compared to CFCs, HCFCs have only 2–10 percent ozone-depletion potential. In contrast to CFCs, a problem with the new foaming agents is that they yield a denser foam, so they are less efficient as an insulating material. Researchers are working on two fronts. They want to improve the effectiveness of the HCFCs and develop foaming agents that contain no chlorine and have no ozone-depleting potential.

Some HCFCs are also scheduled for phaseout. HCFCs are class II ozone-depleting substances. HCFC-141b ended production in December 2002. However, the EPA is not actively working to eliminate all class II ODS because viable substitutes may be unavailable.

The European Union and the Montreal Protocol have developed phaseout schedules for the consumption of HCFCs. According to the schedule of the Montreal Protocol, developed countries were to establish a cap equaling 2.8 percent of the ozone-depletion potential of the CFCs plus the HCFCs consumed in 1989. By January 2004, these countries were to reduce usage by 35 percent, followed in 2010 by a reduction

Figure 7-6. A pelletized azotype blowing agent.

of 65 percent, a 90 percent reduction by 2015, a 99.5 percent reduction by 2020, and a 100 percent reduction by 2030. In contrast, the European Union promotes a schedule that reaches 100 percent reduction of HCFCs by 2026.

Some manufacturers have investigated the use of perfluorocarbons (PFCs) as a substitute for CFCs or HCFCs. However, the PFCs are potent greenhouse gases and have atmospheric lifetimes of 3000 to 5000 years. Consequently, the EPA discourages widespread use of PFCs.

Chemical blowing agents such as azodicarbonamide are widely used to produce cellular HDPE, PP, ABS, PS, PVC, and EVA. This chemical has several advantages, including efficient gas yield, some FDA approvals for food-contact applications, and ease of modification for various plastics (Figure 7-6).

Heat Stabilizers

Heat stabilizers are additives that retard the decomposition of a polymer caused by heat, light energy, oxidation, or mechanical shear. PVC has poor thermal stability and has been the focus of most heat stabilizers. In the past, heat stabilizers were compounds based on lead and cadmium. Lead has been a predominant additive for wire and cable coatings. Because of environmental concerns surrounding heavy metals, noncadmium stabilizers have taken over many applications previously held by cadmium stabilizers (Figure 7-7).

The basic problem is that most of the alternatives do not offer comparable efficiency. Some stabilizers based on calcium/zinc have very low toxicity but poor thermal performance. Stabilizers based on organotin materials are excellent in thermal performance but often cause unpleasant odors. Stabilizers based on barium/zinc have good properties, but barium may be targeted for regulation in the near future. Because of these problems, the use of lead stabilizers continues.

One way to reduce lead use involves utilizing lead-based stabilizers in the inner layers of multilayer PVC pipes. Elimination of all metals can be achieved with the use of

Courtesy of Ciba Specialty Chemicals

Figure 7-7. Plastics used in automotive parts under the hood must withstand high temperatures and hot fluids.

all-organic-based stabilizers. Many of these materials are based on organosulfide compounds. They are still not widely used but have the potential of eliminating the use of heavy metals for stabilization.

Impact Modifiers

One or more monomers (usually elastomers) may be added in varying amounts to rigid plastics to improve (modify) impact properties, melt index, processability, surface finish, and weather resistance. PVC is toughened by modifying it with ABS, CPE, EVA, or other elastomers (see Ethylene-ethyl acrylate and Styrene-butadiene in Appendix E).

Lubricants

Lubricants are needed for making plastics. During the production of polymers, lubricants are added for three basic reasons. First, they help get rid of some of the friction between the resin and manufacturing equipment. Second, lubricants aid in emulsifying other ingredients and provide internal lubrication for the resin. Third, some lubricants prevent plastics from sticking to the mold surface during processing. After the products are taken from the mold, lubricants may exude from the plastics and prevent the products from adhering to each other. They may provide a nonsticking or slippery quality to the plastics surface.

Many different lubricants are used as ingredients in plastics. Some examples are waxes such as montan, carnauba, paraffin, and stearic acid. Metallic soaps such as the stearates of lead, cadmium, barium, calcium, and zinc are also used as lubricants (Table 7-1). During the process of manufacturing the resin,

Table 7-1. Lubricants Chart

Plastics	Alcohol Esters	Amide Waxes	Complex Esters	Comb. Blends	Fatty Acids	Glycerol Esters	Metallic Stearates	Paraffin Waxes	Polyethylene Waxes
ABS		X			X	X			
Acetals	X		X						
Acrylics	X			X					
Alkyd							X		
Cellulosics	X	X			X	X			
Epoxy				X		X			
Ionomers		X							
Melamines			X		X				
Phenolics				X	X	X			
Polyamides	X			X					
Polyester			X	X	X				
Polyethylene		X							
Polypropylene		X			X	X			
Polystyrene		X	X		X			X	
Polyurethanes			X						
Polyvinyl chloride	X	X	X	X	X	X	X	X	X
Sulfones			X						

most of the lubricant is lost. Excess lubricant may slow polymerization or cause a lubrication bloom. This will be seen as an irregular, cloudy patch on the plastics surface.

Some plastics exhibit nonstick and self-lubrication properties. Examples are fluorocarbons, polyamides, polyethylene, and silicone plastics. They are sometimes used as lubricants in other polymers. Remember that all additives must be carefully selected for toxic effects and desired service use.

Nucleating Agents

Nucleating agents are added to a polymer to increase its crystallinity. These agents may shorten cycle times by speeding the transition from melted to solid material. Some common nucleating agents are inert mineral fillers, chalk, clay, talc, titanium dioxide, and potassium stearate. Properties such as density and clarity may be altered by changing the crystallinity of plastics.

Plasticizers

Plasticity refers to the ability of a material to flow or become fluid under force. A plasticizer is a chemical agent added to plastics to increase flexibility, reduce melt temperature, and lower viscosity. All of these properties aid in processing and molding. Plasticizers act much like solvents by lowering viscosity. However, they also act like lubricants by allowing slippage between molecules.

It is important to remember that van der Waals bonds are only physical attractions and not chemical bonds. Plasticizers help neutralize most of these forces. Plasticizers, which are much like solvents, produce a more flexible polymer. However, they are designed not to evaporate from the polymer during normal service life.

Plasticizer leaching or loss is an important consideration. It is undesirable when it occurs in contact with food, pharmaceutical, or other products for consumption. Leaching and degassing may cause PVC hoses, upholstery, and other products to become stiff or brittle and crack. For best results, the plasticizer and polymer must have similar solubility parameters.

Over 500 different plasticizers have been formulated to modify polymers. Plasticizers are vital ingredients in plastics coatings, extrusions, moldings, adhesives, and films. One of the most widely used plasticizers is dioctyl phthalate. Some plasticizers may be hazardous. The EPA found di-2-ethylhexyl phthalate plasticizer as carcinogenic in lab tests done on animals. It is currently labeled as a potential carcinogen. Some plasticizers are listed in Table 7-2.

Table 7-2. Compatibility of Selected Plasticizers and Resins

Plasticizer	Resin											
	Polyvinyl Acetate	Polyvinyl Chloride	Polyvinyl Butyral	Polystyrene	Cellulose Nitrate	Cellulose Acetate	Cellulose Acetate Butyrate	Ethyl Cellulose	Acrylic	Epoxy	Urethane	Polyamide
Butyl benzyl phthalate	C	C	C	C	C	P	C	C	C	C	C	C
Butyl cyclohexyl phthalate	C	C	C	C	C	P	C	C	C	C	C	C
Didecyl phthalate	I	C	C	C	C	C	C	C	C	P	C	P
Butyl octyl phthalate	I	C	P	C	C	I	C	C	C	P	C	C
Dioctyl phthalate	I	C	P	C	C	I	C	C	C	I	C	C
Cresyl diphenyl phosphate	C	C	C	P	C	C	C	C	C	C	C	C
N-Ethyl-o, p-toluenesulfonamide	C	I	C	P	C	C	C	C	C	C	P	C
o, p-Toluenesulfonamide	C	I	C	P	C	C	C	C	C	C	P	C
Chlorinated paraffins	C	P	P	C	P	I	P	C	P	P	C	C
Didecyl adipate	I	C	I	C	C	I	C	C	I	I	P	C
Dioctyl adipate	I	C	C	C	C	I	C	C	I	I	P	C
Dioctyl sebacate	I	C	P	C	C	I	P	C	I	I	P	C

Note:
C–Compatible   I–Incompatible
P–Partially compatible

Preservatives

Elastomers and heavily plasticized PVC are the most susceptible to attack by microorganisms, insects, or rodents. Microbiological deterioration may occur when moisture stays or condenses on shower curtains, automobile tops, pool liners, waterbed liners, cable coatings, tubing, and so on. Antimicrobials, mildewicides, fungicides, and rodenticides may provide the necessary protection in many polymers. The EPA and FDA carefully regulate the use and handling of all antimicrobials.

Processing Aids

There are a variety of additives used to improve processing behavior, increase production rates, or improve surface finish. Antiblocking agents such as waxes exude to the surface and prevent two polymer surfaces from adhering to each other. *Emulsifiers* are used to lower the surface tension between compounds. They act as detergents and wetting agents. Wetting agents used to lower viscosity are called *viscosity depressants*. They are used in plastisol compounds to assist in processing heavily filled materials or those that become too thick due to aging.

Solvents are added to resins for several reasons. Many natural resins are very viscous or hard. Therefore, they must be diluted or dissolved before processing. Resinous varnish and paints must be thinned with solvents for proper application.

Solvents may be considered a processing aid. In solvent molding, the solvent holds the resin in solution while it is being applied to the mold. The solvent rapidly evaporates, leaving a layer of plastics film on the mold surface. Solvents dissolve many thermoplastics. Therefore, they are used for both identification and cementing purposes. Solvents are also useful for cleaning resins from tools and instruments. Benzene, toluene, and other aromatic solvents will dissolve the natural oils of the skin. All chlorinated solvents are potentially toxic. Avoid breathing fumes or allowing skin contact when using plastics additives (see Chapter 4, Health and Safety).

UV Stabilizers

Polyolefins, polystyrene, polyvinyl chloride, ABS polyesters, and polyurethanes are all susceptible to ultraviolet solar breakdown. Solar radiation of polymers may result in crazing, chalking, color changes, or loss of physical, electrical, and chemical properties. This weathering damage is caused when the polymer absorbs light energy. Ultraviolet light is the most destructive portion of the solar radiation striking plastics products. There may be enough energy involved to break chemical bonds between atoms.

To reduce the damage done by exposure to UV light, compounding processes add UV stabilizers to plastics. Carbon black is sometimes used as a UV stabilizer but is limited due to its color. In the past, the most widely used ultraviolet absorbers were 2-hydroxybenzophenones, 2-hydroxyphenylbenzotriazoles, and 2-cyanodiphenyl acrylates. Currently, most developments involve hindered-amine light stabilizers (HALS).

HALS often contain reactive groups that chemically bond onto the backbone of the polymer molecules. This reduces migration and volatility. The combination of a HALS and phosphite or phenolic antioxidant increases the UV resistance.

REINFORCEMENTS

Reinforcements are ingredients added to resins and polymers. These ingredients do not dissolve in the polymer matrix; consequently, the material becomes a composite. There are many reasons for adding reinforcements. One important reason is to produce dramatic improvements in the physical properties of the composite.

Reinforcements are often confused with fillers. However, fillers are small particles and contribute only slightly to strength. On the other hand, reinforcements are ingredients that increase strength, impact resistance, and stiffness. One major reason for confusing the two is that some materials such as glass may act as a filler, reinforcement, or both.

There are six general variables that influence the properties of reinforced composite materials and structures.

1. *Interface bond between matrix and reinforcements.* The matrix functions to transfer most of the stress to the (much stronger) reinforcements. In order to accomplish this task, there must be an excellent adhesion between the matrix and reinforcement.
2. *Properties of the reinforcement.* It is assumed that the reinforcement is much stronger than the matrix. The actual properties of each reinforcement may vary by composition, shape, size, and number of defects. The production, handling, processing, surface enhancement, or hybridization can also determine properties for each type of reinforcement.
3. *Size and shape of the reinforcement.* Some shapes and sizes may help provide superior handling, loading, processing, packing orientation, or adhesion in the matrix. Some fibers are so small that they are handled in bundles, whereas others are woven into cloth. Particulates are more likely to be randomly distributed than long fibers.
4. *Loading of the reinforcement.* Generally, mechanical strength of the composite depends on the amount of reinforcing agent it contains. A part containing 60 percent reinforcement and 10 percent resin matrix is almost six times stronger than a part containing the opposite amounts of these two materials. Some glass filament-wound composites may have up to 80 percent (by weight) loading by unidirectional orientation of the filament. Most

reinforced thermoplastic composites contain less than 40 percent (by weight) reinforcements.

5. *Processing technique.* Some processing techniques allow the reinforcements to be more carefully aligned or oriented. During processing, reinforcements may be broken or damaged, resulting in lower mechanical properties. Depending on the processing technique, particulate reinforcements and short fibers are more likely to have random rather than oriented placement in the matrix.

6. *Alignment or distribution of the reinforcement.* Alignment or distribution of the reinforcement allows versatility in composites. The processor can align or orient the reinforcements to provide directional properties. In Figure 7-8, the parallel (anisotropic) alignment, or continuous **strands**, provides the highest strength; bidirectional (cloth) alignment provides a middle strength range; and random (mat) gives the lowest.

Reinforcements may be divided into two major groups of materials: lamina and fibrous. The basic structural element of laminar composites is the *lamina.*

Lamina

Lamina may consist of unidirectional fibers, woven cloth, or sheets of material. Because individual layers act as a reinforcement, they may be included as an ingredient or additive. These laminar layers are more than just a processing technique (see Chapter 12, Laminating Processes and Materials). The lamina selection, alignment, and composition constitute the performance properties of the laminar composite.

It should become clear that alignment of the reinforcements is the key to designing the composite with anisotropic or isotropic properties. As a rule, if all the reinforcements are placed parallel to each other (0° layup), the composite

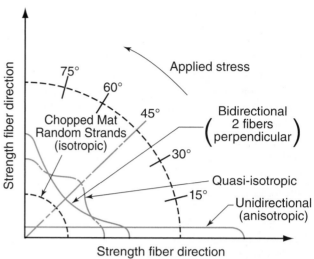

Figure 7-9. The effect of alignment or distribution of the reinforcement.

will be directional (see Pultrusion, Chapter 13). The directional tensile strength properties achieved with different fiber reinforcement alignment is illustrated in Figure 7-9. The random-chopped strand mat provides equal strength properties in all directions. The unidirectional fiber alignment has the highest strength parallel to (or in the direction of) the fiber. As this angle varies from 0° to 90°, the strength varies proportionally. Remember that the matrix must securely adhere to the reinforcement and prevent the reinforcement from buckling in order to transfer the applied stress.

Fibrous Reinforcements

In the fibrous group of reinforcements, there are six subclasses:

1. Glass
2. Carbonaceous
3. Polymer
4. Inorganic
5. Metal
6. Hybrids

Glass fibers. One of the most important reinforcing materials is fibrous glass (Tables 7-3 and 7-4). Because the strength of glass reinforces plastics, many parts previously made from metals have been replaced with plastics.

Glass fibers are produced using several different methods. One common method of production involves pulling a strand of molten glass after it has been formed by a small orifice. The diameter of the strand, as seen in Table 7-5, is controlled by the pulling action.

The major constituent of glass is silica but other ingredients allow the production of many types of fibrous glass. The most common type is **E glass** fiber, which has good electrical (E) properties and high strength. **C glass** is used for chemical

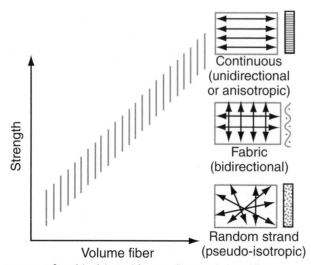

Figure 7-8. Strength in relation to reinforcement alignment and volume of fiber.

Table 7-3. Properties of Thermoplastics: Unreinforced and Reinforced Materials

Property	Polyamide		Polystyrene*		Polycarbonate		Styrene-Acrylonitrile†		Polypropylene		Acetal		Linear Polyethylene	
	U	R	U	R	U	R	U	R	U	R	U	R	U	R
Tensile strength, MPa	82	206	59	97	62	138	76	124	35	46	69	86	23	76
Impact strength, notched, J/mm									0.069					
At 22.8°C	0.048	0.202	0.016	0.133	0.106§	0.213	0.024	0.160	0.112	0.128	3.20	0.160	—	0.240
At −40°C	0.032	0.224	0.010	0.170	0.080§	0.213	—	0.213	—	0.133	—	0.160	—	0.266
Tensile strength, MPa	2.75	—	2.75	8.34	2.2	11.71	3.58	10.34	1.37	3.10	2.75	5.58	0.82	6.20
Shear strength, MPa	66	97	—	62	63	83	—	86	33	34	65	62	—	38
Flexural strength, MPa	79	255	76	138	83	179	117	179	41 to 55	48	96	110	—	83
Compressive strength, MPa	34††	165	96	117	76	130	117	151	59	41	36	90	19 to 24	41
Deformation (27.58 MPa), %	2.5	0.4	1.6	0.6	0.3	0.1	—	0.3	—	6.0	—	1.0	—	0.4‡
Elongation, %	60.0	2.2	2.0	1.1	60–100	1.7	3.2	1.4	>200	3.6	9–15	1.5	60.0	3.5
Water absorption in 24 h, %	1.5	0.6	0.03	0.07	0.3	0.09	0.2	0.15	0.01	0.05	0.20	1.1	0.01	0.04
Hardness, Rockwell	M79	E75 to 80	M70	E53	M70	E57	M83	E65	R101	M50	M94	M90	R64	R60
Relative density	1.14	1.52	1.05	1.28	1.2	1.52	1.07	1.36	0.90	1.05	1.43	1.7	0.96	1.30
Heat distortion temp. (at 1.82 MPa), °C	65.6	261	87.8	104.4	137.8	148.9	93.3	107	68.3	137.8	100	168.6	52.2	126.7
Coef. of thermal expansion, per °C $\times 10^{-6}$	90	15	60	35	60	15	60	30	70	40	65	30	85	25
Dielectric strength (short time), V/mm	15,157	18,898	19,685	15,591	15,748	18,976	17,717	20,276	29,528	—	19,685	—	—	2,362
Volume resistivity, ohm-cm $\times 10^{15}$	450	2.6	10.0	36.0	20.0	1.4	10^{16}	43.5	17.0	15.0	0.6	38.0	10^{15}	29.0
Dielectric constant at 60 Hz	4.1	4.5	2.6	3.1	3.1	3.8	3.0	3.6	2.3	—	—	—	2.3	2.9
Power factor at 60 Hz	0.0140	0.009	0.0030	0.0048	0.0009	0.0030	0.0085	0.005	—	—	—	—	—	0.001
Approximate cost, ¢/cm³	0.256	0.70	0.04	0.21	0.31	0.56	0.08	0.30	0.05	0.18	0.28	0.67	0.06	0.26

Notes: Columns marked "U" unreinforced, "R" reinforced. *Medium-flow, general-purpose grade.
†Heat-resistant grade. §Impact values for polycarbonates are a function of thickness, ‡6.8 MPa load.
††At 1% deformation. Adapted from *Machine Design*, Plastic Reference Issue.

Table 7-4. Properties of Thermosetting Plastics: Glass-Fiber Reinforced Resins

Property	Base Resin				
	Polyester	Phenolic	Epoxy	Melamine	Polyurethane
Molding quality	Excellent	Good	Excellent	Good	Good
Compression molding Temperature, °C	76.7–160	137.8–176.7	148.9–165.6	137.8–171.1	148.9–204.4
Pressure, MPa	1.72–13.78	13.78–27.58	2.06–34.47	13.78–55.15	0.689–34.47
Mold shrinkage, mm/mm	0.0–0.05	0.002–0.025	0.025–0.05	0.025–0.100	0.228–0.762
Relative density	1.35–2.3	1.75–1.95	1.8–2.0	1.8–2.0	1.11–1.25
Tensile strength, MPa	173–206	35–69	97–206	35–69	31–55
Elongation, %	0.5–5.0	0.02	4	–	10–650
Modulus of elasticity, Pa	0.55–1.38	2.28	2.09	1.65	–
Compressive strength, MPa	103–206	117–179	206–262	138–241	138
Flexural strength, MPa	69–276	69–414	138–179	103–159	48–62
Impact, Izod, J/mm	0.1–0.5	0.5–2.5	0.4–0.75	0.2–0.3	No break
Hardness, Rockwell	M70–M120	M95–M100	M100–M108	–	M28–R60
Thermal expansion, per °C	$5-13(\times 10^{-4})$	4×10^{-4}	$2.8-7.6(\times 10^{-4})$	3.8×10^{-4}	$25-51(\times 10^{-4})$
Volume resistivity (at 50% RH, 23°C), ohm-cm	1×10^{14}	7×10^{12}	3.8×10^{15}	2×10^{11}	$2 \times 10^{11}-10^{14}$
Dielectric strength, V/mm	13,780–19,685	5,512–14,567	14,173	6,693–11,811	12,992–35,433
Dielectric constant					
At 60 Hz	3.8–6.0	7.1	5.5	9.7–11.1	5.4–7.6
At 1 kHz	4.0–6.0	6.9	–	–	5.6–7.6
Dissipation factor					
At 60 Hz	0.01–0.04	0.05	0.087	0.14–0.23	0.015–0.048
At 1 kHz	0.01–0.05	0.02	–	–	0.043–0.060
Water absorption, %	0.01–1.0	0.1–1.2	0.05–0.095	0.9–21	0.7–0.9
Sunlight (change)	Slight	Darkens	Slight	Slight	None to slight
Chemical resistance	Fair*	Fair*	Excellent	Very good†	Fair
Machining qualities	Good	–	Good	Good	Good

Notes: *Attacked by strong acids or alkalies. †Attacked by strong acids. *Source: Machine Design*, Plastics Reference Issue.

Table 7-5. Glass Fiber Diameter Designations

Filament Designation	Filament Diameter in µm	(inches)
C	4.50	0.000175
D	5.00	0.000225
DE	6.00	0.000250
E	7.00	0.000275
G	9.10	0.000375
H	11.12	0.000425
K	13.14	0.000525

resistance. Both E and C glass have tensile strength exceeding 3.4 GPa (493,183 psi). D glass is used for low dielectric constant and density. I glass contains lead oxide for radiation protection. S Glass is selected for high-strength uses. It is about 20 percent stronger and stiffer than E glass. S glass has a tensile strength of more than 4.8 gigapascals (GPa) (696,258 psi).

Plastics processors purchase these differing types of glass in several forms. **Rovings** are long strands of fibrous glass that may be easily cut and applied to resins. A roving is made up of many strands of glass and resembles a loosely twisted or stranded rope. *Chopped fibers* (Figure 7-10) are among the least costly forms of glass reinforcement. Chopped strands range in length from 3 to 50 mm (0.125 to 2 in.). Figure 7-11 shows the production of chopped strands from rovings. *Milled fibers* are less than 1.5 mm (0.062 in.) in length and are produced by hammer milling glass strands (Figure 7-12). Milled fibers are added to a resin as a premix to increase viscosity and product strength. **Yarns** resemble rovings but are twisted like a rope (Figures 7-13 and 7-14). Reinforcing yarns are used in the fabrication of large liquid tank containers.

Figure 7-10. Chopped strands of fibrous glass.

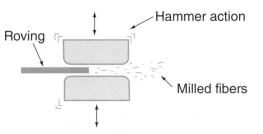

Figure 7-11. Production of chopped glass strands.

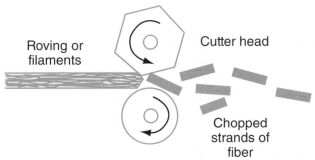

Figure 7-12. Production of milled glass fibers.

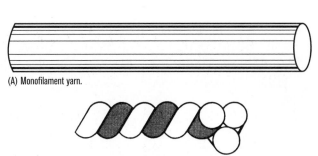

(A) Monofilament yarn.

(B) Multifilament yarn.

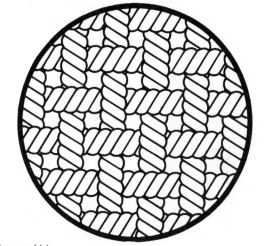

(C) Woven yarn fabric.

Figure 7-13. Yarns.

In Z-twist, the two strands assume an ascending left to right configuration.

In S-twist, these two bobbins of yarn are plied together. The S-twist assumes an ascending right to left configuration.

Figure 7-14. Yarn nomenclature.

Fiberglass yarn product nomenclature is based on a letter-number system. For example, a yarn designated as ECG 150 2/2 2.8 would be:

E = Electrical glass

C = Continuous filament

G = Filament diameter of 9 μm   (see Table 7-5)

150 = 1/100 of the total approximate bare yardage in a pound, or 1500 yards

2/2 = Single strands were twisted, and two of the twisted strands were plied together (S or Z may be used to designate the type of twist)   (Figure 7-14).

2.8 = Number of turns per inch in the twist of final yarn with S-twist

There were two basic strands in the yarn and two twisted strands. Thus,

$$15,000/2 \times 2 = 3.750 \text{ yards per pound of yarn}$$

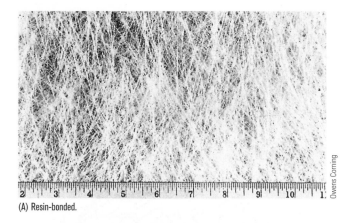

(A) Resin-bonded.

(B) Stitched (needled).

Figure 7-15. Fibrous glass mats.

temperatures and result in a carbon elemental analysis of 99 percent. Once the organic materials have been driven off (pyrolyzed and stretched into filaments), a high-strength, high-modulus, low-density fiber will result (Figure 7-18).

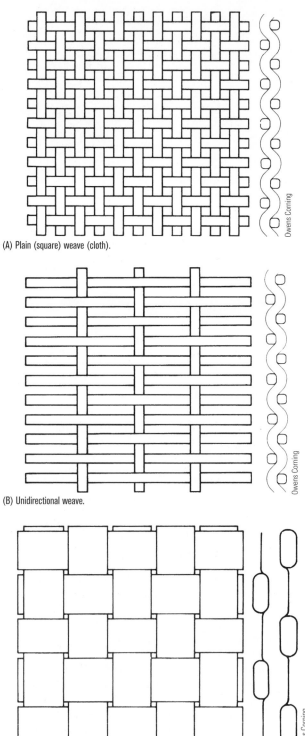

(A) Plain (square) weave (cloth).

(B) Unidirectional weave.

(C) Square-weave woven rovings.

Figure 7-16. Weave patterns and roving.

In addition to fiber and yarn, glass reinforcements are also available in cloth and mat forms. *Mats* consist of nondirectional pieces of chopped strands that are held together by a resinous binder or by mechanical stitching called needling (Figure 7-15).

Woven cloth can provide the greatest physical strength of all the fibrous forms. However, it is about 50 percent more costly than other forms. Standard rovings may be woven into fabric form (known as woven roving) and used for thick reinforcements.

There are several types of woven glass fabrics. Glass fiber yarns are woven into several basic patterns, as shown in Figure 7-16. Figure 7-17 shows three different forms of fibrous glass reinforcements.

Carbonaceous fibers. Carbonaceous fibers are usually made by oxidizing, carbonizing, and graphitizing an organic fiber. Rayon and polyacrylonitrile (PAN) are currently used. Pitch fibers can also be produced directly from oil and coal. Ordinary pitches are isotropic in character and must be oriented to be as useful as reinforcing agents. Although the terms *carbon* and *graphite* are used interchangeably, there is a distinction. Carbon (PAN) fibers are about 95 percent carbon, whereas graphite fibers are graphitized at much higher

(B) Fine strand mat.

(C) Combination woven roving and mat.

Figure 7-17. (*Continued*)

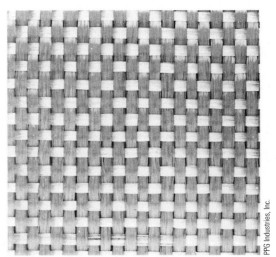

(D) Multifilament wound or twisted roving used in making heavy woven fibrous glass products.

Figure 7-16. (*Continued*)

Polymer fibers. For years, cotton and silk filaments were used as reinforcements in belting, tires, gears, and other products. Currently, synthetic polymers of polyester, polyamide (PA), polyacrilonitrile (PAN), polyvinyl acetate (PVA), cellulose acetate (CA), and others are used. Kevlar® aramid is an aromatic polyamide polymer fiber with nearly twice the stiffness and about half the density of glass. Kevlar is a trademark of DuPont. Aramid is the generic name for a series of Kevlar fibers. Unlike carbon fibers, Kevlar fibers do not conduct electricity, nor are they electrically opaque to radio waves. Kevlar 29 fibers are used for ballistic protection, ropes, army helmets,

Figure 7-18. This reinforcing cloth contains both glass and carbon fibers. The glass is lighter in color, and the carbon is darker.

coated fabrics, and a variety of composite applications. Kevlar 49 is used in boat hulls, flywheels, V-belts, hoses, composite armor, and aircraft structures. It possesses equal strength but a much higher modulus than Kevlar 29.

(A) Woven roving.

Figure 7-17. Some of the many forms of fibrous glass reinforcements.

A common high-strength polymer matrix is epoxy. Polyesters, phenolic polyamide, and other resin and polymer systems are used.

Polyester and polyamide-based thermoplastic fibers find applications in bulk molding compounds (BMCs), sheet molding compounds (SMCs), thick molding compounds (TMCs), layup, pultrusion, filament winding, resin transfer molding (RTM), reinforced reaction injection molding (RRIM), thermal expansion resin transfer molding (TERTM), and injection-molding operations.

Inorganic fibers. Inorganic fibers are a class of short crystalline fibers. They are sometimes called **crystal whisker fibers**. Crystal whisker fibers are made of aluminum oxide, beryllium oxide, magnesium oxide, potassium titanate, silicon carbide, titanium boride, and other materials (Figure 7-19). Potassium titanate whiskers are used in large quantities to strengthen composites in thermoplastic matrices. Inorganic continuous boron fibers are stronger than carbon and may be used in a polymer and aluminum matrix. Boron in an epoxy matrix is used to make many composite parts for military and civilian aircraft.

These fibers are very costly to make with present technologies. However, they display tensile strengths greater than 40 GPa (5,802,146 psi). Research into the use of these reinforcements in dental plastics fillings, turbine compressor blades, and special deep-water equipment has shown encouraging results.

Carbon and graphite fibers may exceed glass in strength. They are used extensively as self-lubricating materials, heat-resistant reentry bodies, blades for turbines and helicopters,

Figure 7-20. Use of high-strength graphite fibers for reinforcement provides this racquetball racquet with strength at a weight of only 200 grams.

and valve-packing compounds. Figure 7-20 shows an injection-molded racquetball racquet that relies on the high strength of graphite fibers.

Ceramic fibers have high tensile strength and low thermal expansion. Some fibers may reach a tensile strength of 14 GPa (2,030,750 psi). Present applications for ceramic fibers include dental fillings, special electronics, and spacecraft research (see tensile strength of whisker fibers in Figure 7-21).

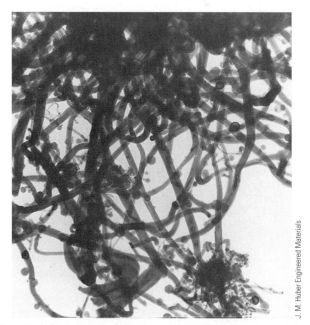

J. M. Huber Engineered Materials

Figure 7-19. Submicron ceramic whiskers grown in a fibrous ball. There is a higher concentration of fibers near the center of the ball. The fibers range from as small as two billionths of a meter up to 50 billionths. The minute diameter and length of these fibers are advantageous in injection molding, permitting greater processing speed with minimal damage.

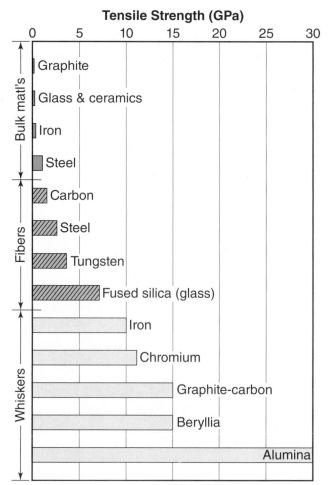

Figure 7-21. Tensile strengths of various materials in bulk, fiber, and whisker forms.

Metal fibers. Steel, aluminum, and other metals are drawn into continuous filaments. They do not compare with the strength, density, and different properties exhibited by other fibers. Metal fibers are used for added strength, heat transfer, and electrical conductivity.

Hybrid fibers. Hybrid fibers refer to a special available form of fibers. Two or more fibers may be combined (hybridization) to tailor the reinforcement to the needs of the designer. Hybrid fibers provide diverse properties and many possible material combinations. They can maximize performance, minimize cost, or improve any deficiency of the other fiber component (a synergistic effect). Glass and carbon fibers are used together to increase impact strength and toughness, prevent galvanic action, and reduce the cost of a 100 percent carbon composite. When these fibers are placed in a matrix, the composite—not the fibers—is called a hybrid. A composite of metal foils or metal composite plies stacked in a specified orientation and sequence is called a super hybrid (see Lamina, this chapter).

Properties of the most commonly used fiber-reinforcing agents are shown in Table 7-6.

FILLERS

The term *filler* is often rather confusing. Filler was originally selected to describe any additive used to *fill* space in the polymer and lower cost. Because some fillers are more expensive than the polymer matrix, the word *extender* can be misleading. The terms *dilutant* and *enhancer* are sometimes used to describe the addition of fillers. Ambiguity of terms and overlapping of function add to the problem. In this book, the term *filler* will mean any minute particle from various sources, functions, composition, and morphology. Fillers can be saucer-, sphere-, or needle-shaped, or irregular (Figure 7-22).

According to the ASTM, a filler is a relatively inert material added to a plastic to modify its strength, permanence, working properties, or other qualities, or to lower costs.

Fillers may be either organic or inorganic ingredients of plastics or resins. They may increase bulk or viscosity, replace more costly ingredients, reduce mold shrinkage, and improve the physical properties of the composite item. The size and shape of the filler greatly influence the composite. The aspect ratio of a filler is the ratio of length to width. Flakes or fibers have

Table 7-6. **Properties of the Most Commonly Used Fiber Reinforcing Agents (Metallic and Nonmetallic)**

Fiber	Relative Density	Tensile Strength Ultimate (MPa)	Tensile Modulus of Elasticity Modulus (GPa)
Aluminum	2.70	620	73.0
Aluminum oxide	3.97	689	323.0
Aluminum silica	3.90	4,130	100.0
Aramid (Kevlar 49)	1.4	276	131.0
Asbestos	2.50	1,380	172.0
Beryllium	1.84	1,310	303.0
Beryllium carbide	2.44	1,030	310.0
Beryllium oxide	3.03	517	352.0
Boron-Tungsten boride	2.30	3,450	441.0
Carbon	1.76	2,760	200.0
Glass, E-glass	2.54	3,450	72.0
S-glass	2.49	4,820	85.0
Graphite	1.50	2,760	345.0
Molybdenum	10.20	1,380	358.0
Polyamide	1.14	827	2.8
Polyester	1.40	689	4.1
Quartz (fused silica)	2.20	900	70.0
Steel	7.87	4,130	200.0
Tantalum	16.60	620	193.0
Titanium	4.72	1,930	114.0
Tungsten	19.30	4,270	400.0

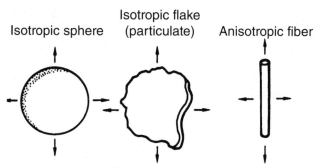

Isotropic sphere   Isotropic flake (particulate)   Anisotropic fiber

Figure 7-22. Spheres are isotropic but have no aspect ratio. Isotropic particulates have uniform mechanical properties in the plane of the flake. Fibers have lower aspect ratios but are anisotropic.

aspect ratios that make them resist movement or realignment, thus improving strength. Spheres have no aspect ratio and produce composites with isotropic properties. Metallic flakes are used in particulate composites to form an electrical barrier or layer in the polymer matrix. The main types of fillers and their functions are shown in Table 7-7. Fillers can be classified by size into two groups: nanoscale and large-scale materials.

Nanocomposites

A strict definition of **nanocomposite** is any composite in which the fillers are submicron in size. Nanocomposites were

Table 7-7. Principle Types of Fillers

Filler	Bulk	Processibility	Thermal Resistance	Electrical Resistance	Stiffness	Chemical Resistance	Hardness	Reinforcement	Electrical Conductivity	Thermal Conductivity	Lubricity	Moisture Resistance	Impact Strength	Tensile Strength	Dimensional Stability
Organic															
Wood flour	x	x												x	x
Shell flour	x	x										x		x	x
Alpha cellulose (wood pulp)	x		x	x										x	
Sisal fibers	x		x	x	x	x	x					x	x	x	x
Macerated paper	x		x										x		
Macerated fabric	x			x									x		
Lignin	x	x													
Keratin (feathers, hair)	x				x								x		
Chopped rayon		x	x	x		x	x	x				x	x	x	x
Chopped Nylon		x	x	x	x	x	x	x			x		x	x	x
Chopped Orlon		x	x	x	x	x	x	x				x	x	x	x
Powdered coal	x		x		x	x						x			
Inorganic															
Mica	x		x	x	x	x	x				x	x			x
Quartz			x	x	x		x					x	x		
Glass flakes		x	x	x	x	x	x	x				x	x	x	
Chopped glass fibers			x	x	x	x	x	x				x	x	x	x
Milled glass fibers	x	x	x	x	x	x	x	x				x	x	x	x
Diatomaceous earth	x	x	x	x	x		x						x		x
Clay	x	x	x	x	x		x						x		x
Calcium silicate			x	x	x		x					x	x		x
Calcium carbonate			x	x	x		x								
Alumina trihydrate			x		x		x					x			
Aluminum powder			x				x		x	x	x			x	
Bronze powder			x				x		x	x	x			x	
Talc	x	x	x	x	x	x	x				x		x		x

This table does not indicate the degree of improvement of function. The prime function will also vary between thermosetting and thermoplastic resins. This table is to be used only as a guide to the selection of fillers.

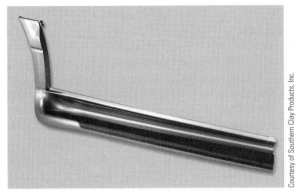

Courtesy of Southern Clay Products, Inc.

Figure 7-23. This nanocomposite step-assist, available on 2002 selected GMC and Chevrolet vans, was the first exterior application of nanocomposites in vehicles.

first developed in the late 1980s. In 1990, Toyota introduced timing belt covers with a clay/nylon nanocomposite. During the 1990s, nylons and other engineering plastics were used in other limited applications. However, in the late 1990s, considerable work occurred on creating nanocomposites using relatively inexpensive commodity plastics. An example of this was the first application of a thermoplastic olefin nanocomposite in selected General Motors vans in 2002 (Figure 7-23). Since then, polymer nanocomposites have grown as an active area of research and development. An indication of this growth appears on the website of the Society of Plastics Engineers (**www.4spe.org/membership/**), which references more than 50 articles, presentations, conferences, and seminars on nanocomposites, most of them no more than one or two years old.

Safe handling of nanofibers and nanoparticles is a growing concern. Because these fibers and particles are so small, they can easily enter airways and lead to health problems. Because of these issues, in December 2004, the EPA created a workgroup to address potential risks from nanotechnologies. An EPA nanotechnology white paper, published in February 2007, was the result. The paper includes a section on toxicity and hazard identification of engineered/manufactured nanomaterials. This section reports that the toxicity of carbon nanotubes is greater than carbon black nanoparticles, although they are very similar in chemistry and size.

Nanocomposites are often categorized by the material or form of the nanomaterials, namely, nanoclays, nanofibers, and nanometals.

Nanoclays. The nanoclay materials often rely on a type of clay called smectite. The smectite group of clay minerals, which includes talc and vermiculite, also includes some unique clays that contain flakes that are only 1 nanometer (one-millionth of a millimeter) thick. Montmorillonite and Hectorite are examples of nanoclays. Montmorrillonite is named for the town of Montmorillon in France, and Hectorite for Hector, California. If uniformly spread out, 5 grams of these nanoclays could cover

the surface of a football field. When properly dispersed in plastics, a nanocomposite of 2.5 percent clay has the properties of a 20 percent talc-filled conventional composite. Consequently, the nanoclay/olefin composite is lighter in weight. Because dispersion is frequently easier in coatings, some applications for nanaclays have opened in packaging. NanoPack Inc. sells coated films under the name NanoSeal™. This company applies a coating that contains nanoclays to polyester, PP, nylon, and PLA films. The advantage is a significant gain in oxygen barrier properties.

The technical difficulty in preparing these nanoclay composites is to achieve a thorough dispersion of the clay in the plastics. Three dispersion techniques are used practically. Solution intercalation involves dissolving the polymer in a solvent, then dispersing the clay in the solution, and finally removing the solvent to leave the clay/plastic composite behind. This is a viable process for water-soluble polymers, but because most plastics are not water-soluble, organic solvents are required. The use of such solvents involves high costs, and health and safety concerns exist, as well as problems with removing the solvents thoroughly. A second technique is to mix the clay with monomers prior to polymerization. Because the reactors used to create the polymers are frequently used for several different plastics, the addition of clay into the process involves serious problems. The third technique is melt mixing of the clays into the plastics, using extruders. Because the many clays are not compatible with the plastics, they will clump together instead of mixing uniformly. To avoid the clumping, clays are pretreated with various organic chemicals to improve their compatibility. With pretreatment, good dispersions of clays in plastics are manageable at low percentages of clay, but low ranges of clay also mean that the gain in properties is limited. As the percentage of clay increases, it becomes increasingly difficult to cover the large surface area of the clay with polymer. Consequently, dispersion problems continue to limit nanoclays.

Nanofibers. The fibers of greatest interest for plastics applications are carbon fibers, or carbon nanotubes. Although carbon nanotubes are grown as single-wall types, multiwall carbon nanotubes (MWCNTs) dominate in plastics applications. Conventional carbon fibers are not nanoscale fibers, but carbon nanofibers are available with diameters from 70 to 200 nanometers (nm). In contrast to nanofibers, MWCNTs are much smaller in diameter, having an inner diameter of about 5 nm and an outer diameter of about 10 nm. The MWCNTs generally have an aspect ratio of 1000:1, which means that they are 1000 times longer than their diameter.

The cost of the MWCNTs has limited their applications, because their price is about $600 per pound. Bayer Material Science plans to increase its capacity to produce MWCNTs under the name Baytubes, with a goal of 200 metric tons per year by 2009 and 3000 metric tons by 2112. The current leader in MWCNTs is Hyperion Catalysis, a company based in Cambridge, Massachusetts. Hyperion Catalysis sells nanotubes, masterbatches containing 15 to 20 percent nanotubes, and ready-to-use composites for antistatic products in polycarbonate, nylon, polyphenylene, and other engineering materials. To compete with nanotubes, some companies are focusing on carbon nanofibers, which are easier to make. Consequently, carbon nanofibers sell for one-half to one-fourth the cost of MWCNTs.

Due to the high cost of MWCNT and carbon nanofibers, most applications in plastics have been restricted to parts that require electrical conductivity or the ability to dissipate static charges. For example, some automotive parts for fuel lines and connectors use MWCNTs to prevent the buildup of static electricity. In computers, both hard drive components and chips need static dissipation. Other applications are in specialty products, particularly sports equipment. MWCNTs are appearing in bicycles, hockey sticks, baseball bats, and golf club shafts.

Nanometals. Metal/polymer nanocomposites are not well commercialized but have received considerable research. The creation of the nanocrystals of metal can occur before polymerization, in which case the crystals are dispersed in a polymer solution or put into a monomer solution. The other approach is to place metal ions in the monomer solution and, after polymerization, cause nanometal particles to develop by chemical or thermal methods. The potential for such materials includes optical devices, color filters, sensors, and magnetic data storage.

Large-Scale Fillers

Fillers may improve processability, product appearance, and other factors. One example of filler is wood flour, which is obtained by grinding waste wood stock into a fine granular state. This powdered filler is often added to phenolic resins to reduce brittleness and cost and improve product finish.

In most molding operations, the volume of fillers does not exceed 40 percent. However, as little as 10 percent resin may be used for molding large desktops, trays, and particleboards. In the foundry industry, as little as 3 percent resin is used to adhere sand together in the shell-molding process.

One product called cultured marble uses inorganic marble dust and polyester resin to produce items that have the appearance of genuine marble. This product has the advantage of being stain resistant and is easily produced in many colors, shapes, or sizes.

Many organic fillers cannot withstand high temperatures, which means they have low heat resistance. To improve heat resistance, a silica filler such as sand, quartz, tripolic, or diatomaceous earth is used.

Diatemaceous earth consists of the fossilized remains of microscopic organisms (diatoms). This filler provides improved compressive strength in rigid polyurethane foam.

A filler as fine as cigarette smoke (0.007 to 0.050 μm) is called fumed silica. This submicroscopic silica is added to resins to achieve thixotropy. **Thixotropy** is a state of a material that is gel-like at rest, but fluid when agitated. Other thixotropic fillers may be made from very fine powders of polyvinyl chloride, china clay, alumina, calcium carbonate, and other silicates. Cab-O-Sil® and Sylodex® are trade names for two commercial thixotropic fillers. Thixotropic fillers may be added to either thermosetting or thermoplastic resins. They are used to thicken the resin, improve strength, suspend other additives, improve the flow properties of powders, and decrease cost (Figure 7-24).

Thixotropic fillers increase viscosity (internal resistance to flow). This makes them desirable in paints, adhesives, and other compounds applied to vertical surfaces. These fillers may be added to polyester resin in fabrication of inclined or vertical surfaces. Thixotropic fillers can also act as emulsifiers to prevent separation of two or more liquids. Both oil- and water-based additives may be added and held in an emulsified state.

Fillers such as steel, brass, graphite, and aluminum are added to resins to produce electrically conductive moldings or give added strength. Plastics with these fillers may be electroplated. Plastics containing powdered lead are used as neutron and gamma-ray shields.

Wax, graphite, brass, or glass is sometimes added to provide self-lubricating qualities to plastics gears, bearings, and slides.

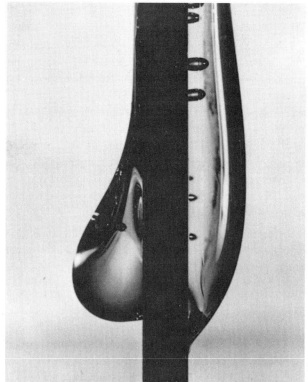

Figure 7-24. The resin on the right contains a thixotropic agent. The one on the left will sag and run without this ingredient. (Cabot Corp.)

Glass is commonly used in plastics for several reasons. It is easily added, relatively inexpensive, improves physical properties, and may be colored. Colored glass has optical advantages (especially color stability) over chemical colorants. Tiny, hollow glass spheres called microballons (also called micro spheres) are used as a filler in producing low-density composites.

RELATED INTERNET SITES

- **www.apsci.com.** Applied Sciences, Inc., produces Pyrograf® products, a line of carbon fibers. Pyrograf® III is a low-cost, carbon nanofiber.

- **www.arap.org.** The Alliance for Responsible Atmospheric Policy monitors trade implications that arise due to the implementation of the Montreal Protocol. In particular, this organization promotes the use of HCFCs instead of CFCs.

- **www.bayermaterialscience.com.** Bayer Material-Science makes CNTs named Baytubes®. Entering "Baytubes" into the search field on the home page leads to the information on nanotubes.

- **www.chemtura.com.** Chemtura Corporation is one of the world's largest producers of additives for plastics, including heat stabilizers, antioxidants, antistatic agents, blowing agents, and coupling agents.

- **www.elementis-specialties.com.** Elementis Specialties offers Bentone® HD, a hectorite clay intended for waterborne coatings.

- **www.epa.gov.** On the EPA home page, an extensive report on nanotechnology can be accessed by entering "Nanotechnology White Paper" in the search field. This paper was published in February 2007 and contains some reports on research on the toxicity of carbon nanotubes.

- **www.fibrils.com.** Hyperion Catalysis International claims to be the world leader in carbon nanotube technology. The website provides good information on both Fibril™, their trade name for carbon multi-walled nanotubes, (CMWNTs) and applications in plastics. This company sells master batches and compounds. The masterbatches contain 15 to 20 percent nanotubes by weight. The letdown process to thoroughly mix the masterbatch into the selected plastics material is not simple and requires a twin-screw extruder with appropriate screw design. To locate the data on compounding, select "masterbatches" under "plastics products" on the home page. Then look for the link "Recommendations for Masterbatch Letdown."

- **www.lockheedmartin.com.** On the home page, select "What We Do," and then find "Emerging Capabilities" and

under that "Nanotechnology" to find data on Lockheed's use of nanomaterials in its products.

- **www.nanoclay.com.** In 2000, Southern Clay Products became part of Rockwood Specialties, Inc. It continues to sell nanoclays as additives for plastics. Its nanoclays, sold under the trade name Cloisite®, are based on montmorillonite.

- **www.nanocor.com.** Nanocor sells Nanomer® nanoclays for use in nylons, epoxies, urethanes, EPDM, and engineering plastics. The company also offers masterbatches of nanoclays in PP, LLDPE, LDPE, and HDPE. It supplies several nanocomposites in pellet form, primarily based on nylon 6.

- **www.nanopackinc.com.** NanoPack Inc. sells coated films. On the website, select "Products & Data" to access information about NanoSeal™ films.

- **www.owenscorning.net.** On the home page you can search for information such as: "Automotive Solutions," "Chopped Strand Mat," "Chopped Strands," "Continuous Filament Mat," "Fabrics," "Milled Fibers," "Multi-End Roving," "Single-End Roving," "Telecommunications Solutions," "Veil & Specialty Non-Woven," "Wet Formed Mat," and "Wet Used Chopped Strands." Each category contains a description of the products, many with a photograph, documents, and questions and answers. Selection of "Processes" from the home page leads to "Centrifugal Casting," "Continuous Casting," "Cold Press Molding," "Compression Molding," "Continuous Lamination," "Filament Winding," "Hand Lay-Up," "Infusion Molding," "Injection Molding," "Preforming," "Pultrusion," "Reaction Injection Molding," "Resin Transfer Molding," "Spray Up," and "Vacuum Bagging."

- **www.tno.nl.** TNO, a company based in the Netherlands, has developed Planomers®, a nanoclay flame retardant.

- **www.zyvexpro.com.** Zyvex Technologies claims "breakthrough results in the fields of nanotechnology." It sells Arovex®, a CNT and grapheme reinforcement that is used in sporting goods, automotive, marine, aerospace, and defense applications.

VOCABULARY

The following vocabulary words are found in this chapter. Use the glossary in Appendix A to look up the definitions of any of these words you do not understand as they apply to plastics.

antiblocking
antioxidants
antistatic agents
aspect ratio
carbonaceous

catalysts (initiators)
char barriers
colorant
coupling agents
crystal whisker fibers
curing agents
exothermic
fillers
flame-retardant
fluorescence
glass, type C
glass, type E
heat stabilizers
inhibitors (stabilizers)

lamina
lubrication bloom
luminescence
microballoons (microspheres)
nanocomposite
phosphorescence
plasticizer
promoters (accelerators)
reinforcements
rovings
strands
thixotropy
yarns

QUESTIONS

7-1. To improve or extend the properties of plastics, _____ are used.

7-2. Static charges may be dissipated by the addition of _____ agents.

7-3. Name four types of colorants.

7-4. The major difference between dyes and pigments is that dyes _____ in the plastics material.

7-5. A major disadvantage of organic dyes is their poor _____ and _____ stability.

7-6. Polyvinyl chloride commonly has plasticizers added to provide _____.

7-7. What name is given to chemical-resistant glass fibers?

7-8. Single crystals used as reinforcements are called _____.

7-9. Thixotropic fillers increase _____.

7-10. One of the most widely used plasticizers is _____.

7-11. Name the type of light that is destructive to plastics.

7-12. Polymerization is initiated by the use of _____.

7-13. Chemical _____ are sometimes used to produce cellular plastics.

7-14. Identify the special stabilizers that retard or inhibit degradation through oxidation.

7-15. Radical initiators such as _____ are curing agents used to cure unsaturated polyester.

7-16. Name three functions fillers provide in plastics.

7-17. A disadvantage of _____ fillers is that they cannot withstand high processing temperatures.

7-18. Long fibrous ingredients that increase strength, impact resistance, and stiffness are called _____.

7-19. Name four reasons why glass fiber is often selected as a reinforcement.

7-20. Long rope-like strands of fibrous glass are called _____.

7-21. To lower viscosity and aid in processing, _____ and _____ are added to resins.

7-22. Zinc stearate is a _____ additive that acts as a lubricant in molding.

7-23. Small particles that contribute only slightly to strength are called _____.

7-24. To help prevent discoloration and decomposition of resins and plastics, _____ are added.

7-25. Crazing, chalking, color changes, and loss of properties may be caused by _____.

7-26. When chemical, electrical, or light energy excites electrons, the radiation of light from the plastics is called _____.

7-27. Lubrication bloom is caused by excess _____.

7-28. Because the color is distributed throughout the product, _____ plastics are superior to painted plastics.

7-29. Which additive will produce the stronger composite, carbon black or graphite fibers?

7-30. Which is stronger, a composite made by curing SMC or a filament-wound structure?

7-31. Pre-promoted polyester resins should not be stored for a period of time in a _____ container or in a _____ place.

ACTIVITIES

Antioxidants

Antioxidants are chemicals that reduce oxidative degradation of plastics. Without antioxidants, many common plastics could not remain useful for long periods of time.

Equipment. Polypropylene reactor flake, natural PP pellets, injection-molding grade, any injection-molding equipment, and an oven (preferably an air-recirculating type).

7-1. Make several parts by injecting any shape or object. Use a PP reactor flake containing no additives, fillers, or reinforcements. Also use pelletized PP, which contains some antioxidant.

7-2. Hang parts in oven at 180°C. Do not use wire or metal clips to hang parts because the metal may promote degradation. Monofilament fishing line may work if the parts are not very heavy.

7-3. Leave parts in the oven until some degradation is apparent on the parts made from the reactor flake. Remove those parts at regular time intervals to document the progression of degradation. Leave the last one or two parts until they are fully degraded and almost crumble when handled. Continue to expose parts made with pelletized material to heat. They should withstand the heat longer before the onset of degradation.

7-4. Write a report summarizing findings.

7-5. To speed up this test, mold PP pellets and reactor flake into thin sheets. Try to produce sheets less then 1 mm thick. Expose samples to heat as in step 7-2. Noticeable degradation should occur in a few hours. Figure 7-25 shows a sample that exhibited degradation after 4 hours at 180°C.

Coupling Agents

Coupling agents promote adhesion between a plastics material and fillers or reinforcements. Fibrous reinforcements cannot provide their strength if they easily slip out of the plastics matrix.

Equipment. Natural PP, extrusion or injection-molding grade, two types of chopped glass fibers—one type compatible

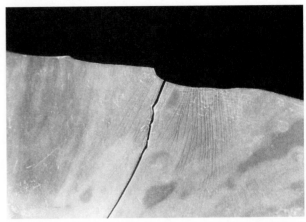

Figure 7-25. This piece of unstabilized polypropylene film exhibited a brittle fracture after 2 hours at 180°C (356°F).

with styrene and another type compatible with PP, an extruder, an injection molder, and a tensile tester.

Procedure

7-1. Compound a batch of PP with a selected loading of PS-compatible glass fibers. Determine the loading percentage to make it feasible for the extruder available. If the extruder does not handle glass well, keep loadings less than 10 percent. Compound a batch of material with the same loading of glass compatible with PP. If the glass fibers are not well dispersed after the first pass through the extruder, run both types through the extruder a second and possibly a third time.

7-2. Injection-mold parts. If available, use a mold that creates tensile test "dog-bones." Pull the parts on the tensile tester. Determine strengths and elongations at yield and break. Calculate means and deviations, and sketch curves comparing PS-compatible glass to PP-compatible glass.

7-3. Write a report summarizing findings.

CHARACTERIZATION AND SELECTION OF COMMERCIAL PLASTICS

CHAPTER 8

INTRODUCTION

Every new plastics product represents a host of decisions made by designers, engineers, processors, and marketing specialists. These professionals determine the shape, color, function, strength, style, appearance, and reliability of the product. One aspect of developing a new product involves identifying the material to use. The purpose of this chapter is to provide an introduction to several facets of material characterization and selection. Here is the content outline for this chapter:

- I. **Basic materials**
  A. Polymerization techniques
  B. Melt index
- II. **Selection of material grade**
- III. **Computerized databases for material selection**

BASIC MATERIALS

Picking among ABS, PC, PS, or PP is often difficult. In addition to cost factors, the decision must involve strength and flexural characteristics, the use/environment, and surface appearance. In a few cases, the choices are well established.

A thermoformed packaging container for a round, decorated cake does not need to be airtight or have oxygen barrier properties. It does not have to be resistant to ultraviolet radiation, but does need to protect the decorations in the frosting and allow some stacking. Such a container will usually consist of polystyrene, which is both transparent and inexpensive. Although polystyrene is rather brittle and will probably crack during use, it should withstand the two or three openings required in its service life.

Plastics lenses for glasses have much more rigorous requirements. They must be highly transparent as well as scratch- and impact-resistant. They must be moldable yet able to be worked on with sanding, grinding, and polishing equipment. A typical material choice is polycarbonate. It will exceed most impact requirements, does not shatter, and can be rather scratch-resistant.

However, many choices of basic plastics material are not obvious. Improvements in commodity resins have made selections even more difficult. For example, reinforced polypropylene has found use in some applications previously reserved for engineering materials such as nylon.

It is far beyond the scope of this chapter to enumerate common uses for a host of plastics. Appendixes E and F list many examples of applications for common plastics. However, there are some considerations that reoccur in material selection processes. One consideration is understanding the characteristics of the basic plastics material.

The determination of basic plastics is not equal to material selection, because even though some commercial products use natural, unmodified, and unreinforced plastics, such products are never 100 percent polymer. They always contain some additives. A key to understanding various plastics is the melt flow index (MFI), often called the *melt index* (see Chapter 6). To clearly understand the importance of the MFI, an introduction to the techniques of polymerization is helpful.

Polymerization Techniques

In polymerization, tiny hydrocarbon molecules (mers) combine to form huge molecules, often called macromolecules.

For this to occur efficiently in a chemical reactor, the monomer needs to be in a form that provides high surface area and low volume. Four differing processes—namely, bulk, solution, suspension, and emulsion polymerization—ensure a high ratio between surface area and volume. In **bulk polymerization**, narrow, tubular reactors guarantee the ratio. **Solution polymerization** requires small drops of monomer to be added to a large bath of solvent. For example, in polymerizing polystyrene from styrene monomer, the monomer, which is approximately 20 percent by weight, is dissolved in about 80 percent benzene. To avoid using solvents, **emulsion and suspension polymerization** processes utilize water to surround the tiny droplet of monomer. Close examination of a polymer that is in the form of reactor flake or sphere can provide a feel for the small physical scale in which polymerization reactions take place.

Regardless of the process, polymerization proceeds according to two major types of chemical reactions: **chain growth** (sometimes called **additive polymerization**) or **stepwise growth** (also called **condensation polymerization**).

In chaining, polymerization begins at one location by the action of a chemical initiator. Almost instantaneously, the complete chain forms without yielding chemical byproducts. One appropriate example would be the image of a train forming when hundreds of boxcars join together. When complete, the train leaves behind no excess parts. The chain stops growing in length due to the effect of chemicals that cause termination of the chain. Termination is also affected by probability, monomer purity, and type.

In stepwise polymerization, monomers combine to form blocks two units long. These in turn combine to form blocks four units long. This sequence continues until the process is terminated. Stepwise polymerization usually requires some chemical alteration of the monomer that results in byproducts. If the byproducts are not removed continuously, they will slow or inhibit the polymerization process. The most common byproducts are water, acetic acid, and hydrogen chloride. For example, in reaction injection molding, some materials polymerize by the stepwise reaction, producing water within the mold. Molds for this manufacturing process must be plated with nickel in order to avoid continuous rusting.

Both chain polymerization and stepwise polymerization are not perfect. Some molecules grow beyond the desired length, and some are too short. This results in a distribution of molecular lengths. Chapter 5 provided instruction on calculating the mean and standard deviation for a distribution of normally distributed values. Distributions of polymer molecules are more complex and cannot be fully described with only the mean and standard deviation. The reason for this is that most distributions of molecular lengths are not

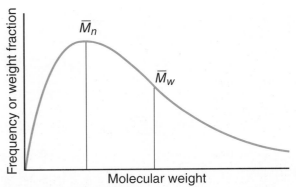

Figure 8-1. This distribution of molecular weights is typical for the polymers in many commercial plastics.

normal. In contrast, they tend to have the shape shown in Figure 8-1.

The shape of this curve (Figure 8-1) is very important. As with a normal distribution, this curve has a peak value identified as M_n. In contrast to normal distributions, the curve is not symmetrical. It has a long tail extending toward the right. That means that there are some very long molecules in the polymer. If these long molecules were rather unimportant, it might be reasonable to pretend that the distribution was normal. However, the long molecules are very significant because they alter the physical properties of the material.

To achieve a more adequate characterization of the distribution of molecular weights in a polymer, it is necessary to use both the number average molecular weight and weight average molecular weight. The **number average molecular weight** (M_n) is based on the frequency of various molecular lengths in a sample. It implies that the short molecules are just as important as the long ones. In contrast, the **weight average molecular weight** (M_w) not only counts the frequency of various molecular lengths but also the contribution of the various molecules to the total weight of the sample. This approach gives the longer molecules greater importance than simply counting their frequency.

The ratio of the M_w to M_n is the **polydispersity index (PI)**. This number is somewhat similar to a standard deviation for normal distribution. Both the standard deviation and the PI indicate the spread of values in the distribution. A PI of 1.0 is a theoretically perfect polymer in which all molecules are exactly the same length. As the PI value increases, the difference between the shortest and longest molecules in a sample increases. Commercial polymers range from a PI of about 2 to 40.

Determining number average molecular weight and weight average molecular weight is not easy. It requires special instruments and trained personnel. Consequently, many manufacturing companies turn to melt index values for a "quick and dirty" estimation of the average molecular lengths in a sample.

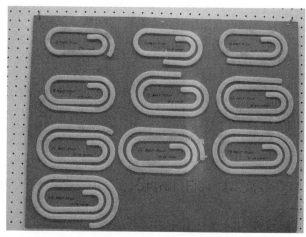

Figure 8-2. Results of spiral flow analysis of polypropylene homopolymer.

Table 8-1. **Relation of MFI to Spiral Flow**

MFI	Flow length, inches
2	19.5
4	23.25
6	25.75
8	27.5
10	34.5
12	36.25
14	37.25
16	39.75
18	41.25
20	43.75

Table 8-2. **Relation of MFI to Molecular Weights**

Melt Flow Range	M_n	M_w	PI(M_w/M_n)
0.3–0.6	90,000	850,000	9.5
1–3	65,000	580,000	9
2.6	61,500	375,000	6.1
3–5	60,000	450,000	8
4.6	57,000	333,000	5.9
5–8	35,000	350,000	10
7.5	51,000	296,000	5.8
8.5	50,000	305,000	6.1
8–16	30,000	300,000	10

Melt Index

Figure 8-2 shows the influence of various melt flow rates on spiral flow. Spiral flow molds for injection molding have very long cavities, resulting in a shot of plastics freezing up long before it reaches the end of the cavity. During spiral flow analysis, the setup on an injection-molding machine provides uniform temperature for the mold and melted plastics. Setup also ensures uniform injection pressure and speed. Each shot has excess material available. This means that the flow does not stop because the material was used up, but because—given the conditions of molding—it would not flow any further.

The material used for the spiral flow shown in Figure 8-2 was natural polypropylene homopolymer. The differences resulted from selecting different melt index materials. The shortest flow was a melt index of 2, whereas the longest was a melt index of 20. The melt index values indicate the number of grams of material extruded through a standard orifice in a 10-minute period. The flow lengths for various melt index values are shown in Table 8-1.

As is apparent, when the melt flow index (MFI) increases, the flow length also increases. The reason the flow length increases is that the average length of the polymer chains decreased as the MFI increased. This means that the higher melt material was "runnier" than the lower melt material. Changes in the MFI are inversely correlated with average chain lengths (the average molecular weight).

Table 8-2 presents these relationships, using data of selected types of polypropylene homopolymer.

The data in Table 8-2 show that as the MFI increases, both the M_n and M_w decrease. Although estimates of molecular weight based on MFI values are not highly accurate, the general trends are very consistent. Understanding these relations can assist in the process of material selection.

SELECTION OF MATERIAL GRADE

The selection of a grade of material is also rather complicated. Because several companies produce comparable materials, the choices are numerous. For example, one major petrochemical company offers five grades of high-heat polycarbonate. These include three general-purpose (GP) grades, three flame-retardant grades, two glass-reinforced grades, one extrusion grade, and one impact-modified grade; and four specialty grades that include medical applications, one lighting grade, three optical types, and three types of automotive lens materials. Another major company offers 12 grades of general-purpose polycarbonate, 3 high-flow grades, 8 health care–product grades, 6 flame-retardant grades, 5 glass-reinforced grades, 7 wear-resistant grades, 7 optical-quality grades, 4 grades for blow molding, and 5 grades of high-heat material.

Most petrochemical companies provide MFI values on material data sheets. These numbers imply the usually unstated values for the M_n, M_w, and PI. Beside variations in polydispersity and molecular weight, other differences stem from an array of fillers and additives. Flame-retardant grades include chemicals that inhibit combustion. High-flow grades often include internal lubricants to promote flow. The search for the optimum material for a specific application can be overwhelming.

COMPUTERIZED DATABASES FOR MATERIAL SELECTION

Because there are hundreds of companies producing plastics and thousands of grades, selecting the *best* material for a product is very difficult. To assist in this task, computerized databases assemble, collect, and organize material data. Typically included is such information as trade name, chemical name, manufacturer name, and most physical properties. Properties include strength, impact, flexural and tensile modules, hardness, filler content, brittleness temp, heat deflection temperature, refractive index, water absorption, melt flow, linear mold shrinkage, and others. The most thorough databases include rheological data, creep curves, aging data, chemical resistance, and weatherability.

There are two basic types of material databases: those produced by software companies and those created by plastics-producing companies. Software companies gather huge numbers of materials into databases and then sell compact discs or online hookups. One such database includes about 17,000 thermoplastic materials. Resin manufacturers, such as GE, DuPont, Bayer, and BASF, generally focus only on their own materials and have databases on about 200 to 600 materials.

One advantage of a company-specific database is that it generally includes more information on the physical properties than that gathered by software companies. Smaller databases may also contain more accurate information.

Databases include sort features that allow a user to search for materials that match a set of specified characteristics. The sort will then return to the user a list of materials that fit the conditions.

Inherent in this process is the assumption that direct comparisons between materials are basically accurate. However, the validity of such comparisons rests on the comparability of the data included. In many cases, tests performed to ASTM standards may not be comparable due to varying size and shape of test specimens, mold designs, and molding conditions. Some of the company-specific databases have moved to ISO tests that may result in more accurate comparisons. However, even within one plastics-producing company the same problem arises. Large companies have multiple testing laboratories in various geographic locations. Often, the machines used to mold test samples are from differing manufacturers; the molds are not identical in size or cooling; and the molding conditions

are not identical from one location to the next. All of these factors introduce variability into the testing results.

SUMMARY

In material selection, awareness of the characteristics of the basic polymer is critical. The average lengths of molecular chains and their distribution impact physical properties and flow characteristics. In addition to the characteristics of the polymeric molecules, the additives, fillers, reinforcing agents, and colors tailor plastics for specific applications. Correct decisions about materials enhance product life and reliability, whereas incorrect decisions often lead to product failures.

RELATED INTERNET SITES

- **www.sabic-ip.com.** Saudi Basic Industries acquired GE Plastics in 2007. A very useful tool is the material selector, which matches materials to desired mechanical, thermal, and electrical properties. To find the material selection tool, choose "Products & Services" on the home page. Then click on the link "Resin and LNP Compounds." That will lead to a link "Engineering Data and Tools," which finally offers "Material Selection" as an option.

- **www2.dupont.com.** To locate CAMPUS (Computer-Aided Material Preselection by Uniform Standards), select "Industries" on the home page. Then go to "Plastics," and then on to "Products and Services," and finally to "Browse Products and Services." That will allow access to a search DuPont box. Type in CAMPUS to reach the database for material selection.

- **www.matweb.com.** MatWeb is one of the generic material databases. It provides mechanical, physical, and electrical properties as well as processing recommendations.

VOCABULARY

The following vocabulary words are found in this chapter. Use the glossary in Appendix A to look up the definitions of any of these words you do not understand as they apply to plastics.

>additive polymerization
>bulk polymerization
>chain growth polymerization
>condensation polymerization
>emulsion polymerization
>number average molecular weight (M_n)
>polydispersity index (PI)
>solution polymerization
>stepwise growth polymerization
>suspension polymerization
>weight average molecular weight (M_w)

QUESTIONS

8-1. Which polymerization processes use water surrounding droplets of monomer?

8-2. How is a high surface area–to–volume ratio created in bulk polymerization?

8-3. What is a common solvent used in the solvent polymerization of styrene?

8-4. How is chain growth different from stepwise growth in polymerization reactions?

8-5. Which type of polymerization reaction produces byproducts?

8-6. What does a polydispersity index of 1.0 mean?

8-7. What is the approximate range of PI values common in commercial plastics?

ACTIVITIES

Introduction

A practical approach to learning about molecular weight distributions involves investigation of melt mixtures. Melt mixtures are blends of plastics in which mixing is done after the initial polymerization. Many compounding companies must provide customers with specific melt flow plastics. To achieve the desired flow rate, compounders mix various percentages of readily available melt flow materials. This can be difficult because predicting a resultant melt flow from flow rates of the ingredients requires a practical understanding of the differences between number average molecular weights and weight average molecular weights.

Equipment. Compounding extruder, pelletizer, extrusion plastometer or other type of melt flow measurement device, selected plastics, and personal safety equipment.

Procedure

8-1. Acquire two or three natural homopolymers with differing melt flow rates. If possible, get information on the number average molecular weight and weight average molecular weight for each type.

8-2. Use melt index equipment to verify the melt flow of the selected materials. For accuracy, run several tests on each type.

8-3. Determine the relative percentages of the ingredients and weigh them. For example, select a 2-melt type and a 20-melt type. For a 500-gram batch, use 50 percent (250 g) of the 2-melt materials and 50 percent (250 g) of the 20 melt. If the differences in flow rates are rather extreme, the difficulty with prediction will be most obvious.

8-4. Melt mix the two melt flow rates with an extruder and pelletize the extrudate.

8-5. Predict the melt flow of the "new" plastics. If the predicted value is 11 based on adding 20 and 2, then divide 22 by 2. Review Figure 8–2. The typical distribution of molecular weights is not symmetrical, but skewed to the right. Because the weight average molecular weight is also to the right of the peak, its influence will be skewed.

8-6. Measure the melt flow of the compounded material.

8-7. Use the following technique to create an estimate of the melt flow of new material:

  a. Determine the log of one melt index value. The log of 2 is 0.3.

  b. Multiply the percentage of the two-melt material by the log value ($0.5 \times 0.3 = 15$).

  c. Determine the log of the other melt index value and multiply by the percentage ($0.5 \times 1.3 = 0.65$).

  d. Add the two values and take the inverse log of the sum. The inverse log of 0.8 is 6.3. This 6.3 value should approximate the measured melt index.

8-8. Melt mix unequal percentages such as 15 percent 2-melt and 85 percent 20-melt material. Predict the resulting melt flow and then determine it experimentally.

8-9. Use more than two ingredients. Does the mathematical procedure accurately predict the mixture of three or more melt index materials?

8-10. Write a report summarizing the results.

MACHINING AND FINISHING

CHAPTER 9

INTRODUCTION

In this chapter, you will learn how plastics and composites are machined and finished. Molded or formed plastics parts often require further processing that includes common operations such as flash removal, slot cutting, polishing, and annealing. Many of the operations are similar to those used in machining and finishing metal or wood products.

The vast number of machines and processes used in the shaping and finishing of plastics do not allow a detailed discussion. However, certain basics apply to all machining and finishing processes. Additives, fillers, and plastics of each family require different shaping and finishing techniques. Few plastics pieces are made solely by machining, even though many molded parts must be finished or fabricated into useful items.

All machining and finishing operations present potential physical hazards. Fine dusts or particles are produced when sawing, laser cutting, or water-jet cutting. Eye protection and a face mask should be worn by the operator to prevent injury or inhalation of particles (see Chapter 4 on Health and Safety).

Processing techniques for plastics are based on those used for wood and metal. Nearly all plastics may be machined (Figure 9-1). As a rule, thermosets are more abrasive to cutting tools than thermoplastics.

Machining techniques for composite materials, such as high-pressure laminates, filament-wound parts, and reinforced plastics, attempt to prevent fraying and delamination of the composite. Reinforcing agents used with various matrix compounds are abrasive. Therefore, most cutting tools must be made from tungsten carbide or coated (with titanium diboride, for example). High-speed steel (M2) or diamond-tipped cutters are also used. Boron/epoxy composites are generally cut using diamond-tipped tools. The lower thermal conductivity and modulus of elasticity (softness, flexibility) of most thermoplastics mean that tools should be kept properly sharp to allow them to cut cleanly without burning, clogging, or causing frictional heat.

Elastic recovery causes drilled or tapped holes to become smaller than the diameter of the drills used, and turned diameters often become larger. The low melting points of some thermoplastic materials tend to make them gum, melt, or craze when machined. Plastics expand more than most materials when heated. The coefficient of thermal expansion for plastics is roughly 10 times greater than that of metals. Cooling agents (liquids or air) may be needed to keep the cutting tool clean and free of chips. Benefits of cooling include increased cutting speed, smoother cuts, longer tool life, and elimination of dust. Because the polymer matrix has a high coefficient of expansion, even small variations of temperature may cause dimensional control problems.

Figure 9-1. Threads on long lengths of pipe are machined, as seen in this sample.

Topics covered in this chapter include these:

I. Sawing

II. Filing

III. Drilling

IV. Stamping, blanking, and die cutting

V. Tapping and threading

VI. Turning, milling, planing, shaping, and routing

VII. Laser cutting

VIII. Induced fracture cutting

IX. Thermal cutting

X. Hydrodynamic cutting

XI. Smoothing and polishing

XII. Tumbling

XIII. Annealing and post-curing

SAWING

Nearly all types of saws have been adapted for cutting plastics. Backsaws, coping saws, hacksaws, saber saws, handsaws, and jeweler's saws may be used for hobby/craft or short-run cutting. For cutting plastics, the shape of the tooth is very important.

Circular blades should have plenty of *set* or be hollow ground. Blades should have a deep, well-rounded *gullet* (Figure 9-2A). The rake (or hook) angle should be zero (or slightly negative). The *back clearance* should be about 30°. The preferred number of teeth per centimeter varies with the thickness of the material to be cut. Four or more teeth per centimeter should be used for cutting thin materials, whereas fewer than four teeth per centimeter are needed for plastics over 25 mm (1 in.) thick.

For materials less than 25 mm (1 in.) thick, precision tooth blades are recommended. For materials over 25 mm (1 in.) thick, buttress tooth (also called skip-tooth) blades are preferred.

A *skip-tooth* (buttress) bandsaw blade (Figure 9-3) features a wide gullet, which provides ample space for plastics chips to be carried out of the kerf (cut made by the saw). For best results, the teeth should have zero front rake and a raker set.

Bandsaw blades may be reversed to achieve a zero or negative rake. Abrasive carbide or diamond grit blades may be used to cut graphite and boron/epoxy composites. In all cutting operations, it is best to back up the work with a solid material to reduce chipping, fraying, and delamination of composites. Table 9-1 suggests the number of teeth per centimeter for various speeds and thicknesses of material.

Note: *Fewer teeth per centimeter are needed for cutting plastics over 6 mm (0.23 in.) thick. Thin or flexible plastics may be cut with shears or blanking dies. Foams require cutting speeds above 8000 feet per minute (fpm) (40 m/s).*

For cutting reinforced or filled plastics and many thermosetting plastics, carbide-tipped blades are recommended

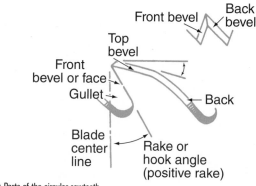

(A) Parts of the circular sawtooth.

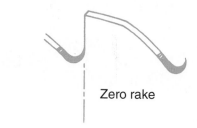

(B) The zero rake of a sawtooth.

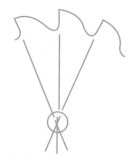

(C) The zero rake where the line of the tooth face crosses the center of the blade.

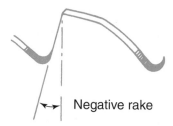

(D) Negative rake.

(E) Positive rake.

Figure 9-2. Tooth characteristics of a circular saw blade.

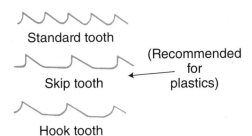

Standard tooth

Skip tooth ← (Recommended for plastics)

Hook tooth

(A) A skip tooth provides a large gullet and a good chip clearance. The hook tooth is sometimes preferred for glass-filled thermosets.

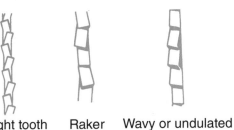

Straight tooth   Raker   Wavy or undulated

(B) Common bandsaw blade teeth (cutting edge view).

Figure 9-3. Bandsaw blade teeth.

Flow International Corp.

Figure 9-4. Cutting plastics. Composites like Kevlar are quickly and cleanly cut by a waterjet.

(Figure 9-4). They provide accurate cuts and have a long blade life. Abrasive- or diamond-tipped blades may also be used. A liquid coolant is advised to prevent clogging or overheating, although jets of CO_2 are often used as a coolant while thermoplastics are being machined. All cutting tools must have protective shields and safety devices.

Table 9-1. Power Saws for Cutting Plastics

Plastics	Circular Saws			Band Saws		
	Teeth per cm		Speed, m/s	Teeth per cm		Speed, m/s (>6 mm)
	(<6 mm)	(>6 mm)		(<6 mm)	(>6 mm)	
Acetal	4	3	40	8	5	7.5–9
Acrylic	3	2	15	6	3	10–20
ABS	4	3	20	4	3	5–15
Cellulose acetate	4	3	15	4	2	7.5–15
Diallyl phthalate	6	4	12.5	10	5	10–12.5
Epoxy	6	4	15	10	5	7.5–10
Ionomer	6	4	30	4	3	7.5–10
Melamine-formaldehyde	6	4	25	10	5	12.5–22.5
Phenol-formaldehyde	6	4	15	10	5	7.5–15
Polyallomer	4	3	45	3	2	5–7.5
Polyamide	6	4	25	3	2	5–7.5
Polycarbonate	4	3	40	3	2	7.5–10
Polyester	6	4	25	10	5	15–20
Polyethylene	6	4	45	3	2	7.5–10
Polyphenylene oxide	6	4	25	3	2	10–15
Polypropylene	6	4	45	3	2	7.5–10
Polystyrene	4	3	10	10	5	10–12.5
Polysulfone	4	3	15	5	3	10–15
Polyurethane	4	3	20	3	2	7.5–10
Polyvinyl chloride	4	3	15	5	3	10–15
Tetrafluoroethylene	4	3	40	4	3	7.5–10

Note: See Appendix G. Useful Tables.

The feed and speed for cutting composites vary greatly with thickness and material but are similar to those used for nonferrous materials (see Table 9-2).

FILING

Thermosetting plastics are quite hard and brittle. Therefore, filing removes material in the form of a light powder. Aluminum type A, shear/tooth, or other files that have coarse, single-cut teeth with an angle of 45° are preferred (Figure 9-5). The deep-angled teeth enable the file to clear itself of plastic chips. Many thermoplastics tend to clog files. Curved-tooth files like those used in auto body shops are good because they clear themselves of plastics chips. Special files designed for plastics should be kept clean and should not be used for filing metals.

DRILLING

Thermoplastic and thermosetting materials may be drilled with any standard twist drill. However, special drills designed for plastics produce better results. Carbide-tipped drills have a long life. Holes drilled in most thermoplastics and some thermosets are usually 0.05 to 0.10 mm (0.002 to 0.004 in.) undersize. Thus, a 6 mm (0.23 in.) drill will not produce a hole wide enough for a 6 mm (0.23 in.) rod. Thermoplastics may need a coolant to reduce frictional heat and gumming during drilling.

For most plastics and fibers, drills should be ground with a 60° to 90° point angle on a low helix drill with a lip clearance angle of 12° to 15° (Figure 9-6). The rake angle on the cutting edge should be zero or several degrees negative. Table 9-3 gives rake and point angles for cutting many

Table 9-2. Machining Composites

Operation	Material	Cutting Tool	Speeds	Feeds (<0.250 thick)
Drilling	GI-Pe	0.250-diamond	20,000 rpm	0.002/rev
	B-Ep	0.250-diamond core (60–120 grit)	100 sfpm	0.002/rev
	B-Ep	0.250 2–4 flute (HSS)	25 sfpm	0.002/rev
	Kv-Ep	Spade-carbide	>25,000 rpm	0.002/rev
	Kv-Ep	Brad point-carbide	>6,000 rpm	0.002/rev
	GI-Ep	Tungsten carbide	<2,000 rpm	<0.5 ipm
	Gr-Ep	Tungsten carbide	>5,000 rpm	<0.5 ipm
Band Saw	Kv-Ep	14 teeth, honed saber	3,000–6,000 sfpm	<30 ipm
	B-Ep	Carbide or 60 gut diamond	2,000–5,000 sfpm	<30 ipm
	Hybrids		3,000–6,000 sfpm	<30 ipm
	G-Pe	14 teeth, honed saber	3,000–6,000 sfpm	<30 ipm
Milling	most	Four-flute carbide	300–800 sfpm	<10 ipm
Circular Saw	Gr-Ep, B-Ep	60 grit diamond	6,000 sfpm	<30 ipm
	GI-Pe	60 tooth carbide or 60 grit diamond	5,000 sfpm 5,000 sfpm	<30 ipm <30 ipm
Lathe	Kv-Ep	carbide	250–300 sfpm	0.002/rev
	GI-Ep	carbide	300–600 sfpm	0.002/rev
Shears	Kv-Ep	HSS or carbide	—	<30 ipm
Countersink or counterbore	most	Diamond grit or carbide	20,000 rpm 6,000 rpm	<0.5 ipm
Laser (10kW)	most <0.250 thickness	CO_2 cooling	—	<30 ipm, depending upon material
Waterjet	most <0.250 thickness	60,000 psi–0.10 in orifice	—	<30 ipm, depending upon material
Abrasive (sanding-grinding)	most	Silicone carbide or alumina grit (wet)	4,000 sfpm	—
Router	most	Carbide or diamond grit	20,000 rpm	—

Note: ipm = inches per minute.

(A) Various file profiles.

(B) Rotary files, sometimes used on plastics.

(C) Comparison of shear-tooth and ordinary mill files.

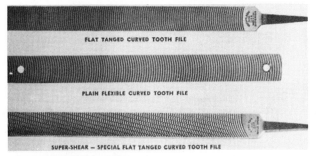

FLAT TANGED CURVED TOOTH FILE

PLAIN FLEXIBLE CURVED TOOTH FILE

SUPER-SHEAR — SPECIAL FLAT TANGED CURVED TOOTH FILE

(D) Curved-tooth files used on plastics.

Figure 9-5. Files used on plastics.

Table 9-3. **Drill Geometry**

Material	Rake Angle	Point Angle	Clearance	Rake
Thermoplastic				
Polyethylene	10°–20°	70°–90°	9°–15°	0°
Rigid polyvinyl chloride	25°	120°	9°–15°	0°
Acrylic (polymethyl methacrylate)	25°	120°	12°–20°	0°
Polystyrene	40°–50°	60°–90°	12°–15°	0° to neg. 5°
Polyamicle resin	17°	70°–90°	9°–15°	0°
Polycarbonate	25°	80°–90°	10°–15°	0°
Acetal resin	10°–20°	60°–90°	10°–15°	0°
Fluorocarbon TFE	10°–20°	70°–90°	91–15°	0°
Thermosetting				
Paper or cotton base	25°	90°–120°	10°–15°	0°
Fibrous glass or other fillers	25°	90°–120°	10°–15°	0°

plastics materials. Highly polished, large, slowly twisting flutes (large helix or helix rake angle) are desirable for good chip removal.

Most plastics are soft enough that conventional drills tend to draw themselves into the material and often cause chipping as the drill exits the backside of a workpiece. The use of a straight flute drill helps reduce this problem. Conventional drills can be modified by grinding a small flat approximately 1.5 mm (1/16 in.) wide on the face of both cutting edges. This will help prevent the drill from drawing itself into the material.

Four factors that affect cutting speeds are as follows:

1. Type of plastics
2. Tool geometry
3. Lubricant or coolant
4. Feed and depth of cut

The cutting speed of plastics is given in surface **feet per minute (fpm)** or **meters per second (m/s)**. Meters per second refers to the distance the cutting edge of the drill travels in 1 second when measured on the circumference of the cutting tool. The following formula is used to determine surface meters per second. This information may be obtained from various handbooks.

$$m/s = \pi D \times rpm$$
$$r/s = (m/s) \div (pD)$$

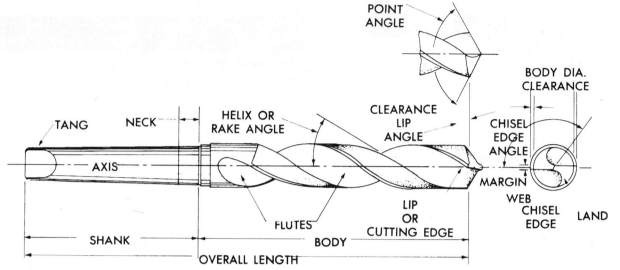

POINT ANGLE

BODY DIA. CLEARANCE

CLEARANCE LIP ANGLE

CHISEL EDGE ANGLE

TANG   NECK

HELIX OR RAKE ANGLE

AXIS

MARGIN

WEB CHISEL EDGE

LAND

FLUTES

LIP OR CUTTING EDGE

SHANK   BODY

OVERALL LENGTH

Morse Twist Drill Machine Co.

(A) Selected nomenclature for tapered-shank twist drills. Drills of 12.5 mm (0.5 in.) or less in diameter usually have straight shanks.

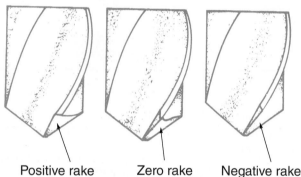

Positive rake   Zero rake   Negative rake

(B) A zero rake is usually preferred for plastics, but a negative rake is sometimes used with polystyrene.

Figure 9-6. Drill nomenclature.

where **rpm = revolutions per minute**
     r/s = revolutions per second
     m/s = meters per second
     D = diameter of cutting tool in meters
     π = 3.14

As a rule of thumb, plastics have a cutting speed of 1 m/s (200 fpm). A guide for drilling thermoplastics and thermosets is shown in Table 9-4.

The rate at which a drill or cutting tool moves into the plastics is crucial. The distance the tool is fed into the work each revolution is called the **feed**. Feed is measured in inches or millimeters. Drill feed ranges from 0.25 to 0.8 mm (0.001 to 0.003 in.) for most plastics, depending on thickness of the material (Table 9-5).

Many of the same principles of speed and feed apply to reaming, countersinking, spotfacing, and counterboring (Figure 9-7). Holes may be made in many thermoplastics with hollow punches. Warming the stock may help the punching operation.

Sintered-diamond core drills, countersinks, reamers, or counterbores may be used with ultrasonic energy to bore

Table 9-4. Guide to Speeds for Drilling Plastics

Drill Size	Speed for Thermoplastics, r/s	Speed for Thermosets, r/s
No. 33 and smaller	85	85
No. 17 through 32	50	40
No. 1 through 16	40	28
1.5 mm	85	85
3 mm	50	50
5 mm	40	40
6 mm	28	28
8 mm	28	20
9.5 mm	20	16
11 mm	16	10
12.5 mm	16	10
A–C	40	28
D–O	20	20
P–Z	20	16

into some composites. Boron/epoxy, graphite/boron/epoxy, and other hybrid composite materials may require ultrasonic techniques.

STAMPING, BLANKING, AND DIE CUTTING

Many thermoplastics and thin pieces of thermosets may be cut using rule, blanking, piercing, or matched molding dies (Figure 9-8). This is done on flat parts less than 6 mm (0.23 in.)

Table 9-5. Drilling Feeds of Plastics

Material	Speed, m/s	Nominal Hole Diameter					
		1.5	3	6	12.5	19	25
Thermoplastics		Feed, mm/revolution					
Polyethylene	0.75–1.0	0.05	0.08	0.13	0.25	0.38	0.5
Polypropylene							
TFE fluorocarbon							
Butyrate							
High-impact styrene	0.75–1.0	0.05	0.1	0.13	0.15	0.15	0.2
Acrylonitrile-butadiene-styrene							
Modified acrylic							
Nylon	0.75–1.0	0.05	0.08	0.13	0.2	0.25	0.3
Acetals							
Polycarbonate							
Acrylics	0.75–1.0	0.02	0.05	0.1	0.2	0.25	0.3
Polystyrenes	0.75–1.0	0.02	0.05	0.08	0.1	0.13	0.15
Thermosets							
Paper or cotton base	1.0–2.0	0.05	0.08	0.13	0.15	0.25	0.3
Homopolymers	0.75–1.5	0.05	0.08	0.1	0.15	0.25	0.3
Fiberglass, graphitized, and asbestos base	1.0–1.25	0.05	0.08	0.13	0.2	0.25	0.3

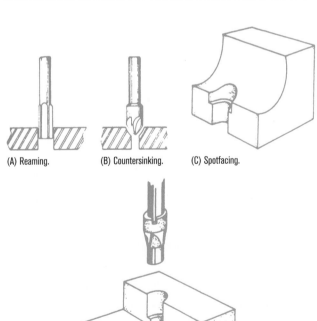

(A) Reaming. (B) Countersinking. (C) Spotfacing.

(D) Counterboring.

Figure 9-7. Drilling operations in plastics.

thick. Holes may either be drilled or die cut, and heating the plastics stock may aid in such operations.

Punching and shearing of laminar composite materials usually results in some delamination, edge raging, or fiber tearing. It is recommended that the part be abrasively ground to the final dimension.

TAPPING AND THREADING

Standard machine-shop tools and methods may be used for tapping and threading. To prevent overheating, taps should be finish ground and have polished flutes. Lubricants may also be used to help clear chips from the hole. If transparency is needed, a wax stick may be inserted in the drilled hole before tapping. The wax lubricates, helps expel chips, and makes a more transparent thread.

Because of the elastic recovery of most plastics, oversized taps should be used. Oversized taps are designated as follows:

H1: Basic size to basic + 0.012 mm
H2: Basic + 0.012 mm to basic + 0.025 mm
H3: Basic + 0.025 mm to basic + 0.038 mm
H4: Basic + 0.038 mm to basic + 0.050 mm

The cutting speed for machine tapping should be less than 0.25 m/s (9.842 in./s). The tap should be backed out often to

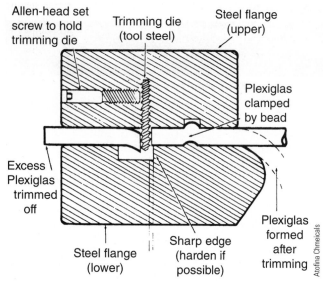

Allen-head set screw to hold trimming die

Trimming die (tool steel)

Steel flange (upper)

Plexiglas clamped by bead

Excess Plexiglas trimmed off

Steel flange (lower)

Sharp edge (harden if possible)

Plexiglas formed after trimming

Atofina Chmeicals

(A) Blank die and clamp for cutting Plexiglas.

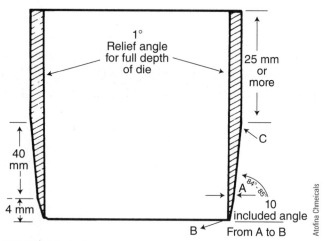

1° Relief angle for full depth of die

25 mm or more

C

40 mm

4 mm

84°–85°

10 included angle From A to B

A

B

Atofina Chmeicals

(B) Modified shoemaker die.

Figure 9-8. Dies used for cutting plastics.

clear chips. Usually, not more than 75 percent of the full thread is cut into plastics. Using sharp V-threads is not advised. Acme threads (Figure 9-9) and ISO metric threads are preferred. Figure 9-10 displays the metric screw thread designation, and selected ISO metric threads are shown in Table 9-6. To obtain tap drill size, subtract the pitch from the diameter. National coarse and fine thread and tap drill sizes are shown in Table 9-7.

Plastics may be tapped and threaded on lathes and screw machines (see Mechanical Fastening in Chapter 18).

TURNING, MILLING, PLANING, SHAPING, AND ROUTING

High-speed steel or carbide cutting tools for machining brass and aluminum should be used for machining plastics (Figure 9-11A). The feeds and speeds are similar. For many

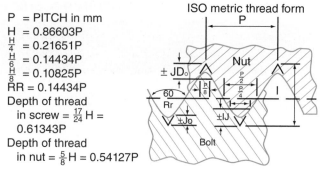

P = PITCH in mm
H = 0.86603P
$\frac{H}{4}$ = 0.21651P
$\frac{H}{6}$ = 0.14434P
$\frac{H}{8}$ = 0.10825P
RR = 0.14434P
Depth of thread
  in screw = $\frac{17}{24}$ H =
  0.61343P
Depth of thread
  in nut = $\frac{5}{8}$ H = 0.54127P

ISO metric thread form

Figure 9-9. Simplified ISO metric thread forms.

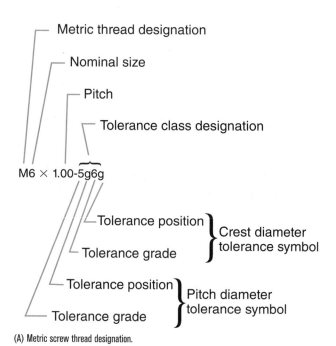

Metric thread designation

Nominal size

Pitch

Tolerance class designation

M6 × 1.00-5g6g

Tolerance position
Tolerance grade
} Crest diameter tolerance symbol

Tolerance position
Tolerance grade
} Pitch diameter tolerance symbol

(A) Metric screw thread designation.

CLASSES OF FIT	INTERNAL THREADS (NUTS)	EXTERNAL THREADS (BOLTS)
CLOSE (close accuracy required)	5H	4h
MEDIUM (general purposes)	6H	6g
FREF (easy assembly)	7H	8g

(B) Classes of fit.

Figure 9-10. Metric screw thread designation and classes of fit.

plastics, a surface speed of 2.5 m/s (492 fpm) with feeds (depth of cut) of 0.12 to 0.5 mm (0.005 to 0.02 in.) per revolution will produce good results. On cylindrical stock, a 1.25 mm (0.049 in.) cut will reduce the diameter by 2.5 mm (0.098 in.).

Climb cutting (or *down-cutting*) refers to a milling operation using lubrication that gives a good machined finish on plastics (Figure 9-11B). In climb milling, the work moves in the same direction as the rotating cutter. The feed rate on multiple-edged milling cutters is expressed in millimeters of cut per cutting edge per second. The feed of a milling machine

Table 9-6. Selected ISO Metric Threads—Coarse Series

Diameter, mm	Pitch, mm	Tap Drill, mm	Depth of Thread, mm	Area of Root, mm²
M 2	0.40	1.60	0.25	1.79
M 2.5	0.45	2.05	0.28	2.98
M 3	0.50	2.50	0.31	4.47
M 4	0.70	3.30	0.43	7.75
M 5	0.80	4.20	0.49	12.7
M 6	1.00	5.00	0.61	17.9
M 8	1.25	6.75	0.77	32.8
M 10	1.50	8.50	0.92	52.3
M 12	1.75	10.25	1.07	76.2
M 16	2.00	14.00	1.23	144

is expressed in millimeters of table movement per second rather than per spindle rotation. The following formula is used to determine the amount of feed in inches per minute or millimeters per second:

$$\text{mm/s} = t \times \text{fpt} \times \text{r/s}$$

where

t = number of teeth
mm/s = feed in millimeters per second
fpt = feed per tooth (chip load)
r/s = revolutions per second (spindle or work)

Table 9-8 gives turning and milling data for various plastics materials. Table 9-9 gives side and end relief angles and back rake angles for tools used to cut different plastics.

Carbide-tipped cutters are advised for all milling, planing, shaping, and routing work. Conventional high-speed steel

Table 9-7. National Coarse and National Fine Threads and Tap Drills

Size	Threads Per Inch	Major Dia.	Minor Dia.	Pitch Dia.	Tap Drill 75% Thread	Decimal Equivalent	Clearance Drill	Decimal Equivalent
2	56	.0860	.0628	.0744	50	.0700	42	.0935
	64	.0860	.0657	.0759	50	.0700	42	.0935
3	48	.099	.0719	.0855	47	.0785	36	.1065
	56	.099	.0758	.0874	45	.0820	36	.1065
4	40	.112	.0795	.0958	43	.0890	31	.1200
	48	.112	.0849	.0985	42	.0935	31	.1200
6	32	.138	.0974	.1177	36	.1065	26	.1470
	40	.138	.1055	.1218	33	.1130	26	.1470
8	32	.164	.1234	.1437	29	.1360	17	.1730
	36	.164	.1279	.1460	29	.1360	17	.1730
10	24	.190	.1359	.1629	25	.1495	8	.1990
	32	.190	.1494	.1697	21	.1590	8	.1990
12	24	.216	.1619	.1889	16	.1770	1	.2880
	28	.216	.1696	.1928	14	.1820	2	.2210
1/4	20	.250	.1850	.2175	7	.2010	G	.2610
	28	.250	.2036	.2268	3	.2130	G	.2610
5/16	18	.3125	.2403	.2764	F	.2570	21/64	.3281
	24	.3125	.2584	.2854	1	.2720	21/64	.3281
3/8	16	.3750	.2938	.3344	5/16	.3125	25/64	.3906
	24	.3750	.3209	.3479	Q	.3320	25/64	.3906
7/16	14	.4375	.3447	.3911	U	.3680	15/32	.4687
	20	.4375	.3725	.4050	25/64	.3906	29/64	.4531
1/2	13	.5000	.4001	.4500	27/64	.4219	17/32	.5312
	20	.5000	.4350	.4675	29/64	.4531	33/64	.5156
9/16	12	.5625	.4542	.5084	31/64	.4844	19/32	.5937
	18	.5625	.4903	.5264	33/64	.5156	37/64	.5781

(Continued)

Table 9-7. National Coarse and National Fine Threads and Tap Drills (*Continued*)

Size	Threads Per Inch	Major Dia.	Minor Dia.	Pitch Dia.	Tap Drill 75% Thread	Decimal Equivalent	Clearance Drill	Decimal Equivalent
5/8	11	.6250	.5069	.5660	17/32	.5312	21/32	.6562
	18	.6250	.5528	.5889	37/64	.5781	41/64	.6406
3/4	10	.7500	.6201	.6850	21/32	.6562	25/32	.7812
	16	.7500	.6688	.7094	11/16	.6875	49/64	.7656
7/8	9	.8750	.7307	.8028	49/64	.7656	29/32	.9062
	14	.8750	.7822	.8286	13/16	.8125	57/64	.8906
1	8	1.0000	.8376	.9188	7/8	.8750	1- 1/32	1.0312
	14	1.0000	.9072	.9536	15/16	.9375	1- 1/64	1.0156
1-1/8	7	1.1250	.9394	1.0322	63/64	.9844	1- 5/32	1.1562
	12	1.1250	1.0167	1.0709	1- 3/64	1.0469	1- 5/32	1.1562
1-1/4	7	1.2500	1.0644	1.1572	1- 7/64	1.1094	1- 9/32	1.2812
	12	1.2500	1.1417	1.1959	1-11/64	1.1719	1- 9/32	1.2812
1-1/2	6	1.5000	1.2835	1.3917	1-11/32	1.3437	1-17/32	1.5312
	12	1.5000	1.3917	1.4459	1-27/64	1.4219	1-17/32	1.5312

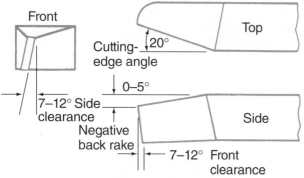

(A) A cutting tool rake and clearance angles for general-purpose turning of plastics. Note the 0–5° back rake angle. A sharp-pointed tool with a +20° rake is used in turning polyamides.

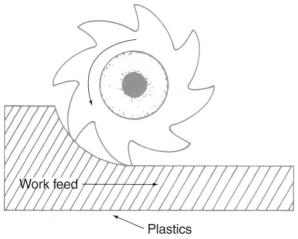

(B) Climb milling, or down milling, a technique in which the work moves in the same direction as the rotating cutter.

Figure 9-11. Machining plastics.

shapers, planers, and routers used for woodworking may be employed with plastics as long as tools are sharpened in a proper manner. Routers and shapers are useful for cutting beads, rabbets, and flutes and for trimming edges. Carbide- or diamond-tipped tools are vital for long runs, uniformity of finish, and accuracy.

LASER CUTTING

A CO_2 laser (light amplification by stimulated emission of radiation) can deliver powerful radiation at a wavelength of 10.6 μm (microns). A laser cutting machine may be used on plastics to make intricate holes and complex patterns (Figure 9-12). Laser power can be controlled to merely etch the plastics surface or actually vaporize and melt it. Holes and cuts made by laser have a slight taper but have a clean, finished appearance. Cuts made by a laser are more precise and tolerances are held more closely than those made with conventional machining operations. There is no physical contact between the plastics and laser equipment. Therefore, no chips are produced. Laser cutting produces a residue of fine dust that is easily removed by vacuum systems. Most polymers and composites may be laser machined, but some laminar composites tend to heat up, bubble, and char.

INDUCED FRACTURE CUTTING

Acrylics and several other plastics, including some composites, may be cut to shape by *induced fracture* methods. These methods are similar to cutting glass. A sharp tool or cutting blade is

Table 9-8. Turning and Milling Plastics

Material	Turning Single Point (H-S Steel)			Milling Tool Per Tooth (H-S Steel)		
	Depth of cut, mm	Speed, m/s	Feed, mm/r	Depth of cut, mm	Speed, m/s	Feed, mm per tooth
Thermoplastics						
Polyethylene	3.8	0.8–1.8	0.25	3.8	2.5–3.8	0.4
Polypropylene	0.6	1.5–2	0.05	3.8	2.5–3.8	0.4
TFE-fluorocarbon				1.5	3.8–5	0.1
Butyrates				3.8	2.5–3.8	0.4
ABS	3.8	1.2–1.8	0.38	3.8	2.5–3.8	0.4
Polyamides	3.8	1.5–2	0.25	3.8	2.5–3.8	0.4
Polycarbonate	0.6	2–2.5	0.05	1.5	3.8–5	0.1
Acrylics	3.8	1.2–1.5	0.05	1.5	3.8–5	0.1
Polystyrenes, low	3.8	0.4–0.5	0.19	3.8	2.5–3.8	0.4
and medium impact	0.6	0.8–1	0.02	3.8	2.5–3.8	0.4
Thermosets						
Paper	3.8	2.5–5	0.3	1.5	2.0–2.5	0.12
and cotton base	0.6	5–10	0.13	1.5	2.0–2.5	0.12
Fiber glass	3.8	1–2.5	0.3	1.5	2.0–2.5	0.12
and graphite base	0.6	2.5–5	0.13	1.5	2.0–2.5	0.12
Asbestos base	3.8	3.2–3.8	0.3	1.5	2.0–2.5	0.12

Table 9-9. Design of Turning Cutting Tool

Work Material	Side Relief Angle, Deg.	End Relief Angle, Deg.	Back Rake Angle, Deg.
Polycarbonate	3	3	0–5
Acetal	4–6	4–6	0–5
Polyamide	5–20	15–25	neg. 5–0
TFE	5–20	0.5–10	0–10
Polyethylene	5–20	0.5–10	0–10
Polypropylene	5–20	0.5–10	0–10
Acrylic	5–10	5–10	10–20
Styrene	0–5	0–5	0
Thermosets: Paper or cloth	13	30–60	neg. 5–0
Glass	13	33	0

used to score or scratch the plastics surface. On thick pieces, both sides are scored. Pressure is applied along the scratch line, causing the plastics to fracture. The fracture will follow the score line (Figure 9-13).

THERMAL CUTTING

Heated wires or dies are used to cut solid and expanded or foamed plastics. Hot dies are used to cut fabrics and silhouette-shaped products, whereas a heated wire or ribbon is commonly used to cut expanded plastics (Figure 9-14). Thermal cutting produces a smooth edge with no chips or dust.

HYDRODYNAMIC CUTTING

High-velocity fluids may be applied to cut many plastics and composites (Figure 9-4). Pressures of 320 MPa (46,417 psi) are used. Foamed or cellular plastics have been reinforced and filled plastics have been successfully machined by this method.

SMOOTHING AND POLISHING

Smoothing and polishing techniques for plastics are similar to those used on woods, metals, and glass.

Because of the elastic and thermal properties of thermoplastics, many are difficult to grind. Abrasive grinding is more easily accomplished on thermosetting materials, reinforced plastics, and most composites. Grinding is not advised unless open-grit wheels are used with a coolant. Hand and machine sanding are important operations. **Open-grit sandpaper** is used

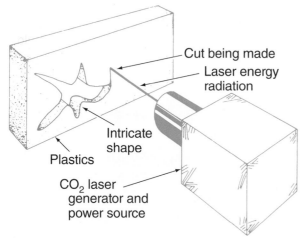

(A) Basic concept.

(B) This 250-watt CO_2 laser can cut acrylic sheet 25 mm (1.0 in.) thick at a rate of 100 mm/min.
Figure 9-12. Light energy from a laser can be used to cut intricate shapes in plastics or to trim plastics into final shape.

Coherent, Inc., USA and EINA SL, Spain

(B) Align the score line with the edge of the table.

(C) Press down on the plastics piece to induce a fracture.

(A) Score plastics with a tool.

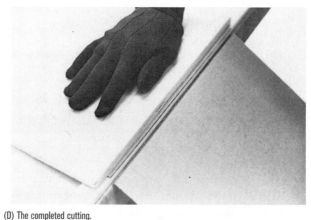

(D) The completed cutting.
Figure 9-13. The induced-fracture cutting method.

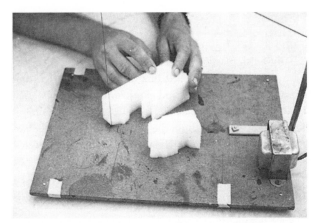

Figure 9-14. Expanded polystyrene is easily cut using a hot Nichrome wire. The wire melts a path through the cellular material.

on machines to prevent clogging (loading). A number 80-grit silicon carbide abrasive is advised for rough sanding. In any machine sanding, light pressure is used to prevent overheating the plastics.

Disc sanders (working at 30 r/s) and belt sanders (working at a surface speed of 18 m/s [59 ft/s]) are used for dry sanding. If water coolants are used, the abrasive lasts longer, and cutting action is increased. Progressively finer abrasives are used. That is, the first rough sanding using 80-grit paper should be followed by 280-grit silicon-carbide wet or dry sandpaper. The final sanding may be with 400- or 600-grit sandpaper. After the sanding is finished and the abrasives removed, further finishing operations are used.

Ashing, buffing, and polishing are done on abrasive-charged wheels. These wheels may be made of cloth, leather, or bristles. A different wheel is used for each abrasive grit. Finishing wheel speeds should not exceed 10 surface m/s (32.8 ft/s). Surface speed may be increased with the use of coolants.

Never finish plastics on wheels used for metals. Small metal particles may be left in the wheel that will damage the plastics surface. Machines should be grounded because static electricity is generated by the movement of the wheels over the plastics. Remove coarse tool marks before finishing wheels are used.

Ashing is a finishing step in which a wet abrasive is applied to a loose muslin wheel. Number 00 pumice is commonly used (Figure 9-15). A hood or shield is placed over the wheel because the operation is wet. Surface speeds of over 20 m/s (65.6 ft/s) may be used. Overheating is avoided in this process, and the loose muslin wheel is fast cutting on irregular surfaces.

Buffing is an operation in which grease- or wax-filled abrasive bars or sticks are applied to a loose or sewn muslin wheel.

Loose buffs are used for more irregular shapes or entering crevices. Hard buffing wheels should be avoided.

Charge a **buffing wheel** by holding the bars or sticks against it as it revolves (Figure 9-16A). This produces frictional heat that leaves the wax-filled abrasive on the wheel. The most common buffing abrasives are **tripoli**, rouge, or other fine silica.

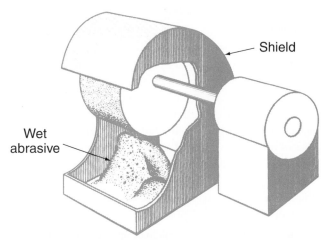

Figure 9-15. Ashing with wet pumice is a faster cutting method than the use of grease- or wax-based compounds. The cooling action is also better.

(A) Charging a wheel with rouge.

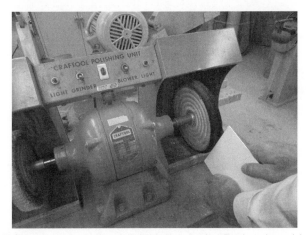

(B) Buffing requires caution. About half of each side is done by pulling the work toward the operator.

Figure 9-16. Buffing plastics materials.

Polishing, sometimes called luster buffing or burnishing, employs wax compounds containing the finest abrasives such as levigated alumina or whiting. Polishing wheels are generally made of loose flannel or chamois. A final polishing is sometimes done with a clean, abrasive-free wax on a flannel or chamois wheel. The wax fills many imperfections and protects the polished surface.

Never let the finishing wheel rotate to the edge of a part because this may cause it to jerk from your hands. The wheel may pass over an edge, but never to it. Always keep the workpiece below the center of the buffing wheel. The preferred procedure is to buff about half the surface and then turn it around and finish the final portion. The part should be moved or pulled toward the operator in rapid, even strokes (Figure 9-16B). Do not spend very much time at the finishing wheels. Move stock around. If you hold a piece in one place, the heat generated by friction between the wheel and the workpiece will melt many thermoplastics.

Solvent-dip polishing of cellulosic and acrylic plastics may be used to dissolve minor surface defects (Figure 9-17A). Parts are either dipped into or sprayed with solvents for about 1 minute. Solvents are sometimes used to polish edges or drilled holes. All solvent-polished parts should be annealed to prevent crazing.

Surface coatings may be used on most plastics to produce a high surface gloss that may cost less than other finishing operations.

Flame polishing with an oxygen–hydrogen flame may also be used to polish some plastics (Figure 9-17B).

TUMBLING

The tumbling-barrel process is one of the least costly ways to rapidly finish plastics molded parts. It produces a smooth finish by rotating plastics parts in a drum with abrasives and lubricants, causing the parts and abrasives to rub against each other with a smoothing effect (Figure 9-18A). The amount of material removed depends on the speed of the tumbling barrel, the abrasive grit size, and the length of the tumbling cycle.

In another tumbling process, abrasive grit is sprayed over the parts as they tumble on an endless rubber belt. Figure 9-18B shows parts being tumbled while being doused with abrasive grit.

Dry ice is sometimes used in tumbling to remove molding flash. The dry ice chills the thin flash, making it brittle. Then tumbling can break it free in a very short time.

ANNEALING AND POST-CURING

During the molding, finishing, and fabrication processes, the plastics or composite part may develop internal stresses. Chemicals may sensitize the plastics and cause crazing.

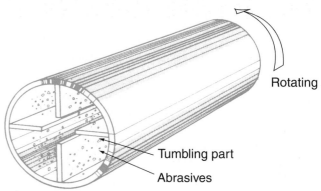

(A) Parts being tumbled in a revolving drum.

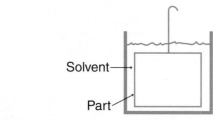

(A) Solvent-dip polishing.

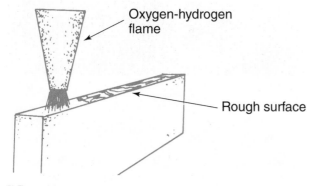

(B) Flame polishing.

Figure 9-17. Two methods of polishing.

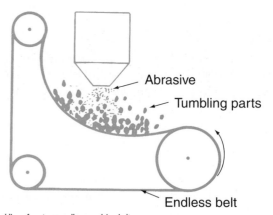

(B) Tumbling of parts on endless revolving belt.

Figure 9-18. Two tumbling methods.

Many parts develop these internal stresses as a result of cooling immediately after molding or post-mold curing because chemical reactions continue during complete polymerization. Composites are sometimes left in the mold or placed in a curing jig until the curing process has been completed and all chemical activity and temperatures are brought to ambient levels. For some plastics and composites, the internal stresses may be reduced or eliminated by annealing. Annealing consists of prolonged heating of the plastics part at a temperature lower than the molding temperature. The parts are then slowly cooled. All machined parts should be annealed before cementing.

Tables 9-10 and 9-11 give heating and cooling times for the annealing of Plexiglas. Figure 9-19 shows a large oven that may be used in the process.

RELATED INTERNET SITES

- **www.beamdynamics.com.** Many companies manufacture laser cutting and engraving equipment, including BEAM Dynamics.

- **www.coherent.com.** To find an extensive brochure on laser machining, including laser cutting of plastics, select "Products" on the home page. Then choose "Laser Machining Tools", and then pick "Laser Cutting Tools," which leads to a choice of "Literature," which contains the brochure of interest.

- **www.coherent.com.** Coherent sells laser cutting and engraving equipment. To locate information about plastics, select "Applications" on the home page. Then choose "Materials Processing and Industrial." Finally select "Non-Metal Cutting."

Table 9-10. Heating Times for Annealing of Plexiglas

Thickness (mm)	Time in a Forced-Circulation Oven at the Indicated Temperature, h									
	Plexiglas G, 11, and 55					Plexiglas I-A				
	110*	100°C*	90°C*	80°C	70°C**	90°C*	80°C*	70°C*	60°C	50°C
1.5 to 3.8	2	3	5	10	24	2	3	5	10	24
4.8 to 9.5	2½	3½	5½	10½	24	2½	3½	5½	10½	24
12.7 to 19	3	4	6	11	24	3	4	6	11	24
22.2 to 28.5	3½	4½	6½	11½	24	3½	4½	6½	11½	24
31.8 to 38	4	5	7	12	24	4	5	7	12	24

Notes: Times include period required to bring part up to annealing temperature, but not cooling time. See Table 9-9.
*Formed parts may show objectional deformation when annealed at these temperatures.
**For Plexiglas G and Plexiglas 11 only. Minimum annealing temperature for Plexiglas 55 is 80°C.
Source: Atofina Chemicals.

Table 9-11. Cooling Times for Annealing of Plexiglas

Thickness (mm)	Rate (°C)/h	Time to Cool from Annealing Temperature to Maximum Removal Temperature							
		Plexiglas G, 11, and 55				Plexiglas I-A			
		230 (110°C)	212 (100°C)	184 (90°C)	170 (80°C)	194 (90°C)	176 (80°C)	158 (70°C)	140 (60°C)
1.5 to 3.8	122 (50)	¾	½	½	¼	¾	½	½	¼
4.8 to 9.5	50 10	1½	1¼	¾	½	1½	1¼	¾	½
12.7 to 19	22–5	3¼	2¼	1½	¾	3	2¼	1½	¾
22.2 to 28.5	18–8	4¼	3	2	1	4	3	2¼	1
31.8 to 38	14–10	5¾	4½	3	1½	5¾	4½	3	1½

Note: Removal temperature is 70°C for Plexiglas G and II, 80°C for Plexiglas 55, and 50°C for Plexiglas 1-A.
Source: Atofina Chemicals.

Figure 9-19. Microprocessor-controlled mechanical convection oven.

- **www.flexa.it.** Flexa is an Italian company that sells automation equipment for sign makers. Click on the British flag for English. Select "Products" on the home page, and then choose "Equipment for the Forming of Thermoplastic Materials." That will lead to a list that includes "Flame Polishing Machines" The unique feature of these machines is that they separate water into hydrogen and oxygen by electrolysis and then use the gases in the polishing torch.

- **www.flowcorp.com.** A leading manufacturer of waterjet cutting equipment is Flow International Corporation. Selection of "Waterjet Technology" on the home page links

to a list of topics including history and function of water jets. Selection of "Waterjet Cutting" leads to a list that contains "Videos, Brochures, Tradeshows." Several videos show water-jet cutting operations.

- **www.thermo.com.** Thermo Scientific sells a number of precision ovens, including gravity convection, mechanical convection, and vacuum ovens. To locate information about these ovens, select "Products" on the home page. Under "Lab Equipment" choose "More," and then "Heating Equipment." Then select "Heating and Drying Ovens."

VOCABULARY

The following vocabulary words are found in this chapter. Use the glossary in Appendix A to look up the definitions of any of these words you do not understand as they apply to plastics.

> ashing
> blanking
> buffing wheel
> feed
> feet per minute (fpm)
> hollow ground
> kerf
> laser cutting
> meters per second (m/s)
> open-grit sandpaper
> rake
> revolutions per minute (rpm)
> revolutions per second (r/s)
> tripoli
> tumbling
> tumbling barrel
> whiting

QUESTIONS

9-1. Name the process of slowly cooling plastics to remove internal stresses.

9-2. Identify the number of teeth per centimeter for cutting 6 mm (0.23 in.) thick polycarbonate on the bandsaw.

9-3. Name the rake angle of a circular saw when the hook angle of the tooth and the center of the blade are in line.

9-4. How many r/s are required for drilling with an 8 mm drill in thermoplastics?

9-5. The distance the cutting tool is fed into the work each revolution is called _____.

9-6. What is the abbreviation or symbol for the International Organization for Standardization?

9-7. Name the slit, or notch, made by a saw or cutting tool.

9-8. Name the burnishing agent sometimes used to polish the edges of acrylic plastics.

9-9. Name the finishing operation in which wet abrasives are used.

9-10. Name the cutting operation that uses high-velocity fluids to cut plastics.

9-11. What type of cutting edges or teeth on tools is essential for long runs, uniform finishing, and accuracy?

9-12. What does the 1.00 show in M6X1.00-5g6g thread designation?

9-13. Name a popular silica abrasive used in some finishing operations.

9-14. Frictional _____ is a major problem in machining most plastics.

9-15. Name the operation that is preferred over sawing for many plastics because it produces a smoother edge.

9-16. How many teeth per centimeter should a circular saw have for cutting thin plastics materials? Thick materials?

9-17. What type of bandsaw blade and what cutting speed should you use to cut 3 mm (0.118 in.) thick acrylic plastics?

9-18. What is a skip-tooth blade?

9-19. Name some saws that may be used to cut plastics. Indicate the kinds of jobs for which each may be used.

9-20. What is drill feed? What is its range in millimeters for most plastics?

9-21. What factors affect the cutting speed of drills in plastics materials?

9-22. What precaution must be observed in drilling a hole of a given size in a plastic rod?

9-23. Why are oversized taps necessary for plastics materials?

9-24. Name the preferred thread forms for taps used on plastics.

9-25. What is climb milling? Why is it used?

9-26. What is laser cutting of plastics? Where is it used?

9-27. What grit number of sandpaper should be used for final finishing of plastics?

9-28. What is ashing? Buffing? Charging a wheel?

9-29. Which abrasives are used in buffing? In burnishing?

9-30. Briefly describe solvent-dip polishing and flame polishing.

9-31. What is tumbling? Why is it so named?

9-32. What does the smoothing in the tumbling operation?

9-33. What is annealing or post-curing of machined or molded parts? Why is this process done?

9-34. What machining operation would you select for shaping or finishing of the following products?

    a. 6 mm (0.23 in.) thick polycarbonate window glazings

    b. Christmas tree decoration shapes made of thin sheet or films

    c. Removing flash from radio cabinet housing

    d. Making the edges of a plastics part smooth and glossy

ACTIVITIES

Drilling

Introduction. Drill geometry dramatically affects the drilling process. Figure 9-20 shows the start of a hole in a piece of acrylic. Notice the surface roughness. A standard twist drill caused this result. Compare Figure 9-20 to Figure 9-21, which shows another drill dimple in acrylic. The surface in Figure 9-21 is much smoother because the drill used had a geometry appropriate for acrylics.

Figure 9-21. The beginning of a drilled hole using a drill with special geometry.

Equipment. Drill press, selected drills, sheet acrylic plastics, and safety glasses.

Procedure

9-1. Acquire a thick acrylic sheet, at least 6 mm (0.23 in. thick). Thicker sheets permit better examination of the walls of the hole.

9-2. Using appropriate safety equipment, safe practices, and a drill at least 12 mm (0.5 in.) in diameter, drill dimples,

Figure 9-20. The beginning of a drilled hole using a standard twist drill.

partial holes, and through holes. (Smaller holes exhibit the same characteristics, but they are harder to thoroughly inspect.) Save all chips and cuttings.

9-3. Examine the start of a hole, as shown in Figures 9-20 and 9-21. Examine the inside walls of the holes. Examine the exit of the drill on the back of the piece.

9-4. Purchase or grind a drill to the geometry recommended for acrylic in Table 9-3.

9-5. Drill a second set of dimples, partial holes, and through holes. Examine them carefully.

9-6. Compare the chips made by the differing drill geometries.

9-7. Did the standard drill cause chips when exiting the piece? Did the special drill also cause chips?

9-8. Drill holes with standard and special drills at various rpms and feed rates.

9-9. At high rpms and/or high feed rates, does the heat generated melt the acrylic? Does the special drill generate more or less frictional heat than the standard bit?

9-10. Record observations and offer explanations for differences caused by drill geometries.

Milling

Introduction. Machining a groove with a ball-end mill or a slot with a flat-end mill provides a direct comparison between climb milling and conventional milling. The effects of speed and feed on the quality of a machined surface are also observable.

Equipment. Milling machine, selected end mills, plastics sheet or moldings thick enough for ease of clamping, and safety glasses.

Procedure

9-1. Use a ball-end mill to cut a groove in a selected material, using the parameters suggested in Table 9-8. Tools ranging from 12 to 25 mm (0.5 to 1 in.) in diameter provide ease of inspection.

9-2. Is the side of the cut that experienced climb milling smoother or rougher than the side that experienced conventional milling?

9-3. Does the surface quality change significantly with changes in speed and feed?

9-4. Record and explain your observations.

Flame Polishing

Introduction. Flame polishing is a common technique used to finish the edges of thermoformed or fabricated acrylic products. Department store displays often exhibit flame-polished edges. Flame polishing is popular because it is several times faster than buffing.

Equipment. Oxyhydrogen torch, selected plastics, gloves, and safety glasses. Oxyacetylene welding or cutting equipment can provide much of the necessary components. A hydrogen regulator is necessary because an acetylene regulator cannot be used on a hydrogen tank.

Procedure

9-1. Cut samples of acrylic or other plastics.

9-2. Use an oxyacetylene flame to polish the edges. Try excess oxygen flames and excess acetylene flames. How do these flames affect the plastics? Can the oxyacetylene be adjusted so that it does not discolor the samples?

9-3. Prepare the oxyhydrogen flame.

! CAUTION

Any oxyhydrogen flame is almost invisible. Exercise great care to avoid burns when manipulating the oxyhydrogen flame.

9-4. Experiment with the rate of travel of the flame along the edge of a sample.

9-5. Does the flame remove the scratches caused by sawing?

9-6. Intentionally scratch a sample and then polish it. How deep must the scratches be to prevent them from being polished out?

9-7. Experiment with the relative amounts of oxygen and hydrogen. Is the neutral flame the best for polishing?

9-8. Record and explain your observations.

MOLDING PROCESSES

CHAPTER 10

INTRODUCTION

Molding processes convert plastics resins, powders, pellets, and other forms into useful products. One characteristic common to all molding processes is that they involve some level of force. In processing powders and pellets, the force required can be enormous. Although driving liquid resins into molds takes much less force than that needed to move melted pellets, some level of pressure is essential.

A host of molding processes is available to processors. This chapter does not attempt to discuss all these techniques. Instead, it focuses on three main areas of molding—namely, injection, compression, and transfer molding and techniques for liquid resins. The outline for this chapter is presented here:

- I. Injection molding
  - A. Injection unit
  - B. Clamping unit
  - C. Injection-molding safety
  - D. Specification of molding machines
  - E. Elements of molding cycles
  - F. Advantages of injection molding
  - G. Disadvantages of injection molding
  - H. Injection-molding thermosets
  - I. Co-injection molding
  - J. Fluid injection
  - K. Overmolding
  - L. Electric and hybrid machines
- II. Molding liquid materials
  - A. Reaction injection molding
  - B. Reinforced reaction injection molding (RRIM)
  - C. Liquid resin molding
- III. Molding granular and sheet thermoset materials
  - A. Compression molding
  - B. Transfer molding
  - C. Sheet molding

INJECTION MOLDING

Injection molding is the principal process for converting plastics into products. The list of injection-molded products that influence daily life is almost endless and includes TV, VCR, and computer housings; CDs; CD players; glasses; toothbrushes; automobile parts; athletic shoes; ballpoint pen barrels; and office furniture.

Injection molding is appropriate for all thermoplastics except polytetraflouroethylene (PTFE) fluoroplastics, polyimides, some aromatic polyesters, and some specialty grades. Injection-molding machines (IMMs) for thermosets provide processing for phenolics, melamine, epoxy, silicone, polyester, and numerous elastomers. In all cases, pelletized or granular materials absorb sufficient heat to make them "flowable." The machine injects hot plastics into a closed mold that creates the desired shape. After cooling or chemical transformation, an ejector system removes the parts from the mold.

Figure 10-1 shows three modern injection-molding machines: one small horizontal model, one medium horizontal model, and one vertical machine. All three machines have the same basic design, namely, the reciprocating screw style. Although there are a number of other types of molding machines, the reciprocating screw type is dominant. In reciprocating screw machines, granular material is quickly made molten by both the heat from the **barrel** and frictional heat created when the screw turns. The screw has the tasks of heating, converting pellets, and acting like a plunger. When the material is uniformly fluid, the screw moves forward, forcing hot melt through the runner system into the mold cavities.

(A) Boy 50m. 55 US tons with "Procan Control®" fully closed loop for all machine functions.

Boy Machines Inc.

(B) An UBEMAX UF-1600 metric ton all-electric injection-molding machine. 1100 US ton all-electric injection-molding machine.

UBEMAX UF-1600 METRIC TON ALL-ELECTRONIC INJECTION MOLDING MACHINE (UBE MACHININERY INC., ANN ARBOR, MI)

Injection-molding machines are also classified as either horizontal or vertical, depending on the direction the clamping unit moves. If the mold opens and closes horizontally, the parts will drop down shoots or onto conveyor belts after ejection. If the mold opens and closes vertically, parts cannot fall out after ejection. The primary use of the vertical press is for insert molding. To increase production, some vertical machines have multiple bottom halves of the mold. If two mold bottoms are used, the press is either a shuttle or rotary type (Figure 10-1C). When one-half is being loaded with fresh inserts, the other half is in the injection cycle. If more than two mold bottoms are used, the presses will be in rotary arrangement. Vertical presses typically require less floor space than horizontal machines and can also have lower tooling costs due to the use of multiple mold buttons.

A closer look at **injection-molding machines (IMMs)** reveals that they contain two major components—the injection unit and the clamp unit.

Injection Unit

The injection unit has the task of melting and injecting materials. The major parts within this unit are the hopper, the barrel, the end cap on the barrel, the nozzle, the screw and nonreturn valve, heater bands, a motor to rotate the screw, and a hydraulic cylinder to move the screw forward and backward. Control systems keep the temperatures at selected levels as well as initiate and time screw rotation and injection strokes. Figure 10-2 is a simple schematic of a typical **plasticating** unit.

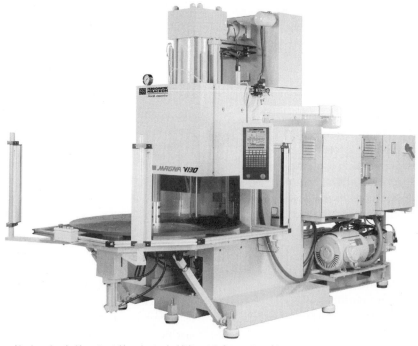

Photo courtesy of Milacron Inc.

(C) This vertical injection-molding machine is equipped with a rotary table and a standard light curtain for operator safety.

Figure 10-1. Three sizes of modern injection-molding machines.

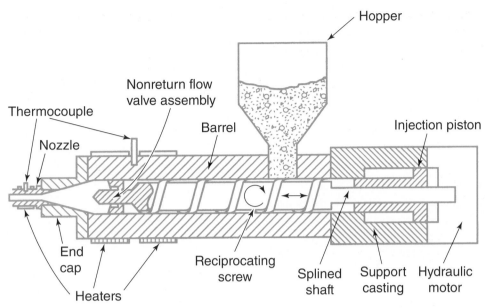

Figure 10-2. The simplified schematic of an injection unit.

The injection cylinder controls movements of the screw. The hydraulic pump generates pressure on hydraulic oil, and as the oil fills the injection cylinder, the screw is driven forward or backward. European and Asian manufacturers generally rate the pressure a machine can exert in a unit called **bar**. A bar is approximately equal to the pressure of 1 atmosphere, which is approximately 100 kPA (14.7 psi). The pressure many injection-molding machines can apply to plastics materials ranges from about 1500 to 2500 bar (20,000 to 30,000 psi).

The action of the screw determines the speed and efficiency of plasticizing pellets. Figure 10-3A shows a small injection screw. Note that the depth of the screw flights near the rear of the screw is greater than near the front end.

A typical screw consists of three major sections: the feed zone, the transition zone, and the metering zone. The feed zone contains deep flights and accounts for approximately half of the total length. The transition zone is about one-fourth of the total length. During transition, the flight depth is reduced, and the resulting compression and friction cause most of the melting of pellets. The melted material and any colorants or additives are mixed in the metering zone, which has a shallow flight depth. The mixing cannot take place until the material is melted. Therefore, this section of the screw is very important. In the metering zone, some screws have special designs to promote mixing. As the material passes through the metering zone, it should reach the desired melt temperature.

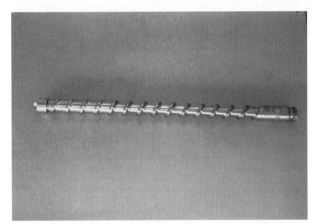

(A) This 30 mm IMM screw shows a ring-type nonreturn valve. Notice the differing flight depths.

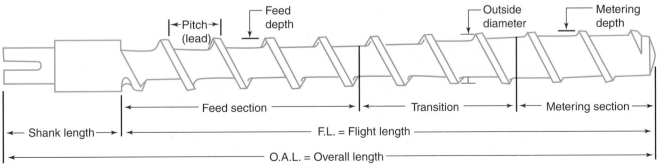

(B) In the metering screw, a great deal of heat is generated as the material is compressed in the transition region.

Figure 10-3. Injection-molding screws.

The two main specifications for an injection screw are the L/D ratio (length/diameter) and the compression ratio (Figure 10-3B). The L/D is usually the length of the flighted section of the screw divided by the outside diameter. Common L/D ratios range from 18:1 to 24:1, with 20:1 being the most common. The compression ratio is the flight depth in the feed zone divided by the flight depth in the metering zone. The compression ratio can range from 1:5:1 to 4:5:1. General-purpose screws typically have compression ratios between 2:5:1 and 3:1.

Nonreturn valves have two common styles, as shown in Figure 10-4. The style shown in Figure 10-3A corresponds to Figure 10-4A. The purpose of the nonreturn valve is to prevent backflow of material during injection. If the nonreturn valve does not function properly, the pressure on the melted plastics may be insufficient to cause the desired flow into the recesses of the mold.

Clamping Unit

The clamping unit has the task of opening and closing the mold and ejecting the parts. The two most common methods used to generate clamping forces are direct hydraulic clamps and toggle clamps actuated by hydraulic cylinders. Figure 10-5 shows the toggle style in both closed and open positions. Figure 10-6 portrays the hydraulic clamp.

Each style is available in a wide range of sizes. Toggle clamps generate force mechanically, thus requiring smaller clamping

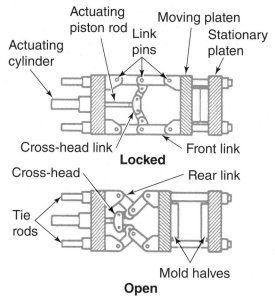

Figure 10-5. The toggle clamp design in open and closed positions.

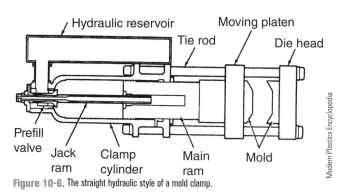

Figure 10-6. The straight hydraulic style of a mold clamp.

cylinders. Hydraulic clamps eliminate mechanical linkages but require much larger clamp cylinders. Very large machines may utilize a combination of hydraulic and mechanical clamping mechanisms, as seen in Figure 10-7.

In addition to plasticating and clamp units, a typical injection-molding machine also includes a hydraulic pump to

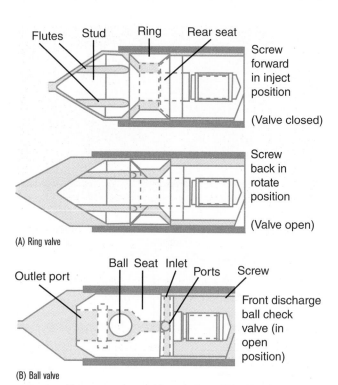

(A) Ring valve

(B) Ball valve

Figure 10-4. Two nonreturn valve assemblies used to prevent the backflow of melted plastics during injection.

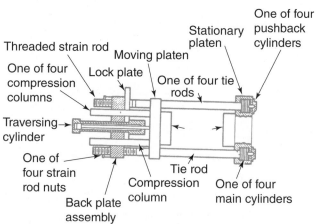

Figure 10-7. The hydromechanical clamp is used most frequently on very large machines.

move and pressurize the hydraulic oil and a reservoir for the oil. There are guards covering the barrel to prevent contact with heaters and electrical connections. There is a guard over the nozzle for protection should molten material fly into the air. The clamping unit is typically guarded on the front, top, and back of the machine. Injection molding is potentially dangerous because it involves molten material and great clamping pressures. To protect operators, several safety systems are triggered by a door that allows operators access to the mold area.

Injection-Molding Safety

Machinery manufacturers use devices to protect both the machine operators and the machine itself. Operator and technician protection relies on guards, doors, mold-closing safety systems, purge guards, and rear-door systems. As required by law, mold-closing safety systems have three separate systems: a mechanical drop bar, electrical interlocks, and hydraulic interlocks.

Mechanical drop bars. The purpose of a drop bar is to prevent the closing of a mold in the event that an operator has hands and arms between the mold halves. If electrical and hydraulic safety systems fail and a mold begins to close on an operator's arm, the drop bar must stop the closing of the mold. All persons working on or around injection molders should recognize drop bars and their proper adjustments.

Drop bars respond to the motion of the front gate of an IMM. If the door opens, the bar must drop. Drop bars come in many sizes and shapes, but two styles predominate: the straight rod and the lobed rod. In the first style, a rod passes through a hole in the stationary **platen** of the machine. The stationary plates hold the half of the mold that does not move. When the door is open, a flap or bar drops in front of the hole. If the mold begins to close, the rod bumps into the bar and prevents mold closing.

The safety rod requires adjustment before every mold change. The distance between the end of the rod and the bar should be adequate to allow the bar to drop easily. It should also be short enough so that the moving platen has little distance to travel before hitting the stop. This will guarantee that the momentum of the moving platen is limited.

Figure 10-8 shows a drop bar in the down position. Figure 10-9 shows the same drop in the up position, allowing the rod to move forward through the hole in the block attached to the platen. Figure 10-10 shows another style of drop bar. The bar is down, meaning the door on the machine was open.

An adjustment on this style of device is either made by a lock nut on the threaded rod or by machined grooves in the rod. The threaded type is visible in Figures 10-8 and 10-10.

Figure 10-8. This drop bar is in the down position, preventing mold closing.

Figure 10-9. With the drop bar in the up position, the threaded rod can move forward and permit mold closing.

Figure 10-10. This safety rod has a threaded portion for length adjustments.

One problem with the straight rod type is the possibility that the bar will not drop and no protection is available. The bar will not drop if the rod is not properly adjusted or the mold does not fully open. Please carefully examine Figure 10-11. It shows that the front door is open, but the bar is still up. This dangerous situation should never be permitted.

To avoid the possible failure of the straight rod type, machine manufacturers developed lobed-style safety rods.

Figure 10-11. Recognize the horrible danger shown here. The front door of the IMM is open, but the bar has not dropped. It is either misadjusted or the mold is not fully open.

These lobes are machined into the rod and allow the mold to open, yet prevent it from closing unless the bar that engages the lobes is up. Figures 10-12, 10-13, and 10-14 show various styles of lobed or notched safety rods. One advantage of the lobed style is that even if the mold is not fully open, the stop will engage after a short distance if the mold begins to close.

Figure 10-12. The drop bar on the left engages the notches in this safety rod.

Figure 10-13 The lobed-style safety rod avoids the danger shown in Figure 10-11.

Figure 10-14. This style of lobed safety rod appears on some medium- and large-size IMMs.

Electrical interlock. The electrical interlock should disable the electrical circuit that controls mold closing when the door is open. Many machines feature a small rod attached to the door. When the door closes, the rod trips a limit switch and allows the machine to initiate an injection cycle. The electrical interlocks can fail if the limit switch does not function or if critical parts become loose and do not meet properly.

Figure 10-15 shows a push rod and the hole that allows the rod to actuate a limit switch.

One way to check on the electrical interlock is to close the door without tripping it. You can achieve this goal by removing or rotating the trip pin for the electrical interlock. Figure 10-16 shows such a setup. The platen should not move as long as the door is closed. If it does, seek maintenance on the electrical interlock immediately.

This procedure of checking the electrical interlock involves tampering with existing safety equipment. Follow this procedure only with great caution, and immediately replace the trip pin when complete.

Hydraulic interlock. Hydraulic interlock devices should prevent the mold from closing when the door is open. These

Figure 10-15. The push rod trips the limit switch, signaling that the door is closed.

Figure 10-16. The push rod shown in Figure 10-15 has been rotated so that the door closes without tripping the limit switch. Return it to the correct position immediately after testing the electrical safety system.

devices often consist of a hydraulic switch and an actuating arm. Figure 10-17 shows an arm that rises when the door is closed and lowers when the door is open. Another style appears in Figure 10-18. Only when the door is closed can hydraulic oil enter the mold-closing cylinder.

Figure 10-17. This arm prevents or allows the flow of hydraulic oil to the mold-closing cylinder.

Figure 10-18. This hydraulic safety arm is directly connected to a switch controlling the flow of hydraulic oil.

In comparison to the electrical interlock, which generally functions correctly or not at all, misadjusted hydraulic safety systems can still partially function. To check the hydraulic safety, open the door, trip the limit electrical system safety switch manually, and attempt to close the mold manually. The platen should not move.

It may be advantageous to manually lift the drop bar before checking hydraulic safety. If the hydraulic safety system fails, the mold will begin to close and hit the mechanical stop. It is not recommended that you stop the platen with the drop bar because the platen may wedge on the tie-rods. Some machines have two drop bars. On these machines, the platen will usually not wedge in the tie-rods. However, this is not true for machines with one safety rod and drop bar. Try to avoid this potential problem.

If the hydraulic safety is misadjusted, the clamping cylinder may receive enough oil to slowly move the mold. If this occurs, adjust the links to prevent any closing motion of the mold.

Purge guards. During purging, hot plastics may spray on nearby personnel. To prevent such accidents, injection-molding machines have purge guards. They are boxlike metal guards that enclose the nozzle of the machine. Figure 10-19 shows one style of purge guard. This guard attaches to the stationary platen with hinges that allow it to swing up. The switch on the front of the guard is a mercury switch. When the guard is up, the switch prevents injection of hot plastics. Another type of purge guard has a hinged panel that swings open like a door. A switch senses when the panel is open and prevents injection.

Rear-door safety systems. Many molding machine operators open and close the front door during every shot. They have three safety systems to provide them protection. However, the rear door does not activate similar redundant safety systems. Most molding machines have a switch attached

Figure 10-19. If this purge is up and the nozzle is exposed, the machine will not initiate an injection cycle.

to the rear door that completely shuts off the machine if the door is open. Figure 10-20 shows a type of rear-door safety device. In order to open the rear door, the cap must come off. When the cap is off, as shown in Figure 10-21, the main motors will not run.

Safe molding practices. Properly guarded machines provide protection for operators and technicians. However, guards cannot replace safe molding practices. Plastics can

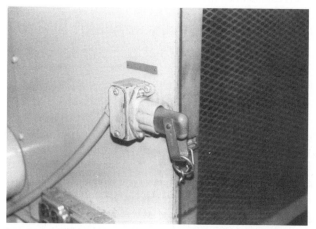

Figure 10-20. This safety switch is attached to the rear door of the molding machine.

Figure 10-21. Removing the cap stops the machine pumps before the rear door can be opened.

degrade in a molding machine, generating immense forces in the barrel and nozzle. Although uncommon, extreme pressures in the barrel have blown off end caps. In order for that to happen, the internal pressure must cause the many bolts that hold the end cap onto the barrel to shear. On a few occasions, barrels have exploded due to internal pressures.

To prevent buildup of pressure inside the barrel, always back off the carriage when a machine will be out of cycle for more than a few minutes. When the nozzle is not in contact with a mold, excess plastics can escape by drooling out the nozzle. This prevents the buildup of pressure.

The throat area of a molding machine should be cool enough not to melt pellets in the throat or the bottom of the hopper. If cooling in the throat area is inadequate or the heater bands in the rear heating zone overheat, pellets may melt in the feed zone of the screw, the throat, and the bottom of the hopper. This can be very dangerous.

Pellets may melt together to form a *bridge* in the bottom of the hopper. This bridge will prevent the flow of fresh pellets into the feed zone. Molding technicians may break through the bridge with a bar or rod. However, this can be extremely dangerous. If hot plastics under pressure are below the bridge, breaking down the bridge may cause them to blow upward. Molding technicians have died from severe burns sustained when hot plastics blew upward from the throat of a machine.

Safety for the machine. Machine safety involves implementing systems to protect the machine from damage. Most IMMs have shear pins or keys connecting the screw motor to the screw. If the barrel is not hot enough or a hard object blocks the screw, the shear pin should break before severe damage occurs to the screw.

If some foreign object gets pinched between the mold halves, low mold pressure protection devices protect the mold from damage. The key to this type of protection is preventing the application of full clamping pressure until the mold is almost fully closed. Only then does the machine exert full pressure against the mold.

A limit switch signals the machine controls when the mold is almost closed. Molding technicians can control the initiation of full clamping pressure by carefully adjusting the limit switch. Some molding technicians use pieces of cardboard as spacers to check the setting of the low mold switches. One or two pieces of cardboard between the mold halves should prevent the application of full clamping pressure. Without the cardboard pieces, the mold should clamp tightly.

When switches controlling the mold pressure are correctly set, the presence of an object between the mold halves will inhibit clamping, thus interrupting the molding cycle. This is extremely important when operators or robots load inserts into a mold. It is possible for metallic inserts to jiggle or fall out of correct position. Unless the mold pressure protection is adjusted correctly, severe mold damage could ensue.

Specification of Molding Machines

Molding machines have dozens of characteristics, but two of them provide a quick method used to describe a machine. These two capabilities are shot size and clamp tonnage.

Shot size. Shot size is the maximum amount of material the machine will inject per cycle. Because of the great variation in densities of commercial plastics, a standard for comparison is needed. The accepted standard for shot size measurements is polystyrene. A small laboratory machine may have a maximum shot size of 20 grams (0.70 oz). Large-capacity machines may have a shot size of more than 9000 grams (g) or 9 kilograms (kg) (19.8 lb).

Clamp tonnage. Clamp tonnage is the maximum force a machine can apply to a mold. One method of categorizing molding machines is to distinguish them as small, medium, or jumbo size. Generally, small machines have clamp tonnages of 99 tons or less; medium-sized machines run from 100 to 2000 tons; and jumbo machines are over 2000 tons. Standard jumbo machines are available up to 10,000 tons. Larger machines require special orders.

Sales figures help determine the utilization of various sizes. During one year in the early 1990s, US sales of small machines were slightly less than 400. Sales of medium-sized machines were about 1200 machines, and the sale of jumbo machines was about 50. It is clear that the medium range was dominant. In that range, the most common press size was about 300 tons.

As a rule of thumb, 3.5 kN (786 pounds-force [lfb]) of force is needed for each square centimeter of mold cavity area. A machine with a clamp force of 3 MN (337 tons-force) should be capable of molding a polystyrene plastics part 250 × 325 mm (9.8 × 12.8 in.). This part would have a surface area of 812.5 cm² (126 in.²). Use the rule of thumb:

$$3000 \text{ kN}/ 3.5 \text{ kN/cm}^2 = 857 \text{ cm}^2$$

Elements of Molding Cycles

Injection molding consists of five basic steps, as shown in Figure 10-22:

1. The mold closes.
2. As the screw begins to move forward, the nonreturn valve on the front end of the screw prevents the plasticated material from moving backward along the screw flights. Consequently, the screw functions as a ram and forces the hot materials into the mold cavity.
3. The screw maintains pressure through the nozzle until the plastics is cooled or set. In thermoplastic molding, timers maintain pressure on the plastics until the gates freeze. Gate freeze effectively separates the molded parts from the injection pressure. Maintaining further pressure on the plastics wastes time.
4. Timers stop injection pressure, and the screw turns to draw fresh material from the feed hopper. The screw backs up until a limit switch signals completion of the shot size. A decompression stroke pulls the screw backward a short distance. The purpose of decompression is to prevent the drooling of hot plastics into the sprue.
5. The mold opens and ejector pins remove the molded part.

It is common to group these basic steps as a *time cycle*. All injection systems have the following four elements in their time cycles:

1. *Fill time* is the time it takes to displace the air in the mold cavity with plastics material.
2. *Pack time* is the time required to maintain enough pressure to fill out the part and achieve gate freeze.
3. *Cooling or dwell time* is the time required for a material to cool or set enough for safe removal from the mold cavity.
4. *Dead time* is the time required to open the mold, remove the molded part, and close the mold.

Advantages of Injection Molding

Injection molding is popular because metal inserts may be used; output rates are high; surface finish can be controlled to produce any desired texture; and dimensional accuracy is good. For thermoplastics, gates, runners, and rejected parts may be ground and reused. The following list enumerates eight advantages of injection molding:

1. It has high output rates.
2. Fillers and inserts may be used.
3. Small, complex parts with close dimensional tolerances can be molded.
4. More than one material may be injection molded (co-injection molding).
5. Parts require little or no finishing.
6. Thermoplastics scrap may be ground and reused.
7. Self-skinning structural foams may be molded (reaction injection molding).
8. Process may be highly **automated**.

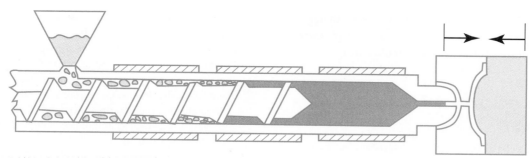

(A) The hot material is ready for injection when the mold closes.

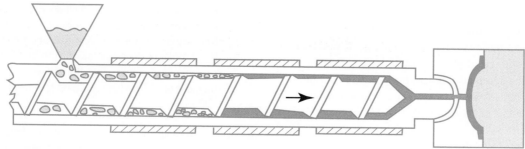

(B) The plastic is injected into the mold.

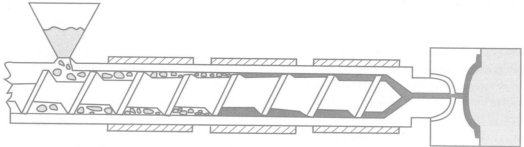

(C) The pressure is maintained to fully pack the mold.

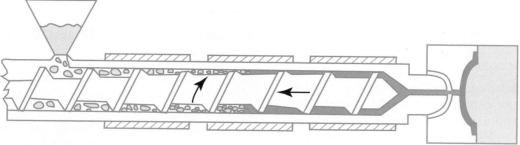

(D) The screw rotation continues until the shot size is achieved.

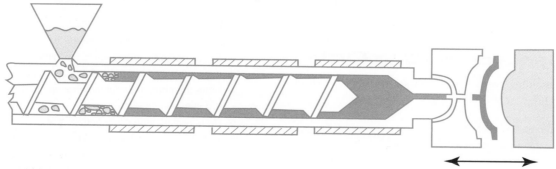

(E) The mold opens and the part is ejected.

Figure 10-22. The steps in the injection cycle.

Disadvantages of Injection Molding

Injection molding is not practical for short production runs. Molding machines are costly. Therefore, hourly costs to operate the machines are considerable. Even small injection molds often cost many thousands of dollars. To make molding cost-effective, the number of parts must be high. Because molding is such a popular process, many companies compete for contracts. Some companies are unable to maintain profits and fall into bankruptcy.

The process of injection molding is complicated. Occasionally, moldmakers find it very difficult to make acceptable parts due to poor part or mold design. When processes are not under control, scrap rates increase, and part rejections by customers may cause major financial losses. Table 10-1 lists some of the problems associated with injection molding.

Table 10-1. Problems in Injection Molding

Difficulty	Cause	Possible Remedy
Black specks, spots, or streaks	Flaking off of burned plastics on cylinder walls	Purge heating cylinder.
	Air trapped in mold causing burning	Vent mold properly.
	Frictional burning of cold granules against cylinder walls	Use lubricated plastics.
Bubbles	Moisture on granules	Dry granules before molding.
Flashing	Material too hot	Reduce temperature.
	Pressure too high	Lower pressure.
	Poor parting line	Reface the parting line.
	Insufficient clamp pressure	Increase clamp pressure.
Poor finish	Mold too cold	Raise mold temperature.
	Injection pressure too low	Raise injection pressure.
	Water on mold face	Clean mold.
	Excess mold lubricant	Clean mold.
	Poor surface on mold	Polish mold.
Short moldings	Cold material	Increase temperature.
	Cold mold	Increase mold temperature.
	Insufficient pressure	Increase pressure.
	Small gates	Enlarge gates.
	Entrapped air	Increase vent size.
	Improper balance of plastics flow in multiple cavity molds	Correct runner system.
Sink marks	Insufficient plastics in mold	Increase injection speed, check gate size.
	Plastics too hot	Reduce cylinder temperature.
	Injection pressure too low	Increase pressure.
Warping	Part ejected too hot	Reduce plastics temperature.
	Plastics too cold	Increase cylinder temperature.
	Too much feed	Reduce feed.
	Unbalanced gates	Change location or reduce gates.
Surface marks	Cold material	Increase plastics temperature.
	Cold mold	Increase mold temperature.
	Slow injection	Increase injection speed.
	Unbalanced flow in gates and runners	Rebalance gates or runners.

Injection-Molding Thermosets

Both machine and mold designs differ when molding thermosetting materials. A nonreturn check valve is not required because the material is extremely viscous and little material is left in the barrel after injection. Screw flights are shallow with a one-to-one compression ratio. Bulk-molding compounds or other heavily filled or reinforced materials generally use a plunger machine. The length-to-diameter (L/D) ratio in these machines generally ranges from 12:1 to 16:1 as compared to higher ratios for injection molding of thermoplastics (Figure 10-23).

For proper injection of most thermosets, temperature control in the machine barrel and the mold is critical. Molding materials are plasticized in a relatively cool cylinder, using compression. However, cross-linking must occur rapidly once the material has filled the mold. Mold temperatures have a great impact on cycle times. Figure 10-24 shows the type of temperature required.

Co-injection Molding

Co-injection molding is a process in which two or more materials are injected into a mold cavity (Figure 10-25). This usually produces a skin of material on the mold surface and a cellular center core. This core material includes blowing agents that produce desired cellular densities. The process is sometimes incorrectly called *sandwich molding* because of the composite layered effect.

Co-injection molding may use different families of plastics for the skin or core layer. Fiber reinforcements provide greater

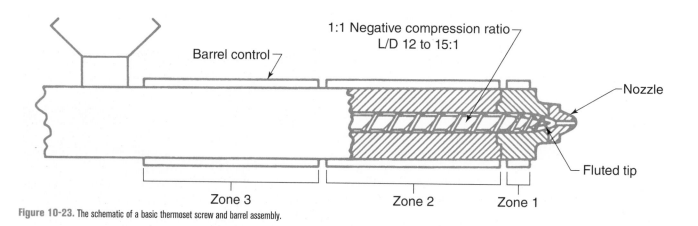

Figure 10-23. The schematic of a basic thermoset screw and barrel assembly.

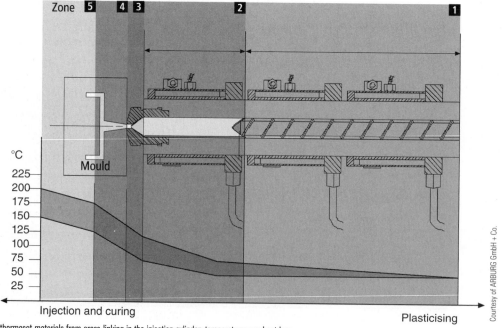

Figure 10-24. To prevent thermoset materials from cross-linking in the injection cylinder, temperatures are kept low.

Plasticated materials from injection machine

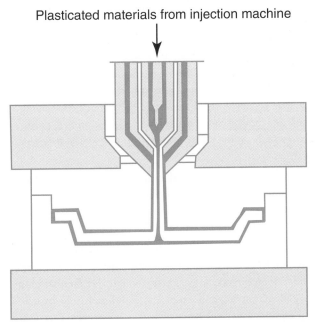

Figure 10-25. This co-injection molding machine feeds three separate melt channels to the mold.

strength, but flow patterns may cause unwanted orientation of fibers. The selection of materials and additives is limited only for exposed skin surfaces. They generally need to be palatable or pigmented.

Items that use the co-injection molding process include automotive parts, office machine housings, furniture components, and appliance housings.

Fluid Injection

Fluid injection involves forcing a fluid into melted plastic during the injection-molding process. The primary goal of fluid injection is to reduce the amount of material used, particularly in thick areas, and consequently to be able to control shrinkage, which can result in excessive sink marks and/or vacuum voids. Two fluids are practical—gas and water.

Gas injection technology (GIT), also called gas-assisted injection molding, was developed more than 20 years ago, but due to legal battles on patents, its widespread application was limited. Because some patents have now expired, its use has increased rapidly. The process injects a compressed gas, usually nitrogen, into the melt. The gas creates a hollow core in the part, helps reduce sink marks, and reduces the amount of material needed. With thick-walled parts, the material saving can reach 40 percent. One of the pioneers in this technology is Gain Technologies based in Michigan. A new version of fluid injection is water injection technology (WIT). Water injection technology is very similar to GIT except that water speeds the cooling of the parts and eliminates the cost of purchasing nitrogen. The process

relies on pressure-generating units that create up to 300 bar (4350 psi). Batttenfeld IMT currently sells both a water injection and a gas-assist system. Data on the Aquamould® and Airmould® systems are available at the Battenfeld website.

Overmolding

In order to achieve products that combine rigidity and softness or flexibility, plastics companies have utilized overmolding. This often involves a tough, stiff component with a skin of softer material. One example, as seen in Figure 10-26, is the blade of a street hockey stick. This blade contains glass-filled polyamide for toughness and shock resistance. To enhance control of the ball, a softer section has been overmolded using a thermoplastic elastomer. The results are both strength and better control.

A similar approach to overmolding is seen in Figure 10-27. The gripping portion of this medical clamp is made of a semirigid base with an elastomeric pad overmolded onto the base.

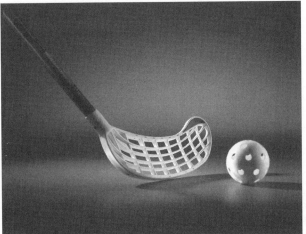

Figure 10-26. This street hockey stick has TPE overmolded on a polyamide base.

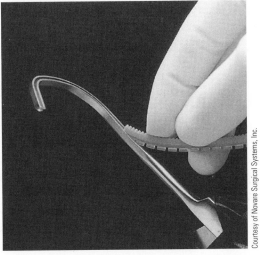

Figure 10-27. This grip section for a medical clamp is overmolded on a semirigid base.

This system provides for secure attachment to a stainless steel clamp plus the advantage of easy removal and disposal of the grip pad. The high flexibility of the pad also helps prevent damage to tissues.

Electric and Hybrid Machines

A number of injection-molding machine manufacturers have predicted a huge switch from hydraulic powered to all-electric machines. Electric IMMs use less energy than hydraulic machines. The major reason for this is that electric machines draw power only when the machine is actually moving. Hydraulic versions require constant energy to drive the hydraulic pumps. In addition, electric machines can provide greater precision of molding cycles.

The manufacturers of electric IMMs feel that they will dominate the small-to-medium range in a few years. Although they offer advantages, the all-electric machines are more expensive—about 15 percent more for small machines and more than 30 percent for midsize and large machines.

Japan leads the United States in turning to all-electric IMMs because of strict laws about disposal of hydraulic fluids. By 2002, all-electric machines were dominant in Japan, whereas they accounted for about 20 to 25 percent of new machines in the United States. Some companies are dropping the production of hydraulic-powered small machines, especially machines using less than 100 tons of clamp force. In 2002, the largest all-electric machine available had a clamp of just over 1500 tons.

Some companies make hybrid machines that use hydraulic power for some functions and electric motors for others. They typically use hydraulic power for linear motions and electric motors for rotary actions. One advantage of such an arrangement is that electric motors improve precision, and hydraulics provide superior speed. All-electric machines do not have the injection speed available in hydraulic machines.

MOLDING LIQUID MATERIALS

Several processes convert liquid resins into finished plastics parts. An important reason liquid materials are used is that they flow into mold cavities with much less force than melted thermoplastic pellets. Liquid materials do not damage delicate inserts because they flow around fibrous reinforcements. The most significant processes relying on liquid materials are reaction injection molding (RIM), reinforced reaction injection molding (RRIM), and liquid resin molding or resin transfer molding (RTM).

Reaction Injection Molding

Reaction injection molding (RIM) is also known as liquid reaction molding or high-pressure impingement mixing. It is a process in which several reactive chemical systems are mixed and forced into the molding cavity, where a polymerization reaction occurs. Although most of the current RIM components are polyols and isocyanates, other modified polyurethanes such as polyester, epoxies, and polyamide monomers are used.

The process involves impingement-atomized mixing of two or more liquids in a mixing chamber. This mixture is immediately injected into a closed mold, resulting in a rigid, structural foamed or cellular product (Figure 10-28).

The automotive and furniture industries are the major users of RIM parts. Bumpers, belts, shock-absorbing parts, fender components, and cabinet elements are familiar examples. Reaction injection molding is limited only by mold and equipment size. Clamp capacity requirements are much lower than those of conventional injection molding. Clamps are often designed to be opened and closed like a book. This allows easy parts removal and operator access to the mold (Figure 10-29).

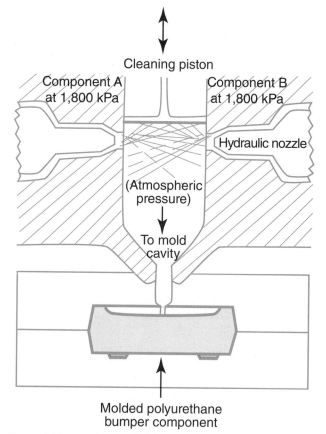

Figure 10-28. Reaction injection molding (RIM) showing impingement mixing. Components are atomized to a fine spray by a pressure drop from 1800 kPa [2500 psi] to atmospheric pressure.

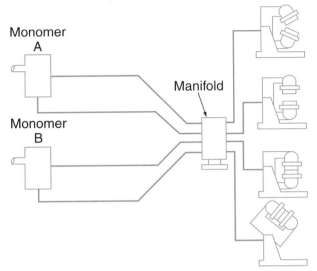

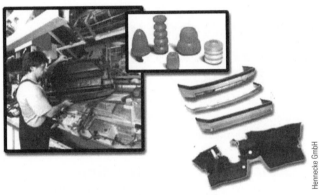

(A) The C-frame stations (clamps) open in book fashion for fast part removal and rock back for venting.

(B) Removal of an automotive bumper from a RIM molding machine.
Figure 10-29. RIM molding systems.

Hennecke GmbH

Reinforced Reaction Injection Molding (RRIM)

When short fibers or flakes (particulates) are used to produce a more isotropic product, the process is called **reinforced reaction injection molding (RRIM)**. Fiber loading increases monomer viscosities and abrasive wear on all flow surfaces.

Polyurethane/urea hybrid, epoxy, polyamide, polyurea, polyurethane/polyester hybrid, polydicyclopentadiene, and other resin systems have been used for RIM and RRIM. RRIM applications include automobile fenders, panels, bumpers, shields, radomes, appliance housings, and furniture components. Seven advantages and four disadvantages of reaction injection molding are listed:

Advantages of Reaction Injection Molding (RIM)

1. Cellular core and integral skin for durable products
2. Fast cycle times for large products
3. Good finishes that are paintable
4. Less costly than castings
5. Polymers may be reinforced
6. Reduced tooling and energy cost (compared with injection molding)
7. Lower equipment cost due to low pressures

Disadvantages of Reaction Injection Molding

1. New technology requiring investment in equipment
2. System requires four or more chemical-component tanks
3. System requires handling of isocyanates
4. Releasing agents required

Liquid Resin Molding

Liquid resin molding (LRM) is a term used to describe products produced by a variety of low-pressure methods in which mixing is often done mechanically rather than by impingement. The term once described a very specialized process for potting and encapsulating components. LRM is used to describe a group of processing methods that include resin transfer molding (RTM), vacuum injection molding (VIM), and thermal expansion resin transfer molding (TERTM), in which resins are forced under low pressure into the molding cavity and rapidly cured. Epoxies, silicones, polyesters, and polyurethanes are often used in liquid resin molding processes.

When liquid thermoset materials are processed in specialty injection machines, the screw design is also unique. These screws must be almost compression free and rather short, with a low L/D ratio. Figure 10-30 shows such a design.

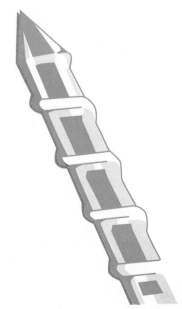

Figure 10-30. To reduce material hand-ups, this screw for liquid thermosets does not have a nonreturn valve.

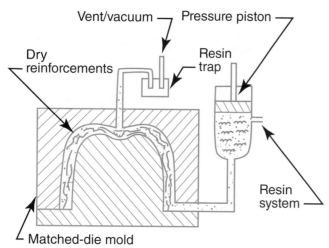

Figure 10-31. A concept of RTM. Preform reinforcements are loaded into a matched-die mold. After mold closing, a liquid resin system is forced into and around the preform.

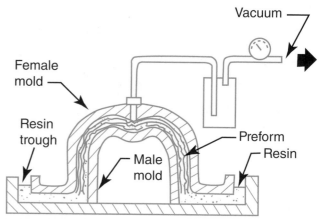

Figure 10-32. A concept of VIM.

Resin transfer molding. Resin transfer molding (RTM), also called **resin injection molding**, is a process whereby catalyzed resin is forced into a mold in which fragile parts or reinforcements have been placed. Low pressure does not distort or move the desired fiber orientation of preforms or other materials. Boat hulls, hatches, computer housings, fan shrouds, or other large composite structures may be produced using this technique. The basic concept of RTM is illustrated in Figure 10-31.

Advantages of RTM

1. Eliminates the plasticizing stage necessary with dry compounds
2. Permits encapsulation of delicate or fragile parts
3. No hand mixing
4. Eliminates preheating and preforming
5. Uses lower pressures
6. Minimum material waste
7. Cures resins rapidly at low temperatures
8. Improves reliability and dimensional stability
9. Reduces material handling

Vacuum injection molding. In a process similar to RTM, preforms are placed on a male mold, and the female mold is closed. A vacuum is drawn, pulling the reactive resin system into the mold cavity. This technique, **vacuum injection molding (VIM)**, is illustrated in Figure 10-32.

Thermal expansion resin transfer molding. Thermal expansion resin transfer molding (TERTM) is a variation of the RTM process. A cellular mandrel of PVC or PU is wound or wrapped with reinforcements and placed into a matched die. Epoxy or other resin systems are injected to impregnate the reinforcements. The heated die causes the cellular material to further expand, forcing the impregnated

reinforcements against the mold walls. The tooling is vented to allow excess matrix or entrapped air to escape.

MOLDING GRANULAR AND SHEET THERMOSET MATERIALS

There are two processes in widespread use for the molding of granular or pelletized thermoset materials: compression molding and transfer molding.

Compression Molding

One of the oldest known molding processes is called **compression molding**. The plastics material is placed in a mold cavity and formed by heat and pressure. As a rule, thermosetting compounds are used for compression molding. However, thermoplastics may also be used. The process is somewhat like making waffles. Heat and pressure force the materials into all areas of the mold. After the heat hardens the substance, the part is removed from the **mold cavity** (Figure 10-33).

To reduce pressure requirements and production (cure) time, the plastics material is usually preheated with infrared, induction, or other heating methods before it is placed in the mold cavity. A screw extruder is sometimes used to reduce cycle time and increase output. The screw extruder is often used to make preformed slugs that are loaded into the molding cavity. The screw-compression process greatly reduces cycle time, eliminating the greatest drawback of compression molding. Compression-molded parts with heavy wall thickness may be produced with up to 400 percent greater product output per mold cavity when this method is used.

Thermosetting bulk-molding compounds (BMCs) are used in forming polyester compounds. BMC is a mixture of fillers,

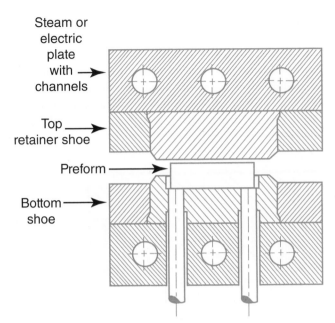

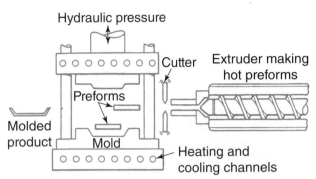

Figure 10-34. A compression molding showing hot preforms being fed into the mold cavity.

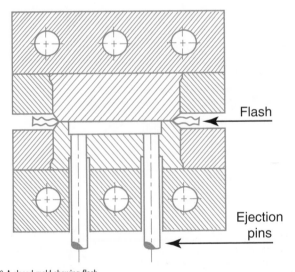

(A) A preform about to be molded

(B) A closed mold showing flash

Figure 10-33 The principle of compression molding.

resins, hardening agents, and other additives. Hot extruded preforms of this material may be loaded straight into the cold cavity or feed chute (Figure 10-34).

Other popular molding materials are phenolic plastics, urea-formaldehyde, and melamine compounds. Like BMCs, they are usually preformed for automation and speed. Heavily filled and reinforced sheet-molding compounds are used. They may be placed in alternating layers to attain more isotropic properties or placed in one direction for more anisotropic properties.

Most compression-molding equipment is sold by the press or platen rating. A force of 20 MPa (2900 psi) is usually required for moldings up to 25 mm (1 in.) thick. An added

5 MPa (725 psi) should be provided for each 25 mm (1 in.) increase. Hydraulic action provides this force.

Steam, electricity, hot oil, and open flame are some of the means used for heating molds, platens, and related equipment. Hot oil is common because it may be heated to high temperatures with little pressure. Electricity is clean but limited by wattage.

During preforming and the actual molding process, heat and various catalysts begin cross-linking the molecules. During the cross-linking reaction, gases, water, or other byproducts may be freed. If they are trapped in the mold cavity, they may affect the plastics part, and the part may be damaged, of poor quality, or marked by surface blisters. Molds are usually vented to allow the escape of these byproducts.

During molding, the thermosetting plastics compound becomes cross-linked and infusible. Therefore, these products may be removed from the molding cavity while hot. Thermoplastic materials must be cooled before removal because they do not cross-link to any extent. Many elastomers are molded by this process.

Long runs of moderately complex parts are often produced by compression molding. Mold maintenance and starting costs are low, materials waste is rather low, and large bulky parts are practical. However, very complex parts are difficult to mold. Inserts, undercuts, side draws, and small holes are not practical with this method when it is necessary to maintain close tolerances.

The sequence of compression molding may include the following six steps:

1. Clean the mold and apply the mold release (if required).
2. Load the preform into the cavity.
3. Close the mold.
4. Open the mold briefly to release trapped gases (**breathing the mold**).
5. Apply heat and pressure until cure is complete (**dwell time**).
6. Open the mold and place the hot part in a cooling fixture.

Six advantages and eight disadvantages of compression molding are given here:

Advantages of Compression Molding

1. There is little waste (no gates, sprues, or runners in many molds).
2. The tooling costs are low.
3. Process may be automated or hand operated.
4. Parts are true and round.
5. Material flow is short—less chance of disturbing inserts, causing product stress, and/or eroding molds.
6. Multiple cavity placement in tooling is not dependent on a balanced feeder system.

Disadvantages of Compression Molding

1. It is hard to mold complex parts.
2. Inserts and fine ejector pins are easily damaged.
3. Complex shapes are sometimes hard to achieve.
4. Long molding cycles may be needed.
5. Unacceptable parts cannot be reprocessed.
6. Trimming flask can be difficult.
7. Some part dimensions are controlled by material charge rather than tooling.
8. It requires external loading and unloading equipment for automation.

Compression-molded products include dinnerware, buttons, buckles, knobs, handles, appliance housings, drawers, parts bins, radio cases, large containers, and many electrical parts.

Two variations of the compression-molding processes deserve special attention. They are cold molding and sintering.

Cold molding. In **cold molding**, plastics compounds (most phenolics) are formed in unheated molds. After forming, a part is hardened into an infusible mass in an oven (Figure 10-35). Electrical insulator parts, utensil handles, battery boxes, and valve wheels are examples of products made in this manner.

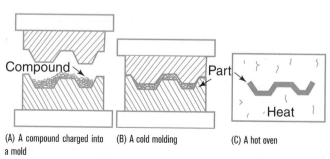

(A) A compound charged into a mold   (B) A cold molding   (C) A hot oven

Figure 10-35. The principle of cold molding.

Sintering. **Sintering** is the process of compressing a powdered plastics into a mold at temperatures just below its melting point for about one-half hour (Figure 10-36). The powdered particles are fused (sintered) together, but the mass as a whole does not melt. Bonding is done through the exchange of atoms between individual particles. After the fusion process, the material may be postformed under heat and pressure to the required dimensions.

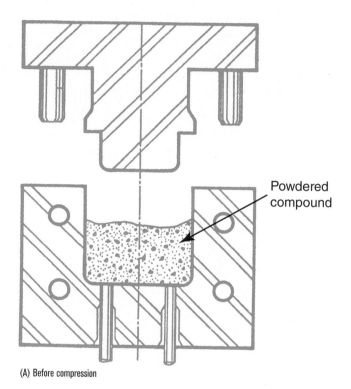

Powdered compound

(A) Before compression

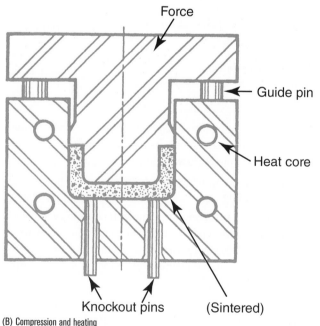

Force

Guide pin

Heat core

Knockout pins   (Sintered)

(B) Compression and heating

Figure 10-36. Sintering plastics parts.

(C) This photo shows PTFE molding powder and a block of sintered powder

Figure 10-36. (*Continued*)

Three major variables governing the sintering process are temperature, time, and the composition of the plastics.

This process is adapted from the sintering operations of powder metallurgy. Sintering can be used to process polytetrafluoroethylene, polyamides, and other specially filled plastics. It is also the primary method by which polytetrafluoroethylene is processed. Dense parts with first-rate electrical and mechanical properties may be produced. The cost of tools and production is high, and parts with thin walls or variations in cross-sectional thickness are hard to form.

Transfer Molding

Transfer molding has been known and practiced since World War II. The process is sometimes called plunger molding, duplex molding, reserve-plunger transfer molding, step molding, injection transfer molding, or impact molding. It is actually a variation of compression molding but differs because the material is loaded in a chamber outside the mold cavity. One advantage of transfer molding is that the molten mass is fluid when entering the mold cavity. Fragile, complex shapes with inserts or pins may be formed with accuracy. Transfer-molding techniques are much like those of injection molding, except thermosetting compounds are normally used.

The American Society of Tool and Manufacturing Engineers recognizes two basic types of transfer molds:

1. Pot or sprue molds
2. Plunger molds

Plunger molds (Figure 10-37) differ from the sprue molds (Figure 10-38) because the plunger or force is pushed to the parting line of the mold cavity when inserting the plastics

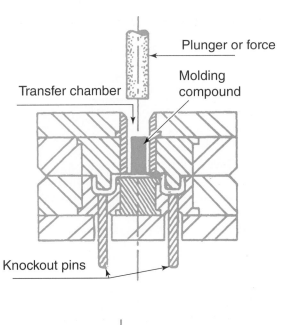

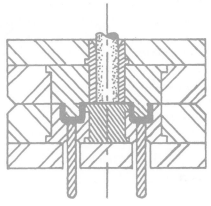

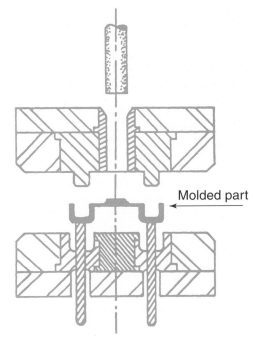

Figure 10-37. A plunger (two-plate) transfer mold.

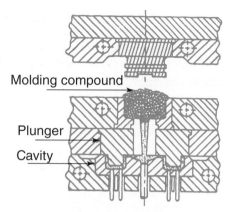

Molding compound

Plunger

Cavity

(A) Open position

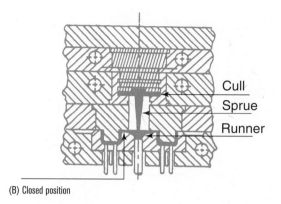

Cull

Sprue

Runner

(B) Closed position

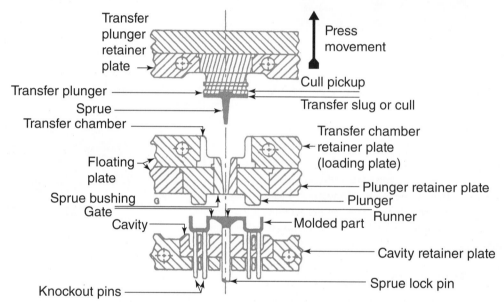

Transfer plunger retainer plate

Press movement

Transfer plunger

Cull pickup

Sprue

Transfer slug or cull

Transfer chamber

Transfer chamber retainer plate (loading plate)

Floating plate

Plunger retainer plate

Sprue bushing

Plunger

Gate

Runner

Cavity

Molded part

Cavity retainer plate

Sprue lock pin

Knockout pins

(C) Release position

Figure 10-38. An integral-transfer (three-plate) mold.

material. In sprue molds, the plastics is fed down a hole (sprue) using gravity. With plunger molds, only runners and gates are left as waste on the molded part.

A third type of mold may also be included in the transfer molding category. In this type, the molding compound is preplasticized by extruder action, and then a plunger forces the melt into the mold (Figure 10-39).

The cost of detailed mold designs and high waste from culls, sprues, and **flash** are two major limitations of transfer molding.

Although most parts are limited in size, there are numerous applications, including distributor caps, camera parts, switch parts, buttons, coil forms, and terminal block insulators. They may also be used for complex shapes such as cups and caps for cosmetic containers.

A variation of compression/transfer molding has found a use in processing comingled recycled scrap plastics. Although the materials are not predominantly thermosets, the process is also not typical for thermoplastics. Dirty, mixed, and unsorted plastics—including films, foams, household waste, and other recycled materials—are ground or densified to make flakes. The flakes then enter a chamber and are accelerated to high speed. Frictional heating from contact with teeth on the inside of the chamber produces a mass of material that is malleable but far too viscous for injection molding. If the part is large, the hot material is transferred manually to a compression mold. If the part is smaller, the material enters a traditional transfer molding process. Figure 10-40 shows wheels that were made with this process.

Injection-transfer plunger

Material hopper

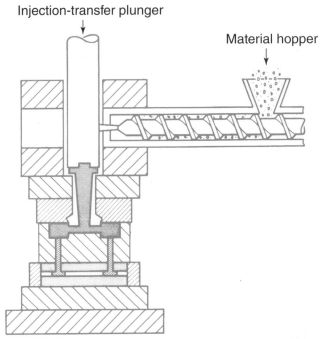

Figure 10-39. An injection-transfer molding showing the hot extruded compound being forced into the mold cavity by the transfer plunger.

Recycled Plastic Products (Utah)

Figure 10-40. These wheels rely on a compression/transfer molding process to convert comingled, unsorted, recycled plastics into useful objects.

Problems in compression and transfer molding are found in Table 10-2. Six advantages and four disadvantages of transfer molding are listed here:

Advantages of Transfer Molding

1. There is less mold erosion or wear.
2. Complex parts (small-diameter holes or thin wall sections) and inserts may be molded and used.
3. Less flash is produced than in compression molding.
4. Densities are even more than in compression molding.
5. Multiple pieces may be molded.
6. It requires shorter mold and loading times than most compression molding processes.

Table 10-2. Problems in Compression and Transfer Molding

Defect	Possible Remedy
Cracks around inserts	Increase wall thickness around inserts.
	Use smaller inserts.
	Use a more flexible material.
Blistering	Decrease cycle and/or mold temperature.
	Vent mold–breathe mold.
	Increase cure–increase pressure.
Short and porous moldings	Increase pressure.
	Preheat material.
	Increase charge weight of material.
	Increase temperature and/or cure time.
	Vent mold–breathe mold.
Burned marks	Reduce preheating and molding temperature.
Mold sticking	Raise mold temperature.
	Preheat to eliminate moisture.
	Clean mold–polish mold.
	Increase cure.
	Check knockout pin adjustments.
Orange peel surface	Use a stiffer grade of molding material.
	Preheat material.
	Close mold slowly before applying high pressure.
	Use more finely ground materials.
	Use lower mold temperatures.
Flow marks	Use stiffer material.
	Close mold slowly before applying high pressure.
	Breathe mold.
	Increase mold temperature.
Warping	Cool on jig or modify design.
	Heat mold more uniformly.
	Use stiffer material.
	Increase cure.
	Lower temperature.
	Anneal in oven.
Thick flash	Reduce mold charge.
	Reduce mold temperature.
	Increase high pressure.
	Close slowly–eliminate breathe.
	Increase temperature.
	Use softer grade materials.
	Increase clamping pressure.

Figure 10-41. This pickup truck box will not rust or corrode like similar steel boxes.

The Budd Company

Disadvantages of Transfer Molding

1. There is more waste from runners and sprues.
2. It requires more costly equipment and molds.
3. Molds must be vented.
4. Runner and gates must be removed.

Sheet Molding

Sheet-molding processes are very similar to compression molding. For more information on sheet-molding compounds, refer to Chapter 13. Sheet-molding equipment tends to be larger than presses for compression molding. Particularly in the automotive industry, sheet moldings are often large parts. Figure 10-41 shows the die in a large sheet-molding press. The product pictured is a full-sized pickup truck box. Notice the ejector pins, which have pushed the part down from the upper mold cavity.

For advantages and disadvantages of sheet molding, see the section on matched die moldings in Chapter 13.

RELATED INTERNET SITES

- **www.milacron.com.** Milacron Inc. provides information about their plastics processing equipment. Select "Products" from the home page, then choose "Extrusion." Each category provides extensive details about the equipment.

- **www.ubemachinery.com.** UBE Machinery Inc. sells injection-molding machines and other specialized process equipment for the plastics industry. Its hydraulic injection-molding machines range in clamp size from 1000 to 7000 tons. Its all-electric line of machines range from 720 to 3300 tons of clamp force.

- Other sites for injection molding equipment:
  www.sumitomopm.com
  www.husky.ca
  www.dr-boy.de
  www.cpm-toyo.com
  www.arburg.com

VOCABULARY

The following vocabulary words are found in this chapter. Use the glossary in Appendix A to look up the definitions of any of these words you do not understand as they apply to plastics.

automated
bar
barrel
breathing the mold
cold molding
compression molding
dwell time
flash
impingement-atomized mixing
injection-molding machine (IMM)
liquid resin molding (LRM)
mold cavity
plasticating
platen
purging
reaction injection molding (RIM)
reinforced reaction injection molding (RRIM)
resin injection molding
resin transfer molding (RTM)
sintering
thermal-expansion resin transfer molding (TERTM)
transfer molding
vacuum injection molding (VIM)

QUESTIONS

10-1. The excess material left on a product after compression molding is called _____.

10-2. The amount of material used to fill a mold during an injection-molding process is called _____.

10-3. The chief advantage of injection molding over other molding processes is _____.

10-4. Name three ways to reduce production time in compression molding.

10-5. Opening the compression mold to allow gases to escape during the molding cycle is called _____.

10-6. The mark on a molding where the halves of the mold meet in closing is called _____.

10-7. The process of molding or forming items from pressed powders at a temperature just below the plastic's melting point is called _____.

10-8. Name three major disadvantages of transfer molding.

10-9. The time it takes to close a mold, form a part, open the mold, and remove the cooled part is called the _____ time in injection molding.

10-10. In _____ molding, the material is loaded in a chamber outside the mold cavity before it is made fluid and forced into the mold cavity.

10-11. The two factors that rate the capacity of an injection-molding machine are _____ and _____.

10-12. Thermosetting materials with fragile, intricate shapes and with inserts or pins may be _____ or _____ molded.

10-13. Name the two molding processes that use thermosetting and selected thermoplastic materials in preform or bulk molding compound forms.

10-14. Name three processes that are similar to transfer-molding processes.

10-15. Name four commonly used polymers in liquid resin molding.

10-16. A common expression for the time during which a molding machine is not operating is _____.

10-17. Reaction injection molding is also known as _____ or high-pressure impingement mixing.

ACTIVITIES

Compression Molding

Introduction. The compression-molding process is one of the oldest and simplest plastics processing techniques. It normally utilizes thermosetting resins.

Equipment. Compression molder, mold, resin, 35 g phenolic, silicone mold release, safety glasses, and gloves to resist high temperatures.

Procedure

10-1. Adjust temperature to 190°C (356°F) and connect cooling hoses to both platens.

10-2. Inspect and clean mold with wooden scraper.

! **CAUTION**

Steel instrument may damage mold surfaces.

10-3. Place all parts of the mold on lower platen and close press platens. Heat for 10 minutes.

Figure 10-42. Carefully measure all phenolic molding materials.

10-4. Carefully measure all phenolic molding materials in a container (Figure 10-42). Too much material will produce a thick product, whereas too little will produce a thin product.

10-5. Remove the mold from the press, using insulated gloves, and set it on a heat-resistant surface, as seen in Figure 10-43.

Figure 10-43. Place a measured amount of phenolic material into the hot mold.

Figure 10-45. A finished phenolic part.

> **! CAUTION**
>
> Mold and platens are hot. Some practice is needed to handle hot molds without wasting time and cooling the molds.

10-6. Assemble the mold and pour measured phenolic charge into it (Figure 10-44).

10-7. Close the mold and return it to the press. Be certain to center the mold on the platen (Figure 10-45).

10-8. Apply 13,500 kPa (1950 psi) on molding surface. After about 10 seconds, release pressure to allow gas to escape, then apply pressure again (Figure 10-46). An MSDS on phenolic molding compounds will indicate that processing releases small amounts of ammonia, formaldehyde, and phenol. Formaldehyde has the lowest TLV of those three gases. It is rated at 0.3 ppm ceiling by the ACGIH. Use adequate ventilation to remove these gases from the work area.

Figure 10-46. The pressure is maintained to complete the compression molding.

10-9. Hold pressure for 5 minutes. Be sure pressure is maintained.

10-10. To cool platens, turn off power and slowly open water valve.

> **! CAUTION**
>
> Steam can cause severe burns.

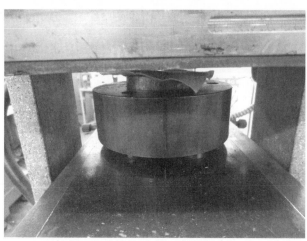

Figure 10-44. Assemble the hot mold and return it to the hot platens.

10-11. Release pressure and remove the mold.

! **CAUTION**

Mold will still be hot.

10-12. Carefully remove part from the mold and allow it to cool. Trim flash from the part. If the part exhibits a dull surface, repeat the process at a higher temperature and/or pressure.

Injection Molding

Introduction. The finished size of injection-molded parts is controllable, within limits, by the processing conditions. To investigate the relation between part size and process parameters, accurate measurement is essential. Figure 10-47 shows a device for rapid and accurate length measurement.

Equipment. Injection-molding machine, mold that produces a rectangular part such as an impact or tensile bar, accurate measuring device, and crystalline thermoplastics (HDPE, LDPE, or PP).

Procedure

10-1. Establish a *control* condition that is capable of producing acceptable parts with little variation in length. This will require holding pressure on the mold until the gate(s) freeze.

To determine this time, step the injection times from a short time to a long time. A short time will be slightly longer than the time needed to fill the part. A long time is over 30 seconds. Carefully cut the parts off the runner system and weigh them individually. Longer injection time will cause weight gain. At some point

in time, the parts will no longer gain weight. That indicates gate freeze.

Some parts with large gates or sprue gated parts will continue to gain weight even after 1 minute of injection hold pressure. Because that will delay the activity, select a mold that has gates small enough to freeze in about 10 seconds.

10-2. Select one process parameter and vary it from one extreme to another. For example, select injection pressure and vary it in increments of 100 psi in the hydraulic pressure. Reduce pressure until the parts are not completely filled, or *short,* then increase incrementally until the mold flashes.

10-3. Number each slot in order and keep track of the parameter under investigation.

10-4. Allow 40 or more hours for part stabilization.

10-5. Measure the parts and create a graph of length on the *Y* axis against the increments of pressure (or other parameter) change on the *X* axis.

10-6. If time allows, investigate changes in injection speed, cooling time, mold temperature, back pressure, screw rpm, and other process parameters.

10-7. Arbitrarily select a length near the high or low extreme and find a process that yields those parts with low variations in length.

Characteristics of Polymer Flow

Introduction. Most new students of plastics do not intuitively grasp polymer flow. The rheological aspects of flow are difficult to observe without instrumentation or a spiral flow mold. However, the nature of fountain flow is easy to observe.

Equipment. Injection-molding machine, plastics pellets, a mold, and safety glasses.

Procedure

10-1. Position a small piece of facial or toilet tissue on a flat cavity surface. A tiny amount of water or grease will gently adhere the tissue to the mold.

10-2. Predict the location of the tissue after the melt fills the cavity.

10-3. Inject the part and observe the location of the tissue. Figure 10-48 shows a tissue piece on a tensile bar.

10-4. Put a piece of tissue at various locations in the cavity or runner system. Do the pieces move?

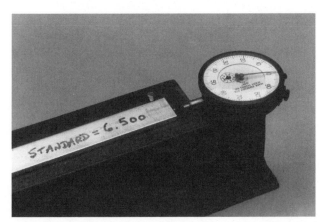

Figure 10-47. Three pins locate the tensile bar exactly and the dial indicator shows its length relative to a standard.

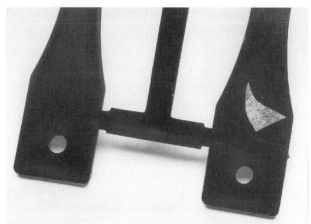

Figure 10-48. This small piece of tissue did not move during the flow of plastics into the mold.

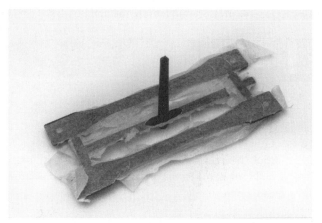

Figure 10-49. The slight wrinkles in the tissue record the shape of the flow front as it advanced through the mold.

10-5. Try clamping a tissue over the entire part surface. Figure 10-49 shows that the force of the flow did not rip or significantly move the tissue in any location.

10-6. Describe the process of mold filling, taking into account the observations that the melt does not slide along the surface of the mold.

EXTRUSION PROCESSES

CHAPTER 11

INTRODUCTION

The word **extrusion** is derived from the Latin word *extrudere* meaning (*ex*) "out" and (*trudere*) "to push." In extrusion, dry powder, granular, or heavily reinforced plastics is heated and forced through an orifice in a die. Although ram extruders are still in use, particularly for products made of ultra-high molecular weight polyethylene (UHMWPE), screw extruders are preferred. The screw plasticates (melts and mixes) the material and forces it through the die.

This chapter deals with extrusion products and processes. These are major topics:

EXTRUSION EQUIPMENT

Figures 11-1 and 11-2 show typical single-screw-type extruders. Although computerized instrumentation has improved process control, the basic design of single-screw extruders has not changed for several decades. Screws are measured by the diameter, ranging from very small machines with 19 mm (0.75 in.) diameter screws, up to very large machines with screws of 300 mm (12 in.) diameter. The most popular machines are the 64 to 76 mm (2.5 to 3 in.) size.

Besides the diameter of the screw, extruders are sold by the amount of material they will plasticate per minute or hour. Extruder capacity with low-density polyethylene may vary from less than 2 kg (4.5 lb) to more than 5000 kg (11,000 lb) per hour.

Screws are characterized by their L/D ratios. A 20:1 screw could be 50 mm (1.96 in.) in diameter and 1000 mm (39.37 in.) long. Short screws, such as one with an L/D ratio of 16:1, would be appropriate for a profile extruder. Long screws, up to 40:1, provide more mixing of the materials than shorter screws. Some screw designs are shown in Figure 11-3.

The channel depth of the screw is greatest in the feed section. That permits the screw to draw in pellets and other forms of material. The channel depth decreases beginning at the transition section. This continuous reduction forces out air and compacts the material (Figure 11-3A). The material is melted as it goes through this transition section. The melted material is then mixed in the final section of the screw, referred to as the metering zone. The metering zone functions to achieve a homogeneous mixture and to regulate the melt. At the end of the **barrel mixer** are a screen pack and a breaker plate. The **breaker plate**, such as the one shown in Figure 11-4, acts as a seal between the barrel and die. It also supports the screen pack and orients the flow of material from the screw. Figure 11-5 shows screens of varying mesh size. Several screens together filter out pieces of foreign material and are called screen packs. As screens become plugged, **back pressure** increases.

(A) The barrel is exposed on this 150 mm (6 in.), L/D 301, single-screw extruder.

Photo courtesy of Davis-Standard, LLC

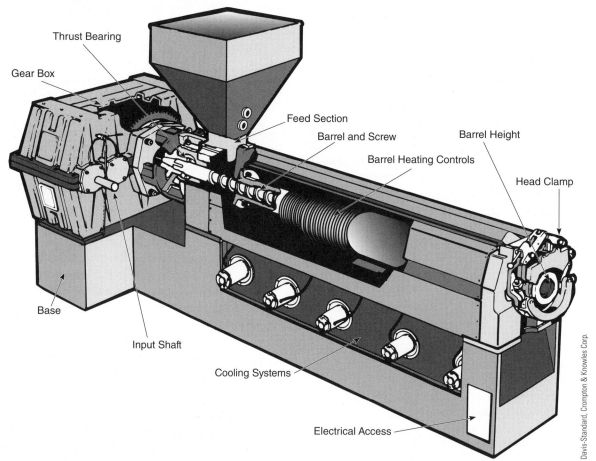

(B) Extruder with parts labeled.

Figure 11-1. Single-screw extruders.

Davis-Standard, Crompton & Knowles Corp.

Most extruders are equipped with a screen changer. The most common is a plate type that moves from one side to the other. Sliding a clean screen pack into place exposes the contaminated screens. Dirty screens are removed, clean ones installed, and the changer is ready. Some machines have a continuous ribbon of screen (sometimes rotary) that can be automatically controlled to maintain a steady head pressure despite varying levels of contamination in the polymer or other flow rate conditions.

After passing through the screen pack and breaker plate, the melt enters the die. The die actually forms molten plastics as it comes out of the extruder. The simplest die is called a single-strand die, which extrudes a strand slightly larger than the diameter of the die itself. Multiple-strand dies create many strands simultaneously.

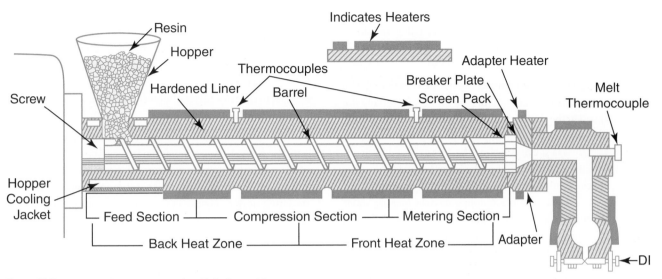

Figure 11-2. Cross section of a typical screw extruder, with the die turned down.

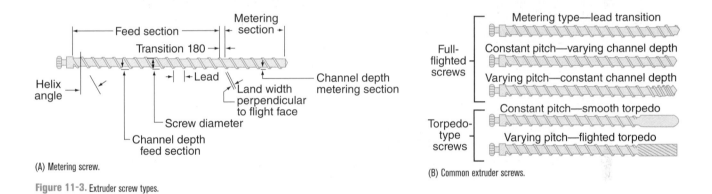

(A) Metering screw.

Figure 11-3. Extruder screw types.

(B) Common extruder screws.

Figure 11-4. The holes through the breaker plates are tapered with the large end toward the extruder screw.

Figure 11-5. Screens in varying mesh size. Mesh numbers refer to the number of openings per inch. 14 mesh is coarse; 200 mesh is very fine.

Sheet, tube, pipe, and profile dies force the material into a desired shape. Although dies may be made of mild steel, they should be made of chromium-molybdenum steel for long runs. Stainless alloys are used with corrosive materials like PVC.

Electric heaters are used around the barrel to help melt the plastics. Once the extruder is mixing, blending, and forcing the material through the die, frictional heat produced by the action of the screw may be enough to partly plasticate the material. Once the process is started, external heaters are used to maintain a fixed temperature. Because the material from the die is molten, special equipment is needed to cool and support the material after it leaves the die. Cooling devices include water baths, water mist, and directed streams of air. Pullers, choppers, and winding devices handle the cooled material and keep the extrudate flowing from the machine at a set rate.

COMPOUNDING

Compounding is the process of blending basic plastics with plasticizers, fillers, colorants, and other ingredients. In the past, compounding companies relied on single-screw extruders, multistrand dies, long water-filled cooling tanks, and pelletizers to chop the strands into pellets. Most compounders are now using twin-screw equipment (Figure 11-6).

Twin-screw compounders consist of two basic types—corotating and counter-rotating. In the corotating design, the material is forced through the gap between two screws, which is also called the nip. The material that goes through the nip experiences intensive mixing. However, some material may move forward down the barrel without passing through the nip.

In the counter-rotating design, the material is transferred from one screw to the other at the point of intermeshing. This creates a figure-8 path for the material, and because this path is rather long, it generally yields more thorough mixing than the corotating design. Figure 11-7 shows the transfer of material from one screw to the other.

Counter-rotating twin-screw compounding extruders are available in L/D ratios from 12:1 to 48:1. Screw diameters range from 25 mm to 300 mm (1 to 12 in.), and output is up to 27,000 kg (about 60,000 lb) per hour. Some companies offer modular screw design. This means that the screw consists of a central splined rod and various mixing and conveying elements that slip over the rod. Kneading and conveying elements can be right- or left-handed. Use of left-handed elements creates back pressure on the melt behind the element and reduces the pressure ahead. With careful arrangement of these elements, effective mixing and conveying of difficult-to-handle materials

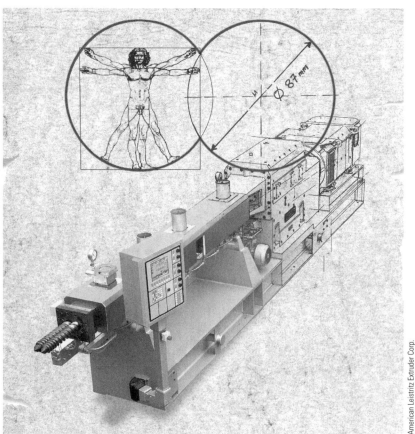

Figure 11-6. The twin screws in this machine are each 87 mm in diameter.

American Leistritz Extruder Corp.

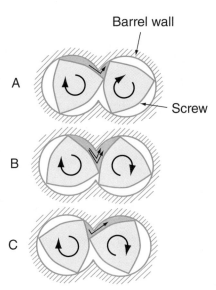

Figure 11-7. In corotating twin-screw machines, one screw wipes the material off the other one, allowing little material to pass between the two screws.

is also possible. Figure 11-8 shows various mixing and conveying elements of a set of screws and barrel sections.

Typically, compounding companies must store, measure, mix, and transport many types of ingredients. Batch blending of ingredients can lead to the problem of segregation, in which heavier ingredients separate from the lighter ones. A homogeneously blended batch may no longer become homogeneous during transportation, particularly when passing through long pipelines.

To avoid problems of segregation due to varying bulk densities, many companies bring the raw materials directly to the extruder and use several weight-loss hoppers to feed precisely measured amounts of ingredients into the extruder feed throat. Well-equipped compounding extruders can handle up to 10 separate feed streams. Some liquids and fibers must be added to already melted plastics. Such **downstream feeding** requires special pumps for liquids and feeders for fibers and powders.

To save space, some compounders purchase underwater pelletizing equipment. Figure 11-9A shows a schematic of an underwater pelletizing system. As compounded plastics material exits the extruder, it passes over a cone and is directed into the holes in the die plate. Figure 11-9B shows a sketch of this portion of an underwater pelletizer. Figure 11-9C provides

a photo of the extruder side of the die plate, and Figure 11-9D shows the cutter assembly on the other side of the die plate.

MAJOR TYPES OF EXTRUSION PRODUCTS

Because extrusion includes so many types, it is useful to break the extrusion category down into major types of products. These are profile, pipe, sheet, film, blown film, filament, and coating and wire covering.

Profile Extrusion

The term **profile extrusions** applies to most extruded products other than pipe, film, sheet, and filaments. Figure 11-10 shows a few of these items. Such profiles are generally extruded horizontally. Achieving a desired shape requires equipment to support and shape the extrudate during cooling. Cooling is done through the use of air jets, water troughs, water sprays, and cooling sleeves. Size control requires sizing dies, hold-down fingers, or sizing plates.

Controlling the size or shape of such profiles can be difficult. When a material exits an extruder die, it changes shape due to a phenomenon called *die swell*. A major factor causing die swell is polymer molecules becoming oriented as they flow through the die. After exiting the die, the oriented molecules return to a tangled, coiled form. This change alters the shape of the extrudate. When the extrudates do not have uniform cross sections, the shrinkage upon cooling will also not be uniform.

To produce exact cross-sectional dimensions after the extrudate has cooled, allowances must be made in the orifice design. In complex cross sections where thin sections or sharp edges are formed, cooling occurs more quickly at these portions. Such areas shrink first, making them smaller than the rest of the section. This means that both the die and shape of the plastics (extrudate) may be different. To correct this problem, the orifice in the die is made larger at these points (Figure 11-11).

To reduce the problems caused by complex geometry of profile dies, some manufacturers choose postforming of simpler shapes. Postforming into different shapes requires the use of sizing plates, shoes, or rollers. A flat-tape shape may be

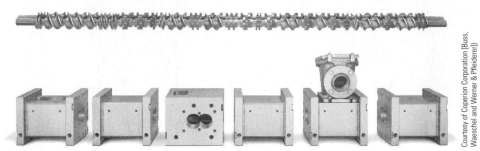

Courtesy of Coperion Corporation (Buss, Waeschel and Werner & Pfleiderer])

Figure 11-8. Twin screws and barrel sections.

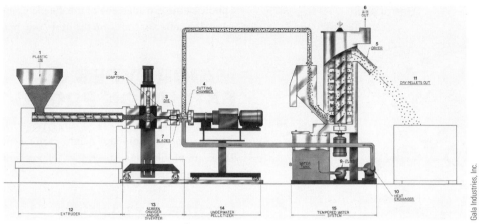

Figure 11-9. (A) Schematic of an underwater pelletizing system.

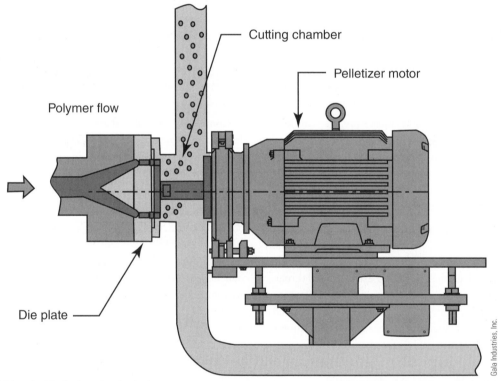

Cutting chamber

Pelletizer motor

Polymer flow

Die plate

Figure 11-9. (B) Schematic of the die plate and cutting chamber.

Figure 11-9. (C) The extruder side of the die plate.

Figure 11-9. (D) The cutter hub, which is removed from the drive shaft and positioned on the exit side of the die place.

Figure 11-10. Some of the many profile extrusions of plastics.

Fellows Corp.

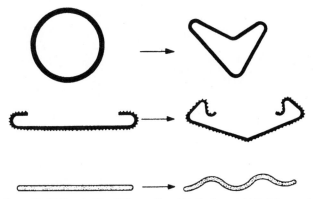

Figure 11-12. The extrudates at left are postformed into the shapes at right by being passed through rollers.

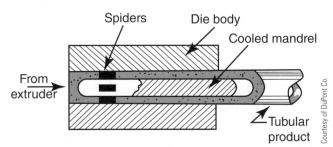

Figure 11-13. In this pipe-forming operation, hot material is extruded around a cold mandrel or pin.

Courtesy of DuPont Co.

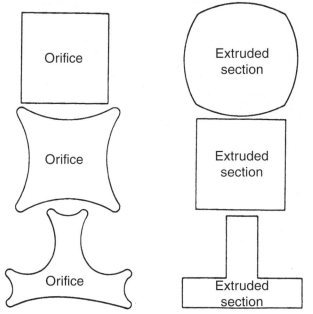

Orifice	Extruded section
Orifice	Extruded section
Orifice	Extruded section

Figure 11-11. Relationships between die orifices and extruded sections.

postformed into corrugated form. While the extrudate is still hot, round rods may be postformed into oval or other new shapes (Figure 11-12).

Pipe Extrusion

Pipes (tubular forms) are shaped by the exterior dimensions of the orifice and by the mandrel (sometimes called a **pin**) that shapes inside dimensions (Figure 11-13). The mandrel is held in place by thin pieces of metal called **spiders**.

The diameter of the pipe or tube is also controlled by the tension of the take-up mechanism. If the tube is pulled faster than

the speed of the extrudate melt, the product will be smaller and thinner than the die.

To prevent the tube from collapsing before cooling, it is pinched shut on the end, and air is forced in through the die. This air pressure expands the pipe slightly. The hot tube may be pulled through a sizing ring or vacuum ring to hold the outside diameter to a close tolerance. The thickness of the pipe wall is controlled by the mandrel and die size.

Sheet Extrusion

The American Society for Testing and Materials (ASTM) has defined **film** as plastics sheeting 0.25 mm (0.01 in.) or less in thickness. Sheet materials thicker than 0.25 mm are considered sheet. Sheet extrusion produces stock for use in most thermoforming operations.

Most sheet forms involve extruding molten thermoplastic materials through dies with long horizontal slots, as seen in Figure 11-14. These dies come in two major styles—the *T-shape* and the *coat-hanger shape*. In both styles, the molten material is fed to the center of the die. It is then formed by the die lands and adjustable jaw. The width may be controlled by the external deckle bars or the actual die width. In Figure 11-15, an adjustable choke or restrictor bar is used in the extrusion of sheeting.

The extruded sheet goes over or through a set of rollers to provide the desired surface finish or texture and accurately size

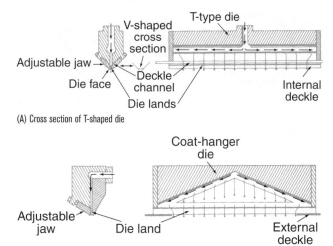

(A) Cross section of T-shaped die

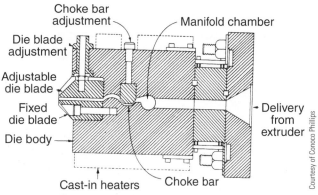

(B) Cross section of coat-hanger shaped die

Figure 11-14. Two types of extrusion die.

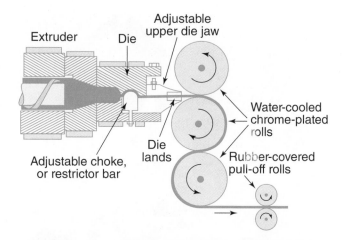

Figure 11-15. Sheet-extrusion die with adjustable choke bars.

Courtesy of Conoco Phillips

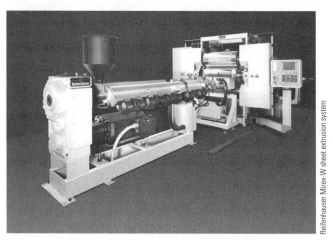

Reifenhauser Mirex-W sheet extrusion system

Figure 11-16. (A) Schematic of sheeting die and takeoff unit. (B) This sheet-extrusion system shows the extruder, die, stack of chill rollers, and controller to match the roll speed to the extrusion rate.

the thickness. Figure 11-16 shows an illustration of sheeting rollers and a photograph of the sheet leaving the die and entering the rollers.

Table 11-1 gives some common sheet-extrusion problems and their remedies.

Film Extrusion

Film extrusion and calendering produce finished products that are rather similar. Although calendering is not technically an extrusion process, it logically deserves discussion along with film extrusion.

Film extrusion is similar to sheet extrusion. Besides the difference in thickness, film-extrusion dies are lighter and possess shorter die lands than sheeting dies.

In Figure 11-17A, the film is extruded into a quench tank. In Figure 11-17B, the film is being drawn over chill rollers in a process sometimes called **film casting**. Both chill-roller and water-tank film extrusions are used commercially. Temperature, vibration, and water currents must be controlled with care when using the water-tank method to allow production of clear,

defect-free film. Heavy-gauge films are currently being made using this high-speed method.

Film extrusion is more costly than blown film, and consequently is used only when the film quality must be higher than blown film. Some plastics are heat sensitive and tend to decompose or degrade due to high temperature. PVC is such a material. To make PVC film, slot extrusion or blown-film techniques are generally not used. PVC film normally requires calendering.

Calendering. In **calendering**, thermoplastic materials are squeezed to final thickness by heated rollers (Figure 11-18). Films and sheet forms with a glossy or embossed finish may be produced by using this method (Figure 11-19). A large amount of calendered film is used in the textile industry. Embossed or textured film is used to produce leather-like apparel, handbags, shoes, and luggage.

The calendering process consists of blending a hot mix of resin, stabilizers, plasticizers, and pigments in a continuous kneader or **Banbury mixer**. This hot mix is led through a two-roll mill to produce a heavy sheet stock. As the sheet passes

Table 11-1. Troubleshooting Sheet Extrusion Equipment

Defect	Possible Remedy
Continuous lines in direction of extrusion	Repair or clean out die.
	Die contamination or polish of rollers scored.
	Reduce die temperatures.
	Use properly dried materials.
Continuous lines across sheet	Jerky operation—adjust tension on sheet.
	Reduce polishing roll temperatures or increase roll temperatures.
	Check back-pressure gauge for surging.
Discoloration	Use proper die and screw design.
	Minimize material contamination.
	Temperature too high—too much regrind.
	Repair and clean out die.
Dimensional variation across sheet	Adjust bead at polishing rolls.
	Balance die heats.
	Reduce polishing roll temperatures.
	Check temperature controllers.
	Repair or clean out die.
Voids in sheet	Balance extruder line conditions.
	Use proper screw design.
	Minimize material contamination.
	Reduce stock temperature.
Dull strip	Die set too narrow at this point.
	Minimize material contamination.
	Increase die temperature.
	Repair or clean out die.
Pits, craters	Balance extruder line conditions.
	Minimize material contamination.
	Use properly dried materials.
	Control stock feed.
	Reduce stock temperature.

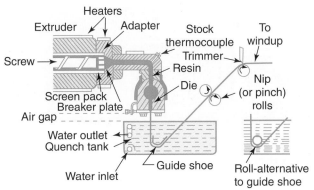

(A) Cross section of the front part of a flat-film extruder, the quench tank, and takeoff equipment.

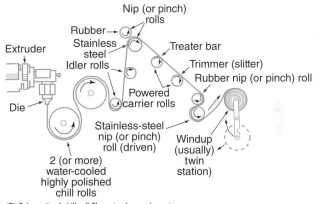

(B) Schematic of chill-roll film extrusion equipment.

Figure 11-17. Types of film takeoff equipment.

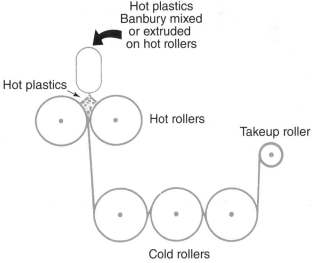

Figure 11-18. The calendering process.

through a series of heated, revolving rollers, it becomes progressively thinner until the desired thickness is reached. A pair of precision, high-pressure finish rollers is used for gauging and embossing. In the end, the hot sheet is cooled on a chill roller and taken off in sheet or film form.

Calender rollers are very costly and easily damaged by metal contaminants. For this reason, metal detectors are often used to scan the sheet before it enters the calender. The calendering equipment, together with accessory controls, is also very costly. Replacement costs for a calender line may exceed $1 million, which has discouraged the installation of calendering equipment. Calendering has some advantage over extrusion

Figure 11-19. A calendar system.

and other methods when producing colored or embossed films and sheets. When colors are changed, a calender requires a minimum of cleaning, whereas an extruder must be purged and cleaned thoroughly.

Though calenders are expensive, calendering remains the preferred method of producing PVC sheets at high rates. Roughly 95 percent of all calendered products are PVC, and only about 15 percent are being used for rigid production.

Calendering equipment may involve multistory complexes, because calender rollers are usually placed in an inverted L or a Z arrangement (Figure 11-20). The rolls and supportive equipment are controlled by many sensing devices and by computers. Calenders are usually rated by the amount (mass) of material that they can produce per unit of time. This rate depends on the material, plasticating rate, required surface finish, and take-up capacity. Large machines have an output rate of nearly 3000 kg/h (6662 lb/h). Most rollers are less than 2 m (6 ft 7 in.) wide. With soft materials, widths over 3 m (10 ft) are possible. Roll forces may approach 350 kN (30.34 tons-force) for thin, rigid materials.

Several methods are used to compensate for roll deflection (bending or bowing): (1) Force is applied to the outboard or inboard main bearings; (2) rolls are produced with slight crowns; or (3) one roll is skewed with respect to the other (roll crossing) (Figure 11-21). Table 11-2 gives some common calendering problems and remedies.

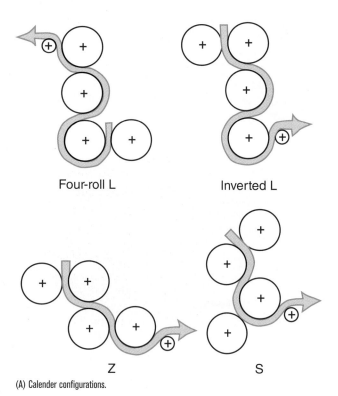

Four-roll L

Inverted L

Z

S

(A) Calender configurations.

From mixer

Metal detector

Two-roll nip mill

135° to 235°C speed 0.5–0.85

Cauging rolls 150° to 260°C speed 0.6–0.9

1–1.5 times thickness of finished gauge

150° to 235°C speed 1.0 finished gauge thickness

Chilled roll optional

Takeoff

Draw ratio 0–300%

(B) A series of rollers arranged in a Z-form is used to calender thermoplastic sheet material.

Figure 11-20. Common calender roll configurations.

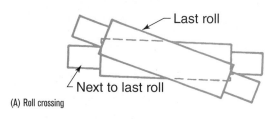

Last roll

Next to last roll

(A) Roll crossing

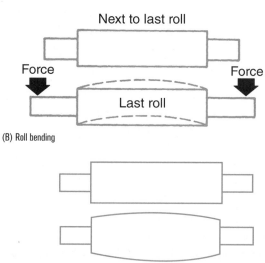

Next to last roll

Force

Force

Last roll

(B) Roll bending

(C) Roll crown

Figure 11-21. Methods used for correcting sheet profile.

Table 11-2. **Troubleshooting Calendering**

Defect	Possible Remedy
Blistering of film or sheet	Reduce melt temperature. Reduce program speed of rolls. Check for resin contamination. Reduce temperature of chill rolls.
Thick section in center, thin edges	Use crowned rolls. Increase nip opening. Check bearing load of rolls.
Cold marks or crow's feet	Increase stock temperature. Decrease program feed.
Pin holes	Check for contamination in resin. Blend plasticizers more thoroughly into resin.
Dull blemishes	Check for lubricant or resin contamination. Check roll surfaces. Increase melt temperature. Increase roll temperature.
Rough finish	Raise roll temperature. Excessive nip between rolls.
Roll bank on sheet	Incorrect stock temperature—increase/decrease. Provide constant takeoff speeds. Reduce roll temperatures and nip clearance.

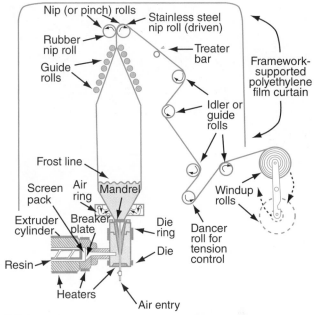

(A) Basic apparatus.

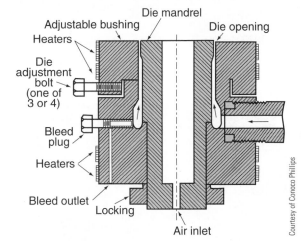

Courtesy of Conoco Phillips

(B) Side-fed manifold blown-film die.

Blown-Film Extrusion

In blown-film extrusion, the film is produced by forcing molten material through a die and around a mandrel. It emerges from the orifice in tube form (Figure 11-22). This is similar to the process used to make pipes or tubes. This tube, or bubble, is expanded by blowing air through the center of the mandrel until the desired film thickness is reached. This process is something like blowing up a balloon. The tube is usually cooled by air from a cooling ring around the die (Figure 11-23). The frost line is the zone where the temperature of the tube has fallen below the softening point of the plastics. In polyethylene or polypropylene film extrusion, the frost zone is evident because it actually appears frosty. The frost zone shows the change taking place as the plastics cools from the melt (an amorphous state) to a crystalline state. However, with some plastics there will be no visible frost line.

The size and thickness of the finished film is controlled by several factors, including extrusion speed, takeoff speed, die (orifice) opening, material temperature, and air pressure inside the bubble or tube. Blow-up ratio is the ratio of the die diameter

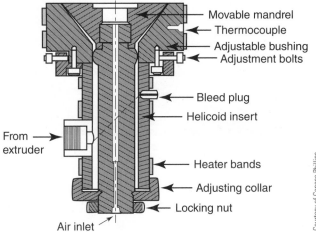

Courtesy of Conoco Phillips

(C) Adjustable-opening blown-film die.

Figure 11-22. Schematic drawings of blown-film extrusion procedures.

(A) This photo shows a microprocessor thickness measurement system that can adjust the extruder to maintain a desired thickness in the blown film.

(C) The bubble case contains the LDPE blown film.

(B) The bubble stalk is clearly visible in this view of HDPE blown film.

(D) The microprocessor control system monitors and adjusts the entire blown film system.

Figure 11-23. Blown-film extrusion.

to the bubble diameter. Blow-extruded film is sold as seamless tubing, flat film, or film molded in a number of ways. Film producers may slit the tubing on one edge during windup. If the tube is blown to a diameter of 2 m (6.5 ft), the flat film will have a width (slit and opened) of over 6 m (19 ft). Slot dies of this

size are not practical. Tubular films are desirable as low-cost packaging for some foods and garments. Only one heat seal is needed in the production of bags from blown tubing.

Blown films are semi-oriented—that is, they have less orientation of molecules in a single direction than film from slot dies. Blown films are stretched as the tube is expanded by air pressure. Such stretching results in a more balanced molecular orientation in two directions. Products are biaxially oriented—one in the direction of length and one across the diameter of the bubble. Improved physical properties are one asset of blown film. However, clarity, surface defects, and film thicknesses are harder to regulate than with slot extrusion. Table 11-3 gives troubleshooting information for blown-extruded film.

Filament Extrusion

Important terms for filament extrusion. A **filament** is a single, long, slender strand of plastics. The fisherman is probably most familiar with monofilament fishing line. This single filament of plastics may be made in any desired length. *Yarns* may be composed of either monofilament or multifilament strands of plastics.

The term **fiber** is used to describe all types of filaments—natural or plastics, and monofilament or multifilament. First, fibers are spun or twisted into yarns. Then, they are woven into finished fabrics, screening, or other products ready for consumer use.

The fineness of a fiber is expressed by a unit called a **denier**. One denier equals the mass in grams of 9000 m (29,527 ft) of fiber. For example, 9000 m of a 10-denier yarn weighs 10 g (0.35 oz). The name of the unit may have evolved from the name for a sixteenth-century French coin used as a standard for measuring the fineness of silk fibers.

Plastics fibers of the same fineness vary in denier because of differences in density. Although two filaments may have the same denier, one may have a larger diameter because of a lower relative density.

To calculate the denier of the filaments in yarn, divide the yarn denier by the number of filaments:

$$\frac{8 - \text{denier yarn}}{40 \text{ filaments}} = 2 \text{ denier for each filament}$$

Remember: 9000 m of 1-denier filament weighs 1 g.

9000 m of 2-denier filament weighs 2 g.

The International Organization for Standardization (ISO) has developed a universal system for designating linear density of textiles called the **tex**. The fabric industry has adopted the tex as a measure of linear density. In the tex system, the yarn count is equal to the mass of yarn in g/km.

Types of filaments. Not all synthetic filaments are used by the textile industry. Some monofilaments are used for bristles in brooms, toothbrushes, and paint brushes. The relative shapes of these filaments vary, as shown in Figure 11-24.

Table 11-3. Troubleshooting Blown Film

Defect	Possible Remedy
Black specks in film	Clean die and extruder.
	Change screen pack.
	Check resin for contamination.
Die lines in film	Lower die pressure.
	Increase melt temperature.
	Polish all rough edges in film path.
	Check nip rolls—make smooth.
Bubble bounces	Increase screw rpm and nip roll speed.
	Enclose tower or stop drafts.
	Adjust cooling ring to obtain constant air velocity around ring.
Poor optical and physical properties	Raise melt temperature.
	Increase blow-up ratio; increase frost line height.
	Clean die lips, extruder, and rollers.
Failures at fold	Decrease nip roll pressure.
Failure at weld lines	If possible, bleed die at weld line.
	Heat die spiders—insulate air lines there.
	Increase melt temperature.
	Check for contamination.
Film won't run continuously	Clean die and extruder.
	Lower melt temperature.
	Increase film thickness.

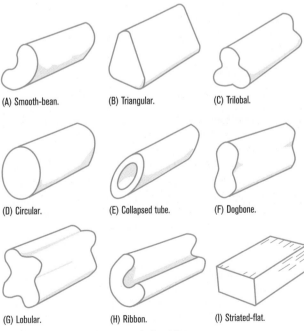

(A) Smooth-bean.   (B) Triangular.   (C) Trilobal.

(D) Circular.   (E) Collapsed tube.   (F) Dogbone.

(G) Lobular.   (H) Ribbon.   (I) Striated-flat.

Figure 11-24. Cross-sectional shapes of selected fibers.

If the fiber is to be pliable and soft, a fine filament is needed. If the fiber is to withstand crushing and be stiff, a thicker filament is used. Clothing fibers range from 2 to 10 denier per filament. Fibers for carpeting range from 15 to 30 denier in fineness per filament.

The cross-sectional shape of the fiber helps determine the texture of the finished product. Triangular and trilobal shapes impart to synthetic fiber many of the properties of the natural fiber silk. Many of the ribbon- and bean-shaped filaments resemble cotton fibers.

Manufacturing of filaments.

Monofilaments are produced much in the same manner as profile shapes, except that a multi-orifice die is used. These dies contain many small openings from which the molten material emerges. Such dies are used to produce granular pellets, monofilaments, and multifilament strands.

Filament shapes are made by forcing plastics through small orifices in a process referred to as *spinning*. The plastics is shaped by the opening in the die or *spinneret*. This process may have been named spinning from the method of spinning natural fibers. The small opening under the jaw of the silkworm is also called the spinneret.

Because the diameter of these orifices is often much finer than human hair, spinnerets are often made of such metals as platinum that will resist acids and orifice wear. In order to be forced or extruded out of these small openings, the plastics must be made fluid.

Acrylic fiber is produced using the spinning process shown in Figure 11-25. A thick chemical solution is extruded into a coagulation bath through the tiny holes of the spinneret. In this bath, the solution coagulates (becomes a solid) and becomes Acrilan® acrylic fiber. The fiber is washed, dried, crimped, and cut into staple lengths. Then it is baled for shipment to textile mills, where it is converted into carpeting, wearing apparel, and many other products.

There are three basic methods of spinning fibers:

1. In **melt spinning**, plastics such as polyethylene, polypropylene, polyvinyl, polyamide, or thermoplastic polyesters are melted and forced out the spinneret. As the filaments hit the air, they solidify and are passed through other conditioners (Figure 11-26A).
2. In **solvent spinning**, plastics such as acrylics, cellulose acetate, and polyvinyl chloride are dissolved by certain solvents. The solution is forced through the spinneret (Figure 11-26B), and the filament then passes through a stream of hot air. The air aids in evaporating the solvents from the slender fibers. For economic reasons, these solvents must be recovered for use again.

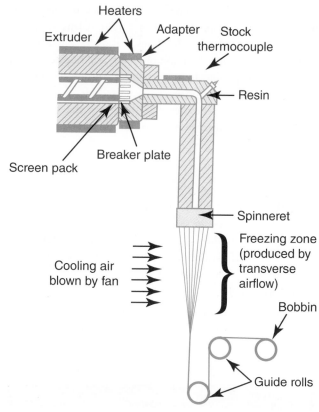

Figure 11-25. Production of acrylic fiber by spinning.

3. The first step of **wet spinning** (Figure 11-26C) is similar to solvent spinning. The plastics is dissolved in chemical solvents. This fluid solution is forced out through the spinneret into a coagulating bath that makes the plastics gel into a solid filament form. Some members of the cellulosic, acrylic, and polyvinyl plastics families may be processed by wet spinning.

All three processes begin by forcing fluid plastics through a spinneret. They all end by solidifying the filament through cooling, evaporation, or coagulation (Table 11-4).

The strength of the single filament may be determined by several factors. Most of the selected filament fibers are linear and crystalline in composition. When groups of molecules lie together in long, parallel molecular chains, additional strong bonding sites are available. When the liquid plastics are forced through the spinneret, many of the molecular chains are forced closer together and parallel to the filament axis. This packing, arranging, and drawing provide increased strength throughout the filament. By mechanically working the filament, further molecular orientation and packing can be accomplished. This mechanical process is called drawing. The drawing of non-crystalline plastics also helps to orient molecular chains, thus improving strength.

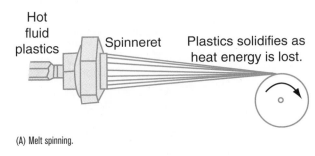

(A) Melt spinning.

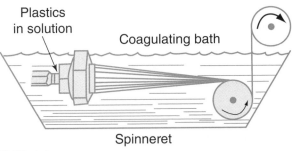

(B) Solvent spinning.

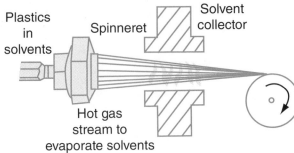

(C) Wet spinning.

Figure 11-26. Three basic methods of spinning plastics fibers.

Table 11-4. Selected Fibers and Production Processes

Fibers	Extrusion Spinning Process
Acrylics and modacrylics	
Acrilan®	Wet
Creslan®	Wet
Dynel® (vinyl-acrylic)	Solvent
Orlon®	Solvent
Verel®	Solvent
Cellulose esters	
Acetate (Acele®, Estron®)	Solvent
Triacetate (Arnel®)	Solvent
Cellulose, regenerated	
Rayon® (viscose, cuprammonium)	Wet
Olefins	
Polyethylene	Melt
Polypropylene (Avisun®, Herculon®)	Melt
Polyamides	
Nylon 6,6®, Nylon 6®, Qiana®	Melt
Polyesters	
Dacron®, Trevira®, Kodel®, Fortrel®	Melt
Polyurethanes	
Glospan®	Wet
Lycra®	Solvent
Numa®	Wet
Vinyls and vinylidines	
Saran®	Melt
Vinyon N®	Solvent

Drawing or stretching of the plastics is accomplished by running the filaments through a series of variable-speed rollers (Figure 11-27). The drawing of crystalline plastics is continuous. As the fiber travels through the drawing process, each roller is rotated at a faster speed. Roller speed determines the amount of stretching or drawing.

Monofilaments range from 0.12 mm to 1.5 mm (0.0047 to 0.0059 in.) in diameter. They may be either handled individually or by takeoff machines.

High-bulk fibers and yarns may be processed from films. This process utilizes a composite film that has been co-extruded or laminated. It is mechanically drawn and cut (fibrillated) into fine strands. The fibrillation is done during the stretching or drawing process while the film passes between serrating rollers rotating at different speeds. The teeth of these rollers cut the film into fibrous form (Figure 11-28). Composite film develops internal stresses as it is extruded, drawn, and fibrillated. This unequal stress orientation in the film layers makes the fibers curl and exhibit properties much like those of natural fibers.

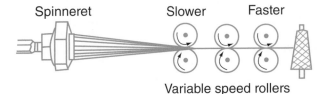

Figure 11-27. Drawing plastics filaments.

Extrusion Coating and Wire Covering

Paper, fabric, cardboard, plastics, and metal foils are common substrates for extrusion coating (Figure 11-29). In extrusion coating, a thin film of molten plastics is applied to the substrate without the use of adhesives while substrate and film are pressed between rollers. For special application, adhesives

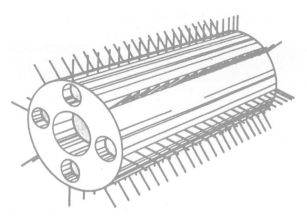

Figure 11-28. This device has rows of pins that fibrillate yarn of very fine denier.

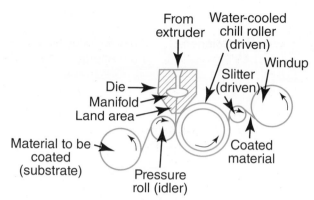

Figure 11-29. Extrusion coating of substrates.

may be needed to ensure proper bonding. Some substrates are preheated and primed with adhesion promoters using slot-extruder dies.

In wire and cable covering, the substrate for extrusion coating is wire. The setup for extrusion coating of wire and cable is shown in Figure 11-30. During this process, a molten plastics is forced around the wire or cable as it passes through the die.

The die actually controls and forms the coating on the wire. Wires and cables are usually heated before coating to remove moisture and ensure adhesion. As the coated wire emerges from the cross-head die, it is cooled in a water bath. Two or more wires may be coated at one time. Television and appliance cords are common examples of this method of coating. Wooden strips, cotton rope, and plastics filaments may be coated by this process.

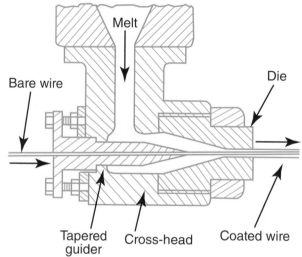

(A) A cross-head holds the wire-coating die and the tapered guide as the soft plastics flows around the moving wire.

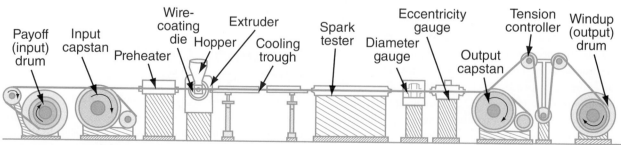

(B) A general component layout in a wire-coating extrusion plant.

Figure 11-30. Extrusion coating of wire and cable.

BLOW MOLDING

This process is sometimes listed as a molding technique because force is used to press hot, soft tubular material against mold walls. In this chapter, it is considered a type of extrusion process because extruders are required to create the tube shape that is later inflated into a hollow container.

Blow molding is a technique adopted from the glass industry and modified for making one-piece containers and other articles. This process has been used for centuries for making glass bottles. However, blow molding of thermoplastics did not develop until the late 1950s. In 1880, blow molding was accomplished by heating and clamping two sheets of celluloid in a mold. Then air was forced in to form a blow-molded baby rattle. This may have been the first blow-molded thermoplastic article produced in the United States.

The basic principle of blow molding is simple (Figure 11-31). A hollow tube (parison) of molten thermoplastic is placed in a female mold, and the mold is closed. It is then forced (blown) by air pressure against the walls of the mold. After a cooling cycle, the mold opens and ejects the finished product. This process is used to produce many containers, toys, packaging units, automobile parts, and appliance housings.

There are two basic blow molding methods:

1. Injection blowing
2. Extrusion blowing

The major difference is in the way that the hot, hollow tube, or **parison**, is produced.

Injection Blow Molding

Injection blow molding can more accurately produce the desired material thickness in specified areas of a part. The major advantage of this type of molding is that any shape with varying wall thickness can be made exactly the same each time. There is no bottom weld or scrap to reprocess. A major disadvantage is that two different molds are required. One is used to mold the preform (Figure 11-32A) and the other for the actual blowing operation (Figure 11-32B). During the

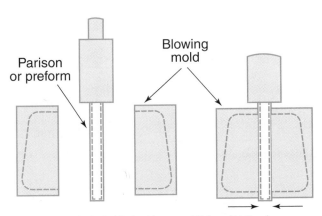

(A) Molded hollow tube (parison) is placed between mold halves, which then close.

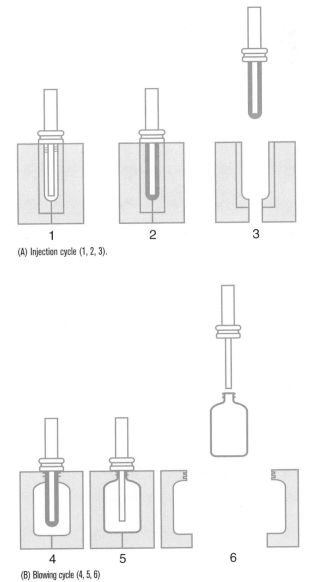

(A) Injection cycle (1, 2, 3).

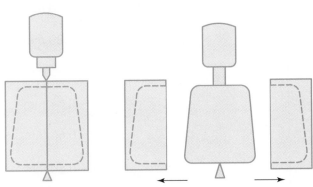

(B) The still molten parison is pinched off and inflated by an air blast. The blast forces the plastics against the cold walls of the mold. Once the product has cooled, the mold opens and the object is ejected.

Figure 11-31. Blow-molding sequence.

(B) Blowing cycle (4, 5, 6)

Figure 11-32. Injection blow-molding sequence.

blow-molding operation, the hot injection-molded preform is placed in the blowing mold. Air is then forced into the preform, making it expand against the walls of the mold. This injection-blow process has been called **transfer blow** because the injected preform must be transferred to the blowing mold (Figure 11-33).

Extrusion Blow Molding

In extrusion blowing, a hot tubular parison is extruded continuously (except when using accumulator or ram systems). The mold halves close, which seals off the open end of the parison (Figure 11-34). Air is then injected and the hot parison

expands against the mold walls. After cooling, the product is ejected. Extrusion blow molding can produce articles large enough to hold 10,000 L (2646 gal.) of water. However, preforms of this size are too costly. Blow extrusion offers strain-free articles at a high production rate, but scrap reprocessing is required.

Controlling wall thickness is the largest disadvantage of extrusion blow molding. By controlling (sometimes called *programming*) the wall thickness of the extruded parison, thinning is reduced. For a part requiring an extremely large body yet needing strength at the corners, a parison could be produced with the corner areas much thicker than the walls (Figure 11-35).

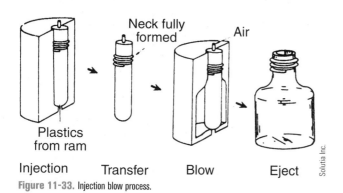

Figure 11-33. Injection blow process.

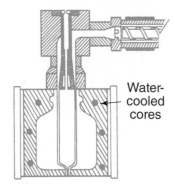

(A) Closing of mold halves.

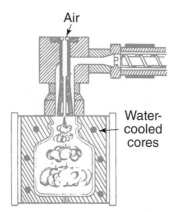

(B) Injection of air.

Figure 11-34. Extrusion blow molding.

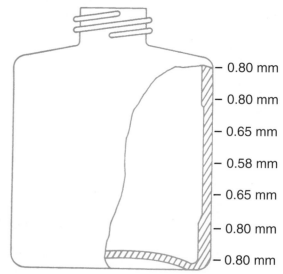

(A) Fixed orifice product.

(B) Programmed orifice product.

Figure 11-35. Parison programming with variable die orifice. Note wall thickness.

In Figure 11-36, the arrangement of the extruder and die parts is shown. By using this method, one or more continuous parisons may be extruded. In Figure 11-37, hot plastics is fed into an accumulator and then forced through the die. A controlled length of parison is produced when the ram or plunger operates. The extruder fills the accumulator and the cycle begins again.

The wall thickness of the tube or parison may be controlled (programmed) to suit the container configuration. This is done by using a die with a variable orifice, as shown in Figure 11-38.

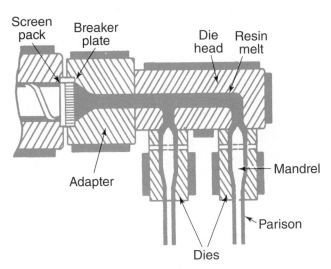

Figure 11-36. Parts found in most extrusion blow molders.

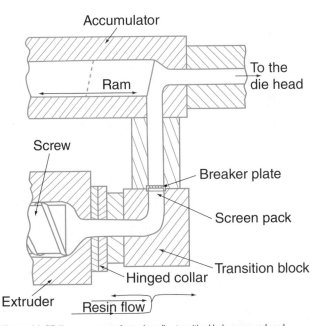

Figure 11-37. The arrangement of extruder collar, transition block, screen pack, and breaker plate found on a blow-molding press with accumulator.

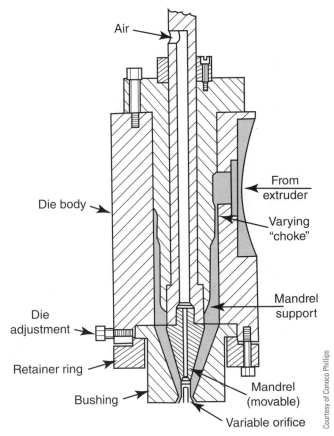

Courtesy of Conoco Phillips

Figure 11-38. Programming die used for blow molding.

Many different ways of forming blow-molded product have been developed (Figure 11-39). Each process may have an advantage in molding a given product. One manufacturer both forms and fills the container in a single operation. Rather than compressed air, the product is forced into the parison.

Extrusion blow molding may be used to produce all forms of composites including fibrous, particulate, and laminar. Short fibers are used to produce a variety of reinforced blow-molded products.

Blow-molded containers that will hold solids or nonreactive liquids are often monolayer, which means that they are made with only one material. Figure 11-40A shows some monolayer products. If products need to contain volatile or reactive liquids, appropriate barrier layers are needed. That requirement necessitates multilayer blow moldings. Figures 11–40B and 11–40D show multilayer products.

Handling equipment for blow-molded products often utilizes pinch-off areas, neck rings, or molded rings. Figure 11-41A shows a large drum being removed from the mold. The handling robot grips the parison tail. Figure 11-41B shows a spin-off bottle trimmer, which rotates round bottles against a cutter to remove handling rings. Table 11-5 shows some common

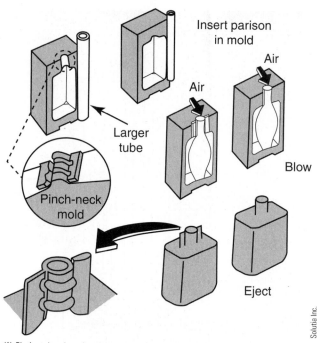

Insert parison in mold

Larger tube

Pinch-neck mold

Air

Air

Blow

Eject

(A) Pinch-neck and regular processes.

Solutia Inc.

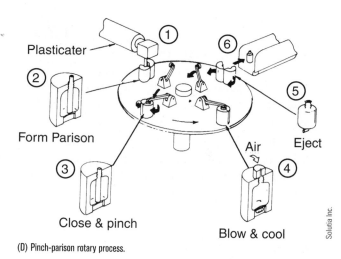

Plasticater

Form Parison

Close & pinch

Blow & cool

Air

Eject

(D) Pinch-parison rotary process.

Solutia Inc.

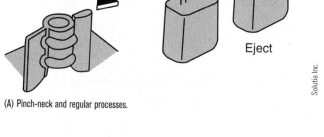

Plasticater

Form parison

Air

Pinch

Blow

Eject

(B) Basic pinch-parison process.

Solutia Inc.

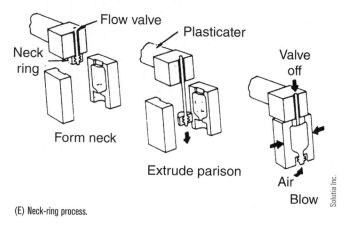

Flow valve

Plasticater

Valve off

Neck ring

Form neck

Extrude parison

Air

Blow

(E) Neck-ring process.

Solutia Inc.

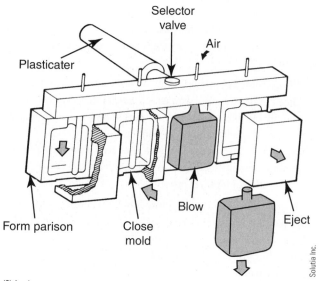

Selector valve

Air

Plasticater

Form parison

Close mold

Blow

Eject

(C) In-place process.

Solutia Inc.

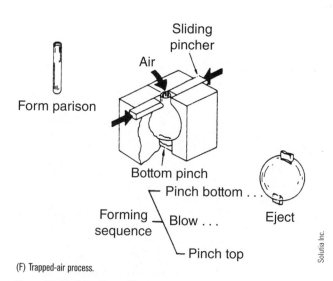

Sliding pincher

Air

Form parison

Bottom pinch

Pinch bottom . . .

Forming sequence

Blow . . .

Eject

Pinch top

(F) Trapped-air process.

Solutia Inc.

Figure 11-39. Various blow-molding processes.

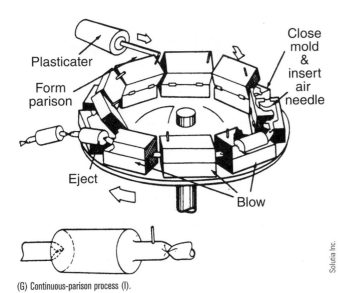

(B) These vehicle gasoline tanks often contain seven layers of material to achieve strength and barrier characteristics that prevent gasoline from evaporating through the walls of the tank.

Courtesy of SIG

Solutia Inc.

(G) Continuous-parison process (I).

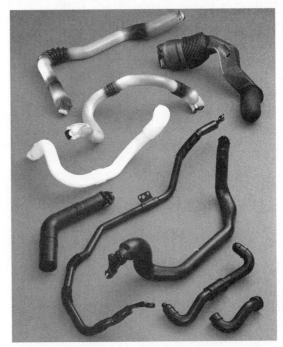

Courtesy of SIG

(C) The complex shapes in these tubes are produced using the suction blow-molding process.

Solutia Inc.

(H) Continuous-parison process (II).

Figure 11-39. (*Continued*)

Courtesy of SIG

(A) These blow-molded containers range in capacities from 50 to 1000 L (13 to 264 gal).

Courtesy of SIG

(D) This trash bin is a three-layered product in which the inner layer or core is made of recycled materials.

Figure 11-40. Selected blow-molded products.

(A) The robot grips the tail of the parison just above the pinch-off during the molding of this 55-gallon drum.

(B) A high-production-rate spin-off trimmer.

Figure 11-41. Blow-molded containers require trimming to remove pinch-off and handling extensions.

problems in blow molding and their remedies; six advantages and five disadvantages of extrusion blow molding are listed below:

Advantages of Extrusion Blow Molding

1. Most thermoplastics and many thermosets may be used.
2. Die costs are lower than those of injection molding.
3. Extruder compounds and blends materials well.
4. Extruder plasticates material efficiently.
5. Extruder is basic to many molding processes.
6. Extrudates may be any practical length.

Table 11-5. Troubleshooting Blow Molding

Defect	Possible Remedy
Excess parison stretch	Reduce stock temperature. Increase extrusion rate. Reduce die tip heat.
Die lines	Die surface is poorly finished or dirty. Blowing air orifice is too small—more air. Extrusion rate is too slow—parison cooling.
Uneven parison thickness	Center mandrel and die. Check heater bands for uneven heating. Increase extrusion rate. Reduce melt temperature. Program parison.
Parison curls up	Excessive temperature difference between mandrel and die body. Increase heating period. Uneven wall thickness or die temperature.
Bubbles (fisheyes) in parison	Check resin for moisture. Reduce extruder temperatures for better melt control. Tighten die tip bolts. Reduce feed-section temperature. Check resin for contamination.
Streaks in parison	Check die for damage. Check melt for contamination. Increase back pressure on extruder. Clean and repair die.
Poor surface	Extrusion temperature is too low. Die temperature is too low. Poor tool finish or dirty tools. Blowing air pressure is too low. Mold temperatures is too low. Blowing speed is too slow.
Parison blowout	Reduce melt temperature. Reduce air pressure or orifice size. Align parison and check for contamination. Check for hot spots in mold and parison.
Poor weld at pinch-off	Parison temperature is too high. Mold temperature is too high. Mold closing speed is too fast. Pinch-off land is too short or improperly designed.
Container breaks on weld lines	Increase melt temperature. Decrease melt temperature. Check pinch-off areas. Check mold temperature and decrease cycle time.
Container sticks in mold	Check mold design—eliminate undercuts. Reduce mold temperature and melt temperature. Increase cycle time.
Part weight too heavy	Parison temperature is too low. Melt index of resin is too low. Annular opening is too large.
Warpage of container	Check mold cooling. Check for proper resin distribution. Lower melt temperature. Reduce cycle time for cooling.
Flashing around container	Lower melt temperature. Check blowing pressure and air start time. Check for molds closing on parison. Check air start time and pressure.

Disadvantages of Extrusion Blow Molding

1. Costly secondary operations are sometimes needed.
2. Machine cost is high.
3. Purging and trimming produce waste.
4. Limited programmed shapes and die configurations are available.
5. Screw design must match material melt and flow characteristics for efficient operation.

Blow-Molding Variations

Four blow-molding variations should be mentioned:

1. Cold parison
2. Sheet
3. Stretched or biaxial
4. Multilayered (co-extrusion or co-injection)

In the cold-parison process, the parison is extruded by normal means (either injection or extrusion), then cooled and stored. Later, the parison is heated and blown to shape. The major advantage of this process is that the parison can be shipped to other locations or stored in case of a breakdown or materials shortage.

Multilayer bottles may be produced by co-injection blow molding or co-extrusion methods. The three-layered product generally contains a barrier layer sandwiched between two body layers.

In the sheet **blow-molding** process, hot extruded sheets are blow formed as they are pinched between mold halves. Edges are fused together by the pinching action of the mold. Two different colored sheets may be extruded and formed into a product with two separate colors (Figure 11-42). However, the pinch-welded seams are a major disadvantage. In addition, two extruders are usually needed, and there is a great deal of scrap to be reprocessed.

In the **stretch**, or **biaxial blow-molding**, process, molded preforms and extruded tubes may be stretched before blowing. This produces a blow-molded product with better clarity, reduced creep, higher impact strength, improved gas and water vapor barrier properties, and lower mass. Homopolymers may be used rather than more costly copolymers.

In the injection-molded preform method, a rod stretches the hot preform during the blow cycle (Figure 11-43). In parison or tube methods, the hot tube is stretched prior to the blow cycle (Figure 11-44).

Co-extrusion blow molding actually produces a laminar bottle product (see Laminating, p. 201). Several extruders are used to extrude the material into the manifold. The multilayered container is then blow molded from the emerging parison. Some food containers may have seven layers for long shelf life. A common food container consists of PP/adhesive/ EVOH/ adhesive/PP (Figure 11-45).

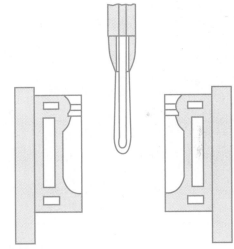

(A) Injection-molded preformed parison.

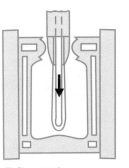

Figure 11-42. Extrudates of two different colors are pinched between mold halves to form a sheet-blown two-color part.

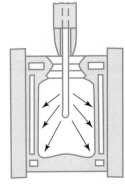

(B) Clamp-stretch.

(C) Blow-cool.

Figure 11-43. The preform is stretched by action of the rod and the air pressure. This biaxial stretching improves properties.

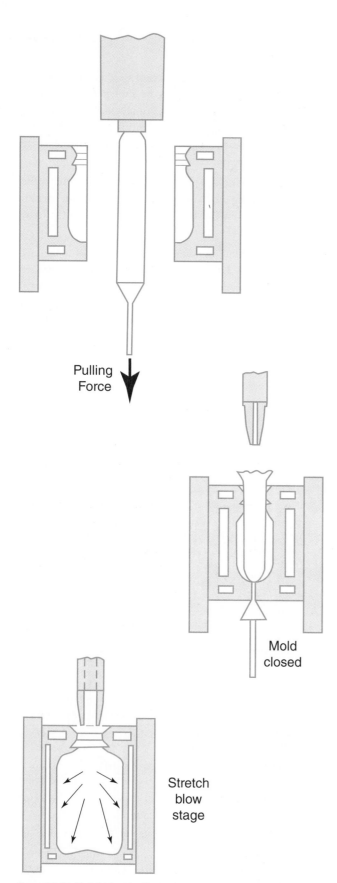

Figure 11-44. Biaxially oriented product is produced by pulling the extruded tube, then blowing.

Pulling Force

Mold closed

Stretch blow stage

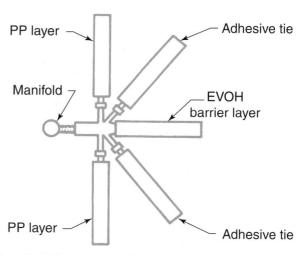

PP layer

Manifold

PP layer

Adhesive tie

EVOH barrier layer

Adhesive tie

Figure 11-45. Five extruders are used to produce a multilayered parison (for blow-molded parts) or bubble (for films).

RELATED INTERNET SITES

- **www.battenfeld-cincinnati.com.** Battenfeld-Cincinnati sells extrusion equipment focused on pipe and tube extrusions. If the website comes up in German, select English on the top of the screen. Selecting "products" leads to a list of the types of extruders available. Choosing "Twin screw extruders for profile, sheet, WPC/NFX." Then select "fiberEX" to a section on wood plastic composite extrusion of WPC. The wood fibers and plastics extruder has a capacity of up to 1000 kg per hour.

- **www.davis-standard.com.** Davis-Standard manufactures extruders and complete extrusion systems. The site leads to information about small extruders for laboratory settings, small to large single-screw extruders, and small to large twin-screw extruders. Entire systems focus on film, pipe and profile, extrusion coating, film blowing, and blow molding.

- **www.dupont.com.** In addition to selling thermoplastics and elastomers, DuPont also makes and sells fibers, woven, and non-woven fabrics. On the home page, enter the name of one of their fiber or non-woven products. These names are Kevlar®, Nomex®, Sorona®, Tychem®, and Tyvek®. The website will provide details on these fibers and materials.

VOCABULARY

The following vocabulary words are found in this chapter. Use the glossary in Appendix A to look up the definitions of any of these words you do not understand as they apply to plastics.

> back pressure
> Banbury mixer
> barrel mixer
> biaxial blow molding
> blow molding
> breaker plate

calender or calendering
compounding
denier
downstream feeding
drawing
extrusion
fiber
filament
film
film casting
frost line

melt spinning
parison
pin
profile extrusions
solvent spinning
spiders
spinneret
stretch blow molding
tex
transfer blow
wet spinning

QUESTIONS

11-1. The hollow plastics tube used in blow molding is called a _____.

11-2. The process that forces hot plastics through machine dies to form continuous shapes is called _____.

11-3. Which molding process would be used to make long sections of plastics pipe?

11-4. Name three extrusion-spinning processes.

11-5. The process of stretching a thermoplastic sheet, rod, or filament to reduce its cross-sectional area and change its physical traits is called _____.

11-6. Because of molecular orientation, _____ extrusion film is generally stronger than _____ film.

11-7. The extruder screw moves the plastics materials through the _____ of the machine.

11-8. The _____ in the extruder helps support the screen pack and creates additional mixing of the plastics before it leaves the die.

11-9. The general term for the product or material delivered by an extruder is _____.

11-10. Name the two methods used to make films by the extrusion process.

11-11. The _____ line is the name of the zone that appears frosty in the extrusion blown-film process.

11-12. Blow molding is a refinement of the ancient _____ processes.

11-13. The two basic blow-molding methods are _____ blowing and _____ blowing.

11-14. In blow molding, a die with a variable orifice may be used to control _____ and thickness.

11-15. In the _____ process, hot material is pressed between two or more rotating rollers.

11-16. Patterns or textures may be placed on calendered films or sheets by passing the soft, hot plastics between _____ rolls.

11-17. Spinnerets are used to produce _____ and _____ plastics forms.

11-18. Films are less than _____ mm in thickness.

11-19. Sheets are greater than _____ mm in thickness.

ACTIVITIES

Extrusion Compounding

Introduction. Extrusion compounding is the process of mixing various ingredients into a base polymer to produce a plastics material with desired properties.

Equipment. An extruder with pelletizer, polypropylene, chopped glass fibers, accurate scale, small furnace, small tongs, Pyrex dish or ceramic crucible, safety glasses, insulated gloves, and special magnifying glass or microscope.

Procedure

11-1. Check the extruder to make sure the rupture disk is good. Most extruders have a rupture disk in a hollow threaded

fastener. The disk is located near the front end of the barrel. Rupture disks blow out if the pressure near the die is too great. This protects the bolts holding the die in place. Extreme pressure can break off the bolts holding the die or cause their threads to shear. This can be extremely dangerous. To provide a release for extreme pressure, the rupture disk fails and the plastics flow out through the hollow fastener. Figure 11-46 shows a ruptured disk. Disks fail at varying pressures. The example shown is a 5000 psi disk.

11-2. If the extruder is dirty, take off the die and push out the extruder screw. With many laboratory-sized machines, it is possible to push out the screw with a rod, provided the barrel is at processing temperature. Use a copper mesh,

Figure 11-46. This rupture disk will fail at 5000 psi.

Figure 11-47. This copper mesh is used to clean extruder screws of hot plastics.

as seen in Figure 11-47, to clean the screw. This kind of mesh is supplied in a roll. To use it, cut a portion about 3 feet long, wrap it around the screw once or twice, and work it back and forth to remove hot plastics. Clean a small portion, then push out another section and clean it. Pushing out a long section may allow enough time for the material in the screw to stiffen before removal.

11-3. Measure the length of the glass fibers before processing.

11-4. Calculate the amount of glass required for batches of 2, 4, 6, 8, 10, and 12 percent glass. A batch of 1.0 kg is usually sufficient to produce many test samples. Measure the glass and propylene and hand mix them to ensure uniform distribution of the glass. Does the mixture tend to separate into glass-rich and glass-poor layers? How can this be avoided?

11-5. Compound the glass into the polypropylene. Select a temperature of about 204°C (400°F) at the die and

front zone. If the L/D ratio of the extruder is low, two passes through the extruder may be needed for adequate dispersion of the glass. Lower percentages should compound rather easily. As the glass content increases, the tendency to surge caused by non-uniform feeding into the screw may increase. If the extruder has a screw no more than 25 mm (1.0 in.) diameter, the glass may hinder the feeding. To reduce this problem, agitate the material in the feed throat with a soft tool.

11-6. After compounding, carefully weigh a crucible or Pyrex container and then measure 5 to 10 g (0.17–0.35 oz) of pellets. Use insulated gloves and small tongs to place the crucible in a furnace. Burn off the propylene in the furnace at a temperature of 482°C to 538°C (900°F to 1000°F). A small heat-treating furnace is appropriate for burnouts. Make sure ventilation is adequate to remove the smoke and fumes produced by the burning polypropylene.

11-7. When the container has cooled, carefully measure the container and glass. Calculate the percentage of glass content. Does the calculated percentage match the percentage of glass in the original batch? If the percents do not match, why is there an error?

11-8. Figure 11-48 shows a specialty magnifying glass with various scales printed on the stages. Such magnifiers are available in many industrial supply catalogs. Select an appropriate scale, and measure the length of the glass fibers. How much shorter are they after compounding than before?

11-9. To compare the relative amounts of glass in various batches, select one pellet from each batch. Arrange them on a microscope slide in order of glass content. See Figure 11-49 for an example of such an arrangement. After burning off the polypropylene, the results give visual comparison of varying glass contents (Figure 11-50). Examine the glass under a microscope.

Figure 11-48. This magnifying glass with graduated stages provides quick measurement of fiber lengths.

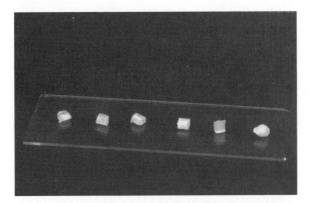

Figure 11-49. These pellets are arranged in order for increasing glass content.

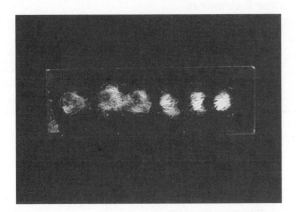

Figure 11-50. After burnout, the remaining glass fibers show changes in percent of glass content.

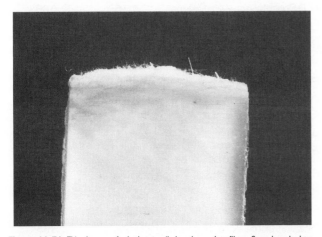

Figure 11-51. This close-up of a broken tensile bar shows glass fibers. Some have broken during testing, others pulled out without breaking.

11-10. Mold test bars and tensile test them. Figure 11-51 shows a close-up of the broken edge of a tensile bar. Notice the glass fibers sticking out of the bar.

11-11. Compare test results to determine how great an effect on strength and elongation the levels of glass have.

11-12. Write a report summarizing the results.

Profile Extrusion

Introduction. Many laboratory extruders do not include required sizing fingers, cooling apparatus, and takeoff equipment needed to support profile extrusion. However, output calculations on strand dies offer the opportunity to determine the efficiency of the extruder under various conditions.

Equipment. Extruder, accurate scale, and selected plastics.

Procedure

11-1. Clean the extruder screw and remove it.

> **! CAUTION**
>
> Use insulated gloves when handling the hot screw. Never use steel tools on the screw. If direct gripping of the screw is required, use brass tools.

11-2. Figure 11-52 shows a single-stage extruder screw. Collect measurements of flighted length, screw diameter, length of feed, transition, and metering zones, as well as flight depth in feed and metering zones.

11-3. Set up the extruder, and extrude enough material for the machine to arrive at a rather stable condition.

11-4. Starting near the top of the rpm capabilities of the equipment, catch the extrudate on a piece of cardboard or sheet metal for a period of 30 seconds.

11-5. Reduce the rpm by 10, and repeat step 2. Continue this process until reaching a low rpm, perhaps 10 or 20.

11-6. When the extrudate samples have cooled, accurately weigh them. Convert the results into output represented as pounds per hour.

11-7. Determine the compression ratio of the screw by dividing the depth of the feed zone by the depth of the metering zone.

11-8. Determine the L/D ratio by dividing the flighted length of the screw by its outside diameter.

11-9. Determine the theoretical output of the screw with this formula:

$$R = 2.2 \, D^2 h g N$$

R = output in pounds per hour
D = screw diameter in inches
h = depth of metering zone in inches
g = melt density
N = screw rpm

11-10. To determine the melt density, refer to Table 11-6.

11-11. Compare the theoretical output to the measured output.

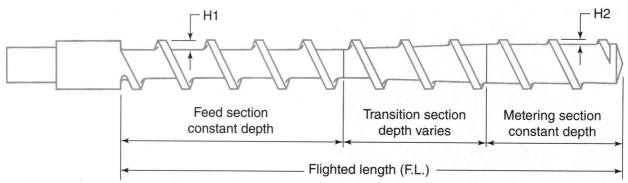

Figure 11-52. Schematic of a single-stage extruder screw.

Table 11-6. Material Data for Extrusion

Material	Melt Density Specific Gravity Grams/cm³ at Processing Temperature	Extrusion Temperature °F	Material	Melt Density Specific Gravity Grams/cm³ at Processing Temperature	Extrusion Temperature °F
ABS—Extrusion	0.88	435	HD Polyethylene—Extrusion	0.72	410
ABS—Injection	0.97		HD Polyethylene—Injection	0.72	
Acetal—Injection	1.17		HD Polyethylene—Blow Molding	0.73	410
Acrylic—Extrusion	1.11	375			
Acrylic—Injection	1.04		LD Polyethylene—Film	0.77	350
CAB	1.07	380	LD Polyethylene—Injection	0.76	
Cellulose Acetate—Extrusion	1.15	380	LD Polyethylene—Wire	0.76	400
Cellulose Acetate—Injection	1.13		LD Polyethylene—Ext. Coating	0.68	600
Cellulose Proprionate—Extrusion	1.10	380	LLD Polyethylene—Extrusion	0.75	500
Cellulose Proprionate—Injection	1.10		LLD Polyethylene—Intrusion	0.70	
CTFE	1.49		Polypropylene—Extrusion	0.75	450
FEP	1.49	600	Polypropylene—Intrusion	0.75	
Ionomer—Extrusion	.73	500	Polystyrene—Impact Sheet	0.96	450
Ionomer—Injection	0.73		Polystyrene—G. P. Crystal	0.97	410
Nylon 6	0.97	520	Polystyrene—Injection Impact	0.96	
Nylon 6/6	0.97	510	Polysulfone	1.16	650
Nylon 6/10	0.97		Polyurethane (non-elastomer)	1.13	400
Nylon 6/12	0.97	475	PVC—Rigid Profiles	1.30	365
Nylon 11	0.97	460	PVC—Pipe	1.32	380
Nylon 12	0.97	450	PVC—Rigid Injection	1.20	
Phenylene Oxide Based	0.90	480	PVC—Flexible Wire	1.27	365
Polyallomer	0.86	405	PVC—Flexible Extruded Shapes	1.14	350
Polyarylene Ether	1.04	460			
Polycarbonate	1.02	550	PVC—Flexible Injection	1.20	
Polyester PBT	1.11		PTFE	1.50	
			SAN	1.00	420
Polyester PET	1.10	480	TFE	1.50	
			Urethane Elastomer (TPE)	0.82	390

LAMINATING PROCESSES AND MATERIALS

CHAPTER 12

INTRODUCTION

This chapter discusses laminates and the processes required for their formation. Before comparing types of laminates and various production techniques, understanding basic definitions is essential.

The verb laminate describes the process of bonding two or more layers of material by cohesion or adhesion. In the plastics industry, laminates contain layers held together by a plastics material. The layers bonded together often provide strength and reinforcement. This makes it difficult to distinguish clearly between laminates and reinforced plastics. One key characteristic of laminates is that they contain layers. In contrast, reinforced plastics generally get their strength from fibers contained within the plastics. Laminates generally consist of flat sheets, whereas reinforced plastics are often molded into complex shapes.

Laminates have numerous uses. They appear as structural materials in automobiles, furniture, bridges, and homes. They are also very important in aircraft structures, helicopter blades, and aerospace applications (Figures 12-1).

Laminates include metallic foils bonded to paper or fabric. In the textile field, layers of cloth, plastics foams, and films are bonded into special-purpose fabrics.

To bring some order to the diverse applications of laminates, the sections in this chapter are organized according to the materials that form the laminate layers:

I.   **Layers of differing plastics**

II. **Layers of paper**

III. **Layers of glass cloth or mat**

IV. **Layers of metal and metal honeycomb**

V.   **Layers of metal and foamed plastics**

Lockheed Martin Aeronautics

Figure 12-1. The exterior skin of the F-35 Lightning II consists of carbon fiber composites.

Table 12-1. **Selected Resins/Plastics and Materials Used in Lamination**

Resin/Plastics	Paper	Cotton Fabric	Fibrous Glass Fabric/Mat	Metallic Foils	Composites, Honeycomb, etc.
Acrylic	LP	—	LP		—
Polyamide	LP	—	LP	LP	—
Polyethylene	LP	—	LP	LP	—
Polypropylene	LP	—	LP	LP	—
Polystyrene	—	—	LP		
Polyvinyl chloride	LP	—	—	LP	LP
Polyester	LP-HP	LP	LP	—	LP
Phenolic	LP-HP	LP-HP	HP-LP	HP	LP-HP
Epoxy	LP	LP-HP	LP-HP	LP	LP
Melamine	HP	HP		HP	
Silicone	—	—	HP	—	LP

Notes: LP = Low-pressure (laminate).
HP = High-pressure (laminate).
— = Only limited amounts manufactured in this category.

Note that this list does not include all laminates. Some laminates, bonded with plastics, do not receive attention. One example is plywood. Table 12-1 provides basic data on the types of plastics used in many laminates.

LAYERS OF DIFFERING PLASTICS

Although it is possible to start with finished film or sheet when making laminates of various plastics, it is usually more economical to extrude the various layers simultaneously. Engraving stock that contains two or more different colors is an example of continuous extrusion laminating. Engraving stock demonstrates the capability of extrusion laminating to make rather thick products.

Extrusion-laminated films are commonly used in the packaging industry. Co-extruded composite films of polyethylene and vinyl acetate produce a tough, durable, two-layer film that may be heat sealed. The plastics layers are brought together in the molten state and extruded through a single die opening to make a multilayered laminate film (Figures 12-2 and 12-3).

A three-ply laminate film is used to wrap bread products. This film laminate is composed of an inner core (ply) or polypropylene with two outer layers (plies) of polyethylene (Figure 12-4).

Multilayer films are used extensively in food packaging. Some foods are sensitive to oxygen and may discolor or spoil rapidly. To reduce these problems, oxygen barrier layers are incorporated into packaging films. Some examples of food packaging are seen in Figures 12-5A, 12-5B, and 12-5C.

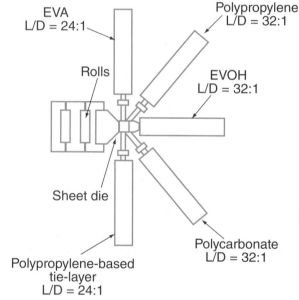

(A) Five extruders are used to produce a true composite laminate.

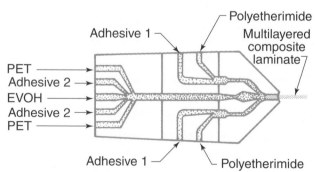

(B) A cross section of this sheet die shows that at least two extruders were used for the adhesives, (1 and 2), one for the outside polyetherimide layers, one for the two PET layers, and one for the EVOH center barrier layer.

Figure 12-2. Multilayer extrusion equipment.

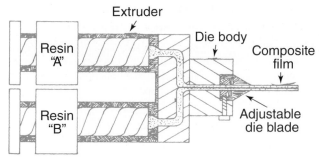

Figure 12-3. Production of co-extruded film.

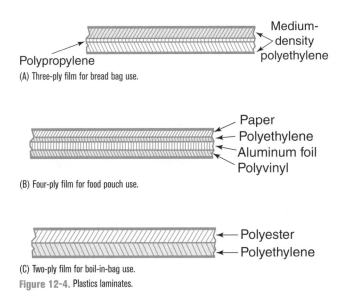

(A) Three-ply film for bread bag use.

(B) Four-ply film for food pouch use.

(C) Two-ply film for boil-in-bag use.

Figure 12-4. Plastics laminates.

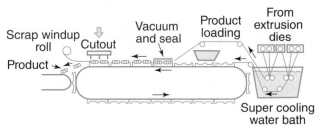

(A) Extrusion and packing.

(B) Bread loaves often require three-ply film.

(C) Packaging for meat and nuts require multilayer films with oxygen barriers.

Figure 12-5. Extrusion technique and product examples.

In another application, two different polymers are blow-film co-extruded with molecular orientation and then squeezed together into a laminate sheet (see Figure 12-6).

There are numerous applications of extruded film laminates. A few examples are listed in Table 12-2.

LAYERS OF PAPER

Paper appears in laminates in two distinct forms—as a non-impregnated layer or as a thoroughly impregnated material.

When paper is not impregnated, one or more layers of plastic—generally an olefinic film—are adhered to the paper with heat and pressure. The goal is to create a glossy water-resistant finish. The cover of this book is an example of this application. Compare the inside of the cover to the outside. The glossy finish comes from a plastics film.

Other products exemplifying this process are ID cards and drivers' licenses. Because the pressure required to bond film to the paper is rather low, these laminates are often called **low-pressure laminates**.

In contrast, impregnated paper laminates are often called **high-pressure laminates**. Some laminates are processed at pressures exceeding 7000 kPa (1015 psi). Many high-pressure laminates contain thermosetting resins—particularly urea, melamine, phenolic, polyester, and epoxy.

J. P. Wright, the founder of the Continental Fiber Company, played a major role in the history of high-pressure laminates. In 1905, he produced one of the earliest phenolic laminates. This was five years before Baekeland had patented the idea of using sheets impregnated with phenolic resins as laminates. Using this basic patent, many high-pressure industrial

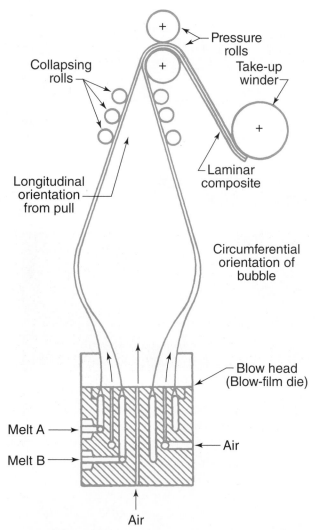

Figure 12-6. Co-extruded film is stretched and oriented in different directions and then pressed together to form a composite laminar sheet.

Table 12-2. Selected Extruded Film Laminates and Applications

Laminate Material	Application
Paper-polyethylene-vinyl	Sealable pouches for dried milk, soups, etc.
Acetate-polyethylene	Tough, heat-sealable packing for nuts
Foil-paper-polyethylene	Moisture barrier pouch for soup mixes, dry milk, etc.
Polycarbonate-polyethylene	Tough, puncture-resistant skin packages
Paper-polyethylene-foil-polyethylene	Strong heat seals for dehydrated soups
Paper-polyethylene-foil-vinyl	Heat-sealable pouches for instant coffee
Polyester-polyethylene	Touch, moistureproof boil-in-bag pouches for foods
Cellulose-polyethylene-foil-polyethylene	Gas and moisture barrier for pouches of ketchup, mustard, jam, etc.
Acetate-foil-vinyl	Opaque, heat-sealing pouches for pharmaceuticals
Paper-acetate	Glossy, scratch-resistant material for record covers, paperback books

laminates were made from a layup of paper, cloth, asbestos, synthetic fiber, or fibrous glass. Today, there are over 50 standard industrial grades of laminates for electrical, chemical, and mechanical uses.

At first, high-pressure industrial laminates took the place of mica as a quality electrical insulation material. Around 1913, the Formica Corporation emerged and made high-pressure laminates that replaced mica (*for mica*) in many electrical and mechanical uses. By 1930, the National Electrical Manufacturers Association (NEMA) saw the potential of decorative laminates. In 1947, a separate NEMA section was established to work with government agencies and associations interested in setting up product standards for decorative laminates.

The basic process used was to fuse layers of paper that had been impregnated with phenolic resin. Decorative and transparent exterior layers were also fused onto the paper. To increase the production rate, many laminates were made at the same time by stacking a press. Figures 12-7 and 12-8 show this procedure.

Impregnating the paper prior to lamination was done by a number of methods. The most common are premix, dipping, coating, or spreading. Figure 12-9 shows these various methods.

After a drying period, the impregnated laminating stock is cut to the desired size and placed in multi-sandwich form between metal plates (separation plates) of the press. Press platens are not smooth enough for the desired finish. These plates may be glossy, matte, or embossed. Metal foils are sometimes used between surface layers to produce a decorative finish. In decorative laminates, a printed pattern layer and a protective overlay sheet are superimposed on the base material. The prepared stock is subjected to heat and high pressure. The combination of heat and pressure causes the resin to flow and the layers to compact into one polymerized mass. Polymerization may also be achieved through chemical or radiation sources. When the thermosetting resins have cured or the thermoplastic resins have cooled, the laminate is removed from the press.

Today, high-pressure laminates enjoy wide use. Formica® and Wilsonart® are brands of laminate often selected for

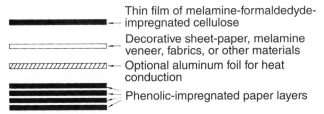

Thin film of melamine-formaldedyde-impregnated cellulose

Decorative sheet-paper, melamine veneer, fabrics, or other materials

Optional aluminum foil for heat conduction

Phenolic-impregnated paper layers

Figure 12-7. Typical arrangement of layers in a decorative highpressure laminate.

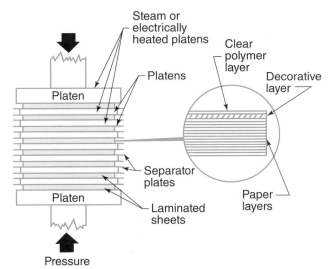

Figure 12-8. Concept of multiple stacking of laminates in press.

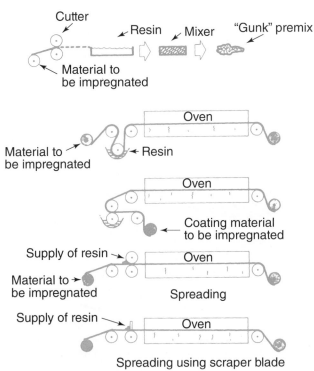

Figure 12-9. Various impregnation methods.

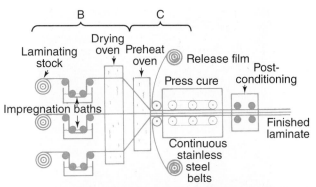

(A) Continuous lamination, in which the (B-stage) resin is changed into an infusible (C-stage) plastics.

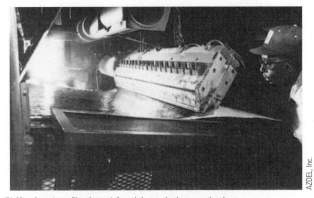

(B) Manufacturing a fiberglass-reinforced thermoplastic composite sheet.

Figure 12-10. Continuous lamination of thermosetting and thermoplastic matrix composite sheets.

countertops. In addition to sheets of laminate, these materials appear as cams, pulleys, gears, fan blades, and printed circuit boards.

A major disadvantage of high-pressure lamination is the slow production rates encountered compared to high-speed injection molding. In order to speed production, some manufacturers have established continuous laminating processes. In *continuous laminating* (Figure 12-10), fabrics or other reinforcements are saturated with resin and passed between two plastics film layers such as cellophane, ethylene, or vinyl.

The thickness of the laminated composite is controlled by the number of layers it consists of and by a set of squeeze rollers. The laminate is then drawn through a heating zone to speed polymerization. Corrugated awnings, skylights, and structural panels are products of continuous laminating.

LAYERS OF GLASS CLOTH OR MAT

Thermosetting resins are used in *hand layup laminates* (Figures 12-11 and 12-12). These products are also called *contact moldings* or *open moldings*. After the mold (either male or female)

(A) A composites specialist cuts pre-impregnated Kevlar with the aid of a laser alignment jig.

(B) A composites specialist applies a pre-impreganted graphite panel with the aid of a laser alignment and then burnishes it into place with a nylon burnishing tool.

Figure 12-11. Hand layup of pre-impregnated Kevlar and graphite.

is coated with a releasing agent, a layer of catalyzed resin is applied and allowed to polymerize to the gel (tacky) state.

This first layer is a specially formulated gel-coat resin used in industry to improve flexibility, blister resistance,

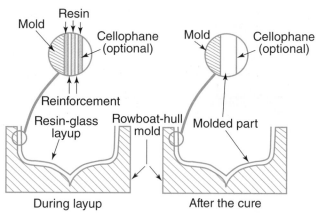

Figure 12-12. In hand layup, reinforcing material in mat or fabric form is applied to the mold, then saturated with a selected thermosetting resin.

surface finish, or color and stain resistance. The gel coat forms a protective surface layer through which fibrous reinforcements do not penetrate. A prime cause of deterioration of fibrous reinforced plastics is the penetration of water that takes place when fibers protrude at the surface. Once the gel coat has partially set, reinforcement is applied. Next, more catalyzed resin is poured, brushed, or sprayed over the reinforcement. This sequence is repeated until the desired thickness is reached. In each layer, the mixture is worked into the mold shape with hand rollers. Then the reinforced composite laminate is allowed to harden or cure (Figure 12-13). External heating is sometimes used to speed polymerization.

Hand layup and sprayed operations are often used alternately. In order to obtain a resin-rich, superior surface finish, a lightweight veil or surface mat is sometimes placed next to the gel layer. Coarser reinforcements are then placed over this layer.

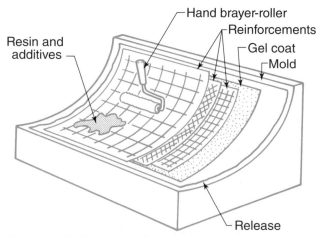

Figure 12-13. Hand layup implies that little equipment is required. In all layup operations, it is important to bray out (remove) any bubbles trapped between the layers.

Some operations and designs use preforms, cloth, mat, and roving materials for directional or additional strength in selected areas of the part.

Among the major advantages of hand layup are low-cost tooling, minimal requirement of equipment, and the ability to mold large components. As for disadvantages, the process is labor intensive and dependent upon the skill of the operator, resulting in a low production rate. Also, this process is messy and exposes workers to hazardous chemicals.

LAYERS OF METAL AND METAL HONEYCOMB

Laminates consisting of metal-face layers with lightweight cores are often called sandwiches. Core types include solid, corrugated paper, cellular plastics, and metallic or plastics honeycombs (Figure 12-14). As a rule, honeycomb, waffle, and cellular sandwiches are isotropic and possess excellent thermal, acoustical, and strength-to-weight ratios. The outer layers must be strong to carry axial and in-plane shear loading. Most of the tensile and compressive forces are transferred to these layers. The core material transfers loads from one facing to the other.

All properties, including thermal and electrical, depend on the selection of facings, cores, and bonding agents. Adhesive bonding is critical if shear and axial loads are to be transmitted to and from the core material. Polyimide, epoxy, and phenolics are commonly used. Resin-impregnated fiber matting, cloth, and paper are adhesive films that may be used in the core-to-face bond.

Honeycomb core materials made of resin-impregnated kraft paper, aluminum, glass-reinforced polymers, titanium, and other materials are among the strongest core structures for their mass. Aluminum is the most commonly used honeycomb core material. All honeycombs are anisotropic, and properties depend on composition, cell size, and geometry. Two major methods of producing honeycomb core materials are illustrated in Figures 12-15 and 12-16.

A unique characteristic of the honeycomb manufactured using the expansion process is that it can be machined prior to expansion. Figure 12-17 shows unexpanded aluminum honeycomb. At this stage, it could be machined into the shape of an airfoil. After machining, it can be stretched or expanded, as shown in Figure 12-18.

The properties of selected aluminum honeycomb materials are shown in Table 12-3. Properties of several glass-reinforced plastics honeycombs are shown in Table 12-4.

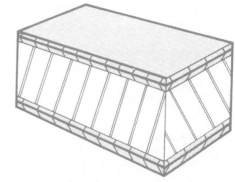

(A) Basic design.

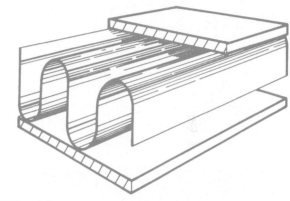

(B) Corrugated paper core.

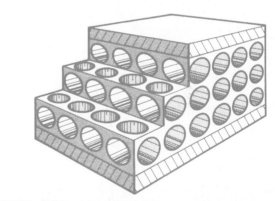

(C) Cellular plastics core.

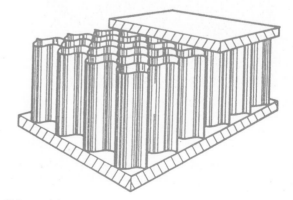

(D) Honeycomb core.

Figure 12-14. Various types of core construction.

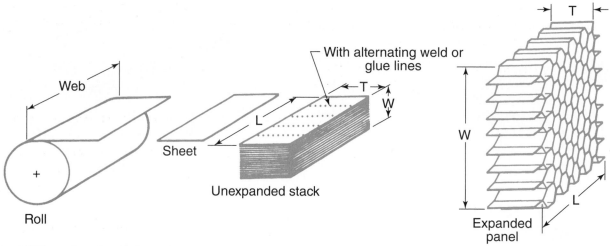

Figure 12-15. Honeycomb manufactured by the expansion process.

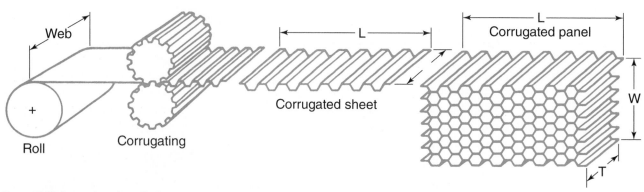

Figure 12-16. Honeycomb manufactured by the corrugation process.

Figure 12-17. Unexpanded aluminum honeycomb.

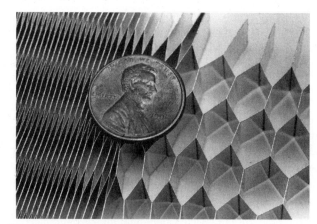

Figure 12-18. Aluminum honeycomb at various stages of expansion.

Table 12-3. **Properties of 5056, 5052, and 2024 Hexagonal Aluminum Honeycomb**

Honeycomb Cell-Material-Gauge	Nominal Density, kg/m³	Compressive			Plate Shear			
		Bare	Stabilized		"L" Direction		"W" Direction	
		Strength, kPa	Strength, kPa	Modulus, MPa	Strength, kPa	Modulus, MPa	Strength, kPa	Modulus, MPa
5056 Hexagonal Aluminum Honeycomb:								
1/16–5056–0.0007	101	6894	7584	2275	4447	655	2551	262
1/16–5056–0.001	144	11,721	12,410	3447	6756	758	4136	344
1/8–5056–0.0007	50	2344	2482	668	1723	310	1068	137
1/8–5056–0.001	72	4343	4619	1275	2930	482	1758	262
5/32–5056–0.001	61	3275	3447	965	2310	393	1413	165
3/16–5056–0.001	50	2344	2482	669	1758	310	1069	138
1/4–5056–0.001	37	1413	1448	400	1172	221	724	103
5056 Alloy Hexagonal Aluminum Honeycomb:								
1/16–5052–0.0007	101	5998	6274	1896	3516	621	2206	276
1/8–5052–0.0007	50	1862	1999	517	1448	310	896	152
1/8–5052–0.001	72	3585	3758	1034	2344	483	1517	214
5/32–5052–0.0007	42	1379	1482	379	1138	255	689	131
5/32–5052–0.001	61	2723	2827	758	1862	386	1207	182
3/16–5052–0.001	50	1862	1999	517	1448	310	896	152
3/16–5052–0.002	91	5309	5585	1517	3172	621	2068	265
1/4–5052–0.001	37	1138	1207	310	965	221	586	113
1/4–5052–0.004	127	9377	9791	2344	4826	896	3034	364
3/8–5052–0.001	26	586	655	138	586	145	345	76

Table 12-4. **Properties of Several Glass-Reinforced Plastics Honeycombs**

Honeycomb Material-Cell-Density	Compressive			Plate Shear			
	Bare	Stabilized		"L" Direction		"W" Direction	
	Strength, kPa	Strength, kPa	Modulus, MPa	Strength, kPa	Modulus, MPa	Strength, kPa	Modulus, MPa
Glass-Reinforced Polyimide Honeycomb:							
HRH 327–3/16–4.0		3033	344	1930	199	896	68
HRH 327–3/16–6.0		5377	599	3171	310	1585	103
HRH 327–3/8 –4.0		3033	344	1930	199	1034	82
Glass-Reinforced Phenolic Honeycomb (Bias Weave Reinforcement):							
HFT–1/8–4.0	2688	3964	310	2068	220	1034	82
HFT–1/8–8.0	9997	11,203	689	3964	331	2344	172
HFT–3/16–3.0	1896	2585	220	1378	165	689	62

(Continued)

Table 12-4. **Properties of Several Glass-Reinforced Plastics Honeycombs** (*Continued*)

Honeycomb Material-Cell-Density	Compressive			Plate Shear			
	Bare	Stabilized		"L" Direction		"W" Direction	
	Strength, kPa	Strength, kPa	Modulus, MPa	Strength, kPa	Modulus, MPa	Strength, kPa	Modulus, MPa
Glass-Reinforced Polyester Honeycomb:							
HRP–3/16–4.0	3447	4137	393	1793	79	965	34
HRP–3/16–8.0	9653	11,032	1131	4551	234	2758	103
HRP–1/4 –4.5	4344	4826	483	2068	97	1172	41
HRP–1/4 –6.5	7076	8136	827	3103	172	793	76
HRP–3/8 –4.5	4205	4757	448	2068	97	1172	41
HRP–3/8 –6.0	6205	6895	689	2758	155	1793	69

LAYERS OF METAL AND FOAMED PLASTICS

Sandwich panels with foamed plastics as the core have wide application in the construction industry. They are used for both insulation and structural components.

Other uses for cellular core sandwiches are refrigerator liners for truck boxes, railcars, and food coolers, and exterior panels for mobile homes. Other applications include doors and construction panels.

In a process called foam reservoir molding or elastic reservoir molding, open-celled polyurethane foam is impregnated with epoxy. The two skin layers are then pressed against the spongy core, forcing some of the epoxy adhesive to adhere to the two face skins. The foam and matrix become a catacomb-like, skeletal structure.

RELATED INTERNET SITES

- **www.thegillcore.com.** Alcore is a part of the M.C. Gill Corporation, which manufactures a variety of metallic honeycombs. Selection of "Products" on the home page will lead to overviews of their main honeycomb products and links to detailed data sheets about the properties of each product.

- **www.hexcel.com.** Hexcel is the largest US producer of carbon fibers. It also sells reinforcements for composites, matrix materials, adhesives, core and honeycomb materials, and structural products.

- **www.kpfilms.com.** The Klöckner Pentaplast Group is one of the world's leading makers of plastics films. Three areas of this site are of particular interest. First, under "KP Advantage" on the home page is a selection "Production Processes" that discusses the various ways to make films. Second, under

"Products & Solutions," look for a selection marked "Food Packaging Films." This will lead to "Pentafood® Food Packaging Films." This provides information on single layer and multi-layer films. Third, under "News" on the home page, select "Case Studies" to find articles on multilayer applications in packaging.

- **www.mastercraft.com.** MasterCraft Boats Company offers a site with elaborate sound and video effects. Selecting " Resources" on the home page leads to "Factory Tour," which provides information about the construction process.

- **www.plascore.com.** Plascore Inc. makes both metallic and nonmetallic honeycomb core materials. Selecting "Products" and then "Honeycomb core" on the home page leads to a choice of "Thermoplastic." Details are then available about polypropylene and polycarbonate cores and their characteristics.

- **www.texasalmet.com.** This site features metallic and nonmetallic honeycomb materials. It also includes information on the machining of honeycomb materials for use in aircraft construction and repair.

VOCABULARY

The following vocabulary words are found in this chapter. Use the glossary in Appendix A to look up the definitions of any of these words you do not understand as they apply to plastics.

elastic reservoir molding
foam reservoir molding
high-pressure laminates
honeycomb
laminate
low-pressure laminates
sandwiches

QUESTIONS

12-1. Two major disadvantages of high-pressure lamination are low _____ rates and high pressures.

12-2. The process in which two or more layers of materials are bonded together is called _____.

12-3. If there are unfavorable interlaminar stresses, what may occur? How can this be effectively prevented?

12-4. What are the major applications for high-pressure laminates?

12-5. How are extruders used to produce laminates? Are calenders used?

12-6. What properties are favorable for applications using honeycomb laminated components?

12-7. Name four honeycomb core materials, and describe the merits of each in a particular application.

12-8. Define a laminated plastics, and describe how several products may be formed.

12-9. Defend the selection of sandwich construction in house doors, airplane components, and cargo containers.

12-10. Describe the process of continuous laminating, and include the type of materials used and typical product applications.

ACTIVITIES

Low-Pressure Thermoplastic Lamination

Introduction. Low-pressure, thermoplastic laminating is a process that bonds two sheets of thermoplastics around a selected paper or card item. Heat softens the plastics, and pressure causes it to flow around the item. The edges of the plastics sheets bond together thermally.

Equipment. Lamination press, polished plates, blotter cushions, and PVC or cellulose acetate sheets.

Procedure

12-1. Connect water-cooling hoses and adjust platen to 175°C (347°F).

12-2. Trim thermoplastic sheets 5 mm larger on all sides of the item to be laminated.

12-3. Assemble a laminating sandwich as follows:

1 top holding plate
2 top cushion blotters
1 polished plate
1 plastics sheet
1 item to be laminated
1 plastics sheet
2 cushion blotters
1 bottom holding plate

12-4. Place the sandwich in the press. Close and apply about 37 MPa (5366 psi) to laminate card items or about 20 MPa (2091 psi) for paper items. Use slightly lower temperature and pressure for PVC sheets.

12-5. To calculate the pressure needed for lamination, use 40,000 kg/m² as a minimum pressure.

12-6. Allow sandwich and plastics laminates to heat for 5 minutes. If multiple sandwiches are pressed, increase the heating time (see Figure 12-19).

12-7. When heating is complete, shut off electric power to heaters, and slowly turn on water for cooling cycle.

> **! CAUTION**
>
> Steam and return hoses are hot. Cool the platens until they reach 40°C (104°F).

12-8. Release the hydraulic pressure and remove the sandwich. Flex sandwich to release the laminate. Do not use a screwdriver to pry the sandwich apart. Any damage to the polished plate will be transferred to the next laminate.

12-9. Examine for bond, color bleeding, air bubbles, delamination, or surface damages.

Figure 12-19. Sandwich assembly ready for removal from compression press.

High-Pressure Lamination with Impregnated Papers

Introduction. In high-pressure laminating, the reinforcing substance is generally paper. However, cloth, wood, or glass fabric may be impregnated with fusible B-stage resin. During the heat and pressure cycle, the impregnated sheets are fused together.

Equipment. Heated platen press; decorative sheet, melamine resin-impregnated; kraft papers, phenolic resin-impregnated; and press plates.

Procedure

12-1. Connect water cooling hoses and adjust platen heater to 175°C.

12-2. Trim sheets to 110 × 110 mm square. After laminating, sheets will be finished trimmed to 100 mm × 100 mm.

12-3. Assemble a sandwich as follows:

   1 top holding plate
   2 cushion blotters
   1 overlay sheet
   1 decorative sheet
   4 or more layers of kraft paper
   1 polished plate
   2 cushion blotters
   1 bottom holding plate

(See Figure 12-20.)

12-4. Place the sandwich in the press and apply about 200 MPa (29,101 psi) of pressure. In calculating pressure requirements, use 245,000 kg/m² (35.6 psi) as a minimum for high-pressure lamination.

12-5. Heat the sandwich for 10 minutes. Pressure will decay as sheets deform. Maintain pressure during the molding cycle. If multiple sandwiches are pressed, increase heating time.

Figure 12-20. Sandwich assembly for high-pressure lamination.

12-6. After the heating cycle, shut off electrical power to the heaters, and slowly turn on water for the cooling cycle.

 CAUTION

Steam and return hoses are hot. Cool platens until they reach 40°C (104°F).

12-7. Release hydraulic pressure and remove the sandwich. Flex the sandwich to release the laminate. Do not use a screwdriver to pry apart.

12-8. Trim laminate and examine for delamination and surface defects.

High-Pressure Lamination without Impregnated Papers

Introduction. If impregnated kraft paper and melamine-impregnated decorative overlays are unavailable, the following exercise can provide experience with thermoset laminates.

Equipment. Heated platen press, pieces of smooth sheet metal (equal in size to the platens), phenolic resin in powder form, mold release, and paper toweling.

Procedure

12-1. Cut several pieces of paper toweling to about the size of the sheet metal plates. Paper toweling is porous enough that the resin can flow in and through it. Less porous papers reduce resin flow and may tear rather than permit resin flow.

12-2. Lay down one layer of paper, and then measure out about ¹/₂ cup of powdered phenolic resin. Cover the resin with another layer of paper, and place a top sheet metal plate on the paper.

12-3. Place the sandwich in a heated platen press and exert pressure. The hydraulic pressure pointer will drop during pressurization. This indicates flow of the resin. Depending on heat, pressure, and volume of resin/paper, the bounce of the hydraulic pressure will stop. This indicates that cross-linking has occurred.

12-4. Release pressure and remove the sandwich. Remove the paper/resin materials. Figure 12-21 shows photos of laminates made with two, four, and six layers of paper.

12-5. Cut a strip out of the laminated paper, and form it into a dog bone for tensile testing. Figure 12-22 shows a strip cut with a band saw and a prepared dog-bone sample.

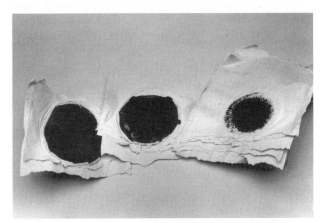

Figure 12-21. Samples of two-, four-, and six-layer laminates.

Figure 12-22. Laminate cut into a strip and prepared as a dog bone.

12-6. Carefully measure the length, width, and thickness of the gage area. Perform a tensile test. Because the cross-linked phenolic is rather hard, it may tend to slip in the jaws of a tensile tester. If this occurs, a piece of emery cloth folded over the grip area should eliminate slippage. The sample shown in Figure 12-22 failed at 37.1 MPa (5384 psi).

12-7. Make a disk of phenolic without any layers of paper and test it. Does the paper increase the strength of the laminate? Does the paper increase the flexibility of the laminate?

12-8. Repeat the procedure with varying amounts of paper or with other fibrous layers such as cloth, glass cloth, glass mat, or Kevlar cloth.

12-9. To see whether the layers of paper or other material affect shrinkage of the laminate, scribe shallow lines on a sheet metal plate. Make sure that spacing between the lines is known and accurate. Figure 12-23 shows the lines transferred to the laminate. These lines were spaced

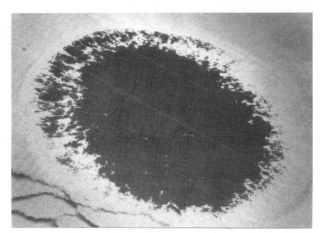

Figure 12-23. Close-up of phenolic laminate.

every 1 inch. The material shrinkage can be calculated by measuring the distance between the lines in a finished, cooled laminate (see the section in Chapter 6 on calculating mold shrinkage).

REINFORCING PROCESSES AND MATERIALS

CHAPTER 13

INTRODUCTION

The term **reinforced plastics** is not very descriptive. It simply implies that an agent has been added to improve or *reinforce* the product. The SPE defines reinforced plastics as "a plastics composition in which reinforcements are embedded with strength properties greatly superior to those of the base resin." Specific terms such as *advanced, high-strength, engineered,* or *structural* composites came into use in the 1960s. With them, a stiffer, higher modulus material of exotic reinforcements in new matrices was used.

Today, *reinforced plastics* is used to describe several forms of composite materials produced by 1 of 10 reinforcing processes. Someday, we may classify all laminating and reinforcing processes as **composite processing**.

Some composite-reinforcing techniques are variations of laminating because they may involve combining two or more different materials in layers. Other techniques are simply modifications of processing methods that produce a new material with specific or unique properties.

The following composite-reinforcing processes will be discussed in this chapter:

I. Matched die
   A. Bulk molding compounds
   B. Sheet molding compounds
II. Hand layup or contact processing
III. Spray-Up
IV. Rigidized vacuum forming
V. Cold-mold thermoforming
VI. Vacuum bag
VII. Pressure bag
VIII. Filament winding
IX. Centrifugal reinforcing and blown-film reinforcing
X. Pultrusion
XI. Cold stamping/forming

In each process, the molds, dies, or rollers must be made with care to ensure proper release of the finished product. Film-, wax-, and silicone-releasing agents are most often used on mold surfaces. **Reinforced molding compounds** should not be confused with laminates, although they occasionally are (Figure 13-1).

In the past, only thermosetting plastics were reinforced in large quantities. Today, the demand for reinforced thermoplastics (RTP) is increasing. Because thermoplastic materials may be processed in many different ways, numerous innovative uses have resulted.

Reinforced molding compounds may be molded by injection, matched die, transfer, compression, or extrusion methods to produce products with complex shapes and a broad range of physical properties. However, some difficulty arises in blow molding small, thick-walled items. Injection molding is the most common method of processing reinforced thermoplastics compounds (see Injection, Extrusion, and Compression molding in the index).

Short fibers of milled or chopped glass are most often used to reinforce molding compounds. See Table 13-1 for a list of properties of fibrous glass reinforced plastics. Plastics fibers in addition to exotic metallic and crystalline whiskers are used as well.

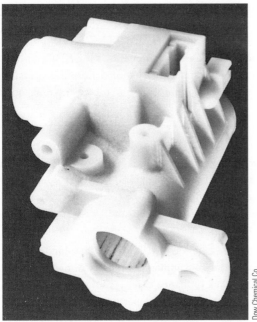

(A) A steering column lock housing made of injection-molded, glass-reinforced nylon instead of the traditional aluminum die-cast and machined housing.

(B) Reinforced parts and housing for a chain saw designed for rugged use.

Figure 13-1. Some examples of reinforced plastics parts.

MATCHED DIE

Bulk molding compounds (BMC) and sheet molding compounds (SMC) are the most common materials used in match die reinforcing.

Bulk Molding Compounds

Bulk molding compounds (BMC) are putty-like mixtures of resin, catalysts, fillers, and short fiber reinforcements. BMC has many names. It is called *gunk, putty, dough,* or *slurry molding.* Dough molding compounds even have their own acronym, DMC. All these names refer to a premix of resins and reinforcements.

Bulk molding compounds are often formed into log or rope shapes to aid molding and handling operations. This fibrous putty may be extruded into H-beam or other profile shapes and automatically fed into the matched die.

Table 13-1. Typical Properties of Fibrous-Glass Reinforced Plastics

Plastics	Relative Density	Tensile Strength, 1000 MPa	Compressive Strength, 1000 MPa	Thermal Expansion $10^{-4}/°C$	Deflection Temperature (at 264 MPa), °C
Acetal	1.54–1.69	62–124	83–86	4.8–4.9	1062–1599
Epoxy	1.8–2.0	10–207	207–262	2.8–8.9	834–1599
Melamine	1.8–2.0	34–69	138–241	3.8–4.3	1406
Phenolic	1.75–1.95	34–69	117–179	2–5.1	1027–2178
Phenylene oxide	1.21–1.36	97–117	124–207	2.8–5.6	910–986
Polycarbonate	1.34–1.58	90–145	117–124	3.6–5.1	965–1000
Polyester (thermoplastic)	1.48–1.63	69–117	124–134	3.6	1379–1586
Polyester (thermosetting)	1.35–2.3	172–207	103–207	3.8–6.4	1406–1792
Polyethylene	1.09–1.28	48–76	34–41	4.3–6.9	800–876
Polypropylene	1.04–1.22	41–62	45–48	4.1–6.1	910–1027
Polystyrene	1.20–1.34	69–103	90–131	4.3–5.6	683–717
Polysulfone	1.31–1.47	76–117	131–145	4.3	1179–1220
Silicone	1.87	28–41	83–138	None	<3323

BMCs are isotropic with fiber lengths usually less than 0.38 mm (0.015 in.). The fiber lengths distinguish BMC from other molding compounds.

Sheet Molding Compounds

Sheet molding compounds (SMC) are leather-like mixtures of resins, catalysts, fillers, and reinforcements.

These mixtures are sometimes called the *flow mat* or *mold mat*. Because they are made in sheet form, fibers may be much longer than in BMC. A typical SMC incorporates about 30% random 25 mm (1 in.) chopped glass fibers, 25% resin, and 45% inorganic filler. SMC-25 indicates a 25% glass content.

SMCs provide greater glass loadings (70% glass) and lighter products compared to BMCs. The longer fibers provide improved mechanical properties. SMCs include many specialized types that have abbreviations such as UMC, TMC, LMC, and XMC. Unidirectional molding compound (UMC) usually has about 30% of its continuous reinforcing fibers aligned in one direction. This provides greater tensile strength in the direction of the fibers. Thick-molding compounds (TMCs) up to 2 inches thick are able to be produced. This thicker sheet allows greater variation in part wall thickness and a wider choice of reinforcements. TMCs are highly filled. Low-pressure molding compounds (LMCs) are SMC formulated to allow use of low-pressure molding techniques. High-strength molding compound (HMC) may contain more than 70% reinforcement for added strength and dimensional stability. Direction ally reinforced molding compound (XMC) is a sheet containing about 75% directionally oriented continuous reinforcements.

SMCs are formed into final shape by the molding operation. Figure 13-2 shows several SMC processes and products. An upper and lower carrier film (usually polyethylene) is used with resins. This film allows the SMCs to be stored neatly at hand and easily handled during processing. The carrier films are removed prior to the molding operation (see Figure 13-3).

SMCs require processing pressures in the range of 3.5 to 14 MPa (507–2030 psi). Processing temperatures vary according to product design and polymer formulation. SMCs are fed into molds and passed through pressing, curing, and demolding in one continuous cycle. This technique eliminates waiting at the press through full cure cycle times.

Major markets for BMC and SMC molded parts are the transportation and appliance industry. Shower floors, heater housings, and appliance cases are all made of BMC. As the name SMC implies, large parts such as automotive body panels, hoods, small boat hulls, furniture, and appliance components are made from this material in matched-die molds.

A variation of this process is called *macerated* reinforced processing. Macerated parts are produced by chopping the reinforcing materials into pieces 0.2 to 10 mm (0.008 to 0.4 in.) long to be processed in the matched molds. Reinforced resin products produced from matched-die molds are strong and may have a superb surface finish both inside and out, but mold and equipment costs are high. Following is a list of five advantages and disadvantages of matched-die processing:

(A) This mold makes SMC pickup truck fenders.
Figure 13-2. SMC processes and products.

(B) Robotic arms handle this tonneau cover.

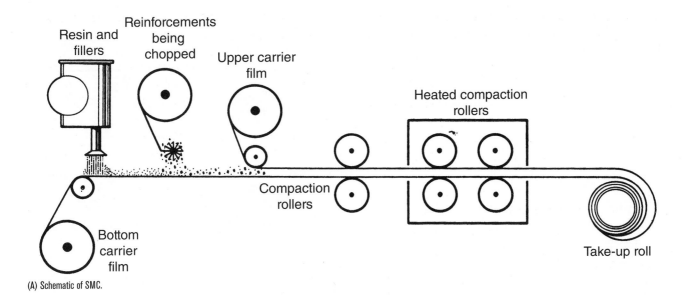

(A) Schematic of SMC.

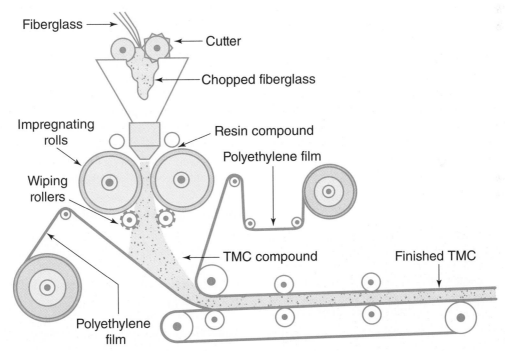

(B) Schematic of TMC.

Figure 13-3. Production method for sheet-molding compound.

Advantages of Matched Die

1. Both interior and exterior surfaces are finished.
2. Complex shapes (including ribs and thin details) are possible.
3. Minimum trimming of parts is needed.
4. Products have good mechanical properties, close part tolerances, and corrosion resistance.
5. Cost and reject rate are relatively low.

Disadvantages of Matched Die

1. Preform, BMC, TMC, XMC, and SMC require more equipment, handling, and storage.
2. Press guides must have good parallelism for close tolerances.
3. Molds and tooling are costly compared to open molds.
4. Surfaces may be porous or wavy.
5. There are no transparent products.

HAND LAYUP OR CONTACT PROCESSING

Thermosetting resins are used in *hand layup moldings*. Because the reinforcement generally consists of a continuous layer, this process was described in Chapter 12 as a type of lamination. However, it can also be considered a reinforcing process (see Chapter 12).

SPRAY-UP

In **spray-up**, catalyst, resin, and chopped roving may be sprayed simultaneously onto mold shapes (Figure 13-4). Although considered a variation of hand layup, this process can be accomplished by hand or machine. After the **gel coat** has been applied, the spray-up of resin and chopped fibers begins. Careful rollout is important in order to avoid damaging the gel coat. Rollout aids in **densifying** (eliminating air pockets and aiding in wetting action) the composite. Poor rollout can induce structural weakness by leaving air bubbles, dislocating the fibers, or causing poor **wet out** (coating of reinforcement). Heat may be applied to speed cure and increase production. This low-cost method allows the production of very complex shapes. Production rates are high compared to hand layup methods because care must be taken to apply uniform layers of materials. Otherwise, mechanical properties may not be consistent throughout the product. Highly contoured or stressed areas can be given additional thickness or metallic tapping plates, stiffeners, or other reinforcing components that can be placed in desired areas and oversprayed.

RIGIDIZED VACUUM FORMING

In a process sometimes called **rigidized shell spray-up**, a thermoplastic sheet is thermoformed into the desired shape, eliminating the gel coat. PVC, PMMA, ABS, and PC are commonly used. This shell is reinforced (by spray or hand layup)

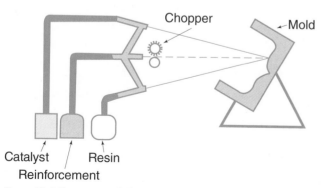

Figure 13-4. The spray-up method can cover either simple or complex shapes easily, an advantage over the hand layup method.

Chopper

Mold

Catalyst

Resin

Reinforcement

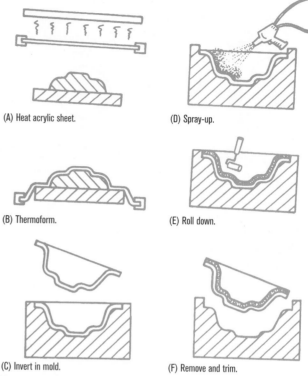

(A) Heat acrylic sheet.

(D) Spray-up.

(B) Thermoform.

(E) Roll down.

(C) Invert in mold.

(F) Remove and trim.

Figure 13-5. The rigidized vacuum-forming process.

on the back side to produce a strong composite bathtub, sink, bathtub-shower combination, small boat, exterior signs, car top carrier, or other similar products. This method is illustrated in Figure 13-5. The following lists give four advantages and three disadvantages of rigidized vacuum forming:

Advantages of Rigidized Vacuum Forming

1. Thermoplastic skin gives smooth finish.
2. It eliminates surface gel coat voids and gel time.
3. It requires only a thermoforming mold.
4. Output rates are faster than those obtained by spray-up methods.

Disadvantages of Rigidized Vacuum Forming

1. It needs thermoforming equipment and sheet storage space.
2. Materials (thermoplastic surface sheets) are costly.
3. Repair of damaged surface sheets is difficult.

COLD-MOLD THERMOFORMING

This process is similar to rigidized vacuum forming except for two major differences. First, the two surfaces are finished, and second, tolerances are more closely controlled. The process involves thermoforming a sheet, then reinforcing the back side by preform, mat, or spray-up methods. Next, the composite is pressed between matched dies until cured. Polymerization is performed using chemical means at room temperature. A major disadvantage is the additional die cost and curing time in the mold.

VACUUM BAG

During vacuum-bag processing, a plastics film (usually polyvinyl alcohol, neoprene, polyethylene, or polyester) is placed over the layup. About 85 kPa of vacuum (25 in. of mercury) is drawn between the film and the mold (Figure 13-6).

Vacuum is usually measured in millimeters of mercury drawn in a graduated tube, or pascals of pressure. Vacuum in millimeters of mercury corresponding to 85 kPa can be calculated using this formula:

$$\frac{101\text{kPa}}{85\text{kPa}} = \frac{760 \text{ mm}}{x}$$

or

$$x = 656 \text{ mm of mercury}$$

where

x = unknown (mm or mercury)
101 kPa = known atmospheric pressure,
760 mm = mm of mercury corresponding to 101 kPa of pressure

The plastics film forces the reinforcing material against the mold surface, producing a high-density product free of air bubbles. Tooling for vacuum-bag processing is costly when large

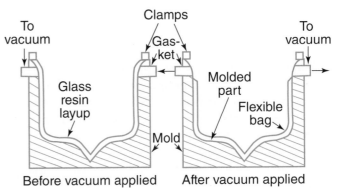

Figure 13-6. In vacuum-bag processing, application of pressure to a layup results in improved strength and a better surface on the unfinished side of the part.

pieces are made. Output is slow compared to high-speed production rates for injection molding.

Both male and female tooling are used. If a smooth surface is required on the exterior of a boat hull, a female mold would be selected, whereas a male mold would probably be selected for a sink. Because heat is required in many operations, ceramic or metal tooling is used. Infrared induction, dielectric, xenon flash, or beam radiation can be used to aid or speed cure.

The mold surface must be protected to allow removal of the finished composite. Plastic films, waxes, silicone resins, PE, PTFE, PVAI, polyester (Mylar®), and polyamide films are used as releasing agents.

Popular **wet layup** resins include epoxies and polyesters. SMC, TMC, and reinforcements pre-impregnated with polysulfone, polyimide, phenolics, diallyl phthalate, silicones, or other resin systems may be used. These materials are often called **prepregs**.

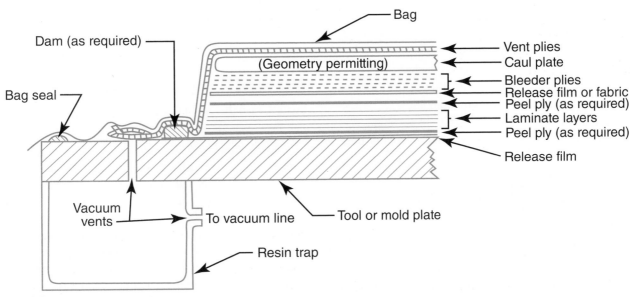

Figure 13-7. Wet layup vacuum-bag processing.

Reinforcements may include honeycomb materials, mats, fabrics, paper, foils, or other pre-impregnated forms.

Wet layup vacuum-bag processing is illustrated in Figure 13-7. After the tooling has been carefully protected with a releasing wax or film (depending on part geometry), a peel ply of finely woven polyester or polyamide fabric is carefully positioned. Sometimes a sacrificial ply (usually a fine, resin-impregnated fabric) is placed on the mold surface. The laminate layers are then placed on the mold surface. A second peel ply is placed on the laminated layers followed by a release film or fabric. Dacron® and Teflon are commonly used as release fabrics. Because perforations allow air and excess resin to escape, this layer is sometimes called the breather ply. Bleeder plies of cloth or mat are laid on the release fabric to collect air and resin that are forced in. On some composite compositions, a caul plate is used to ensure a smooth surface and minimize variations in temperature during the curing process. Next, several vent or breather plies are laid so air can freely pass along the surface of the part inside the bag. The bag can be made of any flexible material that is airtight and won't dissolve in the matrix. Silicone rubber blanket, Neoprene®, natural rubber, PE, PVA®, cellophane, or PA are commonly used. A vacuum of 25 inches of mercury, or about 12 psi of external pressure, is then drawn. A resin trap is used to prevent excess liquid resin from being drawn into the vacuum lines. When additional density or difficult design requirements are needed, pressure-bag, rubber-plunger, rubber-bag, autoclave, and hydroclave forming techniques are used.

Dry, pre-impregnated materials are usually more difficult to form into complex shapes. Additional pressure, plug assistance, and external heat sources are used to soften and aid in shaping the composite against the tooling.

PRESSURE BAG

Pressure-bag processing is also costly and slow but allows the manufacture of large, dense products with good finishes, both inside and out. Pressure-bag processing uses a rubber bag to force the laminating compound against the contours of the mold. About 5.1 psi (35 kPa) of pressure is applied to the bag during the heating and curing cycle (Figure 13-8). Pressures seldom exceed 50.8 psi (350 kPa).

After layup, the mold and compounds may be placed in a steam or heated gas autoclave. Autoclave pressures of 50.8 to 101.5 psi (350 to 700 kPa) will achieve greater glass loading and aid in air removal.

The term hydroclave implies that a hot fluid is used to press the plies against the mold. In all pressure designs, the tooling (including the flexible bag) must be able to withstand molding pressures. Pressure-bag techniques used to force the layup against the mold walls may be appropriate for long hollow pipes, tubes, tanks, or other objects with parallel walls. At least one end of the object must be open in order to insert and remove the bag.

Three advantages and four disadvantages of vacuum- and pressure-bag processing are found in the following list:

Advantages of Vacuum and Pressure Bag

1. There is greater glass loading and fewer voids than hand layup methods.
2. Inside surface has better finish than hand layup methods.
3. There is better adhesion in composites.

Disadvantages of Vacuum and Pressure Bag

1. More equipment is needed than in hand layup methods.
2. Inside surface finish is not as good as matched die molding.
3. Quality depends on skill of operator.
4. Cycle times are long, limiting production with single mold.

FILAMENT WINDING

Filament winding produces strong parts by winding continuous fibrous reinforcements on a mold.

Long continuous filaments are able to carry more load than random, short filaments. Over 80% of all filament winding is accomplished with E-glass roving. Higher modulus fibers of carbon, aramid, or Kevlar may be used. For some applications, boron, wire, beryllium, polyamides, polyimides, polysulfones, bisphenol, polyesters, and other polymers are also used. Specially designed winding machines may lay down these strands in a predetermined pattern to give maximum strength to the desired direction (Figure 13-9).

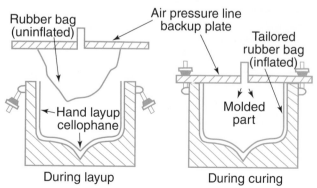

Figure 13-8. Heat and an inflated rubber bag that applies pressure are used in the pressure-bag molding method.

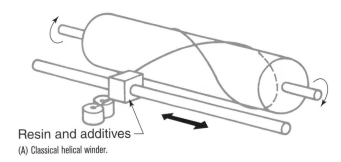

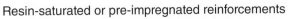

Resin-saturated or pre-impregnated reinforcements

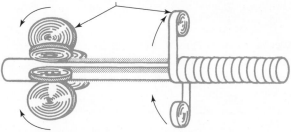

Resin and additives

(A) Classical helical winder.

(E) Continuous normal-axial winder.

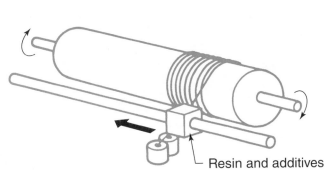

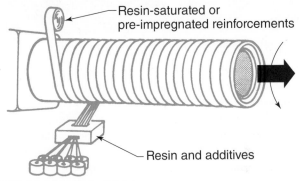

Resin-saturated or pre-impregnated reinforcements

Resin and additives

(B) Circumferential winder.

(F) Selected winding methods and designs.

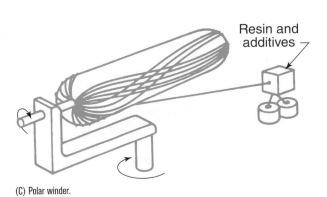

Resin and additives

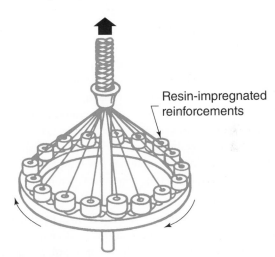

Resin-impregnated reinforcements

(C) Polar winder.

(G) Braid-wrap winder.

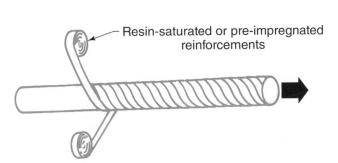

Resin-saturated or pre-impregnated reinforcements

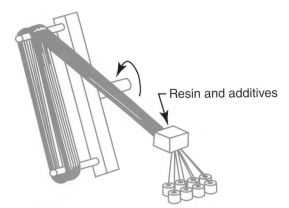

Resin and additives

(D) Continuous helical winder.

Figure 13-9. Selected winding methods and designs.

(H) Loop-wrap winder.

Figure 13-10. Wet filament winding.

During **wet winding**, the fibers must be thoroughly wetted in resin before winding. The strands are drawn through a resin bath, then excess resin matrix and entrapped air are forced (squeezed out from between strands). Filament-winding tension varies from 0.25 to 1 lb per end (a group of filaments) (see Figure 13-10 showing wet filament winding, and note restrictions to shape of filament wound parts).

In **dry winding**, pre-impregnated B-stage reinforcements help to ensure consistency in resin-to-reinforcement content design. These pre-impregnated reinforcements may be machine- or hand-wound on the tooling. Curing may be accelerated by heated **mandrels** (tooling), ambient ovens, chemical hardeners, or other energy sources. Many cylindrical laminated forms are produced using this method. The collapsible mandrel must have the desired shape of the finished product. Soluble or low-temperature-melting mandrels may also be used for special complex shapes or sizes.

An advantage of filament winding is that it allows the designer to place reinforcement in the areas subject to the greatest stress. Containers made by this process usually have a higher strength-to-mass ratio than those made by

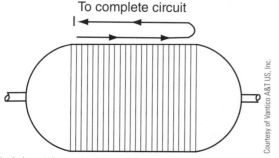

(A) Circular loop windings provide optimum girth or hoop strength in a filament wound structure.

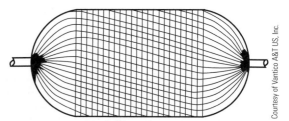

(B) Single-circuit helical winding combined with circular loop windings provide high axial tensile strength.

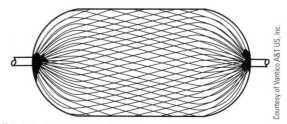

(C) Multiple-circuit helical windings allow optimum use of the glass filament's strain characteristics, without the addition of loop windings.

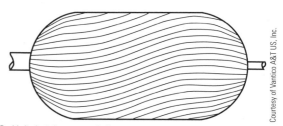

(D) Dual helical windings are used when openings at the ends of the structure are of different diameters.

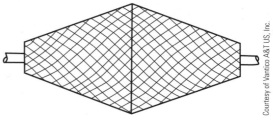

(E) Variable helical windings can produce odd-shaped structures.

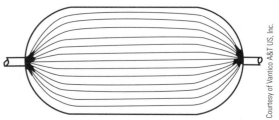

(F) Planar windings provide optimum longitudinal strength (with respect to the winding axis).
Figure 13-11. Advantages of various types of filament winding.

other methods. They may be produced in virtually any size at a lower cost. Figure 13-11 shows various winding patterns used for pressure vessels.

On many pressure vessels, the filament windings are not removed from a mandrel but are overwrapped on thin metal or plastics containers.

Filament-wound applications include rocket engine cases, pressure vessels, underwater buoys, radomes, nose cones, storage tanks, pipes, automotive leaf springs, helicopter blades, spacecraft spars, fuselage, and other aerospace parts.

CENTRIFUGAL REINFORCING AND BLOWN-FILM REINFORCING

In centrifugal reinforcing, resin and reinforced materials are formed against the mold surface as it rotates (Figure 13-12). During this rotation, resin is distributed uniformly through the reinforcement by centrifugal force. Heat is then applied to help polymerize the resin. Tanks and tubing may be produced in this manner.

Another specialty process involves reinforcing blown film. In one proprietary blown-film process, a composite sheet is produced by filament reinforcing the inside of the hot blown film and pressing the fibrous layered film between pinch rollers. This concept is illustrated in Figure 13-13.

PULTRUSION

In **pultrusion**, resin-soaked matting or rovings (along with other fillers) are pulled through a long die heated to between 120°C and 150°C (250°F and 300°F). The product is shaped, and the resin is polymerized as it is drawn through the die. Radio-frequency or microwave heating may also be used to speed production rates. The process appears to resemble extrusion. In the extrusion process, the homogeneous material is *pushed* through the die opening.

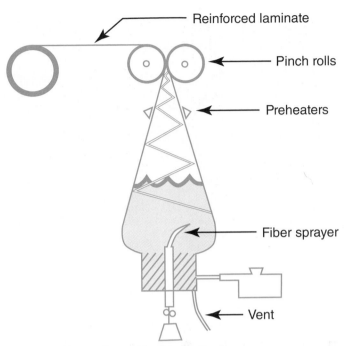

Figure 13-13. Making filament-reinforced sheets by blown-film process.

In pultrusion, however, resin-soaked reinforcements are *pulled* through a heated die where the resin is cured (Figure 13-14).

Pultrusion dies are generally 60 to 150 cm (24 to 60 in.) in length and heated to aid in the polymerization process. Cure must be carefully controlled to prevent cracking, delamination, incomplete curing, or sticking to the die surfaces.

Output ranges from a few millimeters to over 3 m/min. Various resins in use include vinyl esters, polyesters, and epoxies. Fibrous glass is the most widely used reinforcement, although graphite, carbon, boron, polyester, and polyamide fibers may be used. Reinforcements can be positioned in the area of the pultrusion product in which extra strength is needed.

Hot-melt thermoplastic materials and reinforcements may also be used. Parallel orientation of reinforcements produces a strong composite in the direction of the fibers. Some operations may use SMC or wound preforms in combination with other continuous reinforcements to improve omnidirectional properties.

Siding, gutters, I-beams, fishing rods, automotive springs, frames, airfoils, hammer handles, skis, tent poles, golf shafts, ladders, tennis racquets, vaulting poles, and other profile shapes are examples of items produced by pultrusion.

Pulforming is a variation of pultrusion. As materials are pulled from the reinforcement creels and impregnated with

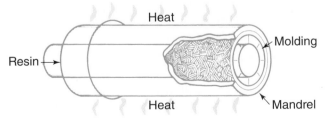

Figure 13-12. In the centrifugal method, chopped reinforcement and resin are evenly distributed on the inner surface of a hollow mandrel. The assembly rotates inside an oven to provide heat for curing.

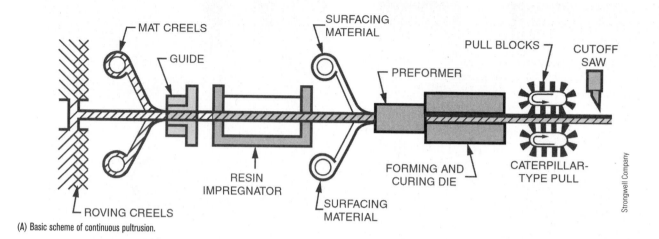

(A) Basic scheme of continuous pultrusion.

(B) Structural supporting members of fiberglass-reinforced polyester for the operating floor of this chemical mixing plant were produced by pultrusion.

Figure 13-14. Pultrusion method and a product application.

resin and other compounds, a number of forming devices (molds) of various cross-sectional shapes form the composite part. In one method, male and female rotary dies are brought together on the pultruded material and the composite is cured. Curved parts can be formed by forcing the pulform into a large circular female mold with a flexible steel belt. The mold and belt are heated to speed cure in a continuous pulform operation (see Figure 13-15.) Applications include hammer handles, bows, curved springs, and other products that do not have a continuous cross-sectional shape.

COLD STAMPING/FORMING

Fibrous-glass-reinforced thermoplastics, available in sheet form, may be cold formed in the same manner as metals (Figure 13-16). Long reinforcements are used to improve the strength-to-mass ratio.

During the forming operation, the sheet is preheated to about 200°C (392°F) and then formed on normal metal stamping presses. It is possible to produce parts with complex designs and varying wall thickness using this method. Production rates may exceed 260 parts per hour. Various products include motor covers, fan guards, wheel covers, battery trays, lamp housings, seat backs, and many interior automotive trim panels.

According to one authority, stampable reinforced thermoplastic composite sheets could become a replacement for stamped steel produced in Detroit. Many of these sheets have a paintable class-A finish from the mold. Non-class-A automotive finish applications account for about 80% of the glass/polypropylene (PP) sheet demand. Polycarbonate/ polybutylene terephthalate (PC/PBT), polyphenylene oxide/ polybutylene terephthalate (PPO/PBT) and polyphenylene oxide/polyamide (PPO/PA) are alloys that are combined with modified glass mat or other special reinforcements in stampable, formable sheets.

Non-impregnated commingled blends of continuous thermoplastic filaments, such as polyetheretherketone (PEEK) and polystyrene (PS) with reinforcing filaments of carbon, glass aramid, or metals, may be made into yarns, fabrics, or felts. Woven, braided, or knitted three-dimensional commingled preforms are then heated under pressure in the mold. The thermoplastic filaments melt and wet out the adjacent reinforcements. These forms are a versatile material for composites.

(A) Glass rovings feeding through wet-out tank.

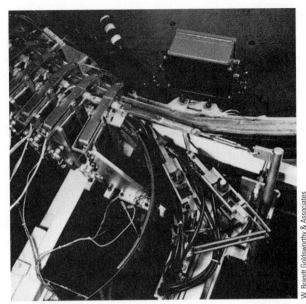

(B) Heated belt die closure on cure section.

(C) Spring stock exiting the die/belt section.

Figure 13-15. Pulforming of a curved composite leaf spring.

(D) Flying spring stock cutoff saw.

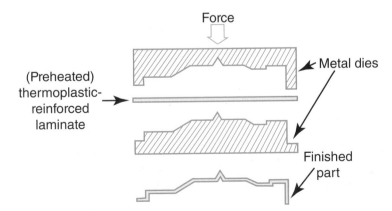

Figure 13-16. Composite sheet material is preheated and formed on cold metal-stamping dies and equipment.

RELATED INTERNET SITES

- **www.acmanet.org.** The American Composites Manufacturers Association (ACMA) provides information about composites. The ACMA publishes *CM, Composites Manufacturing*, bimonthly. This journal is now available online. The ACMA also publishes *Composites Research Journal (CRJ)* quarterly. ACMA also operates a certification program for technician, the CCT (Certified Composites Technician.)

- **www.reichhold.com.** Reichhold Chemicals, Inc., is a leading global supplier of unsaturated polyester resins. These materials go into a variety of SMCs and gel coats. Selecting "read more" under "Composites" on the home page leads to "Literature/Documentation." This section provides numerous documents and brochures about FRPs (fiber reinforced composites.)

- **www.smc-alliance.com.** The SMCBMC Alliance is a European organization of SMC and BMC manufacturers. Selecting "Design" on the home page leads to information on the advantages of SMC components as compared to metal.

VOCABULARY

The following vocabulary words are found in this chapter. Use the glossary in Appendix A to look up the definitions of any of these words you do not understand as they apply to plastics.

breather ply
bulk molding compounds (BMC)
caul plate
commingled
composite processing
densifying
dry winding
filament winding
gel coat
hydroclave
mandrels
prepreg
pulforming
pultrusion
reinforced molding compounds
reinforced plastics
rigidized shell spray-up
sacrificial ply
sheet molding compounds (SMC)
spray-up
wet layup
wet out
wet winding

QUESTIONS

13-1. Plastics with strength properties that increase by adding filler and reinforcing fibers to the base resin are called _____.

13-2. The thin unreinforced layer of resin placed on the surface of a mold in the hand layup process is called a _____.

13-3. The process of using reinforcements to improve selected properties of plastics parts is called _____.

13-4. The two major disadvantages of high-pressure lamination are low _____ rates and high pressures.

13-5. In _____ molding, resin-soaked matting or rovings are pulled through a long heated die.

13-6. Name the simplest processing technique that can be used for making a relatively strong vessel with a complex shape.

CASTING PROCESSES AND MATERIALS

CHAPTER 14

INTRODUCTION

This chapter focuses on plastics casting processes and frequently used casting materials. Casting involves introducing a liquefied plastics into a mold and allowing it to solidify. In contrast to molding and extrusion, casting relies on atmospheric pressure to fill the mold rather than a significant force to push the polymer into the mold cavity.

To fill a mold using atmospheric pressure, the polymer must approach a liquid state. Even at elevated temperatures, many plastics simply do not become liquid enough to flow into molds. Many hot polymers have a viscosity similar to bread dough. Consequently, acetal, PC, PP, and many other plastics are not casting materials. Monomers are typically more liquid than polymers and have viscosities similar to pancake syrup. Because of this property, monomers find considerable use as casting materials.

Casting includes a number of processes in which monomers, modified monomers, powders, or solvent solutions are poured into a mold where they become a solid plastics mass. The transition from liquid to solid may be achieved by evaporation, chemical action, cooling, or external heat. After the cast material solidifies in the mold, the final product is removed from the mold and finished.

Casting techniques listed in the following chapter outline may be placed into six distinct groups:

MATERIAL TYPES

There are four different types of materials used in casting processes. Liquid resins are monomers, syrups, or low-molecular-weight thermosets. Frequently, these materials are short-chain polymers including nylon, acrylic, polyester, and phenolic. These materials typically harden by a chemical reaction to complete the polymerization for thermoplastics or the cross-linking for thermosets. Hot-melt plastics, the second group, are fully polymerized thermoplastics that are melted, then rotationally formed and solidified by cooling. Plastisols and organosols are plastics particles that are suspended in a plasticizing solvent. They harden either by evaporation of the solvent or by drawing the solvent into the particles. Plastisols have a high concentration of solids, usually greater than 90%, which makes them thick and viscous. In contrast, organosols have a lower percentage of solids, often between 50% and 90%. They are generally more fluid than plastisols. The fourth type involves the dissolving of plastics pieces in a volatile solvent and is referred to as dissolved plastics. As the solvent evaporates, the material hardens.

SIMPLE CASTING

In simple casting, liquid resins or molten plastics are poured into molds and allowed to polymerize or cool. Molds may be made of wood, metal, plaster, selected plastics, selected elastomers, or glass. For example, silicones are often cast over patterns to make molds in which plastics or other materials may be cast.

Examples of products made by simple casting include jewelry, billiard balls, cast sheets for windows, furniture parts, watch crystals, sunglass lenses, handles for tools, desk sets, knobs, table tops, sinks, and fancy buttons. Figure 14-1 shows the basic principle of simple casting.

Phenolic castings were part of the early development of the plastics industry. Leo Baekeland introduced numerous articles cast of Bakelite. Today, the most important casting resins are polyester, epoxy, acrylic, polystyrene, silicones, epoxies, ethyl cellulose, cellulose acetate butyrate, and polyurethanes. The most well known is probably polyester resin because it is used in crafts and hobby work.

Polyester casting resins may be filled or unfilled. To reduce the cost of unfilled polyester, casting resins are extended with water. Water-extended polyesters find wide use in casting furniture and cabinet parts.

Many polyester casting resins contain large amounts of fillers and reinforcements. For example, cultured marble is a product that contains marble dust or calcium carbonate (limestone) as the filler and polyester plastics as the binding material. It is used to produce lamp bases, table tops, exterior veneers, statues, and other marble-like products.

Types of clear acrylic plastics are cast rods, sheets, and tubing. Acrylic sheets are often produced by pouring a catalyzed monomer or partially polymerized resin between two parallel plates of glass (Figure 14-2). The glass is sealed with a

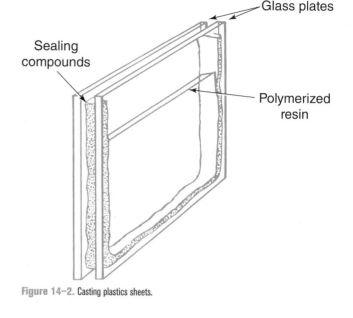

Figure 14–2. Casting plastics sheets.

gasket material to prevent leakage and help control the thickness of the cast sheet. After the resin has fully polymerized in an oven or autoclave, the acrylic sheet is separated from the glass plates and reheated to relieve stresses that occur during the casting process. The faces are covered with masking paper to protect the sheet during shipment, handling, and fabrication. Untrimmed sheets may be purchased with the sealing material still sticking to the edges (Figure 14-3).

Some materials are unsuitable for casting but are utilized in sheet form. In contrast to casting, a cutting process called skiving may be required. Sheets of cellulose nitrate and other plastics may be sliced (skived) from blocks that have been softened by solvents. After the residual solvents have evaporated, the skived piece is pressed between polished plates to improve the surface finish (Figure 14-4).

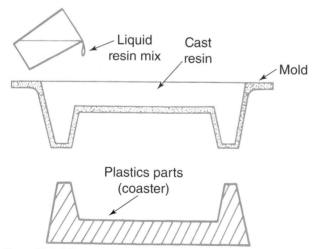

Figure 14–1. Solid casting in an open, one-piece plastics mold.

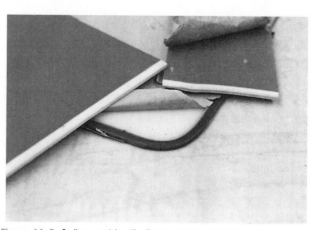

Figure 14–3. Sealing materials still adhere to the edges of these untrimmed acrylic plastics sheets.

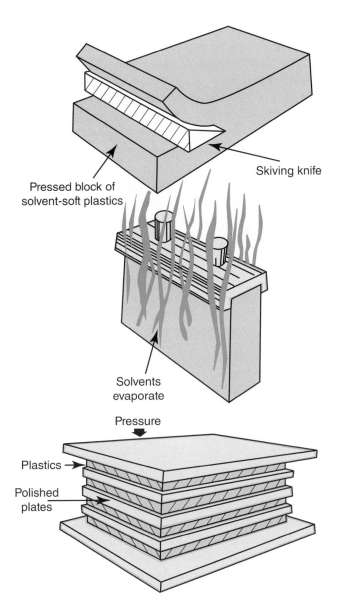

Figure 14-4. Skiving sheets from a block of plastics.

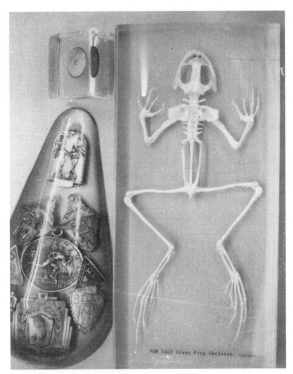

Figure 14-5. Embedding in transparent polyester has wide use for objects such as classroom science samples.

Special Types of Simple Castings

In addition to the common simple castings, three special types of simple castings are popular. These are referred to as embedment, potting, and encapsulation. Although foams may be cast, they are discussed under the foaming processes in Chapter 16.

Embedments. Embedments encase objects completely with a transparent plastics. After polymerization, the casting is removed from the mold and often polished (Figure 14-5).

Objects are embedded for preservation, display, and study. In the biological sciences, animal and plant specimens often are embedded to help preserve them. This allows safe handling of the most fragile samples.

Potting. Potting is used to protect electrical and electronic components from harmful environments. The potting process completely encases the desired components in plastics. In this case, the mold becomes part of the product. Vacuum, pressure, or centrifugal force is frequently applied to ensure that all voids are filled with resin.

Encapsulation. Encapsulation is similar to potting. Encapsulation is a solventless covering on electrical components. This envelope of plastics does not fill all the voids. The process involves dipping the object in the casting resin. After potting, many components are encapsulated.

Listed below are five advantages and four disadvantages of casting processes:

Advantages of Casting

1. Cost of equipment, tooling, and molds is low.
2. It is not a complex forming method.
3. There are a wide number of treatment techniques.
4. Products have little or no internal stress.
5. Material costs are relatively low.

Disadvantages of Casting

1. The output rate is low and cycle time is high.
2. Dimensional accuracy is only fair.
3. Moisture and air bubbles may cause problems.
4. Solvents and other additives may be dangerous.

FILM CASTING

Water-soluble packaging for bleaches and detergents is an example of a cast film. Film casting involves dissolving plastics granules or powder along with plasticizers, colorants, or other additives in a suitable solvent. The solvent solution of plastics is then poured onto a stainless steel belt. The solvents are evaporated by the addition of heat, and the film deposit is left on the moving belt. The film is stripped or removed and wound on a take-up roller (Figure 14-6). This film may be cast as a coating or laminate directly on fabric, paper, or other substrates.

Solvent casting of film offers the following three advantages over other heat-melt processes:

1. Additives for heat stabilization and lubrication are not needed.
2. Films are uniform in thickness and optically clear.
3. No orientation or stress is possible with this method.

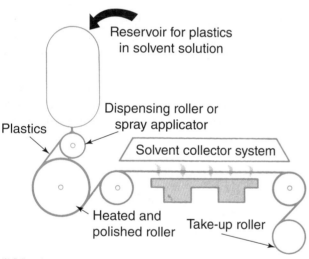

(A) Roller solvent casting.

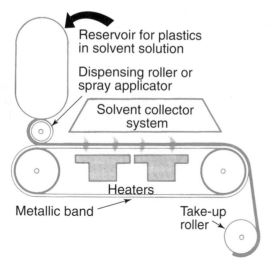

(B) Band solvent casting.

Figure 14-6. Film casting.

To be economically feasible, solvent casting of film requires a solvent recovery system. Plastics that may be solvent cast include cellulose acetate, cellulose butyrate, cellulose propionate, ethyl cellulose, polyvinyl chloride, polymethyl methacrylate, polycarbonate, polyvinyl alcohol, and other copolymers. Casting liquid plastics latexes on Teflon-coated surfaces rather than stainless steel may also be used to produce special films.

Aqueous dispersions of polytetrafluoroethylene and polyvinyl fluoride are cast on heated belts at temperatures below their melting points. This method is handy when producing films and sheets of materials that are hard to process by other means. These films are used as nonstick coatings, gasket material, and sealing components for pipes and joints.

HOT-MELT CASTING

Hot-melt plastics were used for casting as early as World War II. Today, hot-melt formulations may be based on ethyl cellulose, cellulose acetate butyrate, polyamide, butyl methacrylate, polyethylene, and other mixtures. The largest use is the production of strippable coatings and adhesives. Hot-melt resins may be used for making molds for casting other materials. Also, hot-melt resins are used in a casting process for potting and encapsulation (Figure 14-7). Not all potting compounds are thermoplastic and hot melting. Silicone is most often used for coating, sealing, and casting. However, epoxy and polyester resins are also used for these purposes.

Electrical parts may be protected from hostile environments by being placed in molds and having hot resin poured over the components. When cool, the plastics provides protection for wires and vital parts. The encapsulated or potted components may then be assembled with other parts to produce the finished product. Some encapsulations and pottings are not cast in separate molds, but are produced by pouring the molten compound directly over the components inside the case of the finished product. The insulation of parts in a radio chassis or a motor is one well-known example. If the components are

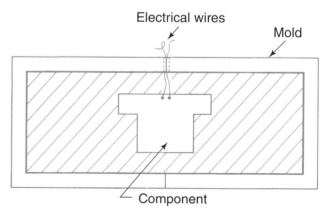

Figure 14-7. Hot-melt encapsulation of an electronic component.

cast in place and not removed from a mold shape, they must be classed as coatings.

SLUSH CASTING AND STATIC CASTING

Slush casting and static castings share the same type of process but often do not share materials. Slush castings rely on liquid casting materials, whereas static castings usually start with powders.

Slush Casting

The major materials for slush castings are plastisols and organosols. Plastisols are mixtures of only finely ground plastics and plasticizers (Figure 14-8). Organosols are vinyl or polyamide dispersions in organic solvents and plasticizers (Figure 14-9). Organosols consist of 50 to 90% solids. The solids are tiny particles of plastic, frequently ground PVC. Organosols contain both plasticizers and varying amounts of solvents. A plastisol may be converted to an organosol by adding selected solvents.

Slush-cast items are hollow but have an opening similar to those found in doll parts, syringe bulbs, and special containers. Any design in the mold will be on the outside of the product.

Slush casting involves pouring dispersions of polyvinyl chloride or other plastics into a heated, hollow, open mold. As the material strikes the walls of the mold, it begins to solidify (Figure 14-10). The wall thickness of the molded part increases as the temperature is increased or solution is left in the hot mold. When the desired wall thickness is reached, the excess material is poured from the mold. The mold is then placed in an oven until the plastics fuses together or evaporation of solvents is complete. After water cooling, the mold is opened and the product is removed.

Commercial molds are usually made from aluminum because this metal allows rapid cycling and lower tooling costs. Ceramic, steel, plaster, or plastics molds may also be used. Vibrating, spinning, or use of vacuum chambers may be necessary to drive out air bubbles in the plastisol product.

Organosols are cast, and the solvents are allowed to escape. The dry, unfused plastics is left on the substrate. Heat is then applied to fuse the plastics.

Static Casting

Thermoplastic powders are also used in a dry process sometimes called static casting. In static casting, a metal mold is filled with powdered plastics and placed in a hot oven (Figure 14-11). As heat penetrates the mold, the powder melts and fuses to the mold wall. When the desired wall thickness is obtained, excess powder is removed from the mold. The mold is then returned to the oven until all powder particles have completely fused together.

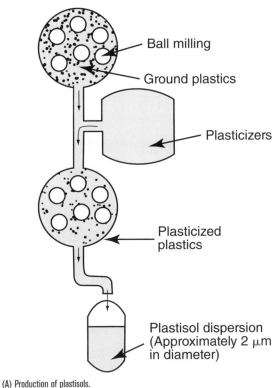

Ball milling

Ground plastics

Plasticizers

Plasticized plastics

Plastisol dispersion (Approximately 2 μm in diameter)

(A) Production of plastisols.

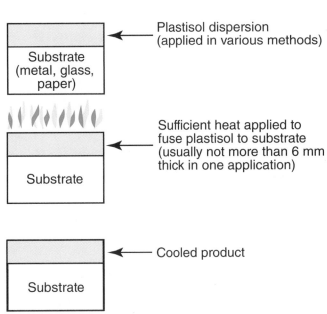

Plastisol dispersion (applied in various methods)

Substrate (metal, glass, paper)

Sufficient heat applied to fuse plastisol to substrate (usually not more than 6 mm thick in one application)

Substrate

Cooled product

Substrate

(B) Fusion of plastisols to substrate.

Figure 14-8. Production and use of plastisols.

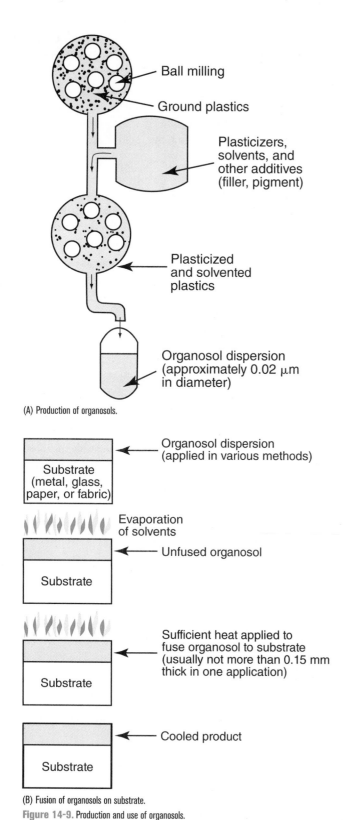

(A) Production of organosols.

(B) Fusion of organosols on substrate.

Figure 14-9. Production and use of organosols.

Huge storage tanks and containers with heavy walls are examples of products made by using this casting method. Cellular polystyrene or polyurethanes may be placed in the remaining space in manufacture of tough-skinned flotation devices.

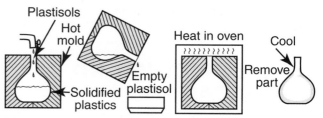

Figure 14-10. Basic slush casting with plastisols.

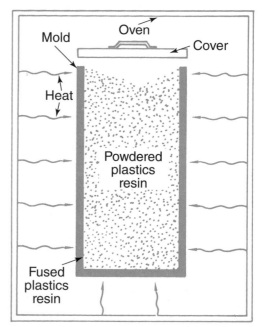

Figure 14-11. Principle of static casting.

In a related process known as **vibrational microlamination (VIM)**, a combination of heat and vibration is used. Thin layers of homopolymer alternating with reinforced layers are utilized to produce huge storage tanks, hollow toys, syringe bulbs, or other containers (Figure 14-12).

ROTATIONAL CASTING

Rotational casting relies on the rotation of a mold to evenly distribute the casting material on its walls. The materials used are plastics powders, monomers, or dispersions. Rotational casting includes two basic categories that stem from the number of axes of rotation. If a mold rotates on only one plane, the process is identified as **centrifugal casting**. If the mold moves on two planes of rotation, the process is **rotational casting**.

Centrifugal Casting

Centrifugal casting commonly produces cylindrical shapes. Large pipes and tubes are examples of typical products. By using specialty processes, inflatable mandrels and reinforced materials may be placed next to the skin layer to produce a

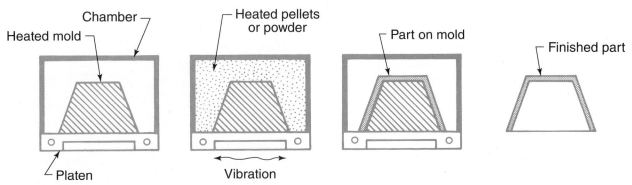

Figure 14-12. In vibrational microlamination (VIM), different formulations, types of plastics, and/or reinforcements may be alternately used to produce a true laminate.

high-density composite containing ribs or other geometric designs. In another operation, wet layup is placed on the mold wall. Centrifugal force causes the reinforcement and matrix to take the shape of the mold.

Rotational Casting

In rotational casting, plastics powders or dispersions are measured and placed in multipiece aluminum molds. The mold is then placed in an oven and rotated in two planes (axes) at the same time (Figure 14-13). This movement spreads the material evenly on the walls of the hot mold. The plastics melts and fuses as it touches the hot mold surfaces, making a one-piece coating. The heating cycle is complete when all powders or dispersions have melted and fused together. However, the mold continues to rotate as it enters a cooling chamber. Cast crystalline polymers are generally air cooled, whereas amorphous polymers may be quickly cooled by a water spray or bath. Finally, the cooled plastics product is removed.

By programming the rotation speed, the wall thickness in different areas may be controlled. If it is desirable to have a thick wall section around the parting line of a ball, the minor axis can be programmed to turn at a faster speed than the major axis. This places more powdered material against the hot mold in that particular area. Figure 14-14 shows how this is done.

Rotational casting may be used for hollow, completely closed objects such as balls, toys, containers, and industrial parts, including armrests, sun visors, fuel tanks, and floats. Figure 14-15 shows rotational casting equipment and Figure 14-16 shows some rotational-cast items.

Foam-filled and double-walled items, including true composites, can also be produced. Short-fiber reinforcements are used, and care must be taken to prevent wicking of protruding reinforcements. Placing a homopolymer layer over the reinforced polymer may overcome this problem. In one operation, a solid outer skin layer is produced, followed by the release of a second charge of material from a *dump box* in the mold.

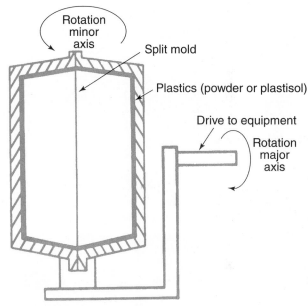

(A) Basic principle of rotation.

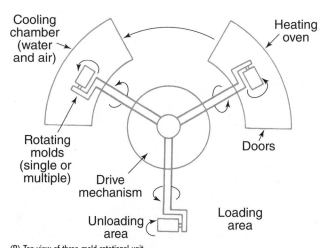

(B) Top view of three-mold rotational unit.

Figure 14-13. Principle of rotational casting.

Figure 14-14. Wall thickness in a ball can be varied by rotating the minor axis faster than the major axis.

(A) Rotationally molded vehicle frontal protection system (VFPS)

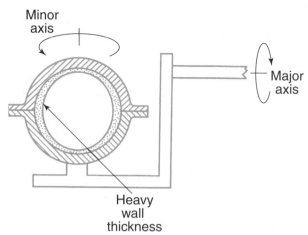

(A) Eight-foot independent-arm style Rotospeed™ rotational molding machine.

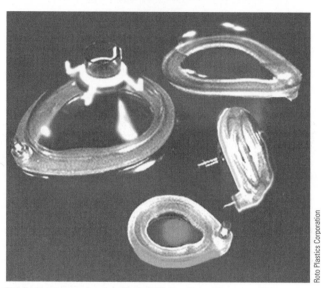

(B) Inflatable oxygen mask seal provides flexibility for patient comfort.

Figure 14-16. Products made by rotational molding equipment.

(B) Sixteen-foot independent-arm style Rotospeed™ rotational molding machine.

DIP CASTING

Dip casting is a simple process in which a heated mold is dipped into a liquid dispersion of plastics. The plastics melts and adheres to the hot metal surface. After the mold is removed from the dispersion, it enters a curing oven to ensure proper fusion of the plastics particles. After curing, the plastics is peeled from the mold.

Dip casting should not be confused with *dip coating*. Coatings are not removed from the substrate. In dip casting, a preheated mandrel, the shape and size of the *inside* of the product, is lowered into a plastisol dispersion (Figure 14-17). As the resin hits the hot mold surface, it begins to melt and fuse. The thickness of the piece continues to increase as it remains in the solution. If additional thickness is desired, the coated piece may be reheated and dipped again. After the desired thickness is obtained, the mold is removed from the oven and cooled, and the part is then stripped from the mold. Thickness of the product is determined by the temperature of the mold and the amount of time the hot mold is in the plastisol.

(C) Six-foot turret-style Rotospeed™ rotational molding machine.

Figure 14-15. Machines for rotational casting.

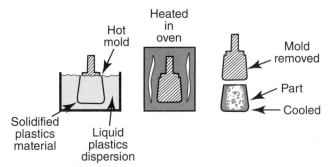

Figure 14-17. Scheme of dip casting.

Several layers or alternate colors and formulations may be applied by alternate heating and dipping. Any design on the mold will appear on the inside of the product. Plastics gloves, overshoes, coin purses, spark plug covers, and toys are examples of dip-cast products.

RELATED INTERNET SITES

- **www.ferryindustries.com.** Ferry Industries, Inc., makes rotational molding machines. Click "Learn More" on RotoSpeed™ to access information about these machines.

- **www.4spe.org.** On the home page of the Society of Plastics Engineers, scroll down to find "Technical Divisions." Then scroll right or left to find "Rotational Molding." Once in the Rotational Molding Division, select "Website" to find information about rotational molding companies and developments.

- **www.rotomolding.org.** The Association of Rotational Molders (ARM) is an international trade association with member companies in almost 60 countries. Its site includes an online bookstore with numerous publications about rotational molding.

VOCABULARY

The following vocabulary words are found in this chapter. Use the glossary in Appendix A to look up the definitions of any of these words you do not understand as they apply to plastics.

> casting
> centrifugal casting
> dip casting
> embedment
> encapsulation
> hot melt
> organosols
> plastisols
> potting
> rotational casting
> simple casting
> slush casting
> static casting
> vibrational microlamination

QUESTIONS

14-1. Identify the amount of pressure required for casting processes.

14-2. List three reasons for taking precautions when using polyester resins and catalyst.

14-3. The process of submerging a hot mold into a resin and removing the plastics from the molds is called _____.

14-4. A major disadvantage of rotational casting is the _____.

14-5. Name three methods that may drive out air bubbles in simple or plastisol castings.

14-6. What materials can be used as molds for dip casting?

14-7. Name the casting compounds that are vinyl dispersions in solvents.

14-8. Identify the plastics sheets that are often cast between two polished sheets of glass. Protective paper is then applied to the surface.

14-9. A process where an object is completely enclosed in transparent plastics is called _____.

14-10. Molds for static casting are normally made of _____.

14-11. What determines the wall thickness of dip castings?

14-12. Name three measures that would help to reduce the problem of air bubbles in simple castings.

14-13. Name a process similar to slush casting where dry thermoplastic powders are used.

14-14. Identify the major difference between dip coating and dip casting.

14-15. Hollow, one-piece objects may be made by the _____ casting process.

14-16. In order to completely fuse the powders or dispersions, a second heating cycle is necessary with _____ casting.

14-17. Because easily shaped tooling materials are used, casting molds are normally _____ expensive than molds for injection molding.

14-18. What two parameters determine the wall thickness of a rotational casting?

14-19. If a plastics is not easily processed by heat methods, what process may be used to make very thin films?

14-20. Plastisol products are popular because inexpensive methods and molds are used with _____ casting.

14-21. Name four reasons why castings processes are less costly than molding operations.

14-22. Name the common cause of pits or pockmarks in or on castings.

ACTIVITIES

Rotational Casting

Introduction. In rotational casting, the mold receives a measured amount of plastics. The mold is closed, heated, and rotated on two axes, producing hollow, one-piece products. Castings may include layers of differing materials.

Equipment. Rotational-casting equipment, rotational casting molds, mold release, powdered polyethylene, heat resistant gloves, and bucket with water.

Procedure

14-1. Preheat molding oven to 200°C (392°F).

14-2. Clean the molds and apply a light coat of mold release. Excessive mold release will damage the surface smoothness of the moldings.

14-3. Place a measured amount of polyethylene powder (precolored or add color pigment if desired) in one half of the mold. Clean all powder from mold lips to prevent flash (see Figure 14-18).

14-4. Close the mold and place it in a rotational device.

! CAUTION

Use protective gear. Oven is hot (see Figure 14-19).

14-5. Close the oven door, and set the time for 10 minutes.

14-6. Set the rotating mechanism at the fastest setting for small molds and slower for larger molds.

Figure 14-18. Mold with unfused powder.

Figure 14-19. Mold closed and placed in oven.

14-7. Turn off electrical heaters, and begin the cooling cycle. Continue rotation because air stream aids cooling; otherwise, the hot plastics in the mold will run or sag. Air cool for 5 minutes. Do not remove from the oven until the mold temperature is below 100°C (212°F).

14-8. Place the mold in water to continue rapid cooling.

! CAUTION

Mold is hot!

14-9. Remove the mold from the water and open it. Do not use metal tools to remove parts because they may damage mold surfaces (see Figure 14-20).

14-10. Trim the flash if necessary. Cut the part in half, and measure the wall thickness. Is the wall thickness uniform?

Figure 14-20. After cooling, part is removed from mold.

14-11. Try to predict the wall thickness that will result from a larger or smaller charge of powder. Make the molding to evaluate the prediction.

Slush Casting

Introduction. Slush-casting variations are used to produce hollow items. Although many industrial products require automatic equipment, some parts must be manually stripped from the mold. Plastisol products range in texture from soft and supple to semirigid. This variation depends on the formulation of the plastisol.

Equipment. Heat-resistant gloves, oven, slush-casting mold, plastisol, mold release, and container with water.

Procedure

14-1. Clean the mold and apply a light coat of mold release.

14-2. Preheat oven to 200°C (392°F).

14-3. Place the mold in a hot oven for 10 minutes. Use protective gear.

14-4. Remove the hot mold from the oven. Place it on a heat-resistant surface, and quickly pour plastisol into the mold (see Figure 14-21).

14-5. After 5 minutes, pour the excess plastisol from the hot mold (see Figure 14-22).

14-6. Set the oven to 175°C (347°F).

14-7. Return the mold to the hot oven for 20 minutes. Steps 4 through 7 may be repeated if a heavier wall thickness is desired.

14-8. Remove the hot mold from the oven and quench it in water (see Figure 14-23).

14-9. Remove the part from the mold. Do not use metal tools for part removal or the mold surface may be scratched (see Figure 14-24). Figure 14-25 shows a slush-cast doorstop.

14-10. Repeat the entire process to find the effect of longer preheat times or greater preheat temperatures.

Figure 14-22. After about 5 minutes, pour excess plastisol from hot mold.

Figure 14-23. After plastisol has cured, quench object in water.

Figure 14-24. Strip plastisol from mold.

Figure 14-21. Fill hot mold with plastisol.

Figure 14-25. Finished plastisol slush-cast item.

Polyester Casting

Introduction. An investigation of polyester casting resins can provide a demonstration of polymerization and an opportunity to observe the polymerization reaction. The catalyst (hardener) used to cure polyester resin is often methyl ethyl ketone peroxide (MEK).

! CAUTION

MEK is an organic peroxide and must be handled carefully.

MEK combines chemically with accelerators in the resin. This reaction releases chemicals called free radicals, which cause the reactive (unsaturated) polyester molecules to begin to polymerize. To document the change from liquid resin to solid, polymerized polyester, complete the following activity.

Equipment. 4-ounce paper cups, polyester casting resin and hardener, stir sticks, vinyl gloves, band saw, milling machine, and Rockwell hardness tester.

Safety Precautions. Wear vinyl gloves when handling liquid resin and catalyst. Work in a well-ventilated area.

Procedure

14-1. Establish a volume for the samples. About half of a 4-ounce cup is convenient. Mark the cups to ensure that all samples are equivalent in volume.

14-2. Select steps of hardener concentration. Figure 14-26 shows cups with 4, 8, 12, 16, 20, and 24 drops of hardener. After placing the hardener in resin, stir thoroughly using stir sticks.

14-3. Allow 24 hours for polymerization.

14-4. Remove the paper cup from the hardened polyester. Figure 14-27 shows hardened samples. The sample with 4 drops did not harden.

Figure 14-26. Polyester samples with various hardener concentrations.

Figure 14-27. One sample as removed from paper cup, one sample as cut with bandsaw, and one machined to create a flat surface.

14-5. Use a band saw to cut two sides to provide clamping surfaces. Also, cut some material off the top and bottom to provide clean surfaces for machining. Figure 14-27 shows one sample ready for machining.

14-6. Mill the top and bottom flat. If these surfaces are not flat, hardness testing results may be erroneous. Figure 14-27 also shows a sample after machining.

14-7. Test the hardness of the samples with a Rockwell hardness tester. The Rockwell M scale should be appropriate.

14-8. Graph the results. If the steps of hardener concentration were too large or too small, repeat with smaller increments. Figure 14-28 shows the results of such a test. Notice that it includes data from samples with 10 and 14 drops of hardener. Notice also that at 8 drops, the material was too soft for reliable measurement. The samples with 10, 12, 14, and 16 drops show the increasing completion of the polymerization reaction.

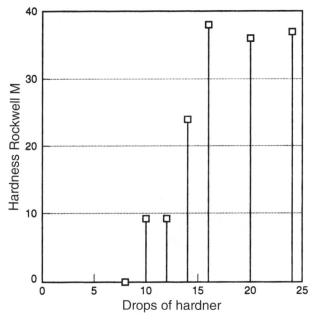

Figure 14-28. Graph of Rockwell hardness test results.

14-9. To gain additional information, select a concentration of hardener that yields fully hardened polyester.

14-10. Make one sample at the selected concentration, and measure the changes in the temperature of the resin as it polymerizes. Figure 14-29 shows a setup used to gather the time/temperature data.

14-11. Graph the results. Figure 14-30 shows a curve created by 16 drops of hardener. The highest temperature reached is called the peak exotherm.

14-12. Determine how changes in hardener concentration affect the time–temperature curve.

Figure 14-29. Setup for measuring temperature of curing resin.

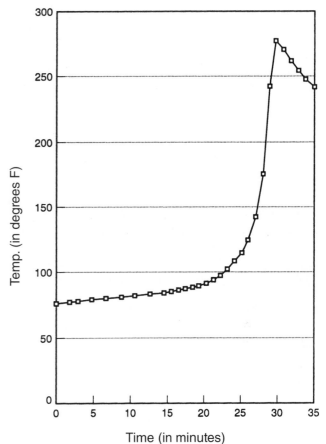

Figure 14-30. Time/temperature graph of curing polyester resin.

THERMOFORMING

CHAPTER 15

INTRODUCTION

Thermoforming is an ancient technique. Ancient Egyptians found that animal horns and tortoise shells could be heated and formed into a variety of vessels and shapes. When synthetic plastics became available, thermoforming was used as an early application. In the United States, John Hyatt thermoformed celluloid sheets over wooden cores, producing piano keys.

Today, thermoformed items surround us. They include signs, light fixtures, ice-cube trays, ducts, drawers, instrument panels, tote trays, housewares, toys, refrigerator panels, transparent aircraft enclosures, and boat windshields (Figure 15-1). The packaging industry relies heavily on thermoforming. Cookies, pills, and other products are commonly packaged by blister packaging. Single portions of butter, jellies, and other foods also appear in blister packs. Replacement parts and hardware are other examples of items that are sometimes skin packaged.

(B) Clear vacuum-formed covers display food products.

(A) This customized car body is made of vacuumed-formed panels that were glued together.

(C) Clear plastics web from the production of meat trays on a roll-fed in-mould-cut GN Thermoformer.

Figure 15-1. Some articles fabricated by thermoforming processes.

The materials used in thermoforming include most thermoplastics—except acetals, polyamides, and fluorocarbons. Usually, thermoforming sheets contain only one basic plastics. However, some thermoplastic composites are also thermoformed.

Thermoforming processes are possible because thermoplastic sheets can be softened and reshaped while retaining the new shape as material is cooled. Because most thermoforming sheets were originally formed by sheet extrusion, considerable energy, time, and space can be saved by directly thermoforming sheets as they leave the extruder. However, many thermoforming industries frequently change materials, colors, and textures. They are not good candidates for immediate thermoforming of extruder sheets.

The type of force required to alter a sheet into the desired product can be mechanical, air, or vacuum pressure. In many cases, thermoforming requires a combination of two or three pressure sources.

Tooling can vary from low-cost plaster molds to expensive water-cooled steel molds. The most common tooling material is cast aluminum. Wood, gypsum, hardboard, pressed wood, cast phenolic resins, filled or unfilled polyester or epoxy resins, sprayed metal, and steel also may be used for molds. Both male (plug) and female (cavity) molds are used. Molds require sufficient draft to assure stress-free parts removal.

Because tooling costs are usually low, parts with large surface areas may be produced economically. Prototypes and short runs are also practical. Although dimensional accuracy is good, thinning can be a problem in some part designs.

In this chapter, 13 basic thermoforming techniques are discussed. They are listed in the following chapter outline:

I. **Straight vacuum forming**

II. **Drape forming**

III. **Matched-mold forming**

IV. **Pressure-bubble plug-assist vacuum forming**

V. **Plug-assist vacuum forming**

VI. **Plug-assist pressure forming**

VII. **Solid phase pressure forming (SPPF)**

VIII. **Vacuum snap-back forming**

IX. **Pressure-bubble vacuum snap-back forming**

X. **Trapped-sheet contact-heat pressure forming**

XI. **Air-slip forming**

XII. **Free forming**

XIII. **Twin sheet thermoforming**

XIV. **Blister pack or skin pack thermoforming**

XV. **Mechanical forming**

Some of the modern industrial thermoforming machines that perform these processes are shown in Figure 15-2.

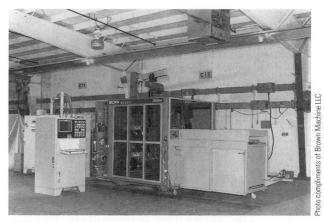

(A) This continuous thermoformer can achieve a draw depth of 127 m (5 in.).

(B) This high speed thermoformer can reach 40 strokes per minute.

(C) The electric index on this thermoformer allows rapid and reliable sheet indexing.

Figure 15-2. Modern industrial thermoforming machines.

Photo compliments of Brown Machine LLC

STRAIGHT VACUUM FORMING

Vacuum forming is the most versatile and widely used thermoforming process. Vacuum equipment costs less than pressure or mechanical processing equipment.

In straight vacuum forming, a plastics sheet is clamped in a frame and heated. While the hot sheet is in a rubbery state, it is placed over a recessed mold cavity. The air is removed from this cavity by vacuum (Figure 15-3), and atmospheric

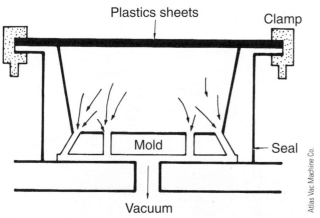

(A) A clamped and heated plastics sheet is forced down into the mold by atmospheric pressure after a vacuum is drawn in the mold.

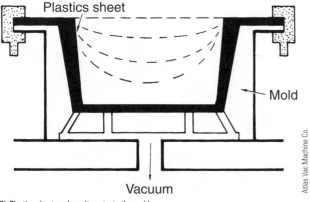

(B) Plastics sheet cools as it contacts the mold.

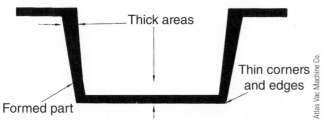

(C) Areas of the sheet that touched the mold last are the thinnest.

Figure 15-3. Straight vacuum forming.

pressure (10 kPa) forces the hot sheet against the walls and contours of the mold. When the plastics has cooled, the formed part is removed if necessary, and final finishing and decorating may be done. Blowers or fans are used to speed cooling. One disadvantage of thermoforming is that formed pieces usually must be trimmed and the scrap must be reprocessed.

Most vacuum systems have a surge tank to ensure a constant vacuum of 500 to 760 mm of mercury. Superior parts are formed by quickly applying the vacuum before any portion of the sheet has cooled. Slots are preferred because they are more efficient than holes in allowing the air to be drawn from the mold. Slots or holes should be smaller than 0.65 mm (0.025 in.) in diameter to avoid surface blemishes on the formed part. A hole or slot should be placed in all low or unconnected portions of the mold. If this is not done, air may be trapped under the hot sheet with no way to escape. Unless they are collapsible, molds should include a 2° to 7° (draft) angle for easy part removal.

Thinning at the edges and corners of a part is one disadvantage of using relatively deep recessed molds. As the sheet material is drawn into the mold, it stretches and thins. Regions that are stretched minimally will remain thicker than regions that are stretched extensively. If preprinted flat sheets are formed, thinning must be kept in mind when trying to compensate for distortion during forming. Straight vacuum forming is limited to simple, shallow designs, and thinning will occur in corners.

The **draw** or **draw ratio** of a recessed mold is the ratio of the maximum cavity depth to the minimum span across the top opening. For high-density polyethylene, the best results are achieved when this ratio does not exceed 0.7:1. Thermoforming equipment and dies are relatively inexpensive.

When the plastics has cooled, it is removed for trimming or post-processing, if needed. Markoff (marks from the mold) appears on the *inside* of the product, whereas such marks are visible on the *outside* of the part in straight vacuum forming.

DRAPE FORMING

Drape forming (incorrectly called mechanical forming) is similar to straight vacuum forming except that after the plastics is framed and heated, it is mechanically stretched over a male mold. A vacuum (actually, a pressure differential) is applied that pushes the hot plastics against all portions of the mold (Figure 15-4). The sheet touching the mold remains close to its original thickness. Side walls are formed by the material draping between the top edges of the mold and the bottom seal at the base.

It is possible to drape form items with a depth-to-diameter ratio of nearly 4:1. Although high draw ratios are possible with

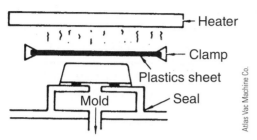

(A) Clamped heated plastics sheet may be pulled over the mold, or the mold may be forced into the sheet.

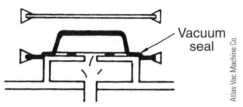

(B) Once the sheet has formed a seal around the mold, a vacuum is drawn to pull the plastics sheet tightly against the mold surface.

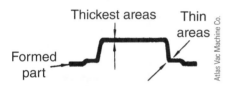

(C) Final wall thickness distribution in the molded part.
Figure 15-4. Principle of drape forming plastics.

drape forming, this technique is also more complex. As a rule, male molds are easy to make and cost less than female molds but are more easily damaged.

Drape forming has also been used to form a hot plastics sheet over male or female molds by gravitational forces alone. Female molds are preferred for multicavity forming because male molds require more spacing.

MATCHED-MOLD FORMING

Matched-mold forming is similar to compression molding. A heated sheet is trapped and formed between male and female dies that may be made of wood, plaster, epoxy, or other materials (Figure 15-5). Accurate parts with close tolerances may be quickly produced in costly water-cooled molds. Very good molded detail and dimensional accuracy, including lettering and grained surfaces, can be obtained with water-cooled molds. Because there is markoff on both sides of the finished product, mold dies *must* be protected from scratches or damage. Such defects would be reproduced by the thermoplastic materials.

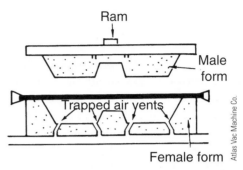

(A) The heated plastics sheet may be clamped over the female die, as shown, or draped over the mold form.

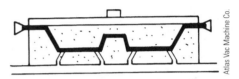

(B) Vents allow trapped air to escape as the mold closes and forms the part.

(C) Distribution of materials in the product depends on the shapes of the two dies.

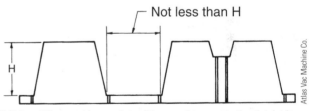

(D) Male mold forms must be spaced at a distance equal to or greater than their height or webbing may occur.
Figure 15-5. Principle of matched-mold forming.

A smooth-surface mold should not be used with polyolefins because air may become trapped between the hot plastics and a highly polished mold. Sandblasted mold surfaces are usually used for these materials.

PRESSURE-BUBBLE PLUG-ASSIST VACUUM FORMING

For deep thermoforming, pressure-bubble plug-assist vacuum forming is an important process. By utilizing this process, it is possible to control the thickness of the formed article. The item may have uniform or varied thickness.

Once the sheet has been placed in the frame and heated, controlled air pressure creates a bubble (Figure 15-6). This bubble stretches the material to a predetermined height usually controlled by a photocell. The male plug assist is then lowered, forcing the stretched stock down into the cavity. The male plug is normally heated to avoid chilling the plastics prematurely. The plug is made as large as possible to enable the plastics to be stretched close to the final shape of the finished product. Plug penetration should range from 70% to 80% of the mold cavity depth. Air pressure is then applied from the plug side while simultaneously a vacuum is drawn on the cavity to help form the hot sheet. For many products, vacuum alone is used to complete the formation of the sheet. In Figure 15-6, both vacuum and pressure are applied during the forming process. The female mold must be vented to allow trapped air to escape from between the plastics and the mold.

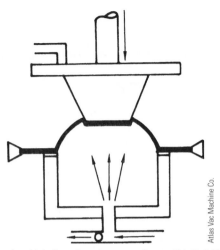

(C) A plug, shaped roughly to the cavity contour presses downward into the bubble, forcing it into the mold.

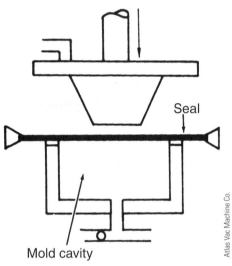

(A) The plastics sheet is heated and sealed across the mold cavity.

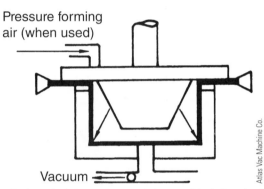

(D) When the plug reaches its lowest point, vacuum is drawn to pull the plastics against the mold walls. Air may be introduced from above to aid forming.

Figure 15-6. Pressure-bubble plug-assist vacuum forming.

PLUG-ASSIST VACUUM FORMING

To help prevent corner or periphery thinning of cup- or box-shaped articles, a **plug assist** is used to mechanically stretch and pull additional plastics stock into the female cavity (Figure 15-7). The plug is normally heated to just below the forming temperature of the sheet stock. The plug should range from 10% to 20% smaller in length and width than the female mold. Once the plug has forced the hot sheet into the cavity, air is drawn from the mold, completing the formation of the part. The plug design or shape determines the wall thickness, as shown in a cross-section in Figure 15-7D.

Plug-assist vacuum and pressure forming allows deep drawing and permits shorter cooling cycles and better control of wall thickness. However, close temperature control is needed, and the equipment is more complex than that used in straight vacuum forming (Figure 15-8).

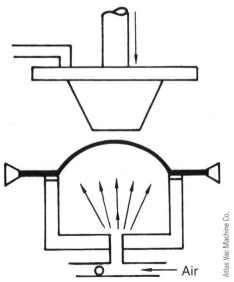

(B) Air is introduced, blowing the sheet upward into an evenly stretched bubble.

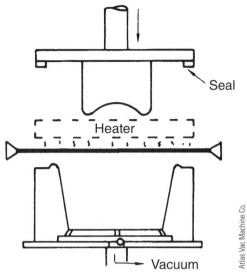

(A) Heated, clamped plastics sheet is positioned over mold cavity.

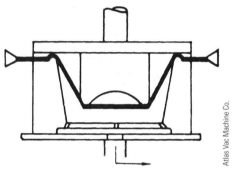

(B) A plug, shaped roughly like the mold cavity, plunges into the plastics sheet to prestretch it.

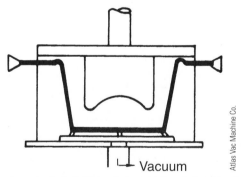

(C) When the plug reaches the limit of its travel, a vacuum is drawn in the mold cavity.

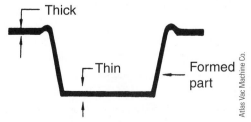

(D) Areas of the plug touching the sheet first form thickened areas due to chilling effect.

Figure 15-7. Plug-assist vacuum forming.

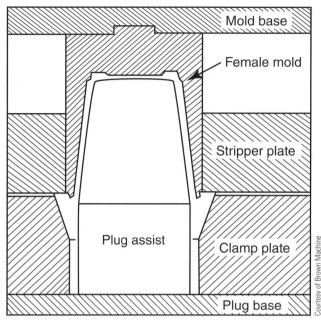

(A) Clamp layout

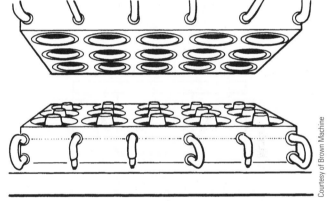

(B) RAM part clamps

Figure 15-8. Restricted-area molding (RAM), with individual part clamps built into the mold. This helps to control material draw and reduces draw ratio.

PLUG-ASSIST PRESSURE FORMING

Plug-assist pressure forming is similar to plug-assist vacuum forming because the plug forces the hot plastics into the female cavity. Air pressure applied from the plug forces the plastics sheet against the walls of the mold (Figure 15-9).

SOLID PHASE PRESSURE FORMING (SPPF)

A process called **solid phase pressure forming (SPPF)** is similar to plug-assist forming. The technique begins with a solid blank (extruded, compression molded, sintered powders)

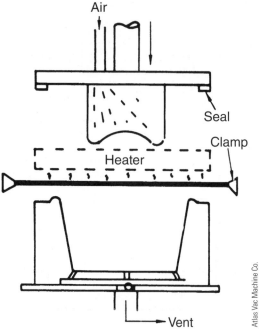

(A) Heated, clamped sheet is positioned over the mold cavity.

Atlas Vac Machine Co.

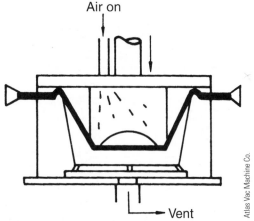

(B) As the plug touches the sheet, air is allowed to vent from beneath the sheet.

Atlas Vac Machine Co.

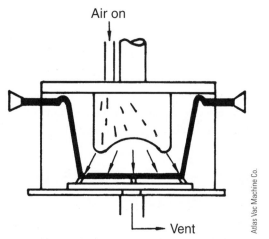

(C) As the plug completes its stroke and seals the mold, air pressure is applied from the plug side, forcing the plastics against the mold.

Atlas Vac Machine Co.

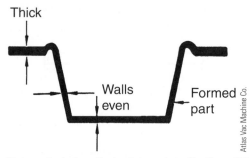

(D) Plug-assist pressure forming is capable of producing products with uniform wall thickness.

Figure 15-9. Plug-assist pressure forming.

that is heated to just below its melting point. Polypropylene or other multilayered PP sheets are used. The blank is then pressed into a sheet form and transferred to the thermoforming press. A plug stretches the hot material further, and air pressure forces it against the mold sides (Figure 15-10). The two (biaxial) stretching operations cause molecular orientation, which

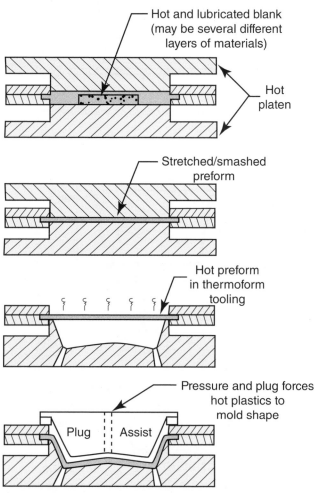

Figure 15-10. Concept of solid phase pressure forming.

enhances the strength, toughness, and environmental-stress crack resistance of the thermoformed product.

VACUUM SNAP-BACK FORMING

In vacuum **snap-back forming**, the hot plastics sheet is placed over a box and a vacuum is drawn, causing a bubble to be forced into the box (Figure 15-11). A male mold is lowered, and the vacuum in the box is released. This causes the plastics to *snap back* around the male mold. A vacuum may also be drawn in the male mold to help pull the plastics into place. Vacuum snap-back forming allows complex parts with recesses to be formed.

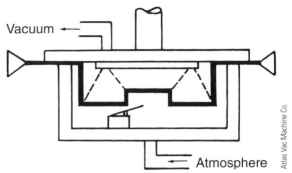

(C) The male plug is lowered and a vacuum drawn through it. At the same time, vacuum beneath the sheet is vented.

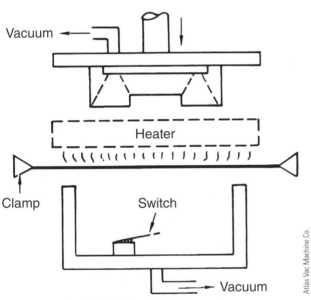

(A) Plastics sheet is heated and sealed over the top of the vacuum box.

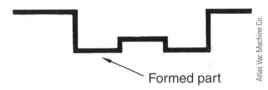

(D) External deep draws can be obtained with this process to form luggage, auto parts, and other items.

Figure 15-11. Vacuum snap-back forming.

PRESSURE-BUBBLE VACUUM SNAP-BACK FORMING

As the name implies, the sheet is heated and then stretched into a bubble shape by air pressure (Figure 15-12). The sheet prestretches about 35% to 40%, and the male mold is then lowered. A vacuum is applied to the male mold while air pressure is forced into the female cavity. This causes the hot sheet to snap back around the male mold. Markoff appears on the male mold side.

Pressure-bubble vacuum snap-back forming allows deep drawing and the formation of complex parts. However, the equipment is complex and costly.

TRAPPED-SHEET CONTACT-HEAT PRESSURE FORMING

This process is like straight vacuum forming except air pressure and a vacuum assist may be used to force the hot plastics into a female mold. Figure 15-13 shows the steps followed during this process.

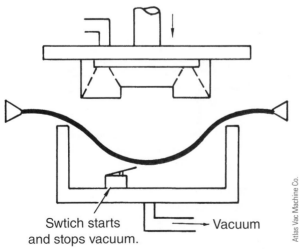

Swtich starts and stops vacuum.

(B) Vacuum is drawn beneath the sheet, pulling it into a concave shape.

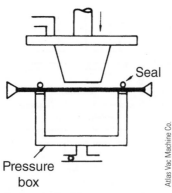

(A) Heated plastics sheet is clamped and sealed across a pressure box.

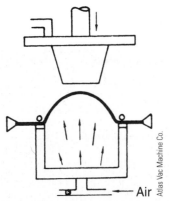

(B) Air pressure is introduced beneath the sheet, causing a large bubble to form.

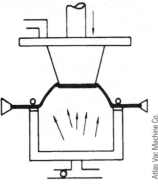

(C) A plug is forced into the bubble, while air pressure is maintained at a constant level.

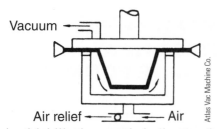

(D) Air pressure beneath the bubble and a vacuum at the plug side create a uniform draw.

Figure 15-12. Pressure-bubble vacuum snap-back forming.

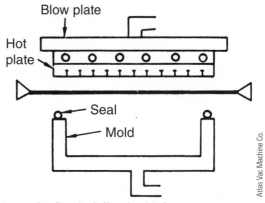

(A) A flat, porous plate allows air to be blown through its face.

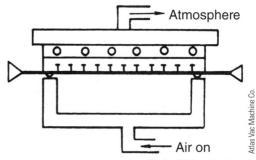

(B) Air pressure from below and a vacuum above force the sheet tightly against the heated plate.

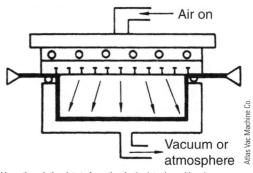

(C) Air is blown through the plate to force the plastics into the mold cavity.

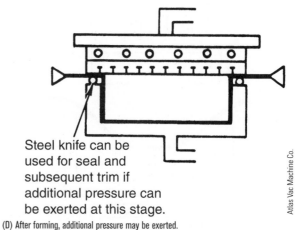

(D) After forming, additional pressure may be exerted.

Figure 15-13. Trapped-sheet contact-heat pressure forming.

AIR-SLIP FORMING

Air-slip forming is similar to snap-back forming with the exception of the method used for creating the stretch bubble. This concept is illustrated in Figure 15-14.

FREE FORMING

In **free forming**, air pressures of over 2.7 MPa (390 psi) may be used to blow a hot plastics sheet through the silhouette of a female mold (Figure 15-15). The air pressure causes the sheet

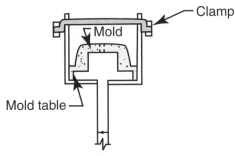

(A) Sheet is clamped to the top of a vertical walled chamber.

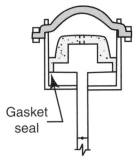

(B) Pre-bellow is achieved by a pressure buildup be-tween the sheet and mold table.

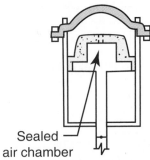

(C) Mold rises in chamber. Mold table is gasketed at edges of chamber wall.

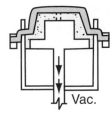

(D) Space between mold and sheet is evacuated as sheet is formed against mold by differential air pressure.

Figure 15-14. Air-slip forming.

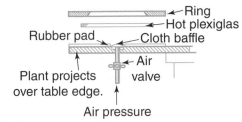

(A) Basic setup

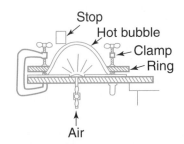

(B) Injection of air

(C) The inspector examines an acrylic bubble-formed side window for Bell Helicopter.

Courtesy of GN Thermoforming Equipment, Canada

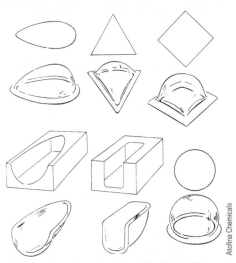

Atofina Chemicals

(D) Examples of free-form shapes that can be obtained with various openings.

Figure 15-15. Free forming of plastics bubbles.

to form a smooth bubble-shaped article. A stop can be utilized to form special contours in the bubble. Skylight panels and aircraft canopies are well-known examples of this technique. Unless a stop is used, there is no markoff. Only air touches each side of the material. There will be markoff from clamping.

TWIN SHEET THERMOFORMING

Twin sheet thermoforming is a method used to make thin-wall hollow products. In this process, two separate sheets are heated until hot and pliable. The sheets are indexed into the molding station, where they rest one just above the other. Vacuum draws the upper sheet into an upper mold cavity and the lower sheet into a lower mold. As the sheets stretch into the mold cavities, the mold halves close, sealing the two sheets together. Sometimes pressure may be introduced between the sheets to aid in forming and cooling. Ductwork, pallets, and plastics cases are items produced in this manner.

BLISTER PACK OR SKIN PACK THERMOFORMING

The packaging industry relies on blister packs or skin packs to hold and display small individual items. Blister packs generally start with a sheet of clear plastics material. This sheet receives forming by using a variety of the methods mentioned earlier in this chapter to create many little wells. After the product is located in the wells, a sheet of cardboard is placed on top. The sheet is reheated and pressed against the cardboard, thereby creating a seal. The full sheet is then cut into individual packages. Skin pack thermoforming is different because the items are placed on a sheet of perforated cardboard. A heated sheet of plastic is placed over the items and cardboard. The perforations allow a vacuum to draw the sheet around the items and seal the sheet to the cardboard. Again, the final step is to cut the sheet into individual packages.

MECHANICAL FORMING

In **mechanical forming**, no vacuum or air pressure is used to form the part. Although it is similar to matched molding, close-fitting matched male and female molds are *not* used. Only the mechanical force of bending, stretching, or holding the hot sheet is used.

This process is sometimes classified as a fabrication or post-forming operation. The forming process may utilize simple wooden forming jigs to give the desired shape, using ovens, a strip heater, or heat guns for the heat source. Flat stock may be heated and wrapped around cylindrical shapes, or stock may be heated in a narrow strip and bent at right angles. Tubes, rods, and other profile shapes may be mechanically formed (Figure 15-16).

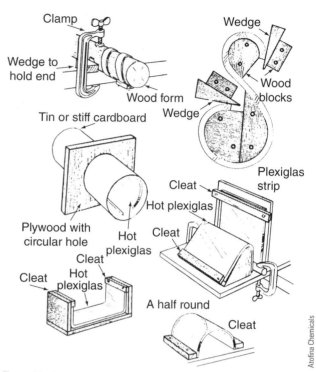

Figure 15-16. Examples of mechanical forming.

Plug-and-ring forming (Figure 15-17) is sometimes considered a separate forming process. No vacuum or air pressure is used. Therefore, it may be classified as a type of mechanical forming.

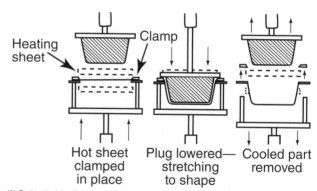

(A) Basic principle of plug-and-ring forming

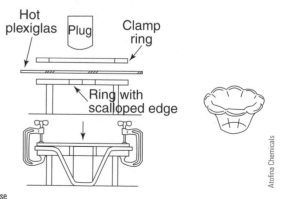

(B) Vase

Figure 15-17. Examples of plug-and-ring forming.

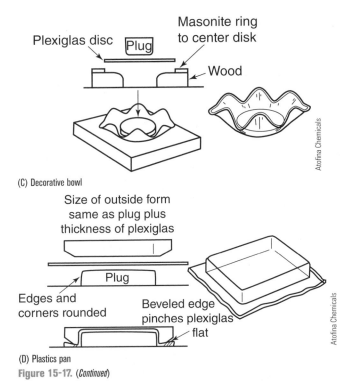

Plexiglas disc | Plug | Masonite ring to center disk
Wood

(C) Decorative bowl

Size of outside form same as plug plus thickness of plexiglas

Plug

Edges and corners rounded

Beveled edge pinches plexiglas flat

(D) Plastics pan

Atofina Chemicals

Figure 15-17. *(Continued)*

The process consists of a male mold shape and a similarly shaped female silhouette mold (not a matched mold). The hot plastics is forced through the *ring* (not necessarily a rough shape) of the female mold by the male. The cooling plastics takes the shape of the male mold that it touches. Table 15-1 lists some common problems encountered in thermoforming plastics.

RELATED INTERNET SITES

- **www.brown-machine.com.** Brown Machine manufactures thermoforming machines and auxiliary equipment. Selecting "Products" on the home page brings up a list of categories of machines, including "Cut sheet," "Continuous," and "Trim Presses." In addition to photographs, the site offers specifications of all machines.

- **www.gnplastics.com.** G.N. Plastics Company Limited manufactures thermoforming equipment. Selecting "Products" on the home page links to "Models." Choosing one of the various models brings up details about the machines and video clips of the machines in operation.

Table 15-1. Troubleshooting Thermoforming

Defect	Possible Remedy
Pinholes or ruptures	Vacuum holes too large, too much vacuum or uneven heating
	Attach baffles to the top clamping frame
Webbing or bridging	Sharp corners on deep draw, change design or mold layout
	Use mechanical drape or plug assists or add vacuum holes
	Check vacuum system and shorten heating cycle
Markoff	Slow draping action may trap air
	Clean mold or remove high surface gloss from mold
	Remove all tool marks or wood grain patterns from mold
	Mold may be chilling plastic sheet too quickly
Excessive post shrinkage	Rotate sheet in relation to mold
	Increase cooling time
Blisters or bubbles	Overheating sheet—lower heater temperature
	Ingredients of sheet formulation incorrect or hygroscopic
Sticking to mold	Smooth mold or increase taper and draft
	Use mechanical releasing tools, air pressure, or mold release
	Mold may be too warm or increase cooling cycle
Incompletely formed pieces	Lengthen heating cycle and increase vacuum
	Add vacuum holes
Distorted pieces	Poor mold design—check tapers and ribs
	Increase cooling cycle or cool molds
	Sheet removed too quickly while still hot
Change in color intensity	Use proper mold design and allow for thinning of piece
	Lengthen heating cycle and warm mold and assists
	Use heavier gauge sheet and add vacuum holes

- **www.plaskolite.com.** Plaskolite, Inc., manufactures a variety of acrylic sheets primarily used for thermoforming. Select "Product Lines" to find descriptions, specifications, and photos of a number of acrylic sheet materials.

- **www.polypack.com.** Polypack, Inc. sells a wide variety of shrink packaging equipment. Select "Equipment" on the home page. The site provides a short video about each of the machines, and a brochure. The video clips are excellent in quality and show the machines in operation.

- **www.tracopackaging.com.** Traco Manufacturing Inc., makes shrink-wrap films and machines. To find information about one of their semi-automatic machines, select "Solutions" on the home page. Then pick "Equipment," and finally "L-Bar & Shrink Tunnel."

VOCABULARY

The following vocabulary words are found in this chapter. Use the glossary in Appendix A to look up the definitions of any of these words you do not understand as they apply to plastics.

air-slip forming
drape forming
draw ratio
free forming
matched-mold forming
mechanical forming
plug assist
plug-and-ring forming
snap-back forming
solid phase pressure forming (SPPF)
thermoforming
twin sheet thermoforming
vacuum forming

QUESTIONS

15-1. Name three materials that vacuum-forming molds can be made from.

15-2. Identify the packaging that makes use of the product as the mold.

15-3. Identify the process that will produce products with the greatest detail.

15-4. Mechanical forming is the name incorrectly used for _____ forming.

15-5. The name of the unit or tool used to heat a small section of plastics so it may be bent at a sharp angle is _____.

15-6. The sides of a thermoforming mold are tapered to aid in the removal of the part. This taper is called _____.

15-7. Name four typical thermoformed products.

15-8. Which thermoforming technique is used to form a part with a very deep draw?

15-9. A _____ or _____ is normally placed in all low or unconnected portions of the thermoforming mold.

15-10. The _____ of a female mold is the ratio of the maximum cavity depth to the minimum span across the top opening.

15-11. What is a major disadvantage of straight vacuum forming using deep cavities?

15-12. In _____ forming, no vacuum or air pressure is used to form the hot plastics sheet.

15-13. Name the term used for marks left on the formed sheet if the mold is not smooth or clean.

15-14. If vacuum holes are too large or there is too much vacuum or uneven heating, _____ or _____ will occur.

15-15. Is a male or female mold used in free forming?

15-16. In plug-assisted methods, plug penetration would normally not exceed _____% of the mold cavity depth.

15-17. In _____-and-_____ forming, a male and similarly shaped female silhouette mold shape the hot plastics.

15-18. Sharp corners on a deep draw may cause _____ or bridging.

15-19. Name three advantages that metal thermoforming molds offer.

15-20. What can you do with the scrap and trim that remain from thermoforming processes?

15-21. One of the most common tooling materials for thermoforming is _____.

15-22. Thermoforming is possible because thermoplastic sheets can be _____ and _____.

15-23. In vacuum forming, _____ forces the hot plastics sheet against the mold contours.

15-24. It is more desirable and efficient to have _____ rather than holes as passages for drawing the air from the mold.

15-25. Male molds are easy to make and generally cost less than _____ molds.

15-26. Accurate, close-tolerance parts with good detail may be made by _____-_____ forming.

15-27. As a general rule, a smooth-surfaced mold should not be used when forming _____.

15-28. It is easier to control the thickness of deep thermoformed products by using the _____ or the pressure-bubble vacuum snap-back forming process.

15-29. Give two advantages of thermoforming.

15-30. Give two disadvantages of thermoforming.

15-31. Name a process where a heated thermoplastic sheet is pulled down into or over a mold surface.

15-32. Is a vacuum used in drape forming?

15-33. What is the major disadvantage of matched-mold thermoforming?

15-34. In vacuum forming, only _____ pressure is used.

15-35. The major difference between plug-assisted pressure forming and plug-assisted vacuum forming is that _____ pressures may be used with pressure forming.

15-36. Describe straight vacuum thermoforming.

15-37. Describe how product thickness is controlled in pressure-bubble plug-assist vacuum forming.

15-38. What determines product wall thickness in plug-assist pressure thermoforming?

15-39. What does the word *snap-back* refer to in vacuum snap-back forming?

15-40. Describe free forming.

15-41. Describe mechanical forming and plug-and-ring forming.

ACTIVITIES

Free-Blow Thermoforming

Introduction. In free-blow thermoforming, a ring or yoke holds down a sheet of hot plastics. The ring must prevent air leakage so that when air is forced below the sheet, it expands upward in a smooth bubble. Acrylic sheets are popular for free-blown products.

Equipment. A thermoforming machine, a ring or yoke, regulated air supply, and personal safety equipment.

Procedure

15-1. Cut a sheet of plastics about 50 mm (2 in.) larger than the opening in the form. Figure 15-18 shows a typical ring for free blowing.

15-2. Set the air pressure to "0" on the regulator.

15-3. Mount the plastics sheet under the ring and heat until formable. If forming acrylic, heat the sheet until the temperature is between 150°C (300°F) and 190°C (375°F).

15-4. Accurately measuring the temperature of a thermoforming sheet with a pyrometer is difficult because the sheet loses temperature very rapidly. If possible, acquire thermoforming temperature strips, as shown in Figure 15-19. The increments on these strips change color at pre-set temperatures. Figure 15-20 shows a piece of plastics that reached a temperature of 143°C (290°F).

15-5. When the thermoformer has been on long enough to stabilize temperature, use the temperature strips to determine the oven time needed to reach the desired temperature. Once the time is known, use this time and eliminate temperature-measuring devices.

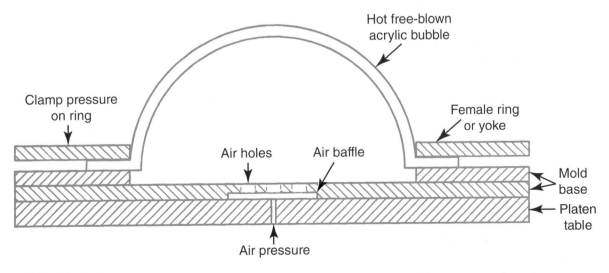

Figure 15-18. Ring for free blowing.

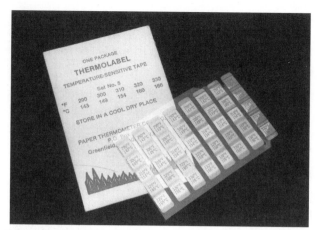

Figure 15-19. Temperature-sensitive tapes come in varying temperature ranges. This sample covers the range 143°C–166°C.

Figure 15-21. Simple thermoforming mold for Taber abrader shells.

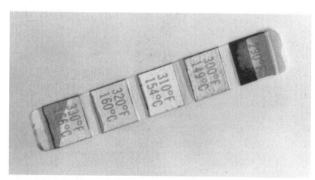

Figure 15-20. This strip reached the temperature of 143°C but did not reach 149°C.

Figure 15-22. Bottom of thermoforming mold, showing air holes and foam tape.

15-6. When the sheet is ready for forming, remove the heat source, and increase the air pressure until the part blows to the desired height.

15-7. Maintain pressure until the part is cool, and remove it from the clamp.

15-8. Cut the part in half, and measure the wall thickness. Is the wall uniform? How does the free-blown wall compare to the wall in straight vacuum-formed items?

Plastics Memory

Introduction. Thermoforming provides a good opportunity to observe plastics memory. The memory can lead to warp if a thermoformed item gets hot enough to permit stress relaxation.

Equipment. Thermoforming machine, thermoforming sheets, and simple molds.

Procedure

15-1. If a laboratory thermoformer is available but an appropriate mold is not, refer to Figure 15-21 and Figure 15-22 for an extremely simple mold. This mold is a piece of 2 × 6 with holes drilled and self-adhesive foam tape used to create both a seal and an opening to the vacuum holes. The aluminum disk permits thermoforming of shells for injection-molded Taber abrader disks. Other forms are equally useful in observing plastics memory.

15-2. After measuring the thickness of a sheet, heat it and form it around a mold, as shown in Figure 15-23.

15-3. Remove the sheet and let it cool. Then remount it in a thermoformer. Heat until the sheet pulls it flat, and then rapidly remove the heat source.

15-4. When the sheet is cool, measure it to see whether any thinning remains. Figure 15-23 shows both a formed sheet and a sheet returned to flat by memory.

15-5. Draw the same sheet further by using a thicker object. Figure 15-24 shows the use of a small C-clamp.

15-6. After removing the clamp or other item, reheat to see whether the sheet will flatten again. If no holes occurred during forming, the sheet should return completely.

15-7. How many times will plastics memory work? How deep can the sheet draw and still retain its memory?

Figure 15-23. Sheet thermoformed around aluminum disk, then flattened using plastics memory.

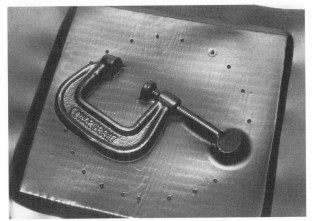

Figure 15-24. Sheet formed around a small C-clamp.

Straight Vacuum Forming and Plug-Assist Forming

Introduction. Localized thinning is a significant problem experienced with straight vacuum forming. The purpose of this exercise is to compare straight vacuum forming with plug-assist forming.

Equipment. A thermoformer, a female mold, a plug for the selected mold, and thermoforming sheets.

Procedure

15-1. If the thermoforming sheets are PS, no problems will arise with moisture. However, if the sheets are ABS, they may be wet and require drying. If not properly dried, these sheets may exhibit bubbles on the surface after forming.

15-2. Determine the time required to reach a chosen thermoforming temperature. If the sheets are PS, select a temperature between 130°C and 180°C (266°F to 356°F). Grid the thermoforming sheets as shown in Figure 15-25. The grid shown is spaced every 1 inch.

15-3. Locate the sheet in the clamp frame, making sure that the grid is parallel to the clamp. The mold used to demonstrate this exercise is seen in Figure 15-26.

15-4. Heat and form the sheet. Figure 15-27 shows straight vacuum forming. Notice how the sheet stretched. Section the formed sheet, and measure the wall thickness

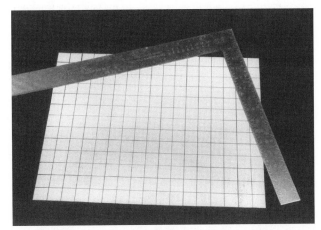

Figure 15-25. Thermoforming sheet with 1-inch-square grid pattern.

Figure 15-26. Mold used to demonstrate thermoforming techniques.

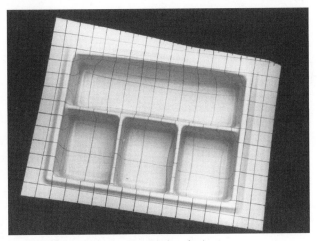

Figure 15-27. Example of draw using straight thermoforming.

at various places. If the elements have stretched in a regular manner, measure the stretched grid. Figure 15-28 shows a regularly stretched element. The original sheet was 0.0625 inch thick. The element stretched to 2.45 in.^{2} and, by calculation, should have a thickness of 0.0256 in. The measured thickness was 0.025 in.

15-5. If the elements are not uniform in stretch, as shown in Figure 15-29, it may be easier to measure the thickness than to calculate it.

15-6. Form a sheet using plug assist. The plug used for this demonstration is seen in Figure 15-26. The depth of plug penetration is critical. If the plug depth is too great, the area of the sheet in contact with the plug will stretch very little. That will force extensive stretch in neighboring elements and may lead to greater localized thinning than found in straight vacuum forming. Figure 15-30 shows a draw with plug depth too great. If the plug depth is too shallow, there will be little change compared to straight vacuum forming. Figure 15-31 shows a draw in which the plug helped reduce localized thinning. Notice the change in the stretch patterns.

15-7. Adjust the plug to optimize the thickness at the corners of the part, and calculate the maximum gain in wall thickness at the corners.

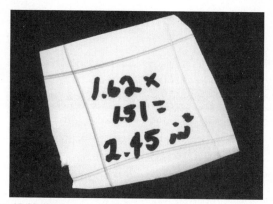

Figure 15-28. This element was 1 in.^{2} before forming. Uniform biaxial stretch would result in a square. This element was almost uniform in stretch.

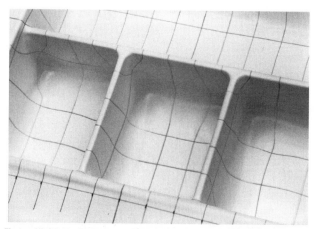

Figure 15-30. Plug depth was too great and prevented the central elements from stretching, causing excessive thinning in neighboring elements.

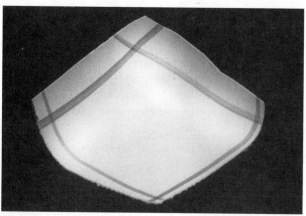

Figure 15-29. This element was not uniform in stretch. Its thickness varies considerably.

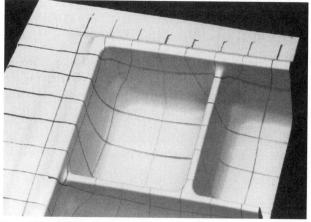

Figure 15-31. This draw shown is an improvement compared to the straight thermoforming shown in Figure 15-27. Blow-molding sequence.

EXPANSION PROCESSES

CHAPTER 16

INTRODUCTION

Methods of expanding plastics are described in this chapter. An **expanded plastics** is similar to a sponge, bread, or whipped cream because all are cellular in structure. Expanded plastics are sometimes called frothed, cellular, blown, foamed, or bubble plastics. They may be classified by cell structure, density, type of plastics, or degree of flexibility including rigid, semirigid, and flexible forms.

These low-density cellular (from Latin *cellula*, meaning "small cell or room") materials can be classified as either **closed cell** or **open cell** (see Figure 16-1). If each cell is a discrete, separate cell, it is a closed-cell material. If the cells are interconnected, with openings between cells (sponge-like), the polymer is an open-celled material. These expanded (cellular) polymers may have densities ranging from that of the solid matrix to less than 9 kg/m³ (0.56 lb/ft³). Nearly all thermoplastic and thermosetting plastics can be expanded, and they may also be made flame retardant. Table 16-1 lists selected properties of some expanded plastics.

Figure 16-1. Examples of closed-cell PS (top) and open-cell PV (bottom) cellular plastics.

Resins are made into expanded plastics using six basic methods:

1. Thermal breakdown of a chemical blowing agent, freeing a gas in a plastics particle (a popular method)

Pentanes, hexanes, halocarbons, or moistures of these materials are forced into the plastics particles under pressure. As the bead or granule of plastics is heated, the polymer becomes soft, which allows the blowing agents to vaporize. This produces an expanded piece sometimes called a pre-puff, preform, or pre-expanded bead. Cooling must be carefully controlled to prevent collapse of the cell or pre-puff. Sudden cooling may create a partial, internal vacuum in the cell. This pre-expansion is facilitated by dry heat, radio-frequency radiation, steam, or boiling water. Pre-expanded materials must be used within a few days to prevent complete loss of all volatile expanding agents. Such materials should be kept in a cool, airtight container until they are ready for molding. PS, SAN, PP, PVC, and PE are made using this method.

Blowing agents vaporize quickly from polyethylene pellets. Therefore, their shelf life is very short. Most molders simply order pre-expanded **pre-puffs** for final processing. To prevent cellular collapse before cooling, polyolefins are sometimes cross-linked by radiation. Prior to molding, the pre-puffs are placed into holding tanks to diffuse air into them. They are pressurized and then molded into a closed-cell product. Pre-puffs are usually stabilized by using thermal drying and annealing for a period of several hours. Pentane and butane are volatile organic compounds.

2. Dissolving a gas that expands at room temperature in the resin (a common method)

Table 16-1. Selected Properties of Expanded Plastics

	Coefficient of Linear Expansion 10°/°C	Water Absorption, Vol %	Flammability, mm/min	Density Range kg/m³	Thermal Conductivity W/m* K	Max Service Temperature, °C	Compression Strength, kPa
Cellulose acetate	6.35	13–17	Slow burning	96–128	0.043	176	862–1034
Epoxy							
Packed in place	38	1–2	Self-extinguishing	210–400	0.028–1.15	260	$13–14 \times 10^3$
Foamed in place	102		Self-extinguishing	80–128	0.035	148	551.5–758
Phenolic							
Reactive type	5–10	15–50	Self-extinguishing	16–1280	0.036–6.48	121	172.3–419
Polyethylene	24.1	1.0	63.5	400–480	0.05–0.058	71	68.9–275.7
Polystyrene							
Extruded	11	0.1–0.5	Self-extinguishing	20–72	0.03–0.05	79	68.9–965
Expanded-beads	10.1	1.0	Self-extinguishing	16–160	0.03–0.039	85	68.9–1375
Self-expanded-beads and others	10.1	0.01	Self-extinguishing	80–160	0.03	85	310.2–838
Polyvinylchloride							
Open cell			Self-extinguishing	48–169		50–107	
Closed cell			Self-extinguishing	64–400		50–107	
Silicone							
Premixed powder		2.1–3.2	Does not burn	192–256	0.043	343	689–2241
Liquid resin, rigid and semirigid		0.28	Self-extinguishing	56–72	0.04–0.43	343	55
Flexible			Self-extinguishing	112–144	0.045–0.052	315	
Urethane							
Rigid	1370		Self-extinguishing	32–640	0.016–0.024	148–176	172–210
Flexible	1650	10	Slow burning	22–320	0.032	107	

Nitrogen and other gases may be forced directly into the polymer melt. By using injection or extruder equipment, special screw-shaft seals prevent the escape of gas from the hot matrix. As the melt leaves the die or enters the mold cavity, the gas vaporizes and causes the polymer to expand. Once the matrix melt falls below the glass transition temperature, the expansion is stabilized.

3. Adding a liquid or solid component that vaporizes when heated in the melt (sometimes done to produce structural foams) (Figure 16-2)

Granular, powder, or liquid blowing agents may be mixed and forced through the melt. They remain compressed in the mold until they are forced into the atmosphere. This will cause rapid decompression (expansion) to occur. PS, CA, PE, PP, ABS, and PVC are expanded by using this method.

4. Whipping air into the resin and rapidly curing or cooling the resin (sometimes used in the production of carpet backing) (Figure 16-3)

Figure 16-3. Mechanically frothed or foamed polyvinyl on thick flooring construction. Magnification 103.

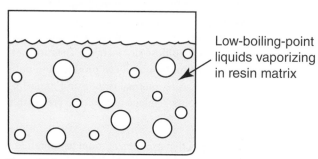

Low-boiling-point liquids vaporizing in resin matrix

Figure 16-2. Bubbles of gas are formed as chemicals change physical states.

Vinyl esters, urea-formaldehyde (UF), phenolics, polyesters, and some dispersion polymers are made cellular by utilizing mechanical methods.

5. Adding components that liberate gas within the resin through a chemical reaction

This is a popular method for producing expanded materials from condensation polymers. Liquid resins, catalysts, and blowing agents are kept separate until they are molded. After mixing, a chemical reaction expands the matrix into a cured cellular material. As the matrix expands, many cell walls rupture, forming a catacomb structure. Polyethers, polyurethane (PU), urea-formaldehyde (UF), epoxy (EP), polyether polyurethane (EU), silicone (SI), isocyanurates, carbodimide, and most elastomers use this expansion method. Polyurethanes and polystyrene account for more than 90% of insulation found in refrigerators, freezers, cryogenic tanks, and cellular construction. Isocyanate is commonly expanded between layers of aluminum foil, felt, steel, wood, or gypsum facings. Flexible polyurethanes are open-celled, whereas rigid polyurethanes are made of closed-cell materials.

The use of chlorofluorocarbons (CFCs) as blowing agents has been severely restricted by the EPA. In order to speed the reduction or elimination of CFCs, a special tax on their use began in 1994. In 1996, the Clean Air Act prohibited the production and import of CFCs. Switching to hydrochlorofluorocarbons (HCFCs), which are less ozone-depleting than traditional CFCs, has not eliminated concerns of the EPA. Production of HCFC-141b, a chemical used as a blowing agent, ended in December 2002. An alternative is the use of perfluorocarbons (PFCs). However, PFCs are potent greenhouse gases, and the EPA also discourages their use.

6. Volatizing moisture (steam left in resins by the heat generated during prior exothermic chemical reaction)

Water-blown systems may be used in some condensation polymers. In addition to the steam from the water condensate, carbon dioxide gas is liberated. Some systems combine water and halocarbon in the resin mixture.

Reinforced cellular materials may be produced with particulate and fibrous reinforcements dispersed throughout the polymer matrix. Fibers tend to orient themselves parallel to cell walls, resulting in improved rigidity. Reinforced epoxies have been mechanically mixed with pre-expanded beads (PS, PVC), then fully expanded and cured in a mold. Cellular materials are also used in the manufacture of sandwich composites.

Syntactic plastics are sometimes classified as a separate expanded plastics group. **Syntactic plastics** are produced by blending microscopically small (0.03 mm) hollow balls of glass or plastics in a resin matrix binder (Figure 16-4). This produces a light, closed-cell material. The putty-like mixture may be molded or applied by hand into spaces not accessible by other means. Syntactic plastics are commonly used in tooling, noise

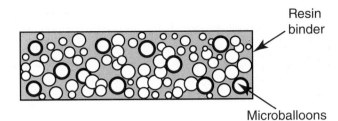

Figure 16-4. Syntactic foam.

alleviation, thermal insulation, and high-compression-strength flotation devices.

Cellular materials have been produced by placing glass spheres in the polymer matrix formulation before sintering. In a process called **leaching**, various salts or other polymers may be sintered together. A solvent solution dissolves the crystals or selected polymer, leaving a porous matrix. Alternate layers of compatible polymers, reinforcements, and soluble crystal mixtures produce a true composite component.

Expanded plastics are used for insulation, packaging, cushioning, and flotation. Some act as an acoustical or thermal insulation. Others are used as moisture barriers in construction. Expanded epoxy materials are used in light tooling fixtures and models. Expanded plastics may also be used in noncorroding, light, shock-absorbing materials for automobiles, aircraft, furniture, boats, and honeycomb structures. In the textile industry, expanded plastics are used in padding and insulation to give garments a special texture or feel.

During World War II, the Dow Chemical Company introduced expanded polystyrene products in the United States and General Electric produced expanded phenolic products. The two main expanded plastics in use today are probably polystyrene and polyurethane. Polystyrene products are rigid, closed cellular structures (Figure 16-5). Polyurethane

Figure 16-5. Closed cellular polystyrene has excellent thermal insulating qualities.

products may be rigid or flexible. PU can also be either closed or open celled. Expanded products that are familiar to the consumer are ceiling tile, Christmas decorations, flotation materials, toys, package liners for fragile items, mattresses, pillows, carpet backing, sponges, and disposable containers.

The following processes are explained in this chapter:

I. Molding
   A. Low-pressure processing
   B. High-pressure processing
   C. Other expanding processes

II. Casting

III. Expanding in place

IV. Spraying

MOLDING

Various processes have been developed for molding expandable plastics. These include injection molding, compression molding, extrusion molding, dri-electric molding (a high-frequency expanding method), steam chamber, and probe molding. Integral skinned cellular or foamed plastics are commonly cast or molded. A solid, dense skin of plastics is formed on the mold surface as it is heated. During the skinning process, a cellular core is formed by forcing blowing agents or gas into the melt, producing the cellular structure (Figure 16-6).

In addition to end products made with expanded plastics, foam patterns for metal castings are also widely used. Foamed polystyrene patterns are popular because polystyrene is rather inexpensive, the patterns have the strength and durability required, and the polystyrene vaporizes readily when contacted by molten metals.

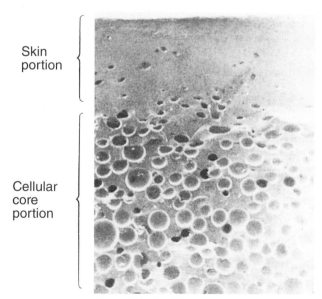

Figure 16-6. In this photograph, the transition from solid skin to cellular core may be seen.

The term **structural foam** includes any cellular plastics with an integral skin. Its stiffness depends largely upon the skin thickness. There are many methods of making this integral skin; however, structural foam products are molded from the melt, by either a low-pressure, or a high-pressure process (Figure 16-7).

Low-Pressure Processing

Low-pressure processing is the simplest as well as the most popular and economical method used for large parts. Typically, the equipment used for this process includes a basic

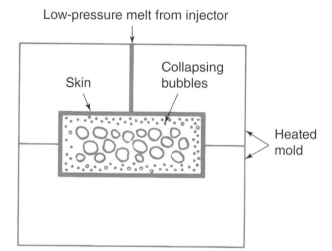

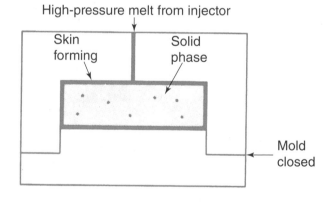

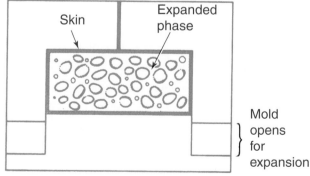

Figure 16-7. Low- and high-pressure methods of expanding products with an integral skin.

injection-molding machine that has been modified to permit addition of the expanding agent. If a pelletized expanding agent is added directly to the hopper along with the base material, no changes may be needed. Modifications may involve special screw designs, the addition of an accumulator, or the installation of multiple injection nozzles in the mold. A mixture of molten plastics and gas is injected into a mold at a pressure of from 1 to 5 MPa (145 to 725 psi). A skin is formed as the gas cells collapse against the sides of the mold. Skin thickness is controlled by the amount of melt forced into the mold as well as mold temperature and pressure. The type of blowing agent used or the amount of gas forced into the melt also helps determine the skin thickness and density of the part.

Low-pressure techniques produce closed-cell parts that are nearly stress free. The collapse of cells on the mold surface sometimes produces a surface swirl pattern on the parts.

High-Pressure Processing

In **high-pressure processing**, the heated melt is forced into the mold with pressures of 30 to 140 MPa (4351 to 20,307 psi). The mold is completely filled with the melt, which allows the melt to become solid against the mold or die surfaces. To allow for expansion, the mold cavity is increased by having the mold open slightly or by withdrawing cores. The melt expands within this increase in volume.

In one method, the expanding agent is added directly to the hot mix. The pressure of the screw does not allow the material to expand until it is forced into a mold (Figure 16-8).

In other injection and extrusion methods, the blowing or expanding agent, colorants, and other additives are melted directly into the molten plastic just before it enters the mold (Figure 16-9). The expansion then takes place in the molding cavity.

In yet another process, the extruded material with expanding agents is fed into an accumulator (Figure 16-10). When the predetermined charge is reached, a plunger forces the material into the mold cavity, where expansion takes place.

Reaction injection molding (RIM), reinforced reaction injection molding (RRIM), and **mat molding/RIM (MM/RIM)**

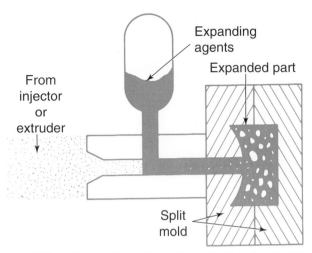

Figure 16-9. A molding method in which the expanding agent is added to the molten plastics just before the plastics enters the mold.

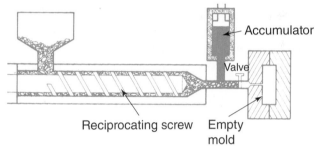

(A) Material enters accumulator.

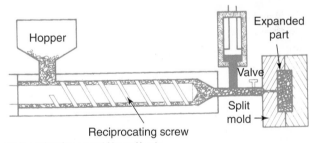

(B) Accumulator forces material into mold cavity.

Figure 16-10. A molding method in which an accumulator is used.

have grown rapidly after extensive use in the automotive industry and numerous structural applications. Reinforcements greatly improve dimensional stability, impact strength, and modulus. In a modified resin transfer molding technique, long fibers or mats are placed in the mold and a mixture of reactive components are forced into the cavity. The resulting MM/RIM product is tough, light, and strong.

Figure 16-11 shows a RIM process for molding polyurethane foam. The liquid materials enter the lower half of the mold cavity. The mold then closes before the chemical reaction takes place, which causes the material to foam. As it expands, the foam fills the mold and, when cured, retains the shape of

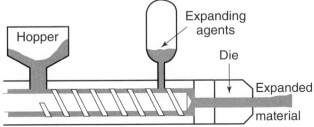

Figure 16-8. Screw method.

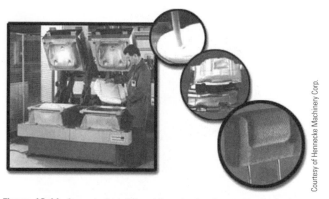

Figure 16-11. Stages in the molding of foamed polyurethane automotive headrests.

the mold. As seen in Figure 16-11, the foam molding requires trimming and upholstering as it leaves the mold before the part is complete.

An expanded plastics with a skin (unexpanded layer) is produced by forcing the hot mixture around a fixed torpedo. The extruded shape is a hollow form the shape of the sizing die. The extrudate then expands, filling the hollow center, and the skin is formed by the cooling action of the sizing and cooling dies (Figure 16-12). Structural profiles are produced by this method.

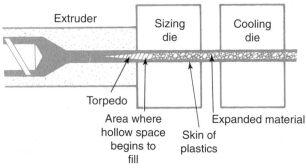

(A) A method of production.

(B) A vinyl foamed mat with a skin on both sides.
Figure 16-12. Forming a skin on expanded plastics.

In another process, two plastics of either the same formulation or different families are injected into a mold one after the other. The first plastics does not contain expanding agents and is partially injected into the mold. The second plastics, containing the expanding agents, is then injected against the first plastics, forcing it against the edges of the mold and forming a shell around the expandable plastics. To close off the shell, more of the first resin is injected into the mold, fully encapsulating the second resin. The resulting part has an outer skin of one type of plastics and an inner core of expanded plastics.

In a unique process called gas counter-pressure, gas is forced into an empty, sealed mold cavity. The melt is then forced into the cavity against the gas pressure. Expansion begins as the mold is vented.

Extruded polyvinyl materials may be expanded as they emerge from the die or stored for future expansion. They are used in the garment industry as single components or given a cloth backing (Figure 16-13).

Foamed materials are used as backing on carpets or other flooring, as shown in Figure 16-14A. Artificial turf for athletic fields has become an extremely complex product. Astroturf®, as seen in Figure 16-14B, consists of a foamed PVC pad beneath the turf. Fillers have been added to the turf to prevent the fibers, which are usually made of nylon or olefin materials, from matting and becoming too hard. Such fillers have consisted of either sand and/or rubber pellets. The difficulty with these systems is that, over time, the sand settles to the bottom. This causes compacting, which can lead to a hard, poorly draining surface. Recent attempts to correct this are shown in Figures 16-14C and 16-14D. These surfaces include foam pads, artificial thatch of nylon 6,6, polyethylene fibers, and rubber fill. The goal is to create a durable surface that does not compact over time.

In compression molding of expandable plastics, the resin formulation is extruded into the molding chamber and the mold is closed. The molten resin quickly expands, filling the mold cavity.

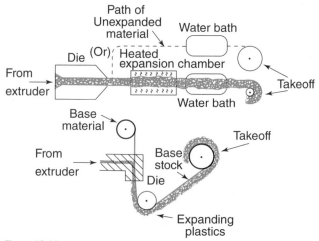

Figure 16-13. Two methods of expanding a plastics as it emerges from the die.

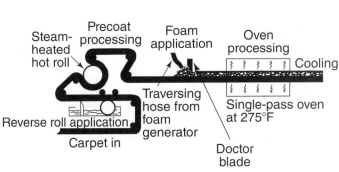

(A) Applying carpet backing.

(C) Astroplay® is a turf system that includes a rubber pad and fiberglass glass backing to provide stability.

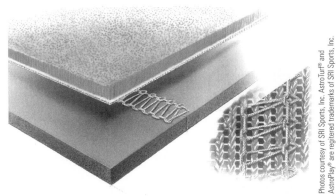

(B) This view of Astroturf® shows a foam pad, woven backing, and nylon pile.

(D) This close-up photo of Astroplay® shows the artificial thatch that supports the fibers and keeps the rubber fill particles at the base of the systems.

Figure 16-14. Foams are used in many flooring and athletic field applications.

One of the largest demands is for extruded polystyrene logs, planks, and sheets. They are produced by extruding the molten plastic containing the expanding agent from a die. Expansion occurs rapidly at the die orifice. Rods, tubes, or other shapes may be produced in this manner.

Table 16-2 gives solutions to problems that may occur during expansion processes.

Other Expanding Processes

Not all expanded plastics make use of the hot-melt method. Polystyrene is commonly produced in the form of small beads containing an expanding agent. These beads may be pre-expanded by heat or radiation and then placed into a mold

Table 16-2. Troubleshooting Expansion Processes

Defect	Possible Causes and Remedies
Mold not filled	Venting, short shot, trapped gas, increase pressure, increase amount of material, use fresh materials.
Pits or holes in surface	Reduce amount of mold release—mold release interfering with blowing agent, increase mold temperature, not enough material in mold, melt temperature incorrect, concentration of blowing agent incorrect, polish die or mold surface.
Distorted pieces	Poor mold design, increase mold time, increase cure cycle, allow thick sections to cool longer, increase mold strength.
Sticking to mold	Increase mold release, select correct release, poor mold design, cool mold, polish molds.
Parts too dense	Increase blowing agent or gas, use fresh pre-expanded beads, lower injection pressures, reduce melt.
Part density varies	Mix compounds thoroughly, check screw design, increase mold temperature, increase melt temperature, increase dwell and cure time.

cavity where they are heated again, often by steam, causing further expansion (Figure 16-15). These beads expand up to 40 times their original size (Figure 16-16), and the resulting pressure packs the beads into a closed cellular structure. A typical bead-expanding system is shown in Figure 16-15. Core box vents are used to allow steam into the mold cavity, and such vents leave marks on the expanded product (Figure 16-15C).

Thermal excitation of molecules using high-frequency radio energy is also used to expand beads. This is sometimes called dielectric molding because it does not require steam lines, moisture, steam vents, or metallic molds. Inlays or decorative

Figure 16-16. Unexpanded (left) and expanded (right) beads of polystyrene.

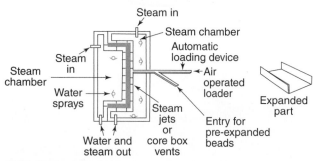

(A) Typical bead-expanding mold that uses steam.

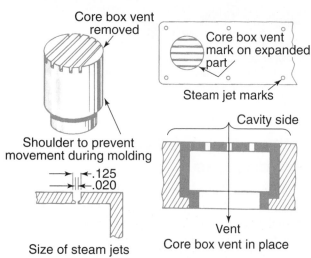

(B) Part showing core box vent and steam jet marks.

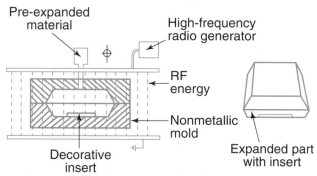

(A) Basic principle of expanding plastics using radiofrequency energy.

(B) Radiation may also be used to expand and cure plastics.
Figure 16-17. Plastics may be expanded by radiation.

substrates of paper, fabric, or plastics may be molded in place by utilizing this method (Figure 16-17).

Well-known examples produced by this method include insulated cups, ice chests, holiday decorations, novelty items, toys, and many flotation and thermal-insulating products.

CASTING

In casting expandable plastics materials, the resin mix containing catalysts and chemical expanding agents is placed in a mold where it expands into a cellular structure (Figure 16-18).

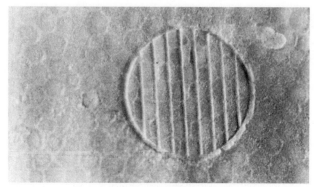

(C) Close-up of core box vent mark left on expanded product.
Figure 16-15. Expanding plastics without using a hot-melt method.

Components

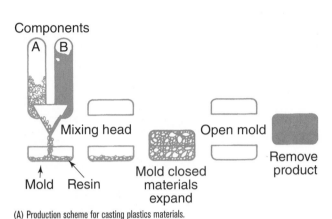

(A) Production scheme for casting plastics materials.

(B) The head and overhead lines deliver the polyurethane components to a mold before expansion occurs.

(C) This unit meters and mixes the components for polyurethane foam.

Figure 16-18. Equipment for expanded polyurethane parts.

Polyurethanes, polyethers, urea-formaldehyde, polyvinyls, and phenolics are often cast expanded. Flotation devices, sponges, mattresses, and safety cushioning materials are often cast. Large slabs or blocks of flexible polyurethanes are cast in open and closed molds. These slabs or blocks are cut into mattress stock or shredded for cushioning. Crash pads and pillow products may be cast in closed molds.

Slab stock is commonly produced using continuous production processes. Some are extruded, whereas many are simply cast or sprayed on a continuous belt. This stock is used as cores for sandwich or laminated composites.

EXPANDING IN PLACE

Expanding in place is similar to casting except that the expanded plastics and the mold together become the finished product. Insulation in truck trailers, rail cars, and refrigerator doors; flotation material in boats; and coatings on fabrics are examples of expanding in place.

In this process, the resin, catalyst, expanding agents, and other ingredients are mixed and poured into the cavity (Figure 16-19). Expansion takes place at room temperature but the mixture may be heated for a greater expanding

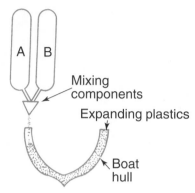

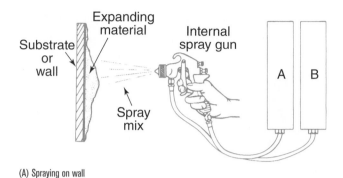

(A) Spraying on wall

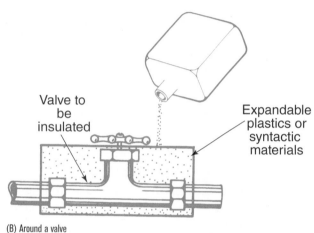

(A) Between inner and outer hulls of a boat

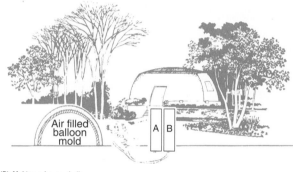

(B) Making a house shell

Figure 16-20. Some examples of spray forming.

(B) Around a valve

Figure 16-19. Expanded-in-place forming.

Disadvantages of Expanded Plastics

1. Process is slow, some need cure cycle.
2. Special equipment is necessary for hot-melt methods.
3. Tool and mold designs are more costly for high-pressure method.
4. Surface finish may be hard to control.
5. Part size is limited in high-pressure method.
6. Some processes emit volatile gases or toxic fumes.

reaction. This method is said to be done *in situ* (in place or position). Syntactic forms of plastics also may be placed in this category.

SPRAYING

A special spraying device is used to place expandable plastics on mold surfaces or walls and roofs for insulation. Figure 16-20 shows two examples of this type of forming.

Five advantages and six disadvantages of expanded plastics are listed here:

Advantages of Expanded Plastics

1. Light, less costly products with low thermal conductivity.
2. There is a wide range of formulations from rigid to flexible.
3. There is a wide range of processing techniques.
4. Less costly molds for low-pressure and casting methods make large products possible.
5. Parts may have high strength-to-mass ratios.

RELATED INTERNET SITES

• **www.dartcontainer.com.** The Dart Container Corporation is an international manufacturer of foam cups and food containers. Selecting "Products" from the home page leads to descriptions and dimensions of numerous containers and cups. Selecting "Environment" leads to information about Dart environmental programs, polystyrene, and many documents concerning recycling and source reduction. In the "Environment" page, there is an area titled "Environmental Media Library," within which is the heading "Scientific Studies." This section contains studies comparing polystyrene packaging to other materials.

• **www.polystyrene.org.** This address leads to the Plastics Foodservice Packaging Group, one of the groups in the American Chemistry Council. Its purpose is to advocate for polystyrene packaging. The site links to extensive materials about polystyrene packaging.

- **www.sierraclub.org.** In contrast to the information from polystyrene manufacturers, the Sierra Club presents information that is very critical of the use of polystyrene. To find this data, enter "styrofoam" in the search box on the home page. This will then lead to a long list of articles about polystyrene and environmental concerns.

VOCABULARY

The following vocabulary words are found in this chapter. Use the glossary in Appendix A to look up the definitions of any of these words you do not understand as they apply to plastics.

blowing agents
closed-cell
expanded (foamed) plastics
expanding in place
high-pressure processing
in situ
leaching
mat molding/RIM (MM/RIM)
open-cell
pre-puffs
structural foam
syntactic plastics

QUESTIONS

16-1. Name three terms used to describe the expanding process.

16-2. The cellular structure of expanded plastics is _____ or _____.

16-3. Cellular plastics have _____ relative densities than solid plastics.

16-4. Can both thermosets and thermoplastics be used for expanding in place?

16-5. Name six methods of forming the cellular structure in the expanding processes.

16-6. Some expanded plastics may be thermoformed. They are commonly made of _____ plastics.

16-7. List four broad product uses of expanded plastics.

16-8. What is used to expand polystyrene beads in the mold?

16-9. Cellular plastics with integral skin are called _____.

16-10. What type of cell structure would a life vest have?

16-11. Supple expanded plastics are used chiefly for _____.

16-12. The two main expanded plastics in use today are probably _____ and _____.

16-13. Name four basic processes of foaming plastics.

16-14. What process would be selected to mold the front of an automobile?

16-15. Styroform is a _____ for polystyrene cellular plastics.

16-16. Which expanding process or processes would be used to produce each of the following products?
  a. Mattress
  b. Padded dashboard
  c. Egg carton
  d. Ice chest or cooler
  e. Insulation in home walls

16-17. As polystyrene beads are heated, the _____ agent causes the beads to swell.

16-18. Pre-expanded beads and unexpanded beads have a limited shelf life because they lose their _____ agent.

16-19. What does the term *in situ* mean?

16-20. As polystyrene beads expand and exert force against the walls of other beads, they form a _____ structure with no ruptured cells.

16-21. Name four items that may be produced by casting of expanded plastics.

16-22. Expanding in place is appropriate when the cellular material will not be _____.

16-23. In casting, the expanded plastics is always _____ from the mold.

16-24. Name the forming process in which the resin and expanding agent are atomized and forced out of a gun to strike a mold or substrate.

16-25. Where is a great amount of flexible polyurethane foam used?

16-26. It is the _____ that determines the physical properties of expanded plastics.

16-27. Name a major application for rigid polyurethane foams.

16-28. Many cellular profile shapes with a surface skin are reproduced by _____ methods.

16-29. Expanded polystyrene parts and molds must be _____ before removing the expanded part because latent heat in the center of the plastics part may continue to cause expansion.

16-30. Small glass or plastics balls are sometimes used to make _____ plastics.

16-31. Name four methods of pre-expanding polystyrene beads.

16-32. In low-pressure forming of structural framed products, pressures from 1 to _____ MPa are common.

16-33. If the product is not completely shaped or the mold is not filled, _____ may be the cause.

16-34. On some expanded polystyrene products, you can see marks left by the _____ used to allow steam into the mold cavity.

16-35. Two major disadvantages of high-pressure methods of expanding products are _____ molds and _____ part size.

16-36. Describe one way of producing an expanded plastics that has a skin.

16-37. Identify one use for syntactic foams and state why the foam would be advantageous for use in the application.

ACTIVITIES

Blowing Agents

Introduction. One type of blowing agent consists of chemicals that decompose at processing temperatures and release gases that form tiny bubbles in a plastics. The amount of the blowing agent used affects the size and number of cells created. A common agent of this type is based on azodicarbonamide. Such materials are often called azo-type blowing agents.

Equipment. Injection molder, HDPE, azo-type blowing agent, razor knife, baking powder, accurate scale, and microscope.

Procedure

16-1. A blowing agent needs to activate at the same temperature range required for processing the desired plastics. Acquire an azo-type blowing agent appropriate for HDPE. Figure 16-21 shows portions of three air shots of HDPE.

　　The top sample contains an azo-type blowing agent, whereas the lower two contain baking powder. The difference between the two samples blown with baking powder was the time in residence in the molding machine. The center sample is discolored compared to the azo material. However, it is not as discolored as the lower sample, which darkened due to longer residence time.

16-2. Mix the blowing agent and the basic plastics at a loading recommended by the manufacturer. If using baking powder in HDPE, use 3% as a starting point.

16-3. Make small air shots. Make sure that adequate ventilation is available to remove the fumes released from the hot plastics. Air shots will provide an unhampered opportunity for the blowing agents to expand. Figure 16-22 is a section about 13 mm (0.5 in.) in diameter cut from the top sample in Figure 16-21.

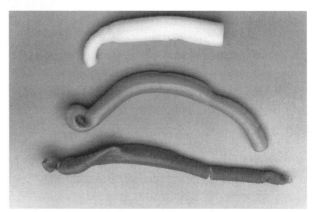

Figure 16-21. Three small expanded air shots.

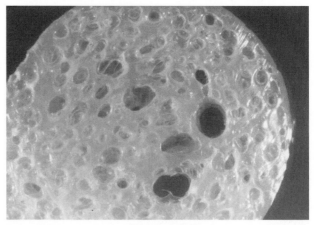

Figure 16-22. Cross section of air shot showing cell structure.

　　Do not confuse vacuum voids with cells caused by blowing agents. Figure 16-23 is the cross section of a HDPE sprue. The hole was not an air bubble but developed during cooling. When the forces caused by shrinkage overcome the strength of the outer surface of the part, the result is a sink mark. When the outer surface is stronger than the contraction forces, the result is often a vacuum void.

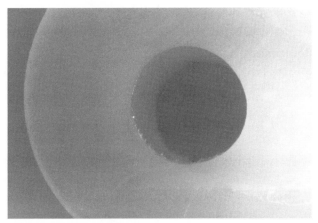

Figure 16-23. A vacuum void in a sprue.

16-4. Measure the specific gravity of a portion of an air shot. Altering the percentage of blowing agent will affect the specific gravity. Visually inspect a slice of the air shot under a microscope. Natural HDPE is sufficiently translucent to permit good inspection in sections easily cut with a sharp razor knife.

16-5. Injection molding with conventional equipment does not yield uniformly dense products. Figure 16-24 shows parts made when the mold opened immediately after complete injection. The hottest regions, those in the center of the parts and near the gates, expanded more than neighboring regions.

16-6. In short shot moldings, material at the end of the flow has the opportunity to expand because it is not pressurized and has one open side in the mold. Figures 16-25A, 16-25B, and 16-25C show cross sections of a short shot—25A near the end of the flow, 16-25B about halfway to the gate, and 16-25C near the gate. The differences in cell size demonstrate the decay in pressure as the flow gets farther from the gate.

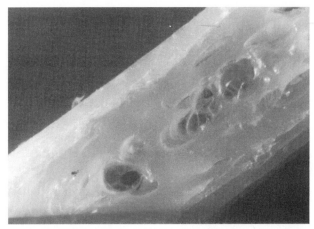

(A) Expansion near the end of flow

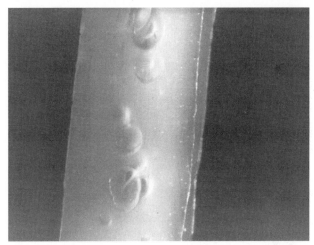

(B) Expansion near the middle of the flow length

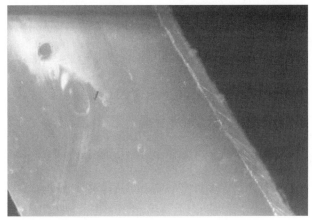

(C) Expansion near the gate

Figure 16-25. Expansion affected by location.

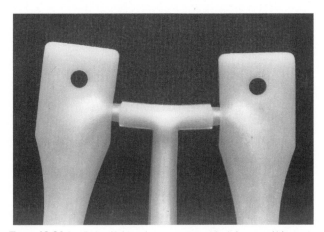

Figure 16-24. Rapidly demolded part shows greatest expansion in hottest or thickest areas.

16-7. Inject full parts under various injection pressures. Measure the specific gravity of sections cut from the same region of the part. What range of specific gravities can be controlled with injection pressure?

16-8. Write a report summarizing your findings.

Micro-balloons

Introduction. Tiny glass spheres have been used for years to reduce weight in SMC and BMC parts. The glass spheres are so small that about 30 spheres can fit in the space occupied by a common grain of salt. However, attempts to use micro-balloons in injection-molded products were largely unsuccessful because the forces needed in compounding and molding crushed the spheres. Recently, new high-strength microspheres have made applications in injection-molded products. One extrusion manufacturer claims survival rates of close to 90%.

Procedure

16-1. Acquire micro-balloons. Industrial grade micro balloons are significantly different from the micro-balloons available in hobby shops. They are generally low-strength spheres and will crush easily. If possible, acquire Scotchlite S60® glass bubbles. This product has a true density of 0.60 g/cc. Its bulk density is about 0.35 g/cc. The bulk density is lower because it includes the voids between the particles in the calculation.

16-2. Examine the balloons under a microscope. Pay particular attention to the frequency of broken balloons.

16-3. Weigh out several batches of balloons and plastics with incremental loadings such as 1%, 2%, and 4%. Determine the specific gravity of the plastics, either experimentally or from manufacturers' material specifications.

16-4. Compound the micro-balloons into the plastics with an extruder.

16-5. Calculate the expected specific gravity of the extrudate based on the loading of micro-balloons. Measure the specific gravity of the extrudate. Burn off the base material, and examine the residual material under a microscope to see whether the extrusion process broke any balloons. If it did break balloons, estimate the percent of breakage. Does the specific gravity data correspond to the breakage estimates?

16-6. Inject mold parts using basic plastics and compounded material. Measure the specific gravity of the parts containing micro-balloons. Burn off the plastics, and examine the ash. Did the injection process cause breakage of the balloons?

16-7. Complete physical tests on parts with and without micro-balloons. Compare the weight reduction to the strength reduction.

16-8. Write up a report summarizing your findings.

Expanded Polystyrene Beads

Introduction. Cellular polystyrene plastics made from beads appear in many products. Small polystyrene beads containing a volatile gas expand when thermal energy (95°C, or 204°F) vaporizes the gas. Under heat and pressure, the beads partially fuse together.

Equipment. PS beads, hot plate or stove-top burner, water container, mold, pigments, pot, pressure cooker, sieve, and mold release.

Procedure

16-1. Boil about 2 L of water (see Figure 16-26).

16-2. Pour 50 ml fresh beads into boiling water. Stir until all beads rise to surface and expand to desired density. Too little pre-expansion will produce a dense, hard product. If beads are pre-expanded too much, they will not fill the mold properly during the expanding operation.

16-3. Using a sieve, remove pre-expanded beads from the water to stop pre-expansion, as shown in Figure 16-27.

! CAUTION

Water and beads are hot. It is best to use pre-expanded beads within 24 hours.

Figure 16-26. Make certain all materials are carefully prepared for expanding polystyrene beads.

Figure 16-27. Pre-expanded beads being removed from water.

16-4. Make certain the mold is clean. Apply a light coat of mold release.

16-5. Fill the mold halfway with pre-expanded beads. Powder pigments may be added and carefully mixed with beads at this time. Beads should be piled high. Remove beads from lips of the mold. Add a teaspoon of unexpanded beads for a harder, denser product. Figure 16-28 shows the mold-filling step.

16-6. Carefully assemble the mold, and shake it to distribute the beads evenly.

16-7. Place the mold in a pressure cooker containing at least 1 liter of water.

16-8. Place the lid on the cooker, making certain it is sealed and the safety valve is functioning.

16-9. Allow pressure to increase to about 100 kPa (14.6 psi). Hold pressure for 5 minutes. Do not allow pressure to exceed 140 kPa (20 psi) (see Figure 16-29).

Figure 16-28. Fill mold half with pre-expanded beads.

Figure 16-29. Place filled mold in cooker, put lid on cooker, and bring pressure up to 100 kPa. DO NOT EXCEED 140 kPa.

16-10. Remove the cooker from the burner, and slowly allow pressure to decease (about 5 minutes).

! CAUTION

Beware of hot escaping steam.

16-11. After all steam has escaped, open the cooker, remove the mold (Figure 16-30), and place it in cooling water (Figure 16-31).

! CAUTION

Wear protective gloves and goggles.

16-12. Open the mold and remove the expanded part. Compressed air may be used to aid in removal of the part. Trim the flash. If the beads were not fully expanded or fused together, increase pressure and/or cycle time. If the product shrinks or has melted areas, reduce pressure, heating cycle, and/or pre-expansion (Figure 16-32).

Figure 16-30. Remove hot mold from cooker.

Figure 16-31. Quench hot mold in water.

Figure 16-32. Carefully remove expanded product and trim.

Optional Boiling Water Method

16-1. Prepare the mold and pre-expanded beads.

16-2. Fill the mold with pre-expanded beads and pack firmly.

16-3. Place the mold in boiling water for 40 to 45 minutes. Times will vary with the size of the mold, amount of pre-expansion, and the age of the bead.

16-4. Cool in water.

16-5. Remove the part.

Optional Dry Oven Method

16-1. Prepare the mold and pre-expanded beads.

16-2. Fill 10% of the volume of the mold with water.

16-3. Fill the mold with a mixture of pre-expanded and unexpanded beads. Use wet beads.

16-4. Place the mold in a 175°C (347°F) oven for about 10 minutes. Time will vary with the size of the mold, amount of pre-expansion, and age of beads.

16-5. Cool in water.

16-6. Remove the part.

COATING PROCESSES

CHAPTER 17

INTRODUCTION

Plastics coatings are found on cars, houses, machinery, and even fingernails. In this chapter, you will learn how various coatings are applied. Many coatings are applied to a substrate to enhance properties of the product by protecting, insulating, lubricating, or adding durable beauty. Coatings may have a combination of properties such as flexibility, texture, color, and transparency that no other material can match.

For a process to be classed as a *coating*, the plastics material must remain on the substrate. Dip-casting and film-casting products are not considered coatings because the plastics is removed from the substrate or mold. Coating may be confused with other processes due to variations in processing and because similar equipment is used.

It is important to select coating materials that are reasonably close to the thermal expansion of the substrate to be coated. This becomes especially important if reinforcements are used. Typically, some reinforced coating processes are better classified as modifications of laminating and reinforcing techniques. Reinforcements can help stabilize the coating matrix. Adhesion is one of the most critical factors in any coating operation. Some substrates must be properly prepared before coating.

In extrusion processing, wire coating exemplifies how more than one material may be put through an extrusion die. In extrusion or calendering of films, hot films are often placed on other substrates coating them. The use of liquid dispersion or solvent solutions is considered a casting method if the film is removed. However, it is considered a coating process if the film remains on the substrate.

There are nine brand (sometimes overlapping) techniques by which plastics are placed on substrates. These are listed in the following chapter outline:

I. **Extrusion coating**

II. **Calender coating**

III. **Powder coating**
   A. Fluidized-bed coating
   B. Electrostatic-bed coating
   C. Electrostatic powder gun coating

IV. **Transfer coating**

V. **Knife or roller coating**

VI. **Dip coating**

VII. **Spray coating**

VIII. **Metal coating**
   A. Adhesives
   B. Electroplating
   C. Vacuum metallizing
   D. Sputter coating

IX. **Brush coating**

EXTRUSION COATING

Single- or co-extrusion of hot melt may be placed on or around the substrate. Extrusion film coating is a technique in which a hot film of plastics is placed on a substrate and allowed to cool. For best adhesion, the hot film should strike the preheated and dried substrate before it reaches the nip of the pressure roller (Figure 17-1). The chill roller is water cooled to speed cooling of the hot film. It is usually chrome plated for durability and high-gloss transfer, and may be

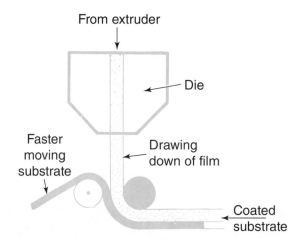

(A) Basic concept of extrusion coating.

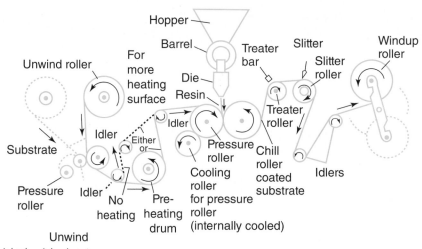

(B) Extrusion coating setup with wind and unwind equipment.

Figure 17-1. Extrusion film coating.

embossed to produce special textures on the film surface. The thickness of the film is controlled by the die orifice and the surface speed of the chill roller. The substrate is moving faster than the hot extrudate as it comes out of the extruder die. The extrudate is drawn out to the desired thickness just before it reaches the nip of the pressure and chill rollers.

Several coating techniques produce a thin laminate composite. Because the primary objective of the coating is to provide protection, some materials are considered more effective for coating than others. Polyolefins, EVA, PET, PVC, PA, and other polymers are commonly used in extrusion coating of various substrates. These materials provide moisture, gas, and liquid barriers and heat-sealable surfaces. The polyethylene coating on paperboard milk cartons is one familiar example of liquid barrier and heat-sealable coatings.

Expanded plastics are also extruded onto various substrates. The substrate may be drawn through the extrusion die, as seen in the coating of wire, cable, rods, and some textiles (Figure 17-2). Five advantages and two disadvantages of extrusion coating are presented here:

Advantages of Extrusion Coating

1. Multilayer plastics may be placed on substrate.
2. No solvents are needed.
3. Thickness applied to substrate may be varied.
4. There is uniform coating thickness on wire and cable.
5. Cellular coatings may be placed on substrate.

Disadvantages of Extrusion Coating

1. Extrudates are hot melts.
2. Equipment is expensive.

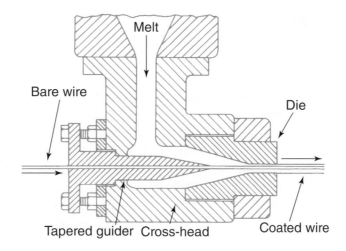

(A) Basic principles of wire coating.

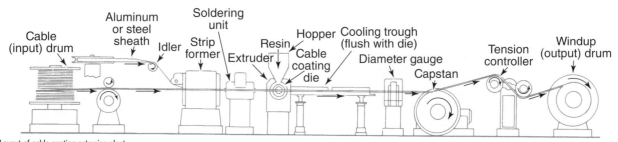

(B) Layout of cable-coating extrusion plant.

Figure 17-2. Process of extrusion wire coating.

CALENDER COATING

Calendered films may be used as a coating on many substrates, through a method similar to extrusion coating. The hot film is squeezed onto the substrate by the pressure of heated gauging rollers (Figure 17-3).

Melt-roll coating is a modification of calendering. In this process, the preheated substrate is pressed into the hot melt by a rubber-covered roller. An embossing roll may also be used. The coated material is cooled and placed on windup rolls.

Pressure-sensitive and heat-reactive hot melts commonly used as adhesives may be coated on a substrate. Paper, plastics, and textiles are coated using this process.

Coating on a paper substrate may add beauty, strength, scuff resistance, moisture, and soil resistance, or provide a sealing system for producing a package.

Five advantages and two disadvantages of calender coating are listed here:

Advantages of Calender Coating

1. It is a high-speed continuous process.
2. It affords precise thickness control.
3. Pressure-sensitive and heat-reactive hot melts may be used.
4. Coatings are stress-free.
5. Short runs are relatively economical.

Disadvantages of Calender Coating

1. Equipment cost is high.
2. Additional equipment is needed for flat stock.

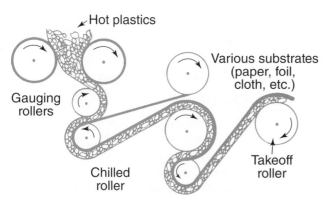

Figure 17-3. Calender coating.

POWDER COATING

There are 10 known techniques for applying plastics powder coatings. However, fluidized-bed, electrostatic-bed, and electrostatic powder gun techniques are the three major processes used today. The process of coating a substrate with a dry plastics powder is sometimes called **dry painting**.

PE, EP, PA, CAB (cellulose acetate-butyrate), PP, PU, ACS (acrylonitrale-chlorinated polyethylene-styrene), PVC, DAIP (diallyl isophthalate resin), AN (acrylonitrile), and PMMA are made into solventless powder formulations for various powder coating techniques. After coating, some techniques require additional heating to guarantee complete fusion or cure.

Fluidized-Bed Coating

In **fluidized-bed coating**, a heated part is suspended in a tank of finely powdered plastics—usually a thermoplastic. The bottom of the tank has a porous base membrane to allow air or inert gas to atomize the powdered plastics into a cloud-like dust storm. Perhaps the term "fog cloud" would be more descriptive, because the air velocity is carefully controlled. This air-solid phase looks and acts like a boiling liquid—hence the term *fluidized bed*.

When the powder hits the hot part, it melts and clings to the part surface. The part is then removed from the coating tank and placed in a heated oven where the heat fuses or cures the powder coating. The part size is limited by the size of the fluidized tank. Epoxy, polyesters, polyethylene, polyamides, polyvinyls, cellulosics, fluoroplastics, polyurethanes, and acrylics are used in powder coating.

The fluidized-bed process originated in Germany in 1953 and has grown into a useful plastics process in the United States.

In a variation of this process, the fluidized powder is sprayed onto preheated parts in a separate chamber. The overspray is collected and reused. (This process is sometimes called **fluidized-bed spray coating**.) The coating on the part is then fused in a heated oven.

The following list shows three advantages and six disadvantages of fluidized-bed coatings:

Advantages of Fluidized-Bed Coating

1. It provides thickness and uniformity.
2. Thermoplastics and some thermosets may be used.
3. No solvents are needed.

Disadvantages of Fluidized-Bed Coating

1. Substrate must be heated above the plastics melt or fusion temperature.

2. Primer may be needed.
3. Thin coatings are hard to control.
4. Continuous automation of line is difficult.
5. Post cure is needed.
6. Surface finish may be uneven (orange peel).

Electrostatic-Bed Coating

In **electrostatic-bed coating**, a fine cloud of negatively charged plastics powders is sprayed and deposited on a positively charged object.

Polarity may be reversed for some operations. Over 100,000 volts with low (less than 100 mA) amperage are used to charge the particles as they are atomized by air or airless equipment. The electrostatic attraction causes the particles to cover all conductive surfaces of the substrate. These parts may or may not require preheating. If preheating is not used, the curing or fusing must take place before the plastics powder loses its charge. The curing is done in a heated oven. Thin foils, screens, pipes, parts for dishwashers, refrigerators, washing machines, and cars as well as marine and farm machines are electrostatic-bed coated. Five advantages and six disadvantages of electrostatic-bed coating are listed here:

Advantages of Electrostatic-Bed Coating

1. Thin, even coats are easily applied.
2. No preheating is needed.
3. Process is readily automated.
4. There is reduced overspray.
5. There is improved finish quality.

Disadvantages of Electrostatic-Bed Coating

1. Thick coatings need preheating of substrate.
2. Small opening or tight angles are difficult to coat.
3. Dust recovery system may be needed.
4. Only ionic resins or plastics can be used.
5. Post-cure is generally needed.
6. Substrates may require special preparation.

Electrostatic Powder Gun Coating

The electrostatic powder gun process is similar to painting with a spray gun. In this process, the dry plastics powder is given a negative electrical charge as it is sprayed on the grounded object to be coated. Fusion or curing must take place in an oven before the powder particles lose their electrical charge, or they will fall from the part. It is possible to coat complex shapes using this method. The fusing oven is the limiting factor relating to size. Automobile manufacturers may replace liquid finishing processes with powder coating methods in the future. Hundreds of products are coated using this process. Examples include outdoor fencing; chemical tanks; plating racks; and dishwasher, refrigerator, and washer parts.

The list below shows five advantages and seven disadvantages of electrostatic gun coating:

Advantages of Electrostatic Powder Gun Coating

1. Thin, even coats are easily applied.
2. No preheating is necessary.
3. Process is readily automated.
4. Short runs and coating of odd-shaped pieces are practical.
5. The equipment cost is lower than for electrostatic-bed coating.

Disadvantages of Electrostatic Powder Gun Coating

1. Thick coatings require preheating of substrate.
2. Small openings or tight angles are hard to coat.
3. Dust recovery system may be needed.
4. Only ionic resins or plastics can be used.
5. Post-cure is needed.
6. Labor cost is high.
7. Thickness is harder to control.

TRANSFER COATING

In transfer coating, a release paper is coated with a plastics solution and dried in an oven. A second coat of plastics is applied over the first coat, and a fabric layer is placed on this wet layer. Next, the coated textile passes through nip rollers and a drying oven. Finally, the release paper is stripped away from the coated fabric. This method produces a tough, leather-like skin on the fabric (Figure 17-4).

Polyurethanes and PVC are commonly used to coat fabrics in the manufacture of awnings, footwear, upholstery, and fashion apparel.

Two advantages and one disadvantage of this process are listed here:

Advantages of Transfer Coating

1. Multicoated and colored substrates are possible.
2. Wide choice of substrates may be coated.

Disadvantage of Transfer Coating

1. Release paper and additional equipment are needed.

KNIFE OR ROLLER COATING

Knife and roller coating methods are other means of spreading a dispersion or solvent mixture of plastics on a substrate. The curing or drying of the plastics coating may be done using heating ovens, evaporating systems, heated rollers, catalysts, or irradiation.

The knife method may involve a simple blade scraper or a narrow jet of air called an air knife. Both sides of the substrate may be coated by utilizing this method.

Paper and fabric are often coated through this procedure. Listed below are four advantages and two disadvantages of knife or roller coating processes:

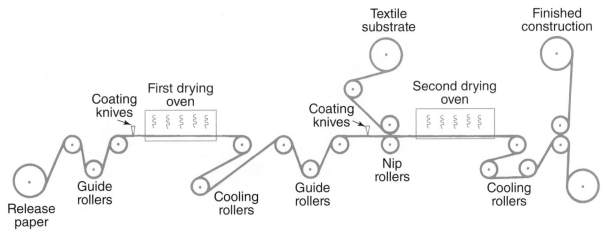

Figure 17-4. Diagram of a transfer coating line.

Advantages of Knife or Roller Coating

1. It is a high-speed continuous process.
2. It provides excellent thickness control.
3. Plastisol coatings are stress- and strain-free.
4. Thick coatings are possible.

Disadvantages of Knife or Roller Coating

1. Equipment and setup time are costly.
2. It is not justified for short runs.

DIP COATING

Dip coatings are applied by dipping a heated object in liquid dispersions or solvent mixtures of plastics. The most common plastics used is polyvinyl chloride. For dispersions, a heating cycle is required to fuse or cure the plastics on the coated object. However, some dip coatings may harden by simple evaporation of solvents. Generally, 10 minutes of heating is needed for each millimeter of coating thickness. Curing temperatures range from 175°C to 190°C (350°F to 375°F). Tool handles and dish-drainer racks are examples of the most common dip-coated products. Objects are limited by the size of the dipping tank (Figure 17-5).

Strippable coatings are often used to ensure that replacement parts arrive in good shape and may be stored under varying conditions. Such coatings are placed on gears, guns, and other hardware. When machined or polished surfaces need protection during fabrication or other operations, strippable coatings may be applied. They have good cohesion, but relatively poor adhesion. As a result, these coatings may be stripped or peeled from the part. Strippable coatings are sometimes used as masking films in electroplating or in applying paints (Figure 17-6).

Wires, cables, woven cords, and tubing may be coated using a modified dipping process in which substrates pass

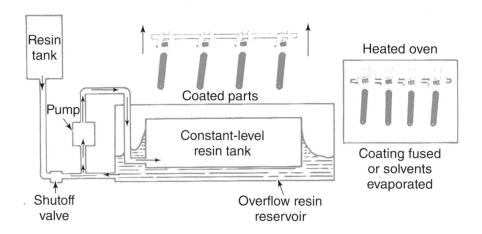

(A) Dip-coating technique.

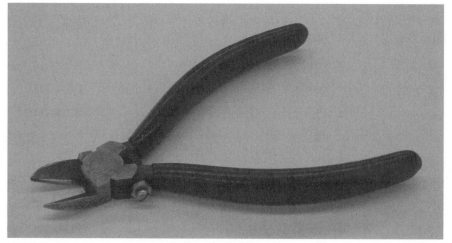

(B) Tool handles dip-coated with PVC.

Figure 17-5. Dip-coating process and products.

Figure 17-6. Strippable coatings protect cutting tools.

through a supply of plastisol or organosol. Both preheating the substrate and post-heating the product speed fusion. Coat thickness is controlled in several ways. The strand may be passed through a die opening, fixing the size and shape of the coating. If no die is used, viscosity and temperature determine the size and shape. Several passes through the plastisol or organosol will increase thickness. This process may be accomplished in a vertical or horizontal position (Figure 17-7). Two advantages and five disadvantages of dip coating are presented here:

Advantages of Dip Coating

1. Light or heavy coatings may be applied on complex shapes.
2. Relatively inexpensive equipment is used.

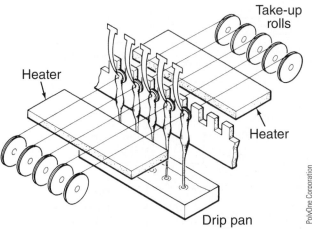

Figure 17-7. This modified dipping process can apply a plastisol coating to wire, cable, woven cord, or tubing at very high speeds.

Disadvantages of Dip Coating

1. Primers may be needed.
2. Plastisols require preheated substrates and post-heating.
3. Organosols require solvent recovery or exhaust.
4. Pot life and viscosity must be controlled.
5. Dip withdrawal rate must be controlled for even coating thickness.

SPRAY COATING

In spray coating, dispersions, solvent solutions, or molten powders are atomized by the action of air, inert gas, or the pressure of the solution itself (airless) and deposited on the substrate. Examples include spray coating of furniture, houses, and vehicles with plastics paints or varnishes. Dispersions of polyvinyl chloride (plastisol) have been spray coated on railroad cars.

In a process sometimes called flame coating, finely ground powders are blown through a specially designed burner nozzle of a spray gun (Figure 17-8). As it passes through this gas or electrically heated nozzle, the powder is quickly melted. The hot molten plastics quickly cools and adheres to the substrate. This process is useful for items that are too large for other coating methods. Three advantages and disadvantages of spray coating are listed here:

Advantages of Spray Coating

1. The equipment cost is low.
2. Short runs are economical.
3. It is fast and adaptable to variation in size.

Disadvantages of Spray Coating

1. It is hard to control coating thickness.
2. Labor costs may be high.
3. Overspray and surface defects (runs and orange peel) may be a problem.

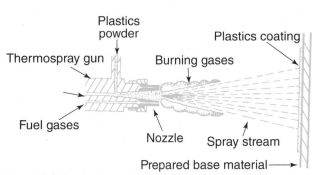

Figure 17-8. Principle of flame coating.

METAL COATING

Perhaps metal coating should not be classified as a basic process of the plastics industry. However, because many plastics are associated with this process, the following information is useful. In addition to being used as a decorative finish, metal coatings may provide an electrically conducting surface, a wear- and corrosion-resistant surface, or added heat deflection. The most common methods for applying a metal coating on a substrate are through the use of adhesives, electroplating, vacuum metallizing, and sputter-coating techniques.

Adhesives

Adhesives are used to apply foils to many surfaces. The textile industry has used this method to adhere metal foils to special garment designs. Complex or irregular parts are difficult to coat, and it is difficult for polyethylene, fluoroplastics, and polyamides to adhere to metals.

Electroplating

Electroplating is done on many plastics. Both the resin and mold design must be considered in producing a metal coating on plastics parts. Ribs, fins, slots, or indentations should be rounded or tapered (Figure 17-9). Phenolic, urea, acetal, ABS, polycarbonate, polyphenylene oxide, acrylics, and polysulfone are often plated.

An electrolysis pre-plating step is accomplished by carefully cleaning the plastics part and etching the surface to ensure adhesion (Figures 17-10 and 17-11). The etched part is cleaned again, and the surfaces are *seeded* with an inactive noble-metal catalyst. An accelerator is added to activate the noble metal,

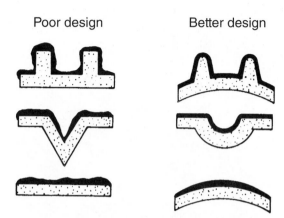

Figure 17-9. When parts are to be plated, large-radius fillets and bends are desirable.

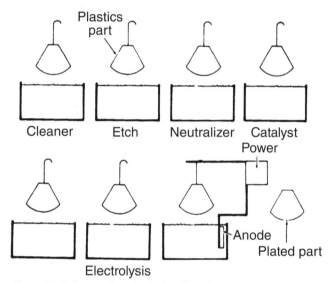

Figure 17-10. Sequence of operations in electroplating plastics.

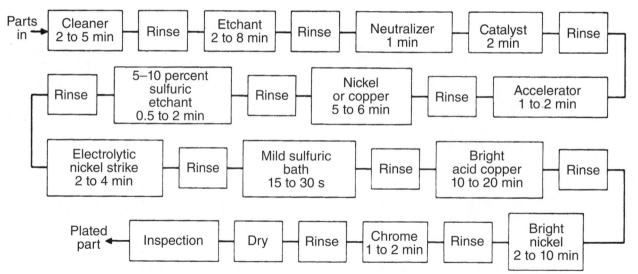

Figure 17-11. Flow chart of a typical plating process.

and the ionic solution of metal reacts autocatalytically in the electrolysis solution. Copper, silver, and nickel electrolysis solutions are used in preparing a deposit ranging from 0.25 to 0.80 micrometers (10 to 30 millionths of an inch) thick. Once a conductive surface has been established, commercial plating solutions such as chrome, nickel, brass, gold, copper, and zinc may be used. Most plated plastics acquire a chrome-like finish. The following list shows two advantages and seven disadvantages of electroplating on plastics:

Advantages of Electroplating

1. It achieves mirror-like finishes.
2. It affords good thickness control.

Disadvantages of Electroplating

1. Holes and sharp angles are hard to plate.
2. Some plastics are not easily plated.
3. Plastics must be cleaned and etched before plating.
4. Cycle time is long.
5. Initial cost is high.
6. It is costly for short runs.
7. Surface finish of plastic must be near perfect in smoothness.

Vacuum Metallizing

In **vacuum metallizing**, plastics parts or films are thoroughly cleaned and given a base coat of lacquer to fill the surface defects and seal the pores of the plastics. Polyolefins and polyamides are chemically etched to ensure good adhesion. The plastics are placed in a vacuum chamber, and small pieces or strips of the coating metal (chromium, gold, silver, zinc, or aluminum) are placed on special heating filaments. The chamber is sealed and the vacuum cycle started. When the desired vacuum is reached (0.5 micrometer Hg or about 0.07 Pa), the filaments are heated. The pieces of metal melt (by high voltage) and vaporize, coating everything the vapor touches in the chamber as well as condensing or solidifying on cooler surfaces (Figure 17-12). Parts must be rotated for full coverage because the vaporized metal travels in a straight path. Once plating is finished, the vacuum is released, and the parts are removed. To help protect the plated surface from oxidization and abrasion, a lacquer coating is applied (Figure 17-13). This finish is best suited for interior applications.

By alternately evaporating two or more metals, it is possible to create a chrome/copper/chrome or stainless/copper/stainless laminate. This process is sometimes called **laminated vapor plating**. The following list shows four advantages and five disadvantages of the vacuum-metallizing coating process.

(A) Cleaning and etching part.

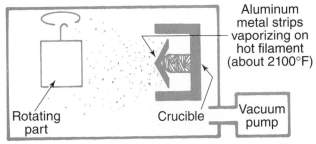

(B) Vaporizing aluminum to coat plastics.

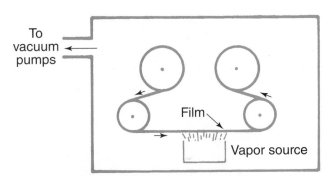

(C) Vacuum metallizing on plastics film.

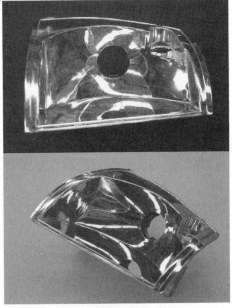

(D) This automotive headlight reflector is vacuum-metallized ABS.

Figure 17-12. Vacuum metallizing of plastics parts.

(A) When electrical current is applied, this filament of stranded tungsten wire becomes incandescent.

(B) These small pieces of pure aluminum, called candy canes, hang on the filaments, melt, and then vaporize. For thicker coatings, heavier pieces are used.

(C) The aluminum melts, spreads in a thin layer over the elements, and vaporizes. Vaporization or flashing the filaments takes only 5 to 10 seconds, with a temperature of 1110°C (2100°F) attained in that time. The metallized products are then removed from the vacuum chamber and dipped in a protective topcoat lacquer. The transparent topcoat can be dyed, allowing a wide choice of colors.

Figure 17-13. Vacuum-metallizing process and supplies.

Advantages of Vacuum-Metallizing Coating

1. Ultrathin, uniform coatings are produced.
2. Nearly all plastics may be used.
3. It produces mirror-like finishes.
4. There is no chemical processing.

Disadvantages of Vacuum-Metallizing Coating

1. Plastics must be coated with lacquer for good results.
2. Vacuum chamber limits part size and output rate.
3. Scratches or flaws are exaggerated.
4. Initial cost is high.
5. It is costly for short runs.

Sputter Coating

Metals or refractories may be deposited by using sputtering systems. Magnetron electronic equipment sprays the metal coating. Chromium atoms fall (sputter) on the plastics surface as argon gas strikes an electrode consisting of the coating metal. Typical thicknesses range from 0.005 mm to 0.07 mm (0.00019 to 0.00275 in.). A clear protective coating of PU, acrylate, or cellulosic is then applied to the metallic coating. This process is used for coating knobs, films, light reflectors, car trim, and plumbing fixtures. Four advantages and three disadvantages of this procedure follow:

Advantages of Sputter Coating

1. Ultrathin coatings
2. It provides excellent adhesion.
3. Automated line systems are possible.
4. Parts are electrically conductive.

Disadvantages of Sputter Coating

1. High technology and capital investment are required.
2. Scratches or flaws are exaggerated.
3. Protective coating is needed over sputter layer.

BRUSH COATING

Solvent and solventless coatings are often brushed onto a substrate by hand. Many paints and finishes are applied in this manner. Solventless finishes consist of two-component systems of resins and curing agents that are mixed and applied. Polyester, epoxy silicone, and some polyurethane resins are used in solventless formulations. Solvent-based coatings may require air drying or heating to cure.

The skill of the applicator and the type and viscosity of the material determine the finish quality.

Protective coatings on large metal tanks are often applied by hand or spray-up methods. If placed underground, the tanks must be protected from corrosion and electrolytic action.

Some familiar examples include coatings on houses, machinery, furniture, and fingernails. Two advantages and three disadvantages of brush coating processes are shown here:

Advantages of Brush Coating

1. Equipment cost is low.
2. Short runs and prototypes are not costly.

Disadvantages of Brush Coating

1. Labor costs are high.
2. There is poor thickness control.
3. Finish is hard to control and reproduce.

RELATED INTERNET SITES

- **www.extrusiondies.com.** Extrusion Dies Industries, LLC manufactures extrusion dies for sheet, film, and coating applications. Selecting "Extrusion & Coextrusion Dies," on the home page leads to "Extrusion Coating Dies," with a discussion of this subject.

- **www.innotekllc.com.** Innotek® manufactures powders for powder coating applications. Select "Products" on the home page, then choose "Thermoplastic," and finally click "Technical Data." This leads to documents on Nylon and vinyl coating, and trouble shooting data for electrostatic and fluid bed coating processes.

- **www.muellercorp.com.** The Mueller Corporation specializes in vacuum metallizing. Its site provides a section on the metallizing process, including details on tooling, loading, basecoating, solvent flashoff, vacuum chamber, topcoating, and inspection. Photographs on the metallizing process are extremely informative.

- **www.southwire.com.** Southwire Company manufactures ROMEX® cable—a widely known cable for electrical wiring. Its site provides information on a wide variety of wire and cable. The section on "Technical Support" contains a wealth of data on the selection of power cables.

- **www.okonite.com.** The Okonite Company manufactures insulated wire and cable. Selecting "Engineering Technical Center" on the home page will lead to a number of topics about conductors and insulation. The site also provides a table on jacket materials selection. This table rates the plastics used for insulation based on a number of criteria.

VOCABULARY

The following vocabulary words are found in this chapter. Use the glossary in Appendix A to look up the definitions of any of these words you do not understand as they apply to plastics.

adhesives
air knife
dip coatings
dry painting
electrostatic-bed coating
extrusion film coating
flame coating
fluidized-bed coating
fluidized-bed spray coating
laminated vapor plating
spray coating
transfer coating
vacuum metalizing

QUESTIONS

17-1. For a process to be classified as a coating, the plastics must remain on the _____.

17-2. The technique by which hot extruded plastics is pressed on a substrate without adhesive is called _____.

17-3. The coating used to protect tools from rusting and damage to cutting edges during shipment is called _____.

17-4. Name two processes used to put a coating on wire.

17-5. Melt-roll coating is a modification of the _____ process.

17-6. What process is used to place coatings on tool handles?

17-7. What is the major advantage of electroplate coating?

17-8. What are two major advantages of brush coating?

17-9. Which coating process would be used to produce reflective window shade films?

17-10. How many minutes of heating time are required for each millimeter of dip-coating thickness.

17-11. Name four methods that may be used to spread dispersion or solvent mixtures on a substrate.

17-12. The ideal wall thickness for a vinyl-dipped coin purse is _____.

17-13. The temperature used to cure plastisols is _____.

17-14. The main element in curing a plastisol coating is _____.

17-15. The main element in curing an organisol coating is _____.

17-16. Name a method that may be used to place a coating on fabric.

17-17. Name two major disadvantages of extrusion coating.

17-18. What causes the powder to atomize in fluidized-bed processing?

17-19. The process of coating a substrate with a dry plastics powder is sometimes called _____.

17-20. Name the major disadvantage of transfer coating.

17-21. In _____ coating processes, the plastics and substrate are given opposite charges.

17-22. The most commonly used plastics for dip coating is _____.

17-23. To speed fusion in dip coating, _____ and _____ the substrate are common practices.

17-24. A coating of _____ is sometimes applied to plated surfaces to minimize oxidization and abrasion.

17-25. Metals or refractories may be deposited by _____ coating systems.

17-26. Name four reasons for coating a substrate.

17-27. Name the three major processes of dry-powder coating.

17-28. Name three commonly transfer-coated materials.

17-29. A narrow jet of air used to spread or disperse resins or plastics on a substrate is called an _____.

17-30. Name four products that are normally spray-coated.

17-31. Name the four major methods of metallizing a substrate.

17-32. Before vacuum metallizing, parts are given a _____ coat to minimize surface defects, provide a reflective surface, and seal the substrate.

17-33. During the vacuum-metallizing cycle, the plastics parts must be _____, because the vaporized metal travels in a line-of-sight path.

17-34. Textiles are coated for moisture and chemical resistance, whereas pots, pans, and tools are coated for _____ and chemical resistance.

17-35. A manufacturer has the following products to be coated. Recommend coating techniques for each.

    a. Plastics grille to look like chrome

    b. Concrete wall slabs for construction

    c. Two-color jewelry piece (black and gold)

    d. Book covers

    e. Textile rainwear

17-36. Describe the fluidized-bed coating process.

17-37. In electrostatic-bed coating, how is the dry powder deposited on the part?

17-38. What is electrostatic powder gun coating? How does it differ from electrostatic-bed coating?

17-39. Briefly describe the electroplating process. Which plastics materials are well suited to this process?

17-40. How would you set up a simple process for coating tool handles with a resilient plastics?

17-41. List the coating processes that require heat for curing. Also, list those that do not require heat for curing.

ACTIVITIES

Fluidized-Bed Coating

Introduction. Fluidized-bed coating is a powder coating process that uses either thermoplastic or thermosetting materials. Figures 17-14, 17-15, and 17-16 show various types of fluidized-bed coaters. All three coaters contain a porous bottom that allows the passage of air or other gases. The air forces its way through the powder, causing the powder to float in the tank somewhat like a fluid. The powder melts and sticks to the surface of preheated substrates when they are dipped into the bed. A post-heating cycle will smooth the surface of thermoplastics and cause final curing of thermosets.

Equipment. Fluidized-bed coater, polyethylene powder, caliper, oven, substrates for dipping, and insulated gloves.

Procedure

17-1. Make substrates for dipping. Figure 17-17 shows a steel substrate, cut from a strip 3 mm × 18 mm. These pieces are about 40 mm long. Other size or shape

Figure 17-14. Small commercial laboratory fluidizer.

Figure 17-15. Small homemade laboratory fluidizer.

Figure 17-16. Laboratory fluidizer with compressor.

Figure 17-17. Uncoated, coated, and peeled substrates.

substrates are also functional. Attach a short piece of wire to each substrate as shown. Measure the thickness of the substrates.

17-2. Hang the substrates in an oven and heat to 177°C (350°F).

17-3. Prepare a fluidized bed by adding polyethylene powder and adjusting the air pressure to thoroughly fluidize the powder. Depending on the type of fluidizer, 20 to 34 kPa (2.9 to 5 psi) may be sufficient.

17-4. Remove one piece of metal from the oven, hold it by the wire, and dip it quickly into the bed. Record the duration of time in the bed. Remove the piece and shake off the excess powder.

17-5. Return the coated piece to the oven to smooth the surface by fusion of powder particles.

17-6. Remove from the oven and cool.

17-7. Measure the thickness of the coated piece and calculate the thickness of the coating. Is the coating thicker at the bottom than at the top of the piece?

17-8. It is often possible to remove the coating by cutting along the edge of the metal and peeling off the coating. One item in Figure 17-17 shows a coating lifted off the substrate. Measuring the thickness on a peeled coating may increase the accuracy of measurement.

17-9. Arbitrarily determine a desired thickness of coating. The coatings in Figure 17-17 were about 0.38 mm (0.015 in.) thick.

17-10. Determine a process that repeatedly yields a smooth, uniform coating of the desired thickness. Here are the parameters of importance:

- Preheat temperature
- Post-heat temperature and time
- Dip time
- Depth substrate is dipped into the bed
- Agitation during dipping
- Cleanness of substrate
- Air pressure

17-11. Write a report summarizing the process that yields optimum coating.

Further investigations. To determine the effects of the substrate material, fabricate substrates from steel, aluminum, and copper or brass. Make sure they are dimensionally identical. Heat to the same temperatures and dip equally. Does the coating thickness change from one type of substrate to another?

Dip Coating

Introduction. A plastisol is a mixture of polyvinyl chloride powder and a plasticizer. When a preheated substrate is dipped into a plastisol, particles of PVC adhere to the substrate. Post-heating completes the fusion of the PVC particles.

Equipment. Vinyl dispersions, substrates for dipping, oven, calipers, tensile tester, and insulated gloves.

Procedure

17-1. Fabricate substrates for dipping. The substrates shown in Figure 17-18 are aluminum and have been cut from a bar 6 mm × 25 mm (0.2 in. × 1 in.) They must be about 100 mm (4 in.) long to fit in a quart can of vinyl dispersion.

17-2. Preheat the substrates to 200°C (392°F). Working rapidly, remove the substrate from the oven and hang it in the dispersion. Figure 17-18 shows a dip coating in progress. Record the dipping time. Remove from the dispersion and allow all drips to fall off.

17-3. Return to the oven and cure at 175°C (347°F).

17-4. When thoroughly fused, remove from oven and cool.

17-5. Slip the coating off the substrate, or cut at the edge of the substrate.

17-6. Figure 17-19 shows a substrate that has been dipped, a piece of vinyl cut into a dog-bone shape, and another piece after tensile testing. To cut the dog bone, use a cutter, as shown in Figure 17-20. If a cutter is unavailable, use scissors to cut the shape. Tensile test the vinyl sample to determine the strength and percentage of elongation. Omitting cutting it into a dog-bone shape

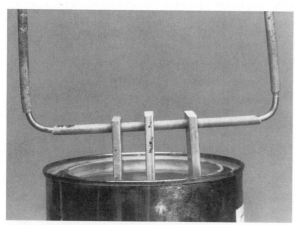

Figure 17-18. Dipping into a quart of plastisol.

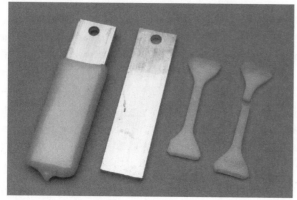

Figure 17-19. Stages in testing of vinyl dip coatings.

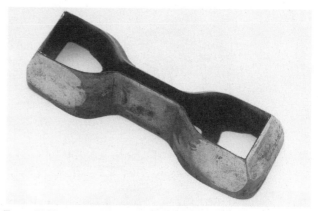

Figure 17-20. Dog-bone cutter showing cutting edges.

will provide a strip of uniform width. This may lead to difficulties if the sample begins to slip out of the jaws of the tensile tester. The broken sample in Figure 17-19 had an ultimate tensile strength of 9.78 MPa (1417 psi) and an elongation of 385%.

17-7. Experimentally determine a process that yields smooth, uniform, strong, and elastic vinyl. The parameters of significance are these:

- Preheat temperature
- Dip time
- Cure temperature
- Cure time
- Surface smoothness of the substrate

! CAUTION

Overheating the plastisol will release toxic hydrogen chloride gas (HCl). Make sure ventilation near the oven is adequate.

17-8. Write a report summarizing the process that yielded optimum vinyl.

Additional activities

a. To observe the effects of substrate smoothness, polish one surface of the substrate and roughen the opposite side. What effect does a rough surface produce?

b. Investigate the effect of post-heating time. Produce several samples, incrementally increase the post-heat time, and test to determine the effects on the physical properties of the vinyl.

c. Do a research paper on latex rubber surgical gloves. Due to concerns about AIDS, the use of latex gloves has increased. How do the manufacturers produce gloves with uniform thickness and no holes?

FABRICATION PROCESSES AND MATERIALS

CHAPTER 18

INTRODUCTION

As with wood, metal, and other materials, plastics components often require assembly or fabrication. This chapter discusses the major fabrication processes and the materials utilized during these processes. There are four broad methods used to join plastics. These are included in the following chapter outline:

MECHANICAL ADHESION

Adhesives are a broad class of substances that adhere materials together with a surface bond. If the adhesives hold parts together by interlocking the surfaces, they are referred to as mechanical or physical adhesives. Mechanical adhesives come in various forms. However, they must be in a liquid or semiliquid state during the bonding operation. This ensures close contact with the adherends (surfaces being adhered). In mechanical adhesion, there is no flow of the adherends.

Animal and some natural plastics have been used as adhesives for centuries. Wax and shellac were once widely used to seal letters and important documents. Many ancient civilizations used pitch to seal cracks in boats and rafts. Archeologists have evidence that more than 30 centuries ago Egyptians used adhesives to attach gold leaf to wooden coffins and crypts.

At one time, the word glue referred to an adhesive obtained from hides, cartilage, bones, and other animal materials. Today, this term is synonymous with adhesives based on plastics and usually refers to bonding wood.

Mechanical adhesives generally fall into three basic categories: (1) thermoplastic resins, (2) thermosetting resins, and (3) elastomeric types. Table 18-1 lists a number of thermosetting and thermoplastic adhesives as well as their available forms.

Thermoplastic Resins

Thermoplastic resins include adhesives based on acrylics, vinyls, cellulosics, and hot-melt materials.

Acrylic adhesives. Acrylic adhesives range from flexible to hard materials. One popular form of acrylic adhesive is a cyanoacrylate adhesive. It is a rapid-setting adhesive that polymerizes when pressure is applied to the joint. The solvent *N,N-dimethylformamide* will thin this adhesive and clean unpolymerized excess. However, this adhesive does not cure by solvent evaporation, but by polymerization.

Vinyl adhesives. Vinyl adhesives encompass a variety of materials. Polyvinyl alcohol is a water-based adhesive used to bond paper, textiles, and leather. The interlayer of safety glass is

Table 18-1. Available Forms of Selected Plastics Adhesives

Plastics Adhesives	Available Forms
Thermosetting	
Casein	Po, F
Epoxy	Pa, D, F
Melamine formaldehyde	Po, F
Phenol formaldehyde	Po, F
Polyester	Po, F
Polyurethane	D, L, Po, F
Resorcinol formaldehyde	D, L, Po, F
Silicone	L, Po, Pa
Urea formaldehyde	D, Po, F
Thermoplastic	
Cellulose acetate	L, H, Po, F
Cellulose butyrate	L, Po, F
Cellulose carboxymethyl	Po, L
Cellulose, ethyl	H, L
Cellulose, hydroxyethyl	Po, L
Cellulose, methyl	Po, L
Cellulose nitrate	Po, L
Polyamide	H, F
Polyethylene	H
Polymethyl methacrylate	L
Polystyrene	Po, H
Polyvinyl acetate	Pt, D, L
Polyvinyl alcohol	Po, D, L
Polyvinyl chloride	Pa, Po, L

Note: Po–Powder; F–Film; D–Dispersion; L–Liquid; Pa–Paste; H–Hot Melt; Pt–Permanently tacky

made of polyvinyl butyral because it has excellent adhesion to glass. The electrical and insulation value of polyvinyl formal makes it ideal for wire enamels. Polyvinyl acetal excels as a bonding adhesive for metals.

One very popular modern polyvinyl acetate adhesive is white glue. This adhesive comes ready to use in a fast-setting liquid form. This familiar adhesive consists of a dispersion of polyvinyl acetate in a solvent. Often the solvent used is water, which must be kept from freezing. Carpenters, artists, secretaries, and many other people make use of the adhesive properties of this material.

Cellulosic adhesives. Cellulosic adhesives are popular and available in solvent, hot-melt, and dry-powder forms. Duco® cement is an example of a general-purpose cellulose nitrate adhesive. It is both waterproof and clear and will adhere to wood, metal, glass, and paper as well as many plastics. Cellulose acetates and butyrates are familiar cements used for plastics models.

Hot-melt adhesives. Hot-melt materials are also popular because they are easy to use, somewhat flexible, and obtain their highest adhesive qualities when cooled. Several thermoplastics are used as hot-melt adhesives, including polyethylene, polystyrene, and polyvinyl acetate.

Small sticks or rods of these plastics are heated in an electric gun. The hot plastics is forced from the gun nozzle onto the gluing surface. Figure 18-1 shows a leather strap being assembled using hot-melt plastics in an electrically heated applicator gun. The shoe industry is presently using this adhesive as an effective means of assembling leather goods.

Small, rapidly assembled articles may be bonded by using this method. Probably the most serious drawback to hot-melt adhesives is the difficulty in making large glue joints. This particular adhesive cools too quickly to ensure adhesion on large bonding surfaces.

Thermosetting Resins

Thermosetting resins gain their strength from polymerization reactions that normally occur after the resin covers the adherends. The polymerization generally occurs due to thermal reactions or catalysts. Major types of thermosetting resins include amino and phenolic resins and epoxies.

Amino resins. Casein and urea-formaldehyde (amino resins) are used in the woodworking industry. Some of the urea resins are sold in liquid form for use in the manufacture of plywood, particle boards, and hardboards.

Shell-molding is an important process used by foundry workers to cast metal parts. Phenolic and amino resins are used to bond the sand mold together.

Phenolic resins. Phenol-formaldehyde (phenolic) resins are sold in liquid, powder, and film forms. The films range from 0.025 mm (0.001 in.) thick and are placed between the

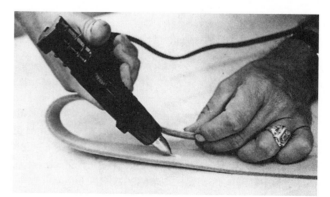

Figure 18-1. This glue gun is used to apply hot-melt adhesives.

materials to be bonded. Moisture in the material or external steam causes the adhesive film to flow and liquefy. The curing reaction takes place at temperatures ranging from 120°C to 150°C (250°F to 300°F). A reinforced film is commonly used. These films are usually thin like tissue paper and saturated with the adhesive. They are used in the same manner as unreinforced films but are easier to handle and apply.

Large quantities of phenolic resin adhesives are used in the manufacture of exterior plywood and tempered or smooth-finished hardboard.

Resorcinol-formaldehyde resins are another type of phenolic-based adhesive that usually comes in liquid form. At the time of use, it is mixed with a powdered catalyst. This resin has the advantages of curing at room temperature and being water and heat resistant. High grades of exterior and marine plywood are bonded with these adhesives (Figure 18-2).

High-frequency heating greatly reduces the curing or polymerization time of many plastics used as adhesives. The high-frequency field excites the molecules of the adhesive, causing heat and rapid polymerization. Wood joints are often assembled using resorcinol-formaldehyde adhesives during this process.

Resin-bonded grinding wheels and sandpapers are made from abrasive grains and a plastics bonding agent. The grinding wheels are made from abrasive grains, powdered resin, and a liquid resin, using a cold-molding process. Figure 18-3 shows some typical sandpapers and grinding wheels bonded with phenolic or other resins.

Epoxy resins. Epoxy resin adhesives are thermosetting plastics available in two-part paste components. Epoxy resin and either a powdered or resinous catalyst are mixed to polymerize the resin. Heat is sometimes used to aid or speed the hardening process. Specially formulated one-part epoxies may be polymerized by the application of heat alone.

Figure 18-3. Many grinding wheels and sandpapers are resin-bonded.

If the surfaces are properly prepared, epoxy adhesives have excellent adhesion to nearly all materials. The superior adhesive properties of epoxies are used to mend broken china, bond copper to phenolic laminates in printed circuits, and bond components in sandwich or skin-type structures. However, even epoxy adhesives have difficulty bonding polyethylene, silicones, and fluorocarbons.

Elastomeric Types

Elastomeric adhesives must bond effectively with the substrate to which they are applied. Their basic purpose is to keep moisture, air, or other agents out of cracks or small openings. To remain effective, these compounds must maintain adhesion yet stretch or compress as the materials they contact expand or contract. For example, the compounds that seal glass windows to aluminum frames must withstand differential expansion, because aluminum expands about 2.5 times more than glass.

Elastomeric adhesives bear many names such as caulk, sealant, glazing compound, and putty. Some of the more common sealants include polysulfides, acrylics, polyurethanes, silicones, and both natural and synthetic rubber compounds.

If the openings and cracks are rather small, caulking and sealing materials are appropriate (Figure 18-4). If the cracks or openings are rather large, the putties or patching compounds are appropriate because they contain a large amount of filler. They are also formulated to minimize shrinkage.

If glass is involved, the materials are usually called glazing compounds. Many glazing compounds and putties contain acrylic compounds. They are easily applied and will not crack, sag, or break down. Figure 18-5 shows automotive glazing applications and Figure 18-6 includes window glazing.

If ceramics need sealing, special formulations of epoxy and silicone compounds are used as sealants around lavatories and bathtubs.

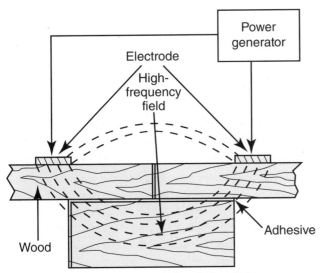

Figure 18-2. High-frequency (radio) waves may be used to heat adhesives.

Figure 18-4. Many caulks, sealants, and glazing materials are available in cartridges for ease of application.

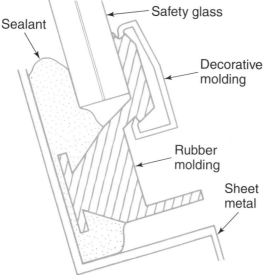

(A) Sealant with rubber molding and decorative metal strip.

(B) Sealant with molding strip.

Figure 18-5. Sealing methods used for automobile windshields.

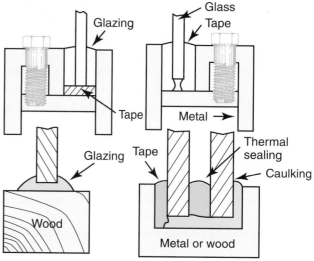

Figure 18-6. Various sealing methods.

The aircraft industry has developed many specialized adhesives that bond aluminum parts. The polymer polysulfide is a very effective sealant with a wide range of applications in the aircraft, electrical, and building industries.

Using compressible tapes or extruder guns is an efficient way to apply sealants. Compressible tapes are rolls of sealant in ribbon form. They are used in the automotive industry for sealing metal joints and as adhesive sealants in window construction.

Extruder guns are convenient applicators that use disposable or refillable containers of sealing compound. These devices push the sealant out of a nozzle during application.

CHEMICAL ADHESION

Chemical or specific adhesion has been defined as adhesion between surfaces held together by valence forces of the same type as those that give rise to cohesion. Forces that hold the molecules of all materials together are referred to as cohesive forces. These forces include strong primary valence bonds and weaker secondary bonds.

In chemical adhesives, there is a strong valence attraction between the materials as molecules flow together. During the welding of metals, for example, molten metal flows and a chemical cohesive action occurs between the pieces.

It should be apparent that chemical bonding can occur only by causing a softening or flow in two materials. If the surfaces are not caused to flow, the pieces are held together only by mechanical or physical forces. In the cohesive bonding of metals, heat must be applied to cause the surface molecules to flow and intermingle. The cohesive bonding of plastics requires either the use of solvents or heat. This type of bonding does not occur with thermoset materials. Heat-techniques cause melting or softening from friction between plastics parts or heat transferred from hot metal.

Solvent Bonding

There are two basic forms of solvent-based adhesives—solvent cements and monomeric cements. Solvent cements dissolve the surfaces of the plastics being joined. This forms a strong intermolecular bond as it evaporates. Monomeric cements are based on a monomer of at least one of the plastics to be joined. It is catalyzed so that a bond is produced by polymerization in the joint. Either cohesive or adhesive bonding will occur, depending on the chemical composition of the materials being joined.

Solvent cements and dope cements are two kinds of cements in common use. The first are solvents or blends of solvents that dissolve the material. When the solvent evaporates, the items are fused together. Dope cements are sometimes called laminating cements or solvent mixes. They are composed of solvents and a small quantity of the plastics to be joined. This cement is a viscous (syrupy) material that leaves a thin film of the parent plastics on the joint when dried.

Solvents with low boiling points evaporate quickly. Therefore, the joint must be positioned before all of the solvent evaporates (Table 18-2). An example of such a material is methylene chloride, with a boiling point of 40°C (104°F). Solvent cements may be applied to plastics joints by any of several methods mentioned below. Regardless of the method used, all joints should be clean and smooth. A V-joint is preferred for making butt joints by many manufacturers and fabricators (Figure 18-7).

By using the soaking method, joints may simply be soaked in a solvent until a soft surface is obtained. The pieces are then placed together at once under slight pressure until all solvents evaporate. If too much pressure is applied, the soft portion may be squeezed out of the joint, resulting in a poor bond.

Large surfaces may be dipped into or sprayed with solvent cements. Cohesive bonds also may be made by allowing the solvent to flow into crack joints using a capillary action. Small paint brushes and bulb applicators are handy cementing tools. Figure 18-8 shows a number of methods used for cementing.

Frictional Heating Techniques

The major chemical adhesion techniques that use frictional heat are spin welding, dielectric bonding, and ultrasonic bonding.

Spin welding (bonding). Spin welding (bonding) is a friction method of joining circular thermoplastic parts. When one or both parts are rotated against each other, frictional heat causes a cohesive melt. Depending on the diameter and the material, joints must spin at 6 m/s (20 ft/s) with less than

Table 18-2. Common Solvent Cements for Thermoplastics

Plastics	Solvent	Boiling Point, °C	°F
ABS	Methyl ethyl ketone	80	[176]
	Methyl isobutyl ketone	117	[243]
	Methylene chloride	40	[104]
Acrylic	Ethylene dichloride	84	[183]
	Methylene chloride	40	[104]
	Vinyl trichloride	87	[189]
Cellulose Plastics:			
Acetate	Chloroform	61	[142]
	Methylene chloride	40	[104]
Butyrate, propionate	Ethylene dichloride	84	[183]
Ethyl acetate	Methyl ethyl ketone	80	[176]
Ethyl cellulose	Acetone	57	[135]
Polycarbonate	Ethylene dichloride	41	[106]
	Methylene chloride	40	[104]
Polyphenylene oxide	Chloroform	61	[142]
	Ethylene dichloride	84	[183]
	Methylene chloride	40	[104]
	Toluene	110	[232]
Polysulfone	Methylene chloride	40	[104]
Polystyrene	Ethylene dichloride	84	[183]
	Methyl ethyl ketone	80	[176]
	Methylene chloride	40	[104]
	Toluene	110	[232]
Polyvinyl chloride and copolymers	Acetone	57	[135]
	Cyclohexane	81	[178]
	Methyl ethyl ketone	80	[176]
	Tetrahydrofuran	65	[149]

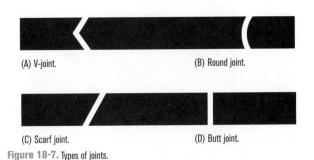

(A) V-joint.   (B) Round joint.

(C) Scarf joint.   (D) Butt joint.

Figure 18-7. Types of joints.

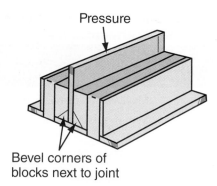

Pressure

Bevel corners of
blocks next to joint

(A) T-joint.

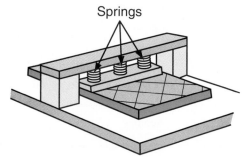

Springs

(B) Cementing a rib on a sheet.

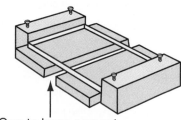

Gap to keep cement
from seeping between
fixture and Plexiglas

(C) Butt joint rig.

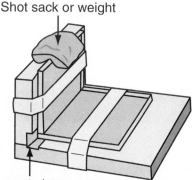

Shot sack or weight

Groove to keep cement
from seeping between
fixture and plastics

(D) Corner cementing.

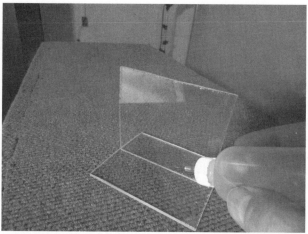

(E) Applying cement with a syringe.

Figure 18-8. Various cementing methods.

138 kPa (20 psi) of contact pressure. When melting takes place, the spinning is stopped and the melt solidifies under pressure.

Joints may also be spin welded by rapidly rotating a filler rod on the joint. A heavy rod consisting of the parent material is rotated at 5000 rpm and moved along the joints as it melts (see Figure 18-9). The plastics weld resembles an arc weld on metal.

Vibration bonding, a variation of spin bonding, is a method by which noncircular parts may be bonded. Vibration frequencies range from 90 to 120 Hz and joint pressures range from about 1300 to 1800 kPa (200 to 250 psi).

Nearly any melt-processable, thermoplastic polymer (even dissimilar polymers with compatible melt temperatures) may be assembled into bottles, tubes, and other containers.

Dielectric or high-frequency bonding. *Dielectric bonding* is used to join plastics films, fabrics, and foams. Only plastics that have a high dielectric loss characteristic (dissipation factor) may be joined using this method. Cellulose acetate, ABS, polyvinyl chloride epoxy, polyether, polyester, polyamide, and polyurethane have sufficiently high dissipation factors that allow dielectric sealing. Polyethylene, polystyrene, and fluoroplastics have very low dissipation factors and cannot be electronically heat sealed. The actual fusion is caused by high-frequency (radiofrequency) waves from transmitters or generators available in several kilowatt sizes. In areas of the parts where the high-frequency waves are directed, molecules try to realign themselves with the oscillations (Figure 18-10). This rapid molecular movement causes frictional heat, causing the areas to become molten.

The Federal Communications Commission (FCC) regulates the use of high-frequency energy. The generated signals are similar to those produced by TV and FM transmitters and operate at frequencies between 20 and 40 MHz.

Ultrasonic bonding. In **ultrasonic welding**, two plastics parts are placed together on a fixture or in a nest. The horn of the ultrasonic welder squeezes the parts together

(A) Plastics rod spin welded to a mirror acrylic sheet. The clarity of the weld zone indicates a complete weld.

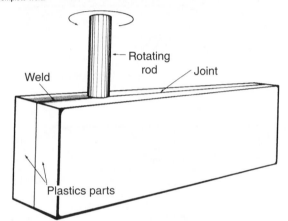

(B) A method of spin welding joints.

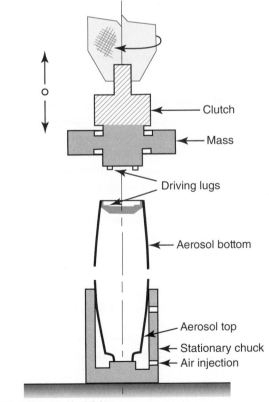

(C) Spin welding aerosol bottle halves.

Figure 18-9. Principle of spin welding (bonding).

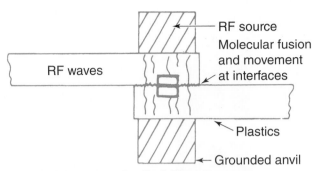

Figure 18-10. Dielectric heat sealing with radio-frequency waves.

using relatively low pressure (see Figures 18-11A and 18-11B). Almost immediately after the horn touches the parts, it begins to vibrate at a high frequency. In the machine, an electronic transducer converts 60 Hz vibrations into frequencies in the horn of either 20 kHz or 40 kHz. The vertical vibrations cause enough frictional heat to melt the plastics. After a short time, usually only a fraction of a second, the vibration stops. The horn continues to clamp the parts until the plastics solidifies.

Ultrasonic techniques are used to activate adhesives to a molten state, spot weld, and sew or stitch films and fabrics together without needles and thread (Figures 18-11C, 18-11D, and 18-11E). Simple joints may be welded in 0.2 to 0.5 seconds.

Staking is a term used to describe the ultrasonic or heated-tool formation of a locking head on a plastics stud similar to forming a head on a metal rivet (see Figure 18-11F). Plastics parts with studs may be assembled by using this technique.

Many adhesives may be melted and cured by ultrasonic vibrations (Figure 18-11G). Ultrasonic systems may be used to cut thermoplastic fabrics and degate parts from runner systems.

Spot bonding is a process similar to metal spot-welding for plastics up to 6 mm (0.25 in.) in thickness, using specially designed horns and high-power equipment (Figure 18-11D). Vibrations from the horn penetrate the first sheet and nearly half the second. Next, the molten material flows into the space between the sheets. Films and fabrics are stitched in a similar fashion.

Insertion bonding uses ultrasonic means to place metallic inserts into plastics parts (Figure 18-11H). The horn is used to hold the insert and direct the high-frequency vibrations into an undersized hole. As the plastics melts, the pressure from the horn forces the insert into the hole. The plastics reforms itself around the insert upon cooling.

Transferred-Heat Techniques

Chemical adhesion may also involve transferring heat from hot metals or gases into plastics parts. Common techniques

(A) The Millennium, model 220, a 20 kHz ultrasonic welder with a dynamic process controller.

Photo courtesy of the Dukane Corporation

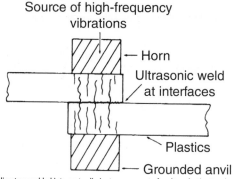

(B) Energy directors molded into parts eliminate movement of molten plastics to edges of joint.

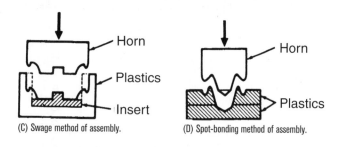

(C) Swage method of assembly.

(D) Spot-bonding method of assembly.

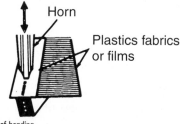

(E) Stitch method of bonding.

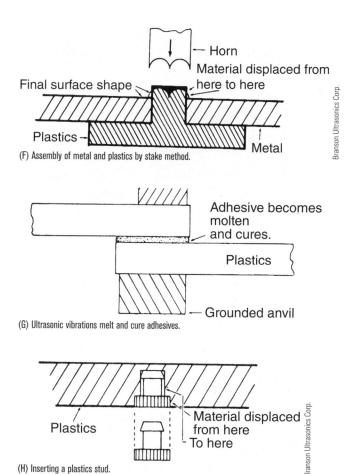

(F) Assembly of metal and plastics by stake method.

(G) Ultrasonic vibrations melt and cure adhesives.

(H) Inserting a plastics stud.

Branson Ultrasonics Corp.

Figure 18-11. Ultrasonic bonding methods.

include hot-gas bonding, heated-tool welding, impulse bonding, and electromagnetic bonding.

Hot-gas welding (bonding). Hot-gas welding consists of directing a heated gas (usually nitrogen) at temperatures ranging from 200°C to 425°C (400°F to 800°F) onto the joints to be melted together. The temperature of the flameless hot-gas torch is controlled by regulating the gas flow or heating source, using electric heating elements with a nitrogen or air pressure of 14 to 28 kPa (2 to 4 psi). This process is similar to open-flame welding of metals (Figure 18-12). Filler rods or materials similar to the parent plastics are used to build up the welded area. Welds may exceed 85% of the tensile strength of the parent material (Table 18-3).

As in any welding technique, the joint area must be properly cleaned and prepared, and butt joints should be beveled to 60°.

Heat staking. Similar to hot-gas welding, this method uses heated air to soften stakes or posts that typically extend through holes in mating parts. When the stakes reach the proper temperature, the hot air is removed, and a cold forming tool presses the molten material into the desired shape, thus fastening the two pieces.

The induction coil must be as close to the joint as possible for rapid bonds to occur. Nonmetallic tooling must be used for alignment.

MECHANICAL FASTENING

A wide choice of mechanical fasteners can be used with plastics. Self-threading screws are used if the fastener will not be removed very often. However, when frequent disassembly is required, threaded metal inserts are placed in the plastics.

Several types of metal inserts are shown in Figure 18-16. These may be molded or placed in the part after molding.

Screws and metallic or plastic rivets provide permanent assembly. Standard nuts, bolts, and machine screws made from both metals and plastics are used in common assembly methods (Figure 18-17).

The spring clips and nuts shown in Figure 18-18 are examples of low-cost, rapid mechanical fasteners. Hinges, knobs, catches, dowels, and other devices are also used in the assembly of plastics.

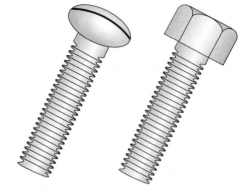

(A) Thread-cutting (self-threading) screws designed for hard plastics.

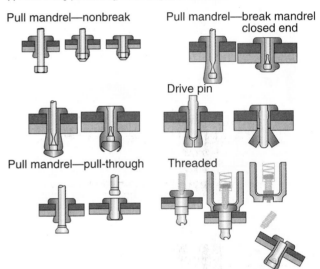

(B) Blind rivets.

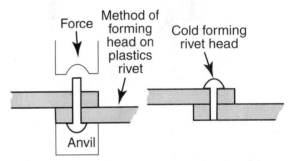

(C) Cold forming of plastics rivets. Heads may be formed by mechanical or explosive means.

Figure 18-17. Selected methods of assembly.

FRICTION FITTING

Friction fitting is a term used to describe a number of pressure-tight joints of permanent or temporary assemblies. The most common are press fits, snap fits, and shrink fits. These techniques may join either like or unlike materials without the use of any mechanical fasteners.

Press Fitting

Press fitting may be used to insert plastic or metallic parts into other plastics components. The parts may be joined while the

Figure 18-16. Inserts designed for use in plastics are installed quickly and easily after molding.

Emhart Technologies/Dodge

plastics are still warm. When a shaft is press fitted into a bearing or sleeve, the outside as well as the inside diameter may be expanded (Figure 18-19).

The difference between a press fit and a snap fit is the undercut and amount of force required for assembly.

Snap Fitting

Snap fitting is a means of assembly in which parts are snapped into place. The plastics is simply forced over a lip or into an undercut retaining ring. Some examples of snap fitting are the simple locks or catches on plastics boxes or the covers on many parts such as automotive dome lenses, flashbulb covers, and instrument panels (Figure 18-20).

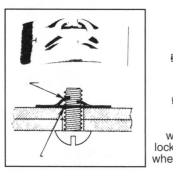

Single-thread locknut, which is speedily applied, locks by grip of arched prongs when bolt or screw is tightened

Tinnerman Palnut Engineered Products, LLC

(A) Single-thread locknut.

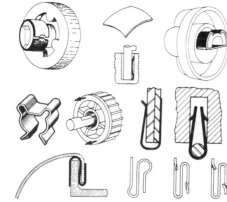

(B) Mechanical fasteners and devices for assembly of plastics parts.

Figure 18-18. Inexpensive mechanical fasteners.

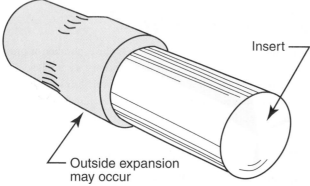

Figure 18-19. Press fitting, showing possible expansion of outside diameter of plastics parts.

- Outside expansion may occur
- Insert

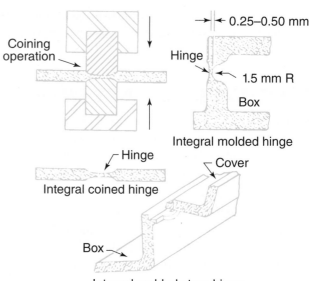

Coining operation

0.25–0.50 mm

Hinge

1.5 mm R

Box

Integral molded hinge

Hinge

Integral coined hinge

Cover

Box

Integral molded strap hinge

(A) Examples of integral molded and coined hinges.

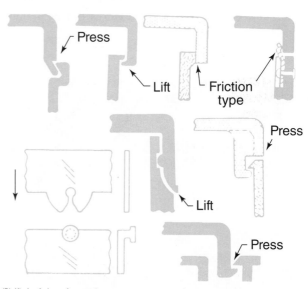

Press

Lift

Friction type

Press

Lift

Press

(B) Kinds of clasps for containers.

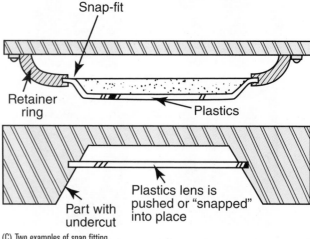

Snap-fit

Retainer ring

Plastics

Part with undercut

Plastics lens is pushed or "snapped" into place

(C) Two examples of snap fitting.

Figure 18-20. Assembly methods using snap-fitting and integral-hinge techniques.

In both press and snap fits, one of two parts is made smaller so that the two cannot fit together without force. The intentional differences in the two part dimensions are called allowances. A negative allowance is called interference, a characteristic necessary for a tight fit. Allowances made for unintentional variations in dimensions are called tolerances. The maximum and the minimum dimensions are referred to as limits. These limits define the tolerance.

Shrink Fitting

Shrink fitting refers to the process of placing inserts in the plastics just after molding and allowing the plastics to cool (Figure 18-21). It also refers to the placement of plastics parts over substrates, where they are heated until the plastics shrinks to its original shape.

RELATED INTERNET SITES

- **www.emersoninduatrial.com/en-us/branson/Pages/home.aspx.** Branson Ultrasonic Corporation, a subsidiary of Emerson Industrial Automation, manufactures a variety of plastics welding and bonding equipment. On the home page, select "Products," and then choose "Plastics Welding." In the section on "Plastics Welding Literature," a number of publications are available on all types of their equipment. Of particular interest is a document in the "Theory" section entitled "Polymer Characteristics and Weldability."

- **www.dukcorp.com.** Near the bottom of the Dukane Corporation's home page, select "Ultrasonics Site Map." Then select "Process Overview," which leads to details on ultrasonic, vibrational, and spin welding as well as heat staking.

- **www.laramyplasticwelders.com.** Laramy Products, LLC, provides hot-gas manual welders for plastics.

- **www.lord.com.** Lord Corporation prepares adhesives and coatings. On the home page, select "Products & Solutions" to reach the section on "Adhesives." Within "Adhesives", choosing "Automotive Applications," and then "Panel Bonding and Sealing" will lead to data sheets on a large number of adhesives specially formulated to bond composites to metals.

Several companies manufacture inserts for plastics. The designs change depending on the method of inserting. The most common methods are molded-in, pressed-in, self-tapping, and ultrasonic or thermal insertion. Some manufacturers are listed here:

- **www.emhart.com**

- **www.yardleyproducts.com**

- **www.tristar-inserts.com**

VOCABULARY

The following vocabulary words are found in this chapter. Use the glossary in Appendix A to look up the definitions of any of these words you do not understand as they apply to plastics.

adherends
allowances
chemical (specific) adhesion
cohesive
cyanoacrylate
dope cements
epoxy resin
friction fitting
glue
hot-gas welding
induction bonding
insertion bonding
interference
monomeric cements
phenol-formaldehyde
polyvinyl acetal
polyvinyl alcohol
polyvinyl butyral
polyvinyl formal
resorcinol-formaldehyde
shrink fitting
solvent cements
spin welding
spot bonding
staking
tolerances
ultrasonic welding

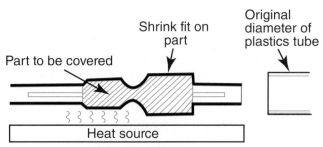

Figure 18-21. Shrink fit into plastics tube over electronic part. Note size of tube before heating.

QUESTIONS

18-1. Identify the type of adhesion created when materials are joined together and an intermingling of molecules occurs.

18-2. Name a bonding process similar to spot welding of metals.

18-3. A _____ dissipation factor is essential for dielectric or high-frequency bonding.

18-4. The bonding process that uses a high-frequency vibration is called _____.

18-5. The percent of bond strength for hot-gas weld bonding of polyethylene is _____.

18-6. Name the bonding that is a method of frictionally joining circular thermoplastics together.

18-7. List three common solvents for solvent bonding acrylic plastics.

18-8. Which solvent from Table 18-2 will evaporate most rapidly?

18-9. Cements composed of solvents combined with a small quantity of the plastics to be joined are called _____.

18-10. Name the term used to describe the ultrasonic forming of a locking head on plastics studs.

18-11. Insertion is a technique of which bonding method?

18-12. Which bonding method would be used in the meat packaging department of a grocery supermarket?

18-13. Name two plastics often bonded by hot-gas welding.

18-14. Name the bonding process in which the plastics film is quickly heated and cooled by the die.

18-15. List two major advantages of impulse bonding.

18-16. Name the four basic ways plastics are joined.

18-17. Solvents or blends of solvents that melt selected plastics joints together are known as _____ cements.

18-18. Into which of the four basic joining methods does spin welding fit?

18-19. In _____, thermoplastic materials are softened by a jet of hot gas.

18-20. Identify the bonding method usually performed on films and fabrics.

18-21. Thermoplastics with _____ factors such as ABS can be sealed by dielectric joining.

18-22. In _____, high frequency causes molecules to move rapidly, thus melting the plastics.

18-23. Thermosetting adhesives are _____ in most solvents once they have cured.

18-24. In solvent cementing, evaporation of the _____ may cause stress cracks.

18-25. Joint preparation for welding thermoplastics resembles that for _____.

18-26. A variety of _____ may be used to join thermoplastics, including knives, soldering irons, strip heaters, and hot plates.

18-27. A special type of nut serving the function of a tapped hole is called an _____.

18-28. Because _____ materials are too brittle to be deformed by a self-tapping screw, a thread-cutter type must be used.

18-29. Name a gas that is used for hot-gas welding.

ACTIVITIES

Solvent Bonding

Introduction. Solvent bonds are readily produced but not often easily evaluated. Tensile testing of solvent bonds often results in failure of the substrate rather than the bond. Shear testing can be successful if the samples are carefully prepared. To maintain a uniform bond area, a fixture is recommended.

Equipment. 6 mm (0.25 in.) thick sheet acrylic, acetone, eyedropper, universal testing equipment, and band or table saw.

Procedure

18-1. Make or acquire an aluminum fixture similar to the one shown in Figure 18-22. This fixture is machined for

strips of 6 mm (0.25 in.) thick acrylic. This fixture also shows three strips of acrylic in the fixture. These strips are 25 mm (1 in.) wide and about 400 mm (16 in.) long.

18-2. After cutting strips of acrylic 25 mm (1 in.) thick, deburr all edges and position the bottom piece on the fixture. Apply acetone to the surface, and lay a second strip into the fixture. Next, apply acetone to the top of the second strip and put the third strip in place. Make sure that the strips are firmly against the fixture and place a weight on the top strip.

18-3. Allow 24 hours for the solvent to evaporate from the bond area. Cut the bonded strips into pieces about

Figure 18-22. Aluminum fixture aligning three strips of acrylic for bond test samples.

Figure 18-23. Bonded sample cut to size for testing.

25 mm (1 in.) long (see Figure 18-23). Measure carefully to determine the bonded area.

18-4. Set up tester for compression test.

18-5. These pieces may shatter during testing. To protect both operators and observers, place a cage around the sample. A one quart can with the bottom removed can provide necessary protection. To hold the can or cage around the sample, hang it from the upper compression plate or place it on supports.

18-6. The sample shown in Figure 18-23 had a bonded area of 450 mm² (0.75 in.²) on each size, a total of 900 mm² (1.5 in.²). That much area requires considerable force to cause failure. Be sure that the load cell in the universal tester has a large enough capacity.

18-7. Run the tester at a low speed, about 5 mm/min. (0.2 in./min.). If the "legs" of the sample are equal in height and the bonded areas are equal, the bonds should fail simultaneously. Examine the broken pieces. If the test was successful, both bonds should fail, yielding three unfractured pieces. The bonds in this sample failed at 16.5 MPa (2400 psi).

Additional activities

a. Compare the strength of solvent bonds to joints made using various mechanical adhesives.

b. Some adhesives involve both cohesive forces and mechanical adhesion. Contact cement is applied to both surfaces and allowed to stiffen. When the two surfaces are pressed together, the resulting bond includes cohesive bonds between the two surfaces of contact cement, as well as mechanical adhesion between the cement and the adherends. Careful examination should indicate whether the failure was mechanical or within the cement.

c. Compare solvent and adhesive bonds to joints made using double-sided adhesive tape.

Spin Welding

In spin welding, frictional heat softens two surfaces, allowing some molecular intertangling. When the heated areas cool, the bond gains considerable strength.

Equipment. Sheet acrylic material, acrylic rod or tube 12 mm (0.5 in.) diameter or less, drill press, and alignment fixtures tensile tester.

Procedure

18-1. Cut an acrylic sheet into squares 25 to 30 mm (1 to 1.2 in.) square.

18-2. Cut an acrylic rod or tube into lengths about 60 to 75 (2.4 to 3 in.) mm long. The rods must be long enough to grip in the chuck of a drill press and fit in the grips of the tensile tester. Figure 18-24 shows materials ready for spin welding.

18-3. Mount the rod in the chuck of the drill press, turn on the drill, and press the spinning rod against the acrylic square. When the heat is sufficient to soften the acrylic surfaces, hold downward pressure on the rod and turn off the press. Hold until the bond begins to cool.

18-4. In order to tensile test the bonds, the rods must be positioned directly opposite one another. To locate

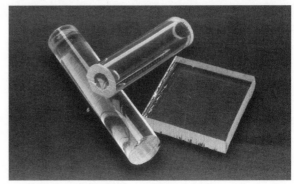

Figure 18-24. Sample materials for spin welding.

Figure 18-25. Fixture for aligning second rod.

the second rod, make a simple fixture that contains a hole slightly larger in diameter than the rods. Position the fixture, using a piece of rod in the chuck. Next, load the square and rod (as seen in Figure 18-25) onto the fixture and bond the second rod.

18-5. Tensile test the resulting sample. It may be necessary to put V-type grips into the tensile tester jaws. The sample seen in Figure 18-25 failed at 4.7 MPa (684 psi).

18-6. Determine the combination of rpms, pressure, and time that resulted in the highest strengths.

Impulse Bonding

Introduction. Heat joining or sealing creates a thermal bond between two or more layers of thermoplastic film. Impulse bonding uses tools that are hot only during the bonding cycle and refers to the quick heating and cooling of the bonding element. Silicone rubber, PTFE, and other antistick agents are used to prevent heated plastics from sticking to the heating element or tool. This type of bonding is common in the packaging industry because it is often used on materials that will not stay bonded unless they are held together while cooling.

Equipment. Impulse bonder and sandwich bags.

Procedure

18-1. Carefully examine a sandwich bag. Was it originally made from sheet material or from a blown tube?

18-2. Cut off the bonded seams.

18-3. Rebond the seams and adjust the dwell and current of the bonder to achieve a good seal and a clean cut. Cutwires are usually coated with PTFE. If the coating is cracked or missing, the cut will be adversely affected.

18-4. The cut-wire should run down the center of the bonded zone. If the cut is too close to one edge of the bonded zone, it may be incomplete.

18-5. A damaged nonstick coating above the heating element results in a poor bond and/or cut.

Hot-Gas Welding

Introduction. Hot-gas welding usually refers to joining thermoplastic materials that are greater than 1 mm in thickness. The hot-gas melts the plastics so that they fuse. Hot air is appropriate for some materials. For other materials, nitrogen gas is essential to prevent oxidation and achieve strong welds. Table 18-4 indicates the temperature, gas, and weld angle for selected plastics.

There are four important variables to consider when making hot-gas welds:

1. Gas temperature
2. Gas pressure
3. Filler rod and torch angle
4. Feed speed

Table 18-4. Welding Data for Thermoplastics

	Welding Temperature, °C	Welding Gas	Butt-Weld Strength, %	Weld Angle, Degrees
ABS	175–200	Nitrogen	50–85	60
Acrylic	315–345	Air	75–85	90
Chlorinated Polyether	315–345	Air	65–90	90
Fluorocarbon	285–345	Air	85–90	90
Polycarbonate	315–345	Nitrogen	65–85	90
Polyethylene	285–315	Nitrogen	50–80	60
Polypropylene	285–315	Nitrogen	65–90	60
Polystyrene	175–400	Air	50–80	60
PVC	260–285	Air	75–90	90

Equipment. Materials to weld, approximately 2 × 20 × 100 mm, rod to match selected plastics, and hot-gas welder.

Procedure

18-1. Protect the table with a temperature-resistant covering.

18-2. Choose a plastics and determine the parameters from Table 18-4.

18-3. Regulate the gas supply to about 25 kPa (3.6 psi). The volume of gas going through the heating element determines the welding temperature. Do not plug in the heating unit until gas is flowing through the torch.

18-4. Check the gas temperature.

18-5. Hold the welding tip about 6 mm (0.25 in.) away from thermometer to determine the temperature of the gas.

18-6. Direct about 60% of the heat on the plastics pieces and 40% on the filler rod. Once the plastics are molten, push the filler rod into the weld joint, using light pressure.

18-7. At the end of the weld, continue to hold light pressure on the filler rod until it has cooled.

18-8. Cut off the filler rod.

18-9. Make several welds before turning off the welder.

18-10. Tensile test the welds. Compare the results with the strengths listed in Table 18-4.

Molded-In Threaded Inserts

Introduction. Threaded inserts enjoy wide use in plastics parts. Inserts of varying lengths, diameters, and thread pitch are available from a number of manufacturers. The holding strength of an insert depends on the surrounding plastics, the temperature of the insert during molding, and several injection-molding process conditions. To systematically examine the effects of varying process conditions on differing materials, it is necessary to prepare molding and testing equipment.

Equipment. Injection-molding machine, mold to position and hold inserts, testing fixture, selected plastics, threaded inserts to match mold.

Procedure

18-1. Prepare a simple injection mold. One significant problem may be removing the part from the mold. A solution would be to use an ejector pin, as shown in Figure 18-26. The pin protruding from the end of the ejector sleeve has a diameter equal to the minor diameter of the threaded brass insert. It also has a length equal to the insert so that no plastic can flash over the end or into the threads. During ejection, the ejector presses on the insert and forces the molding out of the mold. Figure 18-27 shows an insert on the pin in the mold

cavity. Notice the damage on the mold face caused when an insert fell off the pin during mold closing (Figure 18-26 and Figure 18-27).

18-2. The complete part is seen in Figure 18-28. Testing the insert requires a fixture to hold it in a testing machine.

Figure 18-26. Ejector sleeve with protruding rod to position threaded insert.

Figure 18-27. Insert in mold on ejector pin.

Figure 18-28. One trimmed sample.

Figure 18-29 shows a holding fixture. The hole in the fixture is large enough to allow the insert to be pulled through without binding on the fixture. Figure 18-30 shows the part held in the fixture and ready for testing. The threaded rod will go into the upper jaw and the fixture into the lower jaw of the tester.

18-3. Determine molding conditions or materials of interest, and mold a group of parts.

18-4. Wait 40 hours for the parts to stabilize before testing.

18-5. Calculate means and standard deviations to determine the relative magnitude of the effects.

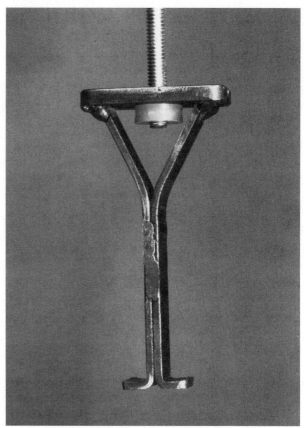

Figure 18-30. Sample positioned in fixture, ready for loading into tensile tester.

Figure 18-29. Holding fixture for testing threaded inserts.

DECORATION PROCESSES

CHAPTER 19

INTRODUCTION

In this chapter, you will learn that plastics may be decorated for some of the same reasons that cloth, ceramics, metals, and other materials are decorated, using many similar processes.

A number of decorating processes are used to produce plastics parts. Decorating may be done during or directly after molding or before final assembly and packaging. The most inexpensive way to produce decorative designs on the product is to include the desired design in the cavity of the mold. These designs may consist of textures, raised or depressed contours, or informative messages such as trademarks, patent numbers, symbols, letters, numbers, or directions.

Embossing, or *texturing*, of hot melts may be referred to as a type of rotary molding. The majority of thermoplastics sheets or films are embossed against a composition roll or matched male and female rolls. To achieve the desired pattern, there must be a proper balance between the pressure of the embossing roll, heat input, and subsequent cooling. Some polyvinyls and polyurethanes are embossed with textured casting paper. After the polymer has cooled or cured, the paper is removed, leaving behind the pattern of the release paper.

When decorating plastics items, in the mold or out, surface treatment and cleanliness are of chief importance. Not only must the molds remain clean and mark-free, but the molded items must be properly prepared to ensure good decorating results. Blushing is the result of applying coatings over items that have not been properly dried in order to eliminate surface moisture. Crazing is due to solvent cutting along lines of stress in the molded plastics. The fine cracks (crazing) may be on or under the surface or extend through a layer of the plastics material.

The problem may be eliminated by changing the mold design to produce a stress-free molding.

The surface of the plastics must be cleaned prior to decoration. All traces of mold release, internal plastics lubricants, and plasticizers must be removed. Plastics parts become electrostatically charged, attracting dust and disrupting the even flow of the coating. Solvent or electronic static eliminators may be used to clean and prepare plastics articles for decorating.

Polyolefins, polyacetals, and polyamides must be treated by one of the methods described below to ensure satisfactory adhesion of the decorating media.

Flame treatment consists of passing the part through a hot oxidizing flame of 1100°C to 2800°C (2012°F to 5072°F). This momentary flame exposure does not cause distortion of the plastics, but makes the surface receptive to decorating methods.

Chemical treatment consists of submerging the part (or portions of the part) in an acid bath. On polyacetals and polymethylpentene polymers, the bath etches the surface, making it receptive to decorating. For many thermoplastics, solvent vapors or baths may be used for the etching treatment.

Corona discharge is a process in which the surface of the plastics is oxidized by an electron discharge (corona). The part or film is oxidized when passing between two discharging electrodes.

Plasma treating subjects plastics to an electrical discharge in a closed vacuum chamber. Atoms on the surface of the plastics are physically changed and rearranged, making excellent adhesion possible.

The nine most widely used decorating treatments for plastics are included in the following chapter outline:

COLORING

The method used to color plastics is to blend the least costly pigments into the base resin. Color matching may be a problem. Therefore, successive batches of the same color of plastics may vary slightly. Most producers of colored resins and plastics encourage the use of stock or standard colors. Plastics parts of an assembly may be produced in different plant locations and at different times. This makes it necessary for color standards to be carefully considered. Plasticizers, fillers, and the molding process may affect the final product color.

Colorants in the form of dry powders, paste concentrates, organic chemicals, and metallic flakes are usually blended with a given resin mix. Banbury, two-roll, and continuous mixers are used to thoroughly disperse the pigments in the resin. The colored resin may then be cast or extruded. Water- and chemical-solvent dyes have been successfully used on many plastics. The procedure consists of dipping the parts in the dye bath and air drying. Three advantages and disadvantages of coloring decorating processes are given here:

Advantages of Coloring Plastics

1. Colored resin control is better in mass production.
2. Dyeing is less costly for short runs.
3. Surface dyeing is better for lenses.

Disadvantages of Coloring Plastics

1. Some colors are hard to produce and match.
2. There may be color migration and varying coloration in pieces with uneven thickness.
3. Pigment mixing in resin is more costly for short runs.

PAINTING

Painting plastics is a popular, low-cost way to decorate parts and provide flexibility in product color design. Transparent, clear, or colored plastics may be painted on the back surface for a striking contrast, variety, or appearance. This effect has not been made possible by other methods. The six painting methods used in decorating plastics include these:

1. Spray painting
2. Electrostatic spraying
3. Dip painting
4. Screen painting
5. Fill-in marking
6. Roller coating

The solvents or curing systems used in the paint must be chosen and controlled with care. As a rule, thermosetting plastics are less subject to swelling, etching, crazing, and deterioration from solvents. Temperature may be a limiting factor for the curing or baking of paints on many plastics. Radiation curing is one method used to cure coatings on plastics (see Chapter 20, Radiation Processes).

Spray Painting

The most versatile and frequently used method of decorating all sizes of plastics articles is spray painting. It is an inexpensive, rapid method of applying coatings. The spray guns may use air pressure (or hydraulic pressure of the paint itself) to atomize the paint.

Masking—needed when areas of the part are not to be painted—may be done with paper-backed masking tape or durable, form-fitting metal masks. Polyvinyl alcohol masks may be sprayed over areas and later removed by stripping or using solvents. Electroformed metal masks are preferred because they are durable and conform to the contour of the item. Four basic types of electroformed masks are shown in Figure 19-1.

Electrostatic Spraying

In electrostatic painting, the plastics surface must first be treated to take an electrical charge. The surface then passes through atomized paint that has an opposite charge. The paint may be atomized by air, hydraulic pressure, or centrifugal force

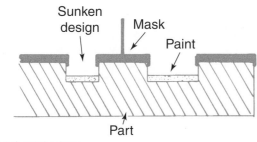

(A) Lip mask on sunken design.

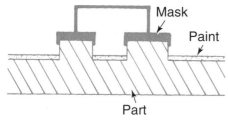

(B) Cap mask or raised design.

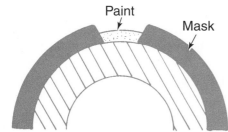

(C) Surface cutout mask.

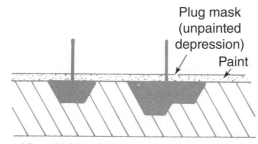

(D) Plug mask for unpainted depressions.

Figure 19-1. Basic types of electroformed masks.

(A) Electrostatic atomization.

(B) Compressed air atomization.

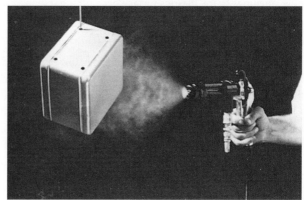

(C) Hydraulic atomization.

Figure 19-2. Methods of atomizing paint in electrostatic painting.

(Figure 19-2). Nearly 95% of atomized paint is attracted to the charged surface. This makes it a highly efficient way of applying paint. However, narrow recesses are hard to coat, and metal masks are not practical. If dry plastics powders are used, the coated substrate must be placed in an oven to fuse the powder to the product. There are no solvents released during the application or curing process. However, the product must be able to withstand curing temperatures.

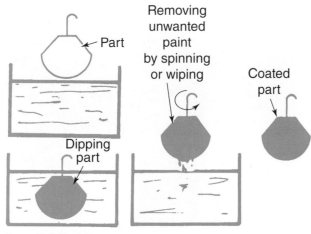

Figure 19-3. Dip painting.

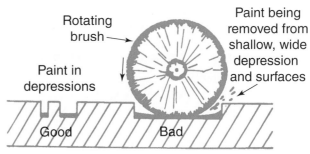

Figure 19-4. Fill-in method of painting.

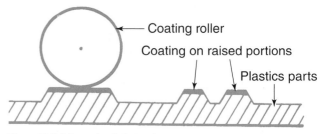

Figure 19-5. Roller coating of raised portions.

Dip Painting

Dip painting (Figure 19-3) is useful when a single or base color is needed. A uniform coating will be achieved if the part is withdrawn from the paint very slowly. Enough time must also be allowed for drainage. Excess paint may be removed by spinning the part, hand wiping, or electrostatic methods.

Screen Painting

Screen painting is a versatile and attractive method of decorating plastics items. It requires forcing a special ink or paint through the small openings of a stenciled screen onto the product surface. The process is sometimes referred to as *silk screen* painting because early screens were made of silk. Screens may be made of metal mesh or finely woven polyamide, polyester, or other plastics. A simple screen stencil is prepared by blocking out the areas where no paint is desired. For intricate designs or lettering, photographic stencils are applied to the screen. When exposed and immersed in a developer bath, the exposed areas will wash away. Paint will be applied through these openings to the plastics surface beneath the screen.

Fill-In Marking

In the fill-in marking process, paint is placed in low or indented portions of the article (Figure 19-4). Letters, figures, or designs on a plastics part are produced as depressions in the molded part. These recesses are filled by spraying or wiping paint into the depressions. To ensure a sharp image, the depression should be deep and narrow. If the depression or design is too wide, the buffing or wiping action may remove the paint. Excess paint around the design may be removed by wiping or buffing operations.

Roller Coating

Raised portions, letters, figures, or other designs may be painted by passing a coating roller over them (Figure 19-5).

In some cases, masking out portions of the article may be required. If edges and corners are sharp and highly raised, good coating details will be obtained. Roller coating may be automated, or small runs may be done by hand with a brayer (hand roller).

Three advantages and six disadvantages of using painting decorating processes follow:

Advantages of Painting

1. Several inexpensive methods are possible.
2. Pretreatment of most plastics is not needed.
3. Variation of methods and designs may hide imperfections.

Disadvantages of Painting

1. Some plastics are solvent sensitive.
2. Hand methods have higher labor costs.
3. Paint reduces cold impact resistance.
4. Fish-eye blemishes may occur from having used silicone or other releases.
5. Solvents may be a health hazard.
6. Oven-drying may be a problem with some thermoplastics.

HOT-LEAF STAMPING

Hot-leaf stamping is sometimes called *roll-leaf stamping* or simply *hot stamping*. Hot stamping is a clean, simple, and economical method for decorating thermoplastic products.

By using this process, a mark or image is transferred from a carrier film to the product surface by heat and pressure. Figure 19-6 shows the multiple layers in hot-stamping foil. A wide range of colors, both pigment and metallic, as well as multicolored patterns and preprinted images are available for use in hot stamping. In contrast to liquid ink methods of decoration, hot foils are less sensitive to humidity and dust in the air. A flat silicone rubber die is often used to highlight or tip raised lettering.

Foil-stamping machinery utilizes two general methods of transferring an image—either by pressure from a stamping press or through the use of roll-on methods. Roll-on presses are preferred for large, flat surfaces because they avoid problems due to air entrapment between the foil and the part.

Using either method, foil stamping can decorate both large and small products. Figure 19-7A shows a large curbside trash cart ready for stamping. Two 10-ton presses simultaneously stamp both sides of the cart. Figure 19-7B shows a rotary indexing table and two 5-ton vertical presses decorating golf bag tags. Figure 19-7C shows a medium-sized part, a container for agricultural pesticide, and the roll-on machine.

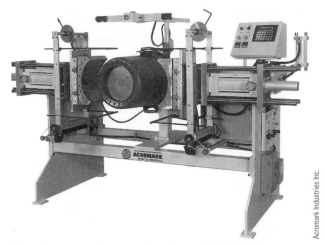

(A) Proper fixtures are needed to support this trash can cart during hot stamping.

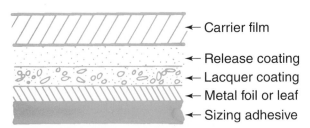

(A) Cross-sectional diagram of a typical metallized hot-stamping foil.

- ← Carrier film
- ← Release coating
- ← Lacquer coating
- ← Metal foil or leaf
- ← Sizing adhesive

(B) This semi-automatic system utilizes a rotary table and a pick-and-place robot to decorate golf bag tags.

(B) This photo of hot-foil-stamping tooling shows metal dies, silicone dies (both flat and curved), silicone rollers, and sheets.

Figure 19-6. Foil-stamping supplies.

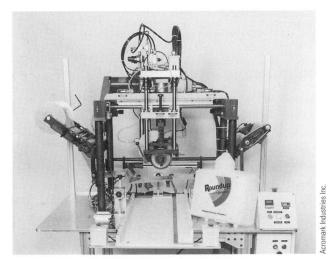

(C) To keep the bottle surface firm during roll-on hot stamping, this press applies air pressure inside the bottle.

Figure 19-7. Various hot-stamping machines.

Advantages of Hot Stamping

1. It is a clean, dry process that does not require pretreatment of part surfaces.
2. No solvent clean up or disposal is needed.
3. It can be automated for high-speed production or direct interface with molding operations.
4. Colors and patterns can be changed quickly.
5. Decorating can be done on patterned or flawed surfaces.

Disadvantages of Hot Stamping

1. Multicolored transfers are relatively costly.
2. It requires a secondary function following the molding of parts.
3. Initial capital investment is significant.

PLATING

Plating and vacuum metallizing have been discussed in Chapter 17 under the topic of metal coatings. There are many functional applications of coating plastics with metal, but they are outnumbered by decorative applications. Metallized foils for dielectrics, electronics items such as semiconductors and resistors, as well as flexible mirrors and plating for corrosion resistance are functional applications. The mirror-like finish on automotive items, appliances, jewelry, and toy parts are examples of decorative applications. Four advantages and five disadvantages of plating are listed here:

Advantages of Plating

1. Metallic finish has a mirror-like quality.
2. Many plastics parts need little or no polishing before plating.
3. Electroplate thickness ranges from 0.00038 to 0.025 mm.
4. Plating is more durable than metallizing.

Disadvantages of Plating

1. Mold finish and design must be considered.
2. Not all plastics are easily plated.

3. Setup is expensive and includes many steps.
4. There are many variables to control for proper adhesion, performance, and finish.
5. Plating is more expensive than metallizing.

ENGRAVING

Engraving is seldom used on a production scale. However, it does provide a durable means of marking and decorating plastics and is often used in engraved tool and die work. Pantographic engraving machines may be automatic or manual and are often used to engrave laminated name tags, door signs, directories, and equipment. They can also place identifying names and marks on bowling balls, golf clubs, and other items. Laminated engraving sheets contain two or more layers of colored plastics. Engraving cuts through the top layer, exposing the contrasting second-color layer (Figure 19-8).

PRINTING

There are over 11 distinct methods and many combinations of these methods used for printing on plastics.

Letterpress is a method in which raised, rigid printing plates are inked and pressed against the plastics part. The raised portion of the plate transfers the image.

Letterflex is similar to letterpress, except flexible printing plates are used. Flexible plates have the ability to transfer their designs to irregular surfaces.

(A) Various sizes of 2D engraving tables.

Figure 19-8. Engraving equipment for flat and curved stock.

(B) This 3D engraved table can handle cylindrical items up to 25 cm [10 in.] in diameter.

Flexographic printing resembles letterflex, but a liquid ink rather than a paste ink is used. The plate is often a rotary type, transferring inks that set or dry rapidly by solvent evaporation.

Dry offset is a method in which a raised, rigid printing plate transfers a paste ink image onto a special roller called an offset blanket. This roller then places the ink image on the plastics part, thus the name *offset*. If multicolor printing is required, a series of offset heads can be used to apply different colors to the blanket roller. The multicolored image is then transferred (offset) to the plastics part in a single printing step.

Offset lithography is similar to dry offset, except the impression on the printing plate is not raised or sunken. This process is based on the principle that oil and water do not mix. The image or message to be printed is placed on the plate by a photographic-chemical process. Images may be placed directly on the plate using special *grease* typewriter ribbons or pencils. The greasy or treated images will be receptive to the type of ink used. Those areas not treated will be receptive to water but will repel ink. A water roller must first pass over the offset plate. Next, the ink roller will deposit ink on the receptive areas. The image is transferred from the printing plate to a rubber offset cylinder (blanket roller) that places the image on the plastics part.

Rotogravure, or *intaglio*, printing involves an image that is depressed or sunken into the printing plate. Ink is applied to the entire surface of the plate, and a device called a doctor blade is used to scrape the plate and remove all excess ink. The ink left in the sunken areas is transferred directly to the product.

Silk screen printing is a process in which ink or paint is forced onto the product through a fine metallic or fabric screen. A rubber squeegee is used to force paint through the screen. The screen is blank or blocked off in areas where no ink is desired.

Stenciling is similar to silk screen printing, except the open areas (those to be printed) do not have a connecting mesh. Stencils may be positive or negative. In positive stencil printing, the image is open, and spray or rollers transfer the ink through these open areas and onto the product. In negative stencil printing, the image is blocked out and the background is inked, leaving no ink in the stencil area. Stencil printing may be considered a masking operation.

Electrostatic printing has been adapted to several well-known printing techniques. In this process, dry inks are attracted to the areas to be printed by a difference in electrical potential. There is no direct contact between the printing plate or screen and the product. There are several methods by which a screen is made conductive in the image areas and nonconductive in other areas. Dry, charged particles are held in these open areas until they are discharged toward an oppositely charged back plate. The object to be printed is placed between the screen and the back plate. When the ink is discharged, it strikes the substrate surface. A fixing agent is then applied to provide a permanent image. The image is faithfully reproduced regardless of the surface configuration of the substrate. Images can be printed on the yolk of an uncooked egg or similar products by using this method. Edible inks are used to identify, decorate, and supply messages on fruits and vegetables.

Heat-transfer printing is used as a decorating process and important printing method. The process is similar to hot-leaf stamping because a carrier film (or paper) supports the release layer and the ink image. The thermoplastic ink is heated and transferred to the product by a heated rubber roll.

Hot-leaf stamping is the process of transferring a colorant or decorative material from a dry carrier film to a product, using heat and pressure. It is sometimes utilized as a printing method.

IN-MOLD DECORATING

During in-mold decorating, an overlay or coated film called a *foil* becomes part of the molded product. Both thermosetting and thermoplastic materials may receive decorative images through this process.

With thermosetting products, the film may be a clear cellulose sheet covered with a partially cured resin-like molding material. The in-mold overlay is placed in the mold cavity while the thermosetting material is only partially cured. The molding cycle then continues, and the decoration becomes an integral part of the product. The bond between the image and the plastics substrate depends on the rear layer of the foil. This layer consists of material that will adhere to the type of thermosetting material selected.

With thermosetting products, an overlay may be held in place by cutting so that it fits snugly in the cavity. Electrostatic methods also are used to hold foils in the proper location.

A polyester film is usually found with thermoplastic products. In injection molding, the foil is placed in the mold before it closes prior to injection. As the melt flows into the cavity, the overlay bonds fully with the plastics substrate. When production runs are long, it is often desirable to automate the loading of the foil. Foil-winding machines hold a fresh roll of foil on one side of the mold and a take-up roll on the other side. When the mold opens and the decorated part is ejected, the foil winder takes up the used foil and locates fresh material.

Similar to hot-stamping foils, in-mold foils contain many layers. The bottom layer adheres to the selected thermoplastic material. The decoration, which often contains several colors or textures, is protected by a transparent topcoat or wear coat.

The wear coat determines the abrasion and scratch resistance of the foil. A heat-activated release coat causes separation of the decorating from the carrier film.

Blow-molded parts are often decorated in the mold. The ink or paint image is placed on a carrier film or paper. As the hot plastics expands and fills the mold cavity, the image is transferred from the carrier to the molded part. The advantages and disadvantages of in-mold decorating are specified here:

Advantages of In-Mold Decorating

1. Full color images, halftones, or combinations may be used.
2. Strong bonding may be achieved.
3. Designs and short runs are economical.
4. Efficiency may reach well over 90%.

Disadvantages of In-Mold Decorating

1. Costs of foil, labor for hand loading, or automated loading machines are high.
2. Mold design must minimize washing and turbulence.
3. Mold design must cause foil to separate cleanly at edges to minimize clean-up.

HEAT-TRANSFER DECORATING

In heat-transfer decorating, the image is transferred from a carrier film onto the plastics part. The structure of heat-transfer decorating stock is shown in Figure 19-9A. The preheated

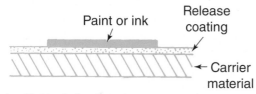

(A) Structure of heat-transfer decorating stock.

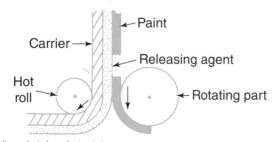

(B) Roller method of transferring design.

Figure 19-9. Transfer decorating supplies and processes.

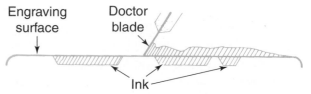

(A) Etched plate with ink held in recessed areas. Surface of the engraving is wiped clean by doctor blade.

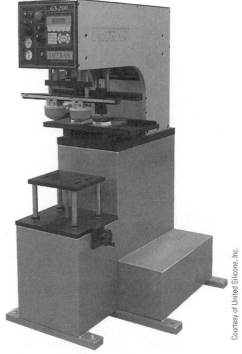

Courtesy of United Silicone, Inc.

(B) This pad printer has a maximum cycle rate of 1800 impressions per hour and has an optional three color configuration.

Figure 19-10. Pad printing equipment.

carrier stock is transferred to the product by a heated rubber roller (Figure 19-9B).

A decorating or printing process that resembles a combination of engraving and offset printing is called the Tampoprint. In this process, a flexible transfer pad picks up the impression from the inked engraving plate (Figure 19-10A) and transfers it to the item to be printed. The entire ink supply carried by the transfer pad is deposited on the part, leaving the pad clean. The flexible pad adapts to rough and uneven surfaces while maintaining absolute reproduction sharpness. Printing heads of various shapes can accommodate a wide variety of objects and textures. In addition, multicolor wet-on-wet printing, including halftones, can be produced (Figure 19-10B). Nearly any type of printing ink or paint may be used by this simple process. Depending on the type of product, up to 20,000 parts per hour can be automatically decorated. Two advantages and disadvantages of the heat-transfer decorating process are listed here:

Advantages of Heat-Transfer Decorating

1. It is similar to hot-leaf stamping except multicolored designs are possible.
2. Many heat-transfer systems are available.

Disadvantages of Heat-Transfer Decorating

1. Carrier film and designs are costly.
2. Secondary operation and equipment are needed.

MISCELLANEOUS DECORATING METHODS

There are many other decorating methods, including pressure-sensitive labels, decalcomanias, flocking, and decorative coatings or clads.

Pressure-sensitive labels are easy to apply. The designs or messages are usually printed on the adhesive-backed foil or film label, and the labels are placed on the finished product by hand or mechanical means.

Decalcomanias, commonly known as decals, are a means of transferring a picture or design to plastics. They generally consist of a decorative film on a paper backing. The decalcomania is moistened in water, and the adhesive-backed film is slipped off the paper and onto the plastics surface. This process is not widely used because decalcomanias are hard to place accurately and quickly on the plastics surface.

Flocking by mechanical or electrostatic means refers to an important method of placing a velvet-like finish on nearly any surface. The process involves coating the product with an adhesive and placing plastics fibers on the adhesive areas. Examples include the velvet-like coatings or designs on wallpapers, toys, and furniture.

There are a number of wood-graining decorative processes. Some are accomplished by rolling engraved or etched wood-graining plates over a contrasting background color. This is actually considered a printing process adaptation. Some decorative laminates and clad coatings are also used to decorate substrates. Polyvinyl-clad metal products are used in store fixtures, partitions, room dividers, furniture, automobiles, kitchen equipment, and bus interiors. These are durable as well as decorative.

Many foils and patterns are thermoformable, permitting the decoration of three-dimensional thermoformed parts. The thermoformed shell may be filled by casting or injection molding. Plastics thinning and pattern distortion must be controlled.

Four advantages and two disadvantages of the pressure-sensitive decorating process are listed here:

Advantages of Pressure-Sensitive Decorating

1. It has variable application rates (relatively high speed to hand dispensed).
2. Multicolored patterns and designs are possible.
3. It may be used on all plastics.
4. Short runs and changes in patterns are economical.

Disadvantages of Pressure-Sensitive Decorating

1. Secondary operation and equipment are required.
2. Surface labels may wear or be removed.

RELATED INTERNET SITES

- **www.ctlaminating.com.** Connecticut Laminating Company is one of many companies that print on plastics. This company utilizes silk screen, offset, and flexographic printing techniques. The site contains numerous examples of products and printing equipment.

- **www.enhancetech.com.** Enhancement Technologies Inc. uses the transfer process to put complex graphics on thermoplastics. On the home page, select "PVC/Vinyl" to find the link "Transfer Process." This page leads to a page entitled "The ETI Manufacturing Process," which provides details on their transfer process. The company claims that its eight color patterns are generally colorfast to light for more than 500 hours QUV (ultraviolet accelerated weathering equipment manufactured by Q-Panel Lab Products). (See Chapter 6 for information on accelerated weathering tests.)

- **www.itwtranstech.com.** ITW Imtran supplies pad print supplies and equipment as well as hot-stamping supplies. This site provides data on the various models of pad printers.

- **www.unitedsilicone.com.** ITW United Silicone provides considerable information on hot stamping, heat transfer, and other decorating processes. Selecting "Hot Stamping & Heat Transfer Equipment" on the home page leads to discussions of the advantages of hot stamping, equipment, and accessories as well as tooling, methods, and theory. The section on heat-transfer processes includes data on several vertical and roll-on transfer machines.

VOCABULARY

The following vocabulary words are found in this chapter. Use the glossary in Appendix A to look up the definitions of any of these words you do not understand as they apply to plastics.

blushing
chemical treatment
corona discharge
crazing
dry offset
electrostatic printing
embossing

flame treatment
flexographic
heat-transfer printing
hot-leaf stamping
in-mold decorating
letterflex
letterpress
offset lithography
plasma treating
rotogravure
silk screen
stenciling
Tampoprint

QUESTIONS

19-1. A _____ may be used to prevent silicone paint from depositing where it is not desired.

19-2. Name two advantages of silicone hot-stamping dies.

19-3. Name the painting process in which surface wear does not easily remove the design.

19-4. What additive may lower the electrical resistance in a plastics?

19-5. Name the result of applying a coating over items that have not been properly dried of surface moisture.

19-6. The name of a well-known mixer used to blend plastics ingredients is _____.

19-7. Hot-leaf stamping is sometimes called roll-leaf stamping or simply _____.

19-8. Name three functional uses of plating.

19-9. Name three decorative uses of plating.

19-10. Electroplate thickness varies from _____ to _____ mm.

19-11. An important method of placing a velvet-like finish on a surface by mechanical or electrostatic means is called _____.

19-12. Name a process in which an image is transferred from the carrier film to the product by stamping with rigid or supple shapes and with heat and pressure.

19-13. It may be less costly to incorporate the desired design in the _____, dies, or rollers.

19-14. Most decorating processes require that the substrate be thoroughly _____ of mold release, lubricants, and plasticizers.

19-15. The best and most inexpensive way to color plastics products is to blend pigments into the basic _____.

19-16. Plasticizers, _____, and molding may affect color.

19-17. An inexpensive and popular method of painting plastics is _____.

19-18. A variety of shapes and different sizes of products may be rapidly painted by _____ methods.

19-19. Name the three methods of atomizing paint.

19-20. Excess paint may be removed in the dip-coating process by spinning the part, hand wiping, or by _____ methods.

19-21. The process of forcing special inks or paint through small openings of a stenciled screen onto a product surface is called _____ or _____.

19-22. In _____, the image depression should be deep and narrow for sharp detail.

19-23. Raised portions, letters, figures, or other designs are easily decorated by _____ methods.

19-24. Name five methods of printing on plastics products.

19-25. The pattern or decoration becomes part of the plastics article as it is fused by heat and pressure of the plastics material during _____ decorating.

19-26. Because of the use of a flexible printing pad, the _____ process is especially useful for decorating irregular surfaces.

19-27. A metal term used to indicate that two or more layers of plastics (or metals) are pressed together under pressure is _____.

19-28. Labels are adhered to the substrate by _____, whereas decals are _____ activated.

19-29. Name four methods for pretreating polyethylene to receive paint or ink.

ACTIVITIES

In-Mold Decoration

Introduction. In-mold decoration uses the heat of hot plastics to adhere decorations or messages to the product. Its major advantage is the elimination of secondary printing or decorating operations.

Equipment. Injection molder, in-mold foil or hot-stamping foil, test-bar mold, and Scotch 3M® transparent tape number 600.

19-1. If possible, acquire in-mold foil. Foil for ABS is popular and available. However, if in-mold foil is not available, use hot-stamping foil as a substitute. A major difference between in-mold foil and hot-stamping foil is the thickness of the polyester carrier film. In-mold foils have much thicker carriers than hot-stamping foils. Figure 19-11 shows an ABS disk covered with woodgrain in-mold material. This disk permits abrasion resistance testing in a Taber abrader.

19-2. If the injection mold used for in-mold decorating has a flat side, tape the foil to that side. Make sure the melt will contact the back side of the foil. If the melt contacts the "wrong" side of the foil, no adhesion will occur.

19-3. If the melt forces the foil into a cavity, the resulting stretch may exceed the strength of the carrier film and cause a rip, especially when using hot-stamping foil. The rip may allow the melt to run on the wrong side of the foil. Figure 19-12 shows an example of this problem.

Notice that the rip began at an edge of the part. To avoid ripping the foil, use a mold with one flat half. Even on the flat half, hot-stamping foil will stretch during injection. Figure 19-13 shows small wrinkles in the foil near the edge of the part. These would not occur if the carrier were thicker.

19-4. Inject at a range of temperatures. If the melt temperature is low, the adhesion will be poor. If the melt temperature is appropriate, adhesion will be good. If the melt is too hot, the foil may stretch and distort the design; loss of gloss may occur; and a spot of foil may be completely removed by *burn through* near the gate. These problems of distortion will be exaggerated when using hot-stamping foil.

19-5. Test the adhesion of the foil using the tape test. First, crosshatch the sample, as shown in Figure 19-14. Next, adhere the tape firmly, taking care to remove all air bubbles under the tape by rubbing firmly with the back of a thumb nail. Pull the tape off at a 90° angle to the

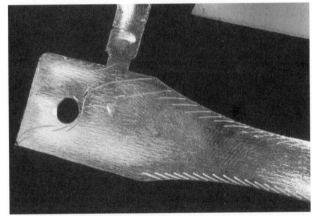

Figure 19-13. Foil wrinkles near edge of part.

Figure 19-11. In-mold foil on a Taber abrader disk.

Figure 19-12. This is an example of melt flow on the "wrong" side of the foil. The foil ripped due to stretching by the injected melt.

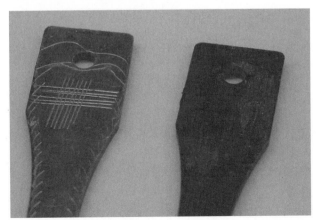

Figure 19-14. One sample is cross-hatched and ready for tape test. The other shows that styrenic foil did not adhere to PP part.

substrate. Check to see whether the adhesion failed. If the tape pulled off pieces of the foil, the adhesion was inadequate. The number 600 tape, manufactured by 3M under the brand name Scotch, is an example of a tape used for foil adhesion testing.

19-6. Use of foils with various compatibilities will show significant differences. Figure 19-14 also shows a hot-foil stamping foil compatible with styrenics used on polypropylene. Almost none of the foil adhered.

19-7. Write a report summarizing your findings.

Hot Stamping

Introduction. Hot-leaf stamping is a decorating process in which metal foils or pigments are placed on carrier films. The word *leaf* is used because thin metal foils of gold are called gold leaf. A heated die is used to press against the foil and plastics, causing the foil or pigment to be released from the carrier film. The heat and pressure of the process bonds the decoration to the substrate.

Hot-foil stamping is a popular process used for decorating plastics. The foil must be compatible with the plastics substrate to be decorated. Foils compatible with styrenics, such as PS, HIPS (high-impact polystyrene), and ABS are commonly used. Another type of foil adheres to olefins. A mismatch of foil type and plastics substrate will dramatically reduce adhesion.

Equipment. Hot-foil stamper, foil for styrenics, foil for olefins, and stamping die or flat silicone rubber stamping plate.

19-1. Stamp words or letters onto various plastics, using both styrenic foil and olefinic foil. Test the adhesion using the eraser on a pencil. If the adhesion is high, the force required to rub off the decoration will also be high.

19-2. Stamp using a flat silicone rubber die onto a flat piece of plastics material. This will yield a surface large enough for tape testing the adhesion of the foil. Crosshatch the surface, and check the adhesion with a tape test. Both the stamping time and the temperature of the die will influence adhesion.

19-3. Stamp onto polyethylene or polypropylene. These materials often resist adhesion. To increase adhesion, flame treat the surface of the substrate. A bunsen burner or propane torch can provide the flame. Do not melt the substrate or char the surface. Vary the time of contact with the flame and test the effects on adhesion.

RADIATION PROCESSES

CHAPTER 20

INTRODUCTION

Radiation processing offers energy savings, low pollution effects, and economic advantages. You will learn how this technology relays new product and manufacturing ideas.

Radiation processing is a growing area of technology in which ionizing and nonionizing systems are used to change and improve the physical properties of materials or components. In 2001, the radiation curing industry in the United States created products worth over $1 billion. Coatings and inks were the largest market, and packaging products were second largest. This industry has been growing at almost 10% annually. The outline for this chapter follows:

I. **Radiation methods**

II. **Radiation sources**
   A. **Ionizing radiation**
   B. **Nonionizing radiation**
   C. **Radiation safety**

III. **Irradiation of polymers**
   A. **Damage by radiation**
   B. **Improvements by radiation**
   C. **Polymerization by radiation**
   D. **Grafting by radiation**
   E. **Advantages of radiation**
   F. **Applications**

RADIATION METHODS

The term **radiation** can refer to energy carried by either waves or particles. The carrier of wave energy is called the **photon**. In radiant energy, the photon is wave-like when in motion (Figure 20-1A). It becomes particle-like when absorbed or emitted by an atom or molecule.

The ordinary electric lightbulb with an element temperature of 2300°C (4172°F) emits radiation waves that are visible. The sun with a surface temperature of 6000°C (10,832°F) emits both visible and invisible radiation. The human eye can see radiation with a wavelength as short as 400 μm (15.7×10^{-6} in.) and as long as 700 μm (27.5×10^{-6} in.). **Ultraviolet** radiations are waves of energy that can burn or tan the exposed parts of the human body, yet are invisible to the human eye. Ultraviolet radiation has wavelengths shorter than 15.7×10^{-6} in. (400 μm).

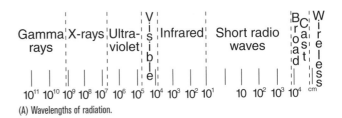

(A) Wavelengths of radiation.

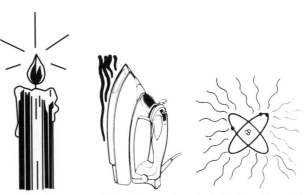

(B) Light (visible) radiation makes things visible.

(C) Heat (infrared) radiation can be felt.

(D) Radioactive radiation can't be seen or felt.

Figure 20-1. Types of radiation.

Photon wavelengths are measured in micrometers. Radiation from the sun, burning fuels, or radioactive elements are considered natural sources of radiation (Figure 20-1B). A few examples of the more important radioactive elements that occur naturally are uranium, radium, thorium, and actinium. These radioactive materials emit photons of energy and/or particles as the nuclei disintegrate and decrease in mass. The earth contains small traces of radioactive materials, whereas the sun is intensely radioactive.

Radiation may be produced by nuclear reactors, accelerators, or natural or artificial radioisotopes. The most important source of controlled radiation is artificial radioisotopes. Scientists have utilized and controlled induced radiation to the point at which it may serve human needs.

When the number of *protons* in the nucleus of an atom changes, a different element is formed. If the number of *neutrons* in the nucleus is changed, a new element is not formed, but the mass of the element is different. Different forms (masses) of the same element are called **isotopes**. Most elements contain several isotopes. For example, the simple element hydrogen consists of three distinct isotopes (Figure 20-2). Most hydrogen atoms have a mass number (the number of protons and neutrons) of 1. This means that they have no neutrons. A very small number of naturally occurring hydrogen atoms have one neutron and one proton, resulting in a mass number of 2. Only when hydrogen has two neutrons and one proton (a mass number of 3) is it considered radioactive.

In 1900, the German physicist Max Planck advanced the idea that photons are bundles or packets of electromagnetic energy. This energy is either absorbed or emitted by atoms or molecules. The unit of energy carried by a single photon is called a **quantum** (*quanta* is the plural). Photons of energy radiation may be classified into two basic groups—electrically neutral and charged radiation.

Alpha particles are heavy, slow-moving masses containing a double positive charge (two protons and two neutrons). When alpha particles strike other atoms, their double positive charge removes one or more electrons, leaving the atom or molecule in a dissociated or ionized state. As discussed in Chapter 2, ionization is the process of changing uncharged atoms or molecules into ions. Atoms in the ionized state have either a positive or negative charge.

Electrons ejected from the nuclei of atoms at a very high speed and energy are called **beta particles**. When a neutron disintegrates, it becomes a proton and an electron. The proton often stays in the nucleus while the electron is emitted as a beta particle. Beta particles are electrons possessing a negative charge. Because beta particles have only 0.000544 times the mass of a proton, they move much faster and have greater penetrating power than alpha particles.

Most of the energy contained in alpha and beta particles is lost when they interact with electrons from other atoms. As the charged particles pass through matter, they lose or transfer all their excess energy to the nuclei or orbital electrons of the atoms they encounter. Because beta particles are negative, they can push or repel electrons, leaving the atom with a positive charge. However, beta particles may also become attached to the atom, giving it a negative charge.

Gamma radiations are short, very high-frequency electromagnetic waves with no electrical charge. **Gamma rays** and *X-rays* are alike except for their origin and penetrating ability. Gamma photons can penetrate even the densest materials. More than 1 meter of concrete is required to stop the radiation effect of gamma rays (Figure 20-3).

The energy of gamma photons is absorbed or lost (Figure 20-4A) in matter in three major ways:

1. Energy may be lost or transferred to an electron it strikes, forcing the electron out of orbit (Figure 20-4B).
2. The gamma photon may strike an orbiting electron a glancing blow, using only a part of its energy while the rest of the energy continues in a new direction (Figure 20-4C).
3. The gamma ray or photon is annihilated when it passes near the powerful electrical field of a nucleus (Figure 20-4D).

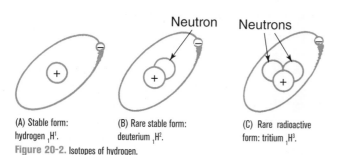

(A) Stable form: hydrogen ₁H¹.

(B) Rare stable form: deuterium ₁H².

(C) Rare radioactive form: tritium ₁H³.

Figure 20-2. Isotopes of hydrogen.

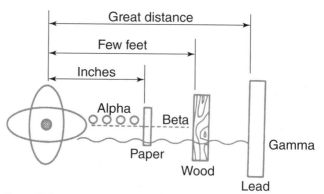

Figure 20-3. Three types of radiation emitted by unstable atoms or radioisotopes: alpha (stopped by paper), beta (stopped by wood), and gamma (stopped by lead).

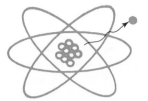

(A) Radioisotope with photon of energy being emitted.

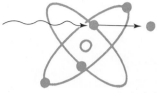

(B) Gamma energy being completely absorbed by forcing an electron from orbit and transferring energy.

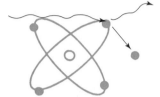

(C) Part of energy continues in new direction, and part is used in ejecting electron from orbit.

(D) Gamma radiation is annihilated. Electron and positron are created and share energy.

Figure 20-4. Interaction of gamma radiation with matter.

In the last method of gamma ray energy loss, the powerful electrical field of an atomic nucleus breaks the gamma photon into two particles of opposite charge—an electron and a positron. The positron quickly loses its energy by colliding with orbital electrons. The net effect of gamma radiation is similar to the effect of alpha and beta radiation. Electrons are knocked from orbit, causing ionization and excitation effects in materials.

Neutrons are uncharged particles that may collide with atomic nuclei, resulting in alpha and gamma radiation as energy is transferred or lost.

RADIATION SOURCES

There are two basic types of radiation sources—those that produce ionizing radiation and those that produce nonionizing radiation.

Ionizing Radiation

Cobalt-60, strontium-90, and cesium-137 are three commercially available *radioisotope* sources that produce ionizing radiation. They are used because of their availability, useful characteristics, reasonably long half-life, and reasonable cost. Another source of ionizing radiation is used or burnt uranium slugs from fission reactor waste.

Gamma radiation, though very penetrating, is not a major source of ionizing radiation for several reasons. It is slow and may require several hours for treatment; its isotope sources are hard to control; the source cannot be turned off; and experienced, well-trained workers are needed.

Electron beam accelerators are the primary ionizing source used for radiation processing. Irradiation processing implies that a controlled or directed treatment of energy is being used on the polymer.

Nonionizing Radiation

Electron accelerators, such as Van de Graaff generators, cyclotrons, synchrotrons, and resonant transformers may be used for producing nonionizing radiation.

Electrons from machines are less penetrating than radioisotope radiation. However, they may be easily controlled and turned off when not required. These machines are capable of delivering 200 kW of power. The dose rating may be expressed in a unit called a gray (Gy). One gray indicates a dose absorption of 1 joule of energy in 1 kilogram of plastics (1 J/kg). As a rule, 1 kW of power is required to deliver a dose of 10 kGy to 360 kg (793 lb) of plastics per hour.

Electron radiation penetrates only a few millimeters into the plastics, but it can irradiate products at a very fast rate. Ultraviolet, infrared, induction, dielectric, and microwave are the most familiar sources of nonionizing radiation. They are generally used to speed up processing by heating, drying, and curing.

Ultraviolet radiation sources such as plasma arcs, tungsten filaments, and carbon arcs produce radiation with enough penetrating power for film and surface treatment of plastics. Prepregs that cure by exposure to sunlight have also been produced.

Infrared radiation is often used in thermoforming, film extrusion, orientation, embossing, coating, laminating, drying, and curing processes.

Induction sources (electromagnetic energy) have been used to produce welds, preheat metal-filled plastics, and cure selected adhesives.

Dielectric sources (radio-frequency energy) have been used to preheat plastics, cure resins, expand polystyrene beads, melt or heat seal plastics, and dry coatings.

Microwave sources are used to speed curing and to heat, melt, and dry compounds.

Radiation Safety

All forms of natural radiation must be considered harmful to polymers and dangerous to work with because they are not easily controlled. However, radiation processes are time-proven production methods used by a wide variety of industries today.

One disadvantage frequently discussed about radiation processing is safety, because an element of emotionalism is often associated with the word *radiation*.

Stray electrons and rays, high voltages, and ozone exposure are all potential hazards of radiation processing. However, safety in radiation processing is possible through understanding the hazards involved and adhering to a sound safety program. Safety standards have been published, and several governmental agencies have set maximum safe exposure levels for different kinds of radiation.

For any company planning to use radiation processing, it is advisable for the manager to hire a qualified consultant to plan, design, and implement a radiation safety program for the plant and its personnel.

IRRADIATION OF POLYMERS

The transfer of energy from the radiation source to the material assists in breaking bonds and rearranging atoms into new structures. The many changes that take place in covalent substances directly affect important physical properties. The effects of radiation on plastics may be divided into four categories:

1. Damage by radiation
2. Improvement by radiation
3. Polymerization by radiation
4. Grafting by radiation

Damage by Radiation

Breaking covalent bonds by using nuclear radiation is called scission. This separation of carbon-to-carbon bonds may lower the molecular mass of the polymer. Figure 20-5 shows that irradiation of polytetrafluorethylene causes long linear plastics to break into short segments. The plastics loses strength as a result of this breaking.

Degradative symptoms include cracking, crazing, discoloration, hardening, embrittlement, softening, and other undesirable physical properties associated with molecular mass, molecular mass distribution, branching, crystallinity, and cross-linking.

With controlled irradiation, polyethylene becomes a cross-linked, insoluble, and non-melting material. Improvements may include increased heat resistance and form stability at elevated temperatures, reduction in cold flow, stress cracking, and thermal cracking.

Effects of radiation on selected polymers are shown in Table 20-1.

The separation of carbon-to-carbon bonds may also form free radicals, which may lead to cross-linking, branching, polymerization, or the formation of gaseous byproducts. In Figure 20-6A, radiation is shown as the cause of polymerization and cross-linking of hydrocarbon radicals. Figure 20-6B shows that a gaseous product is formed during the irradiation process. The radical (R) may be H, F, Cl, and so on, as the gaseous product.

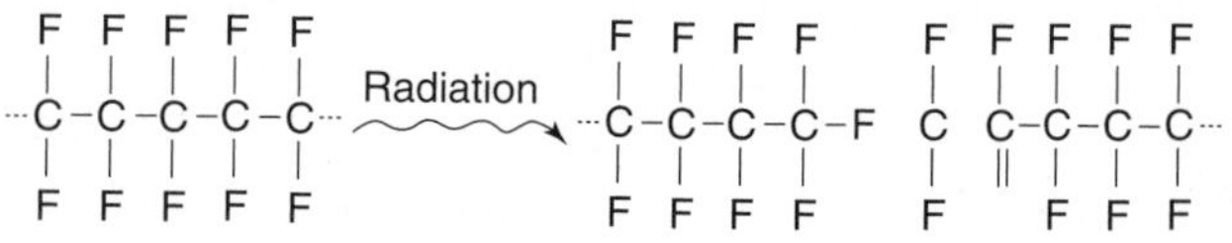

Table 20-1. Effects of Radiation on Selected Polymers

Polymer	Radiation Resistance	Radiation Dose for Significant Damage (Mrads)
ABS	Good	100
EP	Excellent	100–10,000
FEP	Fair	20
PC	Good	100+
PCTFE	Fair	10–20
PE	Good	100
PFV, PFV$_2$, PETFE, PECTFE	Good	100
PI	Excellent	100–10,000
PMMA	Fair	5
Polyesters (aromatic)	Good	100
Polyesters (unsaturated)	Good	1,000
Polymethylpentene	Good	30–50
PP	Fair	10
PS	Excellent	1,000
PSO	Excellent	1,000
PTFE	Poor	2
PU	Excellent	1,000+
PVC	Good	50–100
UF	Good	500

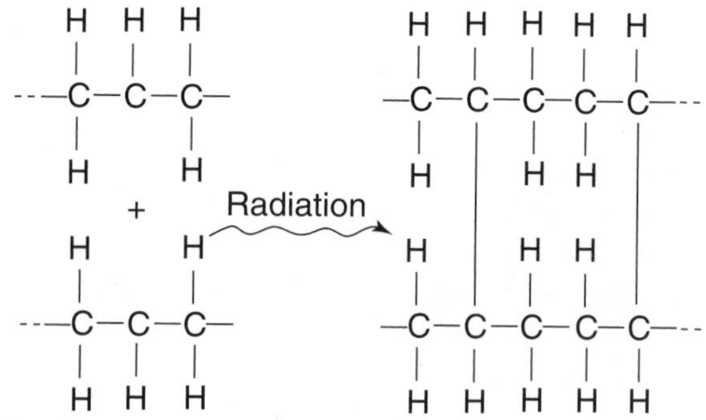

(A) Recombination leading to polymerization or cross-linking of hydrocarbon radicals.

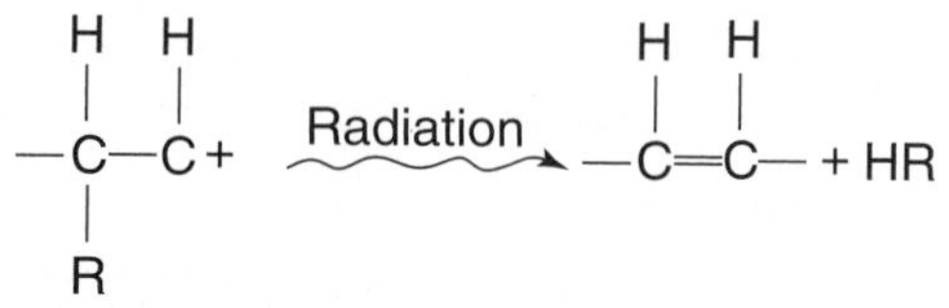

(B) Gaseous product formed by radiation.

Figure 20-6. Formation of free radicals through irradiation.

Figure 20-5. Degradation by irradiation.

Irradiation may result in atoms being knocked from the solid material. This disassociation or displacement of an atom results in a defect in the basic structure of the polymer (Figure 20-7). These vacancies in crystalline structures and other molecular modifications result in changes in the mechanical, chemical, and electrical properties of polymers.

Cross-linkage in elastomers may be considered a form of degradation. For example, natural and synthetic rubbers become hard and brittle in relation to increased cross-linkage or branching (Figure 20-8 and Figure 20-9).

Mineral-, glass-, and asbestos-filled phenolics, epoxies, polyurethane, polystyrene, polyester, silicones, and furan plastics

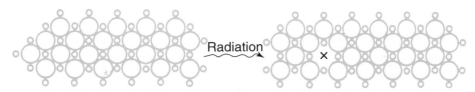

Figure 20-7. Linear structure of plastics with missing atom. The vacancy in the crystalline structure is a potential site for radical attachment.

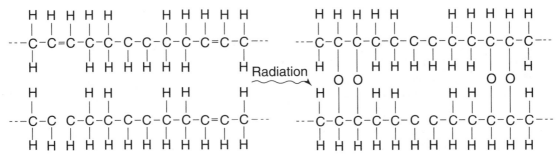

Figure 20-8. Oxidation (radical attachment of oxygen) of polybutadiene. This cross-linking results in a rapid aging effect with loss of elastic strain.

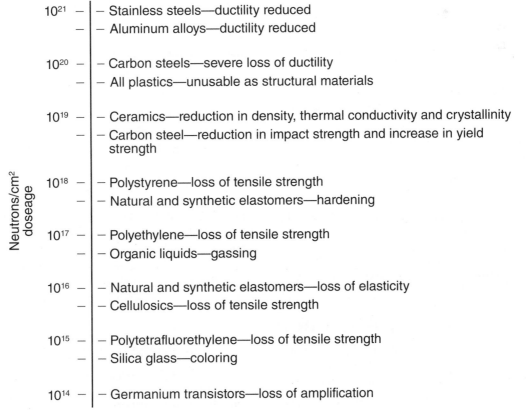

Figure 20-9. Changes in materials' properties caused by radiation. Controlled use of radiation can be beneficial.

have superior radiation resistance. Unfilled methyl methacrylate, vinylidene chloride, polyesters, cellulosics, polyamides, and polytetrafluoroethylene have poor radiation resistance. These plastics become brittle, and their desirable optical properties are affected by discoloring and crazing. Fillers and chemical additives may help to absorb a great deal of radiation energy, whereas heavy pigmentation of the plastics may stop deep penetration of damaging radiation.

Improvements by Radiation

Although some polymers are damaged by radiation, others may actually benefit from controlled amounts. Cross-linkage, grafting, and branching of thermoplastic materials may produce many of the desirable physical properties of thermosetting plastics.

Polyethylene is one example of plastics that benefits from controlled, limited irradiation. Such radiation causes existing bonds to be broken and a rearrangement of the atoms into a branched structure. Branching of the PE chain elevates the softening temperature above that of boiling water. (However, excessive radiation may reverse the effect by rupturing main links in the chains.)

Radiation processing. Today, radiation processing is most often done with electron machines or radioisotope sources such as cobalt-60. This radiation may increase molecular mass by linking molecules of some polymers together or may decrease molecular mass by degrading others. It is this cross-linking and degradation that accounts for most of the property changes in plastics.

The ability of radiation to begin ionization and free-radical formation may prove superior to the ability of other agents such as heat or chemicals.

The main industrial disadvantage of radiation-induced chemical reaction is high cost. With radiation processing integrated directly into processing lines, the cost of radiation systems has been decreasing. Therefore, it may soon be able to compete with chemical processing for some uses.

Ultraviolet treatment may improve such surface characteristics as weather resistance, hardening, penetration, and neutralization of static electricity.

Cross-linking of wire insulation, elastomers, and other plastics parts improves resistance to stress cracking, abrasion, chemicals, and deformation.

Polymerization by Radiation

During the dissociation of a covalent bond by irradiation, a free radical fragment is formed. This radical is available at once for recombinations. The same nuclear energy forces that cause depolymerization of plastics may begin cross-linkage and polymerization of monomer resins (Figure 20-10).

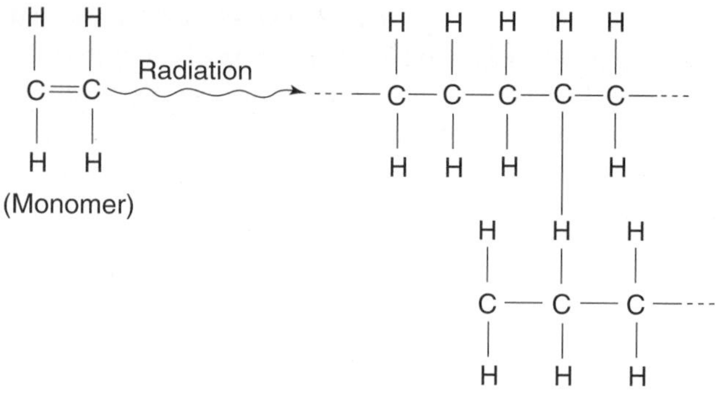

Figure 20-10. Branching of polyethylene.

Polymerization and cross-linking are used to cure polymer coating, adhesives, or monomer layers. Typical doses (Mrads) for cross-linkable polymers range from 20-30 for PE, 5–8 for PVC, 8–16 for PVDF, 10–15 for EVA, and 6–10 for ECTFE.

Grafting by Radiation

When a given type of monomer is polymerized and another kind of monomer is polymerized onto the primary backbone chain, a graft copolymer results. By irradiating a polymer and adding a different monomer and irradiating it again, a graft copolymer is formed. The schematic structure of a graft copolymer is shown in Figure 20-11. The recombining or structuring of two different monomer units (A and B) often yields unique properties. It is possible that graft copolymers with highly specific properties could be combined for optimum product applications. Irradiation may produce a grafting reaction on a thin surface zone or one conducted homogeneously throughout thick sections of a polymer.

Advantages of Radiation

Radiation processing may have numerous broad advantages that could offset the main disadvantage of high cost (Table 20-2).

The first advantage is that reactions can be initiated at lower temperatures than in chemical processing. A second advantage is good penetration, which allows the reaction to occur inside ordinary equipment at a uniform rate. Although gamma radiation from cobalt-60 sources can penetrate more than 300 mm (12 in.), the treatment rate is slow and exposure times are long. Electron radiation sources may react very rapidly with materials less than 10 mm (0.39 in.) thick. For these reasons, over 90% of irradiated products are processed by high-energy electron sources (Figure 20-12A and Figure 20-12B).

AAAAAAAAAAAAAAAAAAAAAAAAAAAAAAAAA
|
BBBBBBBBBB

Figure 20-11. In graft polymerization, a monomer of one type (B) is grated onto a polymer of a different type (A). Because graft copolymers contain long sequences of two different monomer units, some unique properties result.

Table 20-2. **Industrial Applications for Electron Beam Processing**

Product	Product Improvements and Process Advantages	Process
Wire and cable insulation, plastic insulating tubing, plastic packaging film	Shrinkability; impact strength; cut-through, heat, solvent, stress-cracking resistance; low dielectric losses	Cross-linking, vulcanization
Foamed polyethylene	Compression and tensile strength; reduced elongation	Cross-linking, vulcanization
Natural and synthetic rubber	High-temperature stability; abrasion resistance; cold vulcanization; elimination of vulcanizing agents	Cross-linking, vulcanization

Adhesives:
   Pressure-sensitive
   Flock
   Laminate
	Increased bonding; chemical, chipping, abrasion, weathering resistance; elimination of solvent	Curing, polymerization

Coatings, paints, and inks on:
   Woods
   Metals
   Plastics
	100% convertibility of coating; high-speed cure, flexibility in handling techniques; low energy consumption; room-temperature cure; no limitation on colors	Curing, polymerization

Wood and organic impregnates	Mar, scratch, abrasion, warping, swelling, weathering resistance; dimensional stability, surface uniformity; upgrading of softwoods.	Curing, polymerization
Cellulose	Enhanced chemical combination	Depolymerization
Textiles and textile fibers	Soil-release; crease, shrink, weathering resistance; improved dyeability; static dissipation; thermal stability	Grafting
Film and paper	Surface adhesion; improved wettability	Grafting
Medical disposables	Cold sterilization of packages and supplies	Irradiation
Packages and containers	Reduction or elimination of residual monomer.	Polymerization
Polymer	Controlled degradation or modification of melt index	Irradiation, depolymerization, cross-linking

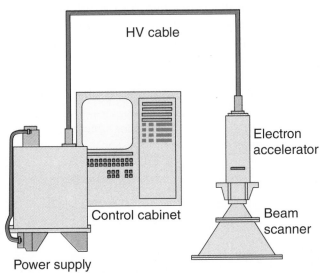

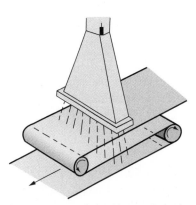

(A) Diagram of the major components of the electron-beam processing system.

(B) Festooning is a method used for processing continuous flexible sheets or webs of material.

Figure 20-12. Principles of electron-beam radiation processing.

A third advantage is that monomers can be polymerized without chemical catalysts, accelerators, and other components that may leave impurities in the polymer. A fourth advantage is that radiation-induced reactions are not greatly affected by the presence of pigments, fillers, antioxidants, and other ingredients in the resin or polymer. A fifth advantage is that cross-linking and grafting may be done on previously shaped parts such as films, tubing, coating, moldings, and other products. Coatings in monomeric form may be applied with radiation processing. This eliminates the use of solvents and the collection or recovery systems for the solvents. Finally, the sixth advantage is that the mixing and storing of chemicals used in chemical processing may be eliminated.

Applications

In addition to the processing advantages just mentioned, radiation processing may provide other marketable features not made possible by other means (Figure 20-13).

Graft and homopolymerization of various monomers on paper and fabrics improves bulkiness, resilience, acid resistance, and tensile strength. Irradiation of some cellulosic textiles has aided in the development of "dura-press" fabrics.

Figure 20-13. The polyethylene container in the center was exposed to controlled radiation to improve heat resistance. Containers at left and right were not treated. They lost shape at 175°C (350°F).

Grafting selected monomers to polyurethane foam, natural fibers, and plastic textiles improves weather resistance as well as eases ironing, bonding, dyeing, and printing. Small radiation dosages, which degrade the surface of some plastics, improve ink adhesion to their surface.

Impregnation of monomers in wood, paper, concrete, and certain composites has increased their hardness, strength, and dimensional stability after irradiation. For example, the hardness of pine has been increased 700% by using this method. Novolacs and resols are soluble and fusible low-molecular-mass resins used in the production of prepregs (reinforcement-impregnated materials). The term *A-stage* is used to refer to novolac and resol resins. While under a high vacuum, wood, fabric, glass fibers, and paper may be saturated with A-stage resins. This supersaturated prepreg or impreg may then be exposed to cobalt radiation, causing the thermosetting, A-stage material to pass through a rubbery stage referred to as the *B-stage*. Further reaction produces a rigid, insoluble, infusible, hard product. This last stage of polymerization is known as the *C-stage*. The terms A-, B-, and C-stage resins are also used to describe analogous states in other thermosetting resins (see Phenolics in Appendix F).

A commercially available *shrinkable* polyethylene film is often used for wrapping food items. This irradiated film is cross-linked by radiation for increased strength. The film can be stretched more than 200% and is usually sold prestressed. When heated to 82°C (180°F) or higher, the film attempts to shrink back to its original dimensions, producing a tight package. Radiation is also being used as a heatless sterilization system for packaged food and surgical supplies.

Radioisotopes are used in many measuring applications. Monomer resins, paints, or other coatings can be measured for thickness without contacting or marking the material's surface, similar to extruded, or blown, films. Using this measuring method may reduce raw material consumption, reduce or eliminate scrap, ensure more uniform thickness, and speed up output (Figure 20-14).

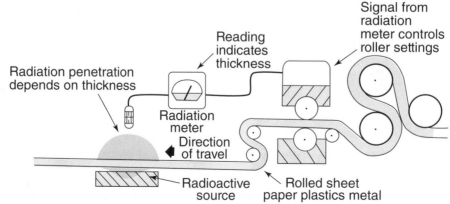

Figure 20-14. Radioisotopes are used to continuously gauge the thickness of a material without being in physical contact with it.

Four advantages of radiation processing and three of its disadvantages are listed here:

Advantages of Radiation Processing

1. It improves many important plastics properties.
2. Many nonionizing radiation processes speed production by heating or initiating polymerization.
3. No physical contact is needed.
4. Machine sources may be controlled with ease and require less shielding.

Disadvantages of Radiation Processing

1. Gamma radiation equipment is somewhat costly; some is specialized.
2. It demands careful handling and trained personnel (especially ionizing radiation).
3. There is potential danger to operator due to ionizing radiation and radioisotopes.

RELATED INTERNET SITES

- **www.iba-industrial.com.** IBA supplies electron beam accelerators for both medical and industrial applications. On the home page, select "Polymer Crosslinking" to reach information on "Easy-E-Beam," their integrated wire and cable crosslinking system. The company claims its system can operate at line speeds of up to 1000 m/minute.

- **www.e-beamservices.com.** E-Beam Services Inc. offers the choice "Polymer Crosslinking" on its home page. This leads to "Technology," which provides details on electron beams.

- **www.electrontech.com.** Electron Technologies Corp. provides irradiation services to several industries, including plastics extrusion and molding. Selecting "About" on the home page leads to an overview of irradiation processing.

VOCABULARY

The following vocabulary words are found in this chapter. Use the glossary in Appendix A to look up the definitions of any of these words you do not understand as they apply to plastics.

alpha particles
beta particles
gamma radiation
gamma rays
gray
irradiation
isotopes
neutrons
photon
quantum
radiation
ultraviolet

QUESTIONS

20-1. Name the term that refers to the bombardment of plastics with a variety of subatomic particles. It may be done to polymerize and change physical properties of plastics.

20-2. The most important source of controlled radiation is _____.

20-3. When one or more different kinds of monomers are attached to the primary backbone of the polymer chain, a _____ results.

20-4. Name five plastics that have poor radiation resistance.

20-5. Name two additives for plastics that may help stop penetration of damaging radiation.

20-6. The carrier of wave energy is called _____.

20-7. Electrons traveling at very high speed and high energy are called _____ particles.

20-8. Different forms of the same element with different atomic masses are called _____.

20-9. Breaking of covalent bonds by nuclear radiation is called _____.

20-10. The two types of radiation systems or sources are _____ and _____.

20-11. Controlled amounts of irradiation may cause _____ of bonds for free-radical formation and cross-linkage.

20-12. Uncontrolled irradiation may break bonds, lowering _____ and _____.

20-13. Name four adverse effects of irradiation.

20-14. Name three possible sources of ionizing radiation for irradiating plastics.

20-15. Name four possible sources of nonionizing radiation for irradiating plastics.

20-16. The _____ is the term often used to describe the dose given in irradiating polymers.

20-17. Radiation sources must be carefully handled only by trained _____.

20-18. Accumulated dosages or exposure to _____ and _____ radiation may cause permanent cell damage.

20-19. Name four major advantages of irradiation processing.

20-20. Energy sources that may be used to preheat molding compounds, heat-seal films, polymerize resins, and expand polystyrene beads are called _____.

20-21. Radiations that sunburn or tan the human body are _____ rays.

20-22. Identify the particles that are heavy, slow-moving masses with a double positive charge.

20-23. Burnt _____ slugs from reactors or fission waste may be a source of radiation.

20-24. As a rule, all forms of natural radiation must be considered _____ to polymers.

20-25. Over _____% of irradiated products are processed by high-energy electron sources.

20-26. For the following products or applications, would you select nonionizing or ionizing processing?

a. Polymerization of selected resins

b. Surface treatment of films

c. Drying plastics granules or preforms

DESIGN CONSIDERATIONS

CHAPTER 21

INTRODUCTION

This chapter summarizes the basic rules for designing products. Due to the diversity of materials, processes, and product uses, designing with plastics demands more experience than designing with other materials. The information presented in this chapter should serve as a fundamental guide and useful starting point in understanding the complexity of designing plastics products. See Appendix H for additional sources of information.

There are many sources that can be used for studying specific design problems, and some are included in the discussion of individual materials and processes.

In the early years of development, plastics were frequently chosen as a substitute for other materials. Some of those early products were very successful due to the consideration and thought given when choosing materials. On the other hand, some of these products failed because the designers did not know enough about the properties of plastics used or were motivated by cost rather than the practical use of the material. The products simply could not stand up to daily wear and tear. The designers' knowledge of the properties of plastics has grown with the industry. Because plastics have combinations of properties that no other materials possess—that is, strength, lightness, flexibility, and transparency (Table 21-1)—they are now chosen as primary materials rather than substitutes.

The design considerations for polymer composites are more complex than those for homopolymers. Most composites vary with time under load, rate of loading, small changes in temperature, matrix composition, material form, reinforcement configuration, and fabric method. Depending upon design requirements, they may be designed to be isotropic, quasi-isotropic, or anisotropic.

In the last few decades, the development of computer-based technologies has had the biggest influence on design. Although an endless array of programs and systems are available, product design and development rely on computer-aided design (CAD), computer-aided engineering (CAE), rapid prototyping (RP), and flow analysis systems such as Moldflow®. CAD and CAE have grown into huge fields, with many companies offering software products. Although CAD and CAE both began using mainframe computers, they migrated to personal computers as well. As the power of personal computers (PCs) grew, more capabilities became available in PC-based CAD and CAE packages. This trend continues, making PCs both more powerful and less expensive.

Computer-Aided Flow Analysis

Until the year 2000, the two biggest companies involved in computer-aided flow analysis were Moldflow and AC Technology. Both companies developed and marketed programs to predict the flow, cooling, and warping of plastics products in simulated injection molding, gas-assisted injection molding, and co-injection molding processes. The programs offered by Moldflow were generally identified as Moldflow, and the software from AC Technology was called C-Mold®. In 2000, Moldflow acquired C-Mold and now holds center stage in the flow analysis world.

Rapid Prototyping

Traditional prototyping, a long-standing industrial activity, involves machining processes such as milling, turning, and grinding. Some prototypes have even been made by casting.

Table 21-1. Plastics vs. Metals

Properties of Plastics Which May Be . . . Favorable

1. Light
2. Better chemical and moisture resistance
3. Better resistance to shock and vibration
4. Transparent or translucent
5. Tendency to absorb vibration and sound
6. Higher abrasion and wear resistance
7. Self-lubricating
8. Often easier to fabricate
9. Can have integral color
10. Cost trend is downward. Today's composite plastics price is approximately 11 percent lower than five years ago.

    However, the long-established, high-volume plastics—phenolics, styrenes, vinyls, for example—appear to have reached a price plateau, and prices change only when demand is out of phase with supply.
11. Often cost less per finished part
12. Consolidation of parts

Unfavorable

1. Lower strength
2. Much higher thermal expansion
3. More susceptible to creep, cold flow, and deformation under load
4. Lower heat resistance—both to thermal degradation and heat distortion
5. More subject to embrittlement at low temperatures
6. Softer
7. Less ductile
8. Dimensions change through absorption of moisture or solvents
9. Flammable
10. Some varieties degraded by ultraviolet radiation
11. Most cost more (per cubic millimeter) than competing metals. Nearly all cost more per kilogram.

Either Favorable or Unfavorable

1. Flexible—even rigid varieties more resilient than metals
2. Electrical nonconductors
3. Thermal insulators
4. Formed through the application of heat and pressure

Exceptions

1. Some reinforced plastics (glass-reinforced epoxies, polyesters, and phenolics) are nearly as rigid and strong (particularly in relation to mass) as most steels; may be even more dimensionally stable.
2. Some oriented films and sheets (oriented polyesters) have greater strength-to-mass ratios than cold-rolled steels.
3. Some plastics are now cheaper than competing metals (Nylons vs. brass, acetal vs. zinc, acrylic vs. stainless steel).
4. Some plastics are tougher at lower than normal temperatures (acrylic has no known brittle point).
5. Many plastics-metal combinations extend the range of useful applications of both (metal-vinyl laminates, leaded vinyls, metallized polyesters, and copper-filled PTFE).
6. Plastics and metal components may be combined to produce a desired balance of properties (plastics parts with molded-in, threaded-metal inserts; gears with cast-iron hubs and Nylon teeth; gear trains with alternate steel and phenolic gears; rotating bearings with metal shaft and housing and Nylon or TFE bearing liner).
7. Metallic fillers in plastics make them electrically or thermally conductive or magnetic.

Source: Adapted from *Machine Design*: Plastics Reference Issue.

In contrast, new prototyping processes are more rapid and can create physical objects directly from CAD drawings. Rapid prototyping models often involve building a shape with the use of many thin layers—sometimes as many as 5 to 10 layers for each millimeter of model. Rapid prototyping began around 1990, and by the year 2002, it had grown to a $1 billion per year market. Since 2002, the greatest growth has occurred in 3D printers, which are lower in cost than other systems and do not require special ventilation. The two companies that have been leading in sales for 3D printers are Stratasys and Z Corp. As an indication of the rapid growth of 3D printers, both companies have been selling systems for more than 10 years, but about half their total sales occurred between 2003 and 2006.

Rapid prototyping equipment can be categorized according to the three common materials used: photopolymers, thermoplastics, and adhesives. The photopolymer systems, often named stereolithography (SLA), use a liquid polymer that is solidified in thin layers by exposure to selected wavelengths of light. Figure 21-1 shows a model created by an SLA system. Often the curing of each layer may require one to two minutes because the light source is laser. During the curing

Photo courtesy of 3D Systems, Inc.

Figure 21-1. This photo shows an SLA® system. Notice the support structure that holds the wheel model as the layers are added.

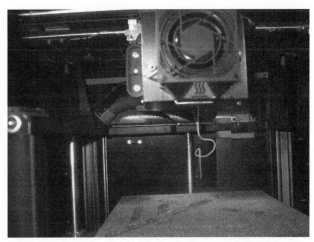

Figure 21-3. Purging the print head guarantees that the model material, white ABS, and the support material, dark water soluble polymer, are both flowing properly.

process, the photopolymer material shrinks and may distort. To minimize distortion, support structures—usually thin walls in a grid pattern—must be added to the models. The thermoplastic systems, often named laser sintering (LS), melt and then apply thin layers of hot plastic to the growing model. The material fuses to the lower layer, creating a continuous object. The adhesive systems, often named paper lamination technology (PLT), use various binders to connect the layers of paper. The paper lamination method is the least expensive of the three approaches and produces models with minimal distortion.

Figure 21-2 shows a model in the process of removal from a bed of powdered plaster in a Z-Corp 3D printer. This type of system does not need a support structure for the model

because the surrounding plaster provides stability. In contrast, the Stratasys 3D printer builds the model with a grade of ABS plastic. Figure 21-3 shows the extrusion head of a Stratasys unit during purging operations. The materials for both the model and the support structure are visible. The white strand (ABS) is for the model, and the darker strand is the support material, which can be dissolved in a hot water solution. Because the base on which the model is built may be uneven, the unit creates a layer of support material first and then begins to build the model on top of that layer. Figure 21-4 shows a layer of support material that provides a flat surface for model construction.

In most designs, a compromise must be made between high performance, good appearance, efficient production, and low cost. Unfortunately, human needs are often deemed less important than factors of cost, processes, or materials used. Three

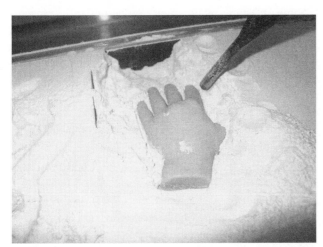

Figure 21-2. A small vacuum removes plaster powder from around a model of a hand in a Z-Corp unit.

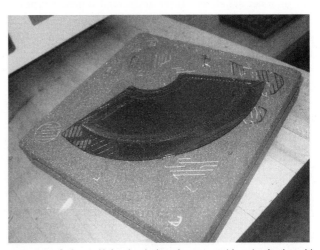

Figure 21-4. On the reusable foam board, a layer of support material remains after the model has been removed. If the model has internal cavities, the support material must be dissolved to remove it.

(sometimes overlapping) major considerations are included in the following chapter outline:

MATERIAL CONSIDERATIONS

Materials with the right properties must be selected to meet design, economic, and service conditions. In the past, the design was commonly changed to compensate for material limitations.

Caution must be exercised when using information about the matrix performance obtained from data sheets or the manufacturer. Much of this data is based on laboratory-controlled evaluations. It is also difficult to compare proprietary data from several different suppliers. However, this does not imply that this data cannot be utilized in the screening of candidate materials.

Customers often use specifications in a document to state all requirements to be satisfied by the proposed product, materials, or other standards. There are several different types of standards, including *physical* standards as kept by the National Bureau of Standards (NBS); *regulatory* standards such as those from the EPA; *voluntary* standards recommended by technical societies, producers, trade associations, or other groups such as Underwriters Laboratories (UL); and *mil* (military) and *public* standards promoted by professional organizations such as the ASTM (American Society for Testing and Materials).

Metrication and internationalization of standards can reduce costs associated with materials, production, inventory, design, testing, engineering, documentation, and quality control. Substantial evidence exists that metrication and standardization lower costs. We must change our attitudes concerning international measurement and standards if we are to significantly lower costs and increase international trade.

With the evolution of the computer, systematic methods of screening and selection of materials are easier. Computer models can predict and anticipate most of the circumstances in which a material can fail. In some computer models, each property is assigned a value according to importance. Next, each property that the part is expected to tolerate is entered. The computer will select the best combination of materials and processes.

Plastics materials must be chosen with care, keeping the final product use in mind. The properties of plastics depend more on temperature than other materials do. Plastics are more sensitive to changes in environment. Therefore, many families of plastics may be limited in use. There is no one material that will possess all the desired qualities. However, undesirable characteristics may be compensated for in the product design.

The final material choice for a product is based on the most favorable balance of design, fabrication, and total cost or selling price of the finished item. Both simple and complicated designs may need fewer processing and fabrication operations using plastics, which may combine with the characteristics of the plastics material to make plastics costs competitive with other materials used for specific parts.

Environment

When designing a plastics product, the physical, chemical, and thermal environments are very important to consider. The useful temperature range of most plastics seldom exceeds 200°C (392°F). Many plastics parts exposed to radiant and ultraviolet energy soon suffer surface breakdown, become brittle, and lose mechanical strength. Fluorocarbons, silicones, polyimides, and filled plastics must be used for products operating above 230°C (450°F). The exotic environments of outer space and the human body are becoming commonplace for plastics materials. The insulation and ablative materials used for space vehicles, as well as artery reinforcements, monofilament sutures, heart regulators, and valves, are only a small portion of these new products.

Some plastics retain their properties at cryogenic (extremely low) temperatures. For instance, containers, self-lubricating bearings, and flexible tubing must function properly in below-zero temperatures. The cold, hostile environments of space and earth are only two examples. Any time refrigeration and food packaging is considered, or taste and odors are a problem, plastics may be chosen. The United States Food and Drug Administration lists acceptable plastics for packaging of foods.

In 1969, the US government passed the Child Protection and Toy Safety Act. It was supplemented in 1973 by the Consumer Product Safety Commission (CPSC). In 1995, these laws were revised in the Child Safety Protection Act, which is still in effect. This act mandates cautionary labeling relating to choking hazards to children less than three years of age. These regulations have helped protect children but, in 1999, almost

70,000 children under the age of five went to the emergency room for toy-related injuries. During that year, 16 children died from toy-related injuries.

In addition to extremes of temperature, humidity, radiation, abrasives, and other environmental factors, the designer must consider fire resistance. There are no fully fire-resistant plastics.

Polyimide-boron fiber composites have high temperature resistance and strength. Polyimide-graphite fiber composites can compete with metals in strength and achieve a significant weight savings at service temperatures up to 316°C (600°F). Boron powder is sometimes added to the matrix to help stabilize the char that forms in thermal oxidation. Other flame-resistant additives or an ablative matrix may also be used. The danger posed by open flames is the most serious disadvantage of the use of plastics in fabrics and architectural structures.

Remember, thermal degradation and cross-linking are not reversible phenomena. Glass transition, melting, and crystallization of most matrixes are reversible.

Moisture may cause deterioration and weaken the reinforcement and matrix bond in composites. Any holes, exposed edges, or machined areas on composite designs should be given a protective coating to prevent moisture infiltration or wicking.

Electrical Characteristics

All plastics have useful electrical insulation characteristics. Although the selection of plastics is usually based on mechanical, thermal, and chemical properties, much of the pioneering in plastics was for electrical uses. Electrical insulating problems such as high altitude, space, undersea, and underground environments are solved by utilizing plastics. All-weather radar and underwater sonar would not be possible without the use of plastics. They are used to insulate, coat, and protect electronic components.

Particulate composites using carbon, graphite, metal, or metal-coated reinforcements provide EMI shielding for many products.

Chemical Characteristics

The chemical and electrical natures of plastics are closely related because of their molecular makeup. There is no general rule for chemical resistance. Plastics must be tested in the chemical environment of their actual use. Fluorocarbons, chlorinated polyethers, and polyolefins are among the most chemical-resistant materials. Some plastics react as semipermeable membranes. They allow selected chemicals or gases to pass while blocking others. The permeability of polyethylene plastics is an asset in packaging fresh fruits and meats. Silicones and other plastics allow oxygen and gases to pass through a thin membrane while stopping water molecules and many chemical ions at the same time. Selective filtration of minerals from water may be done with semipermeable plastics membranes.

Mechanical Factors

Material considerations include the mechanical factors of fatigue, tensile, flexural, impact, compressive strength, hardness, damping, cold flow, thermal expansion, and dimensional stability. All of these properties were discussed in Chapter 6. Although fillers will improve the dimensional stability of all plastics, products that need dimensional stability call for careful choice of materials. A factor sometimes used for evaluation and selection is strength-to-mass ratio—the ratio of tensile strength to the density of the material. Plastics can surpass steels in strength-to-mass ratio.

Example:
Divide the tensile strength of the material by its density.

$$\text{Selected plastics} - \frac{0.70 \text{ Gpa}}{2 \text{ g/cm}^3} = 0.350$$

$$\text{Selected plastics} - \frac{1.665 \text{ Gpa}}{7.7 \text{ g/cm}^3} = 0.214$$

The type and orientation of reinforcements greatly influence the properties of composite products. Some specifications in critical designs will designate a safety factor (SF). A safety factor (sometimes called a design factor) is defined as the ratio of the ultimate strength of the material in relation to the allowable working stress.

$$SF = \frac{\text{Ultimate strength}}{\text{Allowable working stress}}$$

The SF for a composite aircraft landing gear might be 10.0, whereas for an automobile spring it will be only 3.0. With accurate, reliable data, some designs use safety factors of 1.5 to 2.0.

Economics

Economic consideration is the final phase of material selection. It is best not to include material costs in the preliminary screening of candidate materials. Materials with marginal performance properties or expensive materials need not be eliminated at this point. Either material may continue to be a possible candidate, depending on the processing parameters, assembly, finishing, and service conditions. A polymer with minimal performance characteristics may not be the best choice if reliability and quality are important.

Cost is always a major factor in design consideration or material selection. Strength-to-mass ratio or chemical, electrical, and moisture resistance may overcome price disadvantage. Some plastics cost more per kilogram than metals or other materials, but plastics often cost less per finished part. The most meaningful comparison between different plastics is cost per cubic centimeter. In any molding operation, the apparent density and bulk factors are important in cost analysis.

Apparent density, sometimes called *bulk density*, is the mass per unit volume of a material. It is calculated by placing the test sample in a graduated cylinder and taking measurements. The volume (V) of the sample is the product of its height (H) and cross-sectional area (A). Thus $V = HA$.

$$\text{Apparent density} = \frac{W}{V}$$

where

V = volume in cubic centimeters occupied by the material in the measuring cylinder

H = height in centimeters of the material in the cylinder

A = cross-sectional area in square centimeters of the measuring cylinder

W = mass in grams of the material in the cylinder

Many plastics and polymer composite parts cost 10 times more than steel. On a volume basis, some are lower in cost than metals.

Bulk factor is the ratio of the volume of loose molding powder to the volume of the same mass of resin after molding. Bulk factors may be calculated as follows:

$$\text{Bulk factor} = \frac{D_2}{D_1}$$

where

D_2 = average density of the molded or formed specimen

D_1 = average apparent density of the plastics material prior to forming

Economics must also include the method of production and design limitations of the product. One-piece seamless gasoline tanks may be rotationally cast or blow molded. The latter process uses more costly equipment but can produce the products more quickly, thus reducing costs. Conversely, large storage tanks may be produced more economically by rotational casting than blow molding.

Capital investments for new tooling, equipment, or physical space could result in consideration of different materials and/or processes. The labor-intensive operations often associated with wet or open molding cannot hope to compete with the automated facilities of some companies. The number of parts to be produced and the initial production costs may be the decisive factor.

The three overriding factors in plastics design are service, production, and cost. The use and performance of the part or product must be a concern in defining some design factors. For each design, there may be several production or process options. Cost often appears to override most other concerns of design and development and is often based on the production method.

Volume of sales is also very important. If a mold costs $10,000, and only 10,000 parts are to be made, the mold cost would be $1.00 per part. If 1,000,000 parts were to be made, the mold cost would be $0.01 per part.

Many composite components may be more cost-effective in the long term. In many applications, the products may simply outlast metals. One-piece composite components could reduce the number of molds used, tooling, and assembly time in the production of a boat hull, fuselage, or automobile floor pan. Light, corrosion-resistant composites or plastics components could lower energy costs of fuel for the lifetime of the transportation vehicle. Less energy is consumed (including the energy content of raw material) in the production of polymer parts than for metal parts. Plastics are mostly petroleum derived and must continue to compete for depleting resources.

DESIGN CONSIDERATIONS

When considering overall design conditions, the intended application or function, environment, reliability requirements, and specifications must be reviewed. Computer system databases may alert designers that a design is outside the parameters of the material or process selected. See Appendix H for additional sources of information.

Appearance

The consumer is probably most aware of a product's physical appearance and utility. This includes design, color, optical properties, and surface finish. Elements of design and appearance encompass several properties at the same time. Color, texture, shape, and material may influence consumer appeal. One example is the smooth, graceful lines of Danish-style furniture with dark wood and a satin finish. Changing any one of these elements or properties would drastically change the design and appearance of the furniture.

A few outstanding characteristics of plastics are that they may be transparent or colored, as smooth as glass, or as supple and soft as fur. In many cases, plastics may be the only materials with the desired combination of properties to fulfill service needs.

To ensure proper design, there must be close cooperation between mold makers, manufacturers, processors, and fabricators.

Thought must be given to the design of the plastics part before it is molded in order to ensure that the best combination of mechanical, electrical, chemical, and thermal properties will be obtained. Residual stresses develop as a result of forcing the material to conform to a mold shape. These stresses are locked in during cooling or curing and matrix shrinkage. They could cause warpage in flat surfaces. Warpage is somewhat proportional to the amount of shrinkage of the matrix and is generally the result of differential shrinkage.

There are no hard and fast rules to determine the most practical wall thickness of a molded part. Ribs, bosses, flanges, and beads are common methods of adding strength without increasing wall thickness. Large flat areas should be slightly convex or crowned for greater strength and to prevent warpage from stress (Figure 21-5A).

In molding, it is important that all areas of the mold cavity be filled easily to uniformly minimize most of the stress in molding the part. A uniform wall thickness in the design is important to prevent uneven shrinkage of thin and thick sections (Figure 21-5B). If wall thickness is not uniform, the molded part may distort, warp, or have internal stresses or cracks. From 6 to 13 mm (0.24 to 0.51 in.) may be considered heavy wall thickness in molded parts.

In general, the **ribs** at the base should have a width equal to one-half the thickness of the adjacent wall (Figure 21-5C). They should be no higher than three times the wall thickness. **Boss** designs should have an outside diameter equal to twice the inside diameter of the hole. They should be no higher than twice their diameter.

Plastics parts should have liberal fillets and rounds to increase strength, assist molten material flow, and reduce points of stress concentration. All radii should be generous, and the recommended minimum radius is 0.50 mm (0.020 in.). A good design standard is to set the ratio between the radius of fillet and the nominal wall thickness to 0.5. This means that when a rib with a nominal wall of 3 mm (0.125 in.) meets a wall, the radius of the fillet should be 1.5 mm (0.062 in.).

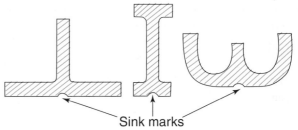

(C) Thickness of walls and ribs in thermoplastic part should be about 60 percent of the thickness of main walls. This will reduce the possibility of sink marks.

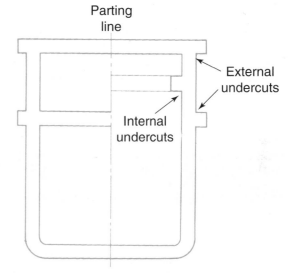

(D) A simple molding with internal and external undercuts.

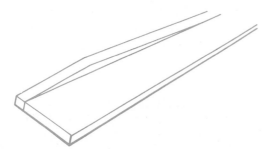

(A) Long, flat strips will warp. Ribs should be added, or the piece crowned in a convex shape.

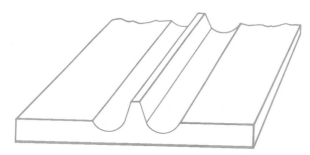

(B) Uneven sections will cause distortion, warpage, cracks, sinks, or other problems because of the difference in shrinkage from section to section.

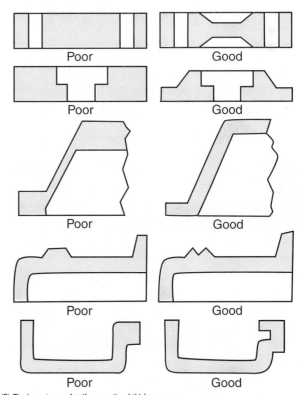

(E) The importance of uniform sectional thickness.

Figure 21-5. Precautions to observe in production of plastics products.

Undercuts (internal or external) in parts should be avoided if possible. **Undercuts** (Figure 21-5D) usually increase tooling costs by requiring techniques for molding, part removal (which usually requires movable parts in the mold), and cooling jigs. A slight undercut could be tolerated in some products when using tough, elastic materials. The molded part can be successfully snapped or stripped out of the cavity while hot. Undercut dimensions should be less than 5 percent of the part diameter. In Table 21-2, the complexity of parts used in several processes is shown.

Decorating may be considered an important function factor in plastics design. The product may include textures, instructions, labels, or letters, and must be decorated in a manner that will not complicate removal from the mold. Decorations should provide durable service to the consumer. Letters are commonly engraved, hobbed, or electrochemically etched into mold cavities (Figure 21-6).

Design Limitations

Next to material selection, tooling and processing have a marked effect on the properties and quality of all plastics products.

The design of the product, and ultimately the mold used to produce the product, is closely related to production. Output rates, parting lines, dimensional tolerances, undercuts, finish, and material shrinkage are among factors that must be kept in mind by the mold maker or tool designer. For example, undercuts and inserts slow output rates and require a more costly mold.

Table 21-2. Plastics Processing Forms—Complexity of Part

Processing Form	Section Thickness, mm		Bosses	Undercuts	Inserts	Holes
	Max.	Min.				
Blow molding	> 6.35	0.254	Possible	Yes—but reduce production rate	Yes	Yes
Injection moldings	> 25.4; normally 6.35	0.381	Yes	Possible—but undesirable; reduce production speed and increase cost	Yes—variety of threaded and non-threaded	Yes—both through and blind
Cut extrusions	12.7	0.254	Yes	Yes—no difficulty	Yes—no difficulty	Yes—in direction of extrusion only; 0.50–1.0 mm min.
Sheet moldings (Thermoforming)	76.2	0.00635	Yes	Yes—but reduce production rate	Yes	No
Slush moldings		0.508	Yes	Yes—flexibility of vinyl allows drastic undercuts	Yes	Yes
Compression moldings		0.889–3.175	Possible	Possible—but not recommended	Yes—but avoid long, slender, delicate inserts	Yes—both through and blind; both should be round, large, and at right angles to surface of part
Transfer moldings		0.889–3.175	Possible	Possible—but should be avoided; reduce production rate	Yes—delicate inserts may be used	Yes—should be round, large, and at right angles to surface of part
Reinforced moldings plastics	Bag: 25.4 matched die: 6.35	Bag: 2.54 matched die: 0.762	Possible	Bag: yes; matched die: no	Bag: yes; matched die: possible	Bag: only large holes; matched die: yes
Castings		3.175–4.762	Yes	Yes—but only with split and cored molds	Yes	Yes

Source: Adapted from *Materials Selector*, Materials Engineering, Reinhold Publishing Corp., Subsidiary of Litton Publications, Inc., Division of Litton Industries.

Figure 21-6. The backward engraved lettering delivers right reading words on the molded part.

The problem of material shrinkage is equally important to both the designer of molds and the designer of molded products. The loss of solvents, plasticizers, or moisture during molding, together with the chemical reaction of polymerization in some materials, results in shrinkage.

During the injection molding of crystalline materials, shrinkage is affected by the rate of cooling.

Thicker sections, which take longer to cool, will experience greater shrinkage than adjacent thinner sections, which cool quickly.

The thermal contraction of the material must also be considered. The thermal values for most plastics are relatively large. This is an asset in removing molded products from mold cavities. If close tolerances are needed, material shrinkage and dimensional stability must be considered. Material shrinkage is sometimes used to ensure snug or shrink fits of metal inserts.

By using a computer model in a CAD system, it is possible to show the stress response of the part with a specific geometry, reinforcement content, reinforcement orientation, and molding orientation (flow). The computer model may require the use of ribs, contours, or other configurations to produce isotropic or anisotropic properties.

After the preliminary design is finished, a physical prototype is often produced. This allows the design engineer and others to view and test a working prototype mold. Simulated performance and service tests may be performed with the prototype. If design or material errors were made, specifications for redesign can be made.

Both material and production considerations are important when designing plastics products. The problems encountered during producing plastics products often demand selection of the production techniques before material considerations are discussed.

PRODUCTION CONSIDERATIONS

In any product design, material behavior and cost are often reflected in molding, fabrication, and assembly techniques. The tooling designer must consider material shrinkage, dimensional tolerance, mold design, inserts, decorations, knockout pins, parting lines, production rates, and other post-processing operations (Table 21-3).

Trimming, cutting, boring holes, or other fabricating or assembly techniques may slow production, lower performance properties, and increase costs.

The part shape, size, matrix formulation, and polymer form will often limit production to one or two possible means.

Processes of Manufacture

With new technology and materials, processing is often the decisive competitive factor. Today, there are fewer limitations in processing thermoplastics and thermosetting materials than in years past. Processes that were unthinkable a few years ago have now become routine. Many thermosets are now injection molded or extruded. Some materials are processed in thermoplastic equipment and later cured. Polyethylene may be cross-linked by chemical or radiation methods after extrusion. Both thermoplastics and thermosets may be made cellular. Because of moldability, output rates, and other material properties, seemingly costly materials become inexpensive products.

A comparison of processing and economic factors is shown in Table 21-4.

Material Shrinkage

Irregularities in wall thickness may create internal stresses in the molded part. Thick sections cool more slowly than thin ones and may create *shrink marks* as well as differential shrinkage in crystalline plastics. As a general rule, injection-molded crystalline plastics have high shrinkage, whereas amorphous plastics shrink less.

A great deal of pressure must be exerted to force the material through thin wall sections in the mold, which creates further problems because of material shrinkage. Polyethylenes, polyacetals, polyamides, polypropylenes, and some polyvinyls shrink between 0.50 and 0.76 mm/mm (0.20 and 0.030 in./in.) after molding. Molds for crystalline and other amorphous plastics must allow for material shrinkage.

Normally, unfilled injection-molded plastics shrink more in the direction of flow as opposed to the axis transverse to flow. This is mainly caused by the orientation pattern developed by the flow direction from the gate or gates. The differential

Table 21-3. **Design Considerations**

Plastics	Linear Mold Shrinkage, mm/mm	Practical Dimensional Tolerances mm/mm (Single Cavity)			Taper Required, Degrees		
		Fine	Standard	Coarse	Fine	Standard	Coarse
ABS	0.127–0.203	0.051	0.102	0.152	0.25	0.50	1.00
Acetal	0.508–0.635	0.102	0.152	0.229	0.50	0.75	1.00
Acrylic	0.025–0.102	0.076	0.127	0.178	0.25	0.75	1.25
Alkyd (filled)	0.102–0.203	0.051	0.102	0.127	0.25	0.50	1.00
Amino (filled)	0.279–0.305	0.051	0.076	0.102	0.125	0.5	1.00
Cellulosic	0.076–0.254	0.076	0.127	0.178	0.125	0.5	1.00
Chlorinated polyether	0.102–0.152	0.102	0.152	0.229	0.25	0.50	1.00
Epoxy	0.025–0.102	0.051	0.102	0.152	0.25	0.50	1.00
Fluoroplastic (CTFE)	0.254–0.381	0.051	0.076	0.127	0.25	0.50	1.00
Ionomer	0.076–0.508	0.076	0.102	0.152	0.50	1.00	2.00
Polyamide (6.6)	0.203–0.381	0.127	0.178	0.279	0.125	0.25	0.50
Phenolic (filled)	0.102–0.229	0.038	0.051	0.064	0.125	0.50	1.00
Phenylene oxide	0.025–0.152	0.051	0.102	0.152	0.25	0.50	1.00
Polyallomer	0.254–0.508	0.051	0.102	0.152	0.25	0.50	1.00
Polycarbonate	0.127–0.178	0.076	0.152	0.203	0.25	0.50	1.00
Polyester (thermoplastic)	0.076–0.457	0.051	0.102	0.152	0.25	0.50	1.00
Polyethylene (high density)	0.508–1.270	0.076	0.127	0.178	0.50	0.75	1.50
Polypropylene	0.254–0.635	0.076	0.127	0.178	1.00	1.50	2.00
Polystyrene	0.025–0.152	0.051	0.102	0.152	0.25	0.50	1.00
Polysulfone	0.152–0.178	0.102	0.127	0.152	0.25	0.50	1.00
Polyurethane	0.254–0.508	0.051	0.102	0.152	0.25	0.50	1.00
Polyvinyl (PVC) (rigid)	0.025–0.127	0.051	0.102	0.152	0.25	0.50	1.00
Silicone (cast)	0.127–0.152	0.051	0.102	0.152	0.125	0.25	0.50

Table 21-4. **Economic Factors Associated with Different Processes**

Production Method	Economic Minimum	Production Rates	Equipment Cost	Tooling Cost
Autoclave	1 – 100	Low	Low	Low
Blow molding	1000 – 10,000	High	Low	Low
Calendering (meters)	1000 – 10,000	High	High	High
Casting processes	100 – 1000	Low-high	Low	Low
Coating processes	1 – 1000	High	Low-high	Low
Compression molding	1000 – 10,000	High	Low	Low
Expanding processes	1000 – 10,000	High	Low-high	Low-high
Extrusion (meters)	1000 – 10,000	High	High	Low
Filament winding	1 – 100	Low	Low-high	Low

Injection molding	10,000 – 100,000	High	High	High
Laminating (continuous)	1000 – 10,000	High	Low	High
Lay-up	1 – 100	Low	Low	Low
Machining	1 – 100	Low	Low	Low
Matched die	1000 – 10,000	High	High	High
Mechanical forming	1 – 100	Low-high	Low	Low
Pressure-bag	1 – 100	Low	Low	Low
Pulforming (meters)	1000 – 10,000	Low-high	Low	High
Pultrusion (meters)	1000 – 10,000	Low-high	High	Low-high
Rotational casting	100 – 1000	Low	Low	Low
Spray-up	1 – 100	Low	Low	Low
Thermoforming	100 – 1000	High	Low	Low
Transfer molding	1000 – 10,000	High	Low	High
Vacuum-bag	1 – 100	Low	Low	Low

shrinkage results because oriented plastics normally have higher shrinkage than non-oriented plastics. Fiber-reinforced polymers are one exception.

Fiber-reinforced polymers will shrink more along the axis transverse to flow than along the axis of material flow. Typical shrinkage of fiber-reinforced polymers is about one-third to one-half that of nonreinforced polymers. The reason is that the fibers that are oriented in the direction of flow prevent the normal free shrinkage of the plastics or polymers.

Tolerances

Maintaining dimensional tolerances is closely related to shrinkage. Molding items with precision tolerances requires careful material selection, and tooling costs are greater. Dimensional tolerances of single-cavity molded articles may be held to ±0.05 mm/mm (±0.002 in./in.) or less with selected plastics. Errors in tooling; variations in shrinkage between multicavity pieces; and differences in temperature, loading, and pressure from cavity to cavity all increase the critical dimensional tolerances of multicavity molds. For example, if the number of cavities is increased to 50, the closest practical tolerance may then be ±0.025 mm/mm (±0.010 in./in.).

Tolerance standards have been established by technical custom molders and the *Standards Committee* of the *Society of the Plastics Industry, Inc.* These standards are to be used only as a guide because the individual plastics material and design must be considered in determining dimensions.

There are three classes of dimensional tolerances for molded plastics parts. They are expressed as plus and minus allowable variations in inches per inch (in./in.) or millimeters per millimeter (mm/mm). Fine tolerance is the narrowest possible limit of variation possible under controlled production. Standard tolerance is the dimensional control that can be maintained under average conditions of manufacture. Coarse tolerance is acceptable on parts where accurate dimensions are not important or critical.

In order for the part to be removed with ease from the molding cavity, draft should be provided (both inside and out). The degree of draft may vary according to molding process, depth of part, type of material, and wall thickness. A draft of 0.25 degrees is sufficient for all shallow molded parts. For textured designs and cores, the draft angles should be increased.

For example, if a part has a depth of 250 mm (10 in.) and a 1.0-degree draft angle, the total draft of the piece will be 4.45 mm (0.175 in.) per side.

Mold Design

Mold design is an important factor in determining molding output. Because mold design is a complex subject, only a broad discussion is possible here (see Chapter 22). A design for a typical two-plate, two-cavity injection mold is shown in Figure 21-7. A two-plate mold has one main parting line. In contrast, a three-plate mold has two main parting lines. One line allows for removal of the molded parts, whereas the other parting line releases the runner system. A three-plate mold, as shown in Figure 21-8, separates the runner from the molded parts.

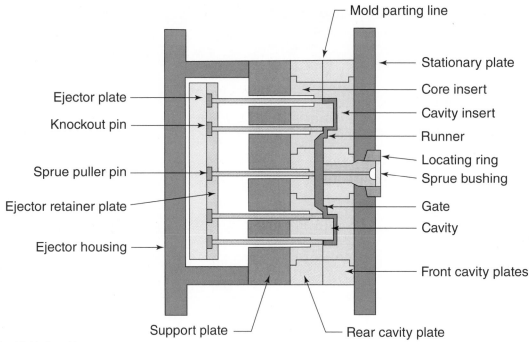

Figure 21-7. Two-plate injection mold.

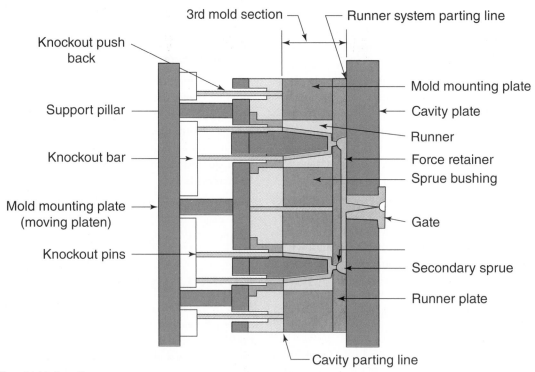

Figure 21-8. Three-plate injection mold.

As the hot, molten material is forced from the nozzle into the mold, it flows through channels or passageways (Figure 21-9A). The terms **sprue**, **runner**, and **gate** are used to designate these channels (Figure 21-9B).

The heavy tapered channel that connects the nozzle with the runners is called the sprue. In a single-cavity mold, the sprue feeds material directly through a gate into the mold cavity. If the sprue feeds directly, the need for a separate runner and gate

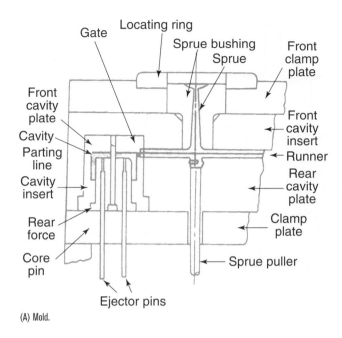

(A) Mold.

(B) Molded part, runner, gate, sprue, and typical part.

Figure 21-9. Construction of an injection mold and a molded part.

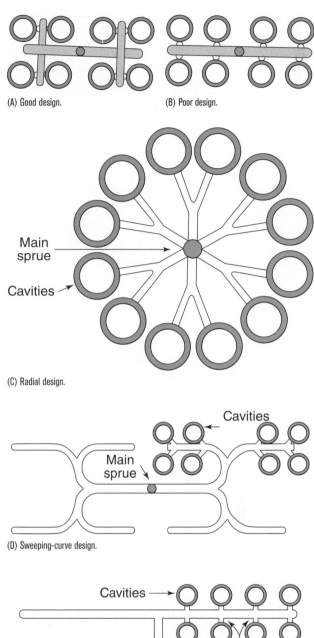

(A) Good design. (B) Poor design.

(C) Radial design.

(D) Sweeping-curve design.

(E) H design.

Figure 21-10. Some typical runner designs.

is eliminated. In most single-cavity molds, there is no need for a runner unless the part is being gated in more than one place.

Runners are narrow channels that convey the molten plastics from the sprue to each cavity. In multicavity molds, the runner system should be designed so that all materials travel equal distances through runners of equal cross sections (Figure 21-10). This ensures that equal pressures are maintained in the runners and cavities. Without equal pressures, the cavities will not fill at the same rate, and part defects may occur.

When a series of runners and gates is used, trapezoidal, half-round, or full-round runners are machined into the mold (Figure 21-11). With trapezoidal and half-round runners and gates, only one-half the mold or die plate is machined. Trapezoidal and half-round runners are easy to machine but generally require more molding pressure. Round runners are advised for transfer molding if extruder-type plasticizers are used. Pinpoint (submarine) gating, as shown in Figure 21-12, may be used, but this system also requires greater molding pressures.

A pinpoint gate is an opening of 0.76 mm (0.030 in.) or less through which the melt flows into a mold cavity. The submarine is a type of edge gate where the opening from the runner into the mold is located below the parting line or mold surface. The part is broken from the runner system or ejection from the mold.

The cost of elaborate mold designs and high waste from culls, sprues, and flash are two major limitations of transfer molding.

In some molds, there is only one cavity, whereas others have many. Regardless of the number of cavities, the *gate* is the point of entry into each mold cavity. In multicavity molds, there is a gate entering each mold cavity. The gates may be any shape or size (Figure 21-13). However, they are usually small so as to leave as small a blemish as possible. Gates must allow a smooth

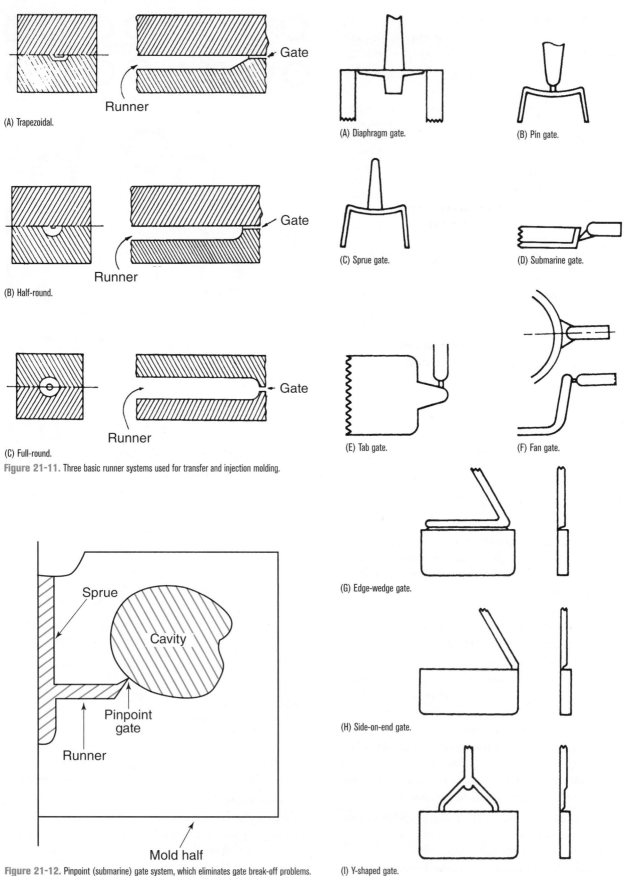

(A) Trapezoidal.

(B) Half-round.

(C) Full-round.

Figure 21-11. Three basic runner systems used for transfer and injection molding.

(A) Diaphragm gate.

(B) Pin gate.

(C) Sprue gate.

(D) Submarine gate.

(E) Tab gate.

(F) Fan gate.

(G) Edge-wedge gate.

(H) Side-on-end gate.

(I) Y-shaped gate.

Figure 21-12. Pinpoint (submarine) gate system, which eliminates gate break-off problems. It also reduces or eliminates finishing problems caused by large gate marks.

Figure 21-13. Some of the many possible gating systems.

flow of molten material into the cavity. A small gate will help the finish item cleanly break away from the sprue and runners (Figure 21-14).

The sprues, gates, and runners are usually cooled and removed from the mold with the parts from each cycle. The sprues, runners, and gates are removed from the parts and then reground for molding. This reprocessing is costly and restricts the mass of molded articles per cycle of the injection-molding machine. In **hot runner molding**, the sprues and runners are kept hot by heating elements built into the mold. As the mold opens, the cured part pulls free from the still molten hot runner system. On the next cycle, the hot material remaining in the sprue and runner is forced into the cavity (Figure 21-15).

A similar system called **insulated runner molding** is used in molding polyethylene or other materials with low thermal transfer (Figure 21-16). This system is also called **runnerless molding**. In this design, large runners are used. As the molten material is forced through these runners, it begins to solidify, forming a plastics lining that serves as insulation for the inner core of the molten material. The hot inner core continues to flow through the tunnel-like runner into the mold cavity.

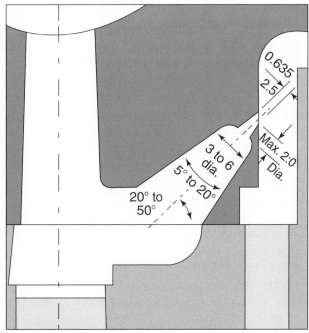

Figure 21-14. A typical tunnel gate, which may be considered a variation of the pinpoint gate design.

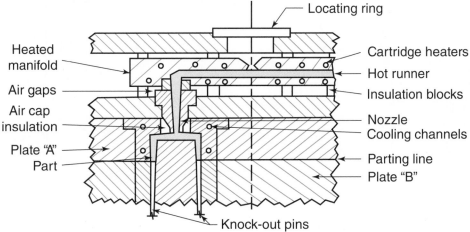

Figure 21-15. Schematic drawing of hot runner mold.

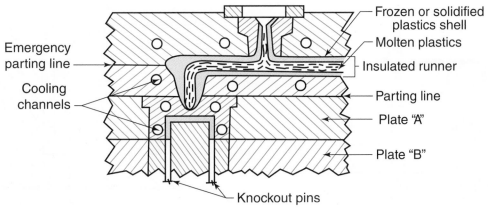

Figure 21-16. Principle of insulated runner molding.

A heated torpedo or probe may be inserted in each gate. This helps control freeze-up and drool.

There are many other mold designs that may include valve-gated molds; single or multiple molds (Figure 21-17); unscrewable molds (for internal or external threads); cam and pin molds (for undercuts and cores); and multicolor, or multimaterial, molds. In multicolor molds, a second color or material is injected around the first molding, leaving portions of the first molding exposed. Examples of multicolor products are special knobs, buttons, and letter or number keys.

Good draft, suitable gating, constant wall thickness, proper cooling, sufficient ejection, proper steels, and ample mold support are all important factors in mold design (see Chapter 22, Tooling and Mold Making).

Compression-mold design. There are three different types of compression-mold designs. Compression molds are usually produced from hardened steel that can withstand the great pressure and abrasive action of the hot plastics compound as it liquefies and flows into all parts of the mold cavity.

The flash mold is the least complex and most economical from the standpoint of original mold cost (Figure 21-18). In this mold, excess material is forced out of the molding cavity to form a flash that becomes waste and must be removed from the molded part.

In positive mold design, vertical or horizontal flash provision is made for excess material in the cavity (Figure 21-19). Vertical flash is easier to remove. The performs must be measured with care if all parts are to be the same density and thickness. If too much material is loaded in the cavity, the mold may not fully close. In positive molds, little or no flash occurs. This design is used for molding laminated, heavily filled, or high-bulk materials.

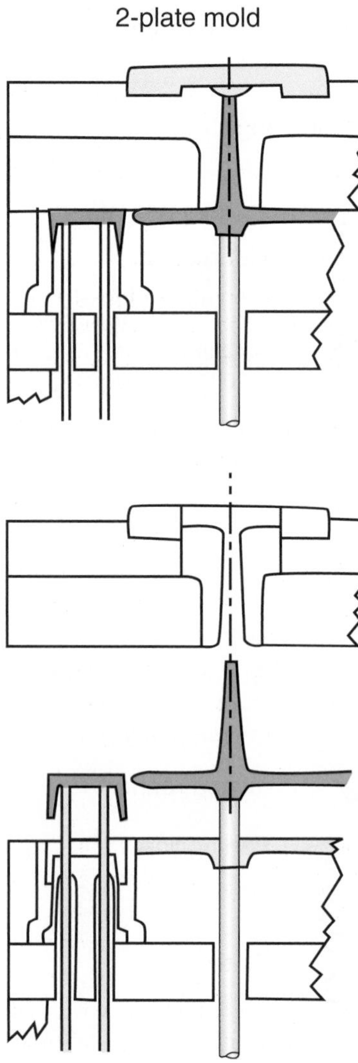

2-plate mold

Figure 21-17. Multiple mold design for injection molding.

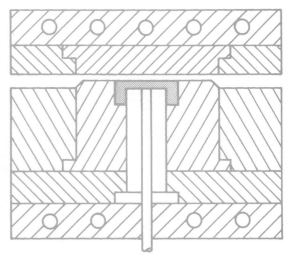

Figure 21-18. Flash mold design.

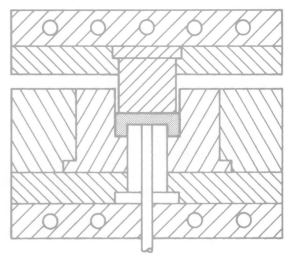

Figure 21-19. Positive mold design.

With fully positive molds, gases freed during the chemical curing of thermosets may be trapped in the mold cavity. The mold may be opened briefly to allow gases to escape. This operation is sometimes called breathing.

Semipositive molds have horizontal and vertical flash waste (Figure 21-20). This design is costly to produce and maintain but is the most practical when many parts or long runs are needed. The design allows for some inaccuracy of charge by allowing flash, thus giving a dense, uniform molded part. As the mold charge is compressed in the cavity, any excess material escapes. As the mold body continues to close, very little material is allowed to flash. When the mold fully closes, the telescoping male half is stopped by the *land*.

Blow-mold design. Construction of a blow mold is less costly. Aluminum, beryllium-copper, or steel is used as basic materials. Aluminum is one of the most inexpensive blow-mold materials. It is light and transfers heat rapidly. Beryllium-copper is harder and more wear resistant, but also more expensive. Steel is used at pinch-off points. If ferrous molds are used, they are plated to prevent rusting or pitting (see Chapter 22).

Parting lines. **Parting lines** are usually placed at the greatest radius of the molded part (Figure 21-21). If the parting line isn't at the plane of greatest dimension, the mold must have movable parts, or flexible molds or materials must be used. If the parting line cannot be concealed or placed on an inconspicuous edge, finishing is generally needed.

Ejector or knockout pins. Knockout or ejector pins push the hardened parts from the mold. They must touch the part in hidden or inconspicuous areas and should avoid contact with a flat surface unless decorative designs can help conceal the marks. Pins should be made as large as possible for longer tool life. Knockout pins, blades, or bushing shapes should not press against thin part areas.

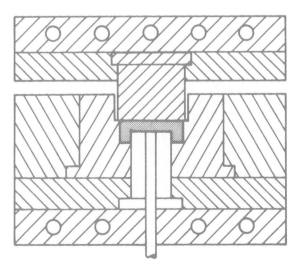

Figure 21-20. Semipositive mold design.

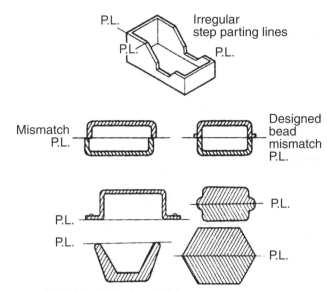

Figure 21-21. Various locations for parting lines.

The pins are usually attached to a master bar or pin plate. They are drawn flush with the surface of the mold by spring action. When the die is opened, a rod attached to the master bar or pin plate strikes a stationary stop, pushing all pins forward, forcing the part from the cavity.

All molds need venting. Ejector pins may be altered to provide venting.

Inserts. Inserts and holes must be designed and placed in the molding cavity or part with care. A liberal draft should be given long pins and plugs. A general rule is never to have a hole depth more than four times the diameter of the pin or plug. Long pins are often made to meet halfway through a hole, allowing twice the hole-depth limitation. Long pins are easily broken and bent by the pressure of flowing plastics.

The placement of molding gates in relation to holes is important. As the molten material is forced into the mold, it must flow around the pins that protrude into the mold cavity. When the mold opens, these pins are withdrawn. If parts are to be assembled on these holes, the material should be made thicker by producing a boss (Figure 21-22). The boss adds strength, which prevents the material from cracking. The pins often restrict the flow of material and may cause flow marks, weld lines, or possible cracking because of molding stress (Figure 21-23). Flow lines and patterns are shown in Figure 21-24.

When metal inserts are molded in place, the molten plastics material is forced around the insert. As the plastics cools, it shrinks around the metal insert substantially, contributing to the holding power of the insert. Inserts may be placed in thermoplastic parts by using ultrasonic techniques. Before molding, inserts may be positioned automatically or by hand on small locating pins in the mold cavity. To avoid cracking, enough material must be provided around all inserts.

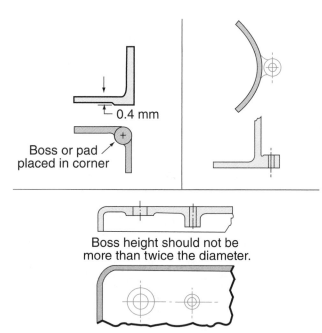

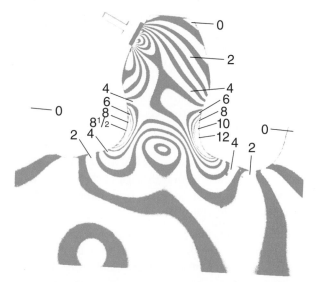

Figure 21-22. Importance of boss design.

Figure 21-23. Molding stress lines in the injection-molded part are visible under polarized light.

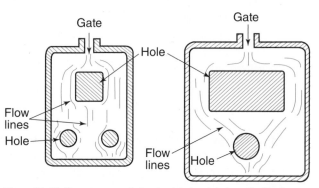

Figure 21-24. Flow patterns must be kept in mind when designing molds. Flow lines may appear around holes and ribs and opposite gates.

Pieces produced with pinpoint gating in the mold may not need any gate or sprue cutting. Many molds are designed so that gates are automatically sheared from the part as the press is opened.

Internal or external threads may be molded into plastics parts. Internal threaded parts may need an unscrewing mechanism to remove the mold. A clearance of 0.08 mm (0.03 in.) should be provided at the end of all threads (Figure 21-25A).

Metallic inserts molded into the part have greater strength. Generally, the ratio of wall thickness around the insert to the outer diameter of the insert should be slightly greater than one. Do not forget that materials have different expansion coefficients.

For blind core pin holes, a minimum 0.04 mm (0.015 in.) of clearance for screws, inserts, or other holding devices should be left (Figure 21-25B).

Matched-mold design. The design parameters are similar to compression molding. Large composite parts, such as sanitary tubs, bathroom shower stalls, and numerous automobile panels, are molded from SMC. Many SMC operations are precut, require no mold pinch-off, and produce no flash. Bosses,

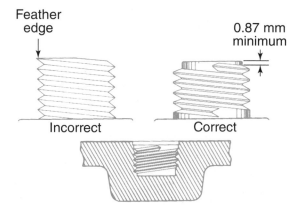

(A) Correct and incorrect threading.

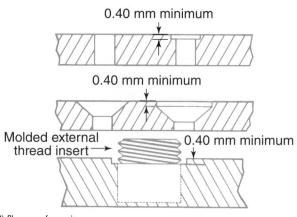

(B) Clearance of core pins.

Figure 21-25. Putting threads and core pins in plastics.

inserts, and ribs are utilized with SMC and TMC methods. If layers or pieces are used, make the overlapping bond as large as possible to prevent stress cracking.

Open-mold design. Layup, spray-up, autoclave, and bag techniques are similar. Careful attention must be given to the matrix formulation and orientation of reinforcements. This may have more influence on the properties of the finished composite than the design. Reinforcements should overlap by 2 inches, and all joints should be staggered. Bosses and ribs are used for added strength but must be liberally tapered. Simple, integrated part designs with gradual changes in thickness are desirable. Blowout holes (pneumatic) may be located in the bottom of the mold to aid in part removal.

Pultrusion design. Bosses, holes, raised numbers, or textured surfaces cannot be used with this continuous reinforcing process. Sharp corners or thickness transitions may result in resin-rich zones with broken fibers.

Filament winding design. In this open-mold, continuous reinforcing process, the fibers are oriented to match the direction and magnitude of stresses. Computer-controlled placement of filaments is designed to compensate for angle, contour reinforcing, bandwidth, equipment backlash, and other design considerations. Designs may call for permanent or removable mandrels (molds) (see Tooling in Chapter 22).

Laminar design. The principle design criterion is concerned with reinforcement orientation in each layer. A design close to optimum that resists all loads may be a laminate consisting of plies at 0°, ±45°, and 90° (Figure 21-26). To avoid distortion of the laminate, there must be a −45° ply for every +45° ply (see Figure 21-27). The lamina should be oriented in the principle direction of anticipated stresses. Fibers arranged in a random manner (isotropic) will have equal strength in all directions. The failure modes and deflection of composites are shown in Figure 21-28.

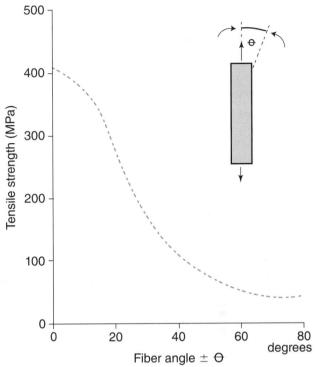

Figure 21-27. Tensile modulus of carbon/epoxy composites drops steeply as the angle between the fibers and the direction of tensile load is increased.

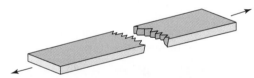

(A) Fiber tensile failure for all fibers in the direction of load (0°).

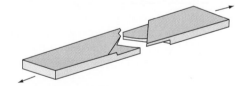

(B) Resin shear failure for fibers at ±45°.

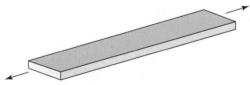

(C) Resin shear failure through the thickness between fibers at 0°. Usually caused by poor fiber-to-resin adhesion.

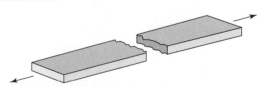

(D) Resin tensile failure between fibers at 90° to load.

Figure 21–28. Failure modes of composites in tension. Tensional strength of a carbon/epoxy structural composite is always related to fiber direction. A simple tensile test shows strikingly different failure behavior in composites having different fiber orientation. In multidirectional composite, single plies can fail with overall structural failure. Recognition of the various failure modes and knowing how composites fail are prerequisites to determining a fix.

(A) All plies at 0°. Axial load results in stretching-shearing behavior.

(B) Two plies at ± (any angle). Opposing shear de formations in the plus and minus plies result in stretching-torsion interaction.

(C) A 0°/90° stacking. This arrangement bends under pure tension because the modulus-weighted centroid is not coincident with the geometric centroid, resulting in an offset load path.

(D) Another 0°/90° stacking. Because of different thermal expansion characteristics in each layer, this stacking deforms into a "saddle" when heated.

Figure 21-26. Symmetry effects on deflection of composites.

For sandwich laminates, the high-density facings must resist most of the applied and bending forces. The lightweight core must resist transverse tension, compression, shear, and buckling.

Table 21-5 shows the advantages and limitations of various processes.

Performance Testing

The true test of any product is performance under actual service conditions. Tests can be used as indicators for product design, redesign, and reliable product service. The term **testing** implies that methods or procedures are employed to determine whether parts meet the required or specified properties. A pressure tank, rocket engine case, or pole vaulter's pole may be subjected to a critical pass/fail (proof) test. **Quality control** procedures must be used to determine whether a product is being manufactured to specifications. It is a technique primarily used by management to achieve quality. **Inspection** ensures that manufacturing personnel check technique procedures and gauge readings as well as detect flaws in processing materials. Inspection is part of quality control. See Appendix G for additional sources of information.

Table 21-5. Plastics Processing Method—Processes, Advantages, Limitations

Processing Method	Process	Advantages	Limitations
Injection molding	Similar to die casting of metals. A thermoplastic molding compound is heated to plasticity in a cylinder at a controlled temperature and then forced under pressure through sprues, runners, and gates into a cool mold; the resin solidifies rapidly, the mold is opened, and the parts ejected; with certain modifications, thermosetting materials can be used for small parts.	Extremely rapid production rate and hence low cost per part; little finishing required; excellent surface finish; good dimensional accuracy; ability to produce variety of relatively complex and intricate shapes	High tool and die costs; high scrap loss; limited to relatively small parts, not practical for small runs
Extrusion	Thermoplastic molding powder is fed through a hopper to a chamber where it is heated to plasticity and then driven, usually by a rotating screw, through a die having the desired cross section; extruded lengths are either used as is or cut into sections; with modifications, thermosetting materials can be used.	Very low tool cost, material can be placed where needed, great variety of complex shapes possible; rapid production rate	Close tolerances difficult to achieve, openings must be in direction of extrusion; limited to shapes of uniform cross section (along length)
Thermoforming	VACUUM FORMING—Heat-softened sheet is placed over a male or female mold; air is evacuated from between sheet and mold, causing sheet to conform to contour of mold. There are many modifications, including vacuum snapback forming, plug-assist, drape forming, etc.	Simple procedure, inexpensive, good dimensional accuracy; ability to produce large parts with thin sections	Limited to parts of low profile
	BLOW OR PRESSURE FORMING—The reverse of vacuum forming in that positive air pressure rather than vacuum is applied to form sheet to mold contour.	Ability to produce deep drawn parts, ability to use sheet too thick for vacuum forming, good dimensional accuracy; rapid production rate	Relatively expensive; molds must be highly polished
	MECHANICAL FORMING—Sheet metal equipment (presses, benders, rollers, creasers, etc.) forms heated sheet by mechanical means. Localized heating is used to bend angles; where several bends are required, heating elements are arranged in series.	Ability to form heavy and/or tough materials: simple, inexpensive; rapid production rate	Limited to relatively simple shapes

Blow moldings	An extruded tube (parison) of heated plastics within the two halves of a female mold is expanded against the sides of the mold by air pressure; the most common method uses injection-molding equipment with a special mold.	Low tool and die cost; rapid production rate; ability to produce relatively complex hollow shapes in one piece	Limited to hollow or tubular parts; wall thickness difficult to control
Slush, rotational, dip castings	Powder (polyethylene) or liquid material (usually vinyl plastisol or organosol) is poured into a closed mold, the mold is heated to fuse a specified thickness of material adjacent to mold surface, excess material is poured out, and the semi-fused part placed in an oven for final curing. A variation, rotational molding, provides completely enclosed hollow parts.	Low-cost molds, relatively high degree of complexity, little shrinkage	Relatively slow production rate, choice of materials limited
Compression moldings	A partially polymerized thermosetting resin, usually preformed, is placed in a heated mold cavity; mold is closed, heat and pressure applied, and the material flows and fills mold cavity; heat completes polymerization and mold is opened to remove hardened part. Method is sometimes used for thermoplastics, e.g., vinyl phonograph records; in this operation, the mold is cooled before it is opened.	Little waste of material and reduced finishing costs due to absence of sprues, runners, gates, etc.; large, bulk parts possible	Extremely intricate parts involving undercuts, side draws, small holes, delicate inserts, etc., not practical; extremely close tolerances difficult to achieve
Transfer moldings	Used primarily for thermosetting materials, this method differs from compression molding in that the plastic is 1) first heated to plasticity in a transfer chamber, and 2) fed, by means of a plunger, through sprues, runners and gates into a closed mold.	Thin sections and delicate inserts easily used; flow of material is more easily controlled than in compression molding, good dimensional accuracy; rapid production rate	Molds more elaborate than compression molds, and hence more expensive; loss of material in cull and sprue; size of parts somewhat limited
Open-mold processing	CONTACT—The lay-up, which consists of a mixture of reinforcement (usually glass cloth or fibers) and resin (usually thermosetting), is placed in mold by hand and allowed to harden without heat or pressure.	Low cost; no limitations on size or shape of part	Parts sometimes erratic in performance and appearance; limited to polyesters, epoxies and some phenolics
	AUTOCLAVE—The vacuum-bag setup is simply placed in an autoclave with hot air at pressures up to 1.38 MPa.	Better quality moldings	Slow rate of production
	FILAMENT WOUND—Glass filaments, usually in the form of rovings, are saturated with resin and machine-wound onto mandrels having the shape of desired finished part; finished part is cured at either room temperature or in an oven, depending on resin used and size of part.	Provides precisely oriented reinforcing filaments; excellent strength-to-mass ratio; good uniformity	Limited to shapes of positive curvature; drilling or cutting reduces strength
	SPRAY MOLDING—Resin systems and chopped fibers are sprayed simultaneously from two guns against a mold: After spraying, layer is rolled flat with a hand roller. Either room temperature or oven cure.	Low cost; relatively high production rate; high degree of complexity possible	Requires skilled workers; lack of reproducibility
Castings	Plastics material (usually thermosetting except for the acrylics) is heated to a fluid mass, poured into mold (without pressure), cured, and removed from mold.	Low mold cost, ability to produce large parts with thick sections; little finishing required; good surface finish	Limited to relatively simple shapes

(Continued)

Table 21-5. Plastics Processing Method—Processes, Advantages, Limitations (*Continued*)

Processing Method	Process	Advantages	Limitations
Cold moldings	Method is similar to compression molding in that material is charged into a split, or open, mold; it differs in that it uses no heat—only pressure. After the part is removed from mold, it is placed in an oven to cure to final state.	Because of special materials used, parts have excellent electrical insulating properties and resistance to moisture and heat; low cost; rapid production rate	Poor surface finish; poor dimensional accuracy; molds wear rapidly; relatively expensive finishing; materials must be mixed and used immediately
Bag molding	VACUUM BAG—Similar to contact except a flexible polyvinyl alcohol film is placed over layup and a vacuum drawn between film and mold (about 82 kPa).	Greater densification allows higher glass contents, resulting in higher strengths	Limited to polyesters, epoxies, and some phenolics
	PRESSURE BAG—A variation of vacuum bag in which a rubber blanket (or bag) is placed against film and inflated to apply about 350 kPa.	Allows greater glass contents	Limited to polyesters, epoxies, and some phenolics
Matched-die molding	MATCHED DIE—A variation of conventional compression molding, this process uses two metal molds that have a close-fitting, telescoping area to seal in the resin and trim the reinforcement; the reinforcement, usually mat or preform, is positioned in the mold, a premeasured quantity of resin is poured in, and the mold is closed and heated; pressures generally vary between 1.04 and 2.75 MPa.	Rapid production rates; good quality and excellent reproducibility; excellent surface finish on both sides; elimination of trimming operations; high strength due to very high glass content	High mold and equipment costs; complexity of part is restricted; size of part limited

Source: Adapted from Materials Selector, *Materials Engineering*, Penton/IPC a subsidiary of Pittway Corp.

RELATED INTERNET SITES

- **www.3dsystems.com.** 3D Systems Corporation offers three main types of rapid prototyping systems. Personal printers are low-cost machines intended for hobby activities and education. The professional printers come in two main lines, the Z-Printers, which use a powder and binder to create the parts, and the ProJet Printers, which use a UV cured acrylic plastic and wax supports. The production printers are either SLA®, stereolithographic, or SLS®, based on laser sintering.

- **www.howstuffworks.com/stereolith.html.** This site provides an introduction to stereolithography (SLA). It contains several close-up photographs of an SLA system and some models. This site contains information about SLA only and does not discuss laser sintering or paper lamination.

- **www.autodesk.com/moldflow.** Autodesk® Simulation Moldflow® is a program that simulates injection molding. The utility of the program is to evaluate temperatures, pressures, shear rates, and shrinkage of designs for plastics parts. Autodesk Simulation Moldflow Insight is the portion of the software that evaluates the flow characteristics of a selected material in a virtual mold. Autodesk Simulation Moldflow Advisor helps solve problems that have been identified by the flow simulations.

- **www.paralleldesign.com.** Select "Design for Moldability" from the home page. This site provides over 30 animations for design problems and solutions for injection molds.

- **www.plastics.dupont.com.** The design and processing guides are some of the most useful publications of DuPont Plastics. However, it is somewhat difficult to locate these guides. On the home page, select "Industries Serviced." Then find "Plastics," and finally type into the search box the plastic of interest. For example, enter "Delrin Design Guide." This will bring up the detailed and instructive guide.

VOCABULARY

The following vocabulary words are found in this chapter. Use the glossary in Appendix A to look up the definitions of any of these words you do not understand as they apply to plastics.

apparent density
ASTM (American Society for Testing and Materials)
boss

bulk factor
computer-aided design (CAD)
computer-aided engineering (CAE)
coarse tolerance
fine tolerance
flash mold
gate
hot runner molding
insulated runner molding
inspection
parting lines
quality control
rapid prototyping

ribs
runner
runnerless molding
safety factor
semipositive molds
specifications
sprue
standard tolerance
standards
testing
tolerances
undercut

QUESTIONS

21-1. In _____, the sprues and runners are kept hot by means of heating elements built into the mold.

21-2. The _____ is the point of entry into the mold cavity.

21-3. Narrow channels that convey the molten plastics from the sprue to each cavity are called _____.

21-4. Parting lines are normally placed at the _____ of the molded part.

21-5. The _____ is the opening in the mold where the product is formed.

21-6. The economic minimum number of pieces produced by hand layup is _____.

21-7. Closely related to shrinkage is dimensional _____.

21-8. The tapered channel connecting the nozzle and runners is called _____.

21-9. In _____ compression mold designs, no provision is made for placing excess material in the cavity.

21-10. Molded parts are pushed from the mold by _____ or _____.

21-11. With _____ gating, pieces may not require any gate or sprue cutting. The dies are designed so that gates are sheared off automatically.

21-12. Name the three overriding requirements in plastics designing.

21-13. If a mold costs $5000 and 10,000 parts are to be made, the mold cost would be _____ per part.

21-14. In nearly every design, a compromise must be made between highest performance, attractive appearance, efficient production, and _____.

21-15. List four favorable properties possessed by most plastics.

21-16. List four unfavorable properties possessed by most plastics.

21-17. In addition to electrical, chemical, mechanical, and economic considerations, name four additional requirements to consider before a product is made.

21-18. The most meaningful comparison in estimating the cost of a plastics product is cost per _____.

21-19. Plastics are lighter per _____ than most materials.

21-20. Problems encountered in producing plastics products often require the selection of the _____ before material or _____ considerations.

21-21. There were many problems associated with early applications of plastics because designers forgot that the finished product must _____ as designed and desired.

21-22. Undercuts in parts usually increase _____ costs.

21-23. With relatively few exceptions, the _____ of plastics developed in the past have been by trial and error.

21-24. Plastics have replaced metals for many applications because of energy and _____ savings.

21-25. If wall thickness is not _____, the molded part may distort, warp, or have internal stresses.

21-26. The mark on a molded piece resulting from the meeting of two or more flow fronts during the molding operation is called _____.

21-27. Wavy surface appearances caused by improper flow of the hot plastics into the mold cavity are known as _____.

21-28. To help prevent flow marks, _____ or change gate location.

21-29. Name three methods of producing an internal thread in a plastics part.

TOOLING AND MOLD MAKING

CHAPTER 22

INTRODUCTION

Toolmaking processes, equipment, and methods for shaping plastics will be discussed in this chapter. Not all of the information will apply to every molding process because some processes are very specialized.

In March 2012, a report from the Congressional Research Service focused on the US tool, die, and industrial mold industry. The report indicated that between 1998 and 2010, the industry lost 45% of its employees and the number of companies dropped by 36%. Although several factors have contributed to this decline, increasing foreign competition, particularly from China, has played a large role. The US industry had sales of slightly under $12 billion in 2010. At the same time, the value of imported tooling was $5.6 billion. In contrast, the value of the tooling exported by the United States was only $1.2 billion. The American Mold Builders Association claims that the Chinese mold makers offer cost savings of 30% to 40% or more. In response, many US-based mold makers have worked to reduce labor costs and institute new computer-based technologies.

Computer-assisted mold making (CAMM) is a result of microprocessor technology and may improve productivity by more than 100%. Computer-assisted design and manufacturing (CAD/CAM) systems are used to design and aid in machining molds. This equipment automatically adjusts cavity dimensions for different resins or special contours. Designing and machining information for making multicavity molds or for the replacement of cores or cavities is stored in the system's memory.

Figure 22-1 shows an extremely complex, multicavity injection mold created by highly skilled machinists using elaborate software programs.

CAMM allows the plastics industry to produce high-quality configurations, close tolerances, and reliable designs at a competitive price. The mold-making and tooling industry is highly labor intensive. Here, computer systems are sophisticated tools that increase productivity. The drafting and design departments can utilize CAD systems to automatically dimension the drawing and allow for material shrinkage. The data base in CAMM programs can select the mold base and recommend the placement of ejector pins, sleeves, locating rings, return pins, pullers, or other details. Although drafting time and changing tooling parameters represent a significant percentage of the total cost of the part, CAMM systems greatly reduce this time and facilitate modifications or changes in tooling parameters.

Figure 22-1. This 144-cavity mold makes PET preforms for soft drink bottles.

Most of the mold-making industry is composed of custom shops that specialize in mold making or offer special services such as plating, polishing, heat treatment, engraving, or machining molds.

Molding information is given in the discussion of each process. However, Chapter 21 should be reviewed for basic design considerations. You should also review the descriptions of the plastics families because they contain information about properties and designs that affect moldability.

The outline for this chapter is presented here:

I. **Planning**

II. **Tooling**
    A. Tooling costs

III. **Machine processing**
    A. Chemical erosion

PLANNING

Most mold designs begin with sketches that allow the mold maker to make decisions about layout and visualize how parts will be made. Final CAD drawings contain notes, dimensions, and tolerances. Some critical dimensions may need surface finishes as low as 0.0025 mm (9.8×10^{-7} in.) for some parts. The design also shows any special requirements. Shrinkage tolerances, finish, engraving, plating, special materials, or other dimensioning factors are noted.

CAM systems allow the user to determine cooling systems, finish, tool path, part geometry, feeds, and inherent equipment limitations (backlash, tool wear) before machining. After the program is verified, it is stored for later use.

CAM systems can utilize the stored information to cut and form the molds. Nearly 80% of the mold maker's time is devoted to setting up the machine tool. Only 20% is actually spent cutting the material. CAD/CAM systems substantially reduce setup, lead, and machining time.

Nearly every phase of the product development process, from concept to completion, can utilize CAMM to save time and reduce costs.

TOOLING

Collectively, jigs, fixtures, molds, dies, gauges, clamping devices, and inspection equipment will be referred to as **tooling**. The terms **jig** and **fixture** are often used interchangeably. Both are devices used to locate and hold a workpiece in the correct position during machining, inspection, or assembly. A *jig* guides the tool during a manufacturing operation such as boring. A *fixture* does not have built-in tool guides. It is primarily used to hold the work securely during machining, cooling, and drying.

Part of the manufacturing cost of plastics is the cost of special fixtures. These are tools used to help measure or load a plastics charge in a molding machine, remove flash, remove molded parts, or hold parts for cooling. Some are used to aid in machining. These include holding blocks, drill jigs, and punch dies.

Tooling Costs

Many factors affect tooling costs: the size of the production run, production technique, reinforcements, additives, fiber orientation, matrix, complexity of design, the tolerance needed, the amount of mold maintenance, and machining.

Multiple-cavity molds, or those with inserts or special surface finishes, add to tooling costs. Designing a mold set to have interchangeable cavities may lower tooling costs because new cavities may be inserted to form other parts, thus extending the use of the original mold. For a small number of parts, a multiple-cavity mold may not be economical because tooling tolerances are much harder to maintain. In addition, mold maintenance and machining are more costly for multiple-cavity molds.

Two remarks often quoted in the mold-making industry are "There is no such thing as a simple plastics part" and "The part is only as good as the mold that makes it." Each of these remarks shows the importance of tooling and mold design (see Figure 22-2).

There are four broad types of tooling: (1) prototype, (2) temporary, (3) short run, and (4) production. These are shown in Table 22-1.

Types of materials used in tooling include gypsum plasters, plastics, wood, and metals.

Figure 22-2. The importance of the tool and die personnel cannot be understated. This tool and die maker puts finishing touches on a steel injection mold.

Bethlehem Steel

Table 22-1. General Types of Tooling

Tool Classification	Number of Parts	Tooling Materials
Prototype	1–10	Plaster, wood, reinforced plasters
Temporary	10–100	Faced plasters, reinforced plasters; backed metal-deposition, cast & machined soft metals
Short-run production	100–1000 >1000	Soft metals, steel Steel, soft metals for some processes

(A) The honeycomb aluminum core is stiff enough to prevent deformation during processing in an autoclave or oven. Machining provides a rough shape to the core.

Courtesy of Vantico A&T US, Inc.

Plastics. Polymer tooling (plastics and elastomers) are used to make master patterns, transfer tools, cores, boxes, templates, draw dies, jigs, fixtures, inspection tools, and prototypes. They are currently replacing wooden and plaster tooling. Laminated, reinforced, and filled plastics are mainly used for making dies, jigs, and foundry patterns.

Polymer molds are generally divided into two groups: those that are backed and those that are not. Foams and honeycomb sheets are often used to provide strong, lightweight support for tools.

Figure 22-3 shows steps in the development of a tool for a composite part that utilizes both a honeycomb aluminum core and a machined epoxy surface. This tool can withstand autoclave temperatures of 177°C (350°F) and pressures of 0.62 MPa (90 psi).

The use of plastics in die fabrication is growing rapidly. Metal-filled and glass-reinforced phenolics, ureas, melamines, polyesters, epoxies, silicones, and polyurethanes are strong, light, and easy to machine. Alumina and steel are common fillers that offer improved thermal conductivity, machinability, strength, and extended service temperature and life. Many may be foam filled. These materials are used in tooling in both the plastics and metals industries. Plastics tools have been used as bending dies, stretch-forming dies, and drop-hammer dies. Acetals, polycarbonates, high-density polyethylene, fluoroplastics, and polyamides are also used as tooling. These materials are used for matched stamping dies, jigs, and fixtures. The acceptability of plastics tools is verified by their wide use in the aerospace, aircraft, and automotive industries.

Large, epoxy-reinforced molds are popular for both layup and spray-up techniques. These toolings are backed and supported by metal frames and bases. In Figure 22-4, polymer tooling is used to form SMC. Polymer tooling must be able to withstand the prolonged exposure of curing temperatures.

Plastics tooling has several advantages over metal or wood tools. Plastics tools may be cast in inexpensive molds,

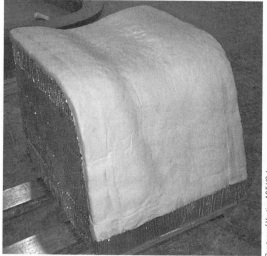

(B) A thick layer of specialized epoxy builds up the surface of the tool face.

Courtesy of Vantico A&T US, Inc.

(C) After the epoxy has cured, machining creates the final contour of the tool.

Figure 22-3. An example of low-cost tooling for composites.

Courtesy of Vantico A&T US, Inc.

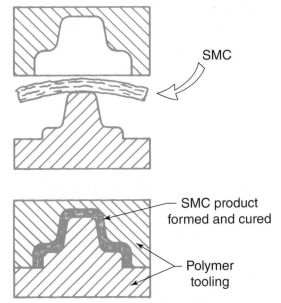

Figure 22-4. Matched-die molding of SMC using polymer tooling.

duplicated easily, and changed frequently in design. Plastics tooling is also light in mass and corrosive resistant. Hot-melt compounds are replacing wood and steel in dies, hammers, mockups, prototypes, and other fixtures used by industry.

Complex furniture parts are sometimes made from flexible polymer molds. Silicones are most popular, but polysulfide and polyurethane elastomers are also used.

Unbacked, flexible molds are used in the furniture industry to faithfully reproduce wood-grain designs. The basic concept of flexible-plunger or elastomer press molding is shown in Figure 22-5.

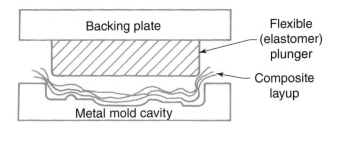
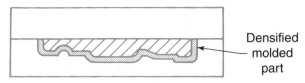

Figure 22-5. Flexible-plunger or elastomeric press molding.

Gypsum plasters

The United States Gypsum Company has developed a number of high-strength plaster materials. These materials have enough strength to produce prototype models, die models, transfer (take-off) tools, patterns, and die molds for forming plastics (Figure 22-4).

Fibers, expanded metal, or other materials are often used to further strengthen the plaster tooling. Metal bases and frames provide secure mountings. In some techniques, templates (loft or templates) help shape the wet plaster. Even a pile of rocks or an expanded bladder (balloon) can be used to help form the general contour. Plaster is easily shaped by hand. Models may be produced by placing plaster over clay, wax, wooden, or wire frame shapes. Rocks, wax, and other model forms are generally removed and replaced by reinforced plastics support. Some plaster molds are designed to be used only once. Hollow designs or some breakaway designs require that the plaster mold be broken and washed away. Typical plaster molds are faced with metal deposition, polymer coating, or composite facing to provide a durable surface for part removal. Metal coatings also improve thermal conductivity necessary in some molding techniques. Cooling coils may be cast during the tooling.

The trade names Ultracal®, Hydrocal®, and Hydrostone® are found on plasters used for tooling. Vacuum-formed pattern molds are often made of low-cost plasters. Hydrostone has an average compressive strength of nearly 11,000 psi (76 MPa).

Plaster is an important master pattern material from which metal or polymer skin master molds are produced.

Wood. Wooden tooling is used for prototypes and occasional short-run work. It is also used for some thermoforming dies and for pattern work.

Metals. Although plastics may become the dominant material in the future, we cannot have plastics products without metals.

For prototype or short-run work, low-melting-point metals may be used. Zinc, lead, tin, bismuth, cadmium, and aluminum are used in thermoforming dies, casting patterns, and duplicating models. Aluminum is a popular metal for many molding processes because it is light, easy to machine, and a good thermal conductor.

Aluminum 7075 is a popular, high-grade, heat-treated alloy. It is sometimes annodized and plated to improve surface hardness and prolong tooling life. Aluminum (7075-T652) is used extensively in blow molding, bag molding, thermoforming, and prototypes. It is soft and lacks sufficient strength and hardness to prevent material wash in injection molds or pinch-off areas in blow molds.

Beryllium-copper (C17200) is used in some injection molds and blow-molding processes. This material can be cast, wrought, and drawn. It can also be pressure cast using a master hob. In other words, many cavities can be produced with one hob. The BeCu is generally supplied to the mold maker tempered to 38–42 Rockwell C. Beryllium-copper will reproduce fine detail. Wood-grain patterns are examples of the possible detail made using BeCu molds.

Kirksite (Zamak®, zinc alloy) molds, made of an alloy of aluminum and zinc, are used because they are inexpensive. This metal will reproduce detail better than aluminum and also last longer. Because of the low pouring temperature of 425°C (800°F), it is possible to cast cooling lines in the mold. These alloys are used for short- and long-run blow-molding operations. Pinch-off edges must be protected against excessive stress concentrations by steel inserts.

Steel is vital to the plastics industry. Carbon, the basic element in plastics, is also an important ingredient of steel, but steel must also have additional alloying elements for mold making. If necessary, mold hardness can vary from 35 to 65 Rockwell C (Figure 22-6).

A popular, non-alloyed carbon steel used for machine bases, frames, and structural components is AISI 1020, 1025, 1030, 1040, and 1045.

Various steel alloys may be used for long production runs. In uses where high compressive strength and wear resistance

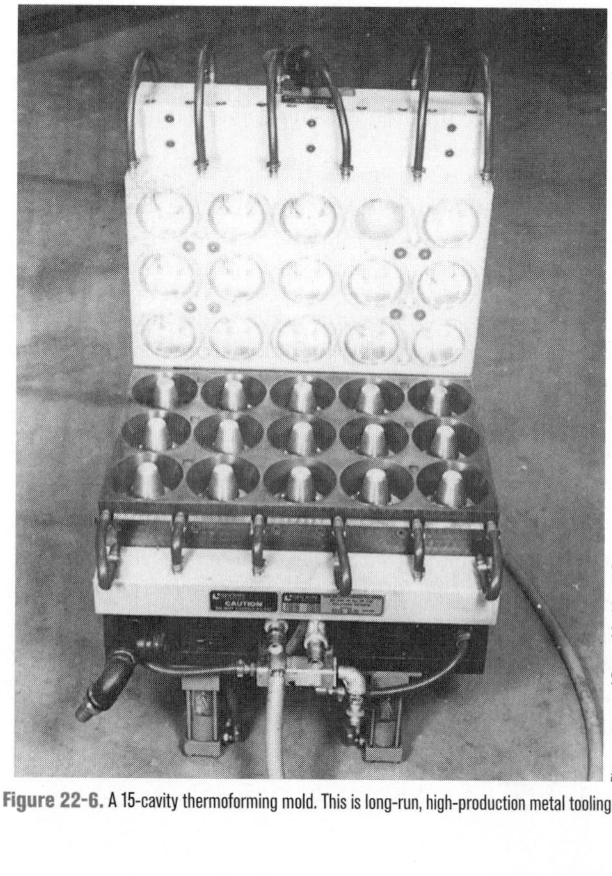

Photo courtesy of Brown Machine LLC

Figure 22-6. A 15-cavity thermoforming mold. This is long-run, high-production metal tooling.

at elevated temperatures are required, an AISI Type H21 steel may be used. The symbol H indicates hot-work tool steels.

> Type H21 steel analysis:
> Carbon, 0.35%
> Manganese, 0.25%
> Silicon, 0.50%
> Chromium, 3.25%
> Tungsten, 9.00%
> Vanadium, 0.40%

AISI Type WI is a high-quality, straight-carbon, water-hardened tool steel used in various tooling applications.

> Type WI steel analysis:
> Carbon, 1.05%
> Manganese, 0.20%
> Silicon, 0.20%
> Alloys, none

Toolmakers have long wished to have a nondeforming die steel. One type of steel that combines the deep-hardening characteristics of air-hardened steels with the simplicity of low-temperature heat treatments possible in many oil-hardened steels is AISI Type A6. The symbol A is used for air-hardened steels. Type A6 steel analysis:

> Carbon, 0.70%
> Manganese, 2.25%
> Silicon, 0.30%
> Chromium, 1.00%
> Molybdenum, 1.35%
> Plus alloy sulfides

Popular alloyed steels, including tool steels, are AISI type A2 and A6, used for injection, transfer, compression, and master hob molds. Fully hardened steels, such as D2 and D3, are also popular. The high-carbon and chromium D3 steels have good wear resistance. Mold cavities, backplates, knives, and die are made from the chromium-molybdenum steel 4140.

The symbol P represents a precipitation-hardened steel. AISI Type P20 is appropriate for all types of injection mold cavities. It may be hardened to a core hardness of 28 Rockwell C.

> Type P20 steel analysis:
> Carbon, 0.30%
> Molybdenum, 0.25%
> Chromium, 0.75%

Mold cavities, holding blocks, dies, and other tooling are made of P20 and P21 steels.

Stainless steel Type 420 and 440C may be used where corrosion resistance is needed or adverse atmospheric conditions exist. It can develop a hardness of 45 to 50 Rockwell C.

Oil-hardened steels are sometimes selected for slides and bushings. Type 02 is most common.

Tungsten carbide is widely used for cutting taps because it has good wear and abrasion resistance. Mold cavities may be made from this ceramic type of material, but it is brittle and harder to shape than tool steels.

Miscellaneous tooling. There are numerous miscellaneous and innovative tooling techniques. Wax and soluble salts are sometimes used. Mandrels, bladders, or other mold shapes have been made of air-inflated elastomers. Even glass shapes are used as molds. After casting, forming, or filament winding, the glass is sometimes broken and removed. In some applications, concrete and ice have been used to produce unique tooling. The advantages and disadvantages of selected tooling materials are shown in Table 22-2.

MACHINE PROCESSING

There are a number of processes used in making tools and dies from steels. Milling, turning, drilling, boring, grinding, hobbing, casting, planing, etching, electroforming, electrical-discharge machining, plating, welding, and heat-treating are only a few.

Toolmaking should be viewed as a tightly controlled process rather than a sequence of discrete tasks. Control of the toolmaking process may be done with the aid of a computer. Computer-integrated manufacturing (CIM) systems may control individual computer numerically controlled (CNC) machine tools. The manufacturing of tooling begins with planning and balancing the workload, specifications, and techniques with machine capabilities. Mistakes in the planning process can be costly and result in excessive production time and even tool failure. Machining of molds and tooling is sometimes separated into two broad areas: initial machining, which involves rough turning and milling, and final machining, which involves grinding, electrical-discharge machining (EDM), and polishing. These processes usually take place after hardening.

Milling, turning, drilling, boring, and grinding are cutting-tool processes. Shapers, planers, lathes, drilling machines, grinding machines, milling machines, and various pantograph duplicating machines are often used to cut metal when making molds and dies.

A duplicating machine is a specially modified vertical milling machine that cuts duplicate molds from a master pattern. The ratio between the pattern and the cavity is usually 1:1.

Table 22-2. Advantages and Disadvantages of Selected Tooling Materials

Tool Material	Advantages	Disadvantages
Aluminum	Low cost; good heat transfer, easily machined; corrosion resistant; lightweight; doesn't rust	Porosity; softness; galling; thermal expansion; easily damaged; limited runs
Copper alloy (brass, bronzes, beryllium)	Easily machined; good surface detail; high thermal conductivity; doesn't rust	Softness; copper may inhibit cure; attacked by some acids; easily damaged; limited runs
Miscellaneous (salts, inflatables, wax, ceramics)	Some are low cost, reusable; designed with undercuts; easily fabricated; light; hard; thermal conductors; ceramics are high-temperature materials	Some are soft; easily damaged; dimensionally unstable; damaged by high temperature or chemicals; poor thermal conductors
Plaster	Low cost; easily shaped; good dimensional stability; doesn't rust	Porosity; softness; poor thermal conductivity; easily damaged; limited runs; limited thermal and strength range
Polymers (laminated, reinforced, filled)	Low cost; easily fabricated; thermal expansion similar to many composites; lightweight; doesn't rust; large designs economical; fewer parts	Limited design; poor thermal conductivity; limited dimensional stability; limited runs; limited thermal range
Steel	Most durable, high thermal resistance; strong and wear resistant; thermal conductivity	Most expensive tooling; machining more difficult; size limitations; many parts; rusts; heavy
Wood	Low cost; easily machined; lightweight; doesn't rust	Porosity; poor dimensional stability; soft; limited runs; poor thermal conductivity and resistance
Zinc alloy (lead, tin)	Low cost; easily machined; good thermal conductivity; good detail; doesn't rust	Soft; easily damaged; limited runs; limited thermal and strength range

Pantograph machines are similar to duplicating machines except they operate on variable ratios of up to 20:1. The large ratio reduction allows the steel to be machined with very delicate detail by coordinated movements of the table and cutting tool.

Hobbing, etching, electroforming, and electrical-discharge machining are *metal-displacement* processes. They are used in making molds in which no cutting tools are involved.

Cold hobbing involves pushing a piece of very hard steel into a blank of unhardened steel (Figure 22-7). The process is performed at room temperature. Pressures vary from 1380 MPa to 2760 MPa (200 to 400 $\times$ 10^3 psi), depending on the hobbing metals and blanking material used. Hobbing machines may need press capacity as high as 2722 tonnes (3000 tons).

In a proprietary procedure called Cavaform, a master hob can supply an unlimited number of impressions distinguished by a remarkable fidelity to the size and finish of the hob. This cold-form swaging procedure is accomplished in most steel in an annealed state. A cavity steel can be chosen to satisfy molding requirements with low heat treat distortion. It may also be vacuum-heat treated to minimize polishing after heat treatment. A major disadvantage is that this process generally requires a through hole at the bottom of the impression.

Hobs are often made from oil-hardened tool steels containing a high percentage of chromium. It may be economical to hob single die cavities, but hobbing is usually used for making large numbers of impressions for multicavity molds. Multicavity molds frequently are numbered to permit instant location of any molding problems.

A slight draft must be provided to allow removal of the hob from the forming blank. The hob must be clean because as much as a pencil mark on the hob may be transferred to the cavity during the hobbing operation.

After hobbing, the blank is machined and hardened before being placed in the mold base. In Figure 22-8, a finished hob (left) has formed the cavity (right) in the blank of steel.

Figure 22-8. Example of hobbing.

Compared with mechanical methods, electrical erosion or **electrical-discharge machining (EDM)** is a fairly slow method of removing metal. Steel is removed at about 4.37 $\times$ 10^{-4} mm³/s (0.016 in.³/min). The workplace may be hardened before the cavity is formed, which removes any problems from heat treatment after machining or forming. During the machining process, a master pattern of copper, zinc, or graphite is made. The pattern is then placed about 0.025 mm (0.00098 in.) from the workpiece, and both the workpiece and the master are submerged in a poor dielectric fluid such as kerosene or light oil. Current is forced across the gap between the master and the workpiece, and each discharge removes minute amounts of substance from both. The loss of material from the tool master must be compensated for in order to obtain accurate cavities in the workpiece.

Most modern EDM machines now incorporate a multi-axis orbital movement in the spindle head that enables one electrode to be used for roughing, final sizing, and finishing.

For inexpensive tool masters made of carbon or zinc, the ratio of material removed from the workpiece to that removed from the tool may be more than 20:1. Accuracy may be within 60.025 mm (60.001 in.) with a finish cut of less than 0.007 mm (30 microinches).

Both wirecut and diesinking EDM techniques often eliminate the need for secondary finishing operations. Wire EDM will produce accurate, intricate cavities that are difficult or impossible to produce with conventional techniques. In Figure 22-9, a CNC-controlled wire EDM cuts the final shape of a die. EDM principles, equipment, and electrodes appear in Figure 22-10.

Chemical Erosion

In **chemical erosion** (etching), an acid or alkaline solution is used to create a depression or cavity. The process usually involves the use of chemically resistant maskants such as wax,

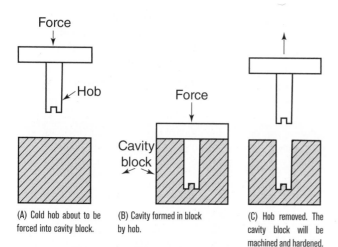

(A) Cold hob about to be forced into cavity block.

(B) Cavity formed in block by hob.

(C) Hob removed. The cavity block will be machined and hardened.

Figure 22-7. Diagram of metal displacement (hobbing) process.

Figure 22-9. This Agie Charmilles wire EMD is in a finishing cut. The workpiece is fully submerged with deionized water to ensure maximum temperature stability, best accuracy, and surface finish.

plastics-based paints, or films. The maskant is removed from the areas where metal is to be chemically removed. Shallow cavities or designs are often reproduced, with textures duplicating fabrics and leather. Photosensitive-resistant materials are commonly used in the printing industry.

Allowances must be made to compensate for the effects of *etching radius*, or *etch factor*. As the etchant acts on the workpiece, it tends to undercut the maskant pattern. In deep cuts, the undercut may be serious. Figure 22-11 depicts the effects of the etch factor in chemical erosion.

Casting and *electroforming* are sometimes called *metal-deposition* processes. They involve depositing a metallic (or sometimes ceramic or plastics) coating on a master form.

In Figure 22-12A, a steel master mandrel is dipped into molten lead compounds until a coating is formed over it. The mandrel may then be removed and used again. Casting resins may be poured into the shell and removed when they are polymerized.

The hot casting of metals by lost wax or sand processes, as well as by permanent metal molds, may be used to produce precision molds. Molten metal may be poured over a hardened steel master to form a cavity, as shown in Figure 22-12B. This process is sometimes called hot hobbing. Molten metal is cast over a hob. Pressure is applied during the cooling.

Spray deposition promises to create molds in a fraction of the time required for traditional machining. The process, as shown in Figure 22-13, begins with a CAD design. A full-size model of the design is used to create a ceramic pattern. Next, molten tool steel is sprayed onto the ceramic pattern. The deposition rate can exceed 200 kilograms per hour and yield a uniform buildup on the pattern. After removing the pattern from the deposit, it is cleaned, polished, and fitted into a standard mold base. The spray machine, shown in Figure 22-13B, can produce an insert in about two hours. Larger machines are still in the design phase.

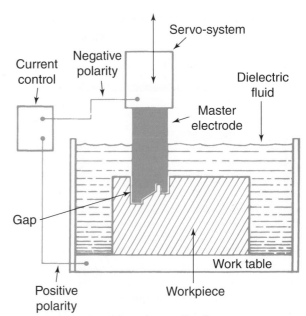

(A) Note the gap between the workpiece and master tool is uniform.

(B) This EDM machine delivers linear travel in three axes and contouring in two axes.

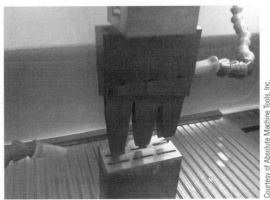

(C) These electrodes have completed a deep, thin cut in tool steel.
Figure 22-10. Electrical erosion (EDM) machining of a die.

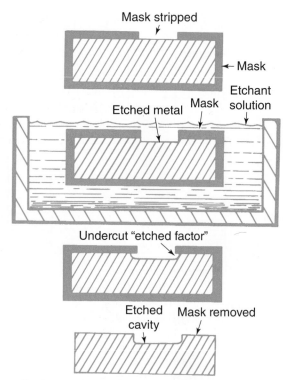

Figure 22-11. Chemical erosion method of producing a die cavity.

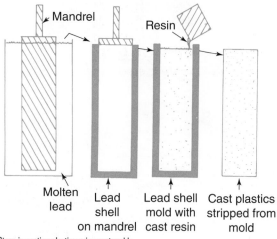

(A) Steps in casting plastics using cast molds.

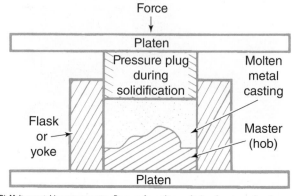

(B) Molten metal is cast on master. Pressure from plug results in a dense, sound casting.
Figure 22-12. Casting of plastics and metal.

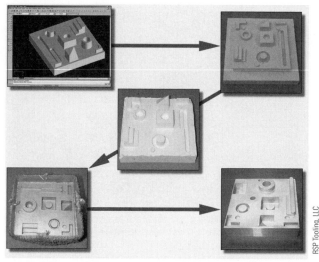

(A) The sprayed tool steel is the reverse image of the ceramic pattern in the center. After removing the over-spray around the edges, the steel is ready to fit into a mold base.

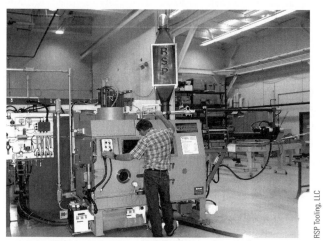

(B) This machine injects molten tool steel into a high-velocity stream of nitrogen. The resulting spray deposits many tiny droplets of metal onto the pattern.
Figure 22-13. Spray deposition processes.

Electroforming is an electroplating process. An accurate mandrel of plastics, glass, wax, or various metals is used as a master to electrically deposit the metallic ions from a chemical solution (Figure 22-14). The molds are thin shelled and may have severe undercuts but usually have a highly polished finish. The cavity may be strengthened by copper plating the back of the shell. Further strength may be provided by placing the die cavity into filled epoxy. The cavities may then be used for thermoforming, blow molding, or injection molding (Figure 22-15).

The major advantages of electroformed molds are their accurate reproduction of detail, zero porosity, zero shrinkage, and lower cost. The major disadvantages include design limitations, relative softness, and difficulty with multiple cavities.

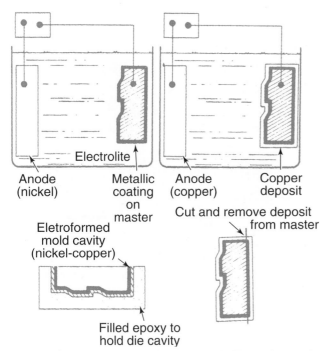

Figure 22-14. Electroformed mold cavity. Copper coating and filled epoxy backing strengthen the nickel cavity.

Figure 22-15. Hard chrome plating of these mold components increases hardness, reduces the coefficient of friction, and provides corrosion resistance.

Sometimes, it is desirable to have the molded piece cling slightly to one-half of the mold, depending on the knockout mechanism. Excessive sticking may be caused by dents or undercuts in the mold or a dirty cavity surface. When cleaning the dies and cavities, a wax, lubricant, or silicone spray is often used. For stubborn spots, a wooden scraper or brass brush may also be used. Never use steel scrapers when cleaning cavities because they may scratch or damage the polished finish of the cavity surface.

A number of other metal-deposition methods may be used for mold making. Flame spraying metals and vacuum metalizing are two such methods (see Chapter 17, Coating Processes).

Electroplating, welding, and heat treatment are also used in making molds. Many finishing operations of molds are done by hand. A steel mold may be electroplated to protect the die cavity from corrosion and provide the desired finish on the plastics product (Figure 22-16).

Mold bases are important to the toolmaker. Bases hold the cavities in place and are made with enough thickness to provide heating and cooling for the cavities. They are made from steel and are available in standard sizes. A standard mold base is shown in Figure 22-17. Bases may be purchased to accept

Figure 22-16. To make the shutoff area on this core pin more wear resistant, it has received a thick deposit of chrome.

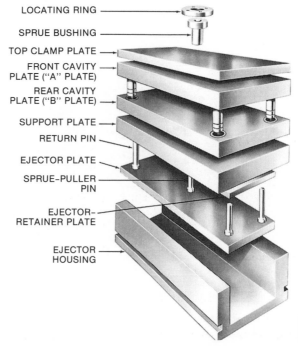

Figure 22-17. Exploded view of a standard mold base, showing the parts.

most custom or proprietary molds (Figure 22-18). Runnerless molds have manifold systems or heated sprue bushings. Mold makers typically purchase the nozzles for the manifold system from a supplier. Figure 22-19 shows a number of nozzle styles for runnerless molds.

Alignment pins ensure proper matching of cavities when the mold base is brought together. If the assembly is not properly aligned and the parting lines are not registered properly, the mold cavities may have to be repositioned. If the molded part sticks in the die cavity, last-minute stoning, hand grinding, and polishing of the mold may be needed.

Mold repair is a significant job for mold makers. Replacing worn, dented, or damaged molds is often very difficult. Typically, worn portions are built up with new metal by welding them together. However, when components are small or delicate, traditional welding may be almost impossible. In most

Photo courtesy of Husky

Figure 22-19. These hot runner system nozzles show many styles of tips.

cases, special laser micro-welding equipment can create tiny welds. Figure 22-20A shows a schematic of the process, which can focus the weld spot to between 0.3 mm and 0.6 mm (0.012 to 0.024 in.). Figure 22-20B shows an operator using the equipment and producing welds such as those seen in Figures 22-20C and 22-20D.

RELATED INTERNET SITES

- **www.amba.org.** The American Mold Builders Association provides announcements about conferences, trade shows, and other issues important to mold makers.

- **www.dme.net.** D-M-E Company sells a wide variety of mold components. The site features photographs and descriptions of standard mold bases, cooling components, ejector pins and sleeves, hot runner systems, temperature controllers, and other mold components. Select "Resources" on the home page, and then choose "CAD Data." This allows a user to download CAD drawings of mold bases and components from an extensive library of drawings.

- **www.makino.com.** Makino is a world leader in EDM technology. On the home page, select "Process," and then click on the photo of an EDM machine. This brings up a good discussion of the process.

- **www.moldmakingtechnology.com.** This site is maintained by the publisher of *MoldMaking Technology* magazine. In addition to accessing the current issue of the magazine, this site also includes "Events," which lists all trade shows and conferences dealing with mold making. This link is found at the bottom of the website.

- **www.pcs-company.com.** PCS Company offers mold bases, hot runner systems, heating and cooling supplies, and other mold components. The site also provides CAD drawings of their standard mold bases.

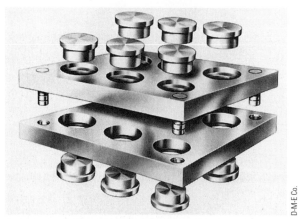

D-M-E Co.

(A) Standard cavity insert rounds showing bored holes in upper and lower cavity plated to receive inserts.

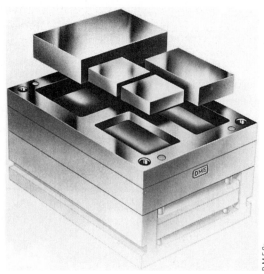

D-M-E Co.

(B) Standard rectangular cavity insert blocks shown with pockets machined in mold base.
Figure 22-18. Standard cavity insert blocks.

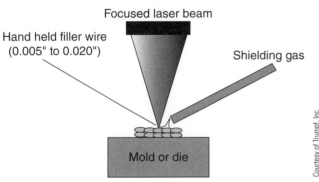

(A) The focused laser creates such a small weld pool that a microscope is needed for accurate welding.

(B) The micro-welding machine accommodates molds up to 150 kilograms (330 lb).

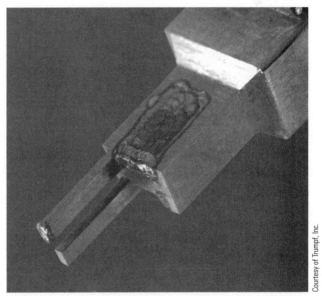

(C) This photo shows the repair of a worn edge on a delicate mold component.

(D) Design changes can be rapidly made if delicate, small parts can be easily altered instead of replaced.

Figure 22-20. Micro-welding for mold repairs.

VOCABULARY

The following vocabulary words are found in this chapter. Use the glossary in Appendix A to look up the definitions of any of these words you do not understand as they apply to plastics.

alignment pins
chemical erosion
electrical-discharge machining (EDM)

electroforming
fixture
hot hobbing
jig
Kirksite
plastics tooling
spray deposition
tooling

QUESTIONS

22-1. Special fixtures, molds, dies, and so on, that enable a manufacturer to produce parts are called _____.

22-2. Name an appliance for accurately guiding and locating tools during the operation involved in producing interchangeable parts.

22-3. Most mold designs begin from preliminary _____.

22-4. A fundamental element in both plastics and steel is _____.

22-5. Hobbing, etching, electroforming, and electrical-discharge machining are generally classed as _____ processes.

22-6. An alloy of aluminum and zinc used for molds that has high thermal conductivity is known as _____.

22-7. Much of the mold-making industry is composed of _____ shops.

22-8. Tooling _____ are more difficult to maintain in multiple-cavity molds.

22-9. Tools for long-run production work are usually made of _____.

22-10. The trade names Ultracal, Hydrocal, and Hydrostone refer to _____ used for tooling.

22-11. Name the tooling material that has several advantages, including lightness, corrosion resistance, and low cost.

22-12. Devices that maintain proper alignment of the cavity as the mold closes are called _____.

22-13. The act of shaping plastics or resins into finished products by heat and/or pressure is called _____.

22-14. The tooling used to hold the cavities in place is called the _____.

22-15. List four methods that may be used to produce a mold where no cutting tools are involved.

22-16. Because _____ are relatively inexpensive and have sufficient strength to produce some types of molds, they are used in prototype or experimental tooling.

22-17. Many metal tools and dies are being replaced by _____ because they are strong, light, and easy to machine.

22-18. A popular material for thermoforming and blow-molding molds is _____ because it is light, easy to machine, and has good thermal conductivity.

22-19. If a blow-molding mold requires fine mold detail and must be easily machined but stronger than aluminum, _____ may be used.

22-20. For high strength, wear resistance, and high-production runs, molds should be made of _____.

22-21. A _____ machine is similar to and functions like a duplicator machine except that it is normally adjusted to operate at ratios as high as 20 to 1.

22-22. No cutting tools are used in machining operations classed as _____ processes.

22-23. The process where a piece of very hard steel is pushed into a blank of unhardened, unheated steel to form a mold cavity is called _____.

22-24. An electrical means of removing metal by electrical erosion is called _____.

22-25. In _____ or etching, an acid or alkaline solution is used to create a cavity.

22-26. Part of the manufacturing cost of plastics products is due to special _____ or tooling.

22-27. Gang or _____ molds and those with inserts require additional tooling costs.

22-28. Steel for molds may vary between 35 and 65 Rockwell C in _____ depending upon requirements.

22-29. Name the type of tooling you would select to produce each of these:

    a. 10 thermoformed serving trays

    b. 10,000 cast woodcut moldings for furniture doors

    c. 10,000 molded polypropylene chair seats

    d. 10 million molded electrical switchplate covers

22-30. Where corrosion resistance is needed or adverse atmospheric conditions exist, _____ steel may be used for molds.

COMMERCIAL CONSIDERATIONS

CHAPTER 23

INTRODUCTION

In this chapter, you will learn that production personnel and management must work together to form a successful plastics business. Financing, equipment, price quotations, plant locality, and other factors must also be considered.

Plastics parts manufacturing is a competitive business. Material selection, processing techniques, production rates, and other variables must be considered in the selling price (see Chapters 21 and 22).

Resin manufacturers and custom molders are the best sources of information concerning the performance of a plastics material. The quantity and compounding of the resin ingredients are important variables. When estimating or planning new items, confer with resin manufacturers, and indicate all specifications. You should be able to answer the following questions:

1. How is the part to be used?
2. What grade of resin is to be used?
3. What physical requirements must the finished part meet?
4. What processing techniques will be used?
5. How many parts are to be made?
6. Will new capital investment for equipment be needed?
7. What are the specifications for reliability and quality of each part?
8. Will we be able to produce what the customer wants at a profit?
9. How and when will the customer pay for our services?

Depending upon quantity, price quotations may vary a great deal. For example, the price of a 0.5 kg (1.1 lb) bag of polyethylene may exceed $2.00 per kg. The same material in 25 kg (55 lb) bags may be about $1.65 per kg. In annual volumes exceeding 100,000 kg (220,000 lb), this material may sell for about $1.10 per kg. This chapter includes the following topics:

I.   Financing
II.   Management and personnel
III.   Plastics molding
IV.   Auxiliary equipment
V.   Molding temperature control
VI.   Pneumatics and hydraulics
VII.   Price quotations
VIII. Plant site
IX.   Shipping

FINANCING

Strong interest and great prospects are not the only prerequisites to starting an industrial enterprise, because no business can succeed and grow without enough financing.

One of the major functions of management is to plan the capital structure of the business with care. In a proprietorship, private capital or loans are used for financing, whereas in a corporation, the sale of stock is used for capitalization.

Some equipment firms provide financing for purchasing machinery through delayed or deferred payments, lease purchase plans, or direct financing. Other sources of capital include insurance companies, commercial banks, mortgages, private lenders, and others. The Small Business Investment Act of 1958 has helped many small firms. In addition, the Small Business Administration (a federal agency) has helped thousands of businesses to secure loans.

MANAGEMENT AND PERSONNEL

It has often been stated, "A business is as strong or successful as its management operations." Many enterprises fail every year, whereas others continue to struggle, barely surviving. Many of the problems of a struggling and failing business may be associated with poor management. Management must coordinate the enterprise, regulating assets, personnel, and time in order to make a profit.

A major concern of the plastics industry is that the labor supply is not keeping pace with its growth rate. There are great employment opportunities for women because they make up nearly half the plastics industry workforce. This area of research in particular needs men and women to work with polymers, processes, and fabrication. Most plastics companies have a need for professional personnel including executives, engineers, and supervisors; technical personnel, such as technicians and para-engineers; skilled workers including machinists, assistant technicians, and machinery setup people; and semiskilled and unskilled workers such as material handlers, equipment operators, and packers.

Most unskilled or semiskilled personnel may be trained by the company. However, professional, technical, and skilled personnel must have college or technical school training.

The plastics industry will continue to compete for a limited supply of skilled designers, engineers, and moldmakers. The use of CAD, CAM, CAE, CAMM, and CIM systems may prove to be one way to meet the challenge of skilled personnel shortages as well as increase productivity.

Management must maintain good labor relations, which may include collective bargaining with labor unions. A good working relationship between labor and management is part of the successful enterprise.

PLASTICS MOLDING

A great deal has been written about the general properties and forming processes of plastics. Molding plastics is a difficult process, and often considerable experience is required to solve production problems. Technology in the plastics industry is constantly changing. Only basic information and precautions about molding plastics may be given here (Table 23-1).

The molding capacity of equipment may limit production output. Limitations include the available pressures of the press,

Table 23-1. Advantages and Disadvantages of Selected Manufacturing Methods

Manufacturing Method	Injection Molding	Filament Winding	Blow Molding	Pultrusion	Rotational Casting	Vacuum Bag	Extrusion	Spray-up	Thermo-forming	Hand Layup
Capital machine cost	high	low-high	low	high	low	low	high	low	low	low
Tool/mold costs	high	low-high	low	low-high	low	low	low	low	low	low
Material costs	high	high	high	high	low	high	high	high	high	high
Cycle times	low-high	high	low-high	high	high	high	low	high	high	high
Output rate	high	low	high	low	low	low	high	low	high	low
Dimensional accuracy	good	fair	fair	fair	fair	fair	fair	fair	poor	fair
Finishing stages	none	some	some	yes	some	some	yes	some	yes	some
Thickness variation	low	fair	fair	fair	low	low	low	low	high	low
Stress in molding	some	some	some	some	some	some	some	some	some	some
Can mold threads	yes	no	yes	no	yes	no	no	no	no	no
Can mold holes	yes	yes	yes	no	yes	yes	no	yes	no	yes
Open-ended components molded	yes	yes	yes	yes	yes	yes	yes	yes	yes	yes
Inserts molded-in	yes	yes	no	no	yes	yes	no	yes	no	yes
Waste material	none	some	some	none	none	some	some	some	some	some

amount of material it will mold, and physical size. For example, compression presses may vary in capacity from less than 5 to more than 1500 metric tons (5.5 to 1653 tons) of pressure. Extruder machines may plasticize less than 8 to more than 5000 kg (17 to 11,000 lb) per hour. Injection machines may range from less than 20 g to more than 20 kg (0.7 oz to 44 lb) per cycle. The clamping pressures vary from less than 2 to more than 1500 metric tons (2.2 to 1653 tons). It is common to run most equipment at 75% capacity rather than maximum capacity.

Many open-mold composite techniques are accomplished by hand. Placing layers of composite tape over specialized tools is slow. Handwork has the added disadvantage of possible incorrect tape orientation, process-induced voids, and/or porosity. One solution to reduce manufacturing costs and assure consistent part quality is to utilize automated tape-laying equipment as part of the molding process. In Figure 23-1 shows part of the process of forming a skin section for the F-35 Lightning II fighter jet. The layers are formed to contour by a specialty tool. When the layup is complete, vacuum bags apply pressure as the parts cure in an autoclave, shown in Figure 23-2.

Figure 23-1. Forming aircraft carbon fiber composite skin components.

Figure 23-2. A large autoclave used to cure composites.

AUXILIARY EQUIPMENT

Plastics materials are poor conductors of heat. Some plastics are hygroscopic, that is, moisture-absorbing. For these reasons, preheating **auxiliary equipment** may be necessary to reduce moisture content and polymerization or forming time. Often, hopper dryers are used on injection and extruders to remove moisture from the molding compounds and help assure consistent molding. Currently, two main types of dryers common in industry are hot air dryers and desiccant dryers. Desiccant dryers utilize a bed of chemicals that remove moisture from the air in order to speed the drying process. Usually, these dryers force hot, dry air through pellets to remove moisture. An alternative approach is to combine heat and vacuum in a drying system. Figure 23-3 illustrates such a system. Preheating of thermosets may be accomplished by using various thermal heating methods that utilize infrared, sonic, or radio-frequency energy. Preheating may reduce cure and cycle time as well as prevent streaking, color segregation, molding stress, and part shrinkage. It also may allow for a more even flow of heavily filled molding compounds.

Often, molders must compound their own additives into resins or other compounds. Both hot and dry mixers may be needed to blend plasticizers, colorants, or other additives. The use of material silos or other conveyor systems may be necessary.

Microprocessors can control many material-handling operations. Through centralized monitoring and control, all settings can be made, and the status of all stations can be checked.

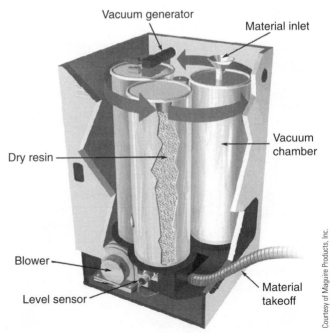

Figure 23-3. This dryer uses heat and vacuum to reduce the time required to dry pelletized plastics.

These systems store molding parameters and actual batch data (formulas) for future use or to prevent the wrong materials from entering a process. Hoppers, loaders, and blenders are controlled to accurately weigh and meter ingredients for each machine. Extrusion compounding operations require precise systems to accurately weigh and dispense several ingredients. Figure 23-4 shows a small weight-loss system for extrusion. Most dryer suppliers agree that microprocessor control is the key to success in their technology. Microprocessors can accurately control energy use in order to create high-performance, dust-free drying systems.

Preform equipment and loaders are important in compression molding and many composite systems.

Injection molding, blow molding, thermoforming, and other molding techniques may require the use of regrinders or granulators to grind sprues, runners, or other cutoff scrap pieces into usable molding materials. Some of these operations are carried out in-line. Processors look for noise and floor-space reduction as well as energy efficiency and ease of maintenance when selecting granulators.

Annealing tanks are used with thermoplastic products to reduce sink marks and distortion. Large housings or similar large pieces are often placed over shrink blocks, in jigs, or in dies to help maintain correct dimensions and minimize warpage during cooling. Many automobile steering wheels once cracked after a length of time due to latent shrinkage of the molded part. Proper annealing and selection of materials have solved such problems.

Closed-loop processing with computer-controlled chillers and cooling towers is helping to reduce energy consumption as well as increase productivity.

The growth of the use of robotics has been phenomenal. The main reason for this growth is increased productivity. In addition, robots may replace human operators in hot, boring, and highly fatiguing jobs. The use of efficient and flexible robots can release humans to carry out the creative and problem-solving tasks of the company (Figure 23-5).

Photo courtesy of Huskey

(A) This photo shows a pick-and-place robot on a small, all-electric, injection-molding machine. The robot can achieve a 4-second cycle time using all three axes of motion.

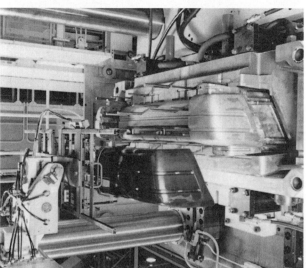

Ford Motor Co.

(B) Robots, shown above removing a car bumper from an injection-molding press, do practically all of the material handling in the production process. Station-to-station movement is by automatically guided vehicles or an automated overhead electric monorail system. Painting also is done by robots. The bumpers, manufactured from a polycarbonate/polyester alloy called Xenoy®, are as strong as steel, but lighter, less expensive, and easier to paint. They meet 5-mile-an-hour crash standards and will not rust.

Figure 23-5. Use of robots in plastics products manufacturing.

Courtesy of Battenfeld-Gloucester

Figure 23-4. This gravimetric system accurately weighs and blends ingredients to ensure a uniform stream of material to the extruder.

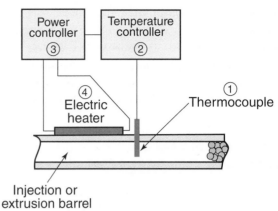

Figure 23-7. The four basic parts of a temperature control system used in the plastics industry.

Figure 23-6. This granulator has a throat opening about 30 cm 3 30 cm–large enough to accommodate rather large parts. For robotic operations, a modified hopper accepts parts dropped from above.

Although parts removal is the main application for robots, they can also perform a variety of secondary operations that improve part quality and reduce labor costs. Robots may be used as parts-handling devices, transporting parts to secondary stations and packaging. This could eliminate nonvalue-added functions to the plant such as assigning employees to package parts. Post-molding operations by robots may include assembling, gluing, sonic welding, decorating, or sprue snipping.

Some manufacturers use bar coding to keep track of material flow, inventory, and part production. Bar coding is used in **flexible manufacturing systems (FMS)**. FMS consists of manufacturing cells that are equipped with a number of computer-controlled machining or molding operations or other automated equipment. As different coded parts travel down the line, FMS or CIM systems determine which operations (assembly, decorate, finish) are to be performed.

Granulators are grinders that are used to reprocess material. These machines spin a series of knives or blades in a chamber to chop sprues, runners, and rejected parts into particles approximately the size of the original pellets (Figure 23-6). Frequently, the reground material is mixed with new pellets and returned to the processing equipment. This method of reusing material is commonly used in extrusion, injection, and blow molding.

MOLDING TEMPERATURE CONTROL

One of the most important factors in efficient molding is controlling temperature. The temperature control system may consist of four basic parts including the thermocouple, temperature controller, power output device, and heaters. A control system using these four parts is shown in Figure 23-7.

The thermocouple is a device made of two unlike metals. Often, combinations of iron with constantan or copper with constantan are used. When heat is applied to the junction of the two metals, electrons are freed, producing an electric current that is measured on a meter calibrated in degrees. Over 98% of all temperature sensors used by the plastics industry are thermocouples. However, resistance temperature detectors (resistance bulbs) have been used for a few installations.

In recent years, the instrumentation for temperature control has progressed through three major designs. The oldest type was based on a millivoltmeter. It is capable of holding a specified temperature at a range of $\pm17°C$ ($\pm30°F$). In contrast, solid-state designs can often hold temperatures within $\pm11°C$ ($\pm20°F$). A microprocessor-based system can often achieve control of $\pm3°C$ ($\pm5°F$). To achieve such accurate control, microprocessors allow PID (proportional, integral, derivative) controls to be used. In addition to accuracy, microprocessor controls are more reliable than the older designs. PID systems tend to be about 12 times more reliable than millivoltmeter designs and six times better than solid-state devices.

Most modern machines are operated by hydraulic power and electrical energy. Vacuum, compressed air, hot water, and chilled water supplies may also be needed. For the most part, cold water is used to cool the mold and reduce the molding cycle time.

It is important in injection molding that the cooling system be able to remove the total heat load generated during each cycle. The system must also be designed to ensure that all molded sections, thick or thin, will cool at the same rate to minimize potential differential shrinkage.

The microprocessor has made major changes in all areas of plastics processing including chillers and mold-temperature controllers. Microprocessor-based mold-temperature controllers may keep molds or chillers within ±1°F. Electromechanical and solid-state controllers usually have ±3°F accuracy ratings.

Balances, scales, pyrometers, clocks, and various timing devices are important accessories. There are many manufacturers of machines and auxiliary equipment for the plastics industry.

PNEUMATICS AND HYDRAULICS

Pneumatic- and hydraulic-actuated accessories and equipment are important in plastics processing (Figure 23-8).

Pneumatics are used to activate air cylinders and provide compact, light, vibrationless power. Filters, air dryers, regulators, and lubricators are accessories needed for pneumatic systems to operate.

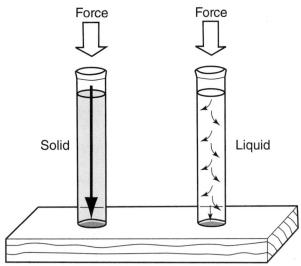

(A) Comparison of transmitting a force through a solid and a liquid.

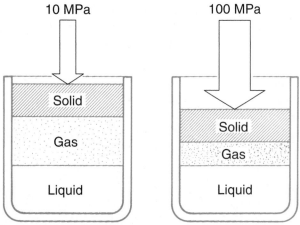

(B) Gas can be compressed, but a liquid resists compression.

Figure 23-8. Force is transmitted differently by solids, liquids, and gases. A solid transmits force only in the direction of the applied force. Liquids (hydraulic systems) and gases (pneumatic systems) transmit force in all directions.

(A) This flow control valve alters the rate that hydraulic oil can pass through the valve.

(B) A directional control valve.

Figure 23-9. Hydraulic components.

Improper oil or the presence of air or moisture may cause noise in hydraulic lines. In addition, chattering valves, pump wear, and high oil temperature may cause noise problems.

Hydraulic power systems may be divided into four basic components:

1. *Pumps* that force the fluid through the system
2. *Motors or cylinders* that convert the fluid pressure into rotation or extension of a shaft
3. *Control valves* that regulate pressure and direction of fluid flow (Figures 23-9A and 23-9B)
4. *Auxiliary components*, which include piping, fittings, reservoirs, filters, heat exchangers, manifolds, lubricators, and instrumentation

Graphic symbols are used to schematically show information about the fluid power system. These symbols do not represent pressures, flow, or compound settings. Some devices have schematic diagrams printed on identification plates.

It is often hard to choose between a hydraulic and a pneumatic system for an application. As a general rule, when a great amount of force is needed, hydraulics should be used; when high speed or a rapid response is needed, pneumatics should be used.

PRICE QUOTATIONS

Because all plastics parts are formed by processes utilizing molds or dies, it is only logical that a great deal of consideration should be given to them.

The cornerstone of any plastics business is tooling. In a tough, competitive market, a business must produce molds faster, cheaper, and better than the competition if the business is to be profitable.

Computer numerically controlled (CNC) machine tools and computer-assisted designs and manufacturing (CAD/CAM) equipment help produce better, more accurate molds at a substantial savings in cost and time.

All price quotations for molded products should be based on the design and general condition of the mold. In the long run, the less costly die is usually not the best buy. Compression, transfer, and injection molds are costly.

It is common for price quotations to be based on **custom molds**. These are molds owned by the customer and turned over to the molder in order to produce parts. The greatest danger in quoting prices is the failure to take into consideration the condition of the custom mold. All quotations should be based on approval of the custom mold because repair or alteration of the mold may be necessary.

If the die is owned or made by the molder, it is called a **proprietary mold**. In quotations for products to be made with proprietary molds, the molder must make enough profit to amortize, or pay for, the mold. Often, a portion of the cost of making the mold is calculated into the price of each piece or each thousand pieces.

Corrosion is an enemy with custom or proprietary molds. Molds should be given a moisture-resistant, noncorrosive coating, and water, air, or steam holes should be dried and given a coating of oil. The mold should then be placed in storage along with all its accessories.

Large tooling (panels, boat hulls, sinks) requires careful consideration of production space and adequate storage facilities. Both add to the cost of the product.

PLANT SITE

The plant site should be determined in relation to the nearness of raw material as well as the potential market. Freight must be reflected in the cost of each plastics product. A quotation must reflect tax rates, labor conditions, and wages. If there is an anticipated increase in taxes or wages, the price quotation must include such an increase in production costs.

Many states and communities encourage the development of new enterprises. They offer reduced taxes and adequate labor force. Labor relations and tax incentives may be major concerns when locating a plant.

Other considerations in site location are an available skilled-labor pool and proximity to educational institutions.

SHIPPING

Shipping plastics, resins, and chemicals falls under the category of special governmental regulations. The United States Postal Service has ruled that any liquid that gives off flammable vapors at or below temperatures of 7°C (20°F) is not mailable (Section 124.2d, Postal Manual). Poisonous materials are generally regarded as nonmailable by the provisions of 124.2d. Caustic or corrosive substances are prohibited in 124.22. The prohibitions mentioned are specified by the law in Section 1716 of Title 18, US Code. Acids, alkalies, oxidizing materials, or highly flammable solids, highly flammable liquids, radioactive materials, or articles emitting objectionable odors are considered to be nonmailable as well.

The Department of Transportation (DOT) regulates the interstate transportation of cellulose nitrate plastics by rail, highway, or water. Special packaging must also be used when shipping this material.

The Interstate Commerce Commission (ICC) regulates the packing, marking, labeling, and transportation of dangerous or restricted materials. Standards for many shipping containers are specified, and regulations also specify the use of labels. For example, flammable liquids require a red label, flammable solids a yellow label, and corrosive liquids a white label. There are also other labels for poisons and shipments of radioactive material. It is the responsibility of the shipping agency to check the label to make certain it is correct and filled in by the shipper.

If there is any doubt about the shipment of flammable or plastics materials, it is best to check local ICC offices. Certain cities have ordinances that prohibit vehicles containing flammable items from driving through tunnels or on bridges. Such Interstate Commerce Commission regulations must always be observed.

RELATED INTERNET SITES

- **www.imscompany.com.** The IMS Company provides molding supplies and equipment for plastics, rubber, and metal processing industries. In addition to supplies, IMS also handles chillers, granulators, dryers, and loaders. On the home page, select "Resources," and then pick "Tech Tips, Product Overviews & Other Services." This leads to a list of topics related to auxiliary equipment. For example, "Sizing a Chiller for Injection Molding" helps identify appropriate sized chillers.

- **www.maguire.com.** Maguire Products Inc. produces metering and blending equipment for plastics molding, extrusion, and compounding. To find information on the function and operation of their products, select "Support" on the home page. Then find "Video and Demonstration Files." Short and longer videos include presentations on gravimetric blending and feeding. Extensive PowerPoint® presentations are also provided.

- **www.parker.com.** The Parker Hannifin Corporation is one of the leading companies producing motion and control technologies. This site includes information about hydraulic components. From the home page, select "Get Support," and then "Catalogs and Literature." This leads to a long list of components and documents about each component.

VOCABULARY

The following vocabulary words are found in this chapter. Use the glossary in Appendix A to look up the definitions of any of these words you do not understand as they apply to plastics.

annealing tanks
auxiliary equipment
custom molds
estimating
flexible manufacturing systems (FMS)
hydraulic
pneumatics
proprietary mold
pyrometer

QUESTIONS

23-1. The greatest danger in quoting custom mold prices is not taking into consideration the _____ of the custom mold.

23-2. Often a portion of the cost of making a mold is calculated into the production of each piece. This is called _____.

23-3. Molds made by the customer and used by the molder are called _____ molds.

23-4. The transmission of power through a controlled flow of liquids is known as _____.

23-5. Molds made and owned by the molder are called _____ molds.

23-6. The _____ is a device made of two dissimilar metals. Combinations of iron and constantan or copper and constantan are commonly used.

23-7. A _____ material tends to absorb moisture.

23-8. The _____ of the molding equipment will limit product size.

23-9. Name four pieces of auxiliary equipment that may be required for injection-molding operations.

23-10. Name four factors that may influence plant locality.

23-11. Name four factors that may influence the cost of each plastics product when estimating the planning of new items.

23-12. The Interstate Commerce Commission regulates the packing, marking, labeling, and transportation of _____ or _____ materials.

23-13. Would you select a hydraulic or a pneumatic power system if great force were required?

23-14. When high speed and rapid response are required, would you select a hydraulic or a pneumatic power system?

23-15. The best sources of information about the performance of a plastics material are _____ manufacturers and _____ molders.

23-16. In a(n) _____ enterprise, private capital or loans are used for financing.

23-17. The sale of stock shares is used for capitalization in a(n) _____ enterprise.

23-18. Planning the capital structure of an enterprise is the major function of _____.

23-19. Successful relations between _____ and _____ are part of the successful enterprise.

23-20. Name the four basic components of hydraulic power systems.

23-21. The cornerstone of any plastics business is _____.

23-22. It is common to run or operate most molding equipment at _____% of maximum capacity.

23-23. Name the four general classifications of plastics personnel.

23-24. Approximately _____% of the plastics industry work force is female.

23-25. If _____ and CNC equipment is used, better, more accurate molds may be produced with a savings in cost and time.

23-26. In addition to increased productivity, _____ may be used to relieve human operators from hot, boring, and highly fatiguing jobs.

GLOSSARY

Additive polymerization—A type of polymerization that adds one mer to another. It usually does not yield any byproducts.

Adherends—Surfaces being adhered.

Adhesive—A substance capable of holding materials together by surface attachment.

Air knife—A narrow jet of air used to control the thickness of liquid coatings.

Air-slip forming—A thermoforming process in which air pressure is used to form a bubble, and a vacuum is then used to form the hot plastics against the mold.

Alignment pins—Devices that maintain proper cavity alignment as a mold closes.

Allowances—The intentional differences in the dimensions of two parts.

Alpha particles—A type of radiation characterized by heavy, slow-moving masses.

Alternating copolymer—A copolymer that has a chemical structure in which the two types of monomers alternate in the polymer chain.

American Conference of Governmental Industrial Hygienists (ACGIH)—This organization publishes guidelines and recommendations on the exposure limits to various chemicals.

Amorphous—A term that means "not crystallized." Plastics that have an amorphous arrangement of molecular chains are often transparent.

Annealing—A process of holding a material at a temperature near, but below, its melting point for a period of time in order to relieve internal stress without shape distortion.

Annealing tanks—Devices to reduce distortion in thermoplastic products.

Antiblocking—Materials that prevent two plastics from undesired adhesion.

Antioxidant—A stabilizer that retards breakdown of the plastics by oxidation.

Antistatic agent—An additive that reduces static charges on a plastics surface.

Apparent density—The mass per unit volume of a material; includes the voids inherent in the material.

Ashing—The use of wet abrasives on wheels to sand and polish plastics.

Aspect ratio—The ratio of the length of a filler to its width.

ASTM (American Society for Testing and Materials)—This organization develops and published testing standards.

Atomic mass units (amu)—A number equal to the number of protons in the nucleus of an atom of the element.

Automated—Mechanical and electronic devices.

Automotive shredder residue (ASR)—The portion of shredded automobiles that is not recycled.

Auxiliary equipment—Equipment needed to help control or form the product. Filters, vents, ovens, and take-up reels are examples.

Backbone—The main chain of a plastics molecule.

Back pressure—The viscous resistance of a material to continued flow when a mold is closed. Resistance to the forward flow of molten material during extrusion.

Bakelite—A phenolic thermosetting plastics invented by Leo Baekeland in 1907.

Banbury mixer—A machine used for compounding materials. The machine contains a pair of counter-rotating rotors that masticate and blend the materials.

Bar—The unit "bar" is not an SI unit, although it is frequently used in conjunction with SI units. One bar is roughly equal to air pressure at sea level, so 1 bar equals 14.5 psi.

Barrel—The cylindrical housing in which the extruder screw rotates.

Barrel mixer—Special mixing device attached to the screw of an extruder.

Bell curve—A graphic representation of a standard, normal distribution.

Benzene ring—A ring of six carbon atoms in a molecule of benzene.

Beta particles—A type of radiation that consists of high-speed and high-energy particles.

Biaxial—Stretching in two directions.

Biaxial blow molding—A blow-molding process that stretches the extrudate in two directions.

Biodegradable—Material that chemically decomposes under the action of bio-organisms.

Blanking—The cutting of flat sheet stock to shape by striking it sharply with a punch while it is supported on a mating die. Punch presses are used.

Block copolymer—A polymer molecule made up of comparatively long sections of one chemical composition, those sections being separated from one another by segments of different chemical character.

Blow molding—A method of fabrication in which a parison is forced into the shape of the mold cavity by internal air pressure.

Blowing agents—Typically, blowing agents are chemicals that decompose to create small nitrogen or carbon dioxide bubbles in melted plastics. This produces various types of foams.

Blushing—A surface defect caused by applying a coating over a surface that was not properly dried.

Boss—A protuberance on a part designed to add strength or to facilitate assembly.

Bottom ash—The residue of combustion that settles in the bottom of an incinerator.

Branched—Side chains attached to the main chain of the polymer. Side chains may be long or short.

Breaker plate—A perforated metal plate located between the end of the screw and the die head.

Breather ply—A layer of fabric that allows air and excess resin to escape during vacuum-bag processing.

Breathing—The opening and closing of a mold to allow gases to escape early in the molding cycle. Also called degassing.

Buffing wheel—An operation that provides a high luster to a surface. The operation, which is not intended to remove much material, usually follows polishing.

Bulk factor—The ratio of the volume of any given weight of loose plastics to the volume of the same weight of the material after molding or forming.

Bulk molding compounds (BMC)—Putty-like mixtures of resins, catalysts, fillers, and short fiber reinforcements.

Bulk polymerization—The polymerization of a monomer without added solvents or water.

CAD—Computer-aided design; a computer system that aids or assists in the creation, modification, and display of a design. It is used to produce three-dimensional designs and illustrations of the proposed part. The term *CAD* is also used to refer to computer-aided drafting.

CAE—Computer-aided engineering; a computer system that assists in the engineering or design cycle. It analyzes the design and calculates the performance predictions of service life and safety design factors.

Calendering—Process of forming a continuous sheet by squeezing the material between two or more parallel rolls to impart the desired finish or ensure uniform thickness.

Carbonaceous—Containing carbon.

Carcinogen—A substance or agent that can cause a growth of abnormal tissue or tumors in humans or animals. A material identified as an animal carcinogen does not necessarily cause cancer in humans. Examples of human carcinogens include coal tar, which can cause skin cancer, and vinyl chloride, which can cause liver cancer.

Casein—A protein material precipitated from skimmed milk by the action of either rennin or dilute acid. Rennet casein is made into plastics.

CAS number—Chemical Abstracts Services registry number; these numbers unambiguously identify all known chemical substances.

Casting—The process of pouring a heated plastics or other fluid resin into a mold to solidify and take the shape of the mold by cooling, loss of solvent, or completing polymerization. No pressure is used. *Casting* should not be used as a synonym for *molding*.

Catalyst—A chemical substance added in minor quantity (as compared to the amounts of primary reactants) that markedly speeds up the cure or polymerization of a compound. Also called initiator.

Caul plate—A smooth metal plate used in contact with the layup during curing to transmit normal pressure and provide a smooth surface to the finished part.

Ceiling—An exposure limit that should never be exceeded, even for short time durations.

Celluloid—A strong, elastic plastics made from nitrocellulose, camphor, and alcohol. Celluloid is used as a trade name for some plastics.

Central tendency—The grouping of data points around one central point.

Centrifugal casting—A process that typically produces large pipes and tubes.

Chain growth polymerization—A type of polymerization in which the chains grow from initiation to completion almost instantaneously.

Char barrier—A layer of charred material that acts as an insulator to prevent a material from burning or continuing to burn.

Chemical Abstracts Services Registry number—A number provided by the Chemical Abstracts Services that identifies each industrial chemical. See *CAS number*.

Chemical erosion—A chemical method of removing metal.

Chemical (specific) adhesion—Adhesion caused by valence forces.

Chemical treatment—Submerging parts in an acid bath.

Closed-cell—Describing the condition of individual cells that make up cellular or foamed plastics when cells are not interconnected.

Coarse tolerance—A tolerance where accurate dimensions are not critical.

Cohesion—The propensity of a substance to adhere to itself; the internal attraction of molecular particles toward each other; the ability to resist partition from the mass.

Cold flow—See *creep*.

Cold molding—A procedure in which a composition is shaped at room temperature and cured by subsequent baking.

Collodion—A thin, transparent film of dried pyroxylin.

Colorants—Dyes or pigments that impart color to plastics.

Combustible—A material that burns easily. Combustible liquids have a flash point of 38°C (100°F) or above.

Commingled—A mixture of many types of filaments.

Composite—A combination of two or more materials (generally a polymer matrix with reinforcements). The structural components of the composites are sometimes subdivided into fibrous, flake, laminar, particulate, and skeletal.

Compound—A substance composed of two or more elements joined together in definite proportions.

Compression molding—A technique in which the molding compound is placed in an open-mold cavity, the mold is closed, and heat and pressure are applied until the material has cured or cooled.

Compression strength—The highest load sustained by a test sample in a compressive test divided by the original area of the sample.

Condensation polymerization—Polymerization by chemical reaction that also produces a byproduct.

Copolymer—A polymer based on the combination of two mers.

Copolymerization—Addition polymerization involving more than one type of mer.

Corona discharge—A method of oxidizing a film of plastics to make it printable. Achieved by passing the film between electrodes and subjecting it to a high-voltage discharge.

Coupling agents—Chemicals that promote the adhesion between reinforcements and the basic plastics material.

Covalent bond—Atomic bonding by sharing electrons.

Crazing—Small cracks caused by solvent cutting along lines of stress.

Creep—The permanent deformation of a material resulting from prolonged application of a stress below the elastic limit. A plastics subjected to a load for a period of time tends to deform more than it would from the same load released immediately after application. The degree of the deformation is dependent on the load duration. Creep at room temperature is sometimes called *cold flow*.

Cross-links—The tying together of adjacent polymer chains.

Crystal whisker fibers—Short crystalline in organic fibers, such as titanium boride.

Crystalline—A basic type of molecular arrangement found in many thermoplastics.

Curing agents—Chemicals that cause thermosetting plastics to cross-link or cure.

Custom molds—Molds owned by the customer and used by the molder.

Cyanoacrylate—A type of thermoplastic adhesive based on acrylics.

Damping—Variations in properties resulting from dynamic loading conditions (vibrations). Damping provides a mechanism for dissipating energy without excessive temperature rise, prevents premature brittle fracture, and is important to fatigue performance.

Degree of polymerization (DP)—Average number of structural units per average molecular mass. In most plastics, the DP must reach several thousand to achieve worthwhile physical properties.

Denier—The mass in grams of 9000 m (29,527 ft) of synthetic fiber in the form of a single continuous filament.

Densifying—The process of removing air from composites.

Density gradient column—A column composed of liquid layers that decrease in density from the bottom to the top. A sample sinks until it reaches the layer that matches its density, where it remains floating.

Depolymerization—A chemical reaction that breaks a polymer down into monomers or other short organic molecules.

Dielectric strength—A measurement of the voltage required to break down or arc through a plastics material.

Dip casting—The process of submerging a hot mold into a resin. After cooling, the product is removed from the mold.

Dip coating—Applying a coating by dipping an item into a tank of melted resin or plastisol, then chilling. The object may be heated and powders may be used for the coating; the powders melt as they strike the hot object.

Dipole interactions—Effects of separated positive and negative charges in molecules.

Distribution—A collection of values.

Dope cements—Cements composed of solvents and a small quantity of the plastics to be joined.

Double covalent bond—A double covalent bond involves two pairs of shared electrons.

Downstream feeding—Causing materials to enter an extruder after the feed throat.

Drape forming—Method of forming a thermoplastic sheet in a movable frame. The sheet is heated and draped over high points of a male mold. Vacuum is then pulled to complete the forming.

Drawing—The process of stretching a thermoplastic sheet, rod, film, or filament to reduce cross-sectional area and change physical properties.

Draw ratio—The ratio of the maximum cavity depth to the minimum span across the top opening of a thermoforming mold.

Dry offset—A printing method that uses a paste ink.

Dry painting—Also known as powder coating, it is a process to coat a substrate with powders.

Dry winding—A type of filament winding that uses pre-impregnated reinforcements.

Dwell time—A pause in the application of pressure to a mold made just before the mold is completely closed. The pause allows gas to escape from the molding material.

Ebonite—A hard, brittle form of vulcanized rubber.

Elastic reservoir molding—A process in which skin layers are attached to open-cell polyurethane foams with epoxy.

Elastomer—A rubberlike substance that can be stretched to several times its original length and that, on release of the stress, returns rapidly to almost its original length.

Electrical-discharge machining (EDM)—A process based on the erosion of metals by discharging sparks.

Electroforming—Making metal objects by electrodeposition.

Electrostatic-bed coating—A type of powder coating that utilizes electrostatic charges to hold the powder on a substrate prior to fusing.

Electrostatic printing—The deposit of ink on a plastics surface where electrostatic potential is used to attract the dry ink through an open area defined by opaquing.

Elutriation—A process that separates contaminants and fines from a stream of chopped plastics materials by controlled updrafts.

Embedment—Enclosing an object in an envelope of transparent plastics by immersing it in a casting resin, allowing the resin to polymerize.

Embossing—A type of rotary molding that places a pattern on plastics films.

Emulsion polymerization—A process where monomers are polymerized by a water-soluble initiator while dispersed in a concentrated soap solution.

Encapsulating—Enclosing an item, usually an electronic component, in an envelope of plastics by immersing it in a casting resin and allowing the resin to solidify by polymerizing or cooling.

Epoxy resin—Material based on ethylene oxide, its derivatives, or homologs. Epoxy resins form straight-chain thermoplastics and thermosetting resins.

Estimating—The act of determining from statistical examples, experience, or other parameters the cost of a product or service.

Exothermic—Evolving heat during reaction (cure).

Expanded (foamed) plastics—Plastics that are cellular or sponge-like.

Expand in place—A process in which resin, catalyst, expanding agents, and other ingredients are mixed and poured at the site where needed. Expansion takes place at room temperature.

Extrusion—The compacting and forcing of a plastics material through an orifice in a more or less continuous fashion.

Extrusion film coating—The resin coating placed on a substrate by extruding a thin film of molten resin and pressing it into or onto the substrate (or both) without use of adhesives.

Fatigue strength—The highest cyclic stress a material can withstand for a given number of cycles before failure occurs.

Feed—The distance the cutting tool moves into the work with each revolution.

Feet per minute (fpm)—A measure of cutting speeds needed for drilling, turning, and milling.

Fibers—This term usually refers to relatively short lengths of various materials with very small cross sections. Fibers can be made by chopping filaments.

Filament—A fiber characterized by extreme length, with little or no twist. A filament is usually produced without the spinning operation required for fibers.

Filament winding—A composite fabrication process that consists of winding a continuous reinforcing fiber (impregnated with resin) around a rotating and removable form (mandrel).

Filler—An inert substance added to a plastic to make it less costly. Fillers may improve physical properties. The percentage of filler used is usually small in contrast to reinforcements.

Film—Plastics material 0.25 mm or less in thickness.

Film casting—A process of film manufacture that converts liquid resins into films.

Fine tolerance—The narrowest possible limit of variation in dimensions.

Fixture—A device used to support the work during processing or manufacturing.

Flame coating—Method of applying a plastics coating in which finely powdered plastics and suitable fluxes are projected through a cone of flame onto a surface.

Flame-retardant—A material that reduces the ability of a plastics to support combustion.

Flame treatment—Passing a part through a hot oxidizing flame.

Flammable—A material that burns readily. Flammable liquids have a flash point of 38°C or below.

Flash—Extra plastics attached to a molding along the parting line. It must be removed to make a finished part.

Flash mold—The least complex and most economical type of compression mold.

Flash point—The temperature at which vapors are given off sufficient to form an ignitable mixture with the air near the surface of the liquid.

Flexible manufacturing systems (FMS)—A series of machines and associated workstations linked by a hierarchical common control and providing for automatic production of a family of workpieces. A transportation system, for both the workpiece and tooling, is as integral to an FMS as is computerized control.

Flexographic—A printing technique that relies on flexible plates and liquid inks.

Flexural strength (modulus of rupture)—The highest stress in the outer fiber of a sample in flexure at the moment of crack or break. In the case of plastics, it is usually higher than tensile strength.

Flotation tank—A tank that promotes the separation of plastics or contaminants based on differing densities.

Fluidized bed coating—Fluidized bed coating uses powdered plastics material made fluid by compressed air to coat heated pieces of metal.

Fluidized-bed spray coating—A method of coating heated items by immersing them in a dense-phase fluidized bed of powdered resin. The objects are usually heated in an oven to provide a smooth coating.

Fluorescence—A property of a substance that causes it to produce light while it is being acted on by radiant energy such as ultraviolet light or X-rays.

Fly ash—A type of ash, produced by incinerators, that is carried by the air. It must be removed before smokestack gases are exhausted into the atmosphere.

Foam reservoir molding—See *Elastic reservoir molding*.

Free forming—Air pressure is used to blow a heated sheet of plastics, the edges of which are being held in a frame, until the desired shape or height is attained.

Friction fitting—Technique to join like or unlike materials without mechanical fasteners.

Frost line—In extrusion, a ring-shaped zone located at the point where the film reaches its final diameter.

Galalith—A plastics made by hardening casein with formaldehyde.

Gamma radiation—A type of radiation with very short, very high-frequency waves.

Gamma ray—Electromagnetic radiation originating in an atomic nucleus.

Gate—In injection and transfer molding, the orifice through which the melt enters the cavity. Sometimes the gate has the same cross section as the runner leading to it.

Gaylord—A large rectangular container with a volume of approximately one cubic yard. A gaylord of plastics pellets weighs about 1000 pounds.

Gel coat—A thin layer of resin that serves as the surface of the product. Reinforced layers may then be built up.

Glass transition temperature (Tg)—A characteristic temperature at which glassy amorphous polymers become flexible or rubberlike because of the motion of molecular segments.

Glass, Type C—Chemically resistant glass fibers.

Glass, Type E— Electrical-grade glass fibers.

Glue—Formerly, an adhesive prepared from animal hides, tendons, and other byproducts by heating with water. In general use, the term is now synonymous with the term *adhesive*.

Graft copolymer—A combination of two or more chains of constitutionally or configurationally different features, one of which serves as a backbone main chain and at least one of which is bonded at some point(s) along the backbone and constitutes a side chain.

Gravity constant—9.807 meters per second squared is the acceleration caused on earth by gravity.

Gray (Gy)—The unit of measurement of the absorbed dose of ionizing radiation, defined as one joule per kilogram (1 Gy = 1 J/kg).

Gum rubber—Non-vulcanized natural latex.

Gutta percha—A rubberlike product obtained from certain tropical trees.

Hardness—The resistance of a material to compression, indentation, and scratching.

Haze—The cloudy or turbid appearance of an otherwise transparent sample caused by the light scattered from within the sample or its surfaces.

Heat stabilizers—Additives that retard polymer decomposition by heat.

Heat-transfer printing—A printing method similar to hot-foil stamping.

High-pressure laminates—Laminates formed and cured at pressures higher than 7000 kPa (1015 psi).

High-pressure processing—High-pressure expansion processes that use pressures up to 140 MPa and then open the mold to allow room for expansion.

Histogram—A vertical bar graph of the frequency of values in a distribution.

Hollow ground—A saw blade that has been specially ground so that the cutting teeth are the thickest portion, to prevent binding in the kerf.

Homopolymer—A polymer consisting of like monomer structures.

Honeycomb—A manufactured product of metal, paper, or other materials that is resin impregnated and has been formed into hexagonal-shaped cells. Used as a core material for sandwich or laminated construction.

Hot-gas welding—A technique for joining thermoplastic materials in which the materials are softened by a jet of hot air from a welding torch and joined together at the softened points. Generally, a thin rod of the same material is used to fill and consolidate the gap.

Hot hobbing—A process in which molten metal is cast over a hob and then pressed during cooling.

Hot-leaf stamping—Decorating operation for marking plastics in which a metal leaf or paint is stamped with heated metal dies onto the face of the plastics. Ink compounds can also be used.

Hot melt—A general term referring to thermoplastic synthetic resins composed of 100 percent solids and used as adhesives at temperatures between 120°C and 200°C (248°F and 392°F).

Hot runner molding—A process using a hot runner mold, which keeps sprues and runners hot.

Hydraulics—The branch of science that deals with liquids in motion; the transmission, control, or flow of energy by liquids.

Hydrocarbon—An organic compound containing only carbon and hydrogen and often occurring in petroleum, natural gas, coal, and bitumens.

Hydroclave—A device that uses hot fluid to press the plies of a composite against a mold.

Hydrogen bond—A strong type of secondary bonding force.

Hydrolysis—A type of depolymerization that yields monomers by chemically attacking polymers.

Hygroscopic—Tending to absorb and retain moisture.

Impact strength—The ability of a material to withstand shock loading.

Impingement-atomized mixing—A method of mixing in which two or more materials collide.

Induction bonding—High-frequency electromagnetic fields are used to excite the molecules of metallic inserts placed in the plastics or in the interfaces, thus fusing the plastics. The inserts remain in the joint.

Infrared (IR)—The part of the light spectrum that contains wave lengths longer than visible red light.

Inhalation—Breathing a material into the lungs. Inhalation is the major route of exposure to toxic materials in the plastics processing industries.

Inhibitor—A substance that slows down a chemical reaction. Inhibitors are sometimes used in certain monomers and resins to prolong storage life.

Injection-molding machine (IMM)—A machine that melts plastics and forces the melt into a mold.

In-mold decorating—Making decorations or patterns on molded products by placing the pattern or image in the mold cavity before the actual molding cycle. The pattern becomes part of the plastics item as it is fused by heat and pressure.

Insertion bonding—Using ultrasonics to place metallic inserts into plastics.

In situ—The technique of depositing a foamable plastics into the place where foaming will occur.

Inspection—A term used to indicate that during manufacture of a part, personnel will conduct visual examinations of materials, placement of plies, gauge readings, and so on.

Insulated runner molding—A type of runner system that relies on large runners to prevent freeze-up.

Interference—The negative allowance used to assure a tight shrink or press fit.

Intermolecular interaction—Effects of molecules on each other because of dipole charges and structure.

Ionic bonds—Atomic bonding by electrical attraction of unlike ions.

Irradiation—A process of exposing polymeric materials to radiation to cause molecular changes.

Isotope—One of a group of nuclides having the same atomic number, but differing atomic mass.

Jig—An appliance for accurately guiding and locating tools during the making of interchangeable parts.

Kerf—The slit or notch made by a saw or cutting tool.

Kirksite—An alloy of aluminum and zinc used for molds. It has high thermal conductivity.

Lac—A dark-red resinous substance deposited by scale insects on the twigs of trees; used in making shellac.

Lamina—A thin layer of material, generally referring to composites.

Laminate—Two or more layers of material bonded together. The term usually applies to preformed layers joined by adhesives or heat and pressure. The term also applies to composites of plastics films with other films, foil, and paper even though they have been made by spread coating or extrusion coating. A reinforced laminate usually refers to superimposed layers of resin-impregnated or resin-coated fabrics or fibrous reinforcements that have been bonded, especially by heat and pressure. When the bonding pressure is at least 7000 kPa (1015 psi), the product is called a high-pressure laminate. Products pressed at pressures under 7000 kPa are called low-pressure laminates. Products produced with little or no pressure, such as hand lay-ups, filament-wound structures, and spray-ups, are sometimes called contact-pressure laminates.

Laminated vapor plating—The process of vacuum metallizing alternate layers of metal coatings on a polymer substrate.

Laser cutting—A means of cutting materials using laser energy.

LD-50—A dose of a material that is lethal to at least 50% of a group of test animals.

Leaching—Removing a soluble component from a polymer mix with solvents.

Letterflex—A type of printing that uses flexible printing plates.

Letterpress—A printing method that inks the top of raised portions of plates.

Liquid resin molding (LRM)—A process that forces liquid resins into mold cavities under low pressure.

Low-pressure laminates—In general, laminates molded and cured at pressures ranging from 2.8 MPa (0.4 psi) to contact pressure.

Lubrication bloom—An irregular, cloudy, greasy film caused by excess lubricants on a plastics surface.

Luminescence—Light emission by the radiation of photons after initial activation. Luminescent pigments are activated by ultraviolet radiation, producing very strong luminescence.

Macromolecules—The large (giant) molecules that make up the high polymers.

Mandrel—A form around which filament-wound and pultruded composite structures are shaped.

Matched-mold forming—Forming hot sheets between matched male and female molds.

Material safety data sheet (MSDS)—A source of information about the health hazards caused by industrial chemicals.

Materials recovery facility (MRF)—A facility for gathering, sorting, and baling recycled materials.

Mean—The arithmetic average of the values in a distribution.

Mechanical forming—Heated sheets of plastics are shaped or formed by hand or with the aid of jigs and fixtures. No mold is used.

Median—A measurement selected so that half the numbers in a series are larger than it and half are smaller.

Melt flow index—Also called MFR, or melt flow rate; the amount of material in grams that is extruded through an orifice in 10 minutes under specified conditions.

Melt spinning—A technique for manufacturing fibers that involves air cooling of fibers.

Mer—The smallest repetitive unit in a polymer.

Metallic bonds—The type of chemical bonding found in metals.

Microballoons—Hollow glass spheres. Also called microspheres.

MM/RIM—Mat molding reaction injection molding.

Mode—A value with the highest frequency in a set of data.

Mohs scale—A hardness scale based on the ability of harder materials to scratch softer ones.

Mold cavity—The cavity or matrix in which plastics are formed. Also, to shape plastics or resins into finished items by using heat or heat and pressure.

Molecular mass—The sum of the atomic mass of all atoms in a molecule. In high polymers, the molecular masses of individual molecules vary widely, so they must be expressed as averages. Average molecular mass of polymers may be expressed as number average molecular weight (M_n) or mass-average molecular weight (M_s). Molecular mass measurement methods include osmotic pressure, light scattering, solution pressure, solution viscosity, and sedimentation equilibrium.

Molecule—The smallest particle of a substance that can exist independently while retaining the chemical identity of the substance.

Monomeric cements—Cements based on monomer of the plastics to be joined.

m/s—meters per second.

Municipal solid waste (MSW)—The waste materials collected from homes and industries. MSW goes into landfills unless recycling programs recover useful materials out of the waste stream.

N—A statistical term that refers to a sample size or the number of observations in a distribution.

Nanocomposite—A composite material in which the fillers are submicron in size.

Natural plastics—Plastics produced by plants, insects, or animals.

Natural rubber—Non-vulcanized natural latex.

Neutron—An atomic particle that is electrically neutral.

Nitrocellulose (cellulose nitrate)—Material formed by the action of a mixture of sulfuric acid and nitric acid on cellulose. The cellulose nitrate used for celluloid manufacture usually contains 10.8 to 11.1 percent nitrogen.

Normal distribution—A symmetrical distribution that has central tendency.

Number average molecular weight (M_n)—A type of molecular weight average based on the frequency of various length molecules in a distribution.

Offset lithography—A printing method that does not have raised or sunken impressions.

Offset yield—For materials that do not exhibit a clear yield point, an offset yield is calculated. It provides a yield point at a determined location on the stress-strain curve.

Open-cell—Referring to the interconnecting of cells in cellular or foamed plastics.

Open-grit sandpaper—Coarse sandpaper (number 80 or less).

Organosol—A dispersion, usually of vinyl or polyamide, in a liquid phase containing one or more organic solvents.

Orientation—Plastics molecules can be oriented in one direction (uniaxially) or in two directions (biaxially). Orientation is caused by flow or stretching and alters the physical properties of the material.

Parison—The hollow plastics tube from which a product is blow molded.

Parkesine—An early plastics made from collodion by Alexander Parkes.

Parting lines—Marks on a molding or casting where halves of the mold meet during closing.

Permissible exposure limit (PEL)—A measure used by OSHA to define the exposure level acceptable for an 8-hour day as part of a 40-hour week.

Phenol-formaldehyde—A type of thermosetting adhesive, commonly called phenolics.

Phenolic—A synthetic resin produced by the condensation of an aromatic alcohol with an aldehyde, particularly of phenol with formaldehyde.

Phosphorescence—Luminescence that lasts for a period after excitation.

Photodegradable—Materials that decompose due to the action of sunlight.

Photon—The least amount of electromagnetic energy that can exist at a given wavelength. A quantum of light energy is analogous to the electron.

Picking line—A type of sorting technology in which operators select types of materials from a moving conveyor belt.

Pin—A hollow tube that supplies the air required to inflate blow-molded objects.

Plasma treating—Subjecting plastics to an electrical discharge in a closed vacuum chamber.

Plastic—An adjective, meaning pliable and capable of being shaped by pressure. Plastic is often incorrectly used as the generic word for the plastics industry and its products.

Plasticate—To make plastic. The plasticating capacity of an injection-molding machine is the maximum weight of material (PS) that it can prepare for injection in 1 hour.

Plasticizer—Chemical agent added to plastics to make them softer and more flexible.

Plastics—Materials based on hydrocarbons that are solid in the finished state.

Plastics tooling—Tools, dies, jigs, or fixtures primarily used for the metal-forming trades and constructed of plastics (usually laminates or casting materials).

Plastisols—Mixtures of finely ground plastics and plasticizers.

Platen—A thick, flat metal plate that supports the mold in an injection-molding machine.

Plug assist—A type of thermoforming that prestretches a sheet with a plug prior to the application of vacuum.

Plug-and-ring forming—A thermoforming technique that requires both a male mold and a female silhouette mold.

Pneumatics—Devices actuated by compressed air.

Polarity—Having a positive or negative electric or magnetic axis.

Polydispersity index (PI)—The ratio of the weight average molecular weight to the number average molecular weight.

Polymer—A compound with high molecular mass (weight), either natural or synthetic, whose structure can be represented by a repeated small unit (the mer). Some polymers are elastic and some are plastics.

Polymerization—The process of growing large molecules from small ones.

Polyvinyl acetal—A type of thermoplastic adhesive based on PVC.

Polyvinyl alcohol—A water-based adhesive for paper and textiles.

Polyvinyl butyral—The adhesive used as the interlayer in safety glass.

Polyvinyl formal—An adhesive used for wire enamels.

Pooled standard deviation—A standard deviation based on two or more standard deviations. It is used in graphic representations of distributions.

Potting—An embedding process for parts, similar to encapsulating except that the object may be simply covered and not surrounded by an envelope of plastics. Normally considered a coating process.

Prepreg—A reinforcement that has already been impregnated with resin.

Pre-puffs—The pre-expanded pieces of polymers used to make cellular polymer parts.

Primary chemical bonds—The basic way atoms attach to each other.

Profile extrusions—Extruded forms with complex shapes.

Promoter—A chemical, itself a feeble catalyst, that greatly speeds up the activity of a given catalyst. Also called an accelerator.

Proprietary molds—Molds made and owned by the molder.

Pulforming—A variation of pultrusion, pulforming uses molds to form various cross-sectional shapes.

Pultrusion—A continuous process for manufacturing composites with a constant cross-sectional shape. The process consists of pulling a fiber-reinforcing material through a resin-impregnation bath and into a shaping die where the resin is subsequently cured.

Purging—Cleaning one color or type of material from the cylinder of a molding machine.

Pyrolysis—Chemical decomposition of a substance by heat and pressure, used to change waste into usable compounds.

Pyrometer—A device used to measure thermal radiation.

Pyroxylin—A moderately nitrated form of cellulose. It was used extensively in early photographic processes.

Quality control—A procedure used to determine whether a product is being manufactured to specifications; a technique of management for achieving quality. Inspection is part of that technique.

Quantum—The smallest amount of energy that exists independently.

Radiation—A form of energy carried by waves or particles. The most common systems for polymers utilize electron beams.

Rake—The angle between a cutting tool and the surface of the workpiece.

Random copolymer—A type of copolymer in which the two types of monomers have random arrangement along the length of the molecular chain.

Rapid prototyping—Creating physical objects directly from CAD drawings without the use of traditional machining processes.

Reaction injection molding (RIM)—The molding process in which two or more liquid polymers are mixed by impingement atomizing in a mixing chamber, then injected into a closed mold.

Recommended exposure limit (REL)—A measure of acceptable exposure to hazardous chemicals.

Reinforced molding compound—A material reinforced with special fillers, fibers, or other materials to meet design needs.

Reinforced plastics—Plastics with strength increased by the addition of filler and reinforcing fibers, fabrics, or mats to the base resin.

Reinforced reaction injection molding (RRIM)—A process that combines reaction injection molding with fibrous reinforcements, usually glass mats.

Reinforcements—Materials that increase strength, stiffness, and impact resistance.

Relative density—The density of any material divided by that of water at a standard temperature, usually 20°C or 23°C (68°F or 73°F). Because water's density is nearly 1.00 g/cm³, density in grams per cubic centimeter and relative density are numerically equal.

Residual stresses—Stresses that remain in objects after manufacturing is complete.

Resin—Gum-like solid or semisolid substance that may be obtained from certain plants and trees or made from synthetic materials.

Resin injection molding—A process that involves injecting liquid resins into a mold.

Resin transfer molding (RTM)—The transfer of catalyzed resin into an enclosed mold in which the fiber reinforcing has been placed. Also called *resin-injection molding* and *liquid resin molding (LRM)*.

Resorcinol-formaldehyde—A type of phenolic adhesive.

Resource Conservation and Recovery Act (RCRA)—A federal act passed in 1976 that promotes reuse, reduction, incineration, and recycling of materials.

Rib—A reinforcing member of a fabricated or molded part.

Rigidized shell spray-up—A process that replaces a gel coat with a thermoformed shell, which is reinforced by spray or hand layup.

Rotational casting—A method used to make hollow objects from plastisols or powders. The mold is charged and rotated in one or more planes. The hot mold fuses the substance into a gel during rotation, covering all surfaces. The mold is then chilled and the product removed.

Rotogravure—A printing method in which ink is applied to a roller that contains recesses to trap the ink.

Roving—A bundle of untwisted strands, usually of fibrous glass.

rpm—revolutions per minute.

r/s—revolutions per second.

Rubber—Natural and synthetic polyisoprene.

Runners—Channels through which plastics flow from the sprue to the gates of mold cavities.

Runnerless molding—A term synonymous with insulated runner molding.

Sacrificial ply—In wet layup, a thin fabric is often used to protect the mold.

SAE code J1344—A code for marking plastics to aid in their identification. This code should assist the recycling of automotive plastics.

Safety factor—The ratio of the ultimate strength of the material to the allowable working stress.

Sandwich—A class of laminar composites composed of a lightweight core material (honeycomb, foamed plastics, etc.) to which two thin, dense, high-strength faces or skins are adhered.

Saturated/unsaturated compounds—Organic compounds that do not contain double or triple bonds and thus cannot add elements or compounds.

Scleroscope—An instrument for measuring impact resilience by dropping a ram with a flattened-cone tip from a given height onto the sample, and then noting the height of rebound.

Semipositive mold—A type of compression mold that has both vertical and horizontal flash.

Shear strength—The maximum load needed to produce a fracture by shearing action.

Sheet-molding compounds (SMC)—Leather-like mixtures of resins, catalysts, fillers, and reinforcements.

Shellac—A natural polymer; refined lac, a resin usually produced in thin, flaky layers or shells and used in varnish and insulating materials.

Short-term exposure limit (STEL)—A level of exposure to hazardous chemicals for 15 minutes.

Shrink fit—A joining method in which an insert is put into a plastics part while that part is hot. Shrink fitting takes advantage of the fact that plastics expand when heated and shrink when cooled. The plastics is normally heated and the insert placed in an undersized hole. Upon cooling, the plastics shrinks around the insert.

Side group—A group of atoms attached to the side of the main chain of plastics molecules.

Silk screen—A type of printing in which ink or paint is forced onto the product through a fine screen material.

Simple casting—Casting in which liquid resins are poured into molds.

Single covalent bond—The type of chemical bond found most frequently in plastics. It allows for rotation, thus permitting twisting and folding of molecular chains.

Sintering—Forming items from fusible powders. The process of holding the pressed powder at a temperature just below its melting point.

Slush casting—Casting in which a resin in liquid or powder form is poured into a hot mold where a viscous skin forms. The excess slush is drained off, the mold is cooled, and the casting removed.

Snap-back forming—A technique in which a plastics sheet is stretched to a bubble shape by vacuum or air pressure, a male mold is inserted into a bubble, and the vacuum or air pressure is released, allowing the plastics to snap back over the mold.

Solid phase pressure forming (SPFF)—A process that presses a blank into a sheet and then thermoforms the sheet.

Solubility parameter—A measure of the reactivity of plastics and organic solvents. A solvent with a lower solubility parameter than that of a selected plastics should dissolve that plastics.

Solution polymerization—A process where inert solvents are used to cause monomer solutions to polymerize.

Solvent cement—An adhesive technique that uses a solvent to dissolve the surfaces being joined.

Solvent resistance—The ability of a plastics material to withstand exposure to a solvent.

Solvent spinning—A technique for manufacturing fibers that involves the evaporation of solvents.

Specification—A statement of a set of requirements to be satisfied by a product, material, process, or system indicating (when appropriate) the procedure by which it may be determined whether the requirements are satisfied. Specifications may cite standards, be expressed in numerical terms, and include contractual agreements or requirements between the buyer and the seller.

Specular gloss—The relative luminous reflectance factor of a plastics sample.

Spiders—Legs that hold a mandrel in the center of an extrusion die.

Spin bonding or welding—A process of fusing two objects by forcing them together while one or both are spinning until frictional heat melts the interface. Spinning is then stopped and pressure held until the parts are frozen together.

Spinneret—A type of extrusion die with many tiny holes. A plastics melt is forced through the holes to make fine fibers and filaments.

Spot bonding—An ultrasonic welding technique similar to spot welding.

Spray coating—A coating process that atomizes a liquid coating with air or pressure.

Spray deposition—A technique for applying a layer of metal onto a mold surface.

Spray-up—A general term covering several processes using a spray gun. In reinforced plastics, the term applies to the simultaneous spraying of resin and chopped reinforcing fibers onto the mold or mandrel.

Sprue—In the mold, the channel or channels through which the plastics is led to the mold cavity.

Staking—Using heat or ultrasonics to form a head on a plastics stud.

Standard—A document or an object for physical comparison to define nomenclature, concepts, processes, materials, dimensions, relationships, interfaces, or test methods.

Standard deviation—A measure of the spread of a distribution.

Standard normal distribution—A distribution with a mean of zero and a standard deviation of 1.0.

Standard tolerance—The limits of dimensional variation that can be maintained in average manufacturing conditions.

Static casting—A casting process that fuses a layer of powdered plastics to the inner surface of a mold.

Stenciling—A printing method similar to silk screen, but without connecting mesh in open areas.

Stepwise growth polymerization—A type of polymerization in which two mers combine to form chains two mers long. Then the two mer pieces combine to form pieces four mers long. The reaction continues in this manner until completed.

Strain—The ratio of the elongation to the gauge length of a test sample; that is, the change in length per unit of original length.

Strand—A bundle of filaments.

Stress—The force producing, or tending to produce, deformation of a substance; expressed as the ratio of applied load to the original cross-sectional area.

Stretch blow molding—A blow-molding process that uses a rod to stretch a heated preform prior to blowing.

Structural foam—Cellular plastics with integral skin.

Suspension polymerization—A process in which liquid monomers are polymerized as liquid droplets suspended in water.

Symmetrical distribution—A distribution that has similar frequencies of values above and below the mean.

Syntactic plastics—Cellular resins or plastics with low-density fillers.

Tampoprint—A process of transferring ink from an engraved ink-filled surface to a product surface by the use of a flexible printing (transfer) pad.

Tensile strength—The maximum load needed to produce a fracture through tension.

Terpolymer—A polymer that consists of three distinct chemical types.

Testing—A term that implies that methods or procedures are used to determine physical, mechanical, chemical, optical, electrical, or other properties of a part.

Tex—An ISO standard unit of linear density used as a measure of yarn count. One tex is the linear density of a fabric that has a mass of 1 g and a length of 1 km and is equal to 10^{-6} kg/m.

Thermal-expansion resin transfer molding (TERTM)—A variation of the RTM process in which, after the resin is injected, heat causes the cellular core material to expand, forcing reinforcements and matrix against the mold walls.

Thermoforming—Any process of forming a thermoplastic sheet that consists of heating the sheet and pulling it down onto a mold surface.

Thermoplastic—(adj.) Capable of being repeatedly softened by heat and hardened by cooling. (n) A linear polymer that will repeatedly soften when heated and harden when cooled.

Thermoplastic elastomers (TPE)—A group of materials that can be processed like plastics but has physical characteristics similar to rubber.

Thermoplastics olefin elastomer (TPO)—A plastics based on polyethylene or polypropylene that exhibits the characteristics of rubber.

Thermoset or thermosetting—A network polymer that will undergo or has undergone a chemical reaction by the action of heat, catalysts, ultraviolet light, and so on, leading to a relatively infusible state.

Thixotropic agents—Materials that when added to a liquid cause it to become gel-like at rest, but fluid when agitated.

Thixotropy—State of materials that are gel-like at rest, but fluid when agitated. Liquids containing suspended solids are apt to be thixotropic.

Threshold limit value (TLV)—A term used by ACGIH to express the airborne concentration of a material to which nearly all persons can be exposed day after day without adverse effects. ACGIH expresses TLVs in three ways: (1) *TLV-TWA*: the allowable time-weighted average concentration for a normal 8-hour workday or 40-hour workweek; (2) *TLV-STEL*: the short-term exposure limit or maximum concentration for a continuous 15-minute exposure period (maximum of four such periods per day, with at least 60 minutes between exposure periods and provided that the daily *TLV-TWA* is not exceeded); (3) *TLV-C*: the ceiling limit—the concentration that should not be exceeded even instantaneously.

Time-weighted average (TWA)—A value that represents an exposure level acceptable for 8 hours per day as part of a 40-hour workweek.

Tolerances—Allowances for unintentional variations in dimensions.

Tooling—A broad term that refers to jigs, fixtures, dies, molds, gauges, and inspection equipment.

Toughness—A term with a wide variety of meanings, no single mechanical definition being generally recognized. Represented by the energy required to break a material equal to the area under the stress-strain curve.

Transfer blow—A type of blow molding that inflates a parison before it cools.

Transfer coating—A coating process that produces a tough, leather-like skin on fabric.

Transfer molding—A method of molding plastics in which the material is softened by heat and pressure in a transfer chamber, then forced by high pressure through the sprues, runners, and gates in a closed mold for final curing.

Triple covalent bond—A type of chemical bond that involves three bonds.

Tripoli—A silica abrasive.

Tumbling barrel—An inexpensive method of deflashing and polishing plastics parts by rotating them in a drum with abrasives and lubricants.

Twin sheet thermoforming—A thermoforming method to make thin-wall hollow objects.

Ultrasonic welding—A bonding process that relies on heat generated by rapid, low-amplitude vibrations.

Ultraviolet—An invisible portion of the light spectrum just beyond violet.

Undercut—Having a protuberance or indention that impedes withdrawal from a two-piece rigid mold. Flexible materials can be ejected intact with slight undercuts.

Uniaxial—Stretching on only one diretion.

Vacuum forming—Method of sheet forming in which the edges of the plastics sheet are clamped in a stationary frame, and the plastics is heated and drawn down by a vacuum into a mold.

Vacuum injection molding (VIM)—A process that utilizes a vacuum that draws reactive liquid resin into a mold cavity.

Vacuum metallizing—A process in which surfaces are thinly coated by exposing them to a metal vapor under vacuum.

Van der Waals forces—Weak secondary interatomic attraction arising from internal dipole effects.

Vehicle Recycling Partnership (VRP)—An organization of automobile companies that seeks to promote the recycling or cars and car parts.

Vibration micro lamination (VIM)—A casting process in which heated molds are vibrated in a bed of polymer pellets or powder.

Vicar softening point—The temperature at which a flat-ended needle of 1 mm² circular or square cross section will penetrate a thermoplastic specimen to a depth of 1 mm under a specified load using a uniform rate of temperature rise (definition from ASTM D1525).

Viscosity—A measure of the internal friction resulting when one layer of fluid is caused to move in relationship to another layer.

Vulcanize—The process of toughening natural rubber by compounding it with powdered sulfur.

Waste-to-energy (WTE)—A term to describe facilities that use municipal solid wastes as fuel in incinerators that produce electricity and steam.

Weight average molecular weight (M_w)—A type of molecular weight based on the contributions of each weight fraction to the total molecular weight of a sample.

Wet layup—A process that involves the addition of liquid resins to layers of reinforcements, usually fibrous mats or cloths.

Wet out—Removing air bubbles from reinforcing materials.

Wet spinning—A technique for manufacturing fibers that involves chemical coagulation of plastics gel.

Wet winding—Filaments pass through a bath of liquid resin prior to winding on a mandrel.

Whiting—Calcium carbonate powder abrasive.

Yarn—Bundle of twisted strands.

ABBREVIATIONS FOR SELECTED MATERIALS

APPENDIX B

Abbreviation	Polymer Term or Generic Name
ABA	Acrylonitrile-butadiene-acrylate
ABS	Acrylonitrile-butadiene-styrene
AES	Acrylonitrile-ethylpropylene-styrene
AI	Amide-imide polymers
AMMA	Acrylonitrile-methyl-methacrylate
AN	Acrylonitrile
AP	Ethylene propylene
APET	Amorphous polyethylene terephthalate
ASA	Acrylic-styrene-acrylonitrile
ATH	Aluminum trihydrate
AU	Polyester polyurethane
BBP	Butyl benzyl phthalate
BDP	Biodegradable polymers
BFK	Boron fiber reinforced plastic
BMC	Bulk molding compounds
BMI	Bismaleimide
BOPP	Biaxially-oriented polypropylene
CA	Cellulose acetate
CAB	Cellulose acetate-butyrate
CAP	Cellulose acetate propionate
CAR	Carbon fiber
CF	Cresol-formaldehyde
CFC	Chlorinated fluorocarbon
CFRP	Carbon fiber reinforced plastics
CMC	Carboxymethyl cellulose
CN	Cellulose nitrate
COPA	Copolyamide
CP	Cellulose propionate
CPE	Chlorinated polyethylene
CPET	Crystallized PET
CPVC	Chlorinated polyvinyl chloride
CS	Casein
CTFE	(Poly) Chlorotrifluoro ethylene
DAIP	Diallyl isophthalate resin

Abbreviation	Polymer Term or Generic Name
DAP	Diallyl phthalate resin
DCHP	Dicyclohexyl phthalate
DCPD	(Poly) Dicyclopentadiene
DGEBA	Diglycidyl ether of bisphenol A (epoxy)
DMAL	Dimethylacetamide
DMC	Dough molding compound
DOPT	Di-2-ethylhexyl terephthalate
DP	Degree of polymerization
EC	Ethyl cellulose
ECTFE	Ethylene-chlorotrifluoroethylene
EEA	Ethylene-ethyl acrylate
EMA	Ethylene-methyl acrylate
EMAC	Polyethylene methyl acrylate
EP	Epoxy
EPDM	Ethylene propylene diene rubber
EPE	Epoxy resin ester
EP-G-G	Prepreg of epoxy resin and glass fabric
EPM	Ethylene propylene copolymer
EPR	Ethylene and propylene copolymer
EPS	Expanded polystyrene
ESI	Ethylene-styrene copolymers
ETFE	Ethylene tetrafluoroethylene
EU	Polyether polyurethane
EVA	Ethylene vinyl acetate
EVOH	Ethylene vinyl alcohol
FEP	Fluorinated ethylene propylene (Also PFEP)
FRP	Glass fiber reinforced polyester
FRTP	Fiberglass-reinforced thermoplastics
GF	Glass fiber reinforced
GF-EP	Glass fiber–reinforced epoxy resin
GR	Glass fiber reinforced
GRP	Glass reinforced plastics
HCl	Hydrogen chloride

Abbreviation	Polymer Term or Generic Name
HCFC	Hydrochlorofluorocarbon
HDPE	High-density polyethylene
HF	Hydrogen fluoride
HIPS	High-impact polystyrene
HMC	High-strength molding compounds
HMW-HDPE	High molecular weight-high density polyethylene
HNP	High nitrile polymer
IPN	Interpenetrating polymer network
LCP	Liquid crystal polymers
LDPE	Low-density polyethylene
LIM	Liquid impingement molding
LLDPE	Linear low-density polyethylene
LMC	Low-cost molding compounds
LRM	Liquid resin molding
MA	Maleic anhydride
MBS	Methacrylate-butadiene-styrene
MC	Methyl cellulose
MDI	Methylene diisocyanate
MEK	Methyl ethyl ketone
MEKP	Methyl ethyl ketone peroxide
MF	Melamine-formaldehyde
MMA	Methyl methylacrylate
NBR	Acrylonitrile-butadiene rubber
OPET	Oriented polyethylene terephthalate
OPP	Oriented polypropylene
OPVC	Oriented polyvinylchloride
OSA	Olefin-modified styrene-acrylonitrile
PA	Polyamide
PAA	Polyacrylic acid
PAI	Polyamide-imide
PAN	Polyacrylonitrile
PAPI	Polymethylene polyphenyl isocyanate
PB	Polybutylene
PBAN	Polybutadiene-acrylonitrile
PBB	Polybrominated biphenyls
PBDE	Polybrominated diphenyl ether
PBS	Polybutadiene-styrene
PBT	Polybutylene terephthalate
PC	Polycarbonate
PCTFE	Polychlorotrifluoroethylene
PDAP	Polydiallyl phthalate
PE	Polyethylene
PEEK	Polyetheretherketone
PEI	Polyetherimide
PEN	Polyethylene naphthalate
PEO	Polyethylene oxide
PES	Polyether sulfone
PET	Polyethylene terephthalate
PETG	Glycol-modified PET
PEVA	Polyethylene vinyl acetate
PF	Phenol-formaldehyde resin
PFA	Perfluoroalkoxy
PFC	Perfluorocarbons

Abbreviation	Polymer Term or Generic Name
PFEP	Polyfluoroethylenepropylene
PHEMA	Polyhydroxyethyl methacrylate
PI	Polyimide
PIBSA	Polyisobutylene
PLA	Polyactic acid
PMA	Polymethyl acrylate
PMCA	Polymethylchloroacrylate
PMMA	Polymethyl methacrylate
PMP	Polymethylpentene
POM	Polyoxymethylene
PP	Polypropylene
PPC	Polyphthalate carbonate
PPE	Polyphenylene ether
PPO	Polyphenylene oxide
PPSO	Polyphenylsulfone
PS	Polystyrene
PSO	Polysulfone
PTFE	Polytetrafluoroethylene
PTMT	Polytetramethylene terephthalate
PU	Polyurethane
PUR	Polyurethane rubber
PVA	Polyvinyl alcohol
PVaC	Polyvinyl acetate
PVB	Polyvinyl butyral
PVC	Polyvinyl chloride
PVDC	Polyvinyl dichloride
PVDF	Polyvinylidene fluoride
PVF	Polyvinyl fluoride
PVF_2	Polyvinylidene fluoride
SAN	Styrene-acrylonitrile
SBP	Styrene-butadiene plastics
SBR	Styrene-butadiene rubber
SI	Silicone
SMA	Styrene-maleic anhydride
SMC	Sheet molding compounds
SRP	Styrene-rubber plastics
TDI	Toluene diisocyanate
TFE	Polytetrafluoroethylene
TMC	Thick molding compounds
TPA	Terephthalic acid
TPE	Thermoplastic elastomer
TPO	Thermoplastic olefin elastomers
TPU	Thermoplastic polyurethane
TPX	Polymethylpentene
UF	Urea-formaldehyde
UHMWPE	Ultra-high molecular weight polyethylene
UF	Urea-formaldehyde
UMC	Unidirectional molding compounds
UP	Urethane plastics
VAE	Vinyl acetate-ethylene
VCP	Vinyl chloride-propylene
VDC	Vinylidene chloride
VLDPE	Very low-density polyethylene

TRADE NAMES AND MANUFACTURERS

Trade Name	Polymer	Manufacturer
Abasfil	Reinforced ABS	AKZO Engineering
Absinol	ABS	Allied Resinous Products, Inc.
Absolac	ABS	Bayer AG
Absolan	SAN	Bayer AG
Abson	ABS resins and compounds	BF Goodrich Chemical Co.
Accord	Polyester amide	Bayer AG
Acelon	Cellulose acetate film	May & Baker, Ltd.
Acetophane	Cellulose acetate film	UCB-Sidac
Achieve	Metallocene PP	Exxon Mobil Chemical Co.
Aclar	CTFE fluorohalocarbon films	Allied Chemical Corp.
Acralen	Ethylene-vinyl acetate polymer	Verona Dyestuffs Div. Verona Corp.
Acrilan	Acrylic (acrylonitrile-vinyl chloride)	Monsanto Co.
Acroleaf	Hot stamping foil	Acromark Co.
Acrylaglas	Fiberglass-reinforced styrene-acrylonitrile	Dart Industries, Inc.
Acrylicomb	Acrylic-sheet-faced honeycomb	Dimensional Plastics Corp.
Acrylite	Acrylic molding compounds; cast acrylic sheets	Cyro
Acryloid	Acrylic modifiers for PVC; coating resins	Rohm & Haas Co.
Acrylux	Acrylic	Westlake Plastics Co.
Adell	Nylon 6,6	Adell
Adflex	Thermoplastic elastomers	Basell Polyolefins B.V.

Trade Name	Polymer	Manufacturer
Aeroflex	Polyethylene extrusions	Anchor Plastics Co.
Aeron	Plastic-coated nylon	Flexfilm Products, Inc.
Aerotuf	Polypropylene extrusions	Anchor Plastics Co.
Afcolene	Polystyrene and SAN copolymers	Pechiney-Saint-Gobain
Afcoryl	ABS copolymers	Pechiney-Saint-Gobain
Affinity	Polyolefin plastomers	Dow Plastics
Akulon	Nylon 6 and 6/6	Schulman
Alathon	Polyethylene resins	E. I. du Pont de Nemours & Co.
Alfane	Thermosetting epoxy resin cement	Atlas Minerals & Chemicals Div., of ESB, Inc.
Algoflon	PTFE	Ausimont
Alpha	Vinyl resins	Alpha Chemical and Plastics
Alpha-Clan	Reactive monomer	Marvon Div., Borg-Warner Corp.
Alphalux	PPO	Marvon Chemical Co.
Alsynite	Reinforced plastic panels	Reichhold Chemicals, Inc.
Amberlac	Modified alkyd resins	Rohm & Haas Co.
Amberol	Phenolic and maleic resins	Rohm & Haas Co.
Amer-Plate	PVC sheet material	Ameron Corrosion Control Div.
Ampol	Cellulose acetates	American Polymers, Inc.

Trade Name	Polymer	Manufacturer
Amres	Cellulose acetates	Pacific Resins & Chemicals, Inc.
Ancorex	ABS extrusions	Anchor Plastics Co.
Anvyl	Vinyl extrusions	Anchor Plastics Co.
APEC	Polycarbonate	Miles
Apogen	Epoxy resin series	Apogee Chemical, Inc.
Araclor	Polychlorinated polyphenyls	Monsanto Co.
Araldite	Epoxy resins and hardeners	CIBA Products Co.
Armorite	Vinyl coating	John L. Armitage & Co.
Arnel	Cellulose triacetate fiber	Celanese Corp.
Arochem	Modified phenolic resins	Ashland Chemical Co.
Arodure	Urea resins	Ashland Chemical Co.
Arofene	Phenolic resins	Ashland Chemical Co.
Aroplaz	Alkyd resins	Ashland Chemical Co.
Aroset	Acrylic resins	Ashland Chemical Co.
Arothane	Polyester resin	Ashland Chemical Co.
Artfoam	Rigid urethane foam	Strux Corp.
Arylon	Polyarl ether compounds	Uniroyal, Inc.
Arylon T	Polyaryl ether	Uniroyal, Inc.
Ascot	Coated spun-bonded polyolefin sheet	Appleton Coated Paper Co.
Ashlene	Nylon 6,6	Ashley
Astralit	Vinyl copolymer sheets	Dynamit Nobel of America, Inc.
Astroplay	Nylon, polyethylene, rubber	Astroturf LLC.
Astroturf	Nylon, polyethylene	Monsanto Co.
Astryn	High-performance compounds	Basell Polyolefins B. V.
Atlac	Polyester resin	Atlas Chemical Industries, Inc.
Aurum	Thermoplastic polyimide	Mitsui Chemicals, Inc.
Averam	Inorganic	FMC Corp.
Avisco	PVC films	FMC Corp.
Avistar	Polyester film	FMC Corp.
Avisun	Polypropylene	Avisun Corp.
Bakelite	Polyethylene, ethylene copolymers, epoxy, phenolic, polystyrene, phenoxy, ABS and vinyl resins and compounds	Union Carbide Corp.

Trade Name	Polymer	Manufacturer
Bayblend	PC/ABS blends	Bayer Corp.
Baydur	Rigid urethane foams	Bayer Corp.
Bayflex	Elastomeric polyurethane	Bayer Corp.
Beetle	Urea molding compounds	American Cyanamid Co.
Betalux	TFE-filled acetal	Westlake Plastics Co.
Biota PLA	Polyactide	Biota
Bio-Flex	PLA blends	FKuR Kunststoff GmbH
Blanex	Cross-linked polyethylene compounds	Reichhold Chemicals, Inc.
Blapol	Polyethylene compounds and color concentrates	Reichhold Chemicals, Inc.
Blapol	Polyethylene molding and extrusion compounds	Blane Chemical Div., Reichhold Chemicals, Inc.
Blendex	ABS resin	Marbon Div., Borg-Warner Corp.
Bolta Flex	Vinyl sheeting and Film	General Tire & Rubber Co., Chemical/ Plastics Div.
Bolta Thene	Rigid olefin sheets	General Tire & Rubber Co., Chemical/ Plastics Div.
Boltaron	ABS or PVC rigid plastic sheets	GenCorp
Boronal	Polyolefins with boron	Allied Resinous Products
Bostik	Epoxy and polyurethane adhesives	Bostik-Finch, Inc.
Bronco	Supported vinyl or pyroxylin	General Tire & Rubber Co., Chemical/ Plastics Div.
Budene	Polybutadiene	Goodyear Tire & Rubber Co., Chemical/ Plastics Div.
Butaprene	Styrene-butadiene latexes	Firestone Plastics Co. Div., Firestone Tire & Rubber
Cadco	Plastics rod, sheet, tubing, and film	Cadillac Plastic & Chemical Co.
Calibre	Polycarbonate	Dow
Capran	Nylon 6 film	Allied Chemical Corp.

Trade Name	Polymer	Manufacturer
Capran	Nylon films and sheet	Allied Chemical Corp.
Capron	Nylon 6,6	Allied Signal
Carbaglas	Fiberglass-reinforced polycarbonate	Fiberfil Div., Dart Industries, Inc.
Carilon	Polyketone	Shell Chemicals
Carolux	Filled urethane foam, flexible	North Carolina Foam Industries, Inc.
Carstan	Urethane foam catalysts	Cincinnati Milacron Chemicals, Inc.
Castcar	Cast polyolefin films	Mobil Chemical Co.
Castethane	Castable molding urethane elastomer system	Upjohn Co., CPR Div.
Castomer	Urethane elastomer system	Baxenden Chemical Co.
Castomer	Urethane elastomer and coatings	Isocyanate Products Div., Witco Chemical Corp.
Celanar	Polyester film	Hoechst Celanese
Celanex	Thermoplastic polyester	Hoechst Celanese
Celcon	Acetal copolymer resins	Hoechst Celanese
Cellasto	Microcellular urethane elastomer parts	North American Urethanes, Inc.
Cellofoam	Polystyrene foam	US Mineral Products Co.
Cellonex	Cellulose acetate	Dyamit Nobel of America, Inc.
Celluliner	Resilient expanded polystyrene foam	Gilman Brothers Co.
Cellulite	Expanded polystyrene foam	Gilman Brothers Co.
Celpak	Rigid polyurethane foam	Dacar Chemical Products Co.
Celthane	Rigid polyurethane foam	Dacar Chemical Products Co.
Chem-o-sol	PVC plastisol	Chemical Products Co.
Chem-o-thane	Polyurethane elastomer casting compounds	Chemical Products Corp.
Chemfluor	Fluorocarbon plastics	Chemplast, Inc.
Chemglaze	Polyurethane-based coating materials	Hughson Chemical Co., Div., Lord Corp.
Chemgrip	Epoxy adhesives for TFE	Chemplast, Inc.

Trade Name	Polymer	Manufacturer
Chevron PE	Polyethylene	Chevron
Cimglas	Fiberglass-reinforced polyester moldings	Cincinnati Milacron, Molded Plastics Div.
Clocel	Rigid urethane foam system	Baxenden Chemical Co.
Clopane	PVC film and tubing	Clopay Corp.
Cloudfoam	Polyurethane foam	International Foam Div., Holiday Inns of America
Co-Rexyn	Polyester resins and gel coats; pigment pastes	Interplastic Corp., Commercial Resins Div.
Cobocell	Cellulose acetate butyrate tubing	Cobon Plastics Corp.
Coboflon	Teflon tubing	Cobon Plastics Corp.
Cobothane	Ethylene-vinyl acetate tubing	Cobon Plastics Corp.
Colorail	Polyvinyl chloride handrails	Blum, Julius & Co.
Colovin	Calendered vinyl sheeting	Columbus Coated Fabrics
Conathane	Polyurethane casting, potting, tooling and adhesive compound	Conap, Inc.
Conolite	Polyester laminate	Woodall Industries, Inc.
Cordo	PVC foam and films	Ferro Corp., Composites Div.
Cordoflex	Polyvinylidene fluoride solutions, etc.	Ferro Corp., Composites Div.
Corlite	Reinforced foam	Snark Products, Inc.
Coror-Foam	Urethane foam systems	Cook Paint & Varnish
Coverlight HTV	Vinyl-coated nylon fabric	Reeves Brothers, Inc.
Creslan	Acrylic	American Cyanamid Co.
Crystic	Unsaturated polyester resins	Scott Bader Co.
Cumar	Coumarone-indene resins	Neville Chemical Co.
Curithane (Series)	Polyaniline polyamine; organo-mercury catalyst	Upjohn Co., Polymer Chemicals Div.
Curon	Polyurethane foam	Reeves Brothers, Inc.
Cycogel	ABS	Bayer AG
Cycolac	ABS resins	Marbon Div., Borg-Warner Corp.
Cycolon	Synthetic resinous compositions	Marbon Div., Borg-Warner Corp.

Trade Name	Polymer	Manufacturer
Cycoloy	Alloys of synthetic polymers w/ABS resins	Marbon Div., Borg-Warner Corp.
Cyclopac 930	ABS and nitrile barrier	Borg-Warner Chemicals
Cyovin	Self-extinguishing ABS graft-polymer blends	Marbon Div., Borg-Warner Corp.
Cyglas	Glass-filled polyester molding compound	American Cyanamid Co.
Cymel	Melamine molding compound	American Cyanamid Co.
Cyrex	Acrylic-PC alloy	Cyro
Dacovin	PVC compounds	Diamond Shamrock Chemical Co.
Dacron	Polyester	E. I. du Pont de Nemours & Co.
Dapon	Diallyl phthalate resin	FMC Corp., Organic Chemicals Div.
Daran	Polyvinyldene chloride emulsion coatings	W. R. Grace & Co., Polymers & Chemicals Div.
Daratak	Polyvinyl acetate homopolymer emulsions	W. R. Grace & Co., Polymers & Chemicals Div.
Darex	Styrene-butadiene latexes	W. R. Grace & Co., Polymers & Chemicals Div.
Davon	TFE resins & reinforced compounds	Davies Nitrate Co.
Delrin	Acetal resin	E. I. du Pont de Nemours & Co.
Densite	Molded flexible urethane foam	General Foam Div., Tenneco Chemical, Inc.
Derakane	Vinyl ester resins	Dow Chemical Co.
Desmopan	Thermoplastic urethane	Bayer Corp.
Dexflex	Thermoplastic olefins	Solvay Engineered Polymers
Dexon	Propylene-acrylic	Exxon Chemical USA
Diaron	Melamine resins	Reichhold Chemicals, Inc.
Dielux	Acetal	Westlake Plastics Co.
Dion-Iso	Isophthalic polyesters	Diamond Shamrock Chemical Co.

Trade Name	Polymer	Manufacturer
Dolphon	Epoxy resin and compounds polyester resins	John C. Dolph Co.
Dorvon	Molded polystyrene foam	Dow Chemical Co.
Dow Corning	Silicones	Dow Corning Corp.
Dowlex	LLDPE	Dow Plastics
Dri-Lite	Expanded polystyrene	Poly Foam, Inc.
Duco	Lacquers	E. I. du Pont de Nemours & Co.
Duracel	Lacquers for cellulose acetate & other plastics	Maas & Waldstein Co.
Duracon	Acetal copolymer	Polyplastics Co.
Duradene	Styrene-butadiene copolymer	Firestone Polymers
Duraflex	Polybutylene	Shell
Dural	Acrylic modified semirigid PVC	Alpha Chemical & Plastics Corp.
Duramac	Oil-modified alkyds	Commercial Solvents Corp.
Durane	Polyurethane	Raffi & Swanson, Inc.
Duraplex	Alkyd resins	Rohm & Haas Co.
Durelene	PVC flexible tubing	Plastic Corp. Warehousing
Durethan	Nylon 6	Miles
Durethene	Polyethylene film	Sinclair-Koppers Co.
Durez	Phenolic and alkyd resins	Chemical/ Occidental/ Plastics Corp.
Duron	Phenolic resins & molding compounds	Firestone Foam Products Co.
Dutral	EPM and EPDM rubber	EniChem SpA
Dyal	Alkyd and styrenated-alkyd resins	Sherwin Williams Chemicals
Dyalon	Urethane elastomer material	Thombert, Inc.
Dyloam	Expanded polystyrene	W. R. Grace & Co.
Dylan	Polyethylene	ARCO/Polymers, Inc.
Dylark	Polystyrene	NOVA Chemicals Corp.
Dylel	ABS plastics	Sinclair-Koppers Co.
Dylene	Polystyrene resin and oriented sheet	ARCO
Dylite	Expandable polystyrene bead, extruded sheets, etc.	ARCO

Trade Name	Polymer	Manufacturer
E-Form	Epoxy molding compounds	Allied Products Corp.
Eastar	Copolyester	Eastman Co.
Easy-Kote	Fluorocarbon release compound	Borco Chemicals, Inc.
Easypoxy	Epoxy adhesive Kits	Conap, Inc.
Ebolan	TFE compounds	Chicago Gasket Co.
Eccosil	Silicone resins	Emerson & Cumming, Inc.
Ecdel	Elastomers	Eastman Co.
Ektar	thermoplastic polyester	Eastman Plastics
El Rexene	Polyethylene, polypropylene, polystyrene and ABS resins	Dart Industries, Inc.
Elastoflex	Flexible polyurethane	BASF Corp.
Elastolit	Urethane engineering thermoplastic	North American Urethanes, Inc.
Elastollyx	Urethane engineering thermoplastic	North American Urethanes, Inc.
Elastolur	Urethane coatings	BASF
Elastonate	Urethane isocyanate prepolymers	BASF
Elastonol	Urethane polyester polyols	North American Urethanes, Inc.
Elastopel	Urethane engineering thermoplastics	North American Urethanes, Inc.
Elastopor	Rigid polyurethane foam	BASF Corp.
Electroglas	Cast acrylic	Gasflex Corp.
Elvace	Acetate-ethylene copolymers	E. I. du Pont de Nemours & Co.
Elvacet	Polyvinyl acetate emulsions	E. I. du Pont de Nemours & Co.
Elvacite	Acrylic resins	E. I. du Pont de Nemours & Co.
Elvamide	Nylon resins	E. I. du Pont de Nemours & Co.
Elvanol	Polyvinyl alcohols	E. I. du Pont de Nemours & Co.
Elvax	Vinyl resins; acid terpolymer resins	E. I. du Pont de Nemours & Co.
Empee	Polyethylene	Monmouth
Ensocote	PVC lacquer coating	Uniroyal, Inc.
Ensolex	Cellular plastic sheet material	Uniroyal, Inc.
Ensolite	Cellular plastic sheet material	Uniroyal, Inc.
Epi-Rez	Basic epoxy resins	Celanese Coatings Co.
Epi-Tex	Epoxy ester resins	Celanese Coatings Co.

Trade Name	Polymer	Manufacturer
Epikote	Epoxy resin	Shell Chemical Co.
Epocap	Two-part epoxy compounds	Hardman, Inc.
Epocast	Epoxies	Furane Plastics, Inc.
Epocrete	Two-part epoxy materials	Hardman, Inc.
Epocryl	Epoxy acrylate resin	Shell Chemical Co.
Epocure	Epoxy curing agents	Hardman, Inc.
Epolast	Two-part epoxy compounds	Hardman, Inc.
Epolene	Low-molecular-weight waxes	Eastman Co.
Epolite	Epoxy compounds	Hexcel Corp. Rezolin Div.
Epomarine	Two-part epoxy compounds	Hardman, Inc.
Epon	Epoxy resin; hardener	Shell Chemical Co.
Eponol	Linear polyether resin	Shell Chemical Co.
Eposet	Two-part epoxy compounds	Hardman, Inc.
Epotuf	Epoxy resins	Reichhold Chemicals, Inc.
Escor	Ethylene acrylic acetate copolymer	ExxonMobil Chemical Co.
Escorene	LDPE and LLDPE	Exxon
Estane	Polyurethane resins and compounds	BF Goodrich Chemical Co.
Estron	Acetate	Eastman Kodak Co.
Ethafoam	Polyethylene foam	Dow Chemical Co.
Ethocel	Ethyl cellulose resin	Dow Chemical Co.
Ethofil	Fiberglass-reinforced polyethylene	AKZO Engineering
Ethoglas	Fiberglass-reinforced polyethylene	Fiberfil Div., Dart Industries, Inc.
Ethosar	Fiberglass-reinforced polyethylene	Fiberfil Div., Dart Industries, Inc.
Ethylux	Polyethylene	Westlake Plastics Co.
Evenglo	Polystyrene resins	Sinclair-Koppers Co.
Everflex	Polyvinyl acetate copolymer emulsion	W. R. Grace & Co., Polymers & Chemicals Div.
Everlon	Urethane foam	Stauffer Chemical Co.
Exact	Metallocene plastomer	ExxonMobil Chemical Co.

Trade Name	Polymer	Manufacturer
Exceed	LLDPE	ExxonMobil Chemical Co.
Excelite	Polyethylene tubing	Thermoplastic Processes
Exon	PVC resins, compounds, and latexes	Firestone Tire & Rubber Co.
Extane	Polyurethane tubing	Pipe Line Service Co.
Extrel	Polyethylene & polyproplylene films	Exxon Chemical, USA
Extren	Fiberglass-reinforced polyester shapes	Morrison Molded Fiber Glass Co.
Fabrikoid	Pyroxylin-coated fabric	Stauffer Chemical Co.
Facilon	Reinforced PVC fabrics	Sun Chemical Corp.
Fassgard	Vinyl coating on nylon	M. J. Fassler & Co.
Fasslon	Vinyl coating	M. J. Fassler & Co.
Felor	Nylon filaments	E. I. du Pont de Nemours & Co.
Ferex	High-gloss PP	Ferro Corp.
Ferrene	Polyolefins	Ferro Corp.
Fiber foam	Polyester-reinforced foam	Weeks Engineered Plastics
Fiberite	Melamine molding compound	ICI Fiberite
Fibro	Rayon	Courtaulds NA, Inc.
Fina	Polypropylene	Fina
Finacene	HDPE and LDPE	ATOFINA
Finaprene	Elastomer SBS	ATOFINA
Finathene	HDPE	ATOFINA
Flexane	Urethanes	Devcon Corp.
Flexathene	TPO	Equistar Chemicals LP
Flexocel	Urethane foam systems	Baxenden Chem Co.
Flexprene	Styrenic TPEs	Teknor Apex
Floranier	Cellulose for esters	ITT Rayonier, Inc.
Fluokem	Teflon spray	Bel-Art Products
Fluon	TFE resin	ICI American, Inc.
Fluorglas	PTFE-coated and impregnated woven glass fabric, laminates, belting	Dodge Industries, Inc.
Fluorocord	Fluorocarbon material	Raybestos Manhattan
Fluorofilm	Cast Teflon films	Dilectrix Corp.
Fluoroglide	Dry-film lubricant of TFE	Chemplast, Inc.
Fluororay	Filled fluorocarbon	Raybestos Manhattan

Trade Name	Polymer	Manufacturer
Fluorored	Compounds of TFE	John L. Dore Co.
Fluorosint	TFE-fluorocarbon base composition	Polymer Corp.
Foamthane	Rigid polyurethane foam	Pittsburgh Corning Corp.
Formadall	Polyester premix compound	Woodall Industries, Inc.
Formaldafil	Fiberglass-reinforced acetal	AKZO Engineering
Formaldaglas	Fiberglass-reinforced acetal	Fiberfil Div., Dart Industries, Inc.
Formaldasar	Fiberglass-reinforced acetal	Fiberfil Div., Dart Industries, Inc.
Formica	High-pressure laminate	American Cyanamid Co.
Formrez	Urethane elastomer chemicals	Witco Chemical Corp., Organics Div.
Formvar	Polyvinyl formal resins	Monsanto Co.
Forticel	Cellulose propionate flake, resins	Hoechst Celanese
Fortiflex	Polyethylene resins	Solvay Polymers
Fortilene	Polypropylene Solvay Polymers	
Fortrel	Polyester, Inc.	Fiber Industries,
Fortron	Polyphenylene sulfide	Ticona
Fosta-Net	Polystyrene foam extruded mesh	Foster Grant Co.
Fosta Tuf-Flex	High-impact polystyrene	Foster Grant Co.
Fostacryl	Thermoplastic polystyrene resins	Foster Grant Co.
Fostafoam	Expandable polystyrene beads	Foster Grant Co.
Fostalite	Light-stable polystyrene molding powder	Foster Grant Co.
Fostarene	Polystyrene molding powder	Foster Grant Co.
Futron	Polyethylene powder	Fusion Rubbermaid Co.
Geloy	Weatherable ASA	GE Plastics
Gelva	Polyvinyl acetate	Monsanto Co.
Genal	Phenolic compounds	General Electric Co.
Genthane	Polyurethane rubber	General Tire & Rubber Co.
Gentro	Styrene butadiene rubber	General Tire & Rubber Co.
Geon	Vinyl resins, compounds, latexes	BF Goodrich Chemical Co.
Gil-Fold	Polyethylene sheet	Gilman Brothers Co.
Glaskyd	Alkyd molding compound	American Cyanamid Co.

Trade Name	Polymer	Manufacturer
Glyptal	Alkyd resins	General Electric Co.
Goldglas	PMMA	ATOFINA
Gordon Superdense	Polystyrene in pellet form	Hammond Plastics, Inc.
Gordon Superflow	Polystyrene in granular or pellet form	Hammond Plastics, Inc.
Gracon	PVC compounds	W. R. Grace & Co.
GravoFLEX	ABS sheets	Hermes Plastics, Inc.
GravoPLY	Acrylic sheets	Hermes Plastics, Inc.
Grilamid	Nylon 12	EMS
Grilon	Nylon 6	EMS
GUR	UHMW-PE	Ticona
Halon	TFE molding compounds	Allied Chemical Corp.
Haylar	CTFE	Ausimont
Haysite	Polyester laminates	Synthane-Taylor Corp.
Herculon	Olefin	Hercules, Inc.
Herox	Nylon filaments	E. I. du Pont de Nemours & Co.
Hetrofoam	Fire-retardant urethane foam systems	Durez Div., Hooker Chemical Corp.
Hetron	Fire-retardant polyester resins	Durez Div., Hooker Chemical Corp.
Hex-One	High-density polyethylene	Gulf Oil Co.
Hicor	Oriented polyethylene	Mobil Plastics
Hi-fax	Polyethylene	Hercules, Inc.
HiGlass	Glass-reinforced PP	Himont
Hipertuf	PEN	Shell Chemicals
Hi-Styrolux	High-impact polystyrene	Westlake Plastics
Hostaform	Acetyl copolymer	Ticona
Hostalen	Polyethylene	Hoechst Celanese
Hostaflon	PTFE	Hoechst Celanese
Hydrepoxy	Water-based epoxies	Acme Chemicals Div., Allied Products Corp.
Hydro Foam	Expanded phenolformaldehyde	Smithers Co.
Hyflon	Perfluoroalkoxy	Ausimont
Hylar	Polyvinylidene fluoride	Solvay Polymers
Impet	Thermoplastic polyester	Ticona
Implex	Acrylic molding powder	Rohm & Haas Co.
Ingeo	Poly(lactic acid)	NatureWorks LLC
Inspire	PP	Dow Plastics

Trade Name	Polymer	Manufacturer
Intamix	Rigid PVC compounds	Diamond Shamrock Chemical Co.
Interpol	Copolymeric resinous systems	Freeman Chemical Corp.
Intol	Styrene-butadiene rubber	EniChem SpA
Iotek	Ionomers	ExxonMobil Chemical Co.
Irvinil	PVC resins and compounds	Great American Chem.
Isoderm	Urethane rigid and flexible integral-skinning foam	Upjohn Co., CPR Div.
Isofoam	Urethane foam systems	Witco Chemical Corp.
Isonate	Diisocyanates and urethane systems	Upjohn Co., CPR Div.
Isoteraglas	Isocyanate elastomer-coated Dacron glass fabric	Natvar Corp.
Isothane	Flexible polyurethane foams	Bernel Foam Products Co.
Iupilon	Polycarbonate engineering	Mitsubishi Plastics Corp.
Jetfoam	Polyurethane foam	International Foam
K-Prene	Urethane cast material	Di-Acro Kaufman
Kadel	Polyketone polymers	BP Amoco
Kalex	Two-part polyurethane elastomers	Hardman, Inc.
Kalidar	PEN	DuPont
Kalspray	Rigid urethane foam system	Baxenden Chemical Co.
Kamax	Acrylic	Rohm & Haas
Kapton	Polyimide	E. I. du Pont de Nemours & Co.
Keltrol	Vinyl toluene copolymer	Spencer Kellogg
Ken U-Thane	Polyurethanes; urethane foam ingredients	Kenrich Petrochemicals, Inc.
Kencolor	Silicone/pigments dispersion	Kenrich Petrochemicals, Inc.
Kevlar	Aramid fiber	E. I. du Pont de Nemours & Co.
Kodacel	Cellulosic film and sheeting	Eastman Chemical Products, Inc.
Kodar	Copolyester thermoplastics	Eastman Chemical Products, Inc.
Kodel	Polyester	Eastman Kodak Co.

Trade Name	Polymer	Manufacturer
Kohinor	Vinyl resins and compounds	Pantasote Co.
Korad	Acrylic film	Rohm & Haas Co.
Koroseal	Vinyl films	BF Goodrich Chemical Co.
Kralastic	ABS high-impact resin	Uniroyal, Inc.
Kralon	High-impact styrene and ABS resins	Uniroyal, Inc.
Kraton	Styrene-butadiene polymers	Shell Chemical Co.
Krene	Plastic film and sheeting	Union Carbide Corp.
Krystal	PVC sheet	Allied Chemical Corp.
Krystaltite	PVC shrink films	Allied Chemical Corp.
Kydene	Acrylic/PVC powder	Rohm & Haas Co.
Kydex	Acrylic PVC sheets	Rohm & Haas Co.
Kynar (Series)	Polyvinylidene Fluoride	Elf Atochem North America
Lacqtene	Polyethylene	ATOFINA
Lamabond	Reinforced Polyethylene	Lamex, Columbian Carbon Co.
Lamar	Mylar vinyl laminate	Morgan Adhesives Co.
Laminac	Polyester resins	American Cyanamid Co.
Last-A-foam	Plastic foam	General Plastics Mfg.
Levapren	EVA	Bayer Corp.
Lexan	Polycarbonate resins, film, sheet	General Electric Co., Plastics Dept.
Lomod	Copolyester elastomer	GE Plastics
Lotryl	EBA and EDA copolymers	ATOFINA
Lucalor	Chlorinated PVC	ATOFINA
Lucite	Acrylic resins	E. I. du Pont de Nemours & Co.
Luflexen	Metallocene MLLDPE	Basell Polyolefins B. V.
Lumasite	Acrylic sheet	American Acrylic Corp.
Lupolen	Polyethylene	Basell Polyolefins B. V.
Luran	Styrene acrylonitrile copolymers	BASF
Lustran	SAN and ABS molding and extrusion resins	Monsanto Co.

Trade Name	Polymer	Manufacturer
Lustrex	Polystyrene molding and extrusion resins	Monsanto Co.
Lycra	Spandex	E. I. du Pont de Nemours & Co.
Lytex	Epoxy	Quantum Composites
Macal	Cast vinyl film	Morgan Adhesives Co.
Maclin	Vinyl resins	Maclin
Magnum	ABS	Dow Plastics
Makrolon	Polycarbonate	Miles
Marafoam	Polyurethane foam resin	Marblette Co.
Maraglas	Epoxy casting resin	Marblette Co.
Maranyl	Nylon 6,6	ICI Americas
Maraset	Epoxy resin	Marblette Co.
Marathane	Urethane compounds	Allied Products Corp.
Maraweld	Epoxy resin	Marblette Co.
Marlex	Polyethylenes, polypropylenes, other polyolefin plastics	Phillips Petroleum
Marnot	Polycarbonate film	Bayer AG
Marvinol	Vinyl resins and compounds	Uniroyal, Inc.
Meldin	Polyimide and reinforced polyimide	Dixon Corp.
Melinar	PET for containers	E. I. du Pont de Nemours & Co.
Merlon	Polycarbonate	Mobay Chemical Co.
Metallex	Cast acrylic sheets	Hermes Plastics, Inc.
Meticone	Silicone rubber dies and sheets	Hermes Plastics, Inc.
Metocene	Metallocene PP	Basell Polyolefins B. V.
Metre-Set	Epoxy adhesives	Metachem Resins Corp.
Micarta	Thermosetting laminates	Westinghouse Electric Corp.
Micro-Matte	Extruded acrylic sheet with matte finish	Extrudaline, Inc.
Micropel	Nylon powders	Nypel, Inc.
Microsol	Vinyl plastisol	Michigan Chrome & Chemical Co.
Microthene	Powdered polyolefins	US Industrial Chemicals Co.
Milmar	Polyester	Morgan Adhesives Co.
Mini-Vaps	Expanded polyethylene	Malge Co., Agile Div.

Trade Name	Polymer	Manufacturer
Minit Man	Epoxy adhesive	Kristal Draft, Inc.
Minlon	Mineral-reinforced nylon	E. I. du Pont de Nemours & Co.
Mipoplast	Flexible PVC sheets	Dynamit Nobel of America, Inc.
Mirasol	Alkyd resins: epoxy ester	C. J. Osborn Chemicals, Inc.
Mirbane	Amino resin	Highpolymer Co.
Mirel	PLA for injection molding	Telles
Mirrex	Calendered rigid PVC	Tenneco Chemicals, Inc., Tenneco Plastics Div.
Mista Foam	Urethane foam systems	M. R. Plastics & Coatings, Inc.
Mod-Epox	Epoxy resin modifier	Monsanto Co.
Molycor	Glass-fiber-reinforced epoxy tubing	A. O. Smith, Inland, Inc.
Mondur	Isocyanates	Mobay Chemical Co.
Monocast	Direct polymerized nylon	Polymer Corp.
Moplen	Isotactic polypropylene	Montecatini Edison S.p.A.
Multrathane	Urethane elastomer chemical	Mobay Chemical Co.
Multron	Polyesters	Mobay Chemical Co.
Mylar	Polyester film	E. I. du Pont de Nemours & Co.
Nakan	Thermoplastic vinyl	ATOFINA
Nanofil	Clay nanocomposite	Süd-Chemie AG
Nanosolve	Carbon nanocomposite	Zyvex Performance Materials
Napryl	Polypropylene	Pechiney-Saint-Gobain
Natene	High-density polyethylene	Pechniney-Saint-Gobain
Natureworks	PLA	NatureWorks LLC
Naugahyde	Vinyl-coated fabrics	Uniroyal, Inc.
NeoCryl	Acrylic resins and resin emulsions	Polyvinyl Chemicals, Inc.
Neopolen	Expanded PE bead	BASF
NeoRez	Styrene emulsions and urethane solutions	Polyvinyl Chemicals, Inc.
NeoVac	PVA emulsions	Polyvinyl Chemicals, Inc.
Nestorite	Phenolic and urea-formaldehyde	James Ferguson & Sons
Nevillac	Modified coumarone-indene resin	Neville Chemical Co.

Trade Name	Polymer	Manufacturer
Nimbus	Polyurethane foam	General Tire & Rubber Co.
Nitrocol	Nitrocellulose base pigment dispersion	C. J. Osborn Chemicals, Inc.
Nob-Lock	PVC sheet material	Ameron Corrosion Control Div.
Nopcofoam	Urethane foam systems	Diamond Shamrock Chemical Co., Resinous Products Div.
Norchem	Low-density polyethylene resin	Northern Petrochemical Co.
Nordel IP	Hydrocarbon rubber Elastomers	DuPont Dow
Noryl	Modified polyethylene oxide	General Electric Co., Plastics Dept.
Novacor	Polystyrene	Novacor
Novamid	Polyamide 6 engineering	Mitsubishi Plastics Corp.
Novarex	Polycarbonate engineering	Mitsubishi Plastics Corp.
Novodur	ABS	Bayer AG
Nupol	Thermosetting acrylic resins	Freeman Chemical Corp.
NYCOA	Nylon 6,6	Nylon Corp. of America
Nyglathane	Glass-filled polyurethane	Nypel, Inc.
Nylafil	Fiberglass-reinforced nylon	AKZO Engineering
Nylaglas	Fiberglass-reinforced nylon	AKZO Engineering
Nylasar	Fiberglass-reinforced nylon	AKZO Engineering
Nylasint	Sintered nylon parts	Polymer Corp.
Nylatron	Filled nylons	Polymer Corp.
Nylo-Seal	Nylon 11 tubing	Imperial-Eastman Corp.
Nylux	Nylon	Westlake Plastics Co.
Nypelube	TFE-filled nylons	Nypel, Inc.
Nyreg	Glass-reinforced nylon molding compounds	Nypel, Inc.
Oasis	Expanded phenol-formaldehyde	Smithers Co.
Oilon Pv 80	Acetal-based resin sheets rods, tubing, profiles	Cadillac Plastic & Chemical Co.
Olefane	Polypropylene film	Amoco Chemicals Corp.
Olefil	Filled polypropylene resin	Amoco Chemicals Corp.

Trade Name	Polymer	Manufacturer
Oleflo	Polypropylene resin	Amoco Chemicals Corp.
Olemer	Copolymer polypropylene	Amoco Chemicals Corp.
Oletac	Amorphous polypropylene	Amoco Chemicals Corp.
Opalon	Flexible PVC materials	Monsanto Co.
Oppanol	Polyisobutylene	BASF Wyandotte Corp.
Optema	Ethylene-methyl acrylate	ExxonMobil Chemical Corp.
Optum	Polyolefin alloy	Ferro Corp.
Orevac	EVA terpolymer	ATOFINA
Orgalacqe	Epoxy and PVC powders	Aquitaine-Organico
Orgamide R	Nylon 6	Aquitaine-Organico
Orlon	Acrylic fiber	E. I. du Pont de Nemours & Co.
Oxy	PVC series	Occidental Chemical Co.
Oxyblend	PVC	Occidental Chemical Co.
Panda	Vinyl and urethane-coated fabric	Pandel-Bradford, Inc.
Papi	Polymethylene polyphenyliso-cyanate	Upjohn Co., Polymer Chemicals Div.
Paradene	Dark coumarone-indene resins	Neville Chemical Co.
Paraplex	Polyester resins and plasticizers	Rohm & Haas Co.
Paxon	Polyethylene	Allied Signal
Pelaspan	Expandable polystyrene	Dow Chemical Co.
Pelaspan-Pac	Expandable polystyrene	Dow Chemical Co.
Pellethane	Thermoplastic urethane	Dow Plastics
Pellon Aire	Nonwoven textile	Pellon Corp.
Pentafood	Co-extruded film	Klöckner Pentaplast Group
Penton	Chlorinated polyether	Hercules, Inc.
PermaRex	Cast epoxy	Permali, Inc.
Permelite	Melamine molding compound	Melamine Plastics, Inc.
Petion	Glass-filled PET	Miles
Petra	Polyester	Allied Signal
Pethrothene	Low-, medium-, and high-density polyethylene	Quantum USI
Petrothene XL	Cross-linkable polyethylene	US Industrial Chemical Co.

Trade Name	Polymer	Manufacturer
Phenoweld	Phenolic adhesive	Hardman, Inc.
Philjo	Polyolefin films	Phillips-Joana Co.
Philprene	Styrene-butadiene	Phillips Chemical Corp.
Piccoflex	Acrylonitrile-styrene resins	Pennsylvania Industrial Chemical Corp.
Piccolastic	Polystyrene resins	Pennsylvania Industrial Chemical Corp.
Piccotex	Vinyl-toluene copolymer	Pennsylvania Industrial Chemical Corp.
Piccournaron	Coumarone-indene resins	Pennsylvania Industrial Chemical Corp.
Piccovar	Alkyl-aromatic: resins	Pennsylvania Industrial Chemical Corp.
Pienco	Polyester resins	Mol-Rex Div., American Petrochemical Corp.
Pinpoly	Reinforced polyurethane foam	Holiday Inns of America, Inc.
Plaskon	Plastic molding	Allied Chemical Corp.
Plastic Steel	Epoxy tooling and repair	Devcon Corp.
Platamid	Hot-melt adhesive	ATOFINA
Platherm	Copolyester hot-melt adhesive	ATOFINA
Plenco	Melamine and phenolic	Plastics Engineering
Pleogen	Polyester resins and gel coats; polyurethane systems	Mol-Rex Div., Whittake Corp.
Plexiglas	Acrylic sheets and molding powders	Rohm & Haas Co.
Plicose	Polyethylene film, sheeting, tubing, bags	Diamond Shamrock Corp.
Pliobond	Adhesive	Goodyear Tire & Rubber Co.
Pliolite	Styrene-butadiene resins	Goodyear Tire & Rubber Co.
Pliothene	Polyethylene-rubber blends	Ametek/ Westchester Plastics
Pliovic	PVC resins	Goodyear Tire & Rubber Co.
Pluracol	Polyethers	BASF Wyandotte Corp.
Pluragard	Urethane foams	BASF Wyandotte Corp.

Trade Name	Polymer	Manufacturer
Pluronic	Polyethers	BASF Wyandotte Corp.
Plyocite	Phenolic-impregnated overlays	Reichhold Chemicals, Inc.
Plyophen	Phenolic resins	Reichhold Chemicals, Inc.
Pocan	Polyester, thermoplastic	Miles
Polex	Oriented acrylic	Southwestern Plastics, Inc.
Pollopas	Urea-formaldehyde compounds	Dynamit Nobel of America, Inc.
Polvonite	Cellular plastic material in sheet form	Voplex Corp.
Polycarbafil	Glass-reinforced PC	Akzo Engineering
Poly-Dap	Diallyl phthalate electrical molding compounds	US Polymeric, Inc.
Poly-Eth	Low-density polyethylene	Gulf Oil Corp.
Poly-Eth-Hi-D	High-density polyethylene	Gulf Oil Corp.
Polycarbafil	Fiberglass-reinforced polycarbonate	Fiberfill Div., Dart Industries, Inc.
Polycure	Cross-linked polyethylene compounds	Crooke Color & Chemical Co.
Polyfoam	Polyurethane foam	General Tire & Rubber Co.
Polyfort	Reinforced polypropylene	Schulman
Polyimidal	Thermoplastic polyimide	Raychem Corp.
Polylite	Polyester resins	Reichhold Chemicals, Inc.
Polymet	Plastic-filled sintered metal	Polymer Corp.
Polymul (series)	Polyethylene emulsions	Diamond Shamrock Chemical Co.
Polyteraglas	Polyester-coated Dacron-glass fabric	Natvar Corp.
Polywrap	Plastic film	Flex-O-Glass, Inc.
Poxy-Gard	Solventless epoxy compounds	Sterling, Div. Reichhold Chemicals, Inc.
PPO	Polyphenylene oxide	Reichhold Chemicals, Inc.
Prevail	Thermoplastics	Dow Plastics
Pro-fax	Polypropylene	Hercules, Inc.
Profil	Fiberglass-reinforced polypropylene	AKZO Engineering
Proglas	Fiberglass-reinforced polypropylene	Fiberfil Div., Dart Industries, Inc.

Trade Name	Polymer	Manufacturer
Prohi	High-density polyethylene	Protective Lining Corp.
Propathene	Polypropylene polymers and compound	Imperical Chemical Ind., Ltd., Plastics Div.
Propylsar	Fiberglass-reinforced polypropylene	Fiberfil Div., Dart Industries, Inc.
Propylux	Polypropylene	Westlake Plastics Co.
Protectolite	Polyethylene film	Protective Lining Corp.
Protron	Ultra-high-strength polyethylene	Protective Lining Corp.
Pulse	PC/ABS blend	Dow Plastics
Purilon	Rayon	FMC Corp.
Quelflam	Urethanes, low surface spread flame	Baxenden Chemical Co.
Radel A	Polyethersulfone polymers	BP Amoco
Radel B	Polyphenylsulfone polymers	BP Amoco
Radilon	Nylon 6,6 International	Polymers
Rayflex	Rayon	FMC Corp.
Regalite	Press-polished clear flexible PVC	Tenneco Advanced Materials, Inc.
REN-Shape	Epoxy material	Ren Plastics, Inc.
Ren-Thane	Urethane elastomers	Ren Plastics, Inc.
Resiglas	Polyester resins, etc.	Kristal Draft, Inc.
Resimene	Malamine resins	Monsanto Co.
Resinoid	Phenolic molding compound	Resinoid
Resinol	Polyolefins	Allied Resinous Products, Inc.
Resinox	Phenolic resins	Monsanto Co.
Resorasa-bond	Resorcinol and phenol-resorcinol	Pacific Resins & Chemicals, Inc.
Respond	Thermoplastic elastomers	Solvay Engineered Polymers
Restfoam	Urethane foam	Stauffer Chemical Co., Plastics Div.
Retain	Recycled ABS and ABS/PC	Dow Plastics
Rexene	PE film	Rexene
Rexolene	Cross-linked polyolefin sheet	Brand-Rex Co.
Rexolite	Polystyrene rod and sheet stock	Brand-Rex Co.
Reynosol	Urethane, PVC	Hoover Ball & Bearing Co.
Rilsan	Nylon 11 and 12 copolymers	ATOFINA
Rhodiod	Cellulose acetate sheet	M & B Plastics, Ltd.

Trade Name	Polymer	Manufacturer
Rhoplex	Acrylic emulsion	Rohm & Haas Co.
Richfoam	Urethane foam	E. R. Carpenter Co.
Rigidite	Modified acrylic and sheet polyester resins	American Cyanamid Co.
Rigidsol	Rigid plastisol	Watson-Standard Co.
Rimtec	Vinyl resins	Rimtec
Rogers	Epoxy molding compound	Rogers
Rolox	Two-part epoxy compounds	Hardman, Inc.
Rotothene	Polyethylene for rotomolding	Rototron Corp.
Rotothon	Polypropylene for rotomolding	Rototron Corp.
Royalex	Structural cellular thermoplastic sheet material	Uniroyal, Inc.
Royalite	Thermoplastic sheet material	Uniroyal, Inc., Uniroyal Plastic Products
Roylar	Polyurethane elastoplastic	Uniroyal, Inc.
Rucoam	Vinyl film and sheeting	Hooker Chemical Corp.
Rucoblend	Vinyl compounds	Hooker Chemical Corp.
Rucon	Vinyl resins	Hooker Chemical Corp.
Rucothane	Polyurethanes	Hooker Chemical Corp.
Rynite	PET	E. I. du Pont de Nemours & Co.
Ryton	Polyphenylene Sulfide	Phillips Chemical Co.
Santolite	Aryl sulfonamide-formaldehyde resin	Monsanto Co.
Saran	Polyvinylidene chloride resin	Dow Chemical Co.
Satin Foam	Extruded polystyrene foam	Dow Chemical Co.
Scotchpak	Heat-sealable polyester film	3M Co.
Scotchpar	Polyester film	3M Co.
Selectrofoam	Urethane foam systems and polyols	PPG Industries, Inc.
Selectron	Polymerizable synthetic resins; polyesters	PPG Industries, Inc.
Sequel	Engineered polyolefins	Solvay Engineered Polymers

Trade Name	Polymer	Manufacturer
Shareen	Nylon	Courtaulds North America, Inc.
Shuvin	Vinyl molding compounds	Blane Chemical Div., Reichhold Chemicals, Inc.
Silastic	Silicone rubber	Dow Corning Corp.
Silly Putty	Silicone	Crayola
Sinkral	ABS	EniChem SpA
Sinvet	Polycarbonate	Bayer AG
Sipon	Alkyl and aryl resin	Alcolac, Inc.
Siponate	Alkyl and aryl sulfonates	Alcolac, Inc.
Skinwich	Urethane rigid and flexible integral-skinning foam	Upjohn Co.
Soarnol	EVA	ATOFINA
Softlite	Ionomer foam	Gilman Brothers Co.
Solarflex	Chlorinated polyethylene	Pantasote Co.
Solef	Polyvinylidene fluoride	Solvay Polymers
Solithane	Urethane prepolymers	Thiokol Chemical Corp.
Sonite	Epoxy resin compound	Smooth-On, Inc.
Spandal	Rigid urethane laminates	Baxenclen Chemical Co.
Spandofoam	Rigid urethane foam board and slab	Baxenden Chemical Co.
Spancloplast	Expanded polystyrene board and slab	Baxenden Chemical Co.
Spectar	Copolyester	Eastman Co.
Spectran	Polyester	Monsanto Textiles Co.
Spenkel	Polyurethane resins	Spencer Kellogg Div., Textron Inc.
Spudware	Starch-based biodegradable Plastic	Excellent Packaging Supply (EPS)
Starez	Polyvinyl acetate resin	Standard Brands Chemical Ind., Inc.
Structoform	Sheet molding compounds	Fiberite Corp.
Stryton	Nylon	Phillips Fibers Corp.
Stylafoam	Coated polystyrene sheet	Gilman Brothers Co.
Stypol	Polyesters	Freeman Chemical Corp., Div., H. H. Robertson Co.
Styrafil	Fiberglass-reinforced polystyrene	AKZO Engineering

Trade Name	Polymer	Manufacturer
Styroblend	Styrene-butadiene blends	BASF
Styrodur C	Rigid polystyrene foam	BASF
Styroflex	Biaxially oriented polystyrene film	Natvar Corp.
Styrofoam	Polystyrene foam	Dow Chemical Co.
Styrolux	Polystyrene	Westlake Plastics Co.
Styropor	Expanded polystyrene	BASF
Styron	Polystyrene resin	Dow Chemical Co.
Styronol	Styrene	Allied Resinous Products, Inc.
Sulfasar	Fiberglass-reinforced polysulfone	Fiberfil Div., Dart Industries, Inc.
Sulfil	Fiberglass-reinforced polysulfone	AKZO Engineering
Sunfrost	Thermoplastic vinyls	ATOFINA
Sunlon	Polyamide resin	Sun Chemical Corp.
Super Aeroflex	Linear polyethylene	Anchor Plastic Co.
Super Coilife	Epoxy potting resin	Westinghouse Electric Corp.
Super Dylan	High-density polyethylene	Sinclair-Koppers Co.
Superflex	Grafted high-impact polystyrene	Gordon Chemical Co.
Superflow	Polystyrene	Gordon Chemical Co.
Sur-Flex	Ionomer film	Flex-O-Glass, Inc.
Surlyn	Ionomer resin	E. I. du Pont de Nemours & Co.
Syn-U-Tex	Urea-formaldehyde and melamine-formaldehyde	Celanese Resins Div., Celanese Coatings Co.
Syntex	Alkyd and polyurethane ester resins	Celanese Resins Div., Celanese Coatings Co.
Syretex	Styrenated alkyd resins	Celanese Resins Div., Celanese Coatings Co.
TanClad	Spray or dip plastisol	Tamite Industries, Inc.
Tedlar	PVF film	E. I. du Pont de Nemours & Co.
Tedur	Reinforced polyphenylene sulfide	Miles
Teflon	FEP and TFE fluorocarbon resins	E. I. du Pont de Nemours & Co.
Tefzel	PE-TFE	DuPont

Trade Name	Polymer	Manufacturer
Tekron	Styrenic TPEs	Teknor Apex Thermoplastic Elastomer Div.
Telcar	Olefin elastomers	Teknor Apex Thermoplastic Elastomer Div.
TempRite	PVC for pipe	BF Goodrich
Tenite	PE and cellulosic compounds	Eastman Chemical Products, Inc.
Tenn Foam	Polyurethane foam	Morristown Foam Corp.
Terblend	ASA/PC alloy	BASF
Tere-Cast	Polyester casting compounds	Sterling Div., Reichhold Chemicals, Inc.
Terluran	ABS	BASF
Terlux	Clear ABS	BASF
Terucello	Carboxymethyl cellulose	Showa Highpolymer Company
Tetra-Phen	Phenolic-type resins	Georgia-Pacific Corp. Chemical Div.
Tetra-Ria	Amino-type resins	Georgia-Pacific Corp. Chemical Div.
Tetraloy	Filled TFE molding compounds	Whitford Chemical Corp.
Tetran	Polytetrafluoroethylene	Pennwalt Corp.
Texalon	Acetal and nylon	Texapol
Texin	Urethane elastomer molding compound	Miles
Textolite	Industrial laminates	General Electric Co., Laminated Products Dept.
Thermaloy	PC/ABS	Bayer AG
Thermalux	Polysulfone	Westlake Plastics Co.
Thermasol	Vinyl plastisols and organosols	Lakeside Plastics International
Thermco	Expanded polystyrene	Holland Plastics Co.
Thermocomp AE	Reinforced ABS	Thermofil
Thermocomp	Reinforced Nylon	LNP
Thorane	Rigid polyurethane foam	Dow Chemical Co.
T-Lock	PVC sheet material	Amercoat Corp.
Torlon	Polyamide-imide	Amoco Performance Products
TPX	Polymethyl pentene	Mitsue Petrochemical Industries
Tran-Stay	Flat polyester film	Transilwrap Co.

Trade Name	Polymer	Manufacturer
Transil GA	Precoated acetate sheets	Transilwrap Co.
Traytuf	PET	Shell Chemicals
Tri-Foil	TFE-coated aluminum foil	Tri-Point Industries, Inc.
Trilon	Polytetrafluoroethylene	Dynamit Nobel of America, Inc.
Triocel	Acetate	Celanese Fibers Marketing Co.
Trolen (series)	Polyethylene and polypropylene sheets	Dynamit Nobel of America, Inc.
Trolitan (series)	Phenol-formaldehyde compounds; boron	Dynamit Nobel of America, Inc.
Trolitrax	Industrial laminates	Dynamit Nobel of America, Inc.
Trosifol	Polyvinyl butyral film	Dynamit Nobel of America, Inc.
Tuffak	Polycarbonate	Rohm & Haas Co.
Tuftane	Polyurethane film and sheet	BF Goodrich Chemical Co.
Tybrene	Acrylonitrile-butadiene-styrene	Dow Chemical Co.
Tynex	Polyamide filaments	E. I. du Pont de Nemours & Co.
Tyril	Styrene-acrylonitrile resin	Dow Chemical Co.
Tyrilfoam	Styrene-acrylonitrile foam	Dow Chemical Co.
Tyrin	Chlorinated polyethylene	Dow Chemical Co.
Udel	Sulfone polymers	Amoco Performance Products
U-Thane	Rigid insulation board stock urethane	Upjohn Co., CPR Div.
Uformite	Urea and melamine resins	Rohm and Haas Co.
Ultem	Polyetherimide	GE Plastics
Ultradur	Polyester, thermoplastic	BASF
Ultraform	Acetal	BASF
Ultramid	Polyamide 6;6,6; and 6,10	BASF Wyandotte Corp.
Ultrapas	Melamine-formaldehyde compounds	Dynamit Nobel of America, Inc.
Ultrason E	Polyethersulfone	BASF
Ultrason S	Polysulfone	BASF
Ultrathene	Ethylene-vinyl acetate resins and copolymers	US Industrial Chemicals Co.
Ultron	PVC film and sheet	Monsanto Co.

Trade Name	Polymer	Manufacturer
Unifoam	Polyurethane foam	William T. Burnett & Co.
Unipoxy	Epoxy resins, adhesives	Kristal Kraft, Inc.
Urafil	Fiberglass-reinforced polyurethane	Fiberfil Div., Dart Industries, Inc.
Urglas	Fiberglass-reinforced polyurethane	Fiberfil Div., Dart Industries, Inc.
Uralite	Urethane compounds	Rezolin Div., Hexcel Corp.
Uramol	Urea-formaldehyde molding compounds	Gordon Chemicals Co.
Urapac	Rigid urethane systems	North American Urethanes, Inc.
Urapol	Urethane elastomeric coating	Poly Resins
Uvex	Cellulose acetate butyrate sheet	Eastman Chemical Products, Inc.
Valite	Phenolic molding compound	Valite
Valox	Thermoplastic polyester	General Electric Co.
Valsof	Polyethylene emulsions	Valchem Div., United Merchants & Mfrs., Inc.
Vandar	PBT alloy	Hoechst Celanese
Varcum	Phenolic resins	Reichhold Chemicals, Inc.
Varex	Polyester resins	McCloskey Varnish Co.
Varkyd	Alkyd and modified alkyd resins	McCloskey Varnish Co.
Varkyclane	Urethane vehicles	McCloskey Varnish Co.
Varsil	Silicone-coated fiberglass	New Jersey Wood Finishing Co.
V del	Polysulfone resins	Union Carbide Corp.
Vectra	Polypropylene fibers	Exxon Chemical USA
Velene	Styrene-foam laminate	Scott Paper Co., Foam Division
Velcro	Nylon, polyester	Velcro Industries B.V.
Velon	Film and sheeting	Firestone Plastics Co., Div., Firestone Tire & Rubber Co.
Versel	Thermoplastic polyester	Allied Chemical Corp.

Trade Name	Polymer	Manufacturer
Versi-Ply	Coextruded films	Pierson Industries, Inc.
Verton	Long fiber composite	LNP Engineering Plastics, Inc.
Vespel	Polyimide	E. I. du Pont de Nemours & Co.
Vestamid	Nylon 6,12	Huels America
Vibrathane	Polyurethane elastomer	Uniroyal, Inc.
Vibrin-Mat	Polyester-glass molding compound	Marco Chemical Div., W. R. Grace & Co.
Vibro-Flo	Epoxy and polyester coating powders	Armstrong Products Co.
Vinidur	Vinyl chloride/acrylate graft copolymers	BASF
Vinoflex	PVC resins	BASF Wyandotte Corp.
Vista	Vinyl resins	Vista
Vistanex	Polyisobutylene	ExxonMobil Chemical Co.
Vitel	Polyester resin	Goodyear Tire & Rubber Co., Chemical Div.
Vithane	Polyurethane resins	Goodyear Tire & Rubber Co., Chemical Div.
Viton	Fluoroelastomer elastomers	DuPont Dow
Vituf	Polyester resin	Goodyear Tire & Rubber Co., Chemical Div.
Volara	Closed-cell, low-density polyethylene foam	Voltek, Inc.
Volaron	Closed-cell, low-density polyethylene foam	Voltek, Inc.
Volasta	Closed-cell, medium-density polyethylene foam	Voltek, Inc.
Voranol	Polyurethane resins	Dow Chemical Co.
Vult-Acet	Polyvinyl acetate latexes	General Latex & Chemical Corp.
Vultafoam	Urethane foam systems	General Latex & Chemical Corp.
Vultathane	Urethane coatings	General Latex & Chemical Corp.
Vycron	Polyester	Beaunit Corp.
Vydyne	Nylon 6,6	Monsanto

Trade Name	Polymer	Manufacturer
Vygen	PVC resin	General Tire & Rubber Co., Chemical/Plastics Div.
Vynaclor	Vinyl chloride emulsion coatings and binders	National Starch & Chemical Corp.
Vynaloy	Vinyl sheet	BF Goodrich Chemical Co.
Vyram	Rigid PVC materials	Monsanto Co.
Weldfast	Epoxy and polyester adhesives	Fibercast Co.
Wellamid (series)	Polyamide 6 and 6,6 molding resins	Wellman, Inc., Plastics Div.
Well-A-Meld	Reinforced nylon resins	Wellman, Inc.
Westcoat	Strippable coatings	Western Coating Co.
Whirlclad	Plastic coatings	Polymer Corp.
Whitcon	Fluoroplastic lubricants	Whitford Chemical Corp.
Wicaloid	Styrene-butadiene emulsions	Wica Chemicals, Div., Ott Chemical Co.
Wicaset	Polyvinyl acetate emulsions	Wica Chemicals, Div., Ott Chemical Co.
Wilfex	Vinyl plastisols	Flexible Products Co.
Xenoy	PBT	GE Plastics
Xylon	Polyamide 6 and 6.6	Fiberfil Div., Dart Industries, Inc.
Zantrel	Rayon	American Enka Co.
Zefran	Acrylic, nylon polyester	Dow Badische Co.
Zelux	Polyethylene films	Union Carbide Corp., Chemicals & Plastics Div.
Zendel	Polyethylene films	Union Carbide Corp., Chemicals & Plastics Div.
Zerion	Copolymer of acrylic and styrene	Dow Chemical Co.
Zetafin	Ethylene copolymer resins	Dow Chemical Co.
ZYLAR	Acrylic terpolymers	NOVA Chemicals Corp.
Zytel	Nylon	E. I. du Pont de Nemours & Co.

MATERIAL IDENTIFICATION

IDENTIFYING PLASTICS

Plastics are complex materials. Identifying them is not easy because they can contain several polymers as well as other ingredients, including fillers. The insolubility of some plastics adds to the problem. Correct identification requires complex tools and techniques.

A student or consumer may have to identify a plastics so that repairs or new parts can be made. Laboratory tests can help to identify ingredients of the unknown material. The methods shown in this appendix are meant for easy identification of basic polymer types. Identification methods that involve complex instruments are not covered.

More sophisticated methods include infrared spectroscopy analysis, which is the only accurate method of obtaining the quantitative identification of unknown polymers. Another highly complex and costly method is X-ray diffraction, which is used for identification of solid crystalline compounds.

IDENTIFICATION METHODS

There are five broad identification methods for plastics:

1. Trade name
2. Appearance
3. Effects of heat
4. Effects of solvents
5. Relative density

Trade Name

The numerous trade names used today identify the product of a fabricator, manufacturer, or processor. Trade names may be associated with the product or with plastics material. In either case, trade names can serve as a guide to identifying plastics. See Appendix C for trade names of plastics materials on the commercial market.

If the trade name is known, the supplier or manufacturer may be the most reliable source of information about types of plastics, ingredients, additives, or physical properties. Batch and lot numbers may vary, but most of the essential information will be known by the supplier. Similar to gasoline brands, additives may vary in each family of plastics manufactured.

Appearance

Many physical or visual clues can be used to help identify plastics materials. Plastics in the raw, uncompounded, or pellet stage are harder to identify than finished products. Thermoplastics are generally produced in powder, granular, or pellet forms. Thermosetting materials are usually made in the form of powders, preforms, or resins.

The method of fabrication as well as the product application are good clues to the identification of plastics. Thermoplastic materials are commonly extruded, injection formed, calendered, blow molded, and vacuum molded. Polyethylene, polystyrene, and cellulosics are used extensively in the container and packaging industry. Harsh chemicals and solvents are likely to be stored in polyethylene containers. Polyethylene, polytetrafluoroethylene, polyacetals, and polyamides have a waxy feel not found in most polymers.

Thermosetting plastics are usually compression molded, transfer molded, cast, or laminated. Some thermosets are not reinforced. Others are reinforced or heavily filled. Some identifiable characteristics of plastics are given in Table D-1.

Table D-1. Identification of Selected Plastics

Plastics	Appearance	Applications
ABS	Styrene-like, tough, metal-like ring when struck, translucent	Appliance and tool housings, instrument panels, luggage, packing crates, sporting goods
Acetal	Tough, hard, metal-like ring when struck, waxy feel, translucent	Aerosol stem valves, cigarette lighters, conveyor belts, plumbing, zippers
Acrylics	Brittle, hard, transparent	Models, glazing, wax
Alkyds	Hard, tough, brittle, usually bulk-filled, opaque	Electrical, paints
Allyl	Hard, filled, reinforced, transparent to opaque	Electrical, sealers
Aminos	Hard, brittle, opaque with some translucence	Appliance knobs, bottle caps, dials, handles
Cellulosics	Varies; tough, transparent	Explosives, fabrics, packaging, pharmaceuticals, handles, toys
Chlorinated Polyesters	Tough, translucent or opaque	Electrical, laboratory equipment, plumbing
Epoxies	Hard, mostly filled, reinforced, transparent	Adhesives, casting, finishes
Fluoroplastics	Tough, waxy feel, translucent	Anti-stick coatings, bearings, gaskets, seals, electrical
Ionomers	Tough, impact resistant, transparent	Containers, paper coatings, safety glasses, shields, toys
Nitrile barrier plastics	Tough, transparent, impact resistant	Packaging
Phenolics	Hard, brittle, filled, reinforced, transparent	Adhesives, billiard balls, handles, molding powders
Phenylene oxide	Tough, hard, often filled, reinforced, opaque	Appliance housings, consoles, electrical, respirators
Polyamides	Tough, waxy feel, translucent	Combs, door catches, castors, gears, valve seats
Polyaryl ether	Impact resistant, polycarbonate-like, translucent to opaque	Appliances, automobile paint, electrical
Polyaryl sulfone	Tough, stiff, opaque, polycarbonate-like	High-temperature uses in aerospace, industry, and consumer goods
Polycarbonate	Styrene-like, tough, metal-like ring when struck, translucent	Beverage dispensers, films, lenses, light fixtures, small appliances, windshields
Polyester, aromatic	Stiff, tough, opaque	Coatings, insulation, transistors
Polyester, thermoplastic	Hard, tough, opaque	Beverage bottles, packaging, photography, tapes, labels
Polyester, unsaturated	Hard, brittle, filled, transparent	Furniture, radar domes, sports equipment, tanks, trays
Polyolefins	Waxy feel, tough, soft, translucent	Carpeting, chairs, dishes, medical syringes, toys
Polyphenylene sulfide	Stiff, hard, opaque	Bearings, gears, coatings
Polystyrene	Brittle, white bend marks, metal-like ring when struck, transparent	Blister packages, bottle caps, dishes, lenses, transparent display boxes
Polysulfone	Rigid, polycarbonate-like, transparent to opaque	Aerospace, distributor caps, hospital equipment, shower heads
Silicones	Tough, hard, filled, reinforced, some flexible, opaque	Artificial organs, grease, inks, molds, polishes, Silly Putty, waterproofing
Urethanes	Tough castings, mostly foams, flexible, opaque	Bumpers, cushions, elastic thread, insulation, sponges, tires
Vinyls	Tough, some flexible, transparent	Balls, dolls, floor coverings, garden hoses, rainwear, wall tiles, wallpaper

Effects of Heat

When plastics specimens are heated in test tubes, distinct odors of specific plastics may be identified (see Table D-2). The actual burning of the sample in an open flame may provide further clues. Polystyrene and its copolymers burn with black (carbon) smoke. Polyethylene burns with a clear flame and drips when molten.

The actual melting point may provide further clues for identification. Thermosetting materials don't melt. Several thermoplastics melt at less than 195°C (203°F). An electric soldering gun can be pressed on the surface of the plastics. If the material softens and the hot tip sinks into it, the material is a thermoplastic. If the material stays hard and merely chars, it is a thermoset.

Table D-2. Identification Test for Selected Plastics

Thermoplastic	Flame-Burn, Smoke, and Flame Danger	Odor and Respiratory Danger	Melting Point, °C
ABS	Yellow flame, black smoke, drips, continues to burn	Rubber, sharp, biting	100
Acetal	Blue flame, no smoke, drips, melts, may burn, continues to burn	Formaldehyde	181
Acrylic	Blue flame, yellow top, white ash, black smoke, popping sound, spurts, continues to burn	Fruit, floral	105
Cellulose acetate	Yellow or yellow-orange-to-green flame, melts, drips, continues to burn, black smoke	Burnt sugar, acetic acid, burning paper	230
Cellulose acetate butyrate	Blue flame, yellow top, sparks, melts, drips, dripping may burn, continues to burn	Camphor, rancid butter	140
Ethyl cellulose	Pale yellow to blue-green flame with blue edge, melts, drips, drips burn	Burnt wood, burnt sugar	135
Fluoronated ethylene propylene	Melts, decomposes, poison gases formed	Slight acid or burned hair. DO NOT INHALE	275
Ionomer	Yellow flame with blue edge, continues to burn, some black smoke, melts, bubbles, drips burn	Paraffin	110
Phenoxy	Burns, no drops	Acid	93
Polyallomer	Yellow or yellow-orange flame, with blue edge, continues to burn, black smoke, melts clear, spurts, drips burn	Paraffin	120
Polyamides 6,6	Blue flame, yellow tip, melts and drips, self-extinguishing, froths	Burned wool or hair	265
Polycarbonate	Decomposes, chars, self-extinguishing, dense black smoke, spurts orange flame	Characteristic, sweet, compare known sample	150
Polychlorotrifluoro-ethylene	Yellow flame, won't support combustion.	Slight, acidic fumes. DO NOT INHALE	220
Polyethylene	Blue flame, yellow top, drippings may burn, transparent hot area, burns rapidly, continues to burn	Paraffin	110
Polyimides	Chars, brittle, blue flame		300
Polyphyenylene oxide	Yellow to yellow-orange flame, no drips, spurts, difficult to ignite, thick black smoke, decomposes	Paraffin, phenol	105
Polypropylene	Blue flame, drips, transparent hot area, burns slowly, trace of white smoke, melts, swells	Heavy, sweet, paraffin, burning asphalt	176
Polysulfones	Yellow or orange flame, black smoke, drips, self-extinguishing, sparkles, decomposes	Acid	200
Polystyrene	Yellow flame, dense smoke, clumps of carbon in air, drips, continues to burn, bubbles	Illuminating gas, sweet, marigold, floral	100

(Continued)

Table D-2. Identification Test for Selected Plastics (*Continued*)

Thermoplastic	Flame-Burn, Smoke, and Flame Danger	Odor and Respiratory Danger	Melting Point, °C
Polytetrafluoroethylene	Yellow flame, slightly green near base, won't support combustion, self-extinguishing, turns clear	None. DO NOT INHALE	327
Polyvinyl acetate	Yellow flame, black smoke, spurts, continues to burn, some soot, green on copper wire test	Vinegar, acetic acid	60
Polyvinyl alcohol	Yellow, smoky	Unpleasant, sweet	105
Polyvinyl chloride	Yellow flame, green at edges, black or gray smoke, chars, self-extinguishing, leaves an ash	Hydrochloric acid	75
Polyvinyl fluoride	Pale yellow	Acid	230
Polyvinylidene chloride	Yellow with green base, spurts green smoke, smoky, self-extinguishing	Pungent	210
Casein	Yellow flame, burns with flame contact, chars	Burnt milk	
Diallyl phthalate	Yellow flame, green-blue edge, self-extinguishing	Acid	
Epoxy	Yellow flame, some soot, spits black smoke, chars, continues to burn	Phenolic phenol, acid	
Melamine formaldehyde	Difficult to burn, self-extinguishing, swells, cracks, yellow flame, blue-green base, turns white	Fish-like, formaldehyde	
Phenolic	Cracks, deforms, difficult to burn, self-extinguishing, yellow flame, little black smoke	Phenolic phenol	
Polyester	Yellow flame, blue edge, burns, ash and black beads, continues to burn, dense black smoke, no drips	Sweet, bittersweet, burning coal	
Polyurethane	Yellow with blue base, thick black smoke, spurts, may melt and drip, continues to burn	Acid	
Silicone	Low, bright yellow-white flame, white smoke, white ash, continues to burn	None	
Urea formaldehyde	Yellow flame with green-blue edge, self-extinguishing	Pancakes	

The melting or softening point may be observed by placing a small piece of the unknown thermoplastic on an electrically heated platen or in an oven. The temperature must be carefully controlled and recorded. When the specimen is within a few degrees of the suspected melting point, the temperature should be increased at a rate of 1°C/min.

A standard method of testing polymers is given in ASTM D-2117. For polymers that have no definite melting point (such as polyethylene, polystyrene, acrylics, and cellulosics) or for those that melt with a broad transition temperature, the Vicat softening point may be of some aid in identification. The Vicat softening point test is described in Chapter 6. Melting points of selected plastics are shown in Table D-2.

Two tests that rely on the effects of heat are the Beilstein test and the Lassaigne test.

The Beilstein test. The Beilstein test is a simple method of determining the presence of a halogen (chlorine, fluorine, bromine, and iodine). For this test, heat a clean copper wire in a Bunsen flame until it glows. Quickly touch the hot wire

to the test sample, and then return the wire to the flame. A green flame shows the presence of halogen. Plastics containing chlorine are polychlorotrifluoroethylene, polyvinyl chloride, polyvinylidene chloride, and others. They give positive results to the halogen test. If the halogen test is negative, the polymer may be composed of only carbon, hydrogen, oxygen, or silicon.

The Lassaigne test. For further chemical analysis, the Lassaigne procedure of sodium fusion may be used.

! CAUTION

Although very useful, this test is dangerous because sodium is highly reactive. It should be performed with great care.

To conduct the Lassaigne test, place 5 grams of the sample in an ignition tube with 0.1 gram of sodium. Heat the tube until the sample decomposes, while keeping the open end of the tube pointing away from you. While the tube is still red-hot, place it in distilled water, and grind it with a mortar and pestle device. While the mixture is still hot, filter out the carbon and glass fragments. Divide the resulting filtrate into four equal portions. Use these portions to perform standard tests for nitrogen, chlorine, fluorine, and sulfur.

Effects of Solvents

Tests for the solubility or insolubility of plastics are easily performed identification methods. Except for polyolefins, acetals, polyamides, and fluoroplastics—thermoplastic materials can be considered soluble at room temperature. Thermosetting plastics may be considered solvent resistant.

! CAUTION

When making solubility tests, remember that solutions may be flammable, give off toxic fumes, be absorbed through the skin, or all three. Appropriate safety precautions should be taken.

Before a solubility test can be made, a chemical solvent must be selected. To help identify polymers and solvents that may react molecularly with each other, solubility parameter numbers have been assigned to selected polymers and solvents (Table D-3).

A polymer should dissolve in a solvent that has a similar or lower solubility parameter. This does not always happen due to crystallization or hydrogen bonding.

When making solubility tests, use a ratio of 1 volume of plastics sample to 20 volumes of boiling or room-temperature solvent. A water-cooled reflux condenser may be used to collect solvent or minimize solvent loss during heating. Table D-4 shows selected solvents with selected plastics.

Relative Density

The presence of fillers or other additives as well as the degree of polymerization (DP) can make identification of plastics by relative density tests difficult. The presence of fillers and additives can cause the relative density to differ greatly from that

Table D-3. Solubility Parameters of Selected Solvents and Plastics

Solvent	Solubility Parameter	Plastics	Solubility Parameter
Water	23.4	Polytetrafluoroethylene	6.2
Methyl alcohol	14.5	Polyethylene	7.9–8.1
Ethyl alcohol	12.7	Polypropylene	7.9
Isopropyl alcohol	11.5	Polystyrene	8.5–9.7
Phenol	14.5	Polyvinyl acetate	9.4
n-Butyl alcohol	11.4	Polymethyl methacrylate	9.0–9.5
Ethyl acetate	9.1	Polyvinyl chloride	9.38–9.5
Chloroform	9.3	Bisphenol A polycarbonate	9.5
Trichloroethylene	9.3	Polyvinylidene chloride	9.8
Methylene chloride	9.7	Polyethylene terephthalate	10.7
Ethylkene dichloride	9.8	Cellulose nitrate	10.56–10.48
Cyclohexanone	9.9	Cellulose acetate	11.35
Acetone	10.0	Epoxide	11.0
Isopropyl acetate	8.4	Polyacetal	11.1
Carbon tetrachloride	8.6	Polyamide 6, 6	13.6
Toluene	9.0	Coumarone indene	8.0–10.6
Xylene	8.9	Alkyd	7.0–11.2
Methyl isopropyl ketone	8.4		
Cyclohexane	8.2		
Turpentine	8.1		
Methyl amyl acetate	8.0		
Methyl cyclohexane	7.8		
Heptane	7.5		

Table D-4. Identification of Selected Plastics by Solvent Test Methods

Plastics	Acetone	Benzene	Furfuryl Alcohol	Toluene	Special Solvents
ABS	Insoluble	Partially soluble	Insoluble	Soluble	Ethylene dichloride
Acrylic	Soluble	Soluble	Partially soluble	Soluble	Ethylene dichloride
Cellulose acetate	Soluble	Partially soluble	Soluble	Partially soluble	Acetic acid
Cellulose acetate, butyrate	Soluble	Partially soluble	Soluble	Partially soluble	Ethyl acetate
Fluorocarbon	Insoluble (most)	Insoluble	Insoluble	Insoluble	Dimethyacetamide (not FEP-TFE)
Polyamide	Insoluble	Insoluble	Insoluble	Insoluble	Hot aqueous ethanol
Polycarbonate	Partially soluble	Partially soluble	Insoluble	Partially soluble	Hot benzene-toluene
Polyethylene	Insoluble	Insoluble	Insoluble	Insoluble	Hot benzene-toluene
Polypropylene	Insoluble	Insoluble	Insoluble	Insoluble	Hot benzene-toluene
Polystyrene	Soluble	Soluble	Partially soluble	Soluble	Methylene dichloride
Vinyl acetate	Soluble	Soluble	Insoluble	Soluble	Cyclohexanol
Vinyl chloride	Soluble	Insoluble	Insoluble	Partially Soluble	Cyclohexanol

of the plastics itself. Polyolefins, ionomers, and low-density polystyrene will float in water (which has a relative density of 1.00).

If a density gradient column is unavailable, a few solutions of known density can provide considerable data. Mix solutions of distilled water and calcium nitrate and measure them with technical-grade hydrometers until a desired relative density is obtained. For densities less than that of water (1.00), isopropyl alcohol may be used. Full-strength isopropyl alcohol has a relative density of 0.92. This value may be raised by adding small amounts of distilled water.

If a plastics floats in a solution with a relative density of 0.94, it may be a medium- or low-density polyethylene plastics. If the sample floats in a solution of 0.92, it must be either low-density polyethylene or polypropylene. If the sample sinks in

all solutions below a relative density of 2.00, the sample is a fluorocarbon plastics.

All of the solutions may be stored in clean containers and reused. However, the density of solutions should be checked during the testing procedure. Factors such as temperature and evaporation may radically change the relative density value.

The strength-to-mass ratio of a plastics can also help in identification. Plastics usually weigh considerably less in relation to volume than metals and most other materials. A reinforced structural foam plastics with a density of 550 kg/m³ (35 lb/ft³) and a tensile strength of 20.0 MPa (2.9 × 103 psi) would have a strength-to-mass ratio of 0.57. In contrast, steel with a tensile strength of 700 MPa (1.0 × 10⁵ psi) and a density of 7750 kg/m³ (484 lb/ft³) has a ratio of 0.09. These ratios are sometimes used in design criteria.

THERMOPLASTICS

APPENDIX E

It is important that you become familiar with thermoplastics. Using this appendix, look for the outstanding properties and applications of each plastics, because plastics properties affect product design, processing, economics, and service.

This appendix treats individual groups of thermoplastics in the following order:

- Acetals (polyacetals)
- Acrylics
- Acrylates
- Acrylonitrile
- Acrylic-styrene-acrylonitrile (ASA)
- Acrylonitrile-butadiene-styrene (ABS)
- Acrylonitrile-chlorinated polyethylene-styrene (ACS)
- Cellulosics
- Chlorinated polyethers
- Coumarone-indene
- Fluoroplastics
- Ionomers
- Nitrile barriers
- Phenoxy
- Polyallomers
- Polyamides
- Polycarbonates
- Polyetheretherketone
- Polyetherimide
- Polyesters (thermoplastic)
- Polyimides (thermoplastic)
- Polymethylpentene
- Polyolefins (PE, PP, PB)
- Polyphenylene oxides
- Polyphpenylene sulfide
- Polyaryl ethers

- Polystyrene (PS)
- Styrene-acrylonitrile (SAN)
- Styrene-butadiene (SBP)
- Polysulfones
- Polyvinyls

POLYACETAL PLASTICS (POMs)

A highly reactive gas, formaldehyde (CH_2O), may be polymerized in a number ways. Formaldehyde, or *methanal*, is the simplest of the aldehyde group of chemicals. The ending for the aldehydes is *al*, derived from the first syllable of aldehyde.

Simple polymers based on formaldehyde have been known since 1859. In 1960, DuPont put the first polyformaldehyde on the market. The polyformaldehyde (polyacetal) polymer is basically a linear, highly crystalline, long molecular structure. The term *acetal* refers to the oxygen atom that joins the repeating units of the polymer structure. Polyoxymethylene (POM) is the correct chemical term for this polymer. Acetal is a generic term.

A number of initiators or catalysts are used to polymerize the basic polyacetal resin including acids, bases, metallic compounds, cobalt, and nickel.

Here is the polyacetal structure:

$$H\text{-}O\text{-}(CH_2\text{-}O\text{-}CH_2\text{-}O)n H : R$$
OR

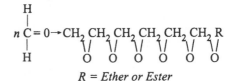

$$R = \textit{Ether or Ester}$$

The best-known polyformaldehyde plastics is the oxymethylene linear structure with attached terminal groups. However, there are many miscellaneous aldehyde-derived polymers.

Thermal and chemical resistance is increased when esters or ethers are attached as terminal groups. Both esters and ethers are relatively inert toward most chemical reagents. They are chemically compatible in organic chemical reactions.

$$\begin{array}{cccc} & & \overset{\displaystyle O}{\overset{\displaystyle \|}{}} & \overset{\displaystyle O}{\overset{\displaystyle \|}{}} \\ \text{H-O-H} & \text{R-O-R} & \text{R-C-OH} & \text{R-C-OR} \\ \textit{Water} & \textit{Ether} & \textit{Carboxylic acid} & \textit{Ester} \end{array}$$

The formulas above show some of the structural relationships among water, ethers, carboxylic acids, and esters.

Close packing and short bond lengths are typically found in polyacetal plastics and provide a hard, rigid, dimensionally stable material. Polyacetals have high resistance to organic chemicals and a wide temperature range.

Polyacetals are easy to fabricate, offer properties not found in metals, and are competitive with many nonferrous metals in cost and performance. They are similar to polyamides in many respects. Acetals provide superior fatigue endurance, creep resistance, stiffness, and water resistance. They are among the strongest and stiffest thermoplastics and may be filled for even greater strength, dimensional stability, abrasion resistance, and improved electrical properties (Figure E-1).

At room temperatures, polyacetals are resistant to most chemicals, stains, and organic solvents. Examples include tea, beet juice, oils, and household detergents. However, hot coffee will usually cause staining. Resistance to strong acids, strong alkalies, and oxidizing agents is poor. Copolymerization and filling improves the chemical resistance of the material.

Moisture and thermal resistance are characteristic of acetal polymers. For this reason, they are used for plumbing fixtures, pump impellers, conveyor belts, aerosol stem valves, and shower heads.

Acetals must be protected from prolonged exposure to ultraviolet light. Such exposure causes surface chalking, reduced molecular mass, and slow degradation. Painting, plating, and/or filling with carbon black or ultraviolet-absorbing chemicals will protect acetal products for outdoor use.

Acetals are available in pellet or powder form for processing in conventional injection molding, blow molding, and extrusion machines. Because of the highly crystalline structure of polyacetals, it is not possible to make optically transparent film. The operator must have adequate ventilation when processing polyacetal materials because upon degrading at high temperature, acetals release a toxic and potentially lethal gas.

Table E-1 gives some of the most important properties of acetal plastics and lists of six advantages and six disadvantages follow:

Advantages of Polyacetals

1. High tensile strength with rigidity and toughness
2. Excellent dimensional stability
3. Glossy molded surfaces
4. Low static and coefficient of friction
5. Retention of electrical and mechanical properties to 120°C (248°F)
6. Low gas and vapor permeability

Disadvantages of Polyacetals

1. Poor resistance to acids and bases
2. Subject to UV degradation
3. Flammable
4. Unsuitable for contact with food
5. Difficult to bond
6. Toxic—releases fumes upon degradation

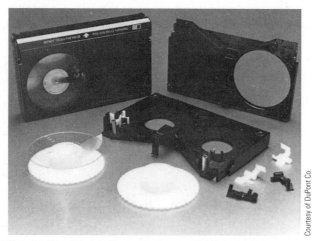

Figure E-1. These high-stress videocassette parts are molded from acetal plastics.

Courtesy of DuPont Co.

ACRYLICS

In 1901, Otto Rohm reported much of the research that later led to the commercial exploitation of acrylics. While pursuing his research in Germany, Dr. Rohm took an active part in the first commercial development of polyacrylates in 1927. By 1931, a Rohm and Haas Company plant was operating in the United States. Most of these early materials were used as coatings or for aircraft components. For example, acrylics were used for windshields and bubble turrets on aircraft during World War II. From these early compounds, an extensive group of monomers has become available, and commercial applications of these polymers have grown steadily.

The term *acrylic* includes acrylic and methacrylic esters, acids, and other derivatives. The principal acid and ester

Table E-1. Properties of Acetals

Property	Acetal (Homopolymer)	Acetal (20% Glass-Filled)
Molding qualities	Excellent	Good to Excellent
Relative density	1.42	1.56
Tensile strength, MPa	68.9	58.6–75.8
(psi)	(10,000)	(8500–11,000)
Compressive strength (10% defl.), MPa	124	124
(psi)	(18,000)	(18,000)
Impact strength, Izod, J/mm	0.07 (Inj) 0.115 (Ext)	0.04
(ft-lb/in.)	1.4 (Inj) 2.3 (Ext)	(0.8)
Hardness, Rockwell	M94, R120	M75–M90
Thermal expansion, (10^{-4}/°C)	20.6	9–20.6
Resistance to heat, °C	90	85–105
(°F)	(195)	(185–220)
Dielectric strength, V/mm	14,960	22,835
Dielectric constant (at 60 Hz)	3.7	3.9
Dissipation factor (at 60 Hz)	3.7	3.9
Arc resistance, s	129	136
Water absorption (24 h), %	0.25	0.25–0.29
Burning rate, mm/min	Slow to 28	20–25.4
(in./min)	(Slow to 1.1)	(0.8–1.0)
Effect of sunlight	Chalks slightly	Chalks slightly
Effect of acids	Resists some	Resists some
Effect of alkalies	Resists some	Resists some
Effect of solvents	Excellent resistance	Excellent resistance
Machining qualities	Excellent	Good to fair
Optical qualities	Translucent to opaque	Opaque

Table E-2. Principal Acid and Ester Monomers

Acrylic acid	Methyl acrylate	Ethyl acrylate	n-Butyl acrylate	Isobutyl acrylate	2-Ethylhexyl acrylate
$CH_2 = CHCOOH$	$CH_2 = CHCOOCH_3$	$CH_2 = CHCOOC_2H_5$	$CH_2 = CHCOOC_4H_9$	$CH_2 = CHCOOCH_2CH(CH_3)_2$	$CH_2 = CHCOOCH_2CH(C_2H_5)C_4H_9$

Methacrylic acid	Methyl methacrylate	Ethyl methacrylate	n-Butyl methacrylate	Lauryl methacrylate
$CH_2 = CCOOH$ $\quad\vert$ $\quad CH_3$	$CH_2 = CCOOCH_3$ $\quad\vert$ $\quad CH_3$	$CH_2 = CCOOC_2H_5$ $\quad\vert$ $\quad CH_3$	$CH_2 = CCOOC_4H_9$ $\quad\vert$ $\quad CH_3$	$CH_2 = CCOO(CH_2)_nCH_3$ $\quad\vert$ $\quad CH_3$

Stearyl methacrylate	2-Hydroxyethyl methacrylate	Hydroxypropyl methacrylate	2-Dimethylaminoethyl methacrylate	2-t-Butylaminoethyl methacrylate
$CH_2 = CCOO(CH_2)_6CH_3$ $\quad\vert$ $\quad CH_3$	$CH_2 = CCOOCH_2CH_2OH$ $\quad\vert$ $\quad CH_3$	$CH_2 = CCOO(C_3H_6)OH_4$ $\quad\vert$ $\quad CH_3$	$CH_2 = CCOOCH_2CH_2N(CH_3)_2$ $\quad\vert$ $\quad CH_3$	$CH_2 = CCOOCH_2CH_2NHC(CH_3)_3$ $\quad\vert$ $\quad CH_3$

monomers are shown in Table E-2. In order to avoid possible confusion, the basic acrylic formula is shown, with possible side groups R_1 and R_2, in Figure E-2.

There are many monomer possibilities and preparation methods. The most important is the commercial preparation of methyl methacrylate from acetone cyanohydrin. These homomonomers and comonomers may be polymerized by using various commercial methods, including bulk, solution, emulsion, suspension, and granulation polymerization. In all cases, an organic peroxide catalyst is used to start

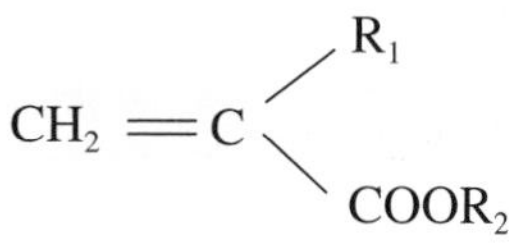

$$CH_2 = C \diagup^{R_1}_{\diagdown COOR_2}$$

(A) Basic acrylic formula.

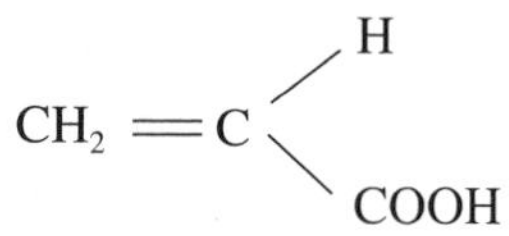

$$CH_2 = C \diagup^{H}_{\diagdown COOH}$$

(B) Hydrogen replaces R_1 and R_2 to produce acrylic acid.

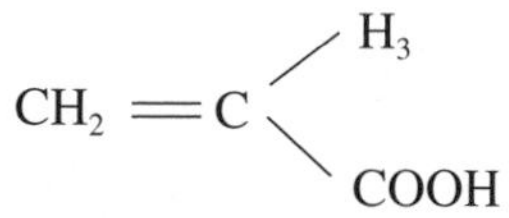

$$CH_2 = C \diagup^{H_3}_{\diagdown COOH}$$

(C) Methyl group replaces R_1 to produce methacrylic acid.

Figure E-2. Acrylic formula, with two possible radical replacements.

polymerization. Many of the molding powders are made by emulsion methods. Bulk polymerization is used for casting sheets and profile shapes.

The versatility of acrylic monomers in processing, copolymerization, and ultimate or final state properties has contributed to their wide use. Table E-3 lists some of the basic properties of acrylics.

Polymethyl methacrylate is an atactic, amorphous, transparent thermoplastic. Because of its high transparency (about 92%), it is used for many optical applications. It is a good electrical insulator for low frequencies and has very good resistance to weathering. Outdoor advertising signs are a familiar use of acrylics.

Polymethyl methacrylate is a standard material used for automobile taillight lenses and covers. This material is used for

Table E-3. Properties of Acrylics

Property	Methyl Methacrylate (Molding)	Acrylic-PVC Copolymer (Molding)	ABS (High Impact)
Molding qualities	Excellent		Good to excellent
Relative density	1.17–1.20	1.30	1.01–1.04
Tensile strength, MPa	48–76	38	30–53
(psi)	(7000–11,000)	(5500)	(4500–7500)
Compressive strength, MPa	83–125	43	30–55
(psi)	(12,000–18,000)	(6200)	(4500–8000)
Impact strength, Izod J/mm	0.015–0.025	0.75	0.25–0.4*
(ft-lb/in.)	(0.3–0.5)	(15)	(5.0–8.0)†
Hardness, Rockwell	M85–M105	R104	R75–R105
Thermal expansion; $10^{-4}/°C$	12–23	12–29	24–33
Resistance to heat, °C	60–94	60–98	60–98
(°F)	(140–200)	(140–210)	(140–210)
Dielectric strength, V/mm	15,800–20,000	15,800	13,800–18,000
Dielectric constant (at 60 Hz)	3.3–3.9	4	2.4–5.0
Dissipation factor (at 60 Hz)	0.04–0.06	0.04	0.003–0.008
Arc resistance, s	No track	25	50–85
Water absorption (24 h), %	0.1–0.4	0.13	0.20–0.45
Burning rate, mm/min	Slow 0.5–30	Nonburning	Slow to self-extinguishing
(in./min)	(Slow 0.6–1.2)		
Effect of sunlight	Nil	Nil	Yellows
Effect of acids	Attacked by strong oxidizing acids	Slight	Attacked by strong oxidizing acids
Effect of alkalies	Attacked	None	None
Effect of solvents	Soluble in ketones, esters, and aromatic and chlorinated hydrocarbons	Attacked by ketones, esters, and aromatic and chlorinated hydrocarbons	Soluble in ketones and esters
Machining qualities	Good to excellent	Excellent	Excellent
Optical qualities	Transparent to opaque	Opaque	Translucent

Notes: *At 23°C, 3 × 12 mmL bar
†At 73°F, 1/8 × 1/2 in bar

aircraft windshields and cockpit covers as well as for bubble bodies on helicopters.

Polymethyl methacrylate may be produced by any typical thermoplastic process. It may be fabricated by solvent cementing. Cast and extruded sheets and profile shapes are popular forms. Sheet forms are widely used for room dividers and skylight domes and as a substitute for glass in windows. These plastics have been widely used in the paint industry in the form of emulsions. Emulsion acrylics are popular as a clear, hard, and glossy "wax" coating for floors. Acrylic-based adhesives are available with a wide range of uses and properties. These adhesives are transparent and available in solvent-based (air-drying), hot-melt, or pressure-sensitive forms. Figure E-3 shows an acrylic sealant being applied directly to an oily aluminum window frame under water.

Glass-reinforced sheets are used to produce sanitary ware, vanities, tubs, and counters. Protective liquid coatings, known as gel coats, may be used with reinforcements as cover stock. Heavily filled and reinforced resins formulated to cross-link into a thermosetting matrix are used to produce marble-like bathroom fixtures and furniture.

Some well-known trade names for polymethyl methacrylate are Plexiglas, Lucite®, and Acrylite®. From the following lists, you can see there are more advantages (11) of acrylics than disadvantages (5).

Advantages of Acrylics

1. Wide range of colors
2. Outstanding optical clarity
3. Slow-burning, releasing little or no smoke
4. Excellent weatherability and ultraviolet resistance
5. Ease of fabrication
6. Excellent electrical properties
7. Unaffected by food and human tissues
8. Rigidity with good impact strength
9. High gloss and good *feel*
10. Excellent dimensional stability and low mold shrinkage
11. Stretch forming improves biaxial toughness

Figure E-3. An acrylic sealant being applied under water.

Disadvantages of Acrylics

1. Poor solvent resistance
2. Possibility of stress cracking
3. Combustibility
4. Limited continuous service temperature of 93°C (200°F)
5. Inflexibility

Polyacrylates

Polyacrylates are transparent, resistant to chemicals and weather, and have a low softening point. Applications include films, adhesives, and surface coatings for paper and textiles. They are usually copolymer compositions. Polyethyl acrylate may be cross-linked to form thermosetting elastomers. Polyacrylate monomers are used as plasticizers for other vinyl polymers.

Acrylic esters may be obtained from the reaction of ethylene cyanohydrin with sulfuric acid and alcohol.

$$HO \cdot CH_2 \cdot CH_2 \cdot CN \xrightarrow[H_2SO_4]{ROH} CH_2 : CH \cdot CO \cdot O \cdot R$$

Acrylic and polyvinyl chloride (PMMA/PVC) may be alloyed to produce a tough, durable sheet that can be easily thermoformed into signs, aircraft trays, and public service seating (see Figure E-4).

Polyacrylonitrile and Polymethacrylonitrile

Before World War II, the elastomers and fibers produced from polyacrylonitrile and polymethacrylonitrile materials were

Figure E-4. A thermoformed acrylic sheet offers sunlight resistance and impact strength for this sign.

merely laboratory curiosities. Since that time, there has been a rapid expansion of the use of acrylonitrile, often as the main ingredient in acrylic fibers. These polymers are copolymerized and stretched to orient the molecular chain. Orlon and Dynel, classified as *modacrylic fibers*, contain less than 85% acrylonitrile. Modacrylic fibers contain at least 35% acrylonitrile units. *Acrylic fibers* such as Acrilan, Creslan, and Zefran® contain more than 85% acrylonitrile.

Unmodified polyacrylonitrile is only slightly thermoplastic and is difficult to mold because its hydrogen bonds resist flow. Copolymers of styrene, ethyl acrylates, methacrylates, and other monomers are extruded into the amorphous fiber form. However, at this point the fiber is too weak to be of value. It is then stretched to produce a greater degree of crystallization. Tensile strength is greatly increased as a result of this molecular orientation (see Nitrile barrier plastics, p. 424).

Monomers of acrylonitrile and methacrylonitrile are shown here:

$$CH_2 = CHCN$$
Acrylonitrile

$$CH_2 = C - CN$$
$$|$$
$$CH_3$$
Methacrylonitrile

Acrylic-Styrene-Acrylonitrile (ASA)

The terpolymer of acrylic-styrene-acrylonitrile may vary in the percentages of each component to enhance or tailor a specific property. The excellent surface gloss makes it attractive as a cover layer for co-extrusion over ABS, PC, or PVC. Exterior applications include signs, downspouts, siding, recreational equipment, camper tops, ATV bodies, and gutters.

This material is also blended with PC (ASA/PC) and alloyed with PVC (ASA/PVC) and PMMA (ASA/PMMA) to enhance specific properties. ASA/PMMA alloys have outstanding weatherability, gloss, and toughness. They are used to produce spas and hot tubs.

Acrylonitrile-Butadiene-Styrene (ABS)

ABS polymers are opaque thermoplastic resins formed by the polymerization of acrylonitrile-butadiene-styrene monomers. Because they possess such a diverse combination of properties, many experts classify them as a family of plastics. However, they are actually terpolymers (*ter* meaning "three") of three monomers and *not* a distinct family. Their development resulted from research efforts on synthetic rubber during and after World War II. The proportions of the three ingredients may vary, which accounts for the great number of possible properties.

The three ingredients are shown here. Acrylonitrile is also known as vinyl cyanide and acrylic nitrile.

$$CH_2 = CHCN$$
Acrylonitrile

$$CH_2 = CH - CH = CH_2$$
Butadiene

$$\bigcirc - CH = CH_2$$
Styrene

Here is a representative structure for acrylonitrile-butadiene-styrene:

Acrylonitrile-Butadiene-Styrene

Graft polymerization techniques are commonly used to make various grades of this material.

The resins are hygroscopic (moisture-absorbing). Pre-drying before molding is advisable. ABS materials can be produced on all thermoplastic processing machines.

ABS materials are characterized by resistance to chemicals, heat, and impact. They are used for appliance housings, light luggage, camera bodies, pipe, power tool housings, automotive trim, battery cases, toolboxes, packing crates, radio cases, cabinets, and various furniture components. They may be electroplates and used in automotive, appliance, and housewares applications.

Table E-3 lists some of the properties of ABS materials, and reference lists of nine advantages and five disadvantages follow:

Advantages of ABS

1. Ease of fabrication and coloring
2. High-impact resistance, with toughness and rigidity
3. Good electrical properties
4. Excellent adhesion by metal coatings
5. Fairly good weather resistance and high gloss
6. Ease of forming by conventional thermoplastic methods
7. Good chemical resistance
8. Lightweight
9. Very low moisture absorption

Disadvantages of ABS

1. Poor solvent resistance
2. Subject to attach by organic materials of low molecular mass
3. Low dielectric strength
4. Only low elongations available
5. Low continuous service temperature

Acrylonitrile-Chlorinated Polyethylene-Styrene (ACS)

Because of the chlorine content, this terpolymer surpasses ABS in flame-retardant properties, weatherability, and service temperatures. Applications include office machine housing, appliance cases, and electrical connectors.

ABS/PA blends have excellent chemical and temperature resistance for under--hood automobile components. ABS/PC alloys fill the price and performance gap between polycarbonate and ABS. Typical applications include typewriter housings, headlight rings, institutional food trays, and appliance housings. ABS/PVC alloys are used for air conditioner fans, grills, luggage shells, and computer housings because of their outstanding impact strength, toughness, and low cost. ABS/EVA alloys have good impact and stress crack resistance. The elastomer content in ABS/EPDM (ethylene-propylene-diene-monomer) improves low-temperature impact and modulus.

CELLULOSICS

Cellulose $(C_6H_{10}O_5)$ is the material that composes the framework, or cell walls, of all plants. Cellulose is our oldest, most familiar, most useful industrial raw material because it is abundant everywhere in one form or another. Plants are also a very inexpensive raw material. We can produce shelter, clothing, and food from them. Cereal straws and grass are composed of nearly 40% cellulose; wood is about 50% cellulose; and cotton may be nearly 98% cellulose. Wood and cotton are major industrial sources of this material. Long-chained molecules of repeating glucose units are referred to as "chemically modified natural plastics."

The chemical structure of cellulose is shown in Figure E-5. Each cellulose molecule contains three hydroxyl (OH) groups to which different groups may attach in order to form various cellulosic plastics. Cellulose can undergo a reaction at the ether linkage between the units.

The term *cellulosics* refers to plastics derived from cellulose, a family that consists of many separate and distinct types of plastics. The relationship of cellulose to many plastics and applications is shown in Figure E-6.

There are three large groups of cellulosic plastics. *Regenerated cellulose* is first chemically changed to a soluble material, then reconverted by chemical means into its original substance.

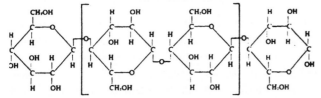

Figure E-5. Chemical structure of cellulose.

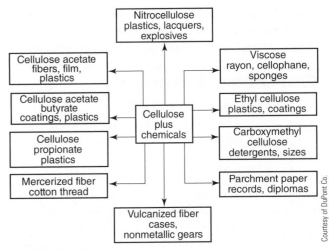

Figure E-6. The cellulosic plastics.

Cellulose esters are formed when various acids react with the hydroxyl (OH) groups of the cellulose. *Cellulose ethers* are compounds derived by the alkylation of cellulose.

Regenerated Cellulose

Examples of regenerated cellulose products are cellophane, viscose rayon, and cuprammonium rayon (no longer of commercial importance).

In its natural form, cellulose is insoluble and incapable of flow by melting. Even in the powdered form, it retains a fibrous structure.

There is evidence that in 1857 cellulose was found to be dissolvable in an ammoniacal solution of copper oxide. By 1897, Germany was commercially producing a fibrous yarn by spinning this solution into an acid or alkaline coagulating bath. Any remaining copper ions were removed by additional acid baths. This expensive process was called the *cuprammonium* process (for copper and ammonia), and the fiber was known as *cuprammonium rayon*. Newer synthetic fibers with equally desirable properties are less costly to produce; therefore, cuprammonium rayon has lost its popularity.

In 1892, C. F. Cross and E. J. Bevan of England produced a different cellulose fiber. They treated alkali cellulose (cellulose treated with caustic soda) with carbon disulfide to form a xanthate. Cellulose xanthate is soluble in water to give a viscous solution (hence the name) called *viscose*. Viscose was then extruded through spinneret openings into a coagulating solution of sulfuric acid and sodium sulfate. The regenerated fiber is called *viscose rayon*. Rayon has become an accepted generic name for fibers composed of regenerated cellulose. It is still found in clothing fabric and has some use as tire cord.

Production patents were granted to J. F. Brandenberger of France in 1908 for an extruded, regenerated cellulose film called *cellophane*. Like viscose rayon, the xanthate solution is regenerated by coagulation in an acid bath. After the cellophane is dried, it is usually given a water-resistant coating of ethyl acetate or cellulose nitrate. Coated and uncoated cellophane films are used for packaging food products and pharmaceuticals.

Cellulose Esters

Among the esters of cellulose are cellulose nitrate, cellulose acetate, cellulose butyrate, and cellulose acetate propionate. In this group of plastics, acids are used to react with the hydroxyl (OH) group to form esters.

Professor Bracconot of France first found the nitration of cellulose in 1832. The discovery that mixing nitric and sulfuric acids with cotton would produce nitrocellulose (or cellulose nitrate) was made by C. F. Schonbein of England. Although this material was a useful military explosive, it found little commercial value as a plastics. Figure E-7 shows the nitration of cellulose.

At the London International Exhibition in 1862, Alexander Parks of England was awarded a bronze medal for his new plastics material called *Parkesine*. It was composed of cellulose nitrate with a plasticizer of castor oil.

In the United States, John Wesley Hyatt created the same material while seeking a substitute for ivory billiard balls. His experiments followed the earlier work of American chemist Maynard. Maynard had dissolved cellulose nitrate in ethyl alcohol and ether to form a product used as a bandage for wounds. He gave this solution the name *collodion*. When the collodion solution was spread over a wound, the solvent evaporated, leaving a thin, protective film. One story of Hyatt's discovery of the material that became known as *celluloid* tells of his accidentally spilling camphor on some pyroxylin (cellulose nitrate) sheets and noticing the improved properties. Another version states that he had been treating a wound covered by collodion with a camphor solution and discovered a change in the cellulose nitrate product.

In 1870, Hyatt and his brother patented the process of treating cellulose nitrate with camphor, and by 1872, celluloid became a commercial success. Products made entirely of nitrocellulose were highly explosive, became brittle, and suffered greatly from shrinkage. The use of camphor as a plasticizer eliminated many of these disadvantages. Celluloid is made from pyroxylin (nitrated cellulose), camphor, and alcohol. It is highly combustible, but not explosive.

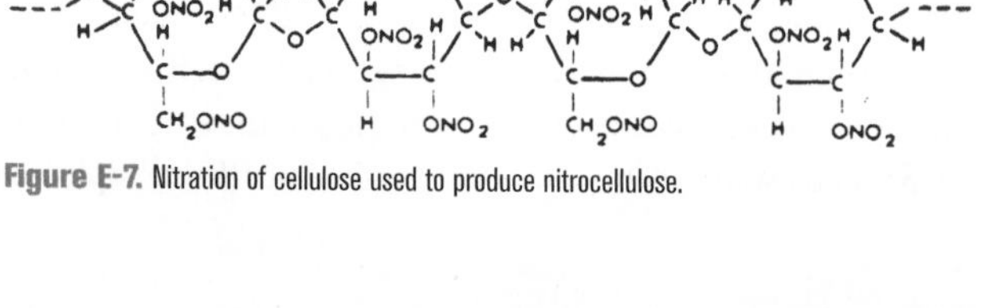

Cellulose Nitrate (CN)

Cellulose nitrate was once used in dental applications, men's clothing, personal hygiene applications, and photographic film (Figure E-8) as well as other applications. Today, it is seldom used because of processing difficulties and high flammability. Cellulose nitrate cannot be injection or compression molded. It is usually extruded or cast into large blocks from which sheets are sliced. Films are made by continuous casting of a cellulose solution on a smooth surface. As the solvents evaporate, the film is removed from the surface and placed on drying rollers. Sheets and films may be vacuum processed. Table-tennis balls and a few novelty items still may be made of cellulose nitrate plastics. Cellulose nitrate esters are found in lacquers for metal and wood finishes, and are common ingredients for aerosol paints and fingernail polishes.

Cellulose Acetate (CA)

Cellulose acetate is the most useful of the cellulosic plastics. During World War I, the British secured the aid of Henry and Camille Dreyfus of Switzerland to start large-scale production of cellulose acetate. Cellulose acetate provided a fire-retardant lacquer for the fabric-covered airplanes used at that time. By 1929, commercial grades of molding powder, fibers, sheets, and tubes were being produced in the United States.

Basic methods for manufacturing this material resemble those used in making cellulose nitrate. Acetylation of cellulose is carried out in a mixture of acetic acid and acetic anhydride, using sulfuric acid as a catalyst. The acetate or acetyl group or radical (CH_3CO) is the source of chemical reaction with the hydroxyl (OH) groups. The structure of cellulose triacetate is as follows:

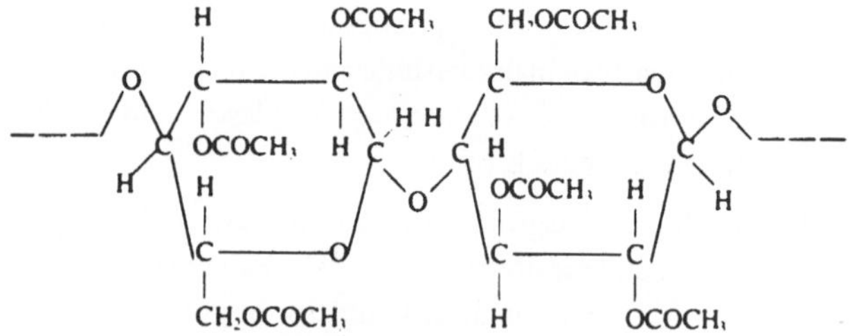

Cellulose triactate

Cellulose acetates exhibit poor heat, electrical, weathering, and chemical resistance. They are fairly inexpensive and may be transparent or colored. Their main uses are as films and sheets in the packaging and display industries. They are fabricated by nearly all the thermoplastic processes and molded into brush handles, combs, and spectacle frames.

Figure E-7. Nitration of cellulose used to produce nitrocellulose.

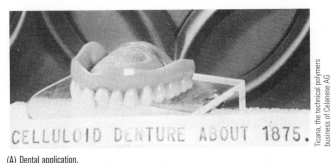

(A) Dental application.

Ticana, the technical polymers business of Celanese AG

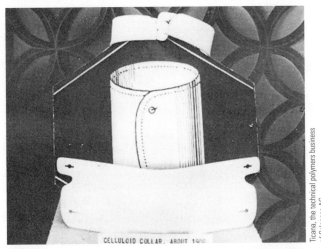

(B) Use in men's clothing.

Ticana, the technical polymers business of Celanese AG

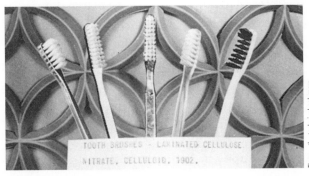

(C) Personal hygiene application.

Ticana, the technical polymers business of Celanese AG

(D) Film base which led to growth of photography as a popular hobby.

Figure E-8. Some early uses of celluloid.

Ticana, the technical polymers business of Celanese AG

Vacuum-formed display containers for hardware or food products are common, and films that permit the passage of moisture and gases are used in commercial packaging of fruits and vegetables. Coated films are used in magnetic recording tapes and photographic film. Cellulose acetate plastics are made into fibers for use in textiles. They are also employed as lacquers in the coating industry. Table E-4 gives some of the properties of cellulose acetate.

Cellulose Acetate Butyrate (CAB)

Cellulose acetate butyrate was developed in the mid-1930s by the Hercules Powder Company and Eastman Chemical. This material is produced by reacting cellulose with a mixture of sulfuric and acetic acids. *Esterification* is completed when the cellulose is reacted with butyric acid and acetic anhydride. The reaction is similar to making cellulose acetate except that butyric acid is also used. The product that results has acetyl groups (CH_3CO) and butyl groups ($CH_3CH_2CH_2CH$) in the repeating cellulose unit. This product has improved dimensional stability, weathers well, and is more chemical and moisture resistant than cellulose acetate.

Cellulose acetate butyrate is used for the tabulator keys of office machines, automobile parts, tool handles, display signs, strippable coatings, steering wheels, tubes, pipes, and packaging components. Probably the most familiar application of this material is in screwdriver handles (Figure E-9).

Cellulose Acetate Propionate

In 1931, cellulose acetate propionate (also simply called cellulose propionate) was developed by the Celanese Plastics Company. It was not widely used until materials shortages developed during World War II. It is made like other acetates, with the addition of propionic acid (CH_3CH_2COOH) in place of acetic anhydride. Its general properties are similar to those of cellulose acetate butyrate. However, it exhibits superior heat resistance and lower moisture absorption. The main uses of cellulose acetate propionate include pens, automotive parts, brush handles, steering wheels, toys, novelties, and film packaging displays. Lists of nine advantages and four disadvantages of cellulosic esters follow:

Advantages of Cellulosic Esters

1. Forms glossy moldings by thermoplastic methods
2. Exceptional clarity (butyrates and propionates)
3. Toughness, even at low temperatures
4. Excellent colorability
5. Non-petrochemical base

Table E-4. Properties of Cellulosics

Property	Ethyl Cellulose (Molding)	Cellulose Acetate (Molding)	Cellulose Propionate (Molding)
Molding qualities	Excellent	Excellent	Excellent
Relative density	1.09–1.17	1.22–1.34	1.17–1.24
Tensile strength, MPa	13.8–41.4	13–62	14–53.8
(psi)	(2000–8000)	(1900–9000)	(2000–22,000)
Compressive strength, MPa	69–241	14–248	165.5–152
(psi)	(10,000–35,000)	(2000–36,000)	(2400–22,000)
Impact strength, Izod J/mm	0.1–0.43	0.02–0.26	0.025–0.58
(ft-lb/in.)	(2.0–8.5)	(0.4–5.2)	(0.05–11.5)
Hardness, Rockwell	R50–R115	R34–R125	R10–R122
Thermal expansion; 10^{-4}/°C	25–50	20–46	28–43
Resistance to heat, °C	46–85	60–105	68–105
(°F)	(115–185)	(140–220)	(155–220)
Dielectric strength, V/mm	13,800–19,685	9840–23,620	11,810–17,715
Dielectric constant (at 60 Hz)	3.0–4.2	3.5–7.5	3.7–4.3
Dissipation factor (at 60 Hz)	0.005–0.020	0.01–0.06	0.01–0.04
Arc resistance, s	60–80	50–310	175–190
Water absorption (24 h), %	0.8–1.8	1.7–6.5	1.2–2.8
Burning rate, mm/min	Slow	Slow to self-extinguishing	Slow 25–33
(in./min)			(1.0–1.3)
Effect of sunlight	Slight	Slight	Slight
Effect of acids	Decomposes	Decomposes	Decomposes
Effect of alkalies	Slight	Decomposes	Decomposes
Effect of solvents	Soluble in ketones, esters, chlorinated hydrocarbons, aromatic hydrocarbons	Soluble in ketones, esters, chlorinated hydrocarbons, aromatic hydrocarbons	Soluble in ketones, esters, chlorinated hydrocarbons, aromatic hydrocarbons
Machining qualities	Good	Excellent	Excellent
Optical qualities	Transparent to opaque	Transparent to opaque	Transparent to opaque

6. Wide range of processing characteristics
7. Resists stress cracking
8. Outstanding weatherability (butyrates)
9. Slow burning (except cellulose nitrate)

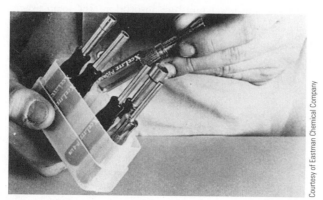

Courtesy of Eastman Chemical Company

Figure E-9. Cellulose butyrate and cellulose acetate are used for the handles of these tools.

Disadvantages of Cellulosic Esters

1. Poor solvent resistance
2. Poor resistance to alkaline materials
3. Relatively low compressive strength
4. Flammable

Cellulose Ethers

Among the ethers of cellulose are ethyl cellulose, methyl cellulose, hydroxymethyl cellulose, carboxymethyl cellulose, and benzyl cellulose. Their manufacture generally involves the preparation of alkali cellulose and other reactants, as shown in Figure E-10.

Ethyl cellulose (EC) is the most important of the cellulose ethers and is the only one used as a plastics material. Basic research and patents were established by Dreyfus in 1912.

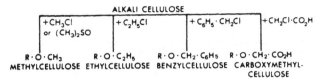

Figure E-10. Cellulose plastics from alkali cellulose and other reactants. The *R* in these formulas represents the cellulose skeleton.

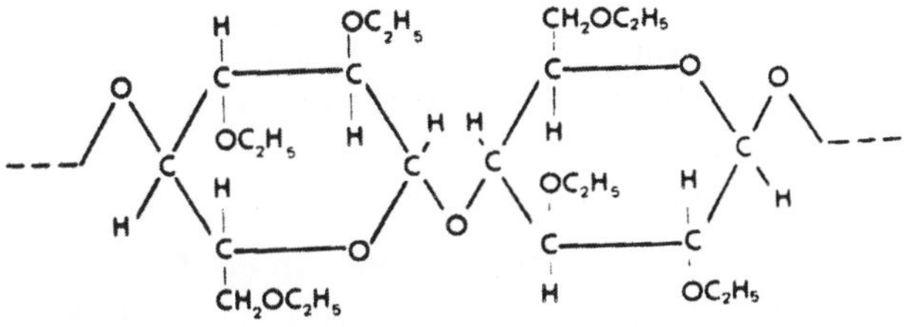

Figure E-11. Ethyl cellulose (fully ethylated).

By 1934, the Hercules Powder Company offered commercial grades in the United States.

Alkali cellulose (soda cellulose) is treated with ethyl chloride to form ethyl cellulose. The radical substitution (*etherification*) of this ethoxy may cause the final product to vary over a wide range of properties. In etherification, the hydrogen atoms of the hydroxyl groups are replaced by ethyl (C_2H_5) groups (Figure E-11). This cellulose plastic is tough, flexible, and moisture resistant. Table E-4 gives some of the properties of ethyl cellulose.

Ethyl cellulose is used in football helmets, flashlight cases, furniture trim, cosmetic packages, tool handles, and blister packages. It has been used for protective coatings on bowling pins and in the formulations of paint, varnish, and lacquers. Ethyl cellulose is also a common ingredient in hair sprays and is often used as a hot melt for strippable coatings. These coatings protect metal parts against corrosion and marring during shipment and storage.

Methyl cellulose is prepared like ethyl cellulose but uses methyl chloride or methyl sulfate instead of ethyl chloride. In etherification, the hydrogen of the OH group is replaced by methyl (CH_3) groups:

$$R(ONa)_{3n} + CH_3Cl \longrightarrow R(OCH)_{3n}$$
Methyl cellulose

Methyl cellulose finds a wide number of uses. It is water soluble and edible. It is used as a thickening emulsifier in cosmetics and adhesives and is a well-known wallpaper adhesive and fabric-sizing material. It is useful for thickening and emulsifying water-based paints, salad dressings, ice cream, cake mixes, pie filling, crackers, and other products. In pharmaceuticals, it is used to coat pills and as contact lens solutions (Figure E-12).

Hydroxyethyl cellulose is produced by reacting the alkali cellulose with ethylene oxide and has many of the same applications as methyl cellulose. In the schematic equation below, R represents the cellulose skeleton.

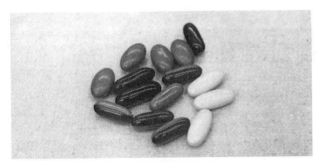

Figure E-12. Methyl cellulose is used as a coating for pharmaceuticals.

$$R(ONa)_{3n} + \overset{O}{\overset{/\backslash}{CH_2CH_2}} \longrightarrow R(OCH_2CH_2OH)_{3n}$$
Hydroxyethyl cellulose

At one time, *benzyl cellulose* was used for moldings and extrusions. However, in the United States it is unable to compete with the other polymers.

Carboxymethyl cellulose (sometimes called sodium carboxymethyl cellulose) is made from alkali cellulose and sodium chloracetate. Like methyl cellulose, it is water soluble and used as a sizing, gum, or emulsifying agent. Carboxymethyl cellulose may be found in foods, pharmaceuticals, and coatings. It is a first-rate water-soluble suspending agent for lotions, jelly bases, ointments, toothpastes, paints, and soaps. It is used to coat pills, paper, and textiles.

$$R(ONa)_{3n} + ClCH_2COONa \longrightarrow R(OCH_2COONa)_{3n} + NaCl$$
Sodium carboxymethyl cellulose

CHLORINATED POLYETHERS

In 1959, the Hercules Powder Company introduced *chlorinated polyethers* with the trade name Penton™. However, in 1972, Hercules discontinued production and sales. This thermoplastic material was produced by chlorinating pentaerythritol. The resulting dichloromethyl oxycyclobutane was polymerized into a crystalline, linear product. This material, penthol, was over 45% chlorine by mass (Figure E-13).

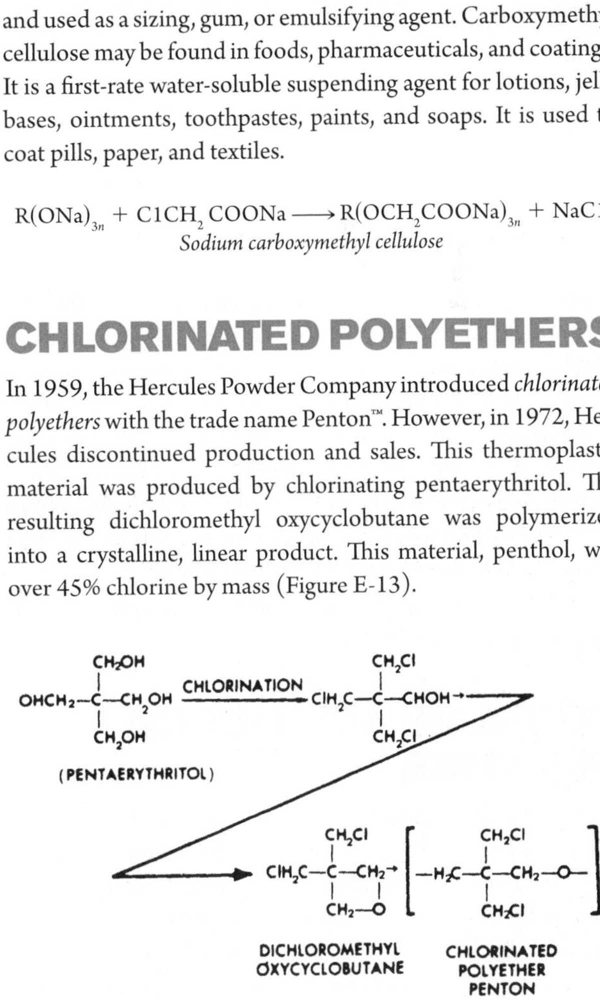

Figure E-13. Production of penthol, a chlorinated polyether.

Table E-5. Properties of Chlorinated Polyether

Property	Chlorinated Polyether
Molding qualities	Excellent
Relative density	1.4
Tensile strength, MPa	41
(psi)	(6000)
Impact strength, Izod, J/mm	0.002
(ft-lb/in.)	(0.4)
Hardness, Rockwell	R100
Thermal expansion, 10^{-4}/°C	20
Resistance to heat, °C	143
(°F)	(290)
Dielectric strength, V/mm	16,000
Dielectric constant (at 60 Hz)	3.1
Dissipation factor (at 60 Hz)	0.01
Water absorption (24 h), %	0.01
Burning rate, mm/min	Self-extinguishing
Effect of sunlight	Slight
Effect of acids	Attacked by oxidizing acids
Effect of alkalies	None
Effect of solvents	Resistant
Machining qualities	Excellent
Optical qualities	Translucent to opaque

Chlorinated polyethers may be processed on thermoplastic equipment. These materials are high-performance, high-priced plastics (Table E-5). They have been coated on metal substrates by using fluidized-bed, flame-spraying, or solvent processes. Molded parts possess high strength, heat resistance, excellent electrical and chemical resistance, and low water absorption. Although the high price restricted their wider use, they were utilized as coatings for valves, pumps, and meters. Molded parts included chemical meter components, pipe lining, valves, laboratory equipment, and electrical insulation. However, at degrading temperatures, lethal chlorine gas is released.

To date, there are no plans for another producer to make this plastics with its distinct chemical nature.

COUMARONE-INDENE PLASTICS

Coumarone and indene are obtained from the fractionation of coal tar but are seldom separated. These inexpensive products resemble styrene in chemical structure. Coumarone and indene may be polymerized by the ionic catalytic action of sulfuric acid. A wide range of products, from sticky resins to brittle plastics, may be obtained by varying the coumarone-indene ratio or by copolymerization with other polymers:

Coumarone   *Indene*   *Styrene*

Although available since long before World War II, they are not available as molding compounds and have found only limited uses as binders, modifiers, or extenders for other polymers and compounds. The largest quantities of coumarone and indene are used in the production of paints or varnishes and as binders in flooring tiles and mats.

The traits of these compounds vary greatly. Coumarone-indene copolymers are good electrical insulators. They are soluble in hydrocarbons, ketones, and esters. They range from light to dark in color and are inexpensive to make. They are true thermoplastic materials, with a softening point of 35°C to 50°C (100°F to 120°F). Applications include printing inks, coatings for paper, adhesives, encapsulation compounds, some battery boxes, brake linings, caulking compounds, chewing gum, concrete-curing compounds, and emulsion binders.

FLUOROPLASTICS

Elements in the seventh column of the Periodic Table are closely related. These elements (fluorine, chlorine, bromine, iodine, and astatine) are called *halogens*, from the Greek word that means "salt producing." Chlorine is found in common table salt. All of these elements are electronegative (they can attract and hold valence electrons) and have only seven electrons in their outermost shell. Fluorine and chlorine are gases not found in a pure or free state. Fluorine is the most reactive element known. Large quantities of fluorine are needed for processes connected with nuclear energy technology, such as the isolation of uranium metal.

Compounds containing fluorine are commonly called *fluorocarbons*, although in the strictest definition fluorocarbon should be used to refer to compounds containing only fluorine and carbon.

The French chemist Moissan isolated pure fluorine in 1886, but it remained a laboratory curiosity until 1930. In 1931, the trade name Freon® was announced. This fluorocarbon is a compound of carbon, chlorine, and fluorine. One example is CCl_2F_2. Freons are used extensively as refrigerants. Fluorocarbons are also used in the formation of polymeric materials.

In 1938, the first polyfluorocarbon was discovered by accident in the DuPont research laboratories. It was discovered that tetrafluoroethylene gas formed an insoluble, waxy, white powder when stored in steel cylinders. A number of fluorocarbon polymers have been developed as a result of this chance discovery.

The term *fluoroplastic* is used to describe alkene-like structures that have some or all of the hydrogen atoms replaced by fluorine.

MONOMER   MONOMER

$$\text{-C-C+C-C+C-C-}$$
Polyethylene

$$\text{-C-C+C-C+C-C-}$$
Polytetrafluoroethylene

It is the presence of the fluorine atoms that provides the unique properties characteristic of the fluoroplastic family. These properties are directly related to the high carbon-to-fluorine bonding energy and the high electronegativity of the fluorine atoms. Thermal stability and solvent, electrical, and chemical resistance are weakened if fluorine (F) atoms are replaced with hydrogen (H) or chlorine (Cl) atoms. The C–H and C–Cl bonds are weaker than C–F bonds and are more vulnerable to chemical attack and thermal decomposition.

The major fluoroplastics are shown in Figure E-14. There are only two types of fluorocarbon plastics: polytetrafluoroethylene (PTFP) and polytetrafluoropropylene (PFEP, FEP, or fluorinated ethylene propylene). The others must be considered copolymers or fluorine-containing polymers.

Polytetrafluoroethylene (PTFE)

Polytetrafluoroethylene $(CF_2=CF_2)_n$ accounts for nearly 90% (by volume) of the fluorinated plastics. The monomer tetrafluoroethylene is obtained by pyrolysis of chlorodifluoromethane. Tetrafluoroethylene is polymerized in the presence of water and

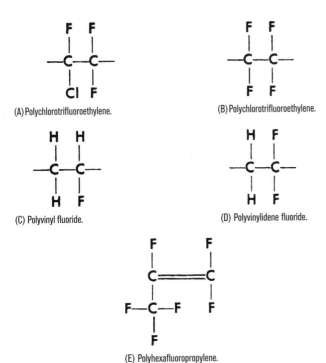

(A) Polychlorotrifluoroethylene.

(B) Polychlorotrifluoroethylene.

(C) Polyvinyl fluoride.

(D) Polyvinylidene fluoride.

(E) Polyhexafluoropropylene.

Figure E-14. Monomers of fluoroplastics and fluorine-containing polymers.

a peroxide catalyst under high pressure. PTFE is a highly crystalline, waxy thermoplastic material with a service temperature of −268°C to +288°C (−450°F to +550°F). The high bonding strength and compact interlocking of fluorine atoms about the carbon backbone prevent PTFE processing by the usual thermoplastic methods. At present, it cannot be plasticized to aid processing. Most of the material is made into preforms and sintered.

Sintering is a special fabricating technique used for metals and plastics. The powdered material is pressed into a mold at a temperature just below its melting or degradation point until the particles are fused (sintered) together. The mass as a whole does not melt in this process. Sintered moldings may be machined. Special formulations may be extruded in the form of rods, tubes, and fibers by using organic dispersions of the polymer. These are later vaporized as the product is sintered. Coagulated suspensoids may be used in much the same way. Presintered grades of this material may be extruded through extremely long compacting and sintering zones of special dies. Many films, tapes, and coatings are cast, dipped, or sprayed from PTFE dispersions, using a drying and sintering process. Films and tapes may also be cut or sliced from sheet stock.

Teflon is a familiar trade name for homopolymers and copolymers of polytetrafluoroethylene. Its antistick property (low coefficient of friction) makes it a very useful coating. Teflon is applied to many metallic substrates, including cookware. There is no known solvent for these materials. They may be chemically etched and adhesive-bonded with contact or epoxy adhesives. Films may be heat sealed together, but not to other materials.

Fluorocarbons have a greater mass than hydrocarbons because fluorine has an atomic mass of 18.9984 and hydrogen, only 1.00797. Fluoroplastics are thus heavier than other plastics. Relative densities range from 2.0 to 2.3.

PTFE requires special fabricating techniques. Its chemical inertness, unique weathering resistance, excellent electrical insulation characteristics, high heat resistance, low coefficient of friction, and nonadhesive properties have led to numerous uses. Parts coated with PTFE have such a low coefficient of friction that they release and slide easily yet require no lubrication. Saw blades, cookware, utensils, snow shovels, bakery equipment, and bearings are common applications (Figure E-15). Aerosol spray dispersions of micron-sized particles of polytetrafluoroethylene are used as a lubricant and antistick agent for metal, glass, or plastics substrates. Many profile shapes are used for chemical, mechanical, and electrical applications (Figure E-16). Shrinkable tubing is used to cover rollers, springs, glass, and electrical parts. Skived or extruded tapes and films may be used for seals, packing, and gasket materials. Bridges, pipes, tunnels, and buildings may rest on slip joints, expansion plates, or bushing or bearing pads of polytetrafluoroethylene (Figure E-17).

The excellent electrical insulating and low dissipation factors of polytetrafluoroethylene make it useful for wire and cable insulation, coaxial wire spacers, laminates for printed circuits,

Figure E-15. Tools coated with Teflon.

Courtesy of DuPont Co.

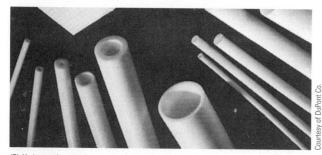

(A) Shrinkable protective tubing.

Courtesy of DuPont Co.

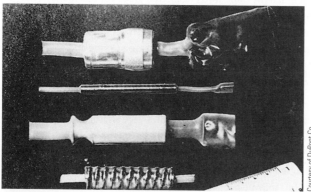

(B) Various rods and tubes .

Figure E-16. Various PTFE applications.

Courtesy of DuPont Co.

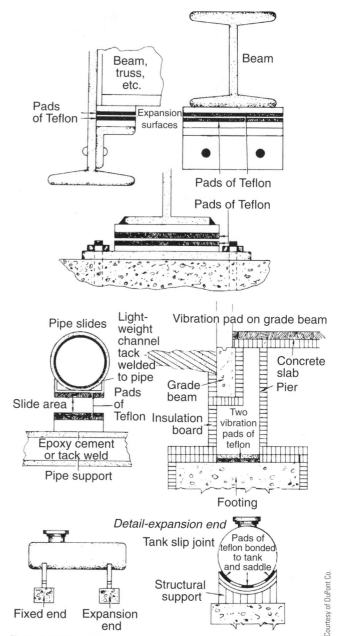

Figure E-17. Teflon pads have many construction uses.

Courtesy of DuPont Co.

and many other electrical applications. Six advantages and eight disadvantages of PTFE follow:

Advantages of PTFE

1. Nonflammable
2. Outstanding chemical and solvent resistance
3. Excellent weatherability
4. Low friction coefficient (antistick property)
5. Wide thermal service range
6. Very good electrical properties

Disadvantages of PTFE

1. Not processable by common thermoplastic methods
2. Toxic in thermal degradation
3. Subject to creep
4. Permeable
5. Requires high processing temperatures
6. Low strength
7. High density
8. Comparatively high cost

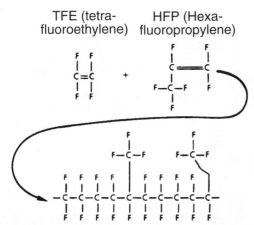

Figure E-18. Manufacture of polyfluoroethylenepropylene.

Figure E-19. Shrinkable PFEP antistick covers applied to rollers.

Polyfluoroethylenepropylene (PFEP) or (FEP)

In 1965, DuPont announced another Teflon fluoroplastic wholly composed of fluorine and carbon atoms. Polyfluoro-ethylenepropylene (PFEP or FEP) is made by copolymerizing tetrafluoroethylene with hexafluoropropylene (Figure E-18).

The partial disruption of the polymer chain by the propylene-like groups $CF_3CF{=}CF_2$ reduces the melting point and viscosity of FEP resins. Polyfluoroethylenepropylene may be processed by using normal thermoplastic methods. This reduces production costs of items previously molded of PTFE. Because of pendant CF_3 groups, this copolymer is less crystalline, more processable, and transparent in films up to 0.25 mm (0.01 in.) thick.

Commercial PFEP plastics possess properties similar to those of PTFE. They are chemically inert and possess very good electrical insulation properties as well as a somewhat greater impact strength. Service temperatures may exceed 205°C (400°F). Polyfluoroethylenepropylene plastics are used extensively by the military and aircraft and aerospace industries for electrical insulation and high reliability at cryogenic temperatures. They are used for lining chutes, pipes, and tubes as well as coating objects where a low coefficient of friction or nonadhesive characteristics is required. PFEP is molded into parts such as gaskets, gears, impellers, printed circuits, pipes, fittings, valves, expansion plates, bearings, and other profile shapes (Figure E-19).

As early as 1933, both Germany and the United States were making a fluoroplastic material in connection with atomic bomb research. It was used in handling uranium fluoride, a compound of uranium with fluorine.

The following lists show six advantages and six disadvantages of polyfluoroethylenepropylene:

Advantages of PFEP

1. Processable by normal thermoplastic methods
2. Resistant to chemicals (including oxidizing agents)
3. Excellent solvent resistance
4. Antistick characteristics
5. Nonflammable
6. Low coefficient of friction, dielectric constant, mold shrinkage, and water absorption

Disadvantages of PFEP

1. Comparatively high cost
2. High density
3. Subject to creep
4. Low compressive and tensile strength
5. Low stiffness
6. Toxic upon thermal decomposition

Polychlorotrifluoroethylene (PCTFE) or (CTFE)

Polychlorotrifluoroethylene is produced in various formulations. Chlorine atoms are substituted for fluorine in the carbon chain:

$$-CF_2-\underset{\underset{Cl}{|}}{CF}$$

Polychlorotrifluoroethylene (PCTFE)

Monomers are obtained by fluorinating hexachloroethane and then dehalogenating (controlled removal of the halogen, chlorine) with zinc in alcohol:

$$CCl_3CCl_3 \xrightarrow[\text{HF}]{\text{Anhydrous}} CCl_2FCClF_2 \xrightarrow[\substack{\text{boiling} \\ \text{ethyl alcohol}}]{\text{Zinc}} CClF$$

Hexachloroethane

$$= CF_2 + Cl_2$$

The polymerization is similar to PTFE because it is accomplished in an aqueous emulsion and suspension. Peroxide or Ziegler-type catalysts are used during bulk polymerization:

$$nCF_2 = CFCl \xrightarrow{\text{polymerize}} (-CF_2CFCl-)_n$$

Chlorotrifluoroethylene   Polychlorotrifluoroethylene

The addition of the chlorine atoms to the carbon chain allows processing by normal thermoplastic equipment. The chlorine presence also allows selected chemicals to attack and break the partially crystalline polymer chain. Depending on the degree of crystallinity, PCTFE may be produced in an optically clear form. Copolymerization with vinylidene fluoride or other fluoroplastics provides varying degrees of chemical inertness, thermal stability, and other unique properties.

Polychlorotrifluoroethylene is harder, more flexible, and possesses a higher tensile strength than PTFE. It is more expensive than PTFE and has a service temperature ranging from $-240°C$ to $+205°C$ ($-400°F$ to $+400°F$). The introduction of the chlorine atom lowers its electrical insulating properties and raises the coefficient of friction. Although it is more expensive than polytetrafluoroethylene, it finds similar uses. Examples include insulation for wire, cable, printed circuit boards, and electronic components (Figure E-20). The property of chemical resistance is best used in producing transparent windows for chemicals, seals, gaskets, O-rings, and pipe lining as well as pharmaceutical and lubricant packaging. Dispersions and films may be used in coating reactors, storage tanks, valve bodies, fittings, and pipes. Films may be sealed by using thermal or ultrasonic techniques. Epoxy adhesives may be used on chemically etched surfaces.

The tough plastics ethylene-chlorotrifluoroethylene (ECTFE) is shown in Figure E-21. The lists that follow give eight advantages and five disadvantages of polychlorotrifluoroethylene:

Advantages of PCTFE (or CTFE)

1. Excellent solvent resistance
2. Optically clear
3. Zero moisture absorption
4. Self-extinguishing
5. Low permeability
6. Creep resistance better than PTFE or PFEP
7. Very low coefficient of friction
8. Good low-temperature capability

Disadvantages of PCTFE (or CTFE)

1. Lower electrical properties than PTFE
2. More difficult to mold than PFEP
3. Crystallizes upon slow cooling
4. Less solvent resistance than PTFE and PFEP
5. Higher coefficient of friction than PTFE and PFEP

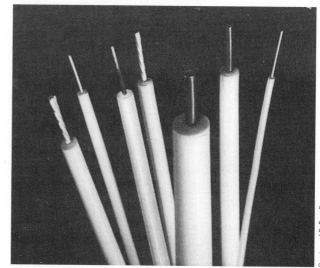

(A) Insulation for wires and cables.

(B) Ball-valve seats.

Figure E-20. Two uses of PCTFE.

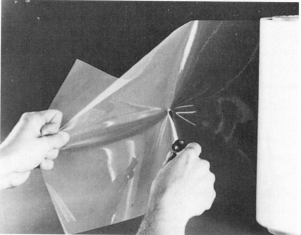

Figure E-21. Ethylene-chlorotrifluoroethylene (E-CTFE) has high-performance properties common to other fluoropolymers. It surpasses them in impermeability, tensile strength, and resistance to abrasion.

Polyvinyl Fluoride (PVF)

In 1900, the earliest preparation of vinyl fluoride gas was considered impossible to polymerize. However, in 1958, DuPont announced the polymerization of vinyl fluoride (PVF). In 1933, monomer resins were prepared in Germany by reacting hydrogen fluoride with acetylene, using selected catalysts:

$$HF + CH \equiv CH \xrightarrow{\text{catalysts}} CH_2 = CHF$$

Although the monomer has been known to chemists for some time, it is hard to manufacture or polymerize. Polymerization is carried out using peroxide catalysts in various aqueous solutions at high pressures.

Polyvinyl fluoride may be processed by using normal thermoplastic methods. These plastics are strong, tough, flexible, and transparent. They have outstanding weather resistance. PVF has good electrical and chemical resistance, with a service temperature near 150°C (300°F).

Uses include protective coatings and surfaces for exterior use, finishes for plywood, sealing tapes, packaging of corrosive chemicals, and many electrical insulating applications. Coatings may be applied to automotive parts, lawn mower housings, house shutters, gutters, and metal siding. Four advantages and four disadvantages follow. (Also see Polyvinyls.)

Advantages of PVF

1. Processable by thermoplastic methods
2. Low permeability
3. Flame retardant
4. Good solvent resistance

Disadvantages of PVF

1. Lower thermal capability than highly fluorinated polymers
2. Toxic (in thermal decomposition)
3. High dipole bond
4. Subject to attack by strong acids

Polyvinylidene Fluoride (PVDF)

Polyvinylidene fluoride (PVF_2) closely resembles polyvinyl fluoride. It became available in 1961 and was produced by the Pennwalt Chemical Corporation under the trade name Kynar. Polyvinylidene fluoride is polymerized thermally by the dehydrohalogenation (removal of hydrogen and chlorine atoms) of chlorodifluoroethane under pressure:

$$CH_3CCIF_2 \xrightarrow{500 - 1700°C} CH_2 = CF_2 + HCl$$

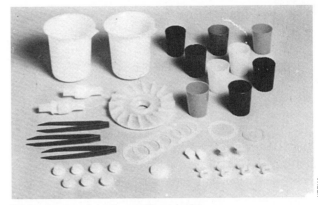

(A) Lab equipment, spools, impellers, containers, and gaskets.

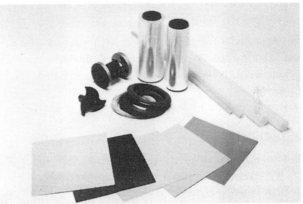

(B) Film, sheet, robs, moldings, and coatings.

Figure E-22. Polyvinylidene fluoride has many uses.

Like polyvinyl fluoride, PVDF does not have the chemical resistance of PTFE or PCTFE. The alternating CH_2 and CF_2 groups in its backbone contribute to its tough, flexible characteristics. The presence of hydrogen atoms reduces chemical resistance and allows solvent cementing and degradation. PVDF materials are processed by utilizing thermoplastic methods, and they are ultrasonically and thermally sealed. PVDF finds wide use in film and coating forms because of its toughness, optical properties, and resistance to abrasion, chemicals, and ultraviolet radiation. Service temperatures range from −62°C to +150°C (−80°F to +300°F). A familiar use is the coating seen on aluminum siding and roofs. Molded parts may include such items as valves, impellers, chemical tubing, ducting, and electronic components (Figure E-22). Six advantages and three disadvantages of polyvinylidene fluoride follow (also see Polyvinyls).

Advantages of PVDF

1. Processable by thermoplastic methods
2. Low creep
3. Excellent weatherability
4. Nonflammable
5. Abrasion resistance better than PTFE
6. Good solvent resistance

Disadvantages of PVDF

1. Lower thermal capability and chemical resistance than PTFE or PCTFE
2. Toxic in thermal decomposition
3. High dipole

Perfluoroalkoxy (PFA)

In 1972, DuPont offered perfluoroalkoxy (PFA) under the trade name of Tefzel®. It is produced from polymerization of perfluoroalkoxyethylene. PFA may be processed by using conventional thermoplastics methods. This plastics has properties similar to PTFE and PFEP. It is available in pellet, film, sheet, rod, and powder forms.

Uses include dielectric and electrical insulators, coatings, and liners for valves, pipes, and pumps. The following lists include seven advantages and six disadvantages of PFA:

Advantages of PFA

1. Higher temperature capability than PFEP
2. Excellent resistance to chemicals (including oxidizing agents)
3. Excellent solvent resistance
4. Antistick characteristics
5. Nonflammable
6. Low coefficient of friction
7. Processable by thermoplastic methods

Disadvantages of PFA

1. Comparatively high cost
2. High density
3. Subject to creep
4. Low compressive and tensile strength
5. Low stiffness
6. Toxic in thermal decomposition

Other Fluoroplastics

There are numerous other fluorine-containing polymers and copolymers. Chlorotrifluoroethylene/vinylidene fluoride is used in fabricating O-rings and gaskets:

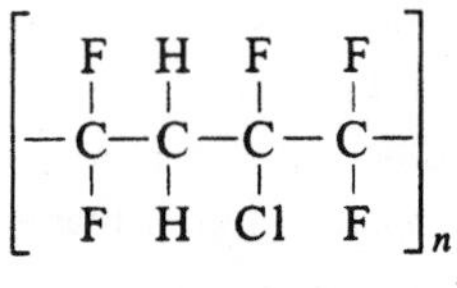

$$\left[\begin{array}{cccc} F & H & F & F \\ | & | & | & | \\ -C & -C & -C & -C- \\ | & | & | & | \\ F & H & Cl & F \end{array} \right]_n$$

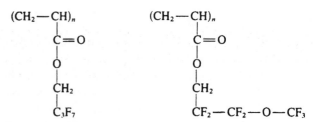

Figure E-23. Two fluoroacrylate elastomers.

Hexafluoropropylene/vinylidene fluoride is an outstanding oil- and grease-resistant elastomer used for O-rings, seals, and gaskets.

The chemical structure of two fluoroacrylate elastometers is shown in Figure E-23. Polytrifluoronitrosomethane, fluorine-containing silicones, and polyesters as well as other fluorine-containing polymers are also being produced.

The linear chained copolymer ethylene-chlorotrifluoroethylene (ECTFE) has high-performance properties similar to other fluoroplastics. It surpasses them in permeability, tensile strength, wear resistance, and creep. Uses include release agents, tank linings, and dielectrics.

The copolymer ethylene tetrafluoroethylene (ETFE) has properties and applications similar to those of ECTFE.

At degradation temperatures, toxic fluorine gas is released (see Chapter 4, Health and Safety). The properties of the basic fluoroplastics are listed in Table E-6.

IONOMERS

In 1964, DuPont introduced a new material, known as *ionomers*, that had characteristics of both thermoplastics and thermosets. Ionic bonding is seldom found in plastics but is characteristic of ionomers. Ionomers possess chains similar to polyethylene, with ion cross-links of sodium, potassium, or similar ions (Figure E-24).

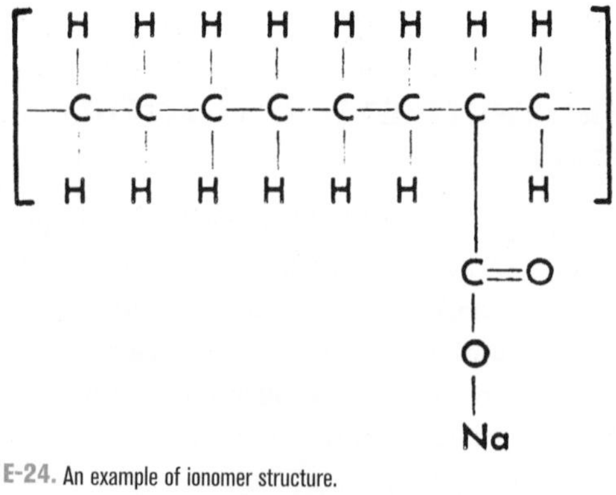

Figure E-24. An example of ionomer structure.

Table E-6. Properties of Fluoroplastics

Property	Polytetra-fluoroethylene (PTFE)	Polyfluoro-ethylenepro-pylene (PFEP)	Polychloro-trifluoroethyl-ene (PCTFE)	Polyvinyl Fluoride (PVF)	Polyvinylidene Fluoride (PVF$_2$)
Molding qualities	Excellent	Excellent	Excellent	Excellent	Excellent
Relative density	2.14–2.2	2.12–2.17	2.1–2.2		1.75–1.78
Tensile strength, MPa	14–35	19–21	30–40	58–124	40–50
(psi)	(2000–5000)	(2700–3100)	(4500–6000)		(5500–7400)
Compressive strength, MPa	12–14		30–50	30–50	60
(psi)	(1700–2000)	(4500–7400)	(4500–7400)	(8680)	(8680)
Impact strength, Izod, J/mm	0.40	No break	0.125–0.135	0.18–0.2	0.18–0.2
(ft-lb/in.)	(8)	(2.5–2.7)	(2.5–2.7)	(3.6–4.0)	(3.6–4.0)
Hardness	Shore D50–D65	Rockwell R25	Rockwell R75–R95	Shore D80	Shore D80
Thermal expansion, 10^{-4}/°C	25	20–27	11–18	70	22
Resistance to heat, °C	287	205	175–199	149	149
(°F)	(550)	(400)	(350–390)	(300)	(300)
Dielectric strength, V/mm	19,000	19,500–23,500	19,500–23,500	10,000	10,000
Dielectric constant (at 60 Hz)	2.1	2.1	2.24–2.8	8.4	8.4
Dissipation factor (at 60 Hz)	0.0002	<0.0003	0.0012	0.049	0.049
Arc resistance, s	300	165+	360	50–70	50–70
Water absorption (24 h), %	0.00	0.01	0.00	0.04	0.04
Burning rate, mm/min	None	None	None	Self-extinguishing	Self-extinguishing
Effect of sunlight	None	None	None	Slight bleach	Slight
Effect of acids	None	None	None	Attacked by fuming sulfuric	Attacked by fuming sulfuric
Effect of alkalies	None	None	None	None	None
Effect of solvents	None	None	None	Resists most	Resists most
Machining qualities	Excellent	Excellent	Excellent	Excellent	Excellent
Optical qualities	Opaque	Transparent	Translucent to opaque	Transparent	Transparent to translucent

In this material, both organic and inorganic compounds are linked together, and because the cross-linking is basically ionic, the weaker bonds are easily broken up on heating. Therefore, this material may be processed as a thermoplastics. At atmospheric temperatures, the plastics has properties usually associated with linked polymers.

The basic ionomer chain is made by polymerizing ethylene and methacrylic acid. Other inexpensive polymer chains may be developed using similar cross-links.

Because they combine ionic and covalent forces in their molecular structure, ionomers can exist in a number of physical states and possess a number of physical properties. They may be processed and reprocessed by using any thermoplastic technique. They are more expensive than polyethylene but possess a higher moisture vapor permeability than polyethylene. Ionomers are available in transparent forms.

Uses of ionomers include safety glasses, shields, bumper guards, toys, containers, packaging films, and electrical insulation as well as coatings for paper, bowling pins, or other substrates (Figure E-25). In the shoe industry, they are used for inner liners, soles, and heels. Ionomers are co-extruded with polyester films to produce a heat-sealable layer while improving package durability.

Ionomers are used for a number of composite market applications. Laminated or co-extruded films are used in tear-open pouches for pharmaceutical and food packaging. Metal, foil laminates, and heat-sealable skin and blister packaging continue to grow. Foamed applications include bumper guards, shoe

Figure E-25. The ionomer surface exhibits high resistance to cutting.

Chris Scredon/E+/Getty Images.

Table E-7. Properties of Ionomers

Property	Ionomer
Molding qualities	Excellent
Relative density	0.93–0.96
Tensile strength, MPa	24–35
(psi)	(3500–5000)
Izod, impact, J/mm	0.3–0.75
(ft-lb/in.)	(6–15)
Hardness, Shore	D50–D65
Thermal expansion, 10^{-4}/°C	30
Resistance to heat, °C	70–105
(°F)	(160–220)
Dielectric strength, V/mm	35,000–40,000
Dielectric constant (at 60 Hz)	2.4–2.5
Dissipation factor (at 60 Hz)	0.001–0.003
Arc resistance, s	90
Water absorption (24 h), %	0.1–1.4
Burning rate, mm/min	Very slow
Effect of sunlight	Requires stabilizers
Effect of acids	Attacked by oxidizing acids
Effect of alkalies	Very resistant
Effect of solvents	Very resistant
Machining qualities	Fair to good
Optical qualities	Transparent

components, wrestling mats, and ski lift seat pads. Ionomer coatings on golf balls and bowling pins extend the service lives of these products.

Table E-7 gives some properties of ionomers, and the following lists show seven advantages and four disadvantages of ionomers:

Advantages of Ionomers

1. Outstanding abrasion resistance
2. Excellent shock resistance, even at low temperatures
3. Good electrical traits
4. High melt strength
5. Abrasion resistance
6. Do not dissolve in common solvents
7. Excellent transparent colors

Disadvantages of Ionomers

1. Some swelling from detergent-alcohol mixtures
2. Must be stabilized for exterior use
3. Service temperature of 72°C (161°F)
4. Must compete with the less expensive polyolefins

NITRILE BARRIER PLASTICS

Formulations of copolymers containing a nitrile (C≡N) functionality of over 50% are called nitrile polymers.

These plastics offer very low permeability and form a barrier against gases and odors. This property is the result of high nitrile content. Their transparency and processibility make them a good candidate for use as containers.

Each manufacturer of nitrile barrier plastics varies the formulation.

Most combinations are based on acrylonitrile (AN or methacrylonitrile). Some formulations may reach 75% AN. All formulations are amorphous with a slight yellow tint. Approximate monomer compositions of nitrile barrier are shown in Table E-8. The Borg-Warner product Cyclopac 930® contains more than 64% acrylonitrile, 6% butadiene, and 21% styrene.

Although they are heat sensitive, nitrile barrier plastics have been used with all thermoplastic processing techniques (Table E-9).

These plastics have been used in beverage containers, food packages, and containers for many non-food fluids.

The AN residual monomers of food containers cannot exceed 0.10 ppm.

Table E-8. Estimated Compositions of Nitrile Barrier Plastic

Composition	Borg-Warner Cycopac 930	Monsanto Lo Pac	Sohlo Barex	DuPont NR-16	ICI LPT*
Acrylonitrile, %	65–75	65–75	65–75	65–75	65–75
Butadiene, %	5–10		5–8		
Methyl acrylate, %			20–25		
Methyl methacrylate, %			3–5		
Styrene, %	20–30	25–35		25–35	25–35

*LPT-Low permeability thermoplastics

PHENOXY

In 1962, a family of resins based on bisphenol A and epichlorohydrin was introduced by Union Carbide.

These resins were called *phenoxy* but could be classified as polyhydroxyethers. The plastics structure resembles polycarbonates, and the material has similar properties.

Table E-9. Properties of Nitrile Barrier Plastics

Property	Nitrile Barrier (Standards)
Molding qualities	Good
Relative density	1.15
Tensile strength, MPa	62
(psi)	(9000)
Impact strength, Izod, J/mm	0.075–0.2
(ft-lb/in.)	(1.5–4)
Hardness, Rockwell	M72–78
Thermal expansion, 10^{-4}/°C	16.89
Resistance to heat, °C	70–100
(°F)	(158–212)
Dielectric strength, V/mm	8660
Dielectric constant (at 60 Hz)	4.55
Dissipation factor (at 60 Hz)	0.07
Water absorption (24 h), %	0.28
Burning rate, mm/min	–
Effect of sunlight	Slight yellowing
Effect of acids	None to attacked
Effect of alkalies	None to attacked
Effect of solvents	Dissolves in acetonitrile
Machining qualities	Good
Optical qualities	Transparent

Phenoxy resins are made and sold as thermoplastic epoxide resins. They may be processed on normal thermoplastic machinery with product service temperatures in excess of 75°C (170°F).

Because of the reactive hydroxyl groups, phenoxy resins may be cross-linked. Cross-linking agents may include diisocyanates, anhydrides, triazines, and melamines.

Homopolymers have good creep resistance, high elongation, low moisture absorption, and low gas transmission, as well as high rigidity, tensile strength, and ductility. They find applications as clear or colored protective coatings, molded electronic parts, pipe for gas and crude oil, sports equipment, appliance housings, cosmetic cases, adhesives, and containers for food or drugs.

Table E-10 gives some of the properties of phenoxy.

POLYALLOMERS

In 1962, a plastics distinctly different from simple copolymers of polyethylene and polypropylene was produced by Eastman. The process, called *allomerism*, is conducted by alternately polymerizing ethylene and propylene monomers. Allomerism refers to varying chemical composition without changing the crystalline form. The plastics exhibits the crystallinity usually associated with the homopolymers of ethylene and propylene. The term *polyallomer* is used to distinguish this alternately segmented plastics from homopolymers and copolymers of ethylene and propylene.

Although highly crystalline, polyallomers may have a relative density as low as 0.896. They are available in various formulations. Properties include high stiffness, impact strength, and abrasion resistance. Their flexibility has been used in the production of hinged boxes, loose-leaf binders, and various other folders. Polyallomers may be used as food containers or films over a wide range of service temperatures. They find limited use where other polyolefins are marginal.

Processing may be done on all normal thermoplastic equipment. Like polyethylene, polyallomers are not cohesively bonded but may be welded.

Table E-10. Properties of Phenoxy

Property	Phenoxy
Molding qualities	Good
Relative density	1.18–1.3
Tensile strength, MPa	62–65
(psi)	(9000–9500)
Impact strength, Izod, J/mm	0.125
(ft-lb/in.)	(2.5)
Resistance to heat, °C	80
(°F)	(175)
Dielectric constant (at 60 Hz)	4.1
Dissipation factor (at 60 Hz)	0.001
Water absorption (24 h), %	0.13
Burning rate	Self-extinguishing
Effect of acids	Resistant
Effect of alkalies	Resistant
Effect of solvents	Soluble in ketones
Machining qualities	Good
Optical qualities	Translucent to opaque

Table E-11. Properties of Polyallomers

Property	Polyallomer (Homopolymer)
Molding qualities	Excellent
Relative density	0.896–0.899
Tensile strength, MPa	20–27
(psi)	(3000–3850)
Impact strength, Izod, J/mm	8.5–12.5
(ft-lb/in.)	(170–250)
Hardness, Rockwell	R50–R85
Thermal expansion, $10^{-4}/°C$	21–25
Resistance to heat, °C	50–95
(°F)	(124–200)
Dielectric strength, V/mm	32,000–36,000
Dielectric constant (at 60 Hz)	2.3–2.8
Dissipation factor (at 60 Hz)	0.0005
Water absorption (24 h), %	0.01
Burning rate	Slow
Effect of sunlight	Slight—should be protected
Effect of acids	Very resistant
Effect of alkalies	Very resistant
Effect of solvents	Very resistant
Machining qualities	Good
Optical qualities	Transparent

Table E-11 lists some of the properties of polyallomers.

POLYAMIDES (PAs)

From research that began in 1928, Wallace Hume Carothers and his colleagues concluded that linear polyesters were not suitable for commercial fiber production. Carothers did succeed in producing polyesters of high molecular mass and oriented them by elongation under tension. However, these fibers were still inadequate and could not be successfully spun. The amino acids present in silk, a natural fiber, prompted Carothers to study synthetic polyamides. Of the many formulations of amino acids, diamines, and dibasic acids, several showed promise as possible fibers. By 1938, the first commercially developed polyamide was introduced by DuPont. It was 6,6 polyamide and was given the trade name *Nylon*. This condensation plastics was called Nylon 6,6 (also written as 66 or 6/6) because both the acid and the amine contain six carbon atoms.

$$NH_2(CH_2)_6NH_2 + COOH(CH_2)_4COOH$$
Hexamethylene   Adipic acid
diamine

$$n[NH_2(CH_2)_6NH \cdot CO(CH_2)_4COOH] \longrightarrow heat$$
Nylon salt

$$[NH(CH_2)_4NH \cdot CO(CH_2)_4CO]_n \longrightarrow + n H_2O$$
Nylon 6,6 polymer chain

The term *nylon* has come to mean any polyamide that can be processed into filaments, fibers, films, and molded parts.

The repeating—CONH—(amide) link is present in a series of linear, thermoplastic Nylons:

- Nylon 6–Polycaprolactam:
$$[NH(CH_2)_5CO]_X$$
- Nylon 6, 6–Polyhexamethyleneadipamide:
$$[NH(CH_2)_6NHCO(CH_2)_4CO]_X$$
- Nylon6,10–Polyhexamethylenesebacamide:
$$[NH(CH_2)_6NHCO(CH_2)_8CO]_X$$
- Nylon11–Poly(11–aminoundecanoic acid)
$$[NH(CH_2)_{10}CO]_X$$
- Nylon12–Poly(12–aminododecanoic acid)
$$[NH(CH_2)_{11}CO]_X$$

There are many other types of nylon currently available, including Nylon 8, 9, and 46 and copolymers from more sophisticated diamines and acids. In the United States, Nylon 6,6 and Nylon 6 are by far the most used. The properties of nylons may also be changed by introducing additives. Amino-containing polyamide resins will react with a number of materials, and cross-linked, thermosetting reactions are possible.

Although developed primarily as a fiber, polyamides find uses as molding compounds, extrusions, coatings, adhesives, and casting materials. Acetals and fluorocarbons share some of the same properties and uses as polyamide. Polyamide resins are costly. They are selected when other resins will not meet service requirements. Acetal resins are superior in fatigue endurance, creep resistance, and water resistance. Nylons have met increased competition from these resins.

Molding compounds were offered in 1941 and have grown in applications. They are among the toughest of the plastics materials. Nylons are self-lubricating, impervious to most chemicals, and highly impermeable to oxygen. They are not attacked by fungi or bacteria. Polyamides may be used in food containers.

The largest applications of homopolymer molding compounds (Nylons 6; 6,6; 6,10; 11; and 12) include gears, earns, bearings, valve seats, combs, furniture casters, and door catches. They are used where wear resistance, quiet operation, and low coefficients of friction are needed.

Because of their crystalline structure, polyamide products have a milky-opaque appearance (Figure E-26). Transparent films may be obtained from Nylon 6 and 6,6 if they are cooled very rapidly. Polyamides are clear amorphous materials when melted. Upon cooling, they crystallize and become cloudy. This crystallinity contributes to stiffness, strength, and heat resistance. Polyamides are harder to process than other thermoplastic materials. All thermoplastic processing equipment may be used, but fairly high processing temperatures are needed. The melting point of a polyamide is abrupt or sharp; that is, polyamides do not soften or melt over a broad range of temperatures (Figure E-27). When sufficient energy has overcome the crystalline and molecular attractions, they suddenly become liquid and may be processed. Because all nylons absorb water, they are dried before molding. This ensures desirable physical properties in the molded product.

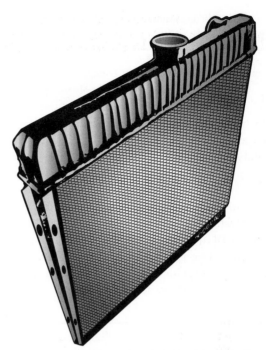

Figure E-27. An automobile radiator can be made of polyamide because it will not melt over a wide temperature range.

Extruded and blown films are used to package oils, grease, cheese, bacon, and other products where low gas permeability is essential. A new application of specialty nylon acts as a barrier layer in containers. Figure E-28 shows several containers that contain a clear nylon barrier to prevent gases from permeating through the walls of the container and entering its contents. The high service temperatures of nylon film are used for boil and bake-in-the-bag food products. Although polyamides are hygroscopic (absorb water), they find many applications as electrical insulators.

Nylon 11 may be used as a protective coating on metal substrates. Polyamides are used in a powder form by using spraying or fluidized-bed processes. Typical uses include rollers, shafts, panel slides, runners, pump impellers, and

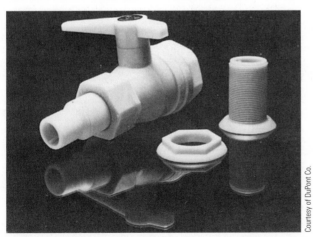

Courtesy of DuPont Co.

Figure E-26. Polyamide products have a milky-opaque appearance as seen in this valve and fitting.

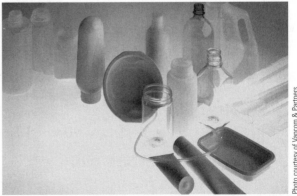

Photo courtesy of Vancom & Partners

Figure E-28. These containers have barrier layers of MXD6, a specialty nylon material made by Mitsubishi Gas Chemical Company.

bearings. Water dispersions and organic solvents of polyamide resin permit certain adhesive and coating applications on paper, wood, and fabrics.

Polyamide-based adhesives may consist of the hot-melt or solution type. Hot-melt adhesives are simply heated above the melting point and applied. Aminopolyamide resins may react with epoxy or phenolic resins to produce a thermosetting adhesive. These adhesives find use in bonding wood, paper laminates, and aluminum, as well as in adhering copper to printed circuit boards. They are used as flexible adhesives for bread wrappers, dried soup packets, cigarette packages, and book-bindings. Polyamide-epoxy combinations are used as two-part systems in casting applications such as potting and encapsulation of electrical components. When combined with pigments and other modifying agents, polyamides may be used as printing inks. The uses of polyamides in textiles and carpets are well known and need little discussion.

Clothing, light tents, shower curtains, and umbrellas are among the products made from nylon. Monofilaments, multi-filaments, and staple fibers are made by melt spinning. This is followed by cold drawing to increase tensile strength and elasticity. Monofilaments are used in fishing lines, surgical sutures, tire cords, rope, sports equipment, brushes, artificial human hair, and synthetic animal fur.

Polyamides are easily machined, but drilled or reamed holes are likely to be slightly undersized due to the resiliency of the material.

Cementing polyamide is difficult because it is solvent resistant. However, phenols and formic acid are specific solvents that are used in cementing polyamides. Epoxy resins are also employed for this purpose.

Some of the basic traits of polyamides are shown in Table E-12. Seven advantages and five disadvantages are listed here:

Table E-12. Properties of Polyamides

Property	Nylon 6,6 (Unfilled)	Nylon 6,10 (Unfilled)	Nylon 6,10 (Glass-Filled)
Molding qualities	Excellent	Excellent	Excellent
Relative density	1.13–1.15	1.09	1.17–1.52
Tensile strength, MPa	62–82	58–60	89–240
(psi)	(9000–12,000)	(8500–8600)	(13,000–35,000)
Compressive strength, MPa	46–86	46–90	90–165
(psi)	(6700–12,500)	(6700–13,000)	(13,000–24,000)
Impact strength, Izod J/mm	0.05–0.1	0.06	0.06–0.3
(ft-lb/in.)	(1.0–2.0)	(1.2)	(1.2–6)
Hardness, Rockwell	R108–R120	R111	M94, E75
Thermal expansion, 10^{-4}/°C	20	23	3–8
Resistance to heat, °C	80–150	80–120	150–205
(°F)	(180–300)	(180–250)	(300–400)
Dielectric strength, V/mm	15,000–18,500	13,500–19,000	16,000–20,000
Dielectric constant (at 60 Hz)	4.0–4.6	3.9	4.0–4.6
Dissipation factor (at 60 Hz)	0.014–0.040	0.04	0.001–0.025
Arc resistance, s	130–140	100–140	92–148
Water absorption (24 h), %	1.5	0.4	0.2–2
Burning rate, mm/min	Self-extinguishing	Self-extinguishing	Self-extinguishing
Effect of sunlight	Discolors slightly	Discolors slightly	Discolors slightly
Effect of acids	Attacked	Attacked	Attacked
Effect of alkalies	Resistant	None	None
Effect of solvents	Dissolved by phenol & formic acid	Dissolved by phenols	Dissolved by phenols
Machining qualities	Excellent	Fair	Fair
Optical qualities	Translucent to opaque	Translucent to opaque	Translucent to opaque

Advantages of Polyamide (Nylon)

1. Tough, strong, and impact resistant
2. Low coefficient of friction
3. Abrasion resistance
4. High temperature resistance
5. Processable by thermoplastic methods
6. Good solvent resistance
7. Resistant to bases

Disadvantages of Polyamide (Nylon)

1. High moisture absorption with related dimensional instability
2. Subject to attack by strong acids and oxidizing agents
3. Requires ultraviolet stabilization
4. High shrinkage in molded sections
5. Electrical and mechanical properties influenced by moisture content

POLYCARBONATES (PCs)

An important material used in the production of plastics is phenol. It is used in producing phenolic, polyamide, epoxy, polyphenylene oxide, and polycarbonate resins.

Phenol is a compound that has one hydroxyl group attached to an aromatic ring. It is sometimes called *monohydroxy benzene*, C_6H_5OH.

Bisphenol A (two phenol and acetone), a vital ingredient in the production of polycarbonates, may be prepared by combining acetone with phenol (Figure E-29). Bisphenol A is sometimes referred to as diphenylol propane or bis-dimethylmethane.

Polycarbonates are linear, amorphous polyesters because they contain esters of carbonic acid and an aromatic bisphenol.

Another important material used in the production of polycarbonate is phosgene. Phosgene, a poisonous gas, was used in World War I.

As early as 1898, A. Einhorn prepared a polycarbonate material from the reaction of resorcinol and phosgene. Both W. H. Carothers and F. J. Natta performed research on a number of polycarbonates, using ester reactions.

Research continued after World War II in Germany by Farbenfabriken Bayer and by General Electric in the United States. By 1957, both had accomplished the production of polycarbonates produced from bisphenol A. Volume production in the United States did not begin until 1959.

There are two general methods of preparing polycarbonates. The most common method is to react purified bisphenol A with phosgene under alkaline conditions (Figure E-30). An alternate method involves the reaction of purified bisphenol A with diphenyl carbonate (meta-carbonate) in the presence of catalysts under a vacuum (Figure E-31).

Purity of the bisphenol A is vital if the plastics must possess high clarity and long linear chains with no cross-linking substances.

The phosgenation process is preferred because it may be carried out at low temperatures using simple technology and equipment. The process requires the recovery of solvents and inorganic salts. However, the product is comparatively high in cost using either method.

Polycarbonates may be processed by using all usual thermoplastic methods. Their resistance to heat and high melt temperatures does require higher processing temperatures. The molding temperature is very critical and must be accurately controlled in order to produce usable products. Polycarbonates are sensitive to hydrolysis at high processing temperatures. Compounds should be dried, or vented barrel equipment used, because water will cause bubbles as well as other blemishes on parts. The unique properties of polycarbonate are due to the carbonate groups and the presence of benzene rings

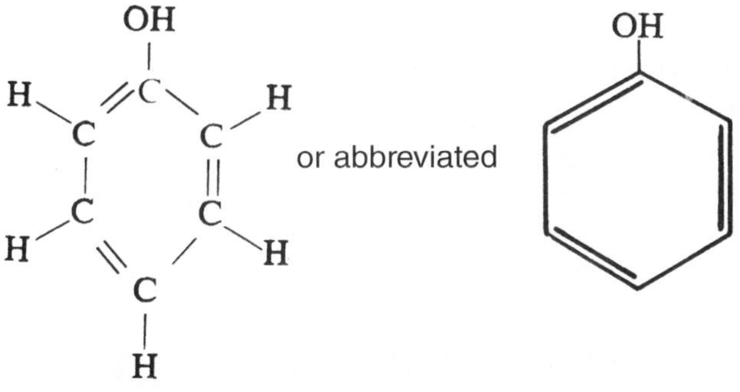

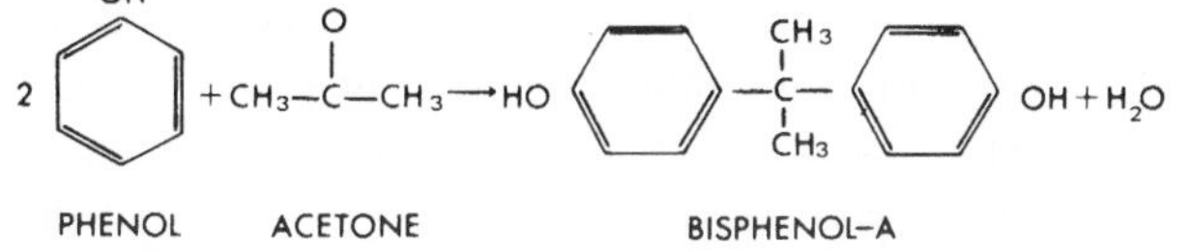

Figure E-29. Preparation of bisphenol-A.

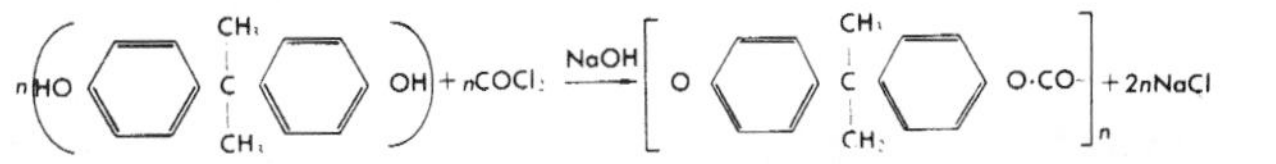

Figure E-30. The first method used to prepare polycarbonates.

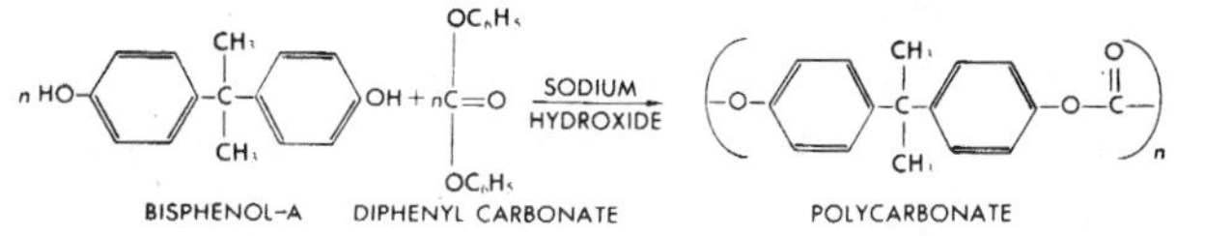

Figure E-31. The second method used to prepare polycarbonates.

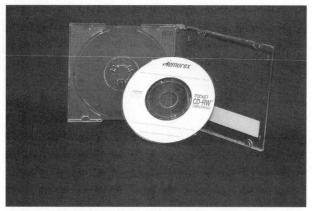

Figure E-32. This small polycarbonate compact disc stores images in a digital camera.

in the long, repeating molecular chain. Polycarbonate properties include high-impact strength, transparency, excellent creep resistance, wide temperature limits, high dimensional stability, good electrical characteristics, and self-extinguishing behavior. Tough transparent grades are used in lenses, films, windshields, light fixtures, containers, appliance components, and compact discs (Figure E-32). Temperature resistance is made use of in hot-dish handles, coffee pots, popcorn popper lids, hair dryers, and appliance housings. These plastics have excellent properties from −170°C to 1132°C (−275°F to 1270°F). Polycarbonates supply the impact and flexural strength needed in pump impellers, safety helmets, beverage dispensers, small appliances, trays, signs, aircraft parts, cameras, and various packaging and film uses. Co-extruded packages are used for oven-safe frozen food trays and microwave pouches. Polycarbonate parts also have very good dimensional stability. Glass-filled grades possess improved impact, moisture, and chemical resistance.

Most aromatic solvents, esters, and ketones will attack polycarbonates. Chlorinated hydrocarbons are used as solvent cements for cohesive bonds.

There are several hundred variations in the polycarbonate structure. The structure may be changed by substituting various radicals as side groups or separating the benzene rings by more than one carbon atom. Some possible structural combinations are shown in Figure E-33.

Some of the properties of polycarbonates are listed in Table E-13. Lists of five advantages and four disadvantages follow:

Advantages of PCs

1. High-impact strength
2. Excellent creep resistance
3. Available in transparent grades
4. Continuous application temperature over 120°C (248°F)
5. Very good dimensional stability

Possible radical side groups

R	R₁
—H	—H
—H	—CH₃
—CH₂	—CH₃
—CH₃	—C₂H₅
—C₂H₅	—C₂H₅
—CH₃	—CH₂—CH₂—CH₃
—CH₂—CH₂—CH₃	—CH₂—CH₂—CH₃

Figure E-33. Possible combination of polycarbonates.

Disadvantages of PCs

1. High processing temperatures
2. Poor resistance to alkalies
3. Subject to solvent crazing
4. Require ultraviolet stabilization

POLYETHERETHERKETONE (PEEK)

The wholly aromatic structure of PEEK contributes to the high temperature resistance of this crystalline thermoplastic. The basic repeating unit is shown in Figure E-34. PEEK can be melt processed in conventional thermoplastic equipment. Applications include coatings on wire and high-temperature composites for aerospace and aircraft components.

POLYETHERIMIDE (PEI)

Polyetherimide (PEI) is an amorphous thermoplastic based on repeating ether and imide units. The general chemical structure of these polymers is shown in Figure E-35. Reinforced and filled grades can improve strength, temperature resistance, and creep. All grades are processed on conventional equipment. Typical applications include jet engine components, oven-safe cookware, flexible circuitry, composite structures for aircraft, and food packaging.

Table E-13. Properties of Polycarbonates

Property	Polycarbonate (Unfilled)	Polycarbonate (10%–40% Glass-Filled)
Molding qualities	Good to excellent	Very good
Relative density	1.2	1.24–1.52
Tensile strength, MPa	55–65	83–172
(psi)	(8000–9500)	(12,000–25,000)
Compressive strength, MPa	71–75	90–145
(psi)	(10,300–10,800)	(13,000–21,000)
Impact strength, Izod (ft-lb/in.)	0.6–0.9	0.06–0.325
J/mm	(12–18)	(1.2–6.5)
Measurement of bar	12.7 × 3.175 mm	6.35 × 12.7 mm
Hardness, Rockwell	M73–78, R115, R125	M88–M95
Thermal expansion, (10^{-4}/°C)	16.8	4.3–10
Resistance to heat, °C	120	135
(°F)	(250)	(275)
Dielectric strength, V/mm	15,500	18,000
Dielectric constant (at 60 Hz)	2.97–3.17	3.0–3.53
Dissipation factor (at 60 Hz)	0.0009	0.0009–0.0013
Arc resistance, s	10–120	5–120
Water absorption (24 h), %	0.15–0.18	0.07–0.20
Burning rate, mm/min	Self-extinguishing	Slow 20–30
(in./min)		(0.8–1.2)
Effect of sunlight	Slight	Slight
Effect of acids	Attacked slowly	Attacked by oxidizing acids
Effect of alkalies	Attacked	Attacked
Effect of solvents	Soluble in aromatic and chlorinated hydrocarbons	Soluble in aromatic and chlorinated hydrocarbons
Machining qualities	Excellent	Fair
Optical qualities	Translucent to opaque	Transparent to opaque

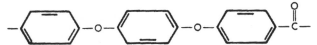

Figure E-34. The general chemical structure of polyetherimide.

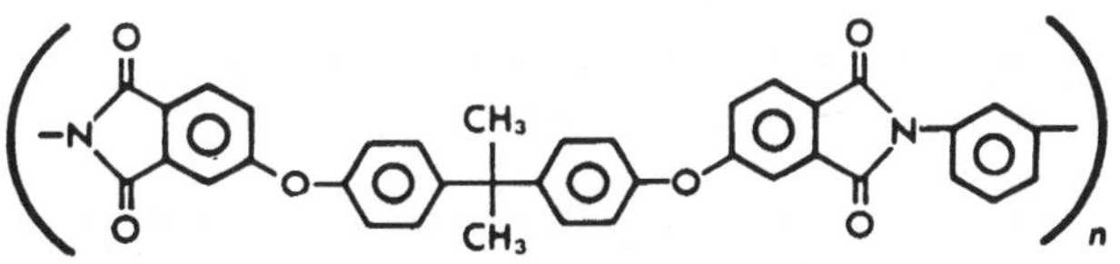

Figure E-35. Chemical structure of PEEK.

THERMOPLASTIC POLYESTERS

The thermoplastic polyester group of plastics includes saturated polyesters and aromatic polyesters. Familiar trade names of saturated polyesters are Dacron and Mylar.

Saturated Polyesters

Saturated polyesters are based on the reaction of terephthalic acid $(C_6H_4(COOH)_2)$ and ethylene glycol $(CH_2)_2(OH)_2)$. They are linear polymers with high molecular mass (Table E-14). Saturated polyesters are used in fiber and film production. W. H. Carothers did his basic research with linear polyesters. After several years, he stopped trying to produce polyester fibers and instead began investigating synthetic polyamides.

Polyethylene terephthalate (PET) may be produced by melt condensation polymerization from terephthalic acid or dimethyl terephthalate and ethylene glycol.

To reduce crystallinity, PET may be copolymerized. Copolyesters of glycol-modified PET are called PETG. Clear shampoo and detergent bottles are familiar applications. PCTA copolyester is produced from cyclohexanedimethanol, terephthalic acid (TPA), and other dibasic acids (Figure E-36).

Table E-14. Properties of Thermoplastic Polyesters: Saturated Polyester

Property	Polyethylene Terephthalate (PETP or PET)	Polybutylene Terephathalate (PBTP) (Unfilled)	Linear Aromatic (Decomposes at 550)	Linear Aromatic (Injection Grade)
Molding qualities	Good	Good	Sinters	Good
Relative density	1.34–1.39	1.31–1.38	1.45	1.39
Tensile strength, MPa	59–72	56	17	20
(psi)	(8550–10,500)	(8100)	(2500)	(2,00)
Compressive strength, MPa	76–128	59–100	76–105	68
(psi)	(11,025–18,130)	(8550–14,500)	(11,025–15,230)	(9860)
Impact strength, Izod (ft-lb/in.)	0.01–0.04	0.04–0.05		0.08
J/mm	(0.2–0.8)	(0.8–1)		(1.6)
Hardness	Rockwell M94–M101	Rockwell M68–M98	Shore D88	–
Thermal expansion, (10^{-4}/°C)	15.2–24	155	7.1	7.36
Resistance to heat, °C	80–120	50–90		280
(°F)	(176–248)	(122–194)		(536)
Dielectric strength, V/mm	13,780–15,750	16,500		13,750
Dielectric constant (at 60 Hz)	3.65	3.29	3.22	
Dissipation factor (at 60 Hz)	0.0055		0.0046	
Arc resistance, s	40–120	75–192		100
Water absorption (24 h), %	0.02	0.08		0.01
Burning rate, mm/min	Slow burning	10		
Effect of sunlight	Discolors slightly	Discolors	None	
Effect of acids	Attacked by oxidizing acids	Attacked	Slight	Slight
Effect of alkalies		Attacked	Attacked	Attacked
Effect of solvents	Attacked by halogen hydrocarbons	Resistant	Resistant	Resistant
Machining qualities	Excellent	Fair	Fair	Good
Optical qualities	Transparent to opaque	Opaque	Opaque	Translucent

Figure E-36. The clarity of PET containers exposes the contents to damage from UV light. Chemical additives can reduce color loss.

Courtesy of Ciba Speciality Chemicals

This plastics has been used for food packaging, clothing fibers, carpeting, and tire cords for nearly 20 years. PET dominates the packaging of carbonated beverages because of its relatively low gas permeability and ease and economy of processing. Most carbonated beverage containers use a two-step process that involves injection molding a preform and then stretch-blowing it to the finished shape. Figure E-37 shows a common preform design.

Most applications require PET to be oriented and crystalline for optimum properties. Orientation processes are accomplished at 100°C to 120°C (212°F to 248°F), or slightly above the glass-transition temperature (T_g).

PET is used for synthetic fibers, photographic film, videotape, dual ovenable containers, computer and magnetic tapes, and numerous beverage bottles, including those for distilled spirits. Reinforced and filled grades are used in gears, cowl vent grills, electrical switches, and sporting goods.

Figure E-37. Wheel housing made out of recycled soda bottles.

Figure E-38. This grill opening reinforcement (GOR) is made of recycled PET that has been compounded with glass fibers to increase strength and stiffness.

Because the volume of recycled PET has increased dramatically in recent years, products using recycled PET have become more familiar. Figure E-38 shows some applications of recycled PET in truck parts.

Polybutylene terephthalate (PBT) or polytetramethylene terephthalate (PTMT) was introduced in 1962. Ethylene glycol, an automobile antifreeze, is also one of the main materials used in producing polyester fibers. The original development began in England by Imperial Chemical Industries (ICI). By 1953, DuPont had purchased the rights to develop Dacron fibers. Extrusion and injection grades were on the market by 1969.

Saturated (unreactive) polyesters do not undergo any cross-linking. These linear polyesters are thermoplastic. Clothing and draperies are common uses of these fibers. Industrial uses may include reinforcements for belting or tires.

Polyester films are used for recording tape, dielectric insulators, photographic film, and boil-in-the-bag food products.

Because of their thermoplastic nature, compounds based on saturated PET and PBT may be injection or extrusion molded.

Well-known trade names include Terylene®, Dacron, Kodel fibers, and Mylar film. Other uses include gears, distributor caps, rotors, appliance housings, pulleys, switch parts, furniture, fender extensions, and packaging. Following are lists of two advantages and three disadvantages of saturated polyesters:

Advantages of Saturated Polyesters

1. Tough and rigid
2. Processable by thermoplastic methods

Disadvantages of Saturated Polyesters

1. Subject to attack by acids and bases
2. Low thermal resistance
3. Poor solvent resistance

Aromatic Polyesters

In 1971 and 1974, oxybenzoyl polyesters were introduced by Carborundum under the trade names Ekonol® and Ekcel®. Both materials are linear chains of *p*-oxybenzoyl units. Because Ekonol does not melt below its decomposition temperature, it must be sintered, compression molded, or plasma sprayed. Ekcel can be processed with injection and extrusion equipment. High-temperature stability, stiffness, and thermal conductivity are important properties.

Some formulations can be melt processed but might require processing temperatures between 300°C and 400°C (572°F and 842°F). Members of this class of materials are sometimes called nematic, anisotropic, liquid crystal polymer (LCP) or self-reinforcing polymers. These terms try to describe the formation of tightly packed fibrous chains during the melt phase. It is the fibrous chain that gives the polymer its self-reinforcing qualities. Parts must be designed to accommodate the anisotropic characteristics of LCP. Applications include chemical pumps, dual ovenable cookware, engine parts, and aerospace components.

Typical uses include bearings, seals, valve seats, rotors, high-performance aerospace and automotive parts, electrical insulation components, and coatings for pans.

In 1978, *polyarylate* was introduced. This light-amber-colored plastics is made from iso- and terephthalic acid and bisphenol A.

Polyarylates are aromatic polyester thermoplastic materials. The term *aryl* refers to a phenyl group derived from an aromatic compound. A number of alloy and filled grades are available.

Polyarylate must be dried before injection or extrusion molding begins. It has excellent ultraviolet, thermal, and heat-deflection resistance. Applications include glazing, appliance housings, electrical connectors and lighting fixtures, exterior glazing, halogen lamp lenses, and selected microwave cookware.

Typical properties are shown in Table E-15.

THERMOPLASTIC POLYIMIDES

Polyimides were developed by DuPont in 1962. They are obtained from condensation polymerization of an aromatic dianhydride and an aromatic diamine (Figure E-39). Aromatic polyimides are linear and thermoplastic and are hard to

Table E-15. Properties of Polyarylate

Property	Polyarylate
Molding qualities	Good
Relative density	1.21
Tensile strength, MPa	48–75
(psi)	(6962–10,879)
Impact strength, Izod (6 mm), J/mm	0.24
(ft-lb/in.)	(4)
Hardness, Rockwell	R105
Deflection temperature (at 1.82 MPa or 264 psi), °C	280
(°F)	(536)
Water absorption (24 h), %	0.01
Optical, refractive index	1.64

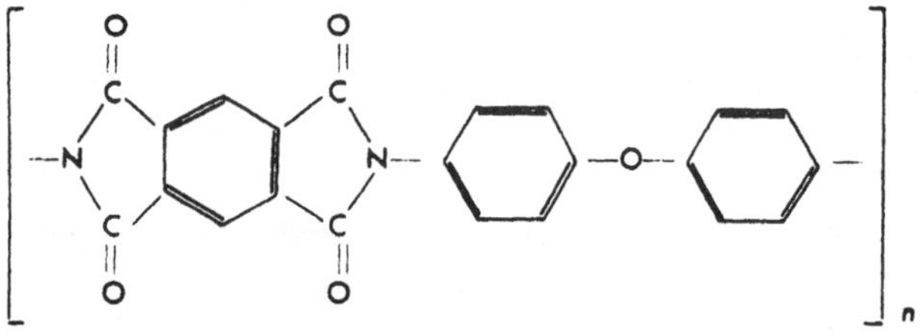

Table E-16. Properties of Polyimides

Property	Polyimide (Unfilled)
Molding qualities	Good
Relative density	1.43
Tensile strength, MPa	70
(psi)	(10,000)
Compressive strength, MPa	>165
(psi)	(>24,000)
Impact Strength, Izod, J/mm	0.045
(ft-lb/in.)	(0.9)
Hardness, Rockwell	E45–E58
Resistance to heat, °C	300
(°F)	(570)
Dielectric strength, V/mm	22,000
Dielectric constant (at 60 Hz)	3.4
Arc resistance, s	230
Water absorption (24 h), %	0.32
Burning rate	Nonburning
Effect of acids	Resistant
Effect of alkalies	Attacked
Effect of solvents	Resistant
Machining qualities	Excellent
Optical qualities	Opaque

Figure E-39. Basic polyimide structure.

process. They can be molded by allowing enough time for flow to occur once glass transition temperature is exceeded. Many polyimides do not melt, but must be fabricated by machining or other forming methods.

Additional polymerization provides plastics with slightly lower heat resistance than condensation polymerization.

Polyimides compete with various fluorocarbons for applications requiring low friction, good strength, toughness, high dielectric strength, and heat resistance. They posses good resistance to radiation but are surpassed in chemical resistance by fluoroplastics. Polyimide is attacked by strong alkaline solutions, hydrazine, nitrogen dioxide, and secondary amine compounds.

Although costly and hard to process, polyimides are used in the making of aerospace, electronics, nuclear power, and office and industrial equipment. Other parts manufactured include valve seats, gaskets, piston rings, thrust washers, and bushings. Films are made by a casting process (usually from prepolymer form) and are used for laminates, dielectrics, and coatings.

Polyimide can be applied as a hot liquid with electrostatic spray equipment. After curing and baking at 290°C (550°F), polyimide forms a hard and flexible glossy finish similar to porcelain.

Prolonged contact with this resin and its reducers may cause serious cracking to workers' skin. The solvents are no more toxic than other aromatics.

Table E-16 gives some properties of polyimides. The following lists give six advantages and six disadvantages of polyimides:

Advantages of Polyimide

1. Short-exposure temperature capability of 315°C to 371°C (600°F to 700°F)
2. Excellent barrier
3. Very good electrical properties
4. Excellent solvent and wear resistance
5. Good adhesion capability
6. Especially suitable for composite fabrication

Disadvantages of Polyimide

1. Difficulty of fabrication
2. Hygroscopic (moisture-absorbing)
3. Subject to attack by alkalies
4. Comparatively high cost
5. Dark color
6. In most types, volatiles or solvents that must be vented during cure

Figure E-40. General structural formula of polyamide-imide.

Polyamide-Imide (PAI)

An amorphous member of the polyimide family is polyamide-imide. It was marketed in 1972 by Amoco Chemicals under the trade name Torlon®. This material contains aromatic rings and a nitrogen linkage, as shown in Figure E-40. Polyamide-imide has striking properties (Table E-17). This material can withstand continuous temperatures of 260°C (500°F). Because of its low coefficient of friction, excellent service temperature, and dimensional stability, polyamide-imide may be melt processed into aerospace equipment, gears, valves, films, laminates, finishes, adhesives, and jet engine components (Figure E-41).

Table E-17. Properties of Polyamide-imide

Property	Poly(amide-imide) (Unfilled)
Molding qualities	Excellent
Relative density	1.41
Tensile strength, MPa	185
(psi)	(26,830)
Compressive strength, MPa	275
(psi)	(39,900)
Impact strength, Izod, J/mm	0.125
(ft-lb/in.)	(2.5)
Hardness, Rockwell	E78
Thermal expansion, $10^{-4}/°C$	9.144
Resistance to heat, °C	260
(°F)	(500)
Dielectric strength, V/mm	>400
Dielectric constant (at 60 Hz)	3.5
Arc resistance, s	125
Water absorption (24 h), %	0.28
Burning rate	Nonburning
Effect of acids	Very resistant
Effect of alkalies	Very resistant
Effect of solvents	Very resistant
Machining qualities	Excellent

(A) These transmission seal rings maintain strength up to 260°C [500°F].

(B) These polyamide imide thrust washers have excellent resistance to creep.
Figure E-41. Applications of polyamide imide.

POLYMETHYLPENTENE

This plastics is reported to be an isotactically arranged aliphatic polyolefin of 4-methylpentene-1. Polymethylpentene was developed in the laboratory as early as 1955. It did not gain commercial value until Imperial Chemicals Industries, Ltd., announced it under the trade name of TPX® in 1965.

Ziegler-type catalysts are used to polymerize 4-methylpentene-1 at atmospheric pressures (Figure E-42).

$$CH_2 - CH$$
$$|$$
$$CH_2$$
$$|$$
$$CH$$
$$/ \quad \backslash$$
$$CH_3 \quad CH_3$$

Figure E-42. Poly(4-methylpentene-1).

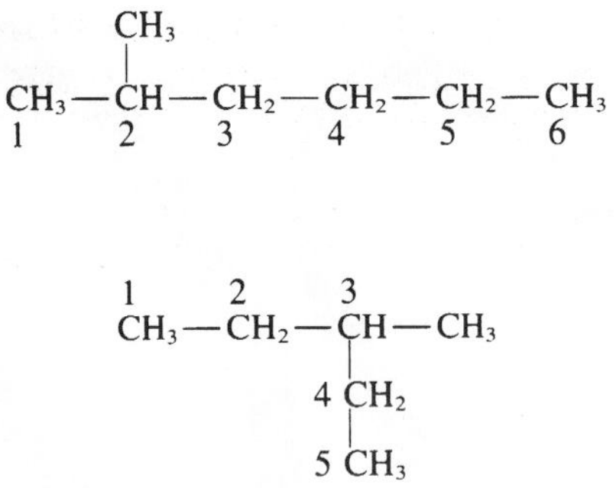

$$CH_3 - \underset{2}{\overset{\overset{\displaystyle CH_3}{|}}{\underset{}{CH}}} - \underset{3}{CH_2} - \underset{4}{CH_2} - \underset{5}{CH_2} - \underset{6}{CH_3}$$
$$_1$$

$$\underset{1}{CH_3} - \underset{2}{CH_2} - \underset{3}{\overset{}{\underset{|}{CH}}} - CH_3$$
$$ \underset{4}{CH_2}$$
$$ \underset{5}{CH_3}$$

Figure E-43. Continuous-chain formulas with carbon atoms numbered.

After polymerization, catalyst residues are removed by washing with methyl alcohol. The material is then compounded into a granular form with stabilizers, pigments, fillers, or other additives.

Formulas for this type of plastics are shown in Figure E-43. To avoid confusion, the carbon atoms of the continuous chain must be numbered. This has been done in the following formulas.

Copolymerization with other olefin units (including hexene-1, octene-1, decene-1, and octadecene-1) can offer enhanced optical and mechanical properties.

Commercial poly (4-methylpentene-1) has a relatively high service temperature that may exceed 160°C (320°F). Although the plastics is nearly 50% crystalline, it has a light-transmission value of 90%. Spherulite growth may be retarded by rapid cooling of the molded mass. The open packing of the crystalline structure gives polymethylpentene a low relative density of 0.83. This is close to the theoretical minimum for thermoplastics.

Polymethylpentene may be processed on normal thermoplastic equipment at processing temperatures that may exceed 245°C (470°F).

In spite of its high cost, this plastics has found uses in chemical plants, autoclavable medical equipment, lighting diffusers, encapsulation of electronic components, lenses, and laboratory items (Figure E-44). A well-known use for this plastics is the packaging of bake-in-the-bag and boil-in-the-bag foods. These packages are used in the home and in catering services for airlines or manufacturing plants. Packaged foods may be boiled in water or cooked in either normal or microwave ovens. Transparency is useful in showing materials in dispensing equipment.

Other side-branched polyolefins are also possible. Three such polymers are shown in Figure E-45. The branched side chains increase stiffness and lead to higher melting points. Polyvinyl cyclohexane melts at about 338°C (640°F). Table E-18 gives some of the properties of polymethylpentene. Lists of five advantages and two disadvantages of polymethylpentene follow:

Figure E-44. The clarity, chemical resistance, and toughness of polymethylpentene make it suitable for laboratory items.

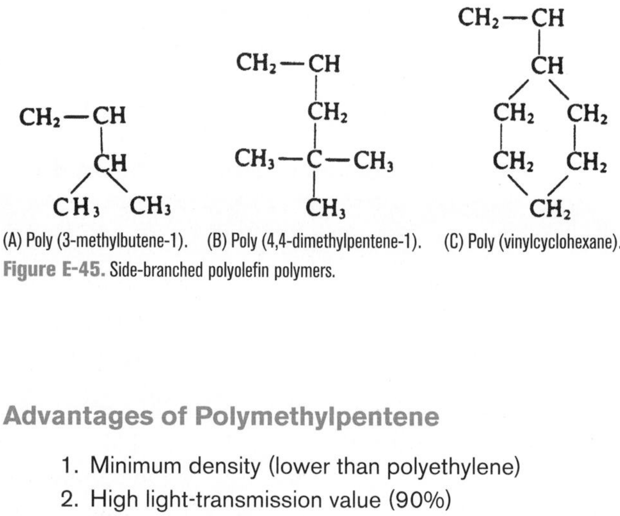

(A) Poly (3-methylbutene-1).   (B) Poly (4,4-dimethylpentene-1).   (C) Poly (vinylcyclohexane).
Figure E-45. Side-branched polyolefin polymers.

Advantages of Polymethylpentene

1. Minimum density (lower than polyethylene)
2. High light-transmission value (90%)
3. Excellent dielectric, volume resistivity, and power factor
4. Higher melting point than polyethylene
5. Good chemical resistance

Disadvantages of Polymethylpentene

1. Must be stabilized against most radiation sources
2. More costly than polyethylene

POLYOLEFINS: POLYETHYLENE (PE)

Ethylene gas is a member of an important group of unsaturated, aliphatic hydrocarbons called *olefins* or *alkenes*. An *ethenic* refers to ethylene materials. The word *olefin* means oil forming.

Table E-18. Properties of Polymethylpentene

Property	Polymethylpentene (Unfilled)
Molding qualities	Excellent
Relative density	0.83
Tensile strength, MPa	25–28
(psi)	(3500–4000)
Impact strength, Izod, J/mm	0.02–0.08
(ft-lb/in.)	(0.4–1.6)
Hardness, Rockwell	L67–74
Thermal expansion, $10^{-4}/°C$	29.7
Resistance to heat, °C	120–160
(°F)	(250–320)
Dielectric strength, V/mm	28,000
Dielectric constant (at 60 Hz)	212
Dissipation factor (at 60 Hz)	0.0007
Water absorption (24 h), %	0.01
Burning rate, mm/min	25
(in./min)	(1.0)
Effect of sunlight	Crazes
Effect of acids	Attacked by oxidizing agents
Effect of alkalies	Resistant
Effect of solvents	Attacked by chlorinated aromatics
Machining qualities	Good
Optical qualities	Transparent to opaque

Table E-19. The Principal Olefin Monomers

Chemical Formula	Olefin Name
$\begin{array}{cc} H & H \\ | & | \\ C & = C \\ | & | \\ H & H \end{array}$	Ethylene
$\begin{array}{cc} H & H \\ | & | \\ C & = C \\ | & | \\ CH_3 & H \end{array}$	Propylene
$\begin{array}{cc} H & H \\ | & | \\ C & = C \\ | & | \\ C_2H_5 & H \end{array}$	Butene-1
$\begin{array}{cc} H & H \\ | & | \\ C & = C \\ | & | \\ H_2C & H \\ | & \\ H-C-CH_3 & \\ | & \\ CH_3 & \end{array}$	4-Methylpentene

The term was originally given to ethylene because oil was formed when ethylene was treated with chlorine. However, olefin now refers to all hydrocarbons with linear carbon-to-carbon double bonds. Due to this bond, olefins are highly reactive. Some of the major olefin monomers are shown in Table E-19.

In the United States, ethylene gas is readily produced by cracking higher hydrocarbons of natural gas or petroleum. The importance and relationship of ethylene to other polymers is shown in Figure E-46.

Between 1879 and 1900, several chemists experimented with linear polyethylene polymers. In 1900, E. Bamberger and F. Tschirner used the expensive material diazomethane to produce a linear polyethylene that they called "polymethylene."

$$2_n \left(\begin{array}{c} CH_2 \\ N \overset{}{=\!=\!=} N \end{array} \right) \longrightarrow -(-CH_2-CH_2-)_n - + 2_n \cdot N_2$$

Diazomethane   *Polyethylene*

In 1930, W. H. Carothers and his coworkers reported producing polyethylene of low molecular mass. The commercial

feasibility of polyethylene resulted from research by Dr. E. W. Fawcett and Dr. R. O. Gibson of the Imperial Church Industries (ICI) in England. In 1933, their discovery was a result of investigating the reaction of benzaldehyde and ethylene (obtained from coal) under high pressure and temperature. In September 1939, ICI began commercial production of polyethylene, and the demands of World War II used all the polyethylene produced

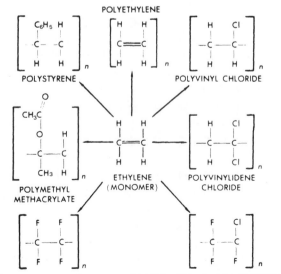

Figure E-46. The ethylene monomer and its relationship to other monomer resins.

to insulate high-frequency radar cables. By 1943, the United States was producing polyethylene by using the high-pressure methods developed by ICI. These early, low-density materials were highly branched, with a disorderly arrangement of molecular chains. Low-density materials are softer, more flexible, and melt at lower temperatures, thus may be more easily processed. By 1954, two new methods were developed for making polyethylene with higher relative densities of 0.91 to 0.97.

One process, developed in Germany by Karl Ziegler and associates, permitted polymerization of ethylene at low pressures and temperatures, in the presence of aluminum triethyl and titanium tetrachloride as catalysts. At the same time, the Phillips Petroleum Company developed a polymerization process using low pressures with a chromium trioxide–promoted silicaalumina catalyst. The conversion of ethylene to polyethylene may also be accomplished with a catalyst of molybdenum oxide on an alumina support and other promoters, a process developed by Standard Oil of Indiana. Only small quantities have been produced in the United States using this process.

The Ziegler process is used more extensively outside the United States, whereas the Phillips Petroleum process is commonly used by US firms.

Polyethylene can be produced with branched or linear chains (Figure E-47) by using either the high-pressure (ICI) or low-pressure (Ziegler, Phillips, Standard Oil) methods. The differentiation of polymer type based on pressures used for polymerization is not employed today. The American Society for Testing and Materials (ASTM) has divided polyethylenes into five groups:

- Type 1   (branched) 0.910–0.925 (low density)
- Type 2   0.926–0.940 (medium density)
- Type 3   0.941–0.959 (high density)
- Type 4   (linear) 0.969 and above (high-density to ultra-high-density homopolymers)
- Type 5   thermoset cross-linked PE

From Figure E-48, it can be seen that the physical properties of low-density (branched) and high-density (linear) polyethylenes are different. Low-density polyethylene has a crystallinity of 60% to 70%. Higher-density polymers may vary in crystallinity from 75% to 90% (Figure E-49).

With increased density, the properties of stiffness, softening point, tensile strength, crystallinity, and creep resistance are increased. Increased density reduces impact strength, elongation, flexibility, and transparency.

Polyethylene properties may be controlled and identified by molecular mass and its distribution. Molecular mass and its distribution may have the effects shown in Table E-20.

Figure E-50 is a schematic representation comparing a polymer with a narrow molecular mass distribution to a polymer with broad molecular-mass distribution. Molecular mass distribution is the ratio of large, medium, and small molecular chains of the resin. If the resin is composed of chains that are all close to the average length, the molecular mass distribution is called *narrow*. Molecular chains that are average length can flow past each other more easily than large ones.

A melt index device is used to measure the melt flow at a specified temperature and pressure. This melt index depends on molecular mass and its distribution. As melt index goes down, the melt viscosity, tensile strength, elongation, and impact strength increase. For many processing methods, it is desirable to have the hot resin flow very easily, indicating use of a resin with a high melt index. Polyethylenes with high molecular mass

MONOMER

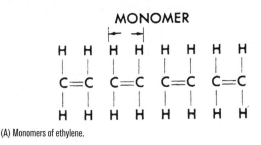

(A) Monomers of ethylene.

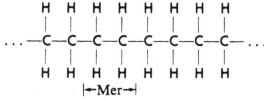

|←Mer→|

(B) Polymer containing many C_2H_4 mers.

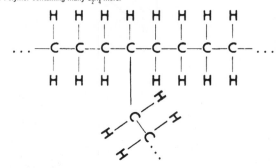

(C) Polymer with branching.

Figure E-47. Addition polymerization of ethylene. The original double bond of the ethylene monomer is broken, forming two bonds to connect adjacent mers.

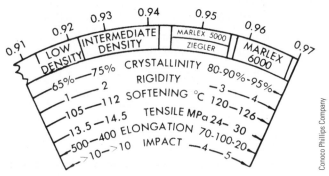

Figure E-48. Density range of polyethylene.

(A) Electron micrograph shows intercrystalline links bridging radial arms of polyethylene spherulite.

Courtesy of Lucent Technologies Bell Laboratories

(B) Small platelet crystals of polyethylenee grown on intercrystalline links may be seen in this electron micrograph.

Courtesy of Lucent Technologies Bell Laboratories

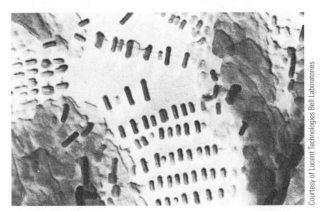

(C) Lamellar crystals of polyethylene were formed by depositing polymer from solution of links.
Figure E-49. Closeup views of polyethylene.

Courtesy of Lucent Technologies Bell Laboratories

Table E-20. Property Changes Caused by Molecular Mass and Distribution

Property	As Average Molecular Mass Increases (Melt Index Decreases)	As Molecular Mass Distribution Broadens
Melt viscosity	Increases	
Tensile strength at rupture	Increases	No significant change
Elongation at rupture	Increases	No significant change
Resistant to creep	Increases	Increases
Impact strength	Increases	Increases
Resistance to low temperature brittleness	Increases	
Environmental stress cracking resistance	Increases	Increases
Softening temperature		Increases

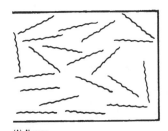

(A) Narrow. (B) Broad.
Figure E-50. Molecular mass (weight) distribution.

There is a growing commercial use of irradiation to cause branching of polyethylene products (see Chapter 20, Radiation Processing). Radiation cross-linkage is rapid and leaves no objectionable residues. Irradiated parts may be exposed to temperatures in excess of 250°C (500°F), as shown in Figure E-51.

Figure E-51. Controlled radiation treatment can improve the heat resistance of polyethylene. The center container was so treated and held its shape at 175°C [350°F].

have a low melt index. By varying density, molecular mass, and molecular mass distribution, polyethylene containing a wide variety of properties may be produced.

Polyethylene may be cross-linked to convert a thermoplastic material to a thermoset. After forming, such conversion opens up many new possibilities. This cross-linking may be accomplished by chemical agents (usually peroxides) or irradiation.

Excessive radiation may reverse the cross-linking effect by breaking the main links in the molecular chain. Stabilizers or pigments of carbon black must be used to absorb or block the damaging effects of ultraviolet radiation on polyethylene.

Because of low price, processing ease, and a broad range of properties, polyethylene has become the most used plastics. As one of the lightest thermoplastics, it may be chosen where costs are based on cubic mass. Very good electrical and chemical resistance has led to wide use of polyethylene in wire coatings and dielectrics. Polyethylene is also used in containers, tanks, pipes, and coatings where chemical agents are present. At room temperature, there is no solvent for polyethylene. However, it is easily welded.

Polyethylene may be easily processed using all thermoplastic methods. Probably the largest use is in the production of containers and film consumed by the packaging industry. Blow-molded containers are seen in every supermarket. These containers replace heavier ones made of glass and metal. Tough plastic bags for packaging foods and films for packaging fresh fruit, frozen foods, and bakery products are examples of the many uses for polyethylene. Colorful rotational molded toys and playground equipment can withstand rough use from active children (Figure E-52).

Lower-density polyethylene films are made with good clarity by quickly chilling the melt as it emerges from the die. As the hot amorphous melt is quickly chilled, widespread crystallization does not have time to occur. Low-density films are used to package new garments, such as shirts and sweaters, as well as sheets and blankets. At the laundry and dry cleaner, very thin films are used for packaging. Higher-density films are used where greater heat resistance is required, such as for boil-in-the-bag food packs. Silage covers, reservoir linings, seedbed covers, moisture barriers, and covering for harvested crops are only a few uses in construction, agriculture, and horticulture.

Although polyethylene is a good moisture barrier, it has a high gas permeability. It should not be used under vacuum or for transporting gaseous materials. Although permeable to oxygen and carbon dioxide, films used for meats and some produce may require small holes to allow ventilation. Oxygen keeps the meat looking red and prevents moisture from condensing on packaged produce. Heat sealing and shrink-wrapping are accomplished using these films. Electronic or radio-frequency heat sealing is difficult because of the low electrical dissipation (power) factor of polyethylene.

Polyethylene is used to coat paper, cardboard, and fabrics in order to improve their wet strength as well as other properties. Coated materials may then be heat sealed. The packaging of milk is one well-known use. Powdered forms are used for dip coating, flame spraying, and fluidized-bed coating when a layer impervious to chemicals and moisture is required.

Injection-molded toys, small appliance housings, garbage cans, freezer containers, and artificial flowers benefit from

(A) Rotationally molded polyethylene toys offer bright colors and high-impact resistance.

Courtesy of Ciba Speciality Chemicals

(B) Playground equipment gets much of its appeal from intense colors.

Figure E-52. Rotationally molded polyethylene products.

Courtesy of Ciba Speciality Chemicals

polyethylene's toughness, chemical inertness, and low service temperatures.

Extruded polyethylene pipe and ducting is used in chemical plants and for some domestic cold water service. Corrugated drain pipe is currently replacing clay or concrete pipe and tiles because it is less costly, faster to install, and lighter. Monofilaments find uses for ropes and fishing nets and are woven for lawn chairs. Polyethylene is widely used as electrical wire and cable covering. Irradiated or chemically cross-linked films are used as dielectrics in winding electrical coils. They also have limited packaging applications.

Polyethylene may be foamed by several methods. A foaming agent that breaks down and releases a gas during the molding operation is preferred for commercial uses. A physical foaming method consists of introducing a gas, such as nitrogen, into the molten resin under pressure. While in the mold and under atmospheric pressure, the gas-filled polyethylene expands. Azodicarbonamide may be used for chemical foaming of either low- or high-density resins. Foams may be selected as gasket material and dielectric materials in coaxial cables. Cross-linked foams are suitable for cushioning, packaging, or flotation, whereas structural foam is used for furniture components and internal panels in automobiles. Low-density foams find uses in wrestling mats, athletic padding, and flotation equipment.

Table E-21 gives properties of low-, medium-, and high-density polyethylene. Lists of six advantages and five disadvantages of polyethylene follow:

Advantages of PE

1. Low cost (except UHMWPE)
2. Excellent dielectric properties
3. Moisture resistance
4. Very good chemical resistance
5. Available in food grades
6. Processable by all thermoplastic methods (except HMWHPE and UHMWPE)

Disadvantages of PE

1. High thermal expansion
2. Poor weathering resistance
3. Subject to stress cracking (except UHMWPE)
4. Difficulty in bonding
5. Flammable

Table E-21. Properties of Polyethylene

Property	Low-Density Polyethylene	Medium-Density Polyethylene	High-Density Polyethylene
Molding qualities	Excellent	Excellent	Excellent
Relative density	0.910–0.925	0.926–0.940	0.941–0.965
Tensile strength, MPa	4–16	8.24	20–38
(psi)	(600–2300)	(1200–3500)	(3100–5500)
Compressive strength, MPa			19–25
(psi)			(2700–3600)
Impact strength, Izod, J/mm	No break	0.025–0.8	0.025–1.0
(ft lb/in)		(.05–16)	(0.5–20)
Hardness, Shore	D41–D46	D50–D60	D60–D70
R10	R15		
Thermal expansion, 10^{-4}/°C	25–50	35–40	28–33
Resistance to heat, °C	80–100	105–120	
(°F)	(180–212)	(220–250)	(250)
Dielectric strength, V/mm	18,000–39,000	18,000–39,000	18,000–20,000
Dielectric constant (at 60 Hz)	2.25–2.35	2.25–2.35	2.30–2.35
Dissipation factor (at 60 Hz)	0.0005	0.0005	0.0005
Arc resistance, s	135–160	200–235	
Water absorption (24 h), %	0.015	0.01	0.01
Burning rate, mm/min	Slow 26	Slow 25–26	Slow 25–26
(in./min)	(1.04)	(1–1.04)	(1–1.04)
Effect of sunlight	Crazes—must be stabilized	Crazes—must be stabilized	Craxes—must be stabilized
Effect of acids	Oxidizing acids	Oxidizing acids	Oxidizing acids
Effect of alkalies	Resistant	Resistant	Resistant
Effect of solvents	Resistant (below 60°C)	Resistant (below 60°C)	Resistant (below 60°C)
Machining qualities	Good	Good	Excellent
Optical qualities	Transparent to opaque	Transparent to opaque	Transparent to opaque

Table E-22. Common Comonomers with Olefins

Formula	Name
H H | | C = C | | C₄H₉ H	1-Hexane
H H | | C = C | | O H | O = C–CH₃	Vinyl acetate
H H | | C = C | | | H | O = C–O–CH₃	Methyl acrylate
H H | | C = C | | | H | O = C – OH	Methyl acrylate

Adding fillers, reinforcements, or other monomers may also change properties. Some of the common comonomers are shown in Table E-22.

A number of new polymerization techniques have expanded the potential application of polyethylene.

Very Low-Density Polyethylene (VLDPE)

This linear, nonpolar polyethylene is produced by copolymerization of ethylene and other alpha olefins. Densities range from 0.890 to 0.915. VLDPE is easily processed into disposable gloves, shrink packages, vacuum-cleaner hoses, tubing, squeeze tubes, bottles, shrink-wrap, diaper film liners, and other health care products.

Linear Low-Density Polyethylene (LLDPE)

Production of linear low-density polyethylene is controlled through catalyst selection and regulation of reactor conditions. Densities range from 0.916 to 0.930. These plastics contain little if any chain branching. As a result, these plastics exhibit good flex life, low warpage, and improved stress-crack resistance. Films for ice, trash, garment, and produce bags are tough as well as puncture and tear resistant.

High-Molecular-Weight, High-Density Polyethylene (HMW-HDPE)

High-molecular-weight, high-density polyethylenes are linear polymers with a molecular mass ranging from 200,000 to 500,000. Propylene, butene, and hexene are common monomers. High molecular mass results in toughness, chemical resistance, impact strength, and high abrasion resistance. High melt viscosities require special attention to equipment and mold designs. Densities are 0.941 or greater. Trash liners, grocery bags, industrial pipe, gas tanks, and shipping containers are familiar applications.

Ultra-High-Molecular-Weight Polyethylene (UHMWPE)

Ultra-high-molecular-weight polyethylenes have molecular weights ranging from 3 to 6 million, which accounts for their high wear resistance, chemical inertness, and low coefficient of friction. These materials do not melt or flow like other polyethylenes. Processing is similar to methods used with polytetraflurorethylene (PTFE).

Sintering produces products with microporosity. Ram extrusion and compression molding are the major forming methods used.

Applications include chemical pump parts, seals, surgical implants, pen tips, and butcher-block cutting surfaces.

Ethylene Acid

A wide variety of properties similar to those of LDPE may be produced by varying the pendent carboxyl groups on the polyethylene chain. These carboxyl groups reduce polymer crystallinity, thus improving clarity, lowering the temperature required to heat seal, and improving adhesion to other substrates.

Molds should be designed to accommodate the adhesive qualities. Processing equipment should be corrosion resistant.

The FDA allows up to 25% acrylic acid and 20% methacrylic acid for ethylene copolymers that come in contact with food.

Most applications of ethylene acid and copolymers are for packaging of foods, coated papers, and composite foil pouches and cans.

Ethylene-Ethyl Acrylate (EEA)

By varying the ethyl acrylate pendent groups in the ethylene chain, properties may vary from rubbery to tough polyethylene-like polymers. The ethyl group on the PE chain lowers crystallinity.

Applications include hot-melt adhesives, shrink-wrap, produce bags, bag-in-box products, and wire coating.

Ethylene-Methyl Acrylate (EMA)

This copolymer is produced by the addition of methyl acrylate (40% by weight) monomer with ethylene gas.

EMA is a tough, thermally stable olefin with good elastomeric characteristics. Typical applications include disposable medical gloves, tough, heat-sealable layers, and coatings for composite packaging.

EMA polymers meet the FDA and USDA requirements for use in food packaging.

Ethylene-Vinyl Acetate (EVA)

A wide variety of properties is available from this family of thermoplastic polymers. The vinyl acetate is copolymerized, in varying amounts ranging from 5% to 50% by weight, onto the ethylene chain. If the vinyl acetate side groups exceed 50%, they are considered vinyl acetate-ethylene (VAE).

Typical EVA applications include hot melts, flexible toys, beverage and medical tubing, shrink-wrap, produce bags, and numerous change coatings.

POLYOLEFINS: POLYPROPYLENE

Until 1954, most attempts to produce plastics from polyolefins had little commercial success, and only the polyethylene family was commercially important. In 1955, Italian scientist F. J. Natta announced the discovery of sterospecific polypropylene. The word *sterospecific* indicates that the molecules are arranged in a definite order in space. This is in contrast to branched or random arrangements. Natta called this regular, arranged material *isotactic polypropylene*. While experimenting with Ziegler-type catalysts, he replaced the titanium tetrachloride in $Al(C_2H_5) + TiCl_4$ with the sterospecific catalyst titanium trichloride. This led to the commercial production of polypropylene.

It is not surprising that polypropylene and polyethylene have many of the same properties. They are similar in origin and manufacture. Polypropylene has become a strong competitor of polyethylene.

Polypropylene gas, $CH_3-CH=CH_2$, is less expensive than ethylene. It is obtained from high-temperature cracking of petroleum hydrocarbons and propane. The basic structural unit of polypropylene is shown here:

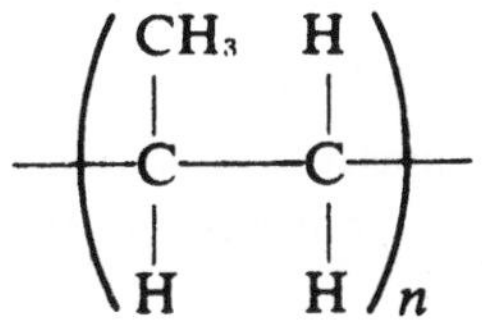

$$\left(\begin{array}{cc} CH_3 & H \\ | & | \\ -C & C- \\ | & | \\ H & H \end{array}\right)_n$$

Figure E-53 shows the stereostatic arrangements of polypropylene. In Figure E-53A, the molecular chains show a high degree of order, with all the CH_3 groups along one side. Atactic polymers are rubbery, transparent materials of limited commercial value. Atactic and syndiotactic plastics grades are more impact resistant than isotactic grades. Both syndiotactic and atactic structure may be present in small quantities in isotactic plastics. Commercially available polypropylene is about 90% to 95% isotactic.

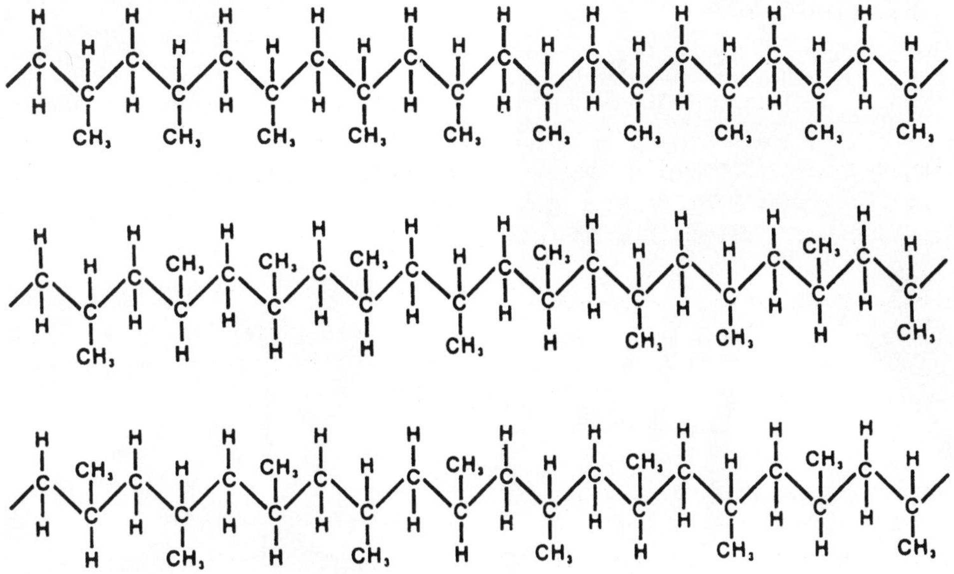

(A) Isotactic.

(B) Atactic.

(C) Syndiotactic.

Figure E-53. Stereotactic arrangements of polypropylene.

The general physical properties of polypropylene are similar to those of high-density polyethylene. However, polyethylene and polypropylene differ in four important respects:

1. Polypropylene has a relative density of 0.90; polyethylene has relative densities of 0.941 to 0.965.
2. The service temperature of polypropylene is higher.
3. Polypropylene is harder, more rigid, and has a higher brittle point.
4. Polypropylene is more resistant to environmental stress cracking (Figure E-54).

The electrical and chemical properties of the two materials are very similar. Polypropylene is more susceptible to oxidation and degrades at elevated temperatures.

Polypropylene may also be made with a variety of properties by adding fillers, reinforcements, or blends of special monomers (Figure E-55). It is easily processed in all conventional thermoplastic equipment. Although it cannot be cemented by cohesive means, it is readily welded.

Polypropylene competes with polyethylene for many uses. It has the advantage of a higher service temperature and is typically used for sterilizable hospital items (Figure E-56), dishes, appliance parts, dishwasher components, containers, items incorporating integral hinges, automotive ducts, and trim. Extruded and cold-drawn monofilaments find use in rot-proof ropes that will float on water. Some fibers are finding increasing uses in textiles and outdoor or automotive carpeting. It may be used as tough packaging film or electrical insulation on wire and cable. Slit film fiber, a process known as filibration, is widely used in producing ropes and fibers from polypropylene. It is co-extrusion-blow-molded into numerous food containers.

Because of first-rate abrasion resistance, high service temperature, and potentially lower cost, foamed polypropylene is finding a growing market. Cellular polypropylene is foamed in much the same manner as polyethylene.

Table E-23 lists some of the properties of polypropylene. Eleven advantages and six disadvantages of polypropylene follow:

Courtesy of Delphi Corp.

Figure E-55. This automotive instrument panel with air-bag doors is made of glass-reinforced polypropylene.

Courtesy of Ciba Speciality Chemicals

Figure E-54. Polypropylene is used in injection-molded chairs because of its resistance to cracking.

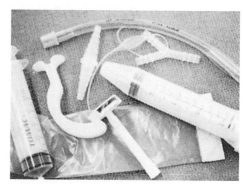

Figure E-56. Gas- and steam-sterilizable hospital items.

Table E-23. Properties of Polypropylene

Property	Polypropylene Homopolymer (Unmodified)	Polypropylene (Glass-Reinforced)
Molding qualities	Excellent	Excellent
Relative density	0.902–0.906	1.05–1.24
Tensile strength, MPa	31–38	42–62
(psi)	(4500–5500)	(6000–9000)
Compressive strength, MPa	38–55	38–48
(psi)	(5500–8000)	(5500–7000)
Impact strength, Izod, J/mm	0.025–0.1	0.05–0.25
(ft-lb/in.)	(0.5–2)	(1–5)
Hardness, Rockwell	R85–R110	R90
Thermal expansion, 10^{-4}/°C	14.7–25.9	7.4–13.2
Resistance to heat, °C	110–150	150–160
(°F)	(225–300)	(300–320)
Dielectric strength, V/mm	20,000–26,000	20,000–25,500
Dielectric constant (at 60 Hz)	2.2–2.6	2.37
Dissipation factor (at 60 Hz)	0.0005	0.0022
Arc resistance, s	138–185	74
Water absorption (24 h), %	0.01	0.01–0.05
Burning rate	Slow	Slow-nonburning
Effect of sunlight	Crazes–must be stabilized	Crazes–must be stabilized
Effect of acids	Oxidizing acids	Slowly attacked by oxidizing acids
Effect of alkalies	Resistant	Resistant
Effect of solvents	Resistant (below 80°C)	Resistant (below 80°C)
Machining qualities	Good	Fair
Optical qualities	Transparent to opaque	Opaque

Advantages of Polypropylene

1. Processable by all thermoplastic methods
2. Low coefficient of friction
3. Excellent electrical insulation
4. Good fatigue resistance
5. Excellent moisture resistance
6. First-rate abrasion resistance
7. Good grade availability
8. Service temperature to 126°C (260°F)
9. Very good chemical resistance
10. Excellent flexural strength
11. Good impact strength

Disadvantages of Polypropylene

1. Broken down by ultraviolet radiation
2. Poor weatherability
3. Flammable (flame-retarded grades are available)
4. Subject to attack by chlorinated solvents and aromatics
5. Difficult to bond
6. Oxidative breakdown accelerated by several metals

POLYOLEFINS: POLYBUTYLENE (PB)

In 1974, a polyolefin, called polybutylene (PB), was introduced by Witco Chemical Corporation. It contains ethyl side groups in the linear backbone. This linear isostatic material may exist in a variety of crystalline forms. Upon cooling, the material is less than 30% crystalline. During aging and before complete crystalline transformation, many postforming techniques may be used. Crystallinity then varies from 50% to 55% after cooling. Polybutylene may be formed by using conventional thermoplastic techniques.

Major uses include high-performance films, tank liners, and pipes. It is also used as hot-melt adhesives and co-extruded as moisture barriers and heat-sealable packages. Table E-24 gives some properties of polybutylene.

Table E-24. Properties of Polybutylene

Property	Polybutylene (Molding grades)
Molding qualities	Good
Relative density	0.908–0.917
Tensile strength MPa	26–30
(psi)	(3770–4350)
Impact strength, Izod, J/mm	No break
Hardness, Shore	D55–D65
Thermal expansion, $10^{-4}/°C$	–
Resistance to heat, °C	<110
(°F)	(<230)
Dielectric constant (at 60 Hz)	2.55
Dissipation factor (at 60 Hz)	0.0005
Water absorption (24 h), %	<0.01–0.026
Burning rate, mm/min	45.7
(in./min)	(1.8)
Effect of sunlight	Crazes
Effect of acids	Attacked by oxidizing acids
Effect of alkalies	Very resistant
Effect of solvents	Resistant
Machining qualities	Good
Optical qualities	Translucent

POLYPHENYLENE OXIDES

This family of materials should probably be called *polyphenylene*. Several plastics have been developed by separating the benzene ring backbone of polyphenylene with other molecules, making these plastics more flexible and moldable by usual thermoplastic methods. Polyphenylene with no benzene-ring separation is very brittle, insoluble, and infusible.

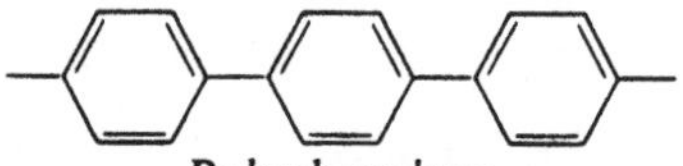

Polyphenylene

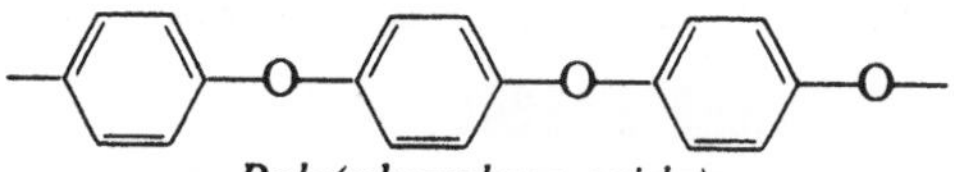

Poly(phenylene oxide)

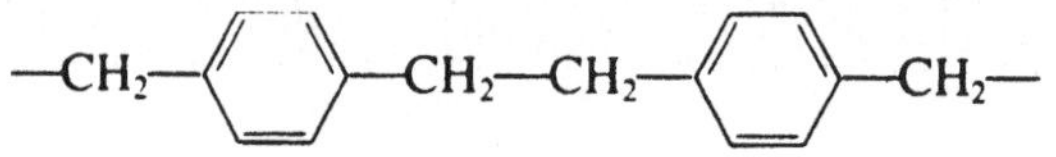

Poly-p-xylylene

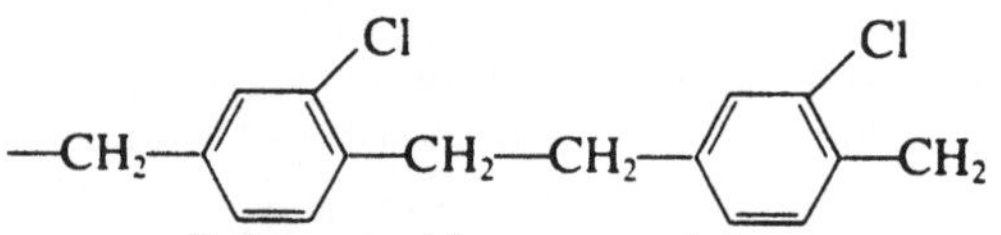

Polymonochloroparaxylylene

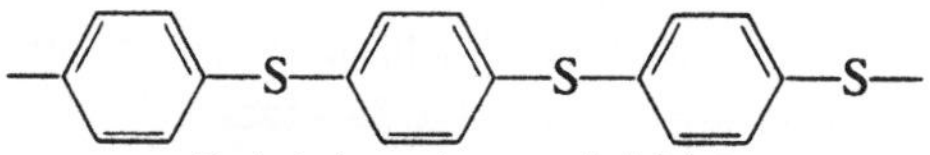

Poly(phenylene sulphide)

Three advantages and three disadvantages of polyphenylenes are listed here:

Advantages of Polyphenylene

1. Excellent solvent resistance
2. Good radiation resistance
3. High thermal and oxidative stability

Disadvantages of Polyphenylene

1. Difficult to process
2. Comparatively expensive
3. Limited availability

Polyphenylene Oxide (PPO)

In 1964, Union Carbide brought out a heat-resisting plastics called polyphenylene oxide. It may be prepared through the catalytic oxidation of 2,6-dimethyl phenol (Figure E-57).

Similar materials have been prepared making use of ethyl, isopropyl, or other alkyl groups. In 1965, the General Electric Co. introduced poly-2, 6-dimethyl-l, 4-phenylene ether as a polyphenylene oxide material. Then in 1966, General Electric announced another similar thermoplastic with the trade name Noryl®. This material is a physical blend of polypheylene oxide and high-impact polystyrene, possessing a large diversity in formulations and properties.

Because Noryl costs less and has properties similar to polyphenylene oxides, many uses are the same. This modified phenylene oxide material (Noryl) may be processed by normal thermoplastic equipment with processing temperatures ranging from 190°C to 300°C (375°F to 575°F). Modified phenylene oxide parts may be welded, heat sealed, or solvent cemented with chloroform and ethylene dichloride. Filled, reinforced, and flame-retardant grades are used as alternatives to die-cast metals, PC, PA, and polyesters. Typical applications include video display terminals, pump impellers, radomes, small appliance housings, and instrument panels. The following lists show five advantages and one disadvantage of polyphenylene oxide:

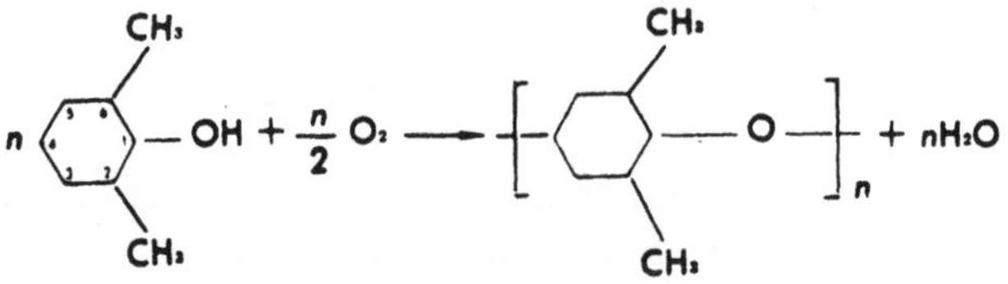

Figure E-57. Preparation of polyphenylene oxide.

Advantages of PPO

1. Good fatigue and impact strength
2. Can be metal plated
3. Thermally and oxidatively stable
4. Resistant to radiation
5. Processable by thermoplastics methods

Disadvantage of PPO

1. Comparatively high cost

Polyphenylene Ether (PPE)

Polyphenylene polymers and alloys belong to a group of aromatic polyethers. To be considered useful, these polyethers are alloyed with PS to lower melt viscosity and allow conventional processing. Without this PS separation, the polymer is very brittle, insoluble, and infusible. These copolymers are used for small appliance housing and electrical components.

Parylenes

In 1965, Union Carbide introduced poly-*p*-xy-lylene under the trade name Parylene® (Figure E-58). Its primary market is in coating and film applications.

Parylene C (polymonochloroparaxlylene) offers improved permeability for moisture and gases. Parylenes are not formed in the same manner as other thermoplastics. They are polymerized as coatings on the surface of the product. The process is

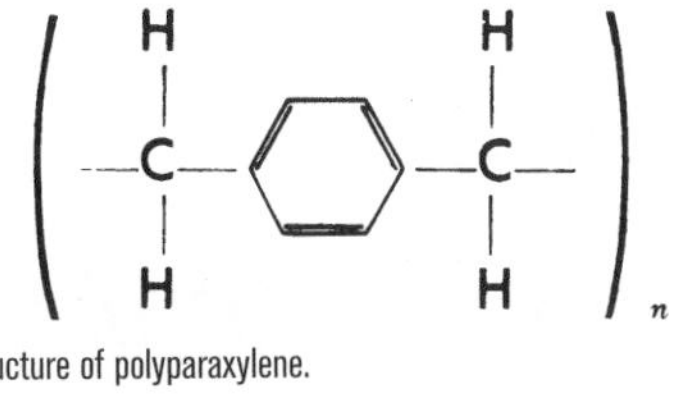

Figure E-58. Structure of polyparaxylene.

similar to vacuum metallizing. Table E-25 lists some properties of parylenes.

Polyphenylene Sulfide (PPS)

In 1968, the Phillips Petroleum Company announced a material known as polyphenylene sulfide with the trade name Ryton®. The material is available as either thermoplastic or thermosetting compounds. Cross-linking is achieved by using thermal or chemical means.

This rigid, crystalline polymer containing benzene rings and sulfur links exhibits outstanding high-temperature stability and chemical and abrasion resistance. Typical uses include computer components, range components, hair dryers, submersible pump enclosures, and small appliance housings. It is also used as an adhesive, laminating resin, and as coatings for electrical parts.

Table E-26 gives some of the traits of three polyphenylene oxides, and the following lists show six advantages plus four disadvantages of polyphenylene sulfide:

Advantages of PPS

1. Capable of extended usage at 232°C (450°F)
2. Good solvent and chemical resistance
3. Good radiation resistance
4. Excellent dimensional stability
5. Nonflammable
6. Low water absorption

Disadvantages of PPS

1. Hard to process (high melt temperature)
2. Comparatively high cost
3. Fillers needed for good impact strength
4. Subject to attack by chlorinated hydrocarbons

Table E-25. Properties of Parylene

Property	Polyparaxlyene	Polymonochloroparaxlyene
Molding qualities	Special process	Special process
Relative density	1.11	1.289
Tensile strength, MPa	44.8	68.9
(psi)	(6500)	(9995)
Thermal expansion, 10^{-4}/°C	17.52	8.89
Resistance to heat, °C	94	116
(°F)	(201)	(240)
Water absorption (24 h), %	0.06	0.01
Effect of solvents	Insoluble in most	Insoluble in most
Optical qualities	Transparent	Transparent

Table E-26. Properties of Polyphenylene Oxide

Property	Polyphenylene oxide (Unfilled)	Noryl SE-1 SE-100	Polyphenylene Sulfides
Molding qualities	Excellent	Excellent	Excellent
Relative density	1.06–1.10	1.06–1.10	1.34
Tensile strength, MPa	54–66	54–66	75
(psi)	(7800–9600)	(7800–9600)	(10,800)
Compressive strength, MPa	110–113	110-113	
(psi)	(16,000–16,400)	(16,000–16,400)	
Impact strength, Izod, J/mm	0.25*	0.25*	0.015 at 24°C 0.5 at 150°C
(ft-lb/in.)	(5.0)*	(5.0)*	(0.3 at 75°F) (1.0 at 300°F)
Hardness, Rockwell	R115–R119	R115–R119	R124
Thermal expansion, 10^{-4}/°C	13.2	8.4–9.4	14
Resistance to heat, °C	80–105	100–130	205–260
(°F)	(175–220)	(212–265)	(400–500)
Dielectric strength, V/mm	15,500–21,500	15,500–21,500	23,500
Dielectric constant (at 60 Hz)	2.64	2.64–2.65	3.11
Dissipation factor (at 60 Hz)	0.0004	0.0006–0.0007	
Arc resistance, s	75		
Water absorption (24 h), %	0.066		0.02
Burning rate	Self-ex., nondrip	Self-ex., nondrip	Nonburning
Effect of sunlight	Colors may fade	Colors may fade	
Effect of acids	None		Attacked by oxidizing acids
Effect of alkalies	None		None
Effect of solvents	Soluble in some aromatics	Soluble in some aromatics	Resistant
Machining qualities	Excellent	Excellent	Excellent
Optical qualities	Opaque	Opaque	Opaque

Polyaryl Ethers

Polyaryl ethers, polyaryl sulfone, and phenylene oxide have good physical and mechanical properties, good heat deflection temperatures, high impact strength, and good chemical resistance.

There are three different chemical groups that link the phenylene structure—isopropylidene, ether, and sulfone. The ether linkage and carbon of the isopropylidene group impart toughness and flexibility to the plastics (see Polyphenylene oxide and Polyphenylene ether, pp. 446, 447).

In 1972, Uniroyal introduced a polyaryl ether plastics under the trade name of Arylon T®.

Polyaryl ethers are prepared from aromatic compounds containing no sulfur links. The resulting polymer is more easily processed, and service temperatures may exceed 75°C (170°F). Uses include business machine parts, helmets, snowmobile parts, pipes, valves, and appliance components.

Table E-27 lists the properties of polyaryl ether.

POLYSTYRENE (PS)

Styrene is one of the oldest known vinyl compounds. However, industrial exploitation of this material did not begin until the late 1920s. This simple, aromatic compound had been isolated as early as 1839 by the German chemist Edward Simon. Early monomer solutions were obtained from such natural resins as storax and dragon's blood (a resin from the fruit of the Malayan rattan palm). In 1851, French chemist M. Berthelot reported the production of styrene monomers by passing benzene and ethylene through a red-hot tube. This dehydrogenation of ethyl benzene is the basis of today's commercial methods.

By 1925, polystyrene (PS) was commercially available in Germany and the United States. For Germany, polystyrene became one of the most vital plastics used during World War II. Germany had already embarked upon large-scale synthetic rubber production. Styrene was an essential ingredient for the production of styrene-butadiene rubber. When the natural sources

Table E-27. **Properties of Polyaryl Ether**

Property	Polyaryl Ether (Unfilled)
Molding qualities	Excellent
Relative density	1.14
Tensile strength, MPa	52
(psi)	(7500)
Compressive strength, MPa	110
(psi)	(16,000)
Impact strength, Izod, J/mm	0.4
(ft-lb/in.)	(8)
Bar size, mm	12.7 × 7.25
(in)	1/2 × 1/4
Hardness, Rockwell	R117
Thermal expansion, 10^{-4}/°C	16.5
Resistance to heat, °C	120–130
(°F)	(250–270)
Dielectric strength, V/mm	16,930
Dielectric constant (at 60 Hz)	3.14
Dissipation factor (at 60 Hz)	0.006
Arc resistance, s	180
Water absorption (24 h), %	0.25
Burning rate	Slow
Effect of sunlight	Slight, yellows
Effect of acids	Resistant
Effect of alkalies	None
Effect of solvents	Soluble in ketones, esters, chlorinated aromatics
Machining qualities	Excellent
Optical qualities	Translucent to opaque

of rubber were cut off in 1941, the United States began a crash program for the production of rubber from butadiene and styrene. This synthetic rubber became known as government rubber styrene (GR-S). There still remains a large demand for styrene-butadiene synthetic rubber.

Styrene is chemically known as vinyl benzene, with the formula:

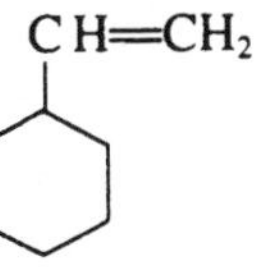

Styrene

In pure form, this aromatic vinyl compound will slowly polymerize by addition at room temperature. The monomer is obtained commercially from ethyl benzene (Figure E-59).

Styrene may be polymerized by using bulk, solvent, emulsion, or suspension polymerization. Organic peroxides are utilized to speed the process.

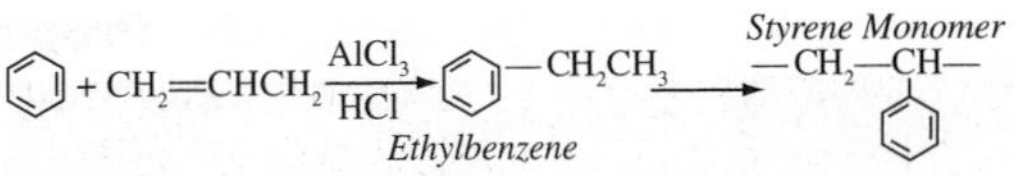

Figure E-59. Production of the vinyl benzene (styrene) monomer.

Polystyrene is an atactic, amorphous thermoplastic with the formula shown in Figure E-60. It is inexpensive, hard, rigid, transparent, easily molded, and possesses good electrical and moisture resistance. Physical properties vary depending on the molecular mass distribution, processing, and additives.

Polystyrene may be processed by all normal thermoplastic processes and may be solvent cemented. Some common uses include wall tile, electrical parts, blister packages, lenses, bottle caps, small jars, vacuum-formed refrigerator liners, containers of all kinds, and transparent display boxes. Thin films and sheet stock have a metallic ring when struck or dropped. These forms are used in packaging foods and other items such as some cigarette packets. Children see the use of polystyrene in model kits and toys. Adults may be aware of this material in inexpensive dishes, utensils, and glasses. Filaments are extruded and deliberately stretched or drawn to orient the molecular chains. This orientation adds tensile strength in the direction of stretching. Filaments may be used for brush bristles.

Expanded or foamed polystyrene is made by heating polystyrene containing a gas-producing or *blowing* agent. The foaming is accomplished by blending a volatile liquid such as methylene chloride, propylene, butylene, or fluorocarbons into the hot melt. As the mixture emerges from the extruder, the blowing agents release gaseous products that result in a low-density cellular material.

Expanded polystyrene (EPS) is produced from polystyrene beads containing an entrapped blowing agent. These agents may be pentane, neopentane, or petroleum ether. Upon either pre-expanding or final molding, the blowing agent volatilizes, causing the individual beads to expand and fuse together. Steam or other heat sources are used to cause this expansion. Both expanded and foamed forms have a closed cellular structure so that they can be used as flotation devices. Because of its low thermal conductivity, this material has found widespread use as thermal insulation (Figure E-61) used in refrigerators, cold storage rooms, freezer display cases, and building walls. EPS has the added advantage of being moisture-proof. It has many packaging uses because of its thermal insulation value and shock absorption characteristics. Packing in cellular polystyrene can save on shipping and breakage costs.

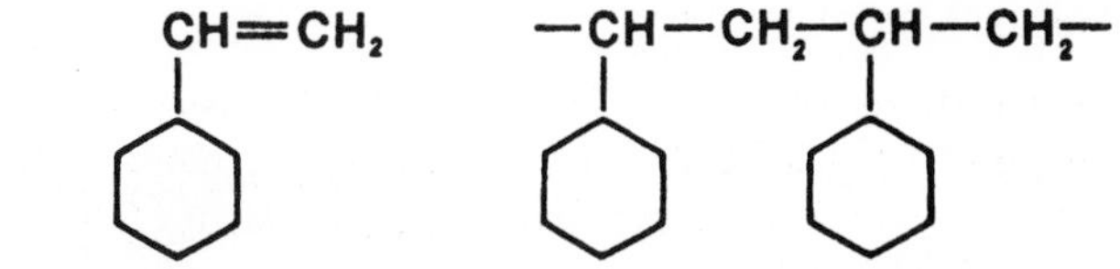

Figure E-60. Polymerization of styrene.

© Sulzer Chemtech Ltd

Figure E-61. Expanded polystyrene beads.

Foamed and expanded polystyrene sheets may be thermoformed. They are made into such familiar packaging items as egg cartons and meat or produce trays. Molded drinking cups, glasses, and ice chests are commonly used items.

Polystyrenes cannot withstand prolonged heat above 65°C (150°F) without distorting. Therefore, they are not good exterior materials. Special grades and additives can be used to correct this problem. Polystyrenes reinforced with glass fiber are used in automotive assemblies, business machines, and appliance housings.

The properties of polystyrene can be considerably varied by copolymerization and other modifications. Styrene-butadiene rubber has been mentioned above. Polystyrene is used in sporting goods, toys, wire and cable sheathing, shoe soles, and tires. Two of the most useful copolymers (terpolymers) are styrene-acrylonitrile and acrylonitrile-butadiene-styrene (ABS).

Table E-28 gives some of the properties of polystyrene, and listed here are nine advantages and six disadvantages of polystyrene:

Table E-28. Properties of Polystyrene

Property	Polystyrene (Unfilled)	Impact- and Heat-Resistant Polystyrene	Polystyrene (20–30% Glass-Filled)
Molding qualities	Excellent	Excellent	Excellent
Relative density	1.04–1.09	1.04–1.10	1.20–1.33
Tensile strength, MPa	35–83	10–48	62–104
(psi)	(5000–12,000)	(1500–7000)	(9000–15,000)
Compressive strength, MPa	80–110	28–62	93–124
(psi)	(11,500–16,000)	(4000–9000)	(13,500–18,000)
Impact strength, Izod, J/mm	0.0125–0.02	0.025–0.55	0.02–0.22
(ft-lb/in.)	(0.25–0.40)	(0.5–11)	(0.4–4.5)
Hardness, Rockwell	M65–M80	M20–M80, R50–R100	M70–M95
Thermal expansion, 10^{-4}/°C	15.2–20	8.5–53	4.5–11
Resistance to heat, °C	65–78	60–80	82–95
(°F)	(150–170)	(140–175)	(180–200)
Dielectric strength, V/mm	19,500–27,500	11,500–23,500	13,500–16,500
Dielectric constant (at 60 Hz)	2.45–2.65	2.45–4.75	
Dissipation factor (at 60 Hz)	0.0001–0.0003	2.45–4.75	0.004–0.014
Arc resistance, s	60–80	10–20	25–40
Water absorption (24 h), %	0.03–0.10	0.05–0.6	0.05–0.10
Burning rate	Slow	Slow	Slow-nonburning
Effect of sunlight	Yellows slightly	Yellows slightly	Yellows slightly
Effect of acids	Oxidizing acids	Oxidizing acids	Oxidizing acids
Effect of alkalies	None	None	Resistant
Effect of solvents	Soluble in aromatic and chlorinated hydrocarbons	Soluble in aromatic and chlorinated hydrocarbons	Soluble in aromatic and chlorinated hydrocarbons
Machining qualities	Good	Good	Good
Optical qualities	Transparent	Translucent to opaque	Translucent to opaque

Advantages of PS

1. Optical clarity
2. Light mass
3. High gloss
4. Excellent electrical properties
5. Good grades available
6. Processable by all thermoplastic methods
7. Low cost
8. Good dimensional stability
9. Good rigidity

Disadvantages of PS

1. Flammable (retarded grades available)
2. Poor weatherability
3. Poor solvent resistance
4. Brittleness of homopolymers
5. Subject to stress and environmental cracking
6. Poor thermal stability

Styrene-Acrylonitrile (SAN)

Acrylonitrile (CH_2=CHCHN) is copolymerized with styrene (C_6H_6), giving products a higher resistance than polystyrene to various solvents, fats, and other compounds (Figure E-62). These products are suitable for components requiring impact strength and chemical resistance and are used in vacuum cleaners and kitchen equipment.

Styrene-acrylonitrile (SAN) copolymers may have about 20% to 30% acrylonitrile content. A wide range of properties and processability may be obtained by varying the proportions of each monomer. A slight yellow cast is typical of SAN due to the copolymerization of acrylonitrile with the styrene member.

This copolymer is easily molded and processed. SAN-type materials inherently absorb more moisture such as silver streaking. Pre-drying is advised.

Methyl ethyl ketone, trichloroethylene, and methylene chloride are among the effective solvents for SAN.

This tough, heat-resistant plastics is used for telephone parts, containers, decorative panels, blender bowls, syringes, refrigerator compartments, food packages, and lenses (Figure E-63).

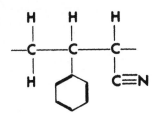

Figure E-62. SAN mer.

(A) Jar for petroleum jelly.

(B) Housing for air precleaner.

Figure E-63. Two uses of transparent SAN.

Table E-29 lists some of the properties of SAN. Three advantages and three disadvantages of styrene-acrylonitrile (SAN) are listed:

Advantages of SAN

1. Processable by thermoplastic methods
2. Rigid and transparent
3. Improved solvent resistance over polystyrene

Disadvantages of SAN

1. Higher water absorption than polystyrene
2. Low thermal capability
3. Low-impact strength

(Olefin-Modified) Styrene-Acrylonitrile (OSA)

A tough, heat and weather resistant polymer is produced by tailoring the molecular weight and monomer ratios of saturated

Table E-29. Properties of SAN

Property	SAN (Unfilled)
Molding qualities	Good
Relative density	1.075–1.1
Tensile strength, MPa	1.075–1.1
(psi)	(9000–12,000)
Compressive strength, MPa	97–117
(psi)	(14,000–17,000)
Impact strength, Izod, J/mm	0.01–0.02
(ft-lb/in.)	(0.35–0.50)
Hardness, Rockwell	M80–M90
Thermal expansion, 10^{-4}/°C	M80–M90
Resistance to heat, °C	60–96
(°F)	(140–205)
Dielectric strength, V/mm	15,750–19,685
Dielectric constant (at 60 Hz)	2.6–3.4
Dissipation factor (at 60 Hz)	0.006–0.008
Arc resistance, s	100–150
Water absorption (24 h), %	0.20–0.30
Burning rate	Slow to self extinguishing
Effect of sunlight	Yellows
Effect of acids	None
Effect of alkalies	Attacked by oxidizing agents
Effect of solvents	Soluble in ketones and esters
Machining qualities	Good
Optical qualities	Transparent

olefinic elastomer with styrene and acrylonitrile. It is used almost exclusively as a co-extrudant over other substrates. Topper covers, boat hulls, and decorative wood and metal construction panels are typical applications.

Styrene-Butadiene Plastics (SBP)

This amorphous copolymer consists of two blocks of styrene repeating units separated by a block of butadiene. This shows a contrast to styrene butadiene (SBR), which is thermosetting.

These polymers are ideally suited for packaging applications including cups, deli containers, meat trays, jars, bottles, skin packages, and overwrap. Reinforced and filled grades are used in tool handles, office equipment housings, medical devices, and toys.

Styrene-Maleic Anhydride (SMA)

This thermoplastic is distinguished from the parent stryenic and ABS families by higher heat resistance. SMA is obtained by the copolymerization of maleic anhydride and styrene. Butadiene is sometimes terpolymerized to produce impact-modified versions. Applications include vacuum cleaner housings, mirror housings, thermoformed headliners, fan blades, heater ducts, and food service trays.

POLYSULFONES

In 1965, Union Carbide introduced a linear, heat-resistant thermoplastic called polysulfone. The basic repeating structure consists of benzene rings joined by a sulfone group (SO_2), an isopropylidene group (CH_3CH_3C), and an ether linkage (O).

One basic polysulfone is made by mixing bisphenol A with chlorobenzene and dimethyl sulfoxide in a caustic soda solution. The resulting condensation polymerization is shown in Figure E-64. The light amber color of the plastics is a result of the addition of methyl chloride, which ends polymerization. The outstanding thermal and oxidation resistance is the result of the benzene-to-sulfone linkages. Polysulfone can be processed using all normal methods. It must be dried before use and may require processing temperatures in excess of 370°C (700°F). Service temperatures range from −100°C to +175°C (−150°F to +345°F).

Polysulfone can be machined, heat sealed, or solvent cemented using dimethyl formamide or dimethyl acetamide.

Polysulfones are competitive with many thermosets. They may be processed in rapid-cycle thermoplastics equipment. The polysulfones have excellent mechanical, electrical, and thermal properties. They are used for hot-water pipes, alkaline battery cases, distributor caps, face shields for astronauts, electrical circuit breakers, appliance housings, dishwasher impellers, autoclavable hospital equipment, interior components for aerospace craft, shower heads, lenses, and numerous electrical insulating components (Figure E-65). When used outdoors, polysulfones should be painted or electroplated to prevent degradation. Five advantages and four disadvantages of polysulfones are found in the following lists:

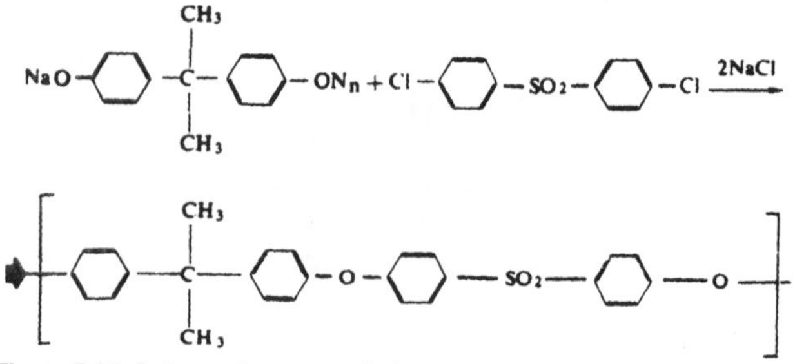

Figure E-64. Basic repeating structure of polysulfone.

(A) This industrial battery case made of polysulfone resists potassium hydroxide electrolyte fluid as well as vibration and high temperatures.

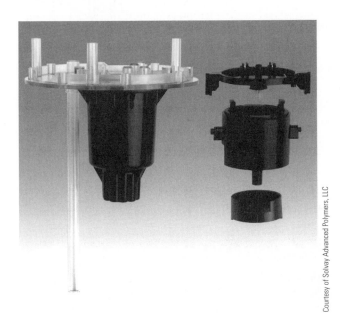

(B) Hot, pressurized water doesn't warp or deflect these polysulfone parts found in a coffee maker.
Figure E-65. Applications of polysulfone.

Advantages of Polysulfones

1. Good thermal stability
2. Excellent high-temperature creep resistance
3. Transparent
4. Tough and rigid
5. Processable by thermoplastic methods

Disadvantages of Polysulfones

1. Subject to attack by many solvents
2. Poor weatherability
3. Subject to stress cracking
4. High processing temperature

Polyarylsulfone

Polyarylsulfone is a high-temperature, amorphous thermoplastic introduced in 1983. It offers properties similar to other aromatic sulfones. Examples of uses include circuit boards, high-temperature bobbins, sight glasses, lamp housings, electrical connectors and housings, and panels of composite materials for numerous transportation components.

Polysulfones have been prepared using a variety of bisphenols with methylene, sulfide, or oxygen linkages. In polyaryl sulfone, the bisphenol groups are linked by ether and sulfone groups. There are no isopropylene (aliphatic) groups present. The term *aryl* refers to a phenyl group derived from an aromatic compound. If more than one hydrogen is substituted in the aryl group when naming these compounds, a numbering system is normally used. Three possible disubstituted benzenes are shown in Figure E-66. The basic properties of polysulfone are given in Table E-30.

Polyethersulfone (PES)

This plastics, with outstanding oxidation and thermal resistance, was introduced in 1973. Polyethersulfone has performed well under creep and stress forces at temperatures above 200°C (390°F). It is characterized by an absence of aliphatic groups and is an amorphous structure. Polyethersulfone is very resistant to both acids and alkalies but is attacked by ketones, esters, and some halogenated and aromatic hydrocarbons. The basic monomer unit is shown in Figure E-67.

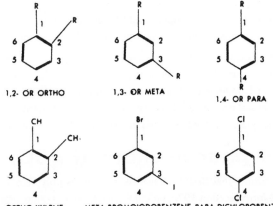

Figure E-66. Three possible disubstituted benzenes.

Table E-30. Properties of Polysulfones

Property	Polysulfone (Unfilled)	Polyary Sulfone (Unfilled)
Molding qualities	Excellent	Excellent
Relative density	1.24	1.36
Tensile strength, MPa	70	90
(psi)	(10,200)	(13,000)
Compressive strength, MPa	96	123
(psi)	(13,900)	(17,900)
Impact strength, Izod, J/mm	0.06; bar 7.25 mm	0.25
(ft-lb/in.)	(1.3); (bar 1/4 in)	(5)
Hardness, Rockwell	M69, R120	M110
Thermal expansion, $10^{-4}/°C$	13.2–14.2	11.9
Resistance to heat, °C	150–175	260
(°F)	(300–345)	(500)
Dielectric strength, V/mm	16,730	13,800
Dielectric constant (at 60 Hz)	3.14	3.94
Dissipation factor (at 60 Hz)	0.0008	0.003
Arc resistance, s	75–122	67
Water absorption (24 h), %	0.22	1.8
Burning rate	Self-extinguishing	Self-extinguishing
Effect of sunlight	Strength loss, yellows slightly	Slight
Effect of acids	None	None
Effect of alkalies	None	None
Effect of solvents	Partly soluble in aromatic hydrocarbons	Soluble in highly polar solvents
Machining qualities	Excellent	Excellent
Optical qualities	Transparent to opaque	Opaque

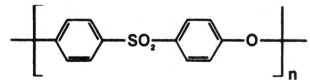

Figure E-67. Basic repeating unit of polyethersulfone.

The distinguishing properties of PES are its high-temperature performance, good mechanical strength, and low flammability.

Polyethersulfone has found uses in aerospace components, sterilizable medical components, and oven windows. Compounded grades extend their useful temperatures and improve mechanical properties. They have been used as adhesives and may be plated. Table E-31 gives properties of these plastics.

Polyphenylsulfone (PPSO)

The sulfone that best resists stress cracking is polyphenylsulfone. Introduced in 1976, polyphenylsulfone is an amorphous structure with very high-impact strength that will withstand a continuous service temperature of 190°C (375°F). Uses

include semiconductor carriers, valves, circuit boards, and aerospace components.

POLYVINYLS

There is a large and varied group of addition polymers that chemists refer to as vinyls. These have the formula:

$$CH_2{=}CH{-}R \text{ or } CH_2 = \overset{\displaystyle R}{\underset{\displaystyle R}{\overset{|}{\underset{|}{C}}}}$$

Radicals (R) may be attached to this repeating vinyl group as side groups to form several polymers related to each other. Additional polymers with the radical side groups attached are shown in Table E-32.

Through common usage, the *vinyl plastics* are those polymers with the vinyl name. Many authorities limit their discussion to include only polyvinyl chloride and polyvinyl acetate. Polyvinyl

Table E-31. Properties of Polyethersulfone

Property	Polyethersulfone (Unfilled)
Molding qualities	Excellent
Relative density	1.37
Tensile strength, MPa	84
(psi)	(12,180)
Impact strength, Izod, J/mm	0.08
(ft-lb/in.)	(1.6)
Hardness, Rockwell	M88
Thermal expansion, $10^{-4}/°C$	13–97
Resistance to heat, °C	150
(°F)	(300)
Dielectric strength, V/mm	15,750
Dielectric constant (at 60 Hz)	3.5
Dissipation factor (at 60 Hz)	0.001
Arc resistance, s	65–75
Water absorption (24 h), %	0.43
Effect of sunlight	Yellows
Effect of acids	None
Effect of alkalies	None
Effect of solvents	Attacked by aromatic hydrocarbons
Machining qualities	Excellent
Optical qualities	Transparent

homopolymers or copolymers may include polyvinyl chloride, polyvinyl acetate, polyvinyl alcohol, polyvinyl butyral, polyvinyl acetal, and polyvinylidene chloride. Fluorinated vinyls are discussed with other fluorine-containing polymers.

The history of polyvinyls may be traced to as early as 1835. French chemist V. Regnault reported that a white residue could be synthesized from ethylene dichloride in an alcohol solution. This tough white residue was again reported in 1872 by E. Baumann. It occurred while reacting acetylene and hydrogen bromide in sunlight. In both cases, sunlight was the polymerizing catalyst that produced the white residue. In 1912, Russian chemist I. Ostromislenski reported the same sunlight polymerization of vinyl chloride and vinyl bromide. By 1930, commercial patents were granted in several countries for the manufacture of vinyl chloride.

In 1933, W. L. Semon of the B. F. Goodrich Company added a plasticizer, tritolyl phosphate, to polyvinyl chloride compounds. The resulting polymer mass could be easily molded and processed without substantial decomposition.

Germany, Great Britain, and the United States commercially produced plasticized polyvinyl chloride (PVC) during World War II. It was largely used as a substitute for rubber.

Today, polyvinyl chloride is the leading plastics produced in Europe, whereas it ranks second after polyethylene in the United States. The polyvinyl chloride molecule (C_2H_3Cl) is similar to polyethylene, as shown in Figure E-68.

Polyvinyl Chloride (PVC)

The basic raw ingredient of polyvinyl chloride, depending on availability, is acetylene or ethylene gas. Ethylene is the chief source in the United States. During its manufacture, polymerization may be initiated by using peroxides, azo compounds, persulfates, ultraviolet light, or radioactive sources. For additional polymerization, the double bonds of the monomers must be broken by the use of heat, light, pressure, or a catalyst system.

The uses of polyvinyl chloride plastics may be expanded by the addition of plasticizers, fillers, reinforcements, lubricants, and stabilizers. They may be formulated into flexible, rigid, elastomeric, or foamed compounds.

Polyvinyl chloride is most widely used in flexible film and sheet forms. These films and sheets compete with other films for use in collapsible containers, drum liners, sacks, and packages. Washable wallpapers and certain clothing such as handbags, rainwear, coats, and dresses are other uses. Sheets are made into chemical tanks and ductwork of all types. They are easily fabricated by welding, heat sealing, or solvent cementing with mixtures of ketones or aromatic hydrocarbons.

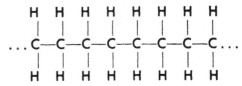

(A) Polyethylene.

(B) Vinyl chloride.

(C) Polyvinyl chloride.

Figure E-68. Similarity of polyethylene and polyvinyl chloride.

Table E-32. Monofunctional Monomers and Their Polymers

Monomer	Polymer
$CH_2=CH_2$ Ethylene	$\rightarrow -CH_2-CH_2-CH_2-CH_2-CH_2-CH_2-CH_2-CH_2-$ Polyethylene
$O-COCH_3$ | $CH_2=CH$ Vinyl acetate	$\rightarrow$ with $O-COCH_3$ groups on alternating carbons $-CH_2-CH-CH_2-CH-CH_2-CH-CH_2-CH-\ldots$ Polyvinyl acetate
Cl | $CH_2=CH$ Vinyl chloride	$\rightarrow -CH_2-CH-CH_2-CH-CH_2-CH-CH_2-CH-\ldots$ with Cl groups Polyvinyl chloride
C_6H_5 | $CH_2=CH$ Styrene (Vinyl benzene)	$\rightarrow -CH_2-CH-CH_2-CH-CH_2-CH-CH_2-CH-\ldots$ with C_6H_5 groups Polystyrene
Cl | $CH_2=C$ | Cl Vinylidene chloride	$\rightarrow -CH_2-CH-CH_2-CH-CH_2-CH-CH_2-CH-$ with Cl above and Cl below Polyvinylidene chloride
$COOH$ | $CH_2=CH$ Acrylic acid	$\rightarrow -CH_2-CH-CH_2-CH-CH_2-CH-CH_2-CH-\ldots$ with $COOH$ groups Polyacrylic acid
$COOH$ | $CH_2=C$ | CH_3 Methacrylic acid	$\rightarrow -CH_2-CH-CH_2-CH-CH_2-CH-CH_2-CH-\ldots$ with $COOH$ above and CH_3 below Polymethacrylic acid
CH_3 | $CH_2=C$ | CH_3 Isobutylene	$\rightarrow -CH_2-CH-CH_2-CH-CH_2-CH-CH_2-CH-\ldots$ with CH_3 above and CH_3 below Polyisobutylene

Extruded profile shapes of both rigid and flexible polyvinyl chloride are used in architectural moldings, seals, gaskets, gutters, exterior siding, garden hose, and moldings for movable partitions (Figure E-69). Injection-molded PVC pipe fittings have been popular for many years. As seen in Figure E-70, a clear version of such pipe fittings has recently become available.

Organosols and plastisols are liquid or paste dispersions or emulsions of polyvinyl chloride. They are used for coating various substrates, including metal, wood, plastics, and fabrics. They may be applied by using dipping, spraying, spreading, or slush and rotational casting. Laminates of polyvinyl film, foam, and fabrics are used for upholstery materials. Dip coatings are found on tool handles, sink drainers, and other substrates as a protective layer. Slush and rotational casting of polyvinyls are used to produce hollow articles such as balls, dolls, and large containers. Heavily filled polyvinyls and their copolymers are used in the production of floor coverings and tiles. Foams have found limited application in the textile and carpeting industries. Large quantities of PVC are used as blow-molded containers and extruded coverings for electrical wire.

Generally, materials in the vinyl family are flame, water, chemical, electrical, and abrasion resistant. They have good

(A) Polyvinyl coating on wood for weatherproof finish and durability.

(B) Solid vinyl gutter and downspout system.

Figure E-69. Polyvinyl coatings have many uses.

weatherability and may be transparent. To aid in processing and provide various properties, polyvinyls are commonly plasticized. Both plasticized and unplasticized PVC compounds are available. Unplasticized grades are used in chemical plants and building industries. Plasticized grades are more flexible and soft. With increased plasticizer, there is more bleeding or migration of the plasticizer chemical into adjacent materials. This is of prime importance when packaging food products and medical supplies.

All thermoplastic processing techniques are employed with vinyls.

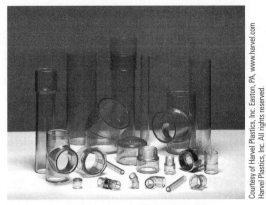

Figure E-70. These clear pipe fittings are made of Harvel Clear™ PVC.

Polyvinyl chloride (PVC) is the most widely used and commonly thought-of vinyl. However, other homopolymer and copolymer polyvinyls are finding increasing use. The following lists show six advantages and five disadvantages of polyvinyl chloride:

Advantages of PVC

1. Processable by thermoplastic methods
2. Wide range of flexibility (by varying levels of plasticizers)
3. Nonflammable
4. Dimensional stability
5. Comparatively low cost
6. Good resistance to weathering

Disadvantages of PVC

1. Subject to attack by several solvents
2. Limited thermal capability
3. Thermal decomposition evolves HCl
4. Stained by sulfur compounds
5. Higher density than many plastics

Polyvinyl Acetate (PVAC)

Vinyl acetate ($CH_2{=}CH{-}O{-}COCH_3$) is industrially prepared from liquid or gaseous reactions of acetic acid and acetylene. Homopolymers find only limited uses due to excessive cold flow and low softening point. They are used in paints, adhesives, and various textile finishing operations. Polyvinyl acetates are usually found in an emulsion form. *White glues* are familiar polyvinyl acetate emulsions. Moisture absorption characteristics are high with chosen alcohols and ketones as solvents. Re-moistenable adhesives and hot-melt formulations are other well-known uses. Polyvinyl acetates are used as binder emulsions in some paint formulations. Their resistance to degradation by sunlight makes them useful for interior or exterior coatings. Other uses may include emulsion binders in paper, cardboard, Portland cements, textiles, and chewing gum bases.

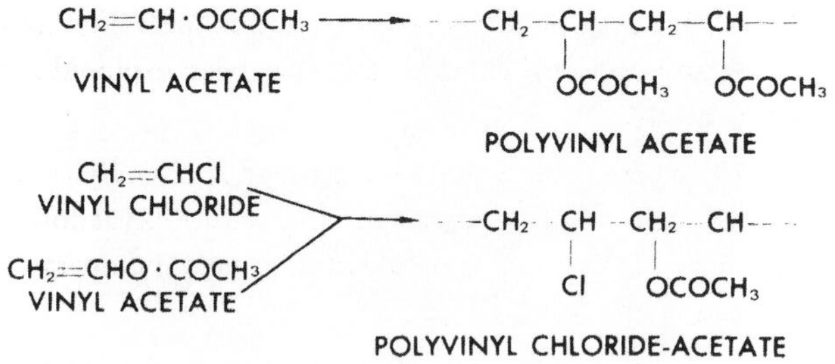

Figure E-71. Production of polyvinyl acetate and polyvinyl chloride-acetate.

Some of the best-known commercial products are copolymers of polyvinyl chloride and polyvinyl acetate (Figure E-71) used for floor coverings and modern phonograph records. These vinyl records have several advantages over polystyrene and the older shellac discs. Two advantages and three disadvantages of polyvinyl acetate follow:

Advantages of PVAC

1. Excellent for forming into films
2. Heat-sealable

Disadvantages of PVAC

1. Low thermal stability
2. Poor solvent resistance
3. Poor chemical resistance

Polyvinyl Formal

Polyvinyl formal is generally produced from polyvinyl acetate, formaldehyde, and other additives that change the alcohol side groups on the chain to *formal* side groups. Polyvinyl formal finds its greatest uses as coatings for metal containers and electrical wire enamels.

Polyvinyl Alcohol (PVA)

Polyvinyl alcohol is a useful derivative produced from the alcoholysis of polyvinyl acetate (Figure E-72). Methyl alcohol (methanol) is used in this process. Polyvinyl alcohol (PVA) is both alcohol- and water-soluble. The properties vary depending on the concentration of polyvinyl acetate that remains in the alcohol solution. Polyvinyl alcohol may be used as a binder and adhesive for paper, ceramics, cosmetics, and textiles. It finds use in water-soluble packages for soap, bleaches, and disinfectants. It is a useful mold-releasing agent used in the manufacture of reinforced plastics products. There has been only limited use of polyvinyl alcohol for moldings and fibers.

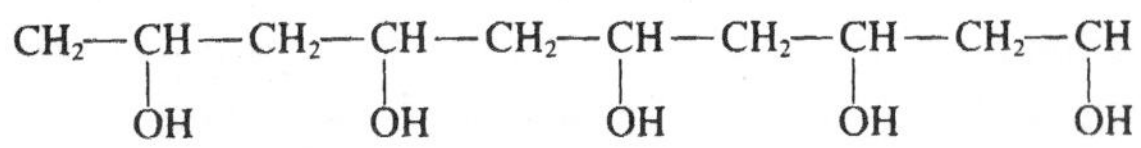

Figure E-72. Polyvinyl alcohol (OH side groups).

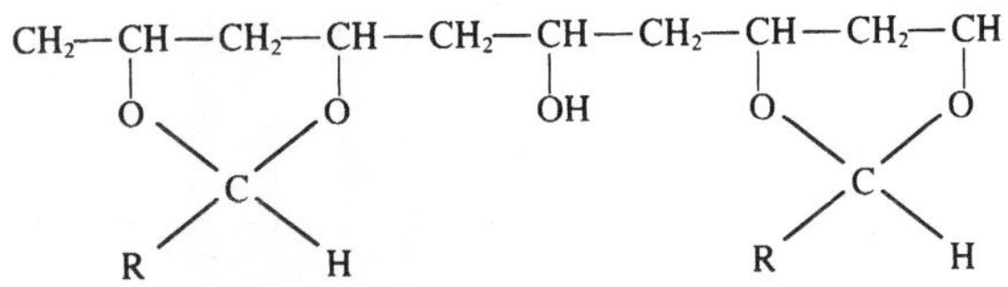

Figure E-73. A polyvinyl acetal.

Polyvinyl Acetate

One more useful derivative of polyvinyl acetate is polyvinyl acetal. It is produced from the treatment of polyvinyl alcohol (from polyvinyl acetate) with an acetaldehyde (Figure E-73). Polyvinyl acetal materials find limited use as adhesives, surface coatings, films, moldings, or textile modifiers.

Polyvinyl Butyral (PVB)

Polyvinyl butyral is produced from polyvinyl alcohol (Figure E-74). This plastics is used as an interlayer film in laminated safety glass.

Polyvinylidene Dichloride (PVDC)

In 1839, a substance similar to vinyl chloride was discovered, but it contained one more chlorine atom (Figure E-75). This material has become commercially important as vinylidene chloride ($H_2C = CCl_2$).

Polyvinylidene dichloride is costly and hard to process. Therefore, it is normally found as a copolymer with vinyl chloride, acrylonitrile, or acrylate esters. A well-known food wrapping film, Saran™, is a copolymer of vinylidene chloride and acrylonitrile. It exhibits clarity and toughness, yet permits little gas or moisture transmission.

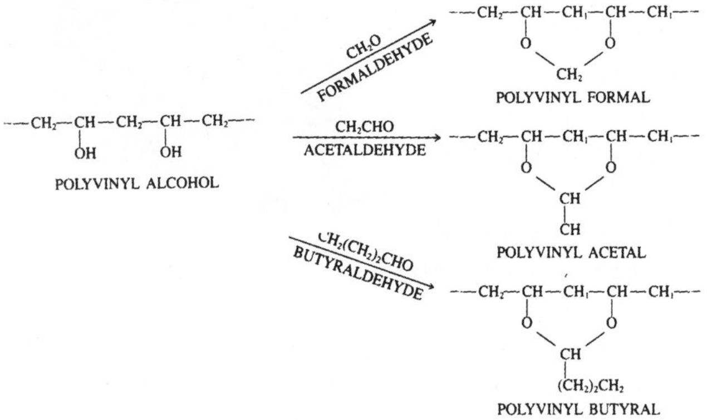

Figure E-74. Production of polyvinyl butyral.

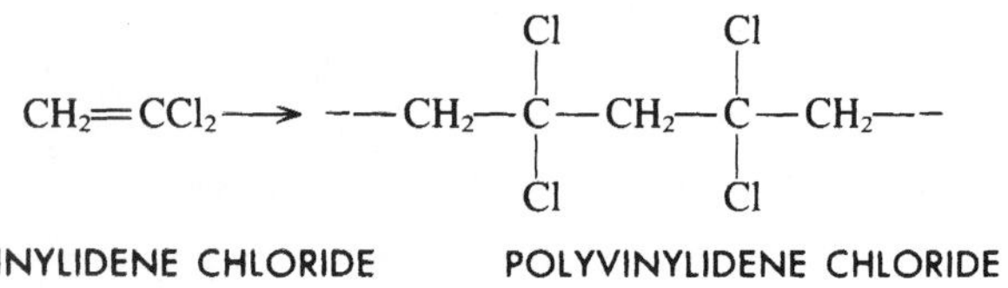

VINYLIDENE CHLORIDE   POLYVINYLIDENE CHLORIDE

Figure E-75. Polymerization of vinylidene chloride.

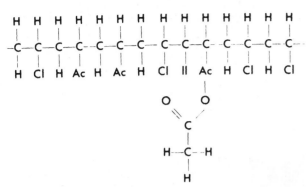

Figure E-76. Copolymerization of vinyl chloride and vinyl acetate.

The chief use of polyvinylidene dichloride copolymers (Figure E-76) is in coating and film packaging. However, they have found some application as fibers for carpeting, automobile seat upholstery, draperies, and awning textiles. Their chemical inertness allows them to be used in pipes, pipe fittings, pipe linings, and filters.

Many other polyvinyl polymers warrant further research and study. Polyvinyl carbazole is used for dielectrics; polyvinyl pyrrolidone is used as a blood plasma substitute; and polyvinyl ethers are used as adhesives. Polyvinyl ureas, polyvinyl isocyanates, and polyvinyl chloroacetate have been explored for commercial use.

All chlorinated or chlorine-containing polymers may emit toxic chlorine gas upon high temperature breakdown. Adequate venting should be provided to protect the operator during processing.

Table E-33 gives some properties of polyvinyl plastics, and lists of four advantages and two disadvantages of polyvinylidene chloride follow:

Advantages of PVDC

1. Low water permeability
2. Approved for food wrap by FDA
3. Processable by thermoplastic methods
4. Nonflammable

Disadvantages of PVDC

1. Lower strength than PVC
2. Subject to creep

Table E-33. Properties of Polyvinyls

Property	Rigid Vinyl Chloride (PVC)	PVC-Acetate (Copolymer)	Vinylidene Chloride Compound
Molding qualities	Good	Good	Excellent
Relative density	1.30–1.45	1.16–1.18	1.65–1.72
Tensile strength, MPa	34–62	17–28	21–34
(psi)	(5000–9000)	(2500–4000)	(3000–5000)
Compressive strength, MPa	55–90		14–19
(psi)	(8000–13,000)		(2000–2700)
Impact strength, Izod, J/mm	0.02–1.0		0.06–0.05
(ft-lb/in.)	(0.4–20.0)		(0.3–1.0)
Hardness, Rockwell	M110–M120	R34–R40	M50–M65
Thermal expansion, 10^{-4}/°C	12.7–47		48.3
Resistance to heat, °C	65–80	55–60	70–90
(°F)	(150–175)	(130–140)	(160–200)
Dielectric strength, V/mm	15,750–19,700	12,000–15,750	15,750–23,500
Dielectric constant (at 60 Hz)	3.2–3.6	3.5–4.5	4.5–6.0
Dissipation factor (at 60 Hz)	0.007–0.020		0.030–0.045
Arc resistance, s	60–80		
Water absorption (24 h), %	0.07–0.4	3.0+	0.1
Burning rate	Self-extinguishing	Self-extinguishing	Self-extinguishing
Effect of sunlight	Needs stabilizer	Needs stabilizer	Slight
Effect of acids	None to slight	None to slight	Resistant
Effect of alkalies	None	None	Resistant
Effect of solvents	Soluble in ketones and esters	Soluble in ketones and esters	None to slight
Machining qualities	Excellent		Good
Optical qualities	Transparent	Transparent	Transparent

THERMOSETTING PLASTICS

Your study of plastics resins is not complete until you become familiar with thermosetting plastics. Look for the outstanding properties and applications of each plastics described in this appendix, because plastics properties affect product design, processing, economics, and service.

You should remember that thermosetting materials undergo a chemical reaction and become "set." In general, most are high-molecular-weight materials resulting in hard, brittle plastics.

This appendix discusses individual groups of thermosetting plastics. These are *alkyds, allylics, amino plastics (urea-formaldehyde and melamine-formaldehyde), casein, epoxy, furan, phenolics, unsaturated polyesters, polyurethane, and silicones.*

ALKYDS

In the past, there was some confusion about ester-based alkyn resins. The term *alkyd* was once used strictly for unsaturated polyesters modified with fatty acids or vegetable oils. These resins are used in paints and other coatings. Today, *alkyd molding compounds* refer to unsaturated polyesters modified by a nonvolatile monomer (such as daillyl phthalate) and various fillers. The compounds are formed into granular, rope, nodular, putty, and log shapes to allow continuous automatic molding.

To obtain a resin suitable for molding compounds, the cross-linking mechanism must be modified so that rapid cure will take place in the mold. The use of initiators in the resin compound also speeds up the polymerization of double bonds. These resins should not be confused with saturated polyester molding compounds, which are linear and thermoplastic.

R. H. Kienle coined the word *alkyd* from the "al" in *alcohol* and the "cid" from a*cid*. It is usually pronounced "al-kid." Alkyd resins may be produced by reacting phthalic acid, ethylene glycol, and the fatty acids of various oils such as linseed, soybean, or tung oil (Figure F-1). In 1927, Kienle combined fatty acids with unsaturated esters while searching for a better electrical insulating resin for General Electric. The need for materials in World War II greatly increased the interest in alkyd finishing resins.

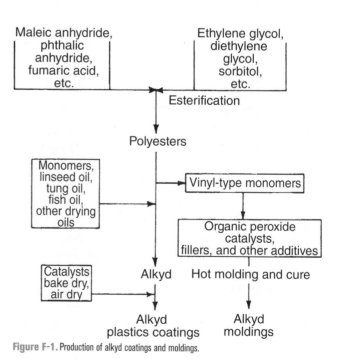

Figure F-1. Production of alkyd coatings and moldings.

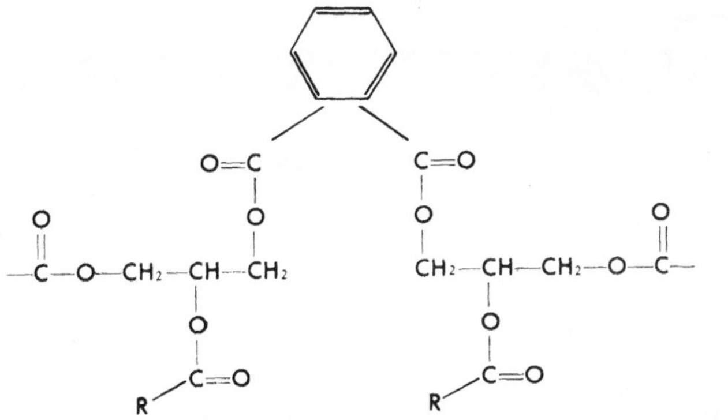

Figure F-2. Long alkyd resin chain with unsaturated groups.

Alkyd resins may have a structure similar to the example shown in Figure F-2. When this resin is applied to a substrate, heat or oxidizing agents are used to begin cross-linking.

One expert estimates that nearly half of the surface coatings used in the United States are alkyd polyester coatings. The increasing use of silicone and acrylic latex coatings may greatly reduce that figure.

Alkyd coatings are valuable because of their relatively low cost, durability, heat resistance, and adaptability. They may be changed to meet special coating needs. To increase durability and abrasion resistance, rosin may be used to modify alkyd resins. Phenolics and epoxy resins may improve hardness and resistance to chemicals and water. Styrene monomers added to the resin may extend flexibility of the finished coating and serve as cross-linking agents.

Alkyd resins are used in *oil-based* house paint, baking enamel, farm implement paint, emulsion paint, porch and deck enamel, spar varnish, and chlorinated rubber paint. Modified alkyd resins may produce special coatings such as *wrinkle* and *hammer* finishes often used on equipment and machinery.

Alkyd baking finishes are referred to as *heat convertible.* They polymerize or harden when heated. Resins that harden when exposed to air are called *air-convertible* resins.

Alkyd resins are used as plasticizers for various plastics; vehicles for printing inks; binders for abrasives and oils; and special adhesives for wood, rubber, glass, leather, and textiles.

New processing technologies and uses for thermosetting alkyd molding compounds are also of current commercial importance.

Alkyd molding compounds are processed in compression, transfer, and reciprocating-screw equipment. By using initiators such a benzoyl peroxide or tertiary butyl hydroperoxide, the unsaturated resins can be made to cross-link at molding temperatures (greater than 50°C, or 120°F). Molding cycles may be less than 20 seconds.

Typical uses of alkyd molding compounds include appliance housings, utensil handles, billiard balls, circuit breakers,

switches, motor cases, and capacitor and commutator parts. They have been used in electrical applications when less costly phenolic or amino resins are not suitable.

Alkyd compounds in putty form are used to encapsulate electronic and electrical components. Different resin formulations, fillers, and processing techniques provide a wide range of physical characteristics.

Table F-1 gives some of the properties of alkyd plastics.

ALLYLICS

Allylic resins usually involve the esterification of allyl alcohol and a dibasic acid (Figure F-3). The pungent odor of allyl alcohol (CH_2=$CHCH_2OH$) has been known to science since 1856. The name *allyl* was coined from the Latin word *allium,* meaning garlic. These resins did not become commercially useful until 1955, although one allyl resin was used in 1941 as a low-pressure laminating resin.

Allylics and polyesters have common applications, properties, and historical backgrounds of development. For this reason, allylics are sometimes erroneously included in the study of polyesters. Allyl monomers are used as cross-linking agents in polyesters, which adds to the confusion. Allylics are a distinct family of plastics based on monohydric alcohols, whereas the chemical basis for polyesters is polyhydric alcohols.

Allylics are unique in that they may form pre-polymers (partly polymerized resins). They may be homopolymerized or copolymerized. Monoallyl esters may be produced as either thermosetting or thermoplastic resins. The saturated monoallyl esters (thermoplastic) are sometimes used as copolymerizing agents in alkyd and vinyl resins. Unsaturated monoallyl esters have been used to produce simple polymers including allyl acrylate, allyl chloroacrylate, allyl methacrylate, allyl crotonate, allyl cinnamate, allyl cinnamalacetate, allyl furoate, and allyl furfuryacrylate.

Simple polymers have been produced from diallyl esters, including diallyl maleate (DAM), diallyl oxalate, diallyl succinate, diallyl sebacate, diallyl pththalate (DAP), diallylcarbonate, diethylene glycol bisallyl carbonate, diallyl isophthalate (DAIP), and others (Figure F-4). The most widely used commercial compounds are DAP (Figure F-4A) and DAIP (Figure F-4B).

Diallyl resins are usually supplied as monomers or prepolymers. Both forms are converted into the fully polymerized thermosetting plastics by the addition of selected peroxide catalysts. Benzoyl peroxide or *tert*butyl perbenzoate are two often-used catalysts for the polymerization of allylic resin compounds. If kept at low ambient temperatures, these resins and compounds may be catalyzed and stored for more than a year. When the material is subjected to normal temperatures of molds, presses, or ovens, complete cure is reached.

Table F-1. **Properties of Alkyds**

Property	Alkyd (Glass-Filled)	Alkyd Molding Compound (Filled)
Molding qualities	Excellent	Excellent
Relative density	2.12–2.15	1.65–2.30
Tensile strength, MPa	28–64	21–62
(psi)	(4000–9500)	(3000–9000)
Compressive strength, MPa	103–221	83–262
(psi)	(15,000–32,000)	(12,000–38,000)
Impact strength, Izod, J/mm	0.03–0.5	0.015–0.025
(ft-lb/in.)	(0.60–10)	(0.30–0.50)
Hardness, Barcol	60–70	55–80
Rockwell	E98	E95
Thermal expansion, (10^{-4}/°C)	3.8–6.35	5.08–12.7
Resistance to heat, °C	230	150–230
(°F)	(450)	(300–450)
Dielectric strength, V/mm	9845–20,870	13,800–17,720
Dielectric constant (at 60 Hz)	5.7	5.1–7.5
Dissipation factor (at 60 Hz)	0.010	0.009–0.06
Arc resistance, s	150–210	75–240
Water absorption (24 h), %	0.05–0.25	0.05–0.50
Burning rate	Slow to nonburning	Slow to nonburning
Effect of sunlight	None	None
Effect of acids	Fair	None
Effect of alkalies	Fair	Attacked
Effect of solvents	Fair to good	Fair to good
Machining qualities	Poor to fair	Poor to fair
Optical qualities	Opaque	Opaque

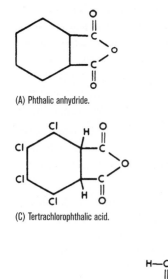

(A) Phthalic anhydride.

(B) Isophthalic acid.

(C) Tertrachlorophthalic acid.

(D) Chlorendic anhydride.

(E) Maleic anhydride.

Figure F-3. Dibasic acid used in the manufacture of allylic monomers.

The diethylene glycol bis-allyl carbonate resin monomer (Figure F-4E) used for laminating and optical castings is illustrated below. This resin could be polymerized with benzoyl peroxide at a temperature of 82°C (180°F).

Diethylene Glycol   Allyl Carbonate

$$CH_2CH_2OCOOCH_2CH=CH_2$$
$$O$$
$$CH_2CH_2OCOOCH_2CH=CH_2$$

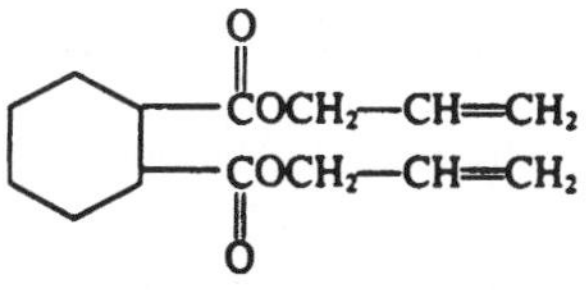

(A) Diallyl phthalate (ortho).

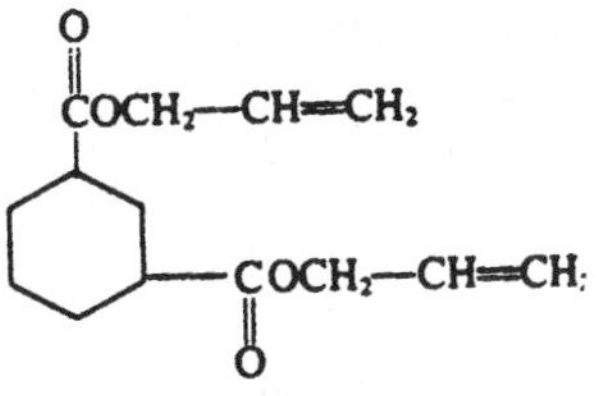

(B) Diallyl isophthalate (meta).

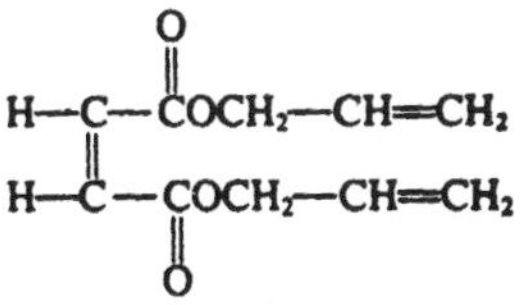

(C) Diallyl maleate.

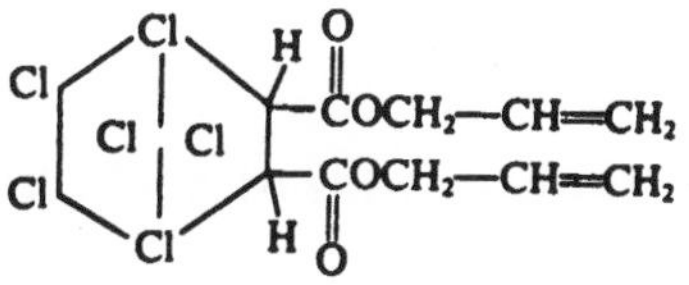

(D) Diallyl chlorendate.

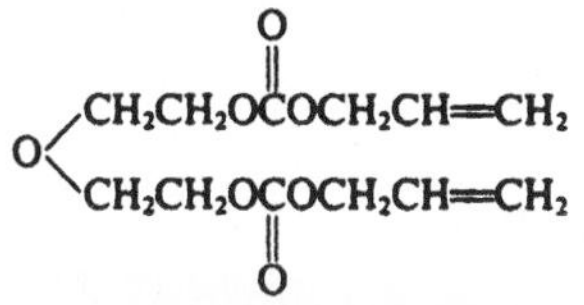

(E) Diethylene glycol bis-(allyl carbonate).

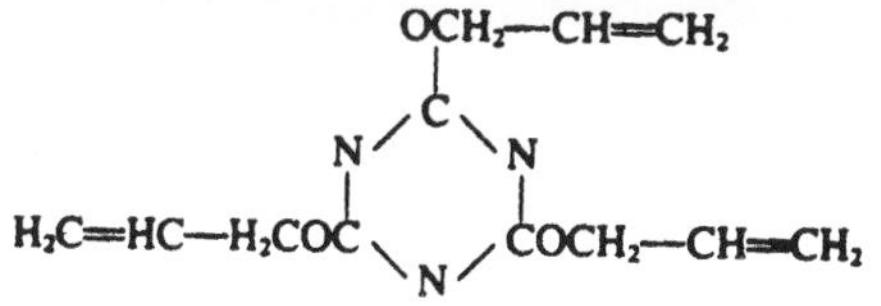

(F) Triallyl cyanurate.

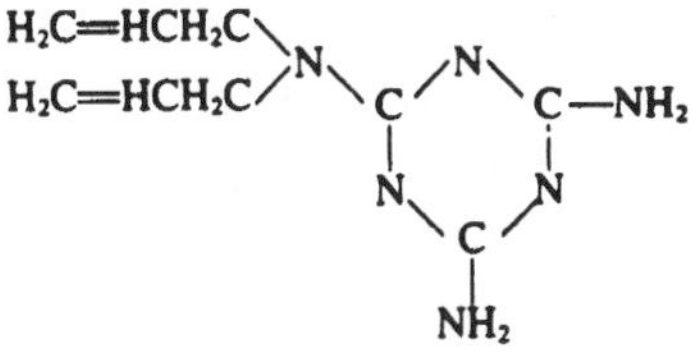

(G) N, N-diallyl melamine.

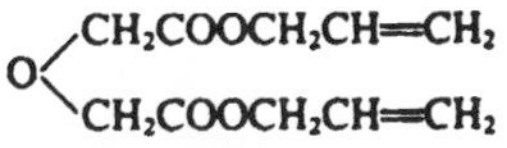

(H) Diallyl diglycollate.

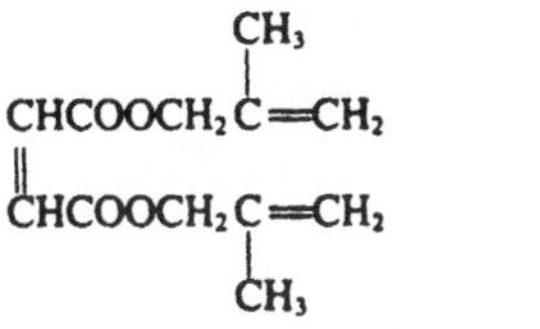

(I) Dimethallyl maleate.

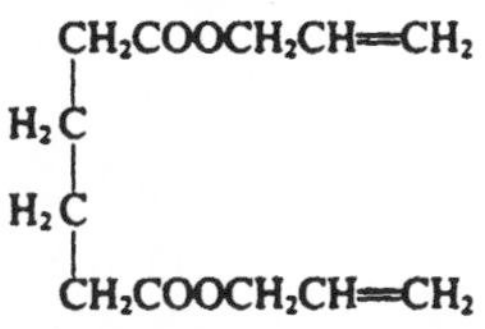

(J) Diallyl adipate.

Figure F-4. Structural formulas of some commercial allylic monomers.

Diallyl phthalate may be used as a coating and laminating material but is probably most noted for its principal use as a molding compound. Diallyl phthalate will polymerize and cross-link because it has two available double bonds (see Figure F-4A).

Good mechanical, chemical, thermal, and electrical properties are only a few of the attractive attributes of diallyl phthalate (Figure F-5). It also offers long shelf life and easy handling.

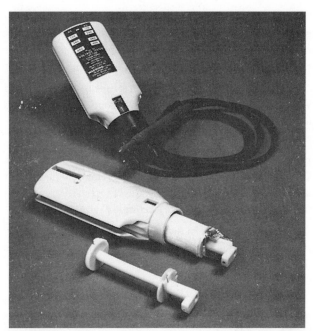

Figure F-5. This handheld voltage tester has a solenoid coil bobbin made of a strong diallyl phthalate molding compound.

Radiation and ablation resistance make these materials useful in space environments.

Nearly all allylic molding compounds include catalysts, fillers, and reinforcements, and are usually blended into putty-like premixes. High cost is a factor that limits the use of allylics in applications where their properties are vital. Compounds may be compression or transfer molded. Some meet the requirements for special high-speed injection machines, low-pressure encapsulation, and extrusion.

Some allylic resin formulations are used to produce laminates. Monomer resins are used to pre-impregnate wood, paper, fabrics, or other materials for lamination. Some improve various properties in paper and garments. Crease, water, and fade resistance may be enhanced by adding diallyl phthalate monomers to selected fabrics. For furniture and paneling, pre-impregnated decorative laminates or overlays are bonded to cores of less costly materials. Melamine and allylic resin-bound laminates have many of the same properties.

Allylic monomers and prepolymers are also used in pre-pregs (wet layup laminates). These prepregs consist of fillers, catalysts, and reinforcements. They are combined just prior to being cured. Some are preshaped for easy molding. Allylic pre-pregs may be prepared in advance and stored until needed. Premix and prepreg molded parts have good flexural and impact strength, and their surface finishes are excellent.

Allylic monomers are used as cross-linking agents for polyesters, alkyds, polyurethane foams, and other unsaturated polymers. They are used because the basic allylic monomer (homopolymer) does not polymerize at room temperature.

This allows materials to be stored for very long periods. At temperatures of 150°C (300°F) and above, diallyl phthalate monomers cause polyesters to cross-link. These compounds may be molded at faster rates than those cross-linked with styrene.

Acrylics (methyl methacrylate) that have been cross-linked with diallyl phthalate monomers have good surface hardness and elasticity.

Allylics have been used in vacuum impregnation to seal the pores of metal castings, ceramics, and other compositions. Other uses include the impregnation of reinforcing tapes that are used to wrap motor armatures. Allylics are used to coat electrical parts and to encapsulate electronic devices.

Table F-2 lists some of the basic properties of allylic (diallyl phthalate) plastics. Four advantages and three disadvantages follow:

Advantages of Allylic Esters (Allyls)

1. Excellent moisture resistance
2. Availability of low-burning and self-extinguishing grades
3. Service temperatures as high as 204°C to 232°C (400°F to 450°F)
4. Good chemical resistance

Disadvantages of Allylic Esters (Allyls)

1. High cost (compared to alkyds)
2. Excessive shrinkage during curing
3. Not usable with phenols and oxidizing acids

Table F-2. Properties of Allylic Plastic

Property	Diallyl Phthalate Compounds		Diallyl Isophthalate
	Glass-Filled	Mineral-Filled	
Molding qualities	Excellent	Excellent	Excellent
Relative density	1.61–1.78	1.65–1.68	1.264
Tensile strength, MPa	41.4–75.8	34.5–60	30
(psi)	(6000–11,000)	(5000–8700)	(4300)
Compressive strength, MPa	172–241	138–221	
(psi)	(25,000–35,000)	(20,000–32,000)	
Impact strength, Izod, J/mm	0.02–0.75	0.015–0.225	0.01–0.015
(ft-lb/in.)	(0.4–15)	(0.3–0.45)	(0.2–0.3)
Hardness, Rockwell	E80–E87	E61	M238
Thermal expansion, $(10^{-4}/°C)$	2.5–9	2.5–10.9	
Resistance to heat, °C	150–205	150–205	150–205
(°F)	(300–400)	(300–400)	(300–400)
Dielectric strength, V/mm	15,000–17,717	15,550–16,535	16,615
Dielectric constant (at 60 Hz)	4.3–4.6	5.2	3.4
Dissipation factor (at 60 Hz)	0.01–0.05	0.03–0.06	0.008
Arc resistance, s	125–180	140–190	123–128
Water absorption (24 h), %	0.12–0.35	0.2–0.5	0.1
Burning rate	Self-extinguishing to nonburning	Self-extinguishing to nonburning	Self-extinguishing to nonburning
Effect of sunlight	None	None	None
Effect of acids	Slight	Slight	Slight
Effect of alkalies	Slight	Slight	Slight
Effect of solvents	None	None	None
Machining qualities	Fair	Fair	Good
Optical qualities	Opaque	Opaque	Transparent

AMINO PLASTICS

Various polymers have been produced by interacting amines or amides with aldehydes. The two most significant and commercially useful amino plastics are produced by the condensation of urea-formaldehyde and melamine-formaldehyde.

Urea-formaldehyde polymers may have been produced as early as 1884. In Germany, Goldschmidt and his colleagues directed their efforts toward making moldable amino plastics.

In 1920, urea-formaldehyde resins were commercially produced in the United States. Thiourea-formaldehyde and melamine-formaldehyde resins were being produced in the period of 1934–1939.

Urea-Formaldehyde (UF)

Reaction between the white crystalline solid, urea (NH_2CONH_2), and aqueous solutions of formaldehyde (formalin) produces urea resins that may be modified by adding other reagents. For complete polymerization and cross-linkage of this thermosetting resin, heat or catalysts and heat are usually needed during the molding operation (Figure F-6).

Urea-formaldehyde polymerization is shown in Figure F-7. The excess water molecule is a result of condensation polymerization.

Many products once produced with urea-formaldehyde plastics are now being made of thermoplastic materials. Faster production cycles and higher output are possible with thermoplastics. Some amino compounds are being processed in specially designed injection-molding equipment. This approach increases the output and makes these products able to compete with most thermoplastics.

Resins containing a urea-formaldehyde base are made into molding compounds that contain several ingredients, including resin, filler, pigment, catalyst, stabilizer, plasticizer, and lubricant. Such compounds speed up curing and production rates.

Molding compounds are produced by adding latent acid catalysts to the resin base. Such catalysts react at molding temperatures. Many urea-formaldehyde molding compounds have prepromoters or catalysts added, causing them to have a limited storage life. Therefore, they should be kept in a cool place. Stabilizers may be added to help control latent catalyst reaction. Lubricants are added to improve molding quality. Plasticizers improve flow properties and help reduce cure shrinkage.

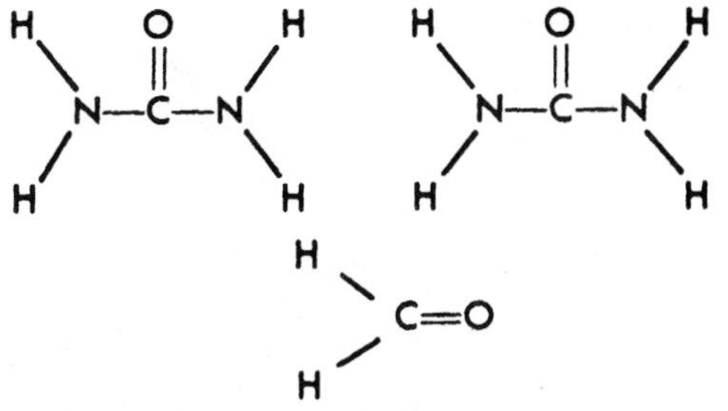

(A) Urea-formaldehyde resin.

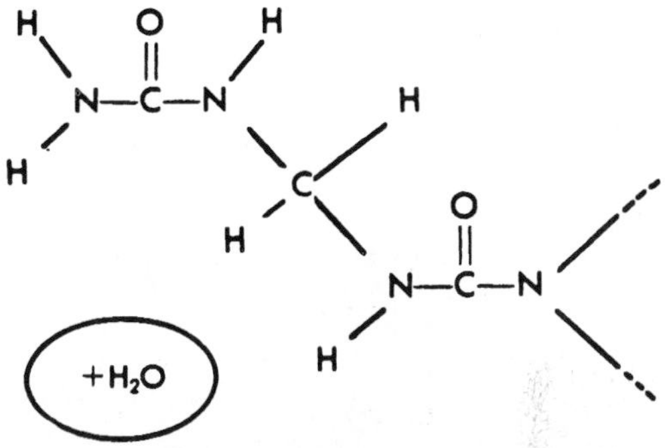

(B) Urea-formaldehyde plastics.

Figure F-7. Formation of urea-formaldehyde resins.

Although the base resin is water-clear, color pigments (transparent, translucent, or opaque) may be used. If color is not needed, fillers such as alpha-cellulose (bleached cellulose) fiber, macerated fabric, or wood flour may be added to improve molding and physical characteristics, as well as lower the cost.

Urea-formaldehyde plastics have many industrial uses because they are inexpensive and have outstanding molding qualities. They are used in electrical and electronic applications where good arc and tracking resistance is needed. They provide good dielectric properties and are unaffected by common organic solvents, greases, oils, weak acids, alkalies, or other hostile chemical environments. Urea compounds do not attract dust by static electricity charges. They will not burn or soften when exposed to an open flame and have very good dimensional stability when filled.

Urea-formaldehyde molding compounds are used to make bottle caps and electrical or thermal insulating materials.

Urea-formaldehyde products do not impart taste or odors to foods and beverages. They are preferred materials for appliance knobs, dials, handles, push buttons, toaster bases, and end plates. Wall plates, switch toggles, receptacles, fixtures, circuit breakers, and switch housings are only a few of the many electrical insulating uses (Figure F-8).

Currently, one of the largest uses of urea-formaldehyde resins is in adhesives for furniture, plywood, and chipboard. Chipboard is made by combining 10 percent resin binder with wood chips, which is then pressed into flat sheets. The product has no grain, so it is free to expand in all directions and does not

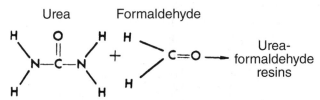

Figure F-6. Polymerization of urea and formaldehyde.

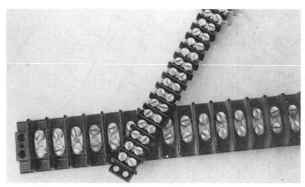

(A) Terminal blocks for electrical connections.

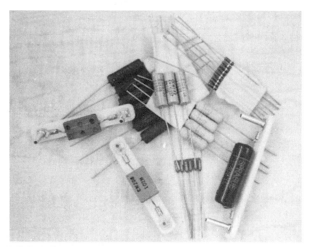

(B) Insulating blocks and packing for electronic components.
Figure F-8. Applications for urea-formaldehyde compounds.

warp. However, water resistance is poor. Plywood bonded with this resin is suitable only for interior use.

Urea-formaldehyde resins may be foamed and cured into a plastics state. Such foams are inexpensive, and various densities may be produced. They are easily produced by vigorously whipping a mixture of resin, catalyst, and a foaming detergent. Another foaming method involves introducing a chemical agent that generates a gas (usually carbon dioxide) while the resin is curing. These foams have been used as thermal insulating materials in buildings and refrigerators as well as low-density cores for structural sandwich construction. Undercured or partially polymerized foams have caused allergy and flu-like symptoms in some people. The degassing of molded furniture parts and wall insulation is a major source of this problem.

Urea-formaldehyde foams may be made flame resistant, but at the expense of foam density. Large amounts of water may be absorbed because these foams are open-celled (sponge-like) structures. This ability to absorb water is used by florists. Stems of cut flowers are inserted into water-soaked urea foam used as a base for flower arrangements. Ground foam has been used as

artificial snow on television and in theatrical productions. The open-celled structure may also be filled with kerosene and used as a lighting agent for fireplaces.

Urea resins find extensive use in the textile, paper, and coating industries. The existence of drip-dry fabrics is largely attributed to these resins.

Urea resins or powders are sometimes used as binders for foundry cores and shell molds.

The coating applications of urea-formaldehyde resins are limited. They can be applied only to substrates that can withstand curing temperatures of 40°C to 175°C (100°F to 350°F). These resins are combined with a compatible polyester (alkyd) resin to produce enamels. From 5 to 50 percent urea resin is added to the alkyd-based resin. These surface coatings are outstanding in hardness, toughness, gloss, color stability, and outdoor durability.

These surface coatings may be seen on refrigerators, washing machines, stoves, signs, venetian blinds, metal cabinets, and many machines. At one time, they were widely applied to automobile bodies. However, the curing or baking time required is not economical for mass production. The baking time depends on the temperature and proportion of amino resin to alkyd resin.

Urea resins modified with furfuryl alcohol have been successfully used in the manufacture of coated abrasive papers (see Furan, p. 472.)

Urea-formaldehyde plastics are easily molded in compression and transfer molding machines. They may also be processed by reciprocating-screw injection-molding machines. Depending on the grade of resin and the filler used, allowances should be made for shrinkage after removal from the mold. Improved dimensional stability can be acquired by post-conditioning the product in an oven.

Table F-3 lists some properties of alpha-cellulose-filled urea-formaldehyde plastics. Five advantages and four disadvantages of urea-formaldehyde follow:

Advantages of Urea-Formaldehyde

1. Good hardness and scratch resistance
2. Comparatively low cost
3. Wide color range
4. Self-extinguishing
5. Good solvent resistance

Disadvantages of Urea-Formaldehyde

1. Must be filled for successful molding
2. Poor long-term oxidation resistance
3. Attacked by strong acids and bases
4. Uncured, partially polymerized plastics and foams degas

Table F-3. Properties of Urea-Formaldehyde

Property	Urea-Formaldehyde (Alpha-Cellulose Filled)
Molding qualities	Excellent
Relative density	1.47–1.52
Tensile strength, MPa	38–90
(psi)	(5500–13,000)
Compressive strength, MPa	172–310
(psi)	(25,000–45,000)
Impact strength, Izod, J/mm	0.0125–0.02
(ft-lb/in.)	(0.25–0.40)
Hardness, Rockwell	M110–M120
Thermal expansion, $10^{-4}/°C$	5.6–9.1
Resistance to heat, °C	80
(°F)	(170)
Dielectric strength, V/mm	11,810–15,750
Dielectric constant (at 60 Hz)	7.0–9.5
Dissipation factor (at 60 Hz)	0.035–0.043
Arc resistance, s	80–150
Water absorption (24 h), %	0.4–0.8
Burning rate	Self-extinguishing
Effect of sunlight	Grays
Effect of acids	None to decomposes
Effect of alkalies	Slight to decomposes
Effect of solvents	None to slight
Machining qualities	Fair
Optical qualities	Transparent to opaque

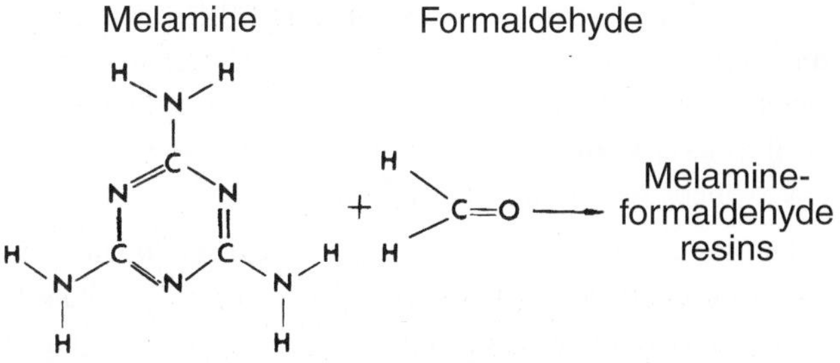

Figure F-9. Formation of melamine-formaldehyde resins.

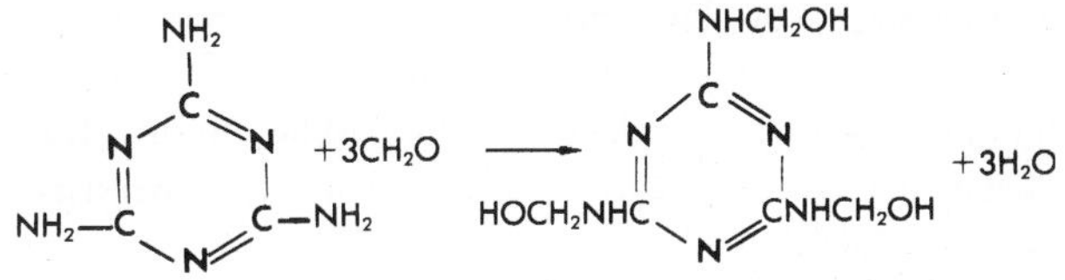

Figure F-10. Formation of trimethylol melamine.

The reaction of formaldehyde with compounds such as aniline, dicyandiamide, ethylene urea, and sulfonamide provide more complex resins for varied applications.

In their pure form, amino *resins* are colorless and soluble in warm solutions of water and methanol.

Amino resins, urea plastics, and melamine plastics are frequently grouped together as one entity. Their chemical structures, properties, and applications are very similar. However, melamine-formaldehyde products are better than the urea-formaldehyde plastics in several respects.

Melamine products are harder and more water resistant. They may be combined with a greater variety of fillers that allow the manufacture of products possessing better heat and scratch, stain, water, and chemical resistance. Melamine products are also much more expensive than urea products.

Probably the largest single use of melamine-formaldehyde is the manufacture of tableware. For this application, molding powders are normally filled with alpha-cellulose. Asbestos or other fillers are sometimes used in handles and utensil housings.

Melamine resin is widely used for surface coatings and decorative laminates. Paper-based laminates are sold under the trade names Formica and Micarta®. Photographic prints on fabrics or paper impregnated with melamine resin are placed on a base or core material and then cured in a large press. Kraft paper impregnated with phenolic resin is commonly used for the base material because it is durable and compatible with melamine resin. This base is also less expensive than multiple layers of melamine-impregnated paper. There is a broad spectrum of uses for these laminates, including surfacing wood, metal, plaster, and hardboard. A familiar use of these laminates is in surfacing for kitchen counters and table tops.

Melamine-Formaldehyde (MF)

Until about 1939, melamine-formaldehyde was an expensive laboratory curiosity. Melamine ($C_3H_6N_6$) is a white crystalline solid. Combining it with formaldehyde results in the formation of a compound referred to as methylol derivative (Figure F-9). With additional formaldehyde, the formulation will react to produce tri-, tetra-, penta-, and hexamethylolmelamine. The formation of trimethylol melamine is shown in Figure F-10.

Commercial melamine resins may be obtained without acid catalysts, but both thermal energy and catalysts are used to speed polymerization and cure. Polymerization of urea and melamine resins is a condensation reaction and produces water, which will evaporate or escape from the molding cavity.

Resins of formaldehyde-plus benzoguanamine ($C_3H_4N_5C_6H_5$) or thiourea ($CS(NH_2)_2$) are of minor commercial importance.

A 3 percent melamine resin solution may be added during the preparation of paper pulp to improve the wet strength of paper. Paper with this resin binder has a wet strength nearly as great as its dry strength. Crush and folding resistance is also greatly increased without adding brittleness.

Finishes resistant to water, chemicals, alkali, grease, and heat are formulated from amino resins. Heat or heat and catalysts are needed to cure the amino resins. These finishes are seen on stoves, washing machines, and other appliances.

Urea or melamine resins may be made compatible with other plastics to produce finishes of outstanding merit. Polyester (alkyd) resins or phenolic resins may be combined to produce finishes possessing the best features of both resins. These finishes are sometimes used when a hard, tough, mar-resistant finish is required.

Melamine resins are often used to make waterproof exterior plywood and marine plywood as well as other adhesive applications requiring a light-colored, nonstaining adhesive. Catalysts, heat, or high-frequency energy are used to cure melamine adhesives in plywood and panel assemblies.

Melamine-formaldehyde resins are commercially employed for textile finishes. The well-known drip-dry fabrics and fabrics with permanent glazing, rot proofing, and shrinkage control largely contribute their existence to such resins. Melamine and silicone resins are used to produce waterproof fabrics.

Melamine-formaldehyde compounds are easily molded in normal compression and transfer molding machines. Special reciprocating-screw injection machines are also used.

Table F-4 lists some of the properties of melamine-formaldehyde plastics, and five advantages and three disadvantages of melamine-formaldehyde follow:

Advantages of Melamine-Formaldehyde

1. Good hardness and scratch resistance
2. Comparatively low cost
3. Wide color range
4. Self-extinguishing
5. Good solvent resistance

Table F-4. Properties of Melamine-Formaldehyde

Property	No Filler	Alpha-Cellulose Filled	Glass-Fiber Filled
Molding qualities	Good	Excellent	Good
Relative density	1.48	1.47–1.52	1.8–2.0
Tensile strength, MPa		48–90	34–69
(psi)		(7000–13,000)	(5000–10,000)
Compressive strength, MPa	276–310	276–310	138–241
(psi)	(40,000–45,000)	(40,000–45,000)	(20,000–35,000)
Impact strength, Izod, J/mm	0.012–0.0175	0.03–0.9	
(ft-lb/in.)	(0.24–0.35)	(0.6–18.0)	
Hardness, Rockwell	M115–M125	M120	
Thermal expansion, $(10^{-4}/°C)$	10	3.8–4.3	
Resistance to heat, °C	99	99	150–205
(°F)	(210)	(210)	(300–400)
Dielectric strength, V/mm		10,630–11,810	6690–11,810
Dielectric constant (at 60 Hz)		6.2–7.6	9.7–11.1
Dissipation factor (at 60 Hz)		0.030–0.083	0.14–0.23
Arc resistance, s	100–145	110–140	180
Water absorption (24 h), %	0.3–0.5	0.1–0.6	0.09–0.21
Burning rate	Self-extinguishing	Nonburning	Self-extinguishing
Effect of sunlight	Color fades	Slight color change	Slight
Effect of acids	None to decomposes	None to decomposes	None to decomposes
Effect of alkalies		Attacked	None to slight
Effect of solvents	None	None	None
Machining qualities		Fair	Good
Optical qualities	Opalescent	Translucent	Opaque

Disadvantages of Melamine-Formaldehyde

1. Must be filled for successful molding
2. Poor long-term oxidation resistance
3. Subject to attack by strong acids and bases

CASEIN

Casein plastics are sometimes classified as natural polymers and referred to by many as *protein plastics*.

Casein is a protein found in a number of sources including animal hair, feathers, bones, and industrial wastes. There has been little interest in these sources, and currently only skimmed milk is of commercial interest in the production of casein.

The history of protein-derived plastics can probably be dated from the work of W. Krische, a German printer, and Adolf Spitteler of Bavaria around 1895. At that time, there was a demand in Germany for what may be described as a *white blackboard*. These boards were thought to possess better optical properties than those with a black surface. In 1897, Spitteler and Krische, while attempting to develop such a product, produced a casein plastics that could be hardened with formaldehyde. Until then, casein plastics had proved unsatisfactory because they were soluble in water. Galaith® (milkstone), Erinoid®, and Ameroid® are trade names of those early protein plastics.

Casein is not coagulated by heat, but must be precipitated from milk by using the action of rennin enzymes or acids. This powerful coagulant causes the milk to separate into solids (curds) and a liquid (whey). After the whey is removed, the curd containing the protein is washed, dried, and made into a powder. When kneaded with water, the doughlike material may be shaped or molded. Next, a simple drying operation then causes a great deal of shrinkage. Casein is thermoplastic while being molded. Molded products may be made water resistant by soaking them in a formalin solution that creates links holding the casein molecules together. The long linear chain of casein molecules is known as the *polypeptide chain*. There are a number of other peptide groups and possible side reactions. No single formula could represent the interaction between casein and formaldehyde. A very simplified reaction is shown in Figure F-11.

It is doubtful that casein will gain popularity, because it is costly to make and the raw material is valuable as food. Casein plastics are seriously affected by humid conditions and cannot be used as electrical insulators. The lengthy hardening process and poor resistance to decomposition by heat make them unsuitable for modern processing rates. There are only limited commercial uses of casein plastics in the United States because they offer no property advantages over synthetic polymers and production costs are high.

Casein plastics are used to a limited extent for buttons, buckles, knitting needles, umbrella handles, and other novelty items. They may be reinforced and filled or obtained in transparent colors. Casein has held some of its appeal because it may be colored to imitate onyx, ivory, and imitation horn. Casein is most widely used in stabilizing rubber latex emulsion, preparing medical compounds, food products, paints, adhesives, and sizing paper and textiles. Casein is also used in insecticides, soaps, pottery, inks, and as modifiers in other plastics.

Films and fibers may be produced from these plastics. The wool-like fibers are warm, soft, and have properties that compare favorably with natural wool. The films are generally of little use except as handy forms for the coating of paper and other materials. Casein glue is a well-known wood adhesive.

Table F-5 lists some of the properties of casein plastics. Two advantages and two disadvantages of casein follow:

Table F-5. Properties of Casein

Property	Casein Formaldehyde (Unfilled)
Molding qualities	Excellent
Relative density	1.33–1.35
Tensile strength, MPa	48–79
(psi)	(7000–10,000)
Compressive strength, MPa	186–344
(psi)	(27,000–50,000)
Impact strength, Izod, J/mm	0.045–0.06
(ft-lb/in.)	(0.9–1.20)
Hardness, Rockwell	M26–M30
Resistance to heat, °C	135–175
(°F)	(275–350)
Dielectric strength, V/mm	15,500–27,500
Dielectric constant (at 60 Hz)	6.1–6.8
Dissipation factor (at 60 Hz)	0.052
Arc resistance, s	Poor
Water absorption (24 h), %	7–14
Burning rate	Slow
Effect of sunlight	Yellows
Effect of acids	Decomposes
Effect of alkalies	Decomposes
Effect of solvents	Slight
Machining qualities	Good
Optical qualities	Translucent to opaque

Figure F-11. Ethyl cellulose (fully ethylated).

Advantages of Casein

1. Produced from nonpetrochemical sources
2. Excellent molding qualities and colorability

Disadvantages of Casein

1. Poor resistance to acids and alkalies, yellows in sunlight
2. High water absorption

EPOXY (EP)

Hundreds of patents have been granted for the commercial uses of epoxide resins. One of the first descriptions of polyepoxides is a German patent by I. G. Farbenindustrie in 1939.

In 1943, the Ciba Company developed an epoxide resin commercially significant in the United States. By 1948, a variety of commercial coating and adhesive applications were discovered.

Epoxy resins are thermosetting plastics. There are several thermoplastic epoxy resins used for coatings and adhesives. Many different epoxy resin structures available today are derived from bisphenol acetate and epichlorohydrin.

Bisphenol A (bisphenol acetate) is made by the condensation of acetone with phenol (Figure F-12). Epichlorohydrin-based epoxies are widely used due to availability and low cost. The epichlorohydrin structure is obtained by chlorination of propylene:

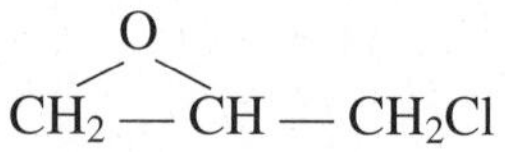

$$CH_2 — CH — CH_2Cl$$

It will become evident that the epoxy group, for which the plastics family is named, has a triangular structure:

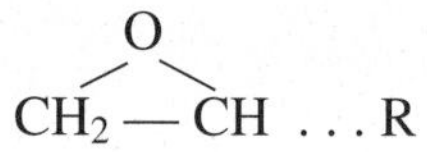

$$CH_2 — CH \ldots R$$

Epoxy structures are usually terminated by this epoxide structure, but many other molecular structures may terminate the long molecular chain. A linear epoxy polymer may be formed when bisphenol A and epichlorohydrin react (Figure F-13). In some literature, these polymers are called polyethers.

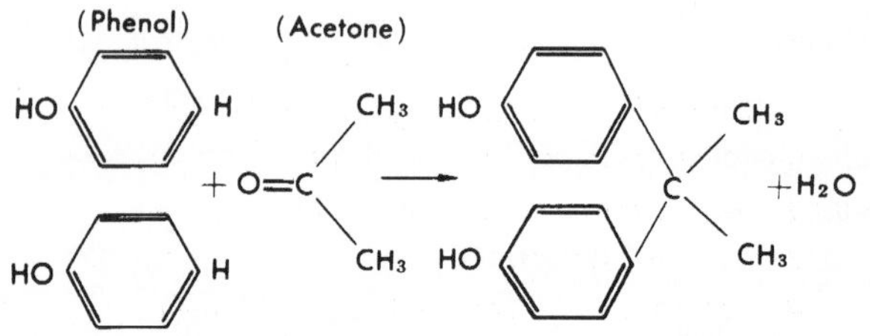

Figure F-12. Production of bisphenol-A.

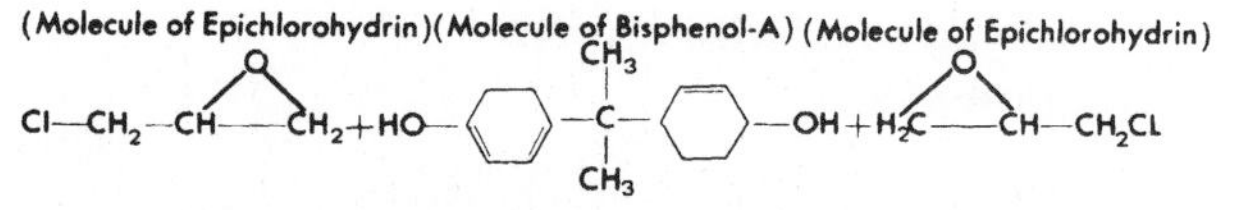

Figure F-13. Formation of linear epoxy polymer.

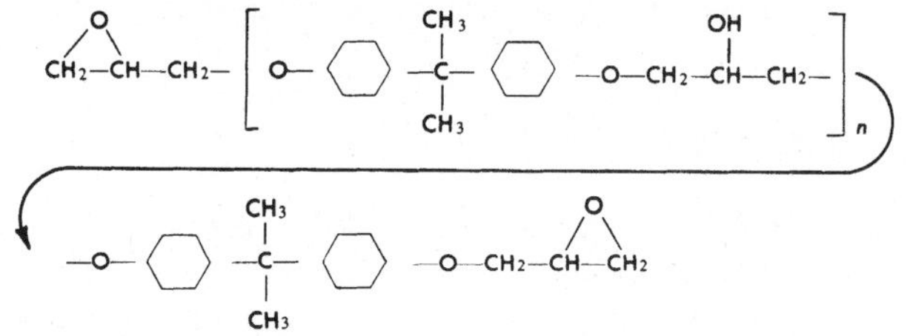

Figure F-14. Bisphenol A-based epoxy resin.

A typical structural formula of epoxy resin based on bisphenol A may be represented as shown in Figure F-14. Other intermediate epoxy-based resins are possible but are too numerous to mention.

As a rule, epoxy resins are cured by adding catalysts or reactive hardeners. Members of the aliphatic and aromatic amine family are commonly used hardening agents. Various acid anhydrides are also used to polymerize the epoxide chain.

Epoxy resins will polymerize and cross-link as thermal energy is added. Catalysts and heat are often used to reach a desired degree of polymerization.

Single-component epoxy resins may contain latent catalysts. These react when enough heat is applied. All epoxy resins share a practical shelf-life expectancy.

Reinforced epoxy resins are very strong. They have good dimensional stability and service temperatures as high as 315°C (600°F). Pre-impregnated reinforcing materials are used to produce products by using hand-layup, vacuum-bag, or filament-winding processes (Figure F-15). Epoxy has very good chemical and fatigue resistance; thus epoxy resins replace the less costly unsaturated polyester resins in many applications. A savings of one-third of the resin mass is realized when epoxy resins are used in place of polyesters.

Figure F-15. Continuous fibrous glass in a matrix of epoxy is used in this composite filament-wound rocket motor case.

Figure F-16. End closures and ribs of this boron slat for aircraft are bonded to the inner skin with epoxy adhesives. High strength makes epoxy resins very useful.

Figure F-17A. This complex carbon fiber composite is a test design for an inlet duct for the F–35 Joint Strike Fighter.

Epoxy-glass laminates are widely used because they have a high strength-to-mass ratio. Superior adhesion to all materials and wide compatibility make epoxy resins desirable (Figure F-16). Laminated circuit boards, radomes, aircraft parts, and filament-wound pipes, tanks, and containers are only a few examples of products made with this plastics material (Figure F-17).

Filled epoxy resins are commonly used for special castings. These strong compounds may be used for low-cost tooling. Epoxies are replacing other tooling materials for dies, jigs, fixtures, and molds for short production runs. Faithful reproduction of details is obtained when epoxy compounds are cast against prototypes or patterns.

A variety of fillers are used in caulking and patching compounds containing epoxy resins. The adhesive qualities and low shrinkage of epoxies during cure make them durable in caulking or patching applications.

Electrical potting is another use for casting. Epoxies are outstanding in protecting electronic parts from moisture, heat, and corrosive chemicals. Electric motor parts, high-voltage transformers, relays, coils, and many other components may be protected from severe environments by being potted in epoxy resins.

Molding compounds of epoxy resins and fibrous reinforcements can be molded by using injection, compression, or transfer processes. They are molded into small electrical items and appliance parts and have many modular uses.

Versatility is achieved by controlling the resin manufacture, the curing agents, and the rate of cure. These resins may be formulated to give results ranging from soft, flexible compounds to hard, chemical-resistant products. By incorporating a blowing agent, low-density epoxy foams may be produced. The qualities offered by the epoxy plastics are adhesion, chemical resistance, toughness, and excellent electrical characteristics.

When the epoxy resins were first introduced in the 1950s, they were recognized as outstanding coating materials. They are more expensive than other coating materials, but their adhesion qualities and chemical inertness make them competitive. Five advantages and three disadvantages of epoxy resins are listed here:

Figure F-17B. This photo shows the EGADS, a robotic drilling system for composites.

Advantages of Epoxy

1. Wide range of cure conditions, from room temperature to 178°C (350°F)
2. No volatiles formed during cure
3. Excellent adhesion
4. Can be cross-linked with other materials
5. Suitable for all thermosetting processing methods

Disadvantages of Epoxy

1. Poor oxidative stability; some moisture sensitivity
2. Thermal stability limited to 178°C to 232°C (350°F to 450°F)
3. Many grades are expensive

Epoxy-based finishes are used on driveways, concrete floors, porches, metal appliances, and wooden furniture. Epoxy finishes on home appliances are a major application of this durable, abrasion-resistant finish. Epoxy coatings have replaced glass enamel finishes for tank car and other container linings that need to resist chemicals. Ship hulls and bulkheads may be coated with epoxy. More durable finishes mean fewer repairs and reduced surface tension between the ship and water. These factors reduce maintenance and fuel costs.

The flexibility of many epoxy coatings makes them popular for postforming coated metal parts. For example, sheets of metal are coated while flat. They are then formed or bent into shallow pans with no damage to the coating.

The ability of epoxy adhesives to bond to dissimilar materials has allowed them to replace soldering, welding, riveting, and other joining methods. The aircraft and automotive industries use these adhesives where heat or other bonding methods might distort the surface. Honeycomb or panel structures also use the superb adhesive and thermal properties of epoxy.

Table F-6 lists some of the properties of various epoxies. Epoxy copolymers are made by cross-linking with phenolics, melamines, polyamide, urea, polyester, and some elastomers.

FURAN

Furan resins are derived from furfurylaldehyde and furfuryl alcohol (Figure F-18). Acid catalysts are used in polymerization. Therefore, it is vital to apply a protective coating to substrates attacked by acids.

Furan plastics have excellent chemical resistance and can withstand temperatures as high as 130°C (265°F). They are

Table F-6. Properties of Epoxies

Property	Epoxy Molding Compounds		
	Glass-Filled	Mineral-Filled	Microballoon-Filled
Molding qualities	Excellent	Excellent	Good
Relative density	1.6–2.0	1.6–2.0	0.75–1.00
Tensile strength, MPa	69–207	34–103	17–28
(psi)	(10,000–30,000)	(5000–15,000)	(2500–4000)
Compressive strength, MPa	172–276	124–276	69–103
(psi)	(25,000–40,000)	(18,000–40,000)	(10,000–15,000)
Impact strength, Izod, J/mm	0.5–1.5	0.015–0.02	0.008–0.013
(ft-lb/in.)	(10–30)	(0.3–0.4)	(0.15–0.25)
Hardness, Rockwell	M100–M110	M100–M110	
Thermal expansion, (10⁻⁴/°C)	2.8–8.9	5.1–12.7	
Resistance to heat, °C	150–260	150–260	
(°F)	(300–500)	(300–500)	
Dielectric strength, V/mm	11,810–15,750	11,810–15,750	14,960–16,535
Dielectric constant (at 60 Hz)	3.5–5	3.5–5	
Dissipation factor (at 60 Hz)	0.01	0.01	
Arc resistance, s	120–180	150–190	120–150
Water absorption (24 h), %	0.05–0.20	0.04	0.10–0.20
Burning rate	Self-extinguishing	Self-extinguishing	Self-extinguishing
Effect of sunlight	Slight	Slight	Slight
Effect of acids	Negligible	None	Slight
Effect of alkalies	None	Slight	Slight
Effect of solvents	None	None	Slight
Machining qualities	Good	Fair	Good
Optical qualities	Opaque	Opaque	Opaque

Furfurylaldehyde   Furfuryl alcohol   Furon

Figure F-18. Acid catalysts will cause condensation of furfurylaldehyde or furfuryl alcohol. Cross-linking occurs between furan rings.

used primarily as additives, binders, or adhesives. Furfurylalde-hyde has been co-reacted with phenolic plastics. Resins based on furfuryl alcohol are used with amino resins to improve wetting. Their wetting and adhesive capabilities make them ideal for impregnating agents. Reinforced and laminated products produced from furan plastics include tanks, pipes, ducting, and construction panels. Furan resins are also used as sand binders in foundries (Table F-7). Two advantages and two disadvantages of furans follow.

Advantages of Furans

1. Produced from non-petrochemical sources
2. Excellent chemical resistance

Disadvantages of Furans

1. Hard to process, limited to fiber-reinforced plastics
2. Subject to attack by halogens

Table F-7. **Properties of Furan**

Property	Furan (Asbestos-Filled)
Molding qualities	Good
Relative density	1.75
Tensile strength, MPa	20–31
(psi)	(2900–4500)
Compressive strength, MPa	68–72
(psi)	(9900–10,450)
Hardness, Rockwell	R110
Resistance to heat, °C	130
(°F)	(266)
Water absorption (24 h), %	0.01–2.0
Burning rate	Slow
Effect of sunlight	None
Effect of acids	Attacked
Effect of alkalies	Little
Effect of solvents	Resistant
Machining qualities	Fair
Optical qualities	Opaque

PHENOLICS (PFs)

Phenolics (phenol-aldehyde) were among the first true synthetic resins produced. They are known chemically as phenol-formaldehyde (PF). Their history reaches back to the work of Adolph Baeyer in 1872. In 1909, the chemist Baekeland invented and patented a technique for combining phenol (C_6H_5OH, also called carbolic acid) and gaseous formaldehyde (H_2CO).

The success of the phenol-formaldehyde resins later encouraged the research of urea- and melamine-formaldehyde resins.

The resin formed from the reaction of phenol with formaldehyde (an aldehyde) is known as a *phenolic*. Figure F-19 shows the reaction of a phenol with formaldehyde. This involves a condensation reaction in which water is formed as a byproduct (A-stage). The first phenol formaldehyde reaction produces a low-molecular-mass resin that is compounded with fillers and other ingredients (B stage). During the molding process, the resin is transformed into a highly cross-linked thermosetting plastics product by heat and pressure (C stage).

Although the monomer solution of phenol is commercially used, cresols, xylenols, resorcinols, or synthetically produced oil-soluble phenols may also be used. Furfural may replace the formaldehyde.

In *one-stage resins*, a *resol* is produced by reacting a phenol with an excess amount of aldehyde in the presence of a catalyst (not acid). Sodium and ammonium hydroxide are common examples of catalysts. This product is soluble and possesses a low molecular mass. It will form large molecules without addition of a hardening agent during the molding cycle.

Two-stage resins are produced when phenol is present in excess with an acid catalyst. The result is a low-molecular mass and soluble *novolac* resin. It will remain a linear thermoplastic

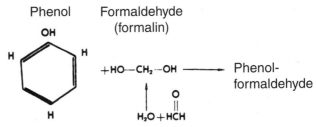

Figure F-19. Reaction of phenol and formaldehyde.

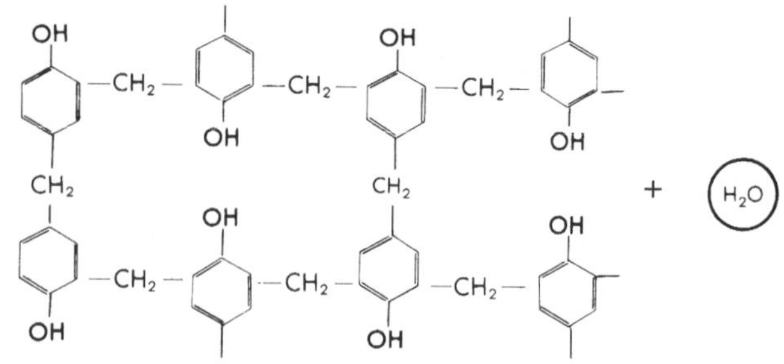

Figure F-20. Final curing or heat-hardening should be considered a further condensation process.

resin unless compounds capable of forming cross-linkage on heating are added. They are called *two-stage resins* because an agent must be added before molding (Figure F-20).

The *A-Stage* novalac resin is a fusible and soluble thermoplastic. The *B-stage* resin is produced by thermally blending the A-stage and hexamethylenetetramine. The B-stage is usually sold in a granular or powder form. Fillers, pigments, lubricants, and other additives are compounded with the resin during this stage. During molding, heat and pressure convert the B-stage resin into an insoluble, infusible *C-stage* thermosetting plastics.

Phenolics are not used as frequently as they once were because so many new plastics have been developed. However, their low cost, mold-ability, and physical properties make phenolics leaders in the thermoset field. These materials are widely used as molding powders, resin binders, coatings, and adhesives.

Molding powders or compounds of novolac resins are rarely used without a filler. The filler is not used simply to reduce cost. It improves physical properties, increases adaptability for processing, and reduces shrinkage. Curing time, shrinkage, and molding pressures may be reduced by preheating phenol-formaldehyde compounds. Advances in equipment and techniques have kept phenolics competitive with many thermoplastics and metals. Phenolics are used in conventional transfer and compression molding operations and in injection and reciprocating-screw machines. Molded phenolic parts are abrasive and hard to machine. Although molded phenolics are widely used as electrical insulation, they exhibit poor tracking resistances and under very humid conditions. In a few uses, they have been replaced by thermoplastics.

Phenolic resins have a major appearance drawback—they are too dark in color for use as surface layers on decorative laminates and as adhesives where glue joints may show. Phenloic-resin-impregnated cotton fabric, wood, or paper is often used in the production of gear wheels, bearings, substrates for electrical circuit boards, and melamine decorative laminates. These laminates are usually made in large presses under controlled heat and pressure. Many methods of impregnation are used, including dipping, coating, and spreading.

Phenol formaldehyde resins may be cast into many profile shapes such as billiard balls, cutlery handles, and novelty items.

Phenolic-based resins are available in liquid, powder, flake, and film forms. The ability of these resins to impregnate and bond with wood and other materials is the reason for their success as adhesives. They improve adhesion and heat resistance and are widely used in the production of plywood and as binder adhesives in wood particle moldings. Wood particleboards are used in many building applications such as sheathing, subflooring, and core stocks.

Phenolic resins are used as binders for abrasive grinding wheels. The abrasive grit and resin are simply molded into the desired shape and cured. Resin binders are an important ingredient in the shell molds and cores used in foundries (Figure F-21). These molds and cores produce very smooth metal castings. As heat-resistant binders, phenolic resins are used in making brake linings and clutch facings.

Due to their high resistance to water, alkalies, chemicals, heat, and abrasion, phenolics are sometimes used in finishes. They are also used for coating appliances, machinery, or other devices requiring maximum heat resistance.

A high-strength, heat- and fire-resistant foam may be produced using phenolic resins. The foam may be produced in the plant or on site by rapidly mixing a blowing agent and catalyst with the resin. As the chemical reaction generates heat and begins the polymerization process, the blowing agent vaporizes. This causes the resin to expand into a multicellular, semipermeable structure. These foams may be used as fill for honeycomb structures in aircraft, flotation materials, acoustic and thermal insulation, and as packing materials for fragile objects.

Micro-balloons (small hollow spheres) may be produced from phenolic plastics filled with nitrogen. These spheres vary

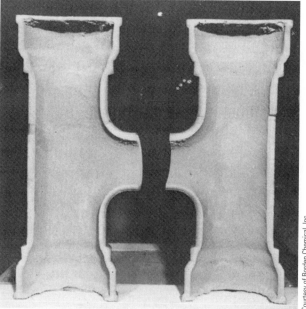

Figure F-21. Phenolic resin was used as a binder for this sand core.

from 0.005 to 0.08 mm (0.0002 to 0.0032 in.) in diameter. They may be mixed with other resins to produce syntactic foams. These foams are used as insulative fillers. They act as vapor barriers when placed on volatile liquids such as petroleum.

Table F-8 lists properties of phenolic materials, and eight advantages and four disadvantages of phenolic resins follow:

Advantages of Phenolics

1. Comparatively low cost
2. Suitable for use at temperatures to 205°C (400°F)
3. Excellent solvent resistance
4. Rigid
5. Good compressive strength
6. High resistivity
7. Self-extinguishing
8. Very good electrical characteristics

Disadvantages of Phenolics

1. Need fillers for moldings
2. Poor resistance to bases and oxidizers
3. Volatiles released during cure (a condensation polymer)
4. Dark color (due to oxidation discoloration)

Phenol-Aralkyl

In 1976, the Ciba-Geigy Corporation introduced a group of resins based on aralkyl ethers and phenols. Two basic prepolymer grades are available. Both are sold as 100 percent prepolymer resin. One grade cures through a condensation reaction. The other undergoes an additional reaction similar to epoxy. Condensation grades are blended with phenolic novolac resins to improve phenolic properties. Additional polymerization

Table F-8. Properties of Phenolics

Property	Phenol-Formaldehyde (Unfilled)	Phenol-Formaldehyde (Macerated Fabric)	Phenolic Casting Resin (Unfilled)
Molding qualities	Fair	Fair to good	
Relative density	1.25–1.30	1.36–1.43	1.236–1.320
Tensile strength, MPa	48–55	21–62	34–62
(psi)	(7000–8000)	(3000–9000)	(5000–9000)
Compressive strength, MPa	69–207	103–207	83–103
(psi)	(10,000–30,000)	(15,000–30,000)	(12,000–15,000)
Impact strength, Izod, J/mm	0.01–0.018	0.038–0.4	0.012–0.02
(ft-lb/in.)	(0.20–0.36)	(0.75–8)	(0.24–0.40)
Hardness, Rockwell	M124–M128	E79–E82	M93–M120
Thermal expansion, (10^{-4}/°C)	6.4–15.2	2.5–10	17.3
Resistance to heat, °C	120	105–120	70
(°F)	(250)	(220–250)	(160)
Dielectric strength, V/mm	11,810–15,750	7875–15,750	9845–15,750
Dielectric constant (at 60 Hz)	5–6.5	5.2–21	6.5–17.5
Dissipation factor (at 60 Hz)	0.06–0.10	0.08–0.64	0.10–0.15
Arc resistance, s	Tracks	Tracks	
Water absorption (24 h), %	0.1–0.2	0.40–0.75	0.2–0.4
Burning rate	Very slow	Very slow	Very slow
Effect of sunlight	Darkens	Darkens	Darkens
Effect of acids	Decomposed by oxidizing acids	Decomposed by oxidizing acids	None
Effect of alkalies	Decomposes	Attacked	Attacked
Effect of solvents	Resistant	Resistant	Resistant
Machining qualities	Fair to good	Good	Excellent
Optical qualities	Transparent to translucent	Opaque	Transparent to opaque

Table F-9. Properties of Phenol-Aralkyl

Property	Phenol-Aralkyl (Glass Filled)
Molding qualities	Good
Relative density	1.70–1.80
Tensile strength, MPa	48–62
(psi)	(6900–9000)
Compressive strength, MPa	206–241
(psi)	(3000–3500)
Impact strength, Izod, J/mm	0.02–0.03
(ft-lb/in.)	(0.4–0.6)
Hardness, Rockwell	
Resistance to heat, °C	250
(°F)	(480)
Dielectric strength, V/mm	
Dielectric constant (at 1 MHz)	2.5–4.0
Dissipation factor (at 1 MHz)	0.02–0.03
Water absorption (24 h), %	0.05
Effect of acids	None to slight
Effect of alkalies	Attacked
Effect of solvents	Resistant
Machining qualities	Fair
Optical qualities	Opaque

grades are used in the fabrication of laminates. These resins are also used as binders for the production of cutting wheels, printed circuit boards, bearings, appliance parts, and engine components. Because of their excellent mechanical properties, processing advantages, and thermal capabilities, these prepolymers will find other uses as well. Table F-9 lists phenol-aralkyl properties.

UNSATURATED POLYESTERS

The term *polyester resin* encompasses a variety of materials. It is often confused with other polyester classifications. A polyester is formed by reacting a polybasic acid and a polyhydric alcohol. Changes with acids, acids and bases, and some unsaturated reactants permit cross-linking, which forms thermosetting plastics.

The term polyester resin should refer to unsaturated resins based on dibasic acids and dihydric alcohols. These resins are capable of cross-linking with unsaturated monomers (often styrene). Alkyds and polyurethanes of the polyester resin group are discussed individually.

Sometimes the term *fiberglass* has been used to refer to unsaturated polyester plastics. However, this term should only refer to fibrous pieces of glass. Various resins may be used with glass fiber acting as a reinforcing agent. The main use for unsaturated polyester resin is in the production of reinforced plastics. Glass fiber is the most used reinforcement.

Credit for the first preparation of polyester resins (alkyd type) is usually attributed to the Swedish chemist Jons Jacob Berzelius in 1847 and to GayLussac and Pelouze in 1833. Further development was conducted by W. H. Carothers and R. H. Kienle. Through the 1930s, most of the work on polyesters was aimed at developing and improving paint and varnish applications. In 1937, further interest in the resin was stimulated by Carleton Ellis. He found that by adding unsaturated monomers to unsaturated polyesters, cross-linking and polymerization time was greatly reduced. Ellis has been called the father of unsaturated polyesters.

Large-scale industrial use of unsaturated polyesters developed quickly as wartime shortages spurred development of many types of resin use. Reinforced polyester structures and parts were widely used during World War II.

The word *polyester* is derived from two chemical processing terms, *poly*merization and *ester*ification. In esterification, an organic acid is combined with an alcohol to form an ester and water. A simple esterification reaction is shown in Figure F-22 (see Alkyds, above).

The reverse of the esterification reaction is called *saponification*. In order to obtain a good yield of ester in a condensation reaction, water must be removed to prevent saponification (Figure F-23). If a polybasic acid (such as maleic acid) is caused to react and the water is removed as it is formed, the result will be an *unsaturated polyester*. *Unsaturated* means that the double-bonded carbon atoms are reactive or possess unused valance bonds. These can be attached to another atom or molecule, making such a polyester capable of cross-linkage. There are many other reactive or unsaturated monomers that can be used to change or tailor the resin in order to meet a special purpose or use. Vinyl toluene, chlorostyrene, methyl methacrylate, and diallyl phthalate are commonly used monomers. Unsaturated

Figure F-22. Examples of esterification reaction.

Figure F-23. To prevent saponification, water should be removed in an esterification reaction.

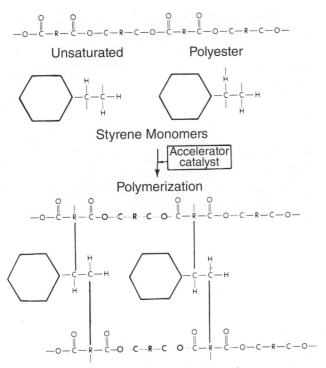

Figure F-24. Polymerization reaction with unsaturated polyester and styrene monomers.

styrene is an ideal, low-cost monomer most often used with polyesters (Figure F-24).

The four main functions of a monomer follow:

1. To act as a solvent carrier for the unsaturated polyester
2. To lower viscosity (thin)
3. To enhance selected properties for specific uses
4. To provide a rapid means of reacting (cross-linking) with the unsaturated linkages in the polyester

As the molecules randomly collide and occasional bonds are completed, a very slow polymerization (cross-linking) process will occur. This process may take days or weeks in simple mixtures of polyesters and monomers.

To speed up polymerization at room temperature, accelerators (promoters) and catalysts (initiators) are added. The accelerators commonly used are cobalt naphthenate, diethyl aniline, and dimethyl aniline. Unless otherwise specified, polyester resins will usually have accelerator added by the manufacturer. Resins that contain an accelerator require only a catalyst to provide rapid polymerization at room temperatures. With the addition of an accelerator, the shelf life of the resin is appreciably shortened. Inhibitors such as hydroquinone may be added to stabilize or retard premature polymerization. These additives do not greatly interfere with the final polymerization. The speed of cure can be influenced by temperature, light, and the amount of additives.

Polyester resins may be formulated without accelerators. All resins should be kept in a cool, dark storage area until used.

Methyl ethyl ketone peroxide, benzoyl peroxide, and cumene hydroperoxide are three common organic peroxides used to catalyze polyester resins. These catalysts break down when they come in contact with accelerators in the resin-releasing free radicals. The free radicals are attracted to the reactive unsaturated molecules, thus beginning the polymerization reaction.

By the strictest definition, the term *catalyst* is incorrectly used when referring to the polymerization mechanism of polyester resins. By the strict definition, a catalyst is a substance that aids a chemical action by its mere presence without being permanently changed itself. However, in polyester resins, the catalyst breaks down and becomes a part of the polymer structure. Because these materials are consumed in initiating the polymerization, the term *initiator* would be more accurate. A true catalyst is recoverable at the end of a chemical process.

Exposure to radiation, ultraviolet light, and heat have also been used to initiate the polymerization of double-bonded molecules. If catalysts are used, the resin mix correspondingly becomes more sensitive to heat and light. On a hot day or in the sunlight, less catalyst is required for polymerization. On a cold day, more catalyst would be needed. The resin and catalyst also could be warmed in order to produce a rapid cure.

The final curing reaction is called *addition polymerization* because no byproducts are present as a result of the reaction. In phenol-formaldehyde reactions, the curing reaction is referred to as *condensation polymerization* because the byproduct water is present (Figure F-25; see Polyallyl esters and Allylics, p. 461.)

Polyester may be specially modified for a wide variety of uses by altering the chemical structure or using additives. More cross-linkage is possible with higher percentages of unsaturated acid. A stiffer, harder product results. The addition of saturated acids will increase toughness and flexibility. Thixotropic fillers, pigments, and lubricants may also be added to the resin.

Polyester resins that contain no wax are susceptible to *air inhibition*. When exposed directly to air, such resins remain undercured, soft, and tacky for some time after setting. This is desirable when multiple layers are to be built up. Resins purchased from the manufacturer without wax are referred to as *air-inhibited* resins. The absence of wax permits better bonds between multiple layers in hand-layup operations.

In some cases, a tack-free cure of air-exposed surfaces is desired. For one-step castings, moldings, or surface coats, such a cure is obtained using a *non-air-inhibited* polyester. Non-air-inhibited resins contain a wax that floats to the surface during the curing operation, blocking out the air and allowing

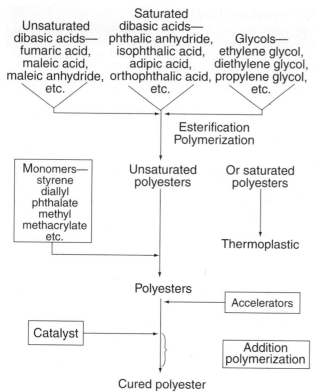

Unsaturated dibasic acids— fumaric acid, maleic acid, maleic anhydride, etc.

Saturated dibasic acids— phthalic anhydride, isophthalic acid, adipic acid, orthophthalic acid, etc.

Glycols— ethylene glycol, diethylene glycol, propylene glycol, etc.

Esterification Polymerization

Monomers— styrene diallyl phthalate methyl methacrylate etc.

Unsaturated polyesters

Or saturated polyesters

Thermoplastic

Polyesters

Accelerators

Catalyst

Addition polymerization

Cured polyester

Figure F-25. Production scheme of cured polyester.

Figure F-26. Classic composite 1953 Chevrolet Corvette was the first composite production body. It was composed of glass fibers in a matrix of polyester.

Figure F-27. This large composite container is highly corrosion resistant.

the surface to cure tack-free. Many waxes may be used in such resins including household paraffin wax, carnauba wax, beeswax, stearic acid, and others. The use of waxes adversely affects adhesion. Therefore, if more layers are to be added, all wax must be removed from the surface by sanding.

By altering the basic combination of raw materials, fillers, reinforcements, cure time, and treatment technique, a wide range of properties is possible.

Polyester is mainly used in making composite products. The primary value of reinforcements is to gain high strength-to-mass ratio. Glass fiber is the most common reinforcing agent. Asbestos, sisal, many plastic fibers, and whisker filaments are also used. The type of reinforcement selected depends on the end use and method of fabrication. Reinforced polyesters are among the strongest materials known. They have been used in automobile bodies (Figure F-26) and boat hulls, and due to their high strength-to-mass ratio, they are used in aircraft and aerospace as well.

Other applications include radar domes, ducts, storage tanks, sports equipment, trays, furniture, luggage, sinks, and various types of vessels (Figure F-27).

Unreinforced casting grades of polyesters are used for embedding, potting, casting, and sealing. Resins filled with wood flour may be cast in silicone molds to produce precise copies of wood carvings and trim. Special resins may be emulsified with water, which further reduces costs. These resins, referred to as

water-extended resins, may contain up to 70 percent water. The castings undergo some shrinkage due to water loss.

Fabrication methods for reinforced polyesters include hand layup, spray-up, matched molding, premix molding, pressure-bag molding, vacuum-bag molding, casting, and continuous laminating. In addition, other molding modifications are sometimes used. Compression molding equipment is sometimes used with doughlike premixes containing all ingredients. Table F-10 lists some of the properties of many polyesters. The following list contains six advantages and two disadvantages of polyesters:

Advantages of Unsaturated Polyesters

1. Wide curing latitude
2. May be used for medical devices (artificial limbs)
3. Accept high filler levels
4. Thermosetting materials
5. Inexpensive tooling
6. Nonburning halogenated grades available

Raymond Boyd/Michael Ochs Archives/Getty Images.

Courtesy of Copmpositesa USA, Inc

Table F-10. Properties of Thermosetting Polyesters

Property	Thermosetting Polyester (Cast)	Thermosetting Polyester (Glass Cloth)
Molding qualities	Excellent	Excellent
Relative density	1.10–1.46	1.50–2.10
Tensile strength, MPa	41–90	207–345
(psi)	(6000–13,000)	(30,000–50,000)
Compressive strength, MPa	90–252	172–345
(psi)	(13,000–36,500)	(25,000–50,000)
Impact strength, Izod, J/mm	0.01–0.02	0.25–1.5
(ft-lb/in.)	(0.2–0.4)	(5.0–30.0)
Hardness, Rockwell	M70–M115	M80–M120
Thermal expansion, $(10^{-4}/°C)$	14–25.4	3.8–7.6
Resistance to heat, °C	120	150–180
(°F)	(250)	(300–350)
Dielectric strength, V/mm	14,960–19,685	13,780–19,690
Dielectric constant (at 60 Hz)	3.0–4.36	4.1–5.5
Dissipation factor (at 60 Hz)	0.003–0.028	0.01–0.04
Arc resistance, s	125	60–120
Water absorption (24 h), %	0.15–0.60	0.05–0.50
Burning rate	Burns to self-extinguishing	Burns to self-extinguishing
Effect of sunlight	Yellows slightly	Slight
Effect of acids	Attacked by oxidizing acids	Attacked by oxidizing acids
Effect of alkalies	Attacked	Attacked
Effect of solvents	Attacked by some	Attacked by some
Machining qualities	Good	Good
Optical qualities	Transparent to opaque	Transparent to opaque

Disadvantages of Unsaturated Polyesters

1. Upper service temperature limited to 93°C (200°F)
2. Poor resistance to solvents

A number of thermoset interpenetrating polymer networks (IPN) that involve a cross-linked unsaturated polyester, vinyl ester, or polyester-urethane copolymer in a urethane network. You may recall that an IPN is a configuration of two or more polymers, each existing in a network. There is a synergistic effect when one of the polymers is synthesized in the presence of the other. Urethane/polyester network resins have good wet-out and may be used in pultrusion, filament winding, RIM, RTM, spray-up, and other composite reinforcing methods. Thermoplastic IPNs do not form chemical crosslinks like thermoset IPNs. There is a physical entanglement that interferes with polymer mobility. Thermoplastic IPN has been produced using PA, PBT, POM, and PP with silicone as the IPN.

THERMOSETTING POLYIMIDE

Polyimides may exist as either thermoplastic or thermosetting materials. Additional polyimides are available as thermosets. During processing, condensation polyimides thermally decompose before they reach their melting point.

Thermosetting polyimides are molded by injection, transfer, extrusion, and compression methods.

Thermosetting polyimides are used in aircraft engine parts, automobile wheels, electrical dielectrics, and coatings.

Table F-11 lists properties of thermosetting polyimides.

Both thermosetting and thermoplastic polyimides are considered high-temperature polymers.

In recent years, some improvements in resin systems have resulted in a high-temperature polymer that is less brittle and more easily processed. Some systems began with a

Table F-11. Properties of Thermosetting Polyimide and Bismalemide

Property	Thermosetting Polyimide (Unfilled)	Bismaleimide 1:1	Bismaleimide 10:0.87
Molding qualities	Good	Good	Good
Relative density	1.43	—	—
Tensile strength, MPa	86	81	92
(psi)	(12,500)	(11,900)	(13,600)
Compressive strength, MPa	275	201	207
(psi)	(39,900)	(29,900)	(30,500)
Impact strength, Izod, J/mm	0.075	—	—
(ft-lb/in.)	(1.5)	—	—
Hardness, Rockwell	E50		
Thermal expansion, (10^{-4}/°C)	13.71		
Resistance to heat, °C	350	272	285
(°F)	(660)	(523)	(545)
Dielectric strength, V/mm	22,050		
Dielectric constant (at 60 Hz)	3.6		
Dissipating factor (at 60 Hz)	0.0018		
Water absorption (24 h), %	0.24		
Effect of acids	Slowly attacked		
Effect of alkalies	Attacked		
Effect of solvents	Very resistant		
Machining qualities	Good		
Optical qualities	Opaque		

thermoplastic polymer based on an aromatic tetracarboxylic dianhydride and an aromatic diamine. The result is a completely imidized powder. Unlike addition polyimides, this product can be processed from many common organic solvents (e.g., cyclohexanone). Although the polyimide is completely imidized, additional cross-linking occurs during processing.

Bismakimides (BMI). These are a class of polyimides with a general structure containing reactive double bonds on each end of the molecule.

Various functional groups such as vinyls, allyls, or amines are used as co-curing agents with bismaleimides to improve the properties of the homopolymer.

In one bismaleimide system, component A (4,49-bismaleimidodiphenylmethane) and component B (o-, o9-diallyl-bisphenol A) react together to provide a BMI molecule with a tougher, flexible backbone.

Tests have shown improved strength and toughness with formulations using a higher ratio (1.0:0.87) of BMI to diallyl-bisphenol A. Typical properties of two bismaleimide systems are shown in Table F-11.

POLYURETHANE (PU)

The term *polyurethane* refers to the reaction of polyisocyanates (-NCO-) and polyhydroxyl (-OH-) groups. A simple reaction between isocyanate and an alcohol is shown below. The reaction product is urethane rather than a polyurethane.

$$R \cdot NCO + HOR_1 — R \cdot NH \cdot COOR_1$$
PolyIsocyanate PolyHydroxyl   PolyUrethane

The German chemists Wurtz, in 1848, and Hentschel, in 1884, produced the first isocyanates.

This later led to the development of polyurethanes. It was Otto Bayer and his coworkers who actually made possible the commercial development of polyurethanes in 1937. Since that time, polyurethanes have developed into many commercially available forms including coatings, elastomers, adhesives, molding compounds, foams, and fibers.

The isocyanates and di-isocyanates are highly reactive with compounds containing reactive hydrogen atoms. For this reason, polyurethane polymers may be reproduced. The recurring link of the polyurethane chain is NHCOO or NHCO.

More complex polyurethanes based on toluene di-isocyanates (TDI) and polyester, diamine, castor oil, or polyether chains have been developed. Other isocyanates used are diphenylmethane diisocyanate (MDI) and polymethylene polyphenyl isocyanate (PAPI).

The first polyurethanes were made in Germany to compete with other polymers produced at that time. Linear aliphatic polyurethanes were used to make fibers. Linear polyurethanes are thermoplastics. They may be processed by using all normal thermoplastics techniques including injection and extrusion. Due to their cost, they have limited use as fibers or filaments.

Polyurethane coatings are noted for their high abrasion resistance, unusual toughness, hardness, good flexibility, chemical resistance, and weatherability (Figure F-28). ASTM has defined five distinct types of polyurethane coatings, as shown in Table F-12.

Polyurethane resins are processed into clear or pigmented finishes for home, industrial, or marine use. They improve the chemical and ozone resistance of rubber and other polymers. These coatings and finishes may be simple solutions for linear polyurethanes or complex systems of polyisocyanate and such OH groups as polyesters, polyethers, and castor oil.

Many polyurethane elastomers (PUR) (rubbers), may be prepared from di-isocyanates, linear polyesters, or polyether resin, and curing agents (Figure F-29). If formulated into a

(A) Foams, insulation, sponges, belts, and gaskets of polyurethane.

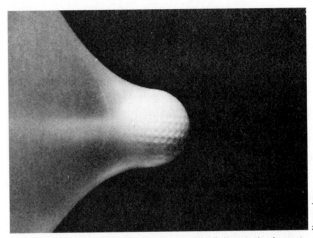

(B) This 0.076 mm (0.003 in.) film stops a hard-driven golf ball, demonstrating the puncture resistance and tensile strength of polyurethane.

Figure F-28. Applications of polyurethane.

$$nHOR\!-\!(OR\!-\!)_x\!-\ OH + n\,OCNR_1NCO \rightarrow (\!-\!OR\!-\!(OR)_x\,OCONHR_1NHCO\!-\!)_n$$

Polyester   Di-isocyanate   Polyurethane

$$nHOR(\!-\!OCOR_2CO\cdot OR\quad)_x OH + nOCNR_1NCO \rightarrow$$

Polyether   Di-isocyanate

$$\!-\!OR(\!-\!OCOR_2CO\cdot OR)_x OCONHR_1NHCO\!-\!)_n$$

Polyurethane

Figure F-29. Production of polyurethane elastomers.

Table F-12. ASTM Designations for Polyurethane Coatings

ASTM Type	Components	Pot Life	Cure	Clear or Pigmented Uses
(I) Oil-modified	One	Unlimited	Air	Interior or exterior wood and marine. Industrial enamels
(II) Prepolymer	One	Extended	Moisture	Interior or exterior. Wood, rubber and leather coatings
(III) Blocked	One	Unlimited	Heat	Wire coatings and baked finishes
(IV) Prepolymer + Catalyst	Two	Limited	Amine/catalyst air	Industrial finishes and leather, rubber products
(V) Polyisocyanate + Polyol	Two	Limited	NCO/OH reaction	Industrial finishes and leather, rubber products

linear thermoplastic urethane, they may be processed by utilizing normal thermoplastic processing equipment. They are used in shock absorbers, bumpers, gears, cable covers, hose jacketing, elastic thread (spandex), and diaphragms. Common uses of cross-linked thermosetting elastometers include industrial tires, shoe heels, gaskets, seals, O-rings, pump impellers, and tread stock for tires. Polyurethane elastomers have extreme resistance to abrasion, ozone aging, and hydrocarbon fluids. These elastometers cost more than conventional rubbers but are tough, elastic, and display a wide range of flexibility at temperature extremes.

Polyurethane foams are widely used and well known. They are available in flexible, semirigid, and rigid forms in a number of different densities. Various flexible foams are used as cushioning for furniture, automobile seating, and mattresses. They are produced by reacting toluene di isocyanate (TDI) with polyester and water in the presence of catalysts. At higher densities, they are cast or molded into drawer fronts, doors, moldings, and complete pieces of furniture. Flexible foams are open-celled structures that may be used as artificial sponges. These foams are used by the garment and textile industries for backing and insulation.

Semirigid foams are used as energy-absorbing materials in crash pads, arm rests, and sun visors.

The three largest uses of rigid polyurethane foam are in the manufacturing of furniture, automotive and construction moldings, and various thermal insulation products. Replicas of woodcarvings, decorative parts, and moldings are produced from high-density, self-skinning foams. The insulation value of these foams makes them an ideal choice for insulating refrigerators as well as refrigerated trucks and railroad cars. They may be foamed in place for many architectural uses. They may be placed on vertical surfaces by spraying the reaction mixture through a nozzle. They are used as flotation devices, packing, and structural reinforcement.

Rigid polyurethane is a closed cellular material produced by the reaction of TDI (prepolymer form) with polyethers and reactive blowing agents such as monofluorotrichloromethane (fluorocarbon). Methylene di isocyanate (MDI) and polymethylene polyphenyl isocyanate (PAPI) are also used in some rigid foams. MDI foams have better dimensional stability, whereas PAPI foams have high-temperature resistance.

Polyurethane-based caulks and sealants are inexpensive polyisocyanate materials used for encapsulation and in construction and manufacturing. Various polyisocyanates are also useful adhesives. They produce strong bonds between flexible fabrics, rubbers, foams, or other materials.

Many blowing or foaming agents are explosive and toxic. When mixing or processing polyurethane foams, make certain that proper ventilation is provided.

Table F-13 lists some properties of urethane plastics. Six advantages and four disadvantages of polyurethane plastics follow:

Advantages of Polyurethane

1. High abrasion resistance
2. Good low-temperature capability
3. Wide variability in molecular structure
4. Possibility of ambient curing
5. Comparatively low cost
6. Prepolymers foam readily

Disadvantages of Polyurethane

1. Poor thermal capability
2. Toxic (isocyanates are used)
3. Poor weatherability
4. Subject to attack by solvents

SILICONES (SIs)

In organic chemistry, carbon is studied because it is capable of forming molecular structures with many other elements. Carbon is considered to be a reactive element. Carbon is capable of entering into more molecular combinations than any other element. Life on earth is based on the element carbon.

The second most abundant element on earth is silicon. It has the same number of available bonding sites as carbon. Some scientists have speculated that life on other planets may be based on silicon. Others find this possibility hard to accept because silicon is an inorganic solid with a metallic appearance. Most of the earth's crust is composed of SiO_2 (silicon dioxide) in the form of sand, quartz, and flint.

As early as 1863, the tetravalent capacity of silicon interested chemists. Friedrich Wohler, C. M. Crafts, Charles Friedel, F. S. Kipping, W. H. Carothers, and many others performed work that led to the development of silicone polymers.

By 1943, Dow Corning Corporation produced the first commercial silicone polymers in the United States. There are thousands of uses for these materials. The word *silicone* should be applied only to polymers containing silicon–oxygen–silicon bonding. However, it is often used to denote any polymer containing silicon atoms.

In many carbon-hydrogen compounds, silicon may replace the element carbon. Methane (CH_4) may be changed to silane or silicomethane (SiH_4). Many structures similar to the aliphatic series of saturated hydrocarbons may be formed.

Table F-13. **Properties of Polyurethane**

Property	Cast Urethane	Urethane Elastomer
Molding qualities	Good	Good to excellent
Relative density	1.10–1.50	1.11–1.25
Tensile strength, MPa	1–69	31–58
(psi)	(175–10,000)	(4500–8400)
Compressive strength, MPa	14	14
(psi)	(2000)	(2000)
Impact strength, Izod, J/mm	0.25 to flexible	Does not break
(ft-lb/in.)	(5)	
Hardness, Shore	10A–90D	30A–70D
Rockwell	M28, R60	
Thermal expansion, $(10^{-4}/°C)$	25.4–50.8	25–50
Resistance to heat, °C	90–120	90
(°F)	(190–250)	(190)
Dielectric strength, V/mm	15,750–19,690	12,990–35,435
Dielectric constant (at 60 Hz)	4–7.5	5.4–7.6
Dissipation factor (at 60 Hz)	0.015–0.017	0.015–0.048
Arc resistance, s	0.1–0.6	0.22
Water absorption (24 h), %	0.02–1.5	0.7–0.9
Burning rate	Slow to self-extinguishing	Slow to self-extinguishing
Effect of sunlight	None to yellows	None to yellows
Effect of acids	Attacked	Dissolves
Effect of alkalies	Slight to attacked	Dissolves
Effect of solvents	None to slight	Resistant
Machining qualities	Excellent	Fair to excellent
Optical qualities	Transparent to opaque	Transparent to opaque

The following general types of bonds may be valuable in understanding the formation of silicone polymers:

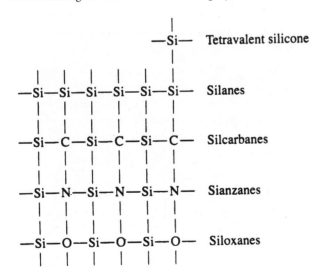

Compounds with only silicon and hydrogen atoms present are called *silanes*. When the silicon atoms are separated by carbon atoms, the structure is called a *silcarbane* (sil-CARB-ane). A *polysiloxane* is produced when more than one oxygen atom separates the silicon atoms in the chain.

A polymerized silicone molecular chain could be based on the structure shown in Figure F-30 modified by radicals (R).

Many silicone polymers are based on chains, rings, or networks of alternating silicon and oxygen atoms. Common examples contain methyl, phenyl, or vinyl groups on the siloxane chain (Figure F-31). A number of polymers are formed by varying the organic radical groups on the silicon chain. Many copolymers are also available.

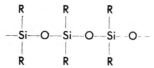

Figure F-30. An example of a polymerized silicone molecular chain.

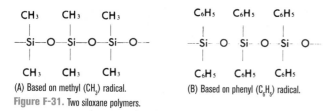

(A) Based on methyl (CH₃) radical.   (B) Based on phenyl (C₆H₅) radical.

Figure F-31. Two siloxane polymers.

The amount of energy needed to produce silicone plastics increase the cost. However, silicone plastics may still be economical due to longer product life, higher service temperatures, and flexibility at temperature extremes.

Silicones are produced in five commercially available categories: fluids, compounds, lubricants, resins, and elastomers (rubber).

Probably the best-known silicon plastics are associated with oils and ingredients for polishes. Examples are lens-cleaning tissues or water-repellent fabrics treated with a thin-film coating of silicone.

Silicone fluids are added to some liquids to prevent foaming (antifoaming), prevent transmission of vibrations (damping), and improve electrical and thermal limits of various liquids. Fluid silicones are used as additives in paints, oils, and inks as well as for mold-release agents, finishes for glass and fabrics, and paper coating.

Silicone compounds are usually granular or fibrous-filled materials. Because of their outstanding electrical and thermal properties, mineral- and glass-filled silicone compounds are used for encapsulating electronic components.

Similar to adhesives and sealants, silicone plastics are limited by their high cost. Their high service temperature and elastic properties make them useful for sealing, gasketing, caulking, and encapsulating as well as for repairing all types of materials (Figure F-32).

The chemical inertness of foamed silicone is useful in breast and facial implants in plastic surgery. Its main uses include electrical and thermal insulation of electrical wires and components.

When used as lubricants, silicones are prized because they do not deteriorate at extreme service temperatures. Silicones are used for lubricating rubber, plastics, ball bearings, valves, and vacuum pumps.

Silicone resins have many uses such as releasing agents for baking dishes. Silicone resins are also found in flexible, tough coatings utilized as high-temperature paints for engine

(A) This is a moisture-cure-type silicone sealant. It cures at room temperature if the relative humidity is between 30% and 80%.

(B) Silicone sealant is used on valve covers because it resists many solvents and has a high service temperature.

(C) Silicone sealant functions as both an adhesive and a sealant.

Figure F-32. Several uses of silicone sealants.

manifolds and mufflers. Their waterproofing capability makes them helpful in treating masonry and concrete walls.

Excellent waterproofing, thermal, and electrical properties make these resins valuable for electrical insulation and generators.

Laminates reinforced with glass cloth are used for structural parts, ducts, radomes, and electronic panel boards. These silicone laminates are characterized by their excellent dielectric and thermal properties and strength-to-mass ratio.

Diatomaceous earth, glass fiber, or asbestos may be used as fillers in preparing a premix or putty for molding small parts from silicone resins.

Some of the best-known silicones are in the form of elastomers. Few industrial rubbers or elastomers can withstand long exposure to ozone (O_3) or hot mineral oils. Silicone "rubbers" are stable at elevated temperatures and remain flexible when exposed to ozone or oils.

Silicone elastomers are used in artificial organs, O-rings, gaskets, and diaphragms. They are also used in flexible molds for casting of plastics and low-melting-point metals.

Room-temperature vulcanizing (RTV) elastomers are used to copy intricate molded parts, seal joints, and adhere parts (Figure F-33).

Silly Putty® and *Crazy Clay*® are novelty silicone products. This bouncing putty is a silicone elastomer that is also used for damping noise and as a sealing and filling compound. A hard bouncing putty will rebound to 80 percent of the height from which it is dropped. Other super-rebounding novelty items are produced from this compound.

Silicone-molding compounds may be processed in the same manner as other thermosetting organic plastics. Silicones are often used in resin form as castings, coatings, adhesives, or laminating compounds.

Research and development is being carried out on various other elements with covalent bonding capacity. Boron, aluminum, titanium, tin, lead, nitrogen, phosphorus, arsenic,

Figure F-34. Possible chemical structure of inorganic or semi-inorganic plastics.

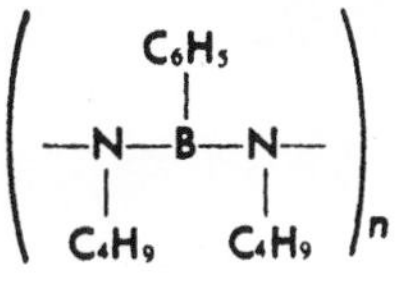

(A) Boron (monomer). 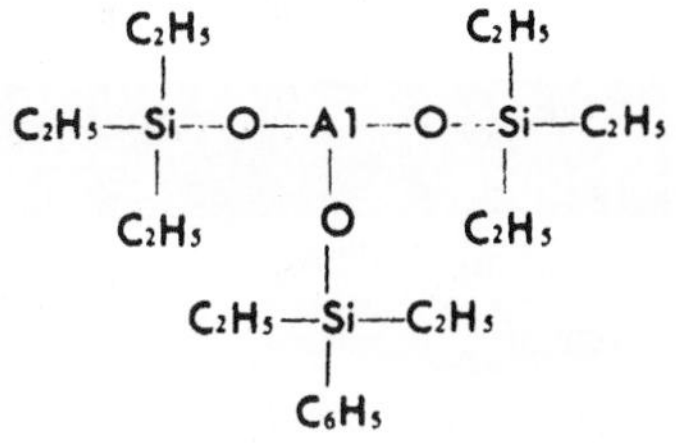 (B) Aluminum. 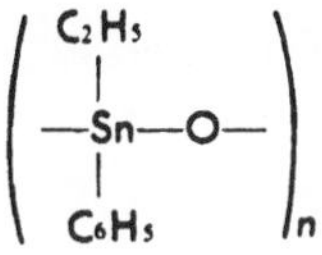 (C) Tin. 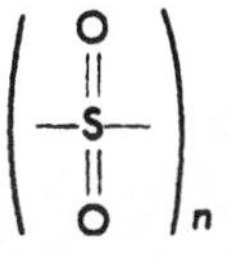 (D) Sulfur (monomer). 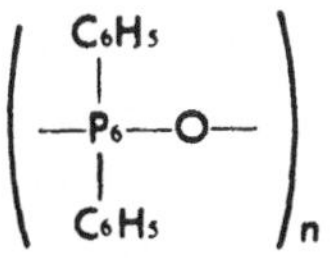 (E) Lead (monomer). 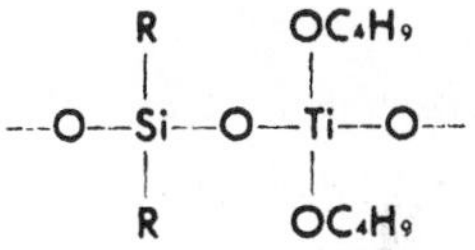 (F) Titanium. 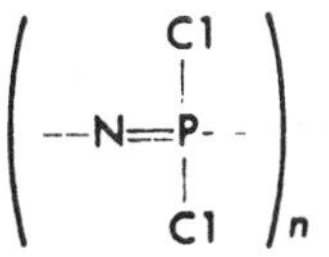 (G) Phosphorous (monomer). 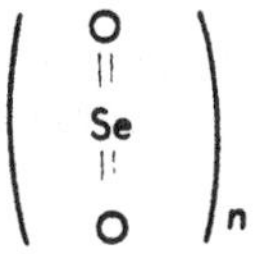 (H) Selenium (monomer).

sulfur, and selenium may be considered for inorganic or semi-organic plastics. The formulas in Figure F-34 show a few of many possible chemical structures of inorganic or semi-organic plastics.

Table F-14 lists the properties of silicone plastics. Seven advantages and three disadvantages of silicone follow:

Advantages of Silicone

1. Wide range of thermal capability, from −73°C to 315°C (−100°F to 600°F)
2. Good electrical traits
3. Wide variation in molecular structure (flexible or rigid forms)
4. Available in transparent grades
5. Low water absorption
6. Available in flame-retardant grades
7. Good chemical resistance

Disadvantages of Silicones

1. Low strength
2. Subject to attack by halogenated solvents
3. Comparatively high cost

Figure F-33. Molds of silicone RTV reproduce fine detail and allow severe undercuts.

Courtesy of Dow Corning Corporation

Table F-14. Properties of Silicones

Property	Cast Resin (including RTV)	Molding Compounds (Mineral-Filled)	Molding Compounds (Glass-Filled)
Molding qualities	Excellent	Excellent	Good
Relative density	0.99–1.50	1.7–2	1.68–2
Tensile strength, MPa	2–7	28–41	28–45
(psi)	(350–1000)	(4000–6000)	(4000–6500)
Compressive strength, MPa	0.7	90–124	69–103
(psi)	(100)	(13,000–18,000)	(10,000–15,000)
Impact strength, Izod, J/mm		0.013–0.018	0.15–0.75
(ft-lb/in.)		(0.26–0.36)	(3–15)
Hardness	Shore A15–A65	Rockwell M71–M95	Rockwell M84
Thermal expansion, $(10^{-4}/°C)$	20–79	5–10	0.61–0.76
Resistance to heat, °C	260	315	315
(°F)	(500)	(600)	(600)
Dielectric strength, V/mm	21,665	7875–15,750	7875–15,750
Dielectric constant (at 60 Hz)	2.75–4.20	3.5–3.6	3.3–5.2
Dissipation factor (at 60 Hz)	0.001–0.025	0.004–0.005	0.004–0.030
Arc resistance, s	115–130	250–420	150–205
Water absorption, %	0.12 (7 days)	0.08–0.13 (24 h)	0.1–0.2 (24 h)
Burning rate	Self-extinguishing	None to slow	None to slow
Effect of sunlight	None	None to slight	None to slight
Effect of acids	Slight to severe	Slight	Slight
Effect of alkalies	Moderate to severe	Slight to marked	Slight to marked
Effect of solvents	Swells in some	Attacked by some	Attacked by some
Machining qualities	None	Fair	Fair
Optical qualities	Transparent to opaque	Opaque	Opaque

USEFUL TABLES

APPENDIX G

English Metric Conversion

	If You Know	You Can Get	If You Multiply by*
Length	Inches (in.)	Millimeters (mm)	25.4
	Millimeters	Inches	0.04
	Inches	Centimeters (cm)	2.54
	Centimeters	Inches	0.4
	Inches	Meters (m)	0.0254
	Meters	Inches	39.37
	Feet (ft)	Centimeters	30.5
	Centimeters	Feet	4.8
	Feet	Meters	0.305
	Meters	Feet	3.28
	Miles (mi)	Kilometers (km)	1.61
	Kilometer	Miles	0.62
Area	Inches^2 (in.^{2})	Millimeters^2 (mm^2)	645.2
	Millimeters^2	Inches^2	0.0016
	Inches^2	Centimeters^2 (cm^2)	6.45
	Centimeters^2	Inches^2	0.16
	Foot^2 (ft^2)	Meters^2 (m^2)	0.093
	Meters^2	Foot^2	10.76
Capacity-Volume	Ounces (oz)	Milliliters (ml)	30
	Milliliters	Ounces	0.034
	Pints (pt)	Liters (l)	0.47
	Liters	Pints	2.1
	Quarts (qt)	Liters	0.95
	Liters	Quarts	1.06
	Gallons (gal.)	Liters	3.8

(Continued)

English Metric Conversion (*Continued*)

	If You Know	You Can Get	If You Multiply by*
	Liters	Gallons	0.26
	Cubic Inches (cu. in.)	Liters	0.0164
	Liters	Cubic Inches	61.03
	Cubic Inches	Cubic Centimeters (cc)	16.39
	Cubic Centimeters	Cubic Inches	0.061
Weight (Mass)	Ounces	Grams (g)	28.4
	Grams	Ounces	0.035
	Pounds (lb)	Kilograms (kg)	0.45
	Kilograms	Pounds	2.2
Force	Ounce	Newtons (N)	0.278
	Newtons	Ounces	35.98
	Pound	Newtons	4.448
	Newtons	Pound	0.225
	Newtons	Kilograms	0.102
	Kilograms	Newtons	9.807
Acceleration	Inch/Sec^2	Meter/Sec^2	.0254
	Meter/Sec^2	Inch/Sec^2	39.37
	Foot/Sec^2 (ft/s^2)	Meter/Sec^2 (m/s^2)	0.3048
	Meter/Sec^2	Foot/Sec^2	3.280
Torque	Pound-inch (Inch-Pound)	Newton-Meters (N-M)	0.113
	Newton-Meters	Pound-Inch	8.857
	Pound-Foot (Foot-Pound)	Newton-Meters	1.356
	Newton-Meters	Pound-Foot	.737
Pressure	Pounds/sq. in. (psi)	Kilopascals (kPa)	6.895
	Kilopascals	Pound/sq. in. (lb/in.^{2})	0.145
	Inches of Mercury (Hg)	Kilopascals	3.377
	Kilopascals	Inches of Mercury (Hg)	0.296
Fuel performance	Miles/gal	Kilometers/litre (km/l)	0.425
	Kilometers/litre	Miles/gal	2.352
Velocity	Miles/hour	Kilometers/hour (km/h)	1.609
	Kilometers/hour	Miles/hour	0.621
Temperature	Fahrenheit Degrees	Celsius Degrees	5/9 (°F −32)
	Celsius Degrees	Fahrenheit Degrees	9/5 (°C +32) = F

*Approximate Conversion Factors to be used where precision calculations are not necessary.

Convert Centigrade Temperature to Fahrenheit and Vice Versa

Centigrade °C = 5/9 (°F − 32)			Temperature Conversion Tables			To Fahrenheit °F = (9/5 × °C) + 32		
°C		°F	°C		°F	°C		°F
−17.8	0	32	−1.67	29	84.2	14.4	58	136.4
−17.2	1	33.8	−1.11	30	86.0	15.0	59	138.2
−16.7	2	35.6	−0.56	31	87.8	15.6	60	140.0
−16.1	3	37.4	−0	32	89.6	16.1	61	141.8
−15.6	4	39.2	0.56	33	91.4	16.7	62	143.6
−15.0	5	41.0	1.11	34	93.2	17.2	63	145.4
−14.4	6	42.8	1.67	35	95.0	17.8	64	147.2
−13.9	7	44.6	2.22	36	96.8	18.3	65	149.0
−13.3	8	46.4	2.78	37	98.6	18.9	66	150.8
−12.8	9	48.2	3.33	38	100.4	19.4	67	152.6
−12.2	10	50.0	3.89	39	102.2	20.0	68	154.4
−11.7	11	51.8	4.44	40	104.0	20.6	69	156.2
−11.1	12	53.6	5.00	41	105.8	21.1	70	158.0
−10.6	13	55.4	5.56	42	107.6	21.7	71	159.8
−10.0	14	57.2	6.11	43	109.4	22.2	72	161.6
−9.44	15	59.0	6.67	44	111.2	22.8	73	163.4
−8.89	16	60.8	7.22	45	113.0	23.3	74	165.2
−8.33	17	62.6	7.78	46	114.8	23.9	75	167.0
−7.78	18	64.4	8.33	47	116.6	24.4	76	168.8
−7.22	19	66.2	8.89	48	118.4	25.0	77	170.6
−6.67	20	68.0	9.44	49	120.2	25.6	78	172.4
−6.11	21	69.8	10.0	50	122.0	26.1	79	174.2
−5.56	22	71.6	10.6	51	123.8	26.7	80	176.0
−5.00	23	73.4	11.1	52	125.6	27.2	81	177.8
−4.44	24	75.2	11.7	53	127.4	27.8	82	179.6
−3.89	25	77.0	12.2	54	129.2	28.3	83	181.4
−3.33	26	78.8	12.8	55	131.0	28.9	84	183.2
−2.78	27	80.6	13.3	56	132.8	29.4	85	185.0
−2.22	28	82.4	13.9	57	134.6	30.0	86	186.8
30.6	87	188.6	149	300	572	343	650	1202
31.1	88	190.4	154	310	590	349	660	1220
31.7	89	192.2	160	320	608	354	670	1238
32.2	90	194.0	166	330	626	360	680	1256
32.8	91	195.8	171	340	644	366	690	1274
33.3	92	196.7	177	350	662	371	700	1292
33.9	93	199.4	182	360	680	377	710	1310
34.4	94	201.2	188	370	698	382	720	1328
35.0	95	203.0	193	380	716	388	730	1346
35.6	96	204.8	199	390	734	393	740	1364
36.1	97	206.6	204	400	752	399	750	1382
36.7	98	208.4	210	410	770	404	760	1400
37.2	99	210.2	216	420	788	410	770	1418
37.8	100	212.0	221	430	806	416	780	1436

(Continued)

Convert Centigrade Temperature to FAhrenheit and Vice Versa (*Continued*)

Centigrade °C = 5/9 (°F − 32)			Temperature Conversion Tables			To Fahrenheit °F = (9/5 × °C) + 32		
°C		°F	°C		°F	°C		°F
38	100	212	227	440	824	421	790	1454
43	110	230	232	450	842	427	800	1472
49	120	248	238	460	860	432	810	1490
54	130	266	243	470	878	438	820	1508
60	140	284	249	480	896	443	830	1526
66	150	302	254	490	914	449	840	1544
71	160	320	260	500	932	454	850	1562
77	170	338	266	510	950	460	860	1580
82	180	356	271	520	968	466	870	1598
88	190	374	277	530	986	471	880	1616
93	200	392	282	540	1004	477	890	1634
99	210	410	288	550	1022	482	900	1652
100	212	413	293	560	1040	488	910	1670
104	220	428	299	570	1058	493	920	1688
110	230	446	304	580	1076	499	930	1706
116	240	464	310	590	1094	504	940	1724
121	250	482	316	600	1112	510	950	1742
127	260	500	321	610	1130	516	960	1760
132	270	518	327	620	1148	521	970	1778
138	280	536	332	630	1166	527	980	1796
143	290	554	338	640	1184	532	990	1814

Decimal Equivalents of Fractions of One Inch

1/64	0.015625	17/64	0.265625	33/64	0.515625	49/64	0.765625
1/32	0.031250	9/32	0.281250	17/32	0.531250	25/32	0.781250
3/64	0.046875	19/64	0.296875	35/64	0.546875	51/64	0.796875
1/16	0.062500	5/16	0.312500	9/16	0.562500	13/16	0.812500
5/64	0.078125	21/64	0.328125	37/64	0.578125	53/64	0.828125
3/32	0.093750	11/32	0.343750	19/32	0.593750	27/32	0.843750
7/64	0.109375	23/64	0.359375	39/64	0.609375	55/64	0.859375
1/8	0.125000	3/8	0.375000	5/8	0.625000	7/8	0.875000
9/64	0.140625	25/64	0.390625	41/64	0.640625	57/64	0.890625
5/32	0.156250	13/32	0.406250	21/32	0.656250	29/32	0.906250
11/64	0.171875	27/64	0.421875	43/64	0.671875	59/64	0.890625
3/16	0.187500	7/16	0.437500	11/16	0.687500	15/16	0.937500
13/64	0.203125	29/64	0.453125	45/64	0.703125	61/64	0.953125
7/32	0.218750	15/32	0.468750	23/32	0.718750	31/32	0.968750
15/64	0.234375	31/64	0.484375	47/64	0.734375	63/64	0.984375
1/4	0.250000	1/2	0.500000	3/4	0.750000	1	1.000000

Standard Draft Angles

Depth	1/4°	1/2°	1°	1¹/2°	2°	2¹/2°	3°	5°	7°	8°	10°	12°	15°	Depth
¹/₃₂	0.0001	0.0003	0.0005	0.0008	0.0011	0.0014	0.0016	0.0027	0.0038	0.0044	0.0055	0.0066	0.0084	¹/₃₂
¹/₁₆	0.0003	0.0006	0.0011	0.0016	0.0022	0.0027	0.0033	0.0055	0.0077	0.0088	0.0110	0.0133	0.0168	¹/₁₆
³/₃₂	0.0004	0.0008	0.0016	0.0025	0.0033	0.0041	0.0049	0.0082	0.0115	0.0132	0.0165	0.0199	0.0251	³/₃₂
¹/₈	0.0005	0.0010	0.0022	0.0033	0.0044	0.0055	0.0066	0.0109	0.0153	0.0176	0.0220	0.0266	0.0335	¹/₈
³/₁₆	0.0008	0.0016	0.0033	0.0049	0.0065	0.0082	0.0098	0.0164	0.0230	0.0263	0.0331	0.0399	0.0502	³/₁₆
¹/₄	0.0011	0.0022	0.0044	0.0066	0.0087	0.0109	0.0131	0.0219	0.0307	0.0351	0.0441	0.0531	0.0670	¹/₄
⁵/₁₆	0.0014	0.0027	0.0055	0.0082	0.0109	0.0137	0.0164	0.0273	0.0384	0.0439	0.0551	0.0664	0.0837	⁵/₁₆
³/₈	0.0016	0.0033	0.0065	0.0098	0.0131	0.0164	0.0197	0.0328	0.0460	0.0527	0.0661	0.0797	0.1005	³/₈
⁷/₁₆	0.0019	0.0038	0.0076	0.0115	0.0153	0.0191	0.0229	0.0383	0.0537	0.0615	0.0771	0.0930	0.1172	⁷/₁₆
¹/₂	0.0022	0.0044	0.0087	0.0131	0.0175	0.0218	0.0262	0.0438	0.0614	0.0703	0.0882	0.1063	0.1340	¹/₂
⁵/₈	0.0027	0.0054	0.0109	0.0164	0.0218	0.0273	0.0328	0.0547	0.0767	0.0878	0.1102	0.1329	0.1675	⁵/₈
³/₄	0.0033	0.0065	0.0131	0.0196	0.0262	0.0328	0.0393	0.0656	0.0921	0.1054	0.1322	0.1595	0.2010	³/₄
⁷/₈	0.0038	0.0076	0.0153	0.0229	0.0306	0.0382	0.0459	0.0766	0.1074	0.1230	0.1543	0.1860	0.2345	⁷/₈
1	0.0044	0.0087	0.0175	0.0262	0.0349	0.0437	0.0524	0.0875	0.1228	0.1405	0.1763	0.2126	0.2680	1
1¹/₄	0.0055	0.0109	0.0218	0.0327	0.0437	0.0546	0.0655	0.1094	0.1535	0.1756	0.2204	0.2657	0.3349	1¹/₄
1¹/₂	0.0064	0.0131	0.0262	0.0393	0.0524	0.0655	0.0786	0.1312	0.1842	0.2108	0.2645	0.3188	0.4019	1¹/₂
1³/₄	0.0076	0.0153	0.0305	0.0458	0.0611	0.0764	0.0917	0.1531	0.2149	0.2460	0.3085	0.3720	0.4689	1³/₄
2	0.0087	0.0175	0.0349	0.0524	0.0698	0.0873	0.1048	0.1750	0.2456	2810	0.3527	0.4251	0.5359	2
Depth	1/4°	¹/₂°	1°	1¹/₂	2°	2¹/₂	3°	5°	7°	8°	10°	12°	15°	Depth

Conversion of Specific Gravity to Grams per Cubic Inch

$$16.39 \times \text{Specific Gravity} = \text{g/in.}^3$$

Specific Gravity	Grams/in.³	Specific Gravity	Grams/in.³	Specific Gravity	Grams/in.³	Specific Gravity	Grams/in.³
1.20	19.7	1.52	24.9	1.84	30.2	2.16	35.4
1.22	20.0	1.54	25.2	1.86	30.5	2.18	35.7
1.24	20.3	1.56	25.6	1.88	30.8	2.20	36.1
1.26	20.7	1.58	25.9	1.90	31.1	2.22	36.4
1.28	21.0	1.60	26.2	1.92	31.5	2.24	36.7
1.30	21.3	1.62	26.6	1.94	31.8	2.26	37.0
1.32	21.6	1.64	26.9	1.96	32.1	2.28	37.4
1.34	22.0	1.66	27.2	1.98	32.5	2.30	37.7
1.36	22.3	1.68	27.5	2.00	32.8	2.32	38.0
1.38	22.6	1.70	27.9	2.02	33.1	2.34	38.4
1.40	22.9	1.72	28.2	2.04	33.4	2.36	38.7
1.42	23.3	1.74	28.5	2.06	33.8	2.38	39.0
1.44	23.6	1.76	28.8	2.08	34.1	2.40	39.3
1.46	23.9	1.78	29.2	2.10	34.4		
1.48	24.3	1.80	29.5	2.12	34.7		
1.50	24.6	1.82	29.8	2.14	35.1		

To determine the cost/cu. in.:

Price/Lb. × Sp. Gravity × 0.03163

$1.32 × 1.76 × 0.03163 = $0.09/cu. in.

Diameters and Areas of Circles

Diameter	Area	Diameter	Area	Diameter	Area	Diameter	Area
1/64"	0.00019	2-"	3.1416	5-"	19.635	3/4	74.662
1/32	0.00077	1/16	3.3410	1/16	20.129	7/8	76.589
3/64	0.00173	1/8	3.5466	1/8	20.629		
1/16	0.00307	3/16	3.7583	3/16	21.125	10-"	78.540
3/32	0.00690	1/4	3.9761	1/4	21.648	1/8	80.516
1/8	0.01227	5/16	4.2000	5/16	22.166	1/4	82.516
5/32	0.01917	3/8	4.4301	3/8	22.691	3/8	84.541
3/16	0.02761	7/16	4.6664	7/16	23.211	1/2	86.590
7/32	0.03758	1/2	4.9087	1/2	23.758	5/8	88.664
1/4	0.04909	9/16	5.1572	9/16	24.301	3/4	90.763
9/32	0.06213	5/8	5.4119	5/8	24.850	7/8	92.886
5/16	0.07670	11/16	5.6727	11/16	25.406		
11/32	0.09281	3/4	5.9396	3/4	25.967	11-"	95.033
3/8	0.11045	13/16	6.2126	13/16	26.535	1/2	103.87
13/32	0.12962	7/8	6.4918	7/8	27.109		
7/16	0.15033	15/16	6.7771	15/16	27.688	12-"	113.10
15/32	0.17257					1/2	122.72
1/2	0.19635	3-"	7.0686	6-"	28.274		
17/32	0.22165	1/16	7.3662	1/8	29.465	13-"	132.73
9/16	0.24850	1/8	7.6699	1/4	30.680	1/2	143.14
19/32	0.27688	3/16	7.9798	3/8	31.919		
5/8	0.30680	1/4	8.2958	1/2	33.183	14-"	153.94
21/32	0.33824	5/16	8.6179	5/8	34.472	1/2	165.13
11/16	0.37122	3/8	8.9462	3/4	35.785		
23/32	0.40574	7/16	9.2806	7/8	37.122	15-"	176.71
3/4	0.44179	1/2	9.6211			1/2	188.69
25/32	0.47937	9/16	9.9678	7-"	38.485		
13/16	0.51849	5/8	10.321	1/8	39.871	16-"	201.06
27/32	0.55914	11/16	10.680	1/4	41.282	1/2	213.82
7/8	0.60132	3/4	11.045	3/8	42.718		
29/32	0.64504	13/16	11.416	1/2	44.179	17-"	226.98
15/16	0.69029	7/8	11.793	5/8	45.664	1/2	240.53
31/32	0.73708	15/16	12.177	3/4	47.173		
				7/8	48.707	18-"	254.47
1-"	0.7854	4-"	12.566			1/2	268.80
1/16	0.8866	1/16	12.962	8-"	50.265		
1/8	0.9940	1/8	13.364	1/8	51.849	19-"	283.53
3/16	1.1075	3/16	13.772	1/4	53.456	1/2	298.65
1/4	1.2272	1/4	14.186	3/8	55.088		
5/16	1.3530	5/16	14.607	1/2	56.745	20-"	314.16
3/8	1.4849	3/8	15.033	5/8	58.426	1/2	330.06
7/16	1.6230	7/16	15.466	3/4	60.132		
1/2	1.7671	1/2	15.904	7/8"	61.862		
9/16	1.9175	9/16	16.349				
5/8	2.0739	5/8	16.800	9-"	63.617		
11/16	2.2465	11/16"	17.257	1/8	65.397		
3/4	2.4053	3/4	17.721	1/4	67.201		
13/16	2.5802	13/16	18.190	3/8	69.029		
7/8"	2.7612	7/8	18.665	1/2	70.882		
15/16	2.9483	15/16	19.147	5/8	72.760		

Steam Temperature Versus Gauge Pressure

Gauge Pressure Lb	Temp. °F
50	297.5
55	302.4
60	307.1
65	311.5
70	315.8
75	319.8
80	323.6
85	327.4
90	331.1
95	334.3
100	337.7
105	341.0
110	344.0
115	347.0
120	350.0
125	353.0
130	356.0
135	358.0
140	361.0
145	363.0
150	365.6
155	368.0
160	370.3
165	372.7
170	374.9
175	377.2
180	379.3
185	381.4
190	383.5
195	385.7
200	387.5

Weight of 1000 Pieces in Pounds Based on Weight of One Piece in Grams

Weight per Piece in Grams	Weight per 1000 Pieces in Pounds	Weight per Piece in Grams	Weight per 1000 Pieces in Pounds
1	2.2	11	24.2
2	4.4	12	26.4
3	6.6	13	28.6
4	8.8	14	30.8
5	11.0	15	33.0
6	13.2	16	35.2
7	15.4	17	37.4
8	17.6	18	39.6
9	19.8	19	41.8
10	22.0	20	44.0

Weight per Piece in Grams	Weight per 1000 Pieces in Pounds	Weight per Piece in Grams	Weight per 1000 Pieces in Pounds
21	46.2	61	134.3
22	48.4	62	136.5
23	50.6	63	138.7
24	52.8	64	140.9
25	55.0	65	143.1
26	57.2	66	145.3
27	59.4	67	147.5
28	61.6	68	149.7
29	63.8	69	151.9
30	66.0	70	154.1
31	68.2	71	156.3
32	70.4	72	158.5
33	72.6	73	160.7
34	74.8	74	162.9
35	77.0	75	165.1
36	79.2	76	167.4
37	81.4	77	169.6
38	83.7	78	171.8
39	85.9	79	174.0
40	88.1	80	176.2
41	90.3	81	178.4
42	92.5	82	180.6
43	94.7	83	182.8
44	96.9	84	185.0
45	99.1	85	187.2
46	101.3	86	189.4
47	103.5	87	191.6
48	105.7	88	193.8
49	107.9	89	196.0
50	110.1	90	198.2
51	112.3	91	200.4
52	114.5	92	202.6
53	116.7	93	204.8
54	118.9	94	207.0
55	121.1	95	209.2
56	123.3	96	211.4
57	125.5	97	213.6
58	127.7	98	215.8
59	129.9	99	218.0
60	132.1	100	220.2

EQUIVALENT WEIGHTS
1 gram = 0.0353 oz.
0.0625 pounds = 1 ounce = 28.3 grams
454 grams = 1 pound.

Length Equivalents Millimeters to Inches

Millimeters	Inches	Millimeters	Inches	Millimeters	Inches
1	0.03937	34	1.33860	67	2.63779
2	0.07874	35	1.37795	68	2.67716
3	0.11811	36	1.41732	69	2.71653
4	0.15748	37	1.45669	70	2.75590
5	0.19685	38	1.49606	71	2.79527
6	0.23622	39	1.53543	72	2.83464
7	0.27559	40	1.57480	73	2.87401
8	0.31496	41	1.61417	74	2.91338
9	0.35433	42	1.65354	75	2.95275
10	0.39370	43	1.69291	76	2.99212
11	0.43307	44	1.73228	77	3.03149
12	0.47244	45	1.77165	78	3.07086
13	0.51181	46	1.81102	79	3.11023
14	0.55118	47	1.85039	80	3.14960
15	0.59055	48	1.88976	81	3.18897
16	0.62992	49	1.92913	82	3.22834
17	0.66929	50	1.96850	83	3.26771
18	0.70866	51	2.00787	84	3.30708
19	0.74803	52	2.04724	85	3.34645
20	0.78740	53	2.08661	86	3.38582
21	0.82677	54	2.12598	87	3.42519
22	0.86614	55	2.16535	88	3.46456
23	0.90551	56	2.20472	89	3.50393
24	0.94488	57	2.24409	90	3.54330
25	0.98425	58	2.28346	91	3.58267
26	1.02362	59	2.32283	92	3.62204
27	1.06299	60	2.36220	93	3.66141
28	1.10236	61	2.40157	94	3.70078
29	1.14173	62	2.44094	95	3.74015
30	1.18110	63	2.48031	96	3.77952
31	1.22047	64	2.51968	97	3.81889
32	1.25984	65	2.55905	98	3.85826
33	1.29921	66	2.59842	99	3.89763
				100	3.93700

VOLUME EQUIVALENTS

1 cc = 0.061 cu. in.

1 cu. in. = 16.387 cc

SOURCES OF HELP
AND BIBLIOGRAPHY

SOURCES OF HELP

The following alphabetical list of service organizations, standards and specifications groups, trade associations, professional societies, references, and US governmental agencies may serve as sources for further information:

American Chemical Society
www.acs.org

American Conference of Governmental Industrial Hygienists (ACGIH)
www.acgih.org

American Industrial Hygiene Association (AIHA)
www.aiha.org

American Insurance Association (AIA)
www.aiadc.org

American Medical Association (AMA)
www.ama-assn.org

American National Standards Institute (ANSI)
www.ansi.org

American Petroleum Institute
www.api.org

American Chemistry Council, Plastics Division
www.plastics.org

(The) American Society for Testing and Materials (ASTM)
www.astm.org

(The) American Society of Mechanical Engineers (ASME)
www.asme.org

The American Society of Safety Engineers
www.asse.org

The Association of Postconsumer Plastic Recyclers (APR)
www.plasticsrecycling.org

Chemical Manufacturers Association
www.cmahq.com

Defense Standardization Program Office (DSPO)
www.dsp.dla.mil

Department of Defense (DOD)
www.dod.gov

Department of Transportation (DOT)
www.dot.gov

Environmental Protection Agency (EPA)
www.epa.gov

Federal Emergency Management Agency (FEMA)
www.fema.gov

Federal Register
federalregister.gov

Food and Drug Administration (FDA)
www.fda.gov

General Services Administration (GSA)
www.gsa.gov

The International Society of Automation
www.isa.org

International Organization for Standardization (ISO)
www.iso.org

National Association of Manufacturers (NAM)
www.nam.org

National Fire Protection Association (NFPA)
www.nfpa.org

National Institute for Occupational Safety
and Health (NIOSH)
www.cdc.gov/niosh/homepage.html

National Institute of Standards and Technology
www.nist.org

(The) National Association for PET Container
Resources (NAPCOR)
www.napcor.com

National Safety Council
www.ncs.org

Occupational Safety & Health Administration (OSHA)
www.osha.gov

Society of Plastics Engineers (SPE)
www.4spe.org

The Plastics Industry Trade Assocation
www.plasticsindustry.org

Underwriters Laboratories (UL)
www.ul.com

US Government Printing Office
www.gpo.gov

BIBLIOGRAPHY

The following bibliography may be useful for further study and more detailed discussion of selected topics presented:

Advanced Composites: Conference Proceedings, American Society for Metals, December 2–4, 1985.

Allegri, Theodore. *Handling and Management of Hazardous Materials and Waste.* New York, NY: Chapman and Hall, 1986.

Beall, Glenn L. *Rotational Molding Design, Materials, Tooling & Processing.* Cincinnati, OH: Hanser Gardner Publications, 1998.

Bernhardt, Ernest. *CAE Computer Aided Engineering for Injection Molding.* New York, NY: Hanser Publishers, 1983.

Billmeyer, Fred W. *Textbook of Polymer Science.* 3rd ed. New York, NY: Wiley, 1984.

Broutman, L., and R. Krock. *Composite Materials.* 6 vols. New York, NY: Academic Press, 1985.

Budinski, Kenneth. *Engineering Materials: Properties and Selection.* 2nd ed. Reston, VA: Reston Publishing Company, Inc., 1983.

Carraher, Charles E., Jr., and James Moore. *Modification of Polymers.* New York, NY: Plenum Press, 1983.

"Chemical Emergency Preparedness Program Interim Guidance," Revision 1, #9223.01A. Washington, DC: United States Environmental Protection Agency, 1985.

Composite Materials Technology, Society of Automotive Engineers, 1986.

"Defense Standardization Manual: Defense Standardization and Specification Program Policies, Procedures and Instruction," DOD 4120. 3-M. August 1978.

Dreger, Donald. "Design Guidelines of Joining Advanced Composites," *Machine Design,* May 8, 1980, pp. 89–93.

Dym, Joseph. *Product Design with Plastics: A Practical Manual.* New York, NY: Industrial Press, 1983.

Ehrenstein, G. W. *Polymeric Materials.* Cincinnati, OH: Hanser Gardner Publications, 2001.

Ehrenstein, G., and G. Erhard. *Designing with Plastics: A Report on the State of the Art.* New York, NY: Hanser Publishers, 1984.

English, Lawrence. "Liquid-Crystal Polymers: In a Class of Their Own," *Manufacturing Engineering,* March 1986, pp. 36–41.

English, Lawrence. "The Expanding World of Composites," *Manufacturing Engineering,* April 1986, pp. 27–31.

Fitts, Bruce. "Fiber Orientation of Glass Fiber-Reinforced Phenolics," *Materials Engineering,* November 1984, pp. 18–22.

Grayson, Martin. *Encyclopedia of Composite Materials and Components.* New York, NY: John Wiley and Sons Inc., 1984.

Johnson, Wayne, and R. Schwed. "Computer Aided Design and Drafting," *Engineered Systems,* March/April 1986, pp. 48–51.

Kliger, Howard. "Customizing Carbon-Fiber Composites: For Strong, Rigid, Lightweight Structures," *Machine Design,* December 6, 1979, pp. 150–157.

Kohan, Melvin I. (Ed.). *Nylon Plastics Handbook.* Cincinnati, OH: Hanser Gardner Publications, 1995.

Lee, Norman C. *Plastic Blow Molding Handbook.* Dordrecht, Netherlands: Kluwer Academic Publishers, 1990.

Levy, S., and J. F. Carley. *Plastics Extrusion Technology Handbook.* 2nd ed. New York, NY: Industrial Press, 1989.

Levy, Sidney, and J. Harry Dubois. *Plastics Product Design Engineering Handbook.* 2nd ed. New York, NY: Chapman and Hall, 1984.

Lubin, George. *Handbook of Composites.* New York, NY: Van Nostrand Reinhold Company, Inc., 1982.

Menges, G., W. Michaeli, and P. Mohren. *How to Make Injection Molds.* 3rd ed. Cincinnati, OH: Hanser Gardner Publications, 2001.

Mohr, G., et. al. *SPI Handbook of Technology and Engineering of Reinforced Plastics/Composites.* 2nd ed. Malabar, FL: Robert Krieger Publishing Company.

Moore, G. R., and D. E. Kline. *Properties and Processing of Polymers for Engineers.* Englewood Cliffs, NJ: Prentice-Hall, Inc., 1984.

Naik, Saurabh, et. al. "Evaluating Coupling Agents for Mica/Glass Reinforcement of Engineering Thermoplastics," *Modern Plastics,* June 1985, pp. 1979–1980.

Plunkett, E. R. *Handbook of Industrial Toxicology.* New York, NY: Chemical Publishing Company, 1987.

Pocius, A. V. *Adhesion and Adhesives Technology: An Introduction.* Cincinnati, OH: Hanser Gardner Publications, 2002.

Powell, Peter C. *Engineering with Polymers.* New York, NY: Chapman and Hall, 1983.

Progelhof, Richard C., and James L. Throne. *Polymer Engineering Principles.* Cincinnati, OH: Hanser Gardner Publications, 1993.

Rauwendaal, Chris. *Polymer Extrusion.* 4th ed. Cincinnati, OH: Hanser Gardner Publications, 2001.

Richardson, Terry. *Composites: A Design Guide.* New York, NY: Industrial Press, 1987.

Rosato, D. V., D. V. Rosato, and M. G. Rosato. *Concise Encyclopedia of Plastics.* Dordrecht, Netherlands: Kluwer Academic Publishers, 2000.

Rotheiser, Jordan I. *Joining of Plastics: Handbook for Designers and Engineers.* Cincinnati, OH: Hanser Gardner Publications, 1999.

Rubin, Irvin I. *Handbook of Plastic Materials and Technology.* New York, NY: John Wiley and Sons, 1990.

Ryntz, Rose. *Plastics and Coatings: Durability, Stabilization, and Testing.* Cincinnati, OH: Hanser Gardner Publications, 2001.

Schwartz, M. M. *Composite Materials Handbook.* New York, NY: McGraw-Hill Book Company, 1984.

Schwartz, Mel. *Fabrication of Composite Materials: Source Book,* American Society for Metals, 1985.

Schultz, Jerome. *Polymer Crystallization: The Development of Crystalline Order in Thermoplastic Polymers.* Oxford University Press, 2001.

Seymour, Ramold B., and Charles Carraher. *Polymer Chemistry.* New York, NY: Marcel Dekker, Inc., 1981.

Shah, Vishu. *Handbook of Plastics Testing Technology.* 2nd ed. New York, NY: John Wiley & Sons, 1998.

Shook, Gerald. *Reinforced Plastics for Commercial Composites: Source Book,* American Society for Metals, 1986.

"Standardization Case Studies: Defense Standardization and Specification Program," Department of Defense, Washington, DC, March 17, 1986.

Stepek, J., and H. Daoust. *Additives for Plastics.* New York, NY: Springer Verlag, 1983, p. 260.

Throne, J. L., and R. J. Crawford. *Rotational Molding Technology.* Plastics Design Library, 2001.

Throne, James L. *Technology of Thermoforming.* Cincinnati, OH: Hanser Gardner Publications, 1996.

Tres, Paul. *Designing Plastic Parts for Assembly.* 4th ed. Cincinnati, OH: Hanser Gardner Publications, 2000.

White, James L. *Twin Screw Extrusion.* Cincinnati, OH: Hanser Gardner Publications, 1991.

Wood, Stuart. "Patience: Key to Big Volume in Advanced Composites," *Modern Plastics,* March 1986, pp. 44–48.

INDEX